Two-Component Signal Transduction

Two-Component Signal Transduction

edited by

James A. Hoch

DIVISION OF CELLULAR BIOLOGY
DEPARTMENT OF MOLECULAR AND EXPERIMENTAL MEDICINE
THE SCRIPPS RESEARCH INSTITUTE, LA JOLLA, CALIFORNIA

and

Thomas J. Silhavy

DEPARTMENT OF MOLECULAR BIOLOGY
PRINCETON UNIVERSITY, PRINCETON, NEW JERSEY

ASM Press • Washington, D.C.

Copyright © 1995 American Society for Microbiology
1325 Massachusetts Avenue, N.W.
Washington, D.C. 20005

Library of Congress Cataloging-in-Publication Data

Two-component signal transduction / edited by James A. Hoch and Thomas J. Silhavy
    p.   cm.
    Includes bibliographical references and index.
    ISBN 1-55581-089-6 (alk. paper)
    1. Bacteria—Physiology.   2. Cellular signal transduction.   3. Bacterial genetics.
4. Bacterial cell walls.   I. Hoch, James A.   II. Silhavy, Thomas J.
    [DNLM: 1. Signal Transduction—physiology.   2. Cell Communication—
physiology.   3. Gene Expression Regulation, Bacterial—physiology. QH 601 T974
1995]
QR84.T96 1995
589.9′01—dc20
DNLM/DLC
for Library of Congress   95-6254
                                                                                                           CIP

Contents

Contributors   *ix*

Preface   *xv*

1. Historical Perspective
Boris Magasanik
1

I   GENERAL PRINCIPLES   7

2. Genetic Approaches for Signaling Pathways and Proteins
John S. Parkinson
9

3. Two-Component Signal Transduction Systems: Structure-Function Relationships and Mechanisms of Catalysis
Jeffry B. Stock, Michael G. Surette, Mikhail Levit, and Peter Park
25

4. Structural and Functional Conservation in Response Regulators
Karl Volz
53

II   PARADIGMS   65

5. Control of Nitrogen Assimilation by the NR_I-NR_{II} Two-Component System of Enteric Bacteria
Alexander J. Ninfa, Mariette R. Atkinson, Emmanuel S. Kamberov, Junli Feng, and Elizabeth G. Ninfa
67

6. Chemotactic Signal Transduction in *Escherichia coli* and *Salmonella typhimurium*
Charles D. Amsler and Philip Matsumura
89

7. **Porin Regulon of** *Escherichia coli*
Leslie A. Pratt and Thomas J. Silhavy
105

8. **Control of Cellular Development in Sporulating Bacteria
by the Phosphorelay Two-Component
Signal Transduction System**
James A. Hoch
129

III   RESPONSE REGULATOR FUNCTIONS   145

9. **Mechanism of Transcriptional Activation by NtrC**
Susan C. Porter, Anne K. North, and Sydney Kustu
147

10. **Transcription Regulation by the** *Bacillus subtilis*
Response Regulator Spo0A
George B. Spiegelman, Terry H. Bird, and Valerie Voon
159

11. **Flagellar Switch**
Robert M. Macnab
181

IV   CELLULAR PHYSIOLOGY   201

12. **Signal Transduction and Cross Regulation in the**
Escherichia coli **Phosphate Regulon by PhoR, CreC, and
Acetyl Phosphate**
Barry L. Wanner
203

13. **Signal Transduction in the Arc System for Control of
Operons Encoding Aerobic Respiratory Enzymes**
Shiro Iuchi and E. C. C. Lin
223

14. **Dual Sensors and Dual Response Regulators Interact to
Control Nitrate- and Nitrite-Responsive Gene Expression in**
Escherichia coli
Valley Stewart and Ross S. Rabin
233

15. **Regulation of Capsule Synthesis: Modification of the
Two-Component Paradigm by an Accessory Unstable Regulator**
Susan Gottesman
253

16. **Expression of the Uhp Sugar-Phosphate Transport
System of** *Escherichia coli*
Robert J. Kadner
263

**17. Symbiotic Expression of *Rhizobium meliloti*
Nitrogen Fixation Genes Is Regulated by Oxygen**
Peter G. Agron and Donald R. Helinski
275

**18. Complex Phosphate Regulation by
Sequential Switches in *Bacillus subtilis***
F. Marion Hulett
289

V   PATHOGENESIS   303

**19. Two-Component Signal Transduction and Its Role in the
Expression of Bacterial Virulence Factors**
Michelle Dziejman and John J. Mekalanos
305

**20. Regulation of *Salmonella* Virulence by
Two-Component Regulatory Systems**
Eduardo A. Groisman and Fred Heffron
319

21. *Bordetella pertussis* BvgAS Virulence Control System
M. Andrew Uhl and Jeff F. Miller
333

**22. Three-Component Regulatory System Controlling
Virulence in *Vibrio cholerae***
Victor J. DiRita
351

**23. Ti Plasmid and Chromosomally Encoded Two-Component
Systems Important in Plant Cell Transformation by
Agrobacterium Species**
Joe Don Heath, Trevor C. Charles, and Eugene W. Nester
367

**24. Regulation of Glycopeptide Resistance Genes of
Enterococcal Transposon Tn*1546* by the VanR-VanS
Two-Component Regulatory System**
*Michel Arthur, Florence Depardieu, Theodore Holman, Zhen Wu,
Gerard Wright, Christopher T. Walsh, and Patrice Courvalin*
387

25. Tetracycline Regulation of Conjugal Transfer Genes
Abigail A. Salyers, Nadja B. Shoemaker, and Ann M. Stevens
393

**VI   CELLULAR COMMUNICATION AND
DEVELOPMENT   401**

**26. Switches and Signal Transduction Networks in the
Caulobacter crescentus Cell Cycle**
Todd Lane, Andrew Benson, Gregory B. Hecht, George J. Burton, and Austin Newton
403

27. The *frz* Signal Transduction System Controls
Multicellular Behavior in *Myxococcus xanthus*
Wenyuan Shi and David R. Zusman
419

28. Intercellular Communication in Marine *Vibrio* Species:
Density-Dependent Regulation of the
Expression of Bioluminescence
Bonnie L. Bassler and Michael R. Silverman
431

29. A Signal Transduction Network in *Bacillus subtilis*
Includes the DegS/DegU and ComP/ComA
Two-Component Systems
Tarek Msadek, Frank Kunst, and Georges Rapoport
447

Index   473

Contributors

Peter G. Agron

Department of Biology and Center for Molecular Genetics, University of California,
San Diego, La Jolla, California 92093-0634
Present address: Division of Infectious Disease, University of California, San Francisco,
San Francisco, California 94143-0654

Charles D. Amsler

Department of Biology, University of Alabama at Birmingham,
Birmingham, Alabama 35294-1170

Michel Arthur

Unité des Agents Antibactériens, Centre National de la Recherche Scientifique,
Institut Pasteur, Paris, France

Mariette R. Atkinson

Department of Biological Chemistry, University of Michigan Medical School,
Ann Arbor, Michigan 48109-0606

Bonnie L. Bassler

Department of Molecular Biology, Princeton University, Princeton, New Jersey 08544

Andrew Benson

Department of Molecular Biology, Princeton University, Princeton, New Jersey 08544

Terry H. Bird

Department of Microbiology and Immunology, University of British Columbia,
Vancouver, Canada V6T 1Z3

George J. Burton

Department of Molecular Biology, Princeton University, Princeton, New Jersey 08544

Trevor C. Charles

Department of Natural Resource Sciences, McGill University, Macdonald Campus,
Ste. Anne de Bellevue, Quebec, Canada H9X 3V9

Patrice Courvalin
Unité des Agents Antibactériens, Centre National de la Recherche Scientifique,
Institut Pasteur, Paris, France

Florence Depardieu
Unité des Agents Antibactériens, Centre National de la Recherche Scientifique,
Institut Pasteur, Paris, France

Victor J. DiRita
Unit for Laboratory Animal Medicine and Department of Microbiology and Immunology,
University of Michigan Medical School, Ann Arbor, Michigan 48109

Michelle Dziejman
Department of Microbiology and Molecular Genetics, Harvard Medical School,
Boston, Massachusetts 02115-5701

Junli Feng
Department of Biological Chemistry, University of Michigan Medical School,
Ann Arbor, Michigan 48109-0606

Susan Gottesman
Laboratory of Molecular Biology, National Cancer Institute,
Bethesda, Maryland 20892-4255

Eduardo A. Groisman
Department of Molecular Microbiology, Washington University School of Medicine,
St. Louis, Missouri 63110

Joe Don Heath
Department of Microbiology, University of Washington, Seattle, Washington 98195

Gregory B. Hecht
Department of Molecular Biology, Princeton University, Princeton, New Jersey 08544

Fred Heffron
Department of Microbiology and Immunology, Oregon Health Sciences University,
Portland, Oregon 97201

Donald R. Helinski
Department of Biology and Center for Molecular Genetics, University of California,
San Diego, La Jolla, California 92093-0634

James A. Hoch
Division of Cellular Biology, Department of Molecular and Experimental Medicine,
The Scripps Research Institute, La Jolla, California 92037

Theodore Holman
Department of Biological Chemistry and Molecular Pharmacology,
Harvard Medical School, Boston, Massachusetts 02115

F. Marion Hulett
Laboratory for Molecular Biology, Department of Biological Sciences,
University of Illinois at Chicago, Chicago, Illinois 60607

Shiro Iuchi
Department of Microbiology and Molecular Genetics, Harvard Medical School,
Boston, Massachusetts 02115

Robert J. Kadner
Department of Microbiology, School of Medicine, University of Virginia,
Charlottesville, Virginia 22908

Emmanuel S. Kamberov
Department of Biological Chemistry, University of Michigan Medical School,
Ann Arbor, Michigan 48109-0606

Frank Kunst
Unité de Biochimie Microbienne, Centre National de la Recherche Scientifique,
Département des Biotechnologies, Institut Pasteur, Paris, France

Sydney Kustu
Departments of Plant Biology and Molecular and Cell Biology, University of California,
Berkeley, Berkeley, California 94270

Todd Lane
Department of Molecular Biology, Princeton University, Princeton, New Jersey 08544

Mikhail Levit
Department of Molecular Biology, Princeton University, Princeton, New Jersey 08544

E. C. C. Lin
Department of Microbiology and Molecular Genetics, Harvard Medical School,
Boston, Massachusetts 02115

Robert M. Macnab
Department of Molecular Biophysics and Biochemistry, Yale University,
New Haven, Connecticut 06520-8114

Boris Magasanik
Department of Biology, Massachusetts Institute of Technology,
Cambridge, Massachusetts 02139

Philip Matsumura
Department of Microbiology and Immunology (M/C 790),
University of Illinois at Chicago, Chicago, Illinois 60612-7344

John J. Mekalanos
Department of Microbiology and Molecular Genetics, Harvard Medical School,
Boston, Massachusetts 02115-5701

Jeff F. Miller
Molecular Biology Institute, University of California, Los Angeles,
Los Angeles, California 90024

Tarek Msadek
Unité de Biochimie Microbienne, Centre National de la Recherche Scientifique,
Département des Biotechnologies, Institut Pasteur, Paris, France
Present address: Division of Cellular Biology, The Scripps Research Institute,
La Jolla, California 92037

Eugene W. Nester
Department of Microbiology, University of Washington, Seattle, Washington 98195

Austin Newton
Department of Molecular Biology, Princeton University, Princeton, New Jersey 08544

Alexander J. Ninfa
Department of Biological Chemistry, University of Michigan Medical School,
Ann Arbor, Michigan 48109-0606

Elizabeth G. Ninfa
Department of Biological Chemistry, University of Michigan Medical School,
Ann Arbor, Michigan 48109-0606

Anne K. North
Departments of Plant Biology and Molecular and Cell Biology, University of California,
Berkeley, Berkeley, California 94270

Peter Park
Departments of Molecular Biology and Chemistry, Princeton University,
Princeton, New Jersey 08544

John S. Parkinson
Biology Department, University of Utah, Salt Lake City, Utah 84112

Susan C. Porter
Departments of Plant Biology and Molecular and Cell Biology, University of California,
Berkeley, Berkeley, California 94270

Leslie A. Pratt
Department of Molecular Biology, Princeton University, Princeton, New Jersey 08544

Ross S. Rabin
NeXagen, Inc., Boulder, Colorado 80301

Georges Rapoport
Unité de Biochimie Microbienne, Centre National de la Recherche Scientifique,
Département des Biotechnologies, Institut Pasteur, Paris, France

Abigail A. Salyers
Department of Microbiology, University of Illinois, Urbana, Illinois 61801

Wenyuan Shi
Department of Molecular and Cell Biology, University of California, Berkeley,
Berkeley, California 94720

Nadja B. Shoemaker
Department of Microbiology, University of Illinois, Urbana, Illinois 61801

Thomas J. Silhavy
Department of Molecular Biology, Princeton University, Princeton, New Jersey 08544

Michael R. Silverman
The Agouron Institute, La Jolla, California 92037

George B. Spiegelman
Department of Microbiology and Immunology, University of British Columbia,
Vancouver, Canada V6T 1Z3

Ann M. Stevens
Department of Microbiology, University of Iowa, Iowa City, Iowa 52242

Valley Stewart
Sections of Microbiology and Genetics and Development, Cornell University,
Ithaca, New York 14853-8101

Jeffry B. Stock
Departments of Molecular Biology and Chemistry, Princeton University, Princeton, New Jersey 08544

Michael G. Surette
Department of Molecular Biology, Princeton University, Princeton, New Jersey 08544

M. Andrew Uhl
Department of Microbiology and Immunology, School of Medicine, University of California, Los Angeles, Los Angeles, California 90024

Karl Volz
Department of Microbiology and Immunology, University of Illinois at Chicago, Chicago, Illinois 60612

Valerie Voon
Department of Microbiology and Immunology, University of British Columbia, Vancouver, Canada V6T 1Z3

Christopher T. Walsh
Department of Biological Chemistry and Molecular Pharmacology, Harvard Medical School, Boston, Massachusetts 02115

Barry L. Wanner
Department of Biological Sciences, Purdue University, West Lafayette, Indiana 47907

Gerard Wright
Department of Biological Chemistry and Molecular Pharmacology, Harvard Medical School, Boston, Massachusetts 02115

Zhen Wu
Department of Biological Chemistry and Molecular Pharmacology, Harvard Medical School, Boston, Massachusetts 02115

David R. Zusman
Department of Molecular and Cell Biology, University of California, Berkeley, Berkeley, California 94720

Preface

Cells must sense and respond to their environment, a process that requires signal transduction across biological membranes. A major mechanism of signal transduction, widespread in bacteria, is the so-called two-component system that has adopted phosphorylation as a means of information transfer. Two-component systems are central to much of the cellular physiology that results from alterations in the environment. Starvation for phosphate or nitrogen, responses to oxygen limitation, and adaptation to new carbon and nitrogen sources are but a few of the environmental insults that cells overcome with modified cellular physiology mediated by two-component systems. Pathogenesis requires two-component modification of cellular physiology as well. There is no doubt that cells sense when they need to express virulence factors, but in most cases what is sensed remains obscure. It is unlikely that any pathogen can survive the varied and changing environments of the human body without involving at least one two-component pathway.

We are only now beginning to understand the bacterial cell cycle and the role of cell-cell communication in population dynamics and development. Yet we can cite examples in which two-component switches process signals required to trigger these events. Because two-component systems form networks that involve more than one system and show dependencies and hierarchies, they are easily adapted for very complex processes. In fact, two-component systems are so widespread, and so important, that without them bacteria would be rendered the equivalent of deaf, dumb, and blind.

In this book we have tried to highlight the global nature of two-component systems and summarize the enormous progress that has been made in less than a decade in our understanding of how these systems work. The book is divided into several sections, each of which deals with a particular aspect of two-component regulation. A few two-component systems have been studied in depth by several investigative groups, and these systems form a reservoir of information about how these systems function. Although some of the systems are complex, the two-component paradigm forms the basis for a common information flow.

Scientists studying microbial physiology, pathogenesis, motility and chemotaxis development, or a variety of other behavioral characteristics of bacteria need to be aware of and understand two-component signal transduction. The functions of two-component systems in eukaryotes such as yeasts and plants are now being appreciated, and astute investigators of these systems will take advantage of the vast knowledge base in bacteria. This book was therefore designed to appeal to the wide variety of disciplines in which signal transduction is a vital component and knowledge of its mechanism is essential.

For those of us who have witnessed the virtual explosion of information on two-component systems in the ten years since we became aware of their existence, the amount of knowledge accumulated seems enormous. Despite this progress, many fundamental issues regarding two-component systems still remain unresolved. We hope that this book will help focus attention on these critical problems and stimulate research to solve them.

The editors would like to thank Susan DiRenzo for service above and beyond the call in keeping the chapters, as well as the authors and editors, organized. We are also grateful to the ASM Press editorial staff—especially Ellie Tupper, Pamela Wilks, and Patrick Fitzgerald—for their continuing help and encouragement.

JAMES A. HOCH
THOMAS J. SILHAVY

Historical Perspective

Boris Magasanik

1

Does a phenomenon, defined a mere 8 years ago, have a history? Perhaps yes, because progress in molecular biology is indeed very rapid. In this short interval of time, two-component systems have been discovered in a very large variety of prokaryotes (see Parkinson and Kofoid, 1992), they appear to exist in eukaryotes (Chang et al., 1993; Ota and Varshavsky, 1993), and they are responsible for the control of a large variety of processes. These facts may serve as an excuse for tracing the development of our current understanding of these intricate control systems.

I shall deal first with the prehistory of this phenomenon. The fact that mutations in either one of two genes could affect the regulation of a biological system was well known before 1986. Mutations in either *crp* or *cya* of *Escherichia coli* resulted in the same phenotype: the inability to grow on many carbon compounds other than glucose (Perlman and Pastan, 1969; Zubay et al., 1970). The contributions of the products of the two genes could be distinguished by the observation that the growth defect of the *cya* mutants but not that of the *crp* mutants could be corrected by the addition of cyclic AMP (cAMP) to the growth medium, indicating that these mutants lack the enzyme necessary for the synthesis of

cAMP. The protein product of the *crp* gene was found to bind cAMP noncovalently to assume the conformation required for the activation of transcription at promoters subject to catabolite repression (Zubay et al., 1970).

Another case was the induction of histidase in *Klebsiella aerogenes* (Schlesinger et al., 1965). The inducible phenotype of the wild-type strain depends on the normal function of two genes, *hutC*, the structural gene for the repressor, and *hutU*, the structural gene for urocanase, the enzyme responsible for the degradation of urocanate, the product of histidase. Mutations causing the loss of either the repressor or urocanase resulted in the constitutive expression of histidase. The explanation for this phenomenon lies in the fact that urocanate, and not histidine, combines noncovalently with the repressor to prevent its binding to the operator (Hagen and Magasanik, 1973). Urocanase thus serves as an inducer-destroying enzyme. In its absence, urocanate, produced by the basal histidase from endogenously synthesized histidine, accumulates in the cell in sufficiently high concentration to neutralize the repressor and to allow the constitutive production of histidase.

Neither of these cases in which two protein components play essential roles in the regulation of gene expression belongs to the category of what now are called two-component sys-

Boris Magasanik, Department of Biology, Massachusetts Institute of Technology, Cambridge, Massachusetts 02139.

Two-Component Signal Transduction, Edited by James A. Hoch and Thomas J. Silhavy, © 1995 American Society for Microbiology, Washington, DC 20005

tems. Yet in these cases, the need for each of the two components is more obvious than in those that constitute the two-component systems, because mutations in the structural genes for the two components of what are now recognized as two-component systems failed to provide evidence that each of the two components plays an essential role in the expression of the regulated genes. In the paradigmatic case, the regulation of gene expression in response to the availability of a source of nitrogen, as well as in two other cases that were used to define the phenomenon, only mutations in one of the two genes, now recognized as the structural gene for the effector (response regulator), result in the loss of the response; mutations destroying the function of the other gene, now recognized as the modulator (sensor), alter the response but do not abolish it (see Albright et al., 1989; Stock et al., 1989).

Nevertheless, a specific role for the modulator was demonstrated when in the case of nitrogen regulation in *E. coli* it became possible to use a system employing only highly purified macromolecules to study the initiation of transcription at the nitrogen-regulated promoter *glnAp2* of *glnA*, the structural gene for glutamine synthetase. In addition to core polymerase and σ^{54}, the product of *glnG(ntrC)* nitrogen regulator I (NR$_I$), the effector, a protein capable of binding to sites in the vicinity of the promoter, as well as the product of *glnL(ntrB)*, NR$_{II}$, the modulator, were required for the initiation of transcription (Hunt and Magasanik, 1985). However, similar experiments using a coupled transcription translation system containing cell extract specifically lacking σ^{54}, NR$_I$, and NR$_{II}$ showed that only the former two, not the latter, had to be added to allow transcription of *glnA* to be initiated (Hirschman et al., 1985).

Together, these observations indicated that NR$_I$ by itself was unable to activate transcription and that it acquired the ability to execute this task under appropriate conditions from NR$_{II}$ but that other cell components could, to a certain extent, substitute for NR$_{II}$. In support

of this view, it could be shown that in intact wild-type cells of *E. coli*, the initiation of transcription at *glnAp2* was immediately activated by shifting the cells from a medium with excess nitrogen to a deficient medium and that it was immediately arrested when the cells were subjected to the opposite shift; but in cells lacking NR$_{II}$, both responses were very slow (Reitzer and Magasanik, 1985).

The recognition that the effectors and the modulators were members of two classes of paired regulatory proteins was the achievement of Ausubel and his coworkers (Nixon et al., 1986). They had cloned and sequenced the *ntrB(glnL)* and *ntrC(glnG)* genes of a species of *Bradyrhizobium* and found that the deduced amino acid sequences of the respective proteins were, not surprisingly, almost identical to those of the corresponding NR$_{II}$ and NR$_I$ proteins of *Klebsiella pneumoniae*. Also, they noticed homology of the N-terminal domains of the product of *ntrC(glnG)* and the products of *phoB, ompR,* and *cheY,* as well as homology of the C-terminal domain of their partners, the products of *ntrB(glnL), phorR, envZ,* and *cheA*. These results suggested a specific interaction of the C-terminal domains of the modulators with the N-terminal domains of the effectors. The nature of this interaction had just been discovered in the case of nitrogen regulation in *E. coli*. It had been shown that NR$_{II}$ activated NR$_I$ by phosphorylating it and inactivated it by removing the phosphate group; the inactivation required, in addition, the small protein P$_{II}$, the product of the *glnB* gene (Ninfa and Magasanik, 1986). It has since been shown that in every case studied so far, the modulator is responsible for the regulated phosphorylation of the effector and, in many but not all cases, also plays a role in its regulated dephosphorylation (see Parkinson and Kofoid, 1992).

The important characteristic of two-component systems is therefore the covalent modification of the effector by the modulator. It is perhaps of some historical interest to ask why such an important and widespread regulatory system was recognized only 8 years ago, particularly when the metabolic regulation of enzyme activity and of gene expression had been intensely investigated for more than 30 years. One reason may be the instability of the phosphorylated response regulators. The other reason may have been the prejudice engendered by the fundamental early studies of feedback inhibition and the induction of β-galactosidase, which showed that regulation in prokaryotes results from the noncovalent interaction of a specific protein with a small molecule (Umbarger, 1956; Pardee et al., 1959). Such an easily reversible mechanism seemed to account in a satisfactory way for the rapid change in the activity of the regulated enzymes and the expression of the regulated genes.

It was the study of the regulation of the activity of glutamine synthetase in *E. coli* that revealed covalent modification of an enzyme as an accurate and rapid mechanism for maintaining the intracellular level of glutamine at exactly the appropriate concentration (Adler et al., 1975). It could be shown that covalent modification allows more accurate control of enzyme activity than can be achieved by noncovalent interactions but at a price: the proper control of glutamine synthetase activity by adenylylation and deadenylylation requires three specific proteins, adenylyltransferase, P_{II}, and uridylyltransferase (Stadtman et al., 1980). The latter two components are also required, together with the modulator NR_{II}, for the control of the activity of the effector NR_I (Bueno et al., 1985; Ninfa and Magasanik, 1986). Indeed, the proper function of many two-component systems depends on the participation of additional proteins to bring about the regulated phosphorylation and dephosphorylation of the effectors.

Turning now to the historical period from 1986 to the present, I shall restrict my comments to the common properties of two-component systems. The mechanism of the phosphorylation reaction is now well understood. Preliminary evidence in the cases of CheA and NR_{II} indicated that these proteins catalyze the transfer of the γ-phosphate of ATP to one of their

histidine residues (Ninfa et al., 1988). This was directly shown in the case of NR_{II}, with the further demonstration of the transfer of the phosphate group from NR_{II}-phosphate to an aspartate residue of NR_I (Weiss and Magasanik, 1988). The specific histidine and aspartate residues have since been identified in these and other cases, and they are located in positions preserved in the appropriate domains of the two classes of proteins (see Parkinson and Kofoid, 1992).

An important advance in our understanding of the mechanism of phosphorylation of the effectors was the discovery that these proteins could become phosphorylated at the correct aspartate residue by incubation with acylphosphates of low molecular weight, such as phosphoramidate and acetylphosphate, but not with ATP (Lukat et al., 1992; Feng et al., 1992). Apparently, the effector has the enzymatic ability for autophosphorylation, and the role of the modulator is to supply the phosphate in a readily accessible form. It was subsequently shown that in the case of phosphate and nitrogen regulation in intact cells, the effector is phosphorylated in the absence of the modulator by acetylphosphate, a normal metabolic product (Wanner and Riesenberg, 1992; Feng et al., 1992). This finding explains why the loss of the modulators does not result in the inability of the effectors to activate the transcription of the dependent genes. Nevertheless, the ability of acetylphosphate to activate the effectors does not interfere with the correct operation of these systems in cells with functional modulators. This is explained by the fact that the modulators are also responsible for the dephosphorylation of the phosphorylated effectors in response to environmental signals.

Another important advance in our understanding of the function of effectors of gene expression was the result of the comparison of the central and C-terminal domains of these proteins (Albright et al., 1989; Stock et al., 1989). The effectors responsible for the activation of transcription at σ^{54}-dependent promoters, such as NR_I, have highly homologous central domains and contain helix-turn-helix motifs characteristic for DNA binding proteins in their C-terminal domains. Another class of effectors responsible for the activation of transcription at σ^{70}-dependent promoters, such as PhoB, share a common domain different from the central domain of the first class. CheY, the effector of the flagellar motor, lacks a C-terminal domain, and still another class of effectors with homologous C-terminal domains of yet undetermined function is known to exist. There is good evidence that the ability of the effectors of the first two classes to activate transcription at σ^{54}- and σ^{70}-dependent promoters resides, respectively, in their central and terminal domains. This conclusion is strengthened by the fact that activators of transcription of both classes are known that have the appropriate central domain but lack the common N-terminal domains that would characterize them as members of the two-component effector class. In the case of NifA, a transcriptional activator of σ^{54}-dependent *nif* genes of *K. pneumoniae,* there is good evidence that its activity does not depend on phosphorylation (Lee et al., 1993).

Another important question is how the phosphorylation of the effector endows it with the ability to exert its effect. An answer to this question comes from a study of the activation of transcription at the *glnAp2* promoter (Weiss et al., 1992). It was found that phosphorylation greatly increases the cooperative interaction of NR$_{\mathrm{I}}$ dimers to form a tetramer or higher oligomer, which in turn was identified as the agent responsible for the activation of transcription. The results of a recent study confirm these conclusions (Porter et al., 1993). It will be of great interest to discover whether oligomerization is generally the result of the phosphorylation of effectors or whether it is a special attribute of NR$_{\mathrm{I}}$.

The history of the development of our understanding of two-component systems is not finished. A great deal remains to be learned about the mechanism of the regulated dephosphorylation of the phosphorylated effectors. Another important area is the evolution of two-component systems. Apparently, the domains of the proteins evolved independently and were then combined. It is likely that effectors dependent on phosphorylation by phosphodonors of low molecular weight existed before the evolution of specific modulators. Eventually, selective advantages resulted in the evolution of the elegant and efficient two-component systems that have been identified in the past 8 years.

ACKNOWLEDGMENTS

I thank Hilda Harris-Ransom for preparing the manuscript.

The work carried out in this laboratory was supported by grant DMB-8817091 from the National Science Foundation.

REFERENCES

Adler, S. P., D. Purich, and E. R. Stadtman. 1975. Cascade control of *Escherichia coli* glutamine synthetase. Properties of the P$_{\mathrm{II}}$ regulatory protein and the uridylyltransferase-uridylyl removing enzyme. *J. Biol. Chem.* **250:**6264–6272.

Albright, L. M., E. Huala, and F. M. Ausubel. 1989. Prokaryotic signal transduction mediated by sensor and regulator protein pairs. *Annu. Rev. Genet.* **23:**311–336.

Bueno, R., G. Pahel, and B. Magasanik. 1985. Role of *glnB* and *glnD* gene products in regulation of the *glnALG* operon of *Escherichia coli. J. Bacteriol.* **164:** 816–822.

Chang, C., S. F. Krook, A. B. Bleecker, and E. M. Meyerowitz. 1993. *Arabidopsis* ethylene response gene *ETR11:* Similarity of product to two-component regulators. *Science* **262:**539–544.

Feng, J., M. R. Atkinson, W. McCleary, J. B. Stock, B. L. Wanner, and A. J. Ninfa. 1992. Role of phosphorylated metabolic intermediates in the regulation of glutamine synthetase synthesis in *Escherichia coli. J. Bacteriol.* **174:**6061–6070.

Hagen, D. C., and B. Magasanik. 1973. Isolation of the self-regulated repressor protein of the *hut* operons of *Salmonella typhimurium. Proc. Natl. Acad. Sci. USA* **70:**808–812.

Hirschman, J., P.-K. Wong, K. Sei, J. Keener, and S. Kustu. 1985. Products of nitrogen regulatory genes *ntrA* and *ntrC* of enteric bacteria activate *glnA* transcription in vitro: evidence that the *ntrA* product is a sigma factor. *Proc. Natl. Acad. Sci. USA* **82:**7525–7529.

Hunt, T. P., and B. Magasanik. 1985. Transcription of *glnA* by purified *Escherichia coli* components: core RNA polymerase and the products of *glnF,* *glnG* and *glnL*. *Proc. Natl. Acad. Sci. USA* **82:**8453–8457.

Lee, H.-S., F. Naberhaus, and S. Kustu. 1993. In vitro activity of NifL, a signal transduction protein for biological nitrogen fixation. *J. Bacteriol.* **175:**7683–7688.

Lukat, G. S., W. R. McCleary, A. M. Stock, and J. B. Stock. 1992. Phosphorylation of bacterial response regulator proteins by low molecular weight phosphodonors. *Proc. Natl. Acad. Sci. USA* **89:**718–722.

Ninfa, A. J., and B. Magasanik. 1986. Covalent modification of the *glnG* product, NR$_I$, by the *glnL* product, NR$_{II}$, regulates the transcription of the *glnALG* operon in *Escherichia coli. Proc. Natl. Acad. Sci. USA* **83:**5909–5913.

Ninfa, A. J., E. B. Ninfa, A. N. Lupas, A. Stock, B. Magasanik, and J. Stock. 1988. Crosstalk between bacterial chemotaxis signal transduction proteins and regulations of transcription of the Ntr regulon: evidence that nitrogen assimilation and chemotaxis are controlled by a common phosphotransferase mechanism. *Proc. Natl. Acad. Sci. USA* **85:**5492–5496.

Nixon, B. C., C. W. Ronson, and F. M. Ausubel. 1986. Two-component regulatory systems responsive to environmental stimuli share strongly conserved domains with the nitrogen assimilation regulatory genes *ntrB* and *ntrC. Proc. Natl. Acad. Sci. USA* **83:**7850–7854.

Ota, I. M., and A. Varshavsky. 1993. A yeast protein similar to two-component regulators. *Science* **262:**566–569.

Pardee, A. B., F. Jacob, and J. Monod. 1959. The genetic control and cytoplasmic expression of "inducibility" in the synthesis of β-galactosidase by *E. coli. J. Mol. Biol.* **1:**165–178.

Parkinson, J. S., and E. C. Kofoid. 1992. Communication modules in bacterial signaling proteins. *Annu. Rev. Genet.* **26:**71–112.

Perlman, R. L., and I. Pastan. 1969. Pleiotropic deficiency of carbohydrate utilization in an adenylylcyclase deficient mutant of *Escherichia coli. Biochem. Biophys. Res. Commun.* **37:**151–157.

Porter, S. C., A. K. North, A. B. Wedel, and S. Kustu. 1993. Oligomerization of NTRC at the *glnA* enhancer is required for transcriptional activation. *Genes Dev.* **7:**2258–2273.

Reitzer, L. J., and B. Magasanik. 1985. Expression of *glnA* in *Escherichia coli* is regulated at tandem promoters. *Proc. Natl. Acad. Sci. USA* **82:**1979–1983.

Schlesinger, S., P. Scotto, and B. Magasanik. 1965. Exogenous and endogenous induction of the histidine-degrading enzymes in *Aerobacter aerogenes. J. Biol. Chem.* **240:**4331–4337.

Stadtman, E. R., E. Mura, P. B. Chock, and S. G. Rhee. 1980. The interconvertible enzyme cascade that regulates glutamine synthetase activity, p. 41–59. *In* J. Mora and R. Palacios (ed.), *Glutamine: metabolism, enzymology and regulation.* Academic Press, Inc., New York.

Stock, J. B., A. J. Ninfa, and A. M. Stock. 1989. Protein phosphorylation and regulation of adaptive responses in bacteria. *Microbiol. Rev.* **53:**450–490.

Umbarger, H. E. 1956. Evidence for a negative feedback mechanism in the biosynthesis of isoleucine. *Science* **123:**848.

Wanner, B. L., and M. R. Riesenberg. 1992. Involvement of phosphotransacetylase, acetate kinase and acetylphosphate synthesis in control of the phosphate regulon in *Escherichia coli. J. Bacteriol.* **174:**2124–2130.

Weiss, V., F. Claverie-Martin, and B. Magasanik. 1992. Phosphorylation of nitrogen regulator I (NR$_I$) of *Escherichia coli* induces strong cooperative binding to DNA essential for the activation of transcription. *Proc. Natl. Acad. Sci. USA* **89:**5088–5092.

Weiss, V., and B. Magasanik. 1988. Phosphorylation of nitrogen regulator I (NR$_I$) of *Escherichia coli. Proc. Natl. Acad. Sci. USA* **85:**8919–8923.

Zubay, G., D. Schwartz, and J. Beckwith. 1970. Mechanism of activation of catabolite-sensitive genes: a positive control system. *Proc. Natl. Acad. Sci. USA* **66:**104–110.

GENERAL PRINCIPLES

Genetic Approaches for Signaling Pathways and Proteins

John S. Parkinson

2

Bacterial signaling proteins are built from modular components: input sensing domains; output effector domains; and transmitter and receiver domains for promoting protein-protein communication. Signaling circuits are assembled by "wiring" these elements in various configurations. This chapter discusses some genetic approaches for studying signaling pathways and for elucidating the molecular mechanisms of information processing by modular signaling proteins.

The components of a signaling pathway can be identified through genetic dissection, much like a conventional biochemical pathway. Given an appropriate phenotypic handle, brute-force screens will yield a more representative spectrum of mutants than will elegant, but less general, selection schemes. The multifunctional nature of signaling proteins can lead to complex mutant phenotypes and complementation properties, but information about the nature of the mutational lesion will help in relating mutant behaviors to functional defects. Finally, epistasis tests among signaling mutants can establish the sequence of steps in the signaling pathway.

Domain surgeries (ablation, scission, and transplantation) can provide clues to the mechanisms of interdomain communication in signaling

proteins. Liberated communication modules can disrupt normal signaling by quenching input signals, by jamming output elements with inappropriate signals, or by shielding output elements from their input signals. These pathological effects can be exploited to identify the targets of inhibition and the binding determinants that mediate signaling interactions within and between proteins.

Comparative sequence analyses and site-directed mutation can provide initial clues to the importance of particular structural features in transmitter and receiver modules. Screens and selections can be devised to look for mutants with defects in the various functional activities of communication modules (phosphorylation, dephosphorylation, and input or output control). Reversion analyses of such mutants, either through bypass or conformational suppression, can also provide valuable insight into the structure-function organization of signaling proteins.

SENSORY SIGNALING IN BACTERIA

Bacteria live in precarious environments. Nutrient and toxin levels, acidity, temperature, osmolarity, humidity, and many other conditions can change rapidly and unexpectedly. To survive, the cells must constantly monitor external conditions and adjust their structure, physiol-

John S. Parkinson, Biology Department, University of Utah, Salt Lake City, Utah 84112.

Two-Component Signal Transduction, Edited by James A. Hoch and Thomas J. Silhavy,
© 1995 American Society for Microbiology, Washington, DC 20005

ogy, and behavior accordingly. Given strong selective pressures such as these, it is no surprise that bacteria have devised sophisticated signaling systems for eliciting adaptive responses to their environment. (For recent reviews, see Bourret et al., 1991; Parkinson, 1993; Parkinson and Kofoid, 1992; Stock et al., 1990.) They readily detect minute fluctuations in many chemical and physical conditions, which in turn trigger changes in gene expression or motility that enhance survival prospects. The sensory machinery underlying these behaviors handles signaling tasks fundamental to all cell sensory systems: stimulus detection; signal processing, including amplification and integration of sensory inputs; and production of appropriate output responses. The sensory systems of prokaryotes provide tractable models for exploring these events in molecular detail and have begun to reveal general principles of cellular signaling mechanisms.

Bacterial signaling systems are amenable to detailed genetic and biochemical analyses. This chapter focuses on genetic methods; biochemical studies are covered separately (see Chapter 3). Readers should appreciate that these tools are not mutually exclusive but rather complementary. A combination of genetic and biochemical approaches undoubtedly offers the most incisive experimental strategy for elucidating sensory pathways and signaling mechanisms. Genetic methods are uniquely valuable, however, for identifying the components of a signaling pathway and for determining the sequence in which they act. Moreover, simple genetic tests can shed considerable light on the information-processing mechanisms of signaling proteins. In this chapter, rather than reviewing genetic studies of specific signaling systems, which are amply covered elsewhere in this book, some general strategies for using genetic methods to study sensory pathways and signaling proteins are discussed.

TWO-COMPONENT PARADIGM

Many signaling proteins, from both gram-positive and gram-negative bacteria, contain characteristic "transmitters" and "receivers," domains that promote information transfer within and between proteins (Parkinson and Kofoid, 1992). Similar communication modules are now turning up in eukaryotic signaling proteins, indicating that this could be a fundamental and widespread strategy for building signaling circuits. Transmitters and receivers function in combination with a variety of input and output domains and can be arranged in different configurations to build signaling circuits of many types. The simplest circuits have two protein components—a sensor, often located in the cytoplasmic membrane, that monitors some environmental parameter; and a cytoplasmic response regulator that mediates an adaptive response, usually a change in gene expression (Fig. 1). Sensors typically contain a C-terminal transmitter module coupled to an N-terminal input domain. Response regulators typically contain an N-terminal receiver module coupled to one or more C-terminal output domains. On detecting a stimulus, the input

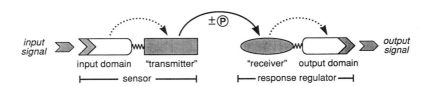

FIGURE 1 "Two-component" paradigm for sensory signaling via communication modules. Sensory information flows through noncovalent controls exerted by one domain on another (dashed arrows) and through phosphorylation reactions between transmitter and receiver domains. The convention of representing transmitters by rectangles and receivers by ovals is used in all subsequent figures.

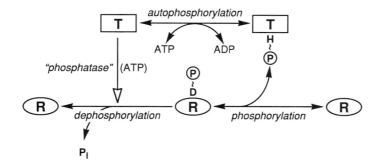

FIGURE 2 Phosphorylation activities of transmitters and receivers. Abbreviations: T, transmitter; R, receiver; H, histidine; D, aspartic acid; P_i, inorganic phosphate. Details of the phosphorylation reactions are discussed in the text. ATP is required for the "phosphatase" activity exhibited by some transmitters but is not hydrolyzed in the reaction.

domain of a sensor modulates the signaling activity of its associated transmitter to communicate with its response regulator partner. The receiver domain of the response regulator detects the incoming sensor signal and then alters the activity of its associated output domain to trigger the response.

Phosphorylation Activities of Transmitters and Receivers

The only demonstrated mechanisms of transmitter-receiver communication involve phosphorylation and dephosphorylation reactions (Fig. 2). Transmitters have an autokinase activity that attaches phosphate groups from ATP to a histidine residue. This reaction is readily reversible. The product phosphohistidine serves as a high-energy intermediate for subsequent transfer of the phosphate group to an aspartate residue in the receiver. Although formally reversible, the phosphotransfer step is effectively unidirectional. The receiver probably catalyzes the transfer reaction, with the transmitter phosphohistidines simply serving as substrates. Receivers also catalyze hydrolytic loss of their phosphate groups, with half-lives ranging from a few seconds to many minutes. Finally, transmitters can also have an apparent phosphatase activity toward their cognate receivers. It is not yet clear whether this reaction is catalyzed by the transmitter or whether the transmitter acts as an allosteric effector to enhance the intrinsic dephosphorylation ability of its target receiver.

Signaling Properties of Transmitters and Receivers

Most transmitter-containing proteins are located in the cytoplasmic membrane, with their transmitters projecting into the cell. They typically have two membrane-spanning segments flanking their input domain, which is consequently deployed in the periplasmic space between the inner membrane and cell wall. Sensor input domains differ broadly in primary structure, reflecting the variety of chemical and physical stimuli they detect. Some have demonstrable ligand binding functions, but most are still poorly characterized, often because the exact nature of the stimulus is unknown. Communication with the cytoplasmic transmitter domain involves propagation of sensory information across the cytoplasmic membrane, presumably via stimulus-induced conformational changes. A few sensor proteins are soluble and contain N-terminal domains that may serve comparable input roles.

Receiver-containing proteins are generally cytoplasmic. In most cases, their output domains have DNA binding or other regulatory functions that provide transcriptional control over one or more target genes. The receiver and output domains in response regulators are often joined by flexible linkers, suggesting that pliable connections may be important in enabling a receiver to exert control over its adjoining output domain.

Transmitters and receivers are ideally suited as circuit elements for assembling signaling pathways. The interplay of kinase and phosphatase activities in a transmitter is subject to input control, enabling the transmitter to regulate the phosphorylation state of its cognate receiver in response to sensory signals. The phosphorylation state of the receiver in turn controls the activity of an adjoining output domain to trigger the behavioral response. The signaling characteristics of module-based circuits thus depend on several parameters. The most critical ones include the basal and stimulated phosphotransfer rates between transmitter and receiver; the lifetimes of activated transmitters and receivers; the ways in which these elements are wired together; and the extent of crosstalk from other signaling pathways.

Although the signaling attributes of transmitters and receivers are now apparent, their activities are still poorly understood at the molecular level. Consequently, the overall logic and information-processing properties of many module-based signaling circuits remain mysterious. Even the simplest sensor-response regulator pathways must carry out four discrete communication transactions (Fig. 3): stimulus detection by the input domain (Fig. 3a); input control over transmitter activity (Fig. 3b); transmitter-receiver communication (Fig. 3c); and receiver control over output activity (Fig. 3d). None of these events is well understood:

- How does stimulus detection alter the conformation of the input domain? How does the stimulated input domain communicate with the adjoining transmitter? Is the segment linking the input and transmitter domains important in this process?

- How do transmitter conformational changes alter its kinase, phosphotransfer, and phosphatase activities? Does the receiver catalyze the phosphotransfer reaction? Which component catalyzes the "phosphatase" reaction?

- What confers signaling specificity to transmitter-receiver transactions? How do matching modules recognize one another? How do they avoid unwanted crosstalk? Does phosphorylation modulate their binding interactions?

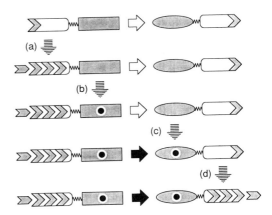

FIGURE 3 Signaling transactions in sensor-response regulator circuit. (a) Generation of a conformational change in the sensor input domain on detection of an input signal; (b) modulation of the autophosphorylation-phosphatase activities of the transmitter by the stimulated input domain; (c) communication between transmitter and receiver via specific docking and phosphotransfer; (d) stimulation or inhibition of the response regulator output domain on a change in phosphorylation state of the receiver.

- How does phosphorylation alter the conformation of the receiver domain? How does this conformational change regulate the activity of the adjoining output domain? Does the receiver directly contact the output domain? What structural features determine the lifetime of the phosphorylated receiver?

Genetic approaches can provide mechanistic answers to these sorts of questions. First, general strategies for identifying and ordering the components of a signaling pathway are described. Then, some specific experimental schemes for elucidating the communication transactions within and between signaling proteins are outlined. The ensuing discussion and illustrative examples deal almost exclusively with an idealized two-component signaling circuit such as that shown in Fig. 3, in which the sensor detects a small-molecule stimulus and the response regulator controls the expression of a nonessential gene.

GENETIC ANALYSIS OF SIGNALING PATHWAYS

Signaling circuits are amenable to genetic dissection in much the same manner as conventional biochemical pathways. In both cases, gene products participate in an ordered series of reactions whose end product, an essential metabolite on the one hand or a behavioral response on the other, influences the organism's phenotype. A defect at any step in the sequence should block the pathway and alter its phenotypic outcome, permitting identification of the responsible gene products through mutants. However, as discussed below, the inherent complexity of biological signaling processes poses some special challenges to genetic analysis.

Identifying the Pathway Elements

GETTING A GRIP ON BEHAVIOR

The phenotypic consequences of a signaling process are often far removed from the underlying molecular events. Some signaling systems elicit discrete regulatory responses to relatively simple stimuli (e.g., a change in porin synthesis on osmolarity shift [see Chapter 7]), but others trigger elaborate developmental programs in response to complex stimuli (e.g., induction of sporulation by starvation conditions [see Chapter 8]). The apparent speed of the overall signaling process can range from fractions of a second (e.g., locomotor responses to chemotactic signals [see Chapter 6]) to many hours (e.g., fruiting body formation in *Myxococcus* [see Chapter 27]). Simple, easily assayed phenotypes can greatly facilitate isolation and characterization of mutants with behavioral defects. For example, gene expression reporters (e.g., promoter fusions to β-galactosidase) provide especially convenient phenotypic handles for following the consequences of regulatory signaling transactions. Unfortunately, these tools are not often applicable to tactic behaviors, which depend on a locomotor apparatus for phenotypic expression.

MUTANT SCREENS OR SELECTIONS?

Selection schemes simplify the process of isolating mutants but are not necessarily the most effective way to dissect a signaling pathway.

Selections based on a special attribute of the desired mutant phenotype could easily bias the kinds of mutants obtained. For example, *Escherichia coli* mutants that cannot tumble while swimming are nonchemotactic. Because they always swim forward, such mutants move faster down a vertical race course than wild-type cells, which tumble fairly frequently. However, taking the winners does not yield a representative spectrum of nonchemotactic mutants because those with other swimming patterns (e.g., excessively tumbly) get overlooked. Brute-force screens based on a more general phenotype (e.g., colony morphologies on motility plates) afford more comprehensive mutant hunts. As a general rule, mutant screens are preferable to selections whenever the desired mutants are reasonably frequent or there is an abundant supply of labor.

Most adaptive behaviors are not essential for viability, at least in the laboratory. In general, then, mutations that block a response pathway should not be lethal, unless the signaling components play other vital roles. Therefore, it should be possible to identify many of the key elements in a signaling pathway through loss-of-function mutants induced by transposon insertions or other knockout mutations. However, to reconstruct the pathway through epistatic analysis (see below), some gain-of-function mutants are needed. If null mutants fail to respond to the stimulus, gain-of-function mutants would show constitutive, stimulus-independent responses. Mutations that activate a signaling pathway will probably be relatively uncommon, but their potential value justifies devising special screens or selections to get them.

Determining the Functional Defects of Signaling Mutants

CAUTION: MULTIFUNCTIONAL PROTEINS

The phenotypes of signaling mutants may provide misleading clues about their underlying functional defects. Signaling proteins must perform several different functions to serve as information-processing devices. These include recognition and docking with other signaling components, catalysis of phosphorylation or dephosphorylation reactions, presentation of the substrate sites involved in those reactions, and control of these activities in response to input signals. Because signaling proteins are inherently multifunctional, genetic lesions can cause a variety of functional defects. Knockout or null mutants would be expected to exhibit complete loss-of-function phenotypes, but missense mutants may well have residual activities that confound their defective phenotypes. In attempting to relate mutant phenotypes to functional defects, it is extremely helpful to know the nature of the structural lesion in the mutant protein. This can be easily determined by DNA sequencing or sometimes inferred from the mutagenic agent used. Without this information, it is foolhardy to offer more than a superficial interpretation of mutant phenotypes when dealing with signaling proteins.

COMPLEMENTATION ANALYSIS OF SIGNALING MUTANTS

The complementation behavior of a mutant reflects the severity of its functional defect. Null mutants should be recessive; gain-of-function mutants should be dominant. The multifunctional nature of signaling proteins is likely to complicate the complementation properties of signaling mutants. Defects in any one subfunction could lead to dominant negative behavior caused by the residual activities in the mutant protein. A catalytic defect, for example, might not prevent docking interactions with signaling partners, enabling the mutant protein to act as a "spoiler" that interferes or competes with its normal counterpart in complementation tests. A similar spoiling effect can occur on interaction of mutant and wild-type protomers to form inactive oligomers. The severity of dominance should depend on the relative stoichiometry of the mutant subunits, so it may be possible to alleviate much of the effect by adjusting gene dosages.

Partially dominant mutants can provide unique subjects for subsequent mutational studies. If a mutant protein disrupts signaling in wild-type cells, it should be possible to look for second-step mutants that are more or less dominant. In the example above, a signaling protein with a catalytic defect might block signal propagation by titrating a target protein. Mutant proteins that block signaling at lower expression levels might have enhanced affinity for their target. Conversely, mutant proteins with diminished effectiveness might have reduced affinities. Affinity mutants could serve to identify the structural determinants involved in docking interactions and signaling specificity (see below).

Reconstructing the Signaling Pathway

LITTLE HOPE FOR PATHWAY INTERMEDIATES

It is often possible to elucidate the stepwise reactions in a conventional biochemical pathway by identifying the precursor compounds that accumulate in different mutants. In principle, pathway intermediates should also accumulate in a blocked signaling circuit. For example, a lesion in the receiver module of a response regulator should, in the presence of an activating stimulus, cause a buildup of the autophosphorylated form of the transmitter module in the sensor. Such intermediates might be ephemeral and, in any case, would be difficult to detect unless they conferred an aberrant phenotype, for example, through inappropriate crosstalk. So, unlike biosynthetic pathways, signaling intermediates would probably be of little use in deducing the sequence of transactions in a communication pathway.

EXPLOITING EPISTASIS

Epistasis is a dominance relationship between nonallelic genes, assessed at the level of their mutant phenotypes. The order in which gene products act in a signaling pathway can be determined from their epistatic interactions. The test is simple: if mutations in two different genes produce two different phenotypes, which phenotype prevails in a double mutant? In signaling pathways, where the scored phenotype is an output response, the test is typi-

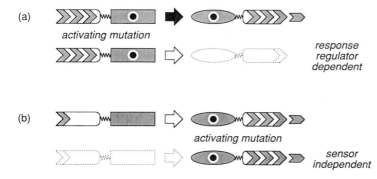

FIGURE 4 Ordering components of signaling pathways by epistasis tests. (a) An activating mutation in the sensor requires a functional response regulator to produce a constitutive output signal. (b) An activating mutation in the response regulator does not require a functional sensor to produce a constitutive output signal.

cally performed by combining gain-of-function and loss-of-function defects (Fig. 4). First, consider an activating mutation in the sensor that enables it to generate transmitter signals with no stimulus input, causing a constitutive output response. When combined with a null defect in the cognate response regulator, the aberrant output ceases, demonstrating that the null mutation affects a component required for the constitutive response (i.e., a later step in the signaling pathway) (Fig. 4a). By contrast, an activating mutation in the response regulator should enable it to generate a constitutive output signal independent of its cognate sensor, demonstrating that the sensor defect blocks an earlier step in the signaling pathway (Fig. 4b).

With caution, the logic of epistatic analysis can be extended to more elaborate signaling pathways that have branches, feedback loops, and so on. However, it is essential at the outset to know the general nature of the mutant defects (gain- or loss-of-function). If loss-of-function mutations in different genes produce dissimilar phenotypes, either the pathway or the phenotypic handle must be more complicated than the simple example shown in Fig. 4.

DOMAIN INTERACTIONS IN SIGNALING PROTEINS

Input-Output Communication within Signaling Proteins

The primary input and output functions of sensors and response regulators are carried out by different domains joined through flexible linkers. How do the input and output modules within a signaling protein communicate with one another? On the one hand, the input domain might make specific direct contact with the output domain to stimulate or inhibit its activity. On the other hand, the input domain might control output activity indirectly by manipulating the subunit organization or overall conformation of the protein. Indirect control mechanisms would not require specific contacts between domains, although the segment connecting them might play an important role.

The modular design of signaling proteins suggests three conceptually simple genetic tests for distinguishing between direct and indirect mechanisms of interdomain communication (Fig. 5). All three approaches involve wholesale surgery on the signaling protein: ablation of the input domain (Fig. 5a); scission of the input and output domains (Fig. 5b); and transplantation of foreign domains (Fig. 5c). These genetic alterations can be readily accomplished by in vitro methods but may lead to postoperative complications. In domain ablation and scission experiments, a change in expression level or stability of the modified proteins could confound interpretation of their signaling properties. In domain transplantation experiments, the length and flexibility of the linker segment might prove critically important for proper domain interactions. Because such experiments could fail for a variety of reasons, it would be unwise to draw conclusions from any surgical operations that produce negative results.

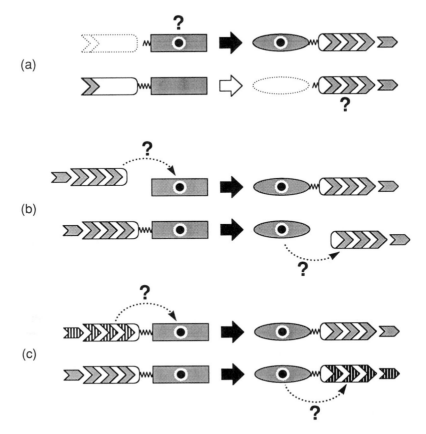

FIGURE 5 Analyzing intraprotein communication mechanisms by domain surgery. (a) Does ablation of the input domain activate the adjoining output domain? (b) Does scission of input and output domains disrupt signal propagation? (c) Does transplantation of foreign domains disrupt signal propagation?

ABLATIONS

Ablation experiments can determine the general manner in which an input domain exerts control over an adjoining output domain (Fig. 5a). If the control mechanism is negative, removal of the input domain might lead to a stimulus-independent output response. This would imply that the input domain normally inhibits output activity and that interaction of the stimulus with the input domain alleviates the inhibition. Negative controls appear to be common in sensor proteins because removal of their input domain often leads to activation of transmitter kinase activity. However, this control strategy seems to be less

prevalent in response regulators, where removal of the receiver seldom leads to activation of the output domain.

SCISSIONS

Scission experiments can establish whether the control mechanism requires a physical connection between the input and output domains (Fig. 5b). If the input domain makes direct specific contact with the output domain, those binding interactions might still occur when the linker is severed, provided that the concentrations of the untethered domains can be adjusted to compensate for the loss of a covalent connection. By contrast, domain scission would definitely abrogate control mediated through nonspecific mechanisms such as monomer-dimer transitions. Indirect controls are probably common in sensors, many of which are transmembrane proteins whose input domain is in the periplasmic space, presumably incapable of contacting the cytoplasmic transmitter. Few scission experiments have been reported with response regulators, and those have been negative. Whether receivers can communicate with adjoining domains through specific contacts is still an open issue.

TRANSPLANTATIONS

Transplantation experiments also address the mechanism of communication between input and output domains but can provide information complementary to that from scission approaches (Fig. 5c). Input and output domains that fail to communicate when physically disconnected could conceivably interact through nonspecific mechanisms. If so, combinations of heterologous domains that use the same control strategy might communicate properly. This approach has succeeded with the EnvZ sensor, which modulates OmpR phosphorylation state in response to changes in medium osmolarity. When coupled to the sensing domain of the aspartate-maltose (Tar) or ribose-galactose (Trg) chemoreceptor, the EnvZ transmitter is able to modulate OmpR activity in response to the appropriate chemoeffector (Baumgartner et al., 1994; Utsumi et al., 1989). Because EnvZ, Tar, and Trg are transmembrane proteins, their input and output domains cannot contact one another directly but instead must communicate through the membrane-spanning segments of the protein. The signaling properties of the hybrid proteins imply that the chemosensing domains of Tar and Trg use the same conformational control mechanisms as the osmosensing input domain of EnvZ. There have been several transplantation attempts with response regulators, all with negative outcomes.

Signaling Transactions between Transmitters and Receivers

E. coli probably contains at least 50 transmitter-receiver pairs and nearly as many signaling circuits. Inappropriate crosstalk between them is minimal, implying that receivers are precisely tuned to their cognate transmitters. High-fidelity signaling presumably derives from specific binding interactions between transmitters and receivers, but their structural basis is not yet understood. Domain liberation approaches can provide useful experimental subjects for exploring the process of target recognition and the ensuing phosphotransfer reactions in molecular detail.

Domain liberation is a general method of identifying functional subdomains within proteins (Morrison and Parkinson, 1994). The approach is based on the premise that protein domains invariably function through specific interactions with some partner, either a small molecule, another macromolecule, or another part of the same protein. When subcloned and overexpressed, a liberated domain should compete with its counterpart in the intact protein, disrupting its activity. This could happen in several ways, for example, through formation of nonfunctional hetero-oligomers with the parent protein, through stoichiometric titration of a common interaction target, or through creation of an aberrant or unregulated catalytic activity. In the case of two-component signaling pathways, liberated transmitter or receiver domains could conceivably disrupt communication in three different ways (Fig. 6).

QUENCHING

Liberated receiver domains could attenuate communication between a sensor and its response regulator target by intercepting incoming transmitter signals (Fig. 6a). To quench signal flow in this manner, the freed receivers must be incapable of exerting control over the activity of the output domain in the response regulator. This appears to be the case in the EnvZ-OmpR system, where expression of the OmpR receiver leads to disruption of the

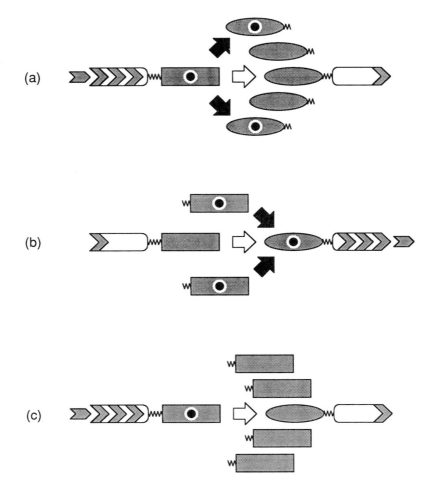

FIGURE 6 Analyzing transmitter-receiver interactions by domain liberation. (a) Quenching of transmitter signals by liberated receivers; (b) jamming signals from liberated, constitutively active transmitters; (c) shielding of receivers by liberated quiescent transmitters.

wild-type regulatory responses (Nakashima et al., 1991).

JAMMING
Removal of the input domain from a transmitter may lead to activation of its output signal. If so, the liberated transmitter should interfere by overloading its response regulator target with inappropriate signals, causing a constitutive behavioral response (Fig. 6b). The resultant phenotype should resemble that of sensor or

response regulator mutants with an activating mutation. It should be possible to identify liberated transmitters that are capable of signal jamming by simply looking for transmitter clones that cause a constitutive output phenotype in wild-type cells. This approach has been used to identify the signaling domain of Tsr, the serine chemoreceptor in *E. coli* (Ames and Parkinson, 1994). When liberated, the wild-type Tsr signaling domain exhibits constitutive signaling activity that jams chemotactic responses.

SHIELDING

Transmitters that are inactive on liberation could still block signal flow by docking with target receivers in response regulators, thereby shielding them from their sensor partners (Fig. 6c). Liberated transmitters need only retain the ability to recognize and bind to their receiver targets to disrupt communication by this mechanism. For example, mutant signaling domains from the Tsr chemoreceptor that cannot activate the signaling pathway still block chemotaxis by shielding intracellular signaling components from interactions with other chemoreceptors (Ames and Parkinson, 1994).

USING LIBERATED DOMAINS

The pathological effects produced by a liberated signaling domain provide a genetic handle for exploring its normal signaling role. The clone that expresses the liberated domain can be treated much like any other gene, except that its function is to disrupt normal lines of communication in the cell. The component that serves as the target of the inhibitory effect can be identified through gene dosage and epistasis tests. Also, mutations that alter the inhibitory properties of the liberated domain could be isolated to identify the structural determinants involved in its interaction with target proteins. If the inhibitor is under regulatable control, its expression could be reduced to look for mutations that enhance potency, or increased to look for mutations that reduce potency. Several uses for such docking affinity mutants are described in the next section. Clones that express functional domains from a signaling protein also provide useful material for biochemical studies (e.g., measurements of binding and catalytic activities and structural determinations).

STRUCTURE-FUNCTION STUDIES OF COMMUNICATION MODULES

Although their structures are undoubtedly very different, transmitters and receivers have remarkably similar functional properties that enable them to serve as signal-processing devices. First, both types of modules have enzymatic activities, primarily autophosphorylation but sometimes dephosphorylation as well. To carry out these reactions, they must have determinants for substrate binding and catalysis and an autophosphorylation site composed of the target residue and the determinants that promote its interaction with the catalytic center. Second, these enzymatic activities are subject to input control and, in turn, regulate output activities. The input determinants of a transmitter modulate its autophosphorylation activity in response to stimulus-induced conformational changes in an adjoining input domain. Its output determinants enable it to communicate with its receiver partner through docking and phosphotransfer transactions. The input determinants of a receiver promote recognition and docking with its transmitter partner. On phosphorylation, its output determinants trigger conformational changes that enable it to control an adjoining output domain. Thus, both transmitters and receivers possess determinants for modulating autophosphorylation activity in response to input signals and for producing output signals that are regulated by phosphorylation state.

The communication functions of transmitters and receivers are still poorly understood in molecular terms. The genetic approaches outlined below can serve to identify structural features important for signal input, processing, and output activities.

Sequence Analyses

The amino acid sequence of a transmitter or receiver, inferred from DNA sequence data, can provide useful clues to its structural and functional organization. The sequence data base of "two-component" systems contains hundreds of transmitter- and receiver-containing proteins. Sequence comparisons within these extended module families have revealed conserved residues and other motifs that probably play important roles in their signaling functions (Fig. 7).

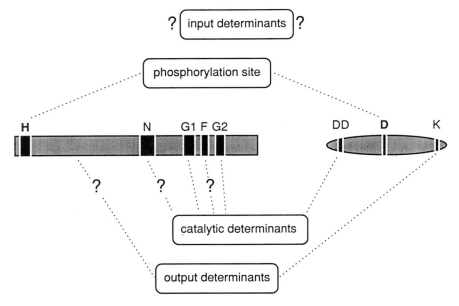

FIGURE 7 Structure-function relationships in transmitters and receivers. Sequence motifs characteristic of transmitters and receivers are indicated by black bars whose widths are proportional to the lengths of the motifs. Each sequence tract is labeled with a letter or two indicating their most prominent amino acid residue. Possible functions for some of these structural features are discussed in the text.

Transmitters are about 240 amino acids in length. Essentially nothing is known about their secondary or tertiary structures, but their primary structures exhibit several blocks of nearly invariant residues (Fig. 7). The histidine phosphorylation site is typically located near the N terminus of the module. The residues flanking the phosphorylation site are not highly conserved, suggesting they may not play important roles in interactions with the catalytic center. Rather, these flanking sequences and other variable blocks in the amino half of the transmitter define transmitter subfamilies whose receiver partners are also similar to one another. These regions might contain the specificity determinants for properly identifying receiver targets, but there is no experimental evidence bearing on this idea. Four blocks of residues in the carboxyl half of the transmitter may comprise the catalytic center. Two of these segments (G1 and G2) are glycine-rich and resemble nucleotide binding motifs seen in other proteins. The sequences of the other two segments (N and F) reveal little about their possible structures or functions.

Receiver modules are roughly 120 amino acids in length. Unlike transmitters, receiver structure is well understood (see Chapter 4). Extrapolating from X-ray and nuclear magnetic resonance studies of CheY, all receiver modules are probably α/β-barrels with five sets of alternating β-strands and α-helices. The β-strands align to form a hydrophobic inner core with the α-helical segments wrapped around the outside of the molecule. Prominent sequence features of receivers include a pair of aspartates near the amino terminus, a lysine near the carboxyl terminus, and a centrally located aspartate (Fig. 7). The three aspartate residues comprise an acid pocket at one end of the barrel, into which the side chain of the conserved lysine protrudes. The central aspartate is the site of phosphorylation, whereas the amino-terminal pair are probably important for catalysis. The conserved lysine may be involved in effecting the phosphorylation-induced conformational changes that regulate output activity.

The primary structures of communication modules have furnished few clues about their input determinants. In receivers, it seems likely that residues in or near the acid pocket could be involved in docking interactions with transmitters, but distal residues might also play roles in input specificity. The external α-helices, for example, could contain additional contact sites for transmitter binding. The input determinants in transmitters are even less apparent but could be involved in controlling interactions between transmitter subunits. Transmitters probably function as dimers, with the catalytic site of one subunit phosphorylating the acceptor site in the other. The fact that many sensors are transmembrane proteins suggests that transmitters are designed to be controlled by long-distance conformational changes rather than by direct contact with an input domain. Conformational changes that alter the spatial orientation of the two subunits could provide the basis for input control of transmitter autophosphorylation.

Mutant Modules

To ascribe communication functions to particular structural features in a transmitter or receiver, mutants are needed that affect specific signaling activities while leaving others intact. They are likely to be partially dominant, possibly leaky, and relatively rare, necessitating a variety of isolation schemes to obtain a representative spectrum of functional defects. Single amino acid replacements, generated by site-directed or random mutagenesis, are best suited for this purpose, but in special circumstances, more drastic structural changes (e.g., deletions and substitution chimeras) might also be informative. The ensuing discussion pertains mainly to missense mutants and assumes the existence of phenotypic or biochemical assays for the various functions under study (input control, catalysis, output control, and so on).

Site-directed mutation, often the genetic equivalent of turning gold into lead, is nevertheless a gratifying way to obtain an initial collection of signaling mutants. For example, alterations can be engineered to test the functional importance of particular structural features in a transmitter or receiver. Popular targets for amino acid replacements are the conserved residues believed to comprise the substrate or catalytic sites (Fig. 7). Such mutant proteins invariably fail to function in vivo and, if examined in vitro, cannot be phosphorylated. This demonstrates that the target residue is important but not what its functional role might be. The amino acid change could conceivably affect the synthesis, folding, stability, or activity of the protein. Distinguishing these possibilities calls for additional biochemical tests. Perversely, site-directed mutation reveals more about a protein when it fails to yield an expected loss-of-function result. If the mutant protein still supports normal in vivo behavior, then the altered residue must not be important for any of its signaling functions. In principle, one could identify all the protein's functionally important residues in this manner, but there are better ways to go about this.

It should be possible to devise isolation schemes for obtaining particular kinds of signaling mutants. When seeking mutants with a specific functional change, try to exploit the anticipated properties and residual activities of the mutant protein to eliminate unwanted types (e.g., complete loss-of-function). Or start the hunt with a mutant that limits the spectrum of possible functional changes. A few general examples will illustrate the possibilities. (i) To identify structural determinants involved in protein-protein interactions, first look for mutants with increased, rather than reduced, target binding affinity. Lesions of this sort should enhance the inhibitory potency of liberated modules or module parts (e.g., phosphorylation site peptides) and might cause dominant signaling defects in intact proteins. They can provide a structural foothold for exploring the interaction surface by conformational suppression (see below). (ii) To identify structural determinants involved in input control, look for dominant mutants with stimulus-independent constitutive output signals. (iii) To identify structural determinants involved in autophosphorylation or phosphotransfer activity, begin with a constitutively active mutant and look for mutations that amplify its output signals. If the constitutive alteration fully short-circuits input control, any increases in output activity should come through enhanced efficiency of a subsequent signaling function.

Reversion Analysis

The ways in which a mutant protein regains function can reveal a lot about its structure-function organization. In addition to back mutation and informational suppression, which are irrelevant in the present context, a variety of functional suppression mechanisms could conceivably alleviate a missense defect in a signaling protein. The suppressors either restore activity to the mutant protein or create an alternative function that bypasses the defective signaling step. Here only the most salient features of these two general suppression mechanisms and some specific examples of each are discussed.

BYPASS SUPPRESSION

Bypass suppressors are gene-specific but not allele-specific. They act on knockout mutations (e.g., deletions) as well as less drastic lesions because they make no use of the defective protein. Consider, for example, a null defect in a sensor that mediates a regulatory response to a particular stimulus, say osmolarity changes. "Revertants" that exhibit appropriate behavioral responses to osmolarity shifts could conceivably arise through bypass mutations in other signaling proteins. Another osmosensor might acquire the ability to communicate with the response regulator partner of the missing sensor, or the response regulator of another osmosensing circuit might acquire the ability to control the signal output of the defective pathway. The mechanisms of bypass suppression are as varied as the imagina-

tion. Except for general principles, lessons learned with one signaling pathway will not apply to another. However, understanding how a particular bypass suppressor operates can provide fresh insights into a signaling pathway and its elements.

CONFORMATIONAL SUPPRESSION

Conformational suppressors restore activity to the mutant protein through compensatory structural alterations. They cannot correct null defects because the mutant gene product is essential to the suppression mechanism. Conformational suppressors act through stereospecific contacts within and between proteins and, consequently, are highly allele-specific. These types of compensatory changes can arise at secondary sites within the mutant protein or in another protein that interacts with it. The example in Fig. 8 shows the kinds of conformational suppressors that might be obtained by reverting a mutant with a missense defect in a transmitter module.

Transmitter defects in the catalytic center, phosphorylation site, or other regions important for autokinase activity might be most easily corrected by additional structural changes within the transmitter domain (Fig. 8a). For example, a mutation that distorts the catalytic center might regain autophosphorylation ability through a nearby change that alleviates the distortion, through alterations of the phosphorylation site that enable it to access the mutant catalytic center, or through other conformational changes that influence the interaction between the phosphorylation site and catalytic center. Transmitter defects in input or output determinants should also be correctable by structural changes within the domain. However, alteration of the communication partner might also compensate for such defects. Input defects might be suppressed by structural changes in the adjoining input domain (Fig. 8b); output defects might be suppressed by changes in the receiver domain of the response regulator target (Fig. 8c).

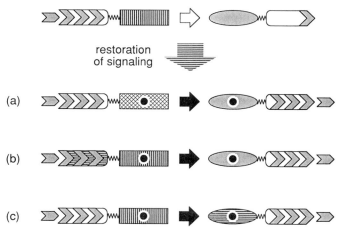

FIGURE 8 Functional suppression of signaling defects. The starting circuit has a missense mutation in the transmitter that interrupts the signal pathway. Three different suppression mechanisms that may be represented among phenotypic revertants are shown: (a) mutations within the mutant domain that compensate for a defect in folding, stability, or signaling function; (b) mutations in another domain of the same protein that restore normal communication with the mutant domain; (c) mutations in another signaling component that restore normal communication with the mutant protein.

TAKE-HOME LESSON

Genetic approaches can provide considerable insight into the operation of signaling pathways and proteins. Even though actual signaling circuits are unlikely to be as simple as the two-component examples in this chapter, the same basic principles should apply. Extension of these ideas to real-world situations is left as an exercise for the reader.

REFERENCES

Ames, P., and J. S. Parkinson. 1994. Constitutively signaling fragments of Tsr, the *E. coli* serine chemoreceptor. *J. Bacteriol.* **176:**6340–6348.

Baumgartner, J. W., C. Kim, R. R. Brissette, M. Inouye, C. Park, and G. L. Hazelbauer. 1994. Transmembrane signalling by a hybrid protein: communication from the domain of chemoreceptor Trg that recognizes sugar-binding proteins to kinase/phosphatase domain of osmosensor EnvZ. *J. Bacteriol.* **176:**1157–1163.

Bourret, R. B., K. A. Borkovich, and M. I. Simon. 1991. Signal transduction pathways involving protein phosphorylation in prokaryotes. *Annu. Rev. Biochem.* **60:**401–441.

Morrison, T. B., and J. S. Parkinson. 1994. Liberation of an interaction domain from the phosphotransfer region of CheA, a signaling kinase of *E. coli. Proc. Natl. Acad. Sci. USA* **91:**5485–5489.

Nakashima, K., K. Kanamaru, H. Aiba, and T. Mizuno. 1991. Osmoregulatory expression of the porin genes in *Escherichia coli:* evidence for signal titration in the signal transduction through EnvZ-OmpR phosphotransfer. *FEMS Microbiol. Lett.* **66:**43–47.

Parkinson, J. S. 1993. Signal transduction schemes of bacteria. *Cell* **73:**857–871.

Parkinson, J. S., and E. C. Kofoid. 1992. Communication modules in bacterial signaling proteins. *Annu. Rev. Genet.* **26:**71–112.

Stock, J. B., A. M. Stock, and J. M. Mottonen. 1990. Signal transduction in bacteria. *Nature* **344:**395–400.

Utsumi, R., R. E. Brissette, A. Rampersaud, S. A. Forst, K. Oosawa, and M. Inouye. 1989. Activation of bacterial porin gene expression by a chimeric signal transducer in response to aspartate. *Science* **245:**1246–1249.

Two-Component Signal Transduction Systems: Structure-Function Relationships and Mechanisms of Catalysis

Jeffry B. Stock, Michael G. Surette, Mikhail Levit, and Peter Park

3

Two-component systems are signal-transducing ATPases that use energy released from ATP hydrolysis to effect responses to changing environmental conditions (for reviews, see J. B. Stock et al., 1989; Parkinson, 1993; Bourret et al., 1991). The chemistry involves three phosphotransfer reactions and two phosphoprotein intermediates:

(i) $\quad$ ATP + His $\leftrightarrow$ ADP + His~P

(ii) $\quad$ His~P + Asp $\leftrightarrow$ His + Asp~P

(iii) $\quad$ Asp~P + H_2O $\leftrightarrow$ Asp + Pi

    (i) The γ-phosphoryl group in ATP is first transferred to a histidine side chain.

   (ii) The phosphoryl group is then transferred from the phosphohistidine residue to an aspartate side chain.

  (iii) Finally, the phosphoryl group is transferred from the phosphoaspartate residue to water.

These reactions work to transduce information. The ATP-dependent phosphorylation of histidine is generally regulated in response to environmental signals by a family of histidine protein kinases. The phosphorylated aspartate is contained within another type of protein termed a *response regulator*, which undergoes a phosphorylation-induced conformational change that serves to elicit a response. Thus, there are two essential families of proteins that function in pairs, the sensory kinases and their associated response regulators.

HISTIDINE PROTEIN KINASES

The histidine protein kinase superfamily is characterized by conserved sequences that generally extend for approximately 200 residues and correspond to a domain that binds ATP and catalyzes the phosphorylation of a histidine side chain (Fig. 1). This histidine kinase domain is invariably flanked by sequences that are not conserved within the family and supply specific regulatory functions. Most commonly, there is an N-terminal extension with stretches of hydrophobic amino acids that traverse the membrane. These histidine kinases function as membrane receptors with sensory domains at the external surface of the membrane that serve to regulate the activity of the conserved histidine kinase domain at the membrane-cytoplasm interface.

The site of histidine phosphorylation is generally located within the kinase protein itself, N-proximal to the conserved kinase domain

Jeffry B. Stock and Peter Park, Departments of Molecular Biology and Chemistry, Princeton University, Princeton, New Jersey 08544. *Michael G. Surette and Mikhail Levit,* Department of Molecular Biology, Princeton University, Princeton, New Jersey 08544.

Two-Component Signal Transduction, Edited by James A. Hoch and Thomas J. Silhavy,
© 1995 American Society for Microbiology, Washington, DC 20005

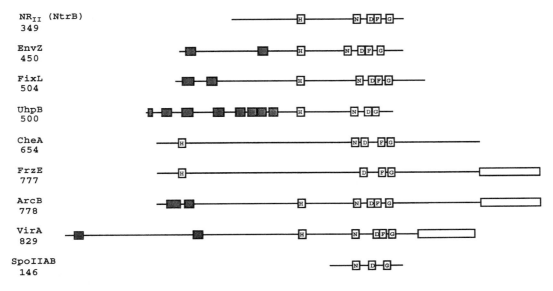

FIGURE 1 Organization of sequence motifs in histidine kinases. Examples of different sequence organizations among representative members of the histidine kinase superfamily are shown with the corresponding total number of residues. Conserved sequences are indicated with the number of the first conserved residue given below. As described in the text and Fig. 2, these are designated as H, N, D, F, and G boxes. Attached response regulator domains are indicated by open boxes, and hydrophobic sequences are indicated by closed boxes. Sequences were chosen from a library of more than 75 different histidine kinase homologs obtained from searching a variety of standard protein sequence data bases. For an extensive delineation of histidine kinase sequences, see Parkinson and Kofoid (1992).

(Hess et al., 1988a; Ninfa and Bennett, 1991; Roberts et al., 1993). Histidine phosphorylation does not proceed by an intramolecular mechanism, however. In all cases that have been studied, a bimolecular reaction is involved with one kinase monomer catalyzing the phosphorylation of a histidine residue in a second monomer (Yang and Inouye, 1992; Ninfa et al., 1993; Wolfe and Stewart, 1993; Swanson et al., 1993a; Pan et al., 1993). This mechanism requires formation of a homodimer with the kinase domain of one subunit positioned such that it can catalyze the phosphorylation of a second subunit.

Structure-Function Relationships between Histidine Kinase Domains

Almost 100 different histidine protein kinases have now been sequenced. The average degree of sequence identity between any pair of ho-

mologous kinase domains is approximately 25%, with only a few specific residues being conserved in almost all members of the family (Fig. 2). The kinase domain appears to be an independently folding unit that functions to bind MgATP and catalyze the phosphorylation of a histidine residue. From an analysis of sequence conservation between kinase domains, Parkinson and Kofoid (1992) concluded that the conserved histidine kinase domain was divided into two subdomains, with a relatively variable connecting linker. The second subdomain contains several highly conserved sequences that are termed the N, D, F, and G boxes (Fig. 2). Presumably, these motifs are arranged in the tertiary structure of the protein to form a nucleotide binding surface within the active site. The G box is a glycine-rich sequence reminiscent of the glycine-rich sequences found in many kinases and other nu-

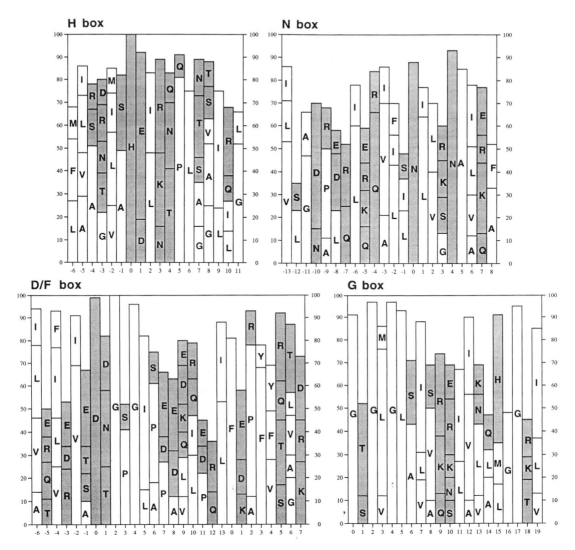

FIGURE 2 Amino acid preference plot for the boxes of most conserved residues among members of the histidine protein kinase superfamily. A histogram of the relative occurrence of each amino acid is plotted against the number of the amino acid within the box, starting with the first signature amino acid (H, N, D, F, or G) designated as 0. The D and F boxes are generally contiguous except in CheA and FrzE, which have intervening inserts of 25 to 28 residues. Amino acid frequencies are derived from an analysis of 68 histidine kinases from GenEMBL and SwissProt data bases. Amino acids with less than 10% frequencies of occurrence at a given position are not shown. The pattern code is light, nonpolar (GAVLMIPFYW) and dark, polar (STCNQKRHDE).

cleotide binding proteins. In the cyclic AMP-dependent protein kinase, a glycine-rich loop serves as a clamp whose backbone amide nitrogens hydrogen-bond to the β-phosphoryl of ATP (Taylor et al., 1992). The D box contains a conserved DXG motif. In serine/threonine/tyrosine protein kinases, the aspartate β-carboxylate in a highly conserved DXG sequence functions to coordinate Mg(II) and the backbone amide of the glycine hydrogen-bonds to the backbone carbonyl of the aspartate to hold it in position (Taylor et al., 1992).

Although lysines are commonly involved in kinase function, there are no conserved lysine residues in histidine kinase domains. One might suppose that an essential lysine might not have been detected because it was flanked by sequences that are relatively divergent, but there are several examples of histidine kinases that have no lysine residues either within the entire first subdomain leading up to the N box, or within the regions between the N and D boxes, or within the region between the D and G boxes, or in the entire C terminus beyond the G box. The lack of a conserved catalytic lysine distinguishes the histidine protein kinase superfamily from other protein kinases that have been characterized.

In terms of function, histidine protein kinases are closely related to serine/threonine/tyrosine protein kinases, but even a cursory examination of the conserved sequences in the two proteins clearly shows that they have different structures and catalytic mechanisms. The serine/threonine/tyrosine kinases have two distinct subdomains, one primarily involved in binding protein substrates and the other functioning to bind MgATP (Parkinson and Kofoid, 1992; Taylor et al., 1992). The two subdomains of the histidine kinases may supply similar functions, but in the histidine kinases, the putative substrate binding domain is N-terminal to the nucleotide binding domain, whereas in serine/threonine/tyrosine kinases, it is C-terminal. Thus, although both kinases have G-rich sequences, in the serine/threonine/tyrosine kinases, this motif is located near the N terminus, whereas in histidine kinases, the G box is near the C terminus. The only other common motif, the DXG sequence, is located about 20 residues to the N terminus of the G-rich region in histidine kinases within the putative ATP binding subdomain, whereas it is C-terminal to the G-rich region in serine/threonine/tyrosine kinases in a region associated with the substrate binding subdomain.

Serine/threonine/tyrosine kinases also differ from histidine kinases in that the product is a phosphoester rather than a phosphoramidate. The phosphotransfer potential of serine and threonine phosphoesters is several kilocalories per mole lower than that of MgATP, so unlike histidine phosphorylation, ATP-dependent phosphorylation at serines and threonines is essentially an irreversible reaction (Shizuta et al., 1975). Phosphotyrosines have a phosphotransfer potential that is closer to MgATP (Fukami and Lipmann, 1983; Hubler et al., 1989), but these phosphorylations also appear to be thermodynamically favored.

There are several well-characterized enzymes whose modes of action involve transfer of a γ-phosphoryl group from a nucleotide triphosphate to a histidine nitrogen. Among the best understood of these is nucleotide diphosphate kinase (NDPK). NDPK is highly conserved from bacteria to vertebrates. The enzyme serves an essential housekeeping function, catalyzing the interconversion of nucleotide di- and triphosphates via a phosphohistidine enzyme intermediate. The protein is composed of approximately 150 amino acids folded in an α/β-structure that is very similar to the regulatory nucleotide binding domain of *Escherichia coli* aspartate transcarbamoylase but quite different from other kinase structures (Dumas et al., 1992; Williams et al., 1993). The histidine that is phosphorylated is at the base of a deep cleft in close proximity to the bound nucleotide, and phosphotransfer appears to proceed by an intrasubunit mechanism. None of the highly conserved sequence motifs in the bacterial histidine kinases are found in NDPK. The distinctive feature of NDPK is tight, low-specificity, nucleotide binding, with phosphotransfer not going beyond the bound nucleotide. The bacterial histidine kinases are generally specific for ATP and have affinities for this nucleotide that are approximately 100-fold lower than those exhibited by NDPK. Moreover, unlike NDPK, they must accommodate transfer from the phosphorylated histidine to a second type of substrate, the aspartate residue in cognate response regulator proteins.

Succinyl–CoA synthetase provides another well-characterized example of an enzyme that catalyzes the transfer of a phosphoryl group from ATP/GTP to a protein histidine side chain. The phosphoryl group is then transferred to a carboxylate in succinate to make succinyl phosphate, which goes on to react with CoA to form succinyl CoA and P_i. This series of reactions works in reverse in the tricarboxylic acid cycle to convert succinyl-CoA to succinate with concomitant production of ATP or GTP. Although this chemistry is very analogous to that catalyzed by histidine protein kinases, there is no indication of any relationship between the sequences of the two types of enzymes. Succinyl-CoA synthetase from *E. coli* is an $\alpha_2\beta_2$-tetramer (Wolodko et al., 1994). Each subunit is, in turn, composed of two distinct domains. In the α-subunit, both domains are doubly wound α/β-structures with Rossmann Folds characteristic of nucleotide binding sites (Rao and Rossmann, 1973). The N-terminal domain binds CoA. The site of histidine phosphorylation is in the C-terminal domain of the α-subunit. The β-subunit also participates in CoA binding and interacts with the phosphorylated histidine. The X-ray crystal structure of succinyl-CoA synthetase has been obtained in the phosphohistidine form (Wolodko et al., 1994). This is currently the only example of a protein structure with phosphohistidine. The N-3 phosphohistidinyl group appears to be stabilized through interactions with two α-helical dipoles. Also, a glutamyl side chain carboxylate makes a salt bridge to the protonated N-1 position, where it could serve a role in the phosphotransfer mechanism.

Sites of Phosphorylation and Dimerization

In most histidine protein kinases, the phosphorylated histidine is located about 110 amino acids N-proximal to the N box at the N-terminal border of the conserved kinase domain (see Fig. 1). This residue is nested in a highly conserved sequence motif termed the H box (Fig. 2). Mutations in this region frequently have a deleterious effect on kinase activity. Evidence has been reported for several kinases, including EnvZ (Yang and Inouye, 1992), NR_{II} (Ninfa et al., 1993), and CheA (Wolfe and Stewart, 1993; Swanson et al., 1993a), that phosphorylation occurs in *trans*. Thus, mutants in the kinase domain that have a histidine phosphorylation site intact can generally be complemented by mutations with an intact kinase domain that lacks a site for histidine phosphorylation. Data with purified NR_{II} and CheA indicate that intramolecular phosphorylation does not occur at a significant rate. In the case of NR_{II}, a wild-type protein in association with a mutant protein exhibits no autophosphorylation (Ninfa et al., 1993). In the case of CheA, the kinetics of phosphorylation exhibit a second-order dependence on CheA concentration that is consistent with an obligatory bimolecular reaction (Ninfa, 1992). From these results, it is clear that the kinases function as dimers and that at least one type of interaction between monomers is between substrate histidine residues and kinase domains.

Insights concerning the mechanism of dimerization come primarily from studies of CheA. CheA has been regarded in the literature as a "stable" dimer because of physical (Gegner and Dahlquist, 1991) and enzymatic (Hess et al., 1988b) studies showing that it is a dimer at concentrations in the micromolar range. From the kinetics of CheA autophosphorylation (Ninfa, 1992) and subunit exchange studies (Wolfe and Stewart, 1993; Swanson et al., 1993a), it is clear that CheA dimerization is a much more dynamic process than had originally been appreciated. CheA is unusual in that the phosphorylated histidine is located several hundred residues N-proximal to the kinase domain (see Fig. 1) within a distinct 15-kDa phosphotransfer domain that can function independently of the remainder of the CheA protein (Hess et al., 1988a). The kinase domain can be isolated as an independently folded unit as well and shown to phosphorylate the phosphotransfer domain (Swanson et al., 1993b).

If the kinase and phosphotransfer domains of each CheA monomer within a dimer bind to one another at independent and nonoverlapping sites, then the minimum K_d expected for the dimer would be the square of the K_d of a single phosphotransfer domain-kinase domain complex. Conversely, the minimum K_d for dissociation of the complex between isolated domains would be the square root of the K_d for the CheA dimer. Kinetic data from an analysis of the autophosphorylation activities of mixtures of various mutants and fragments of CheA are consistent with this idea but only if one assumes that portions of the phosphotransfer domain that are at least 50 residues distal to the phosphorylated histidine, His48, rather than the histidine phosphorylation site per se, are involved in binding to the kinase. This conclusion derives from the observation that a naturally occurring CheA variant called CheAs, produced from an alternative translational initiation site at Met-98 (Kofoid and Parkinson, 1991), forms dimers as well as full-length CheA (Wolfe and Stewart, 1993). A mutant CheA with a defective kinase domain can form a heterodimer with CheAs just as well as with wild type. Thus, a 1:1 mixture of these two proteins exhibits about 25% the activity of an equivalent concentration of wild-type CheA. Recent studies indicate an effect of mutations in the first subdomain of the conserved kinase region of CheA in the kinase activity of an associated wild-type CheA monomer as if the site of dimerization involved this region (Tawa and Stewart, 1994a). This makes sense in terms of the structure of most other kinases, where the H box is located at the N terminus of the first kinase subdomain (see Fig. 1).

The stabilities of the phosphohistidine groups in the kinases CheA (Wylie et al., 1988), NR$_{II}$ (Ninfa et al., 1988), PhoR (Makino et al., 1989), and PhoM (Amemura et al., 1990) are all consistent with a 3-phosphohistidine as opposed to a 1-phosphohistidine, which is significantly more labile. This has been confirmed by chromatographic analysis of the phosphohisti-

dine in PhoM (Amemura et al., 1990) and phosphorus nuclear magnetic resonance analysis of the intact phosphorylated CheA protein (Lukat and Stock, unpublished data). A report of cochromatography of phosphohistidine from CheA with 1-phosphohistidine involved a procedure that would not have resolved the 1- and 3-phosphohistidine isomers (Hess et al., 1988a).

Phosphohistidines in proteins have a large standard free energy of hydrolysis (Weigel et al., 1982), and this appears to be the case for phospho-CheA, which has a phosphotransfer potential of 1 to 3 kcal/mol greater than that of MgATP (Wylie et al., 1988; Tawa and Stewart, 1994b). The forward reaction in vivo is presumably driven by a high ratio of ATP to ADP and by rapid transfer of the phosphoryl group from the phosphohistidine to a response regulator protein.

An example of a bacterial histidine kinase homolog that phosphorylates a target protein at a serine residue has been identified in *Bacillus subtilis* (Min et al., 1993). Endospore development in these cells requires activation of a specific sigma factor, σ^F. This factor is present in vegetative cells in an inactive complex with a 16-kDa protein, SpoIIAB. The sequence of SpoIIAB contains a histidine kinase domain, but the H box is missing. In the presence of SpoIIAB, another protein, SpoIIAA, is phosphorylated at a serine residue. This impairs the ability of SpoIIAA to bind SpoIIAB and thereby releases SpoIIAB so that it can interact with σ^F (Alper et al., 1994). The binding of ATP or nonhydrolyzable analogs of ATP to SpoIIAB also facilitates binding to σ^F (Alper et al., 1994). SpoIIAB does not appear to phosphorylate itself.

Histidine kinases that are indistinguishable from the bacterial enzymes have been sequenced in eukaryotic microorganisms such as *Saccharomyces cerevisiae* (Ota and Varshavsky, 1993), as well as in higher plants (Chang et al., 1993). The kinases that have been identified in *S. cerevisiae* (Maeda et al., 1994) and *Arabidopsis* (Chang et al., 1993) have cognate response regulator domains that presumably accept phosphoryl groups from the phosphorylated histidine. The *Arabidopsis* kinase functions to sense ethylene (Chang et al., 1993), and the yeast kinase is an osmosensor (Maeda et al., 1994). Another example from plants are the phytochromes, which serve important signaling functions in response to light (Schneider-Poetsch, 1992). In this case, no phospho-accepting response regulator domain has been identified, and autophosphorylation has not been demonstrated. In vertebrate tissues, histidine kinase homologs function in mitochondria to phosphorylate and inactivate α-keto acid dehydrogenases such as pyruvate dehydrogenase and branched-chain α-keto acid dehydrogenase (Popov et al., 1992). No response regulator domain has been associated with the mitochondrial kinases, and the site of phosphorylation in the dehydrogenases is a serine residue.

Autophosphorylation Kinetics

The autophosphorylation reaction catalyzed by histidine protein kinases is relatively complex and involves several intermediate reactions. The essential features are kinase dimerization, nucleotide binding, and histidine phosphorylation. There are potentially six forms of each monomer that need to be considered: kinase alone, kinase with ATP bound, kinase with ADP bound, phosphorylated kinase, phosphorylated kinase with ATP bound, and phosphorylated kinase with ADP bound. Allowing all combinations of these within a dimer gives an additional 21 forms, for a total of 27. By using different kinetic methods, one can eliminate many of these to make an analysis experimentally tractable.

The simplest interaction to examine is kinase dimerization. There are two examples of kinases whose states of oligomerization have been characterized, NR_{II} (Ninfa et al., 1986) and CheA (Gegner and Dahlquist, 1991). Both have been purified as dimers. By mixing genetically tagged variants of NR_{II}, it has been demonstrated that the rate of subunit exchange is

negligible over a period of hours (Ninfa et al., 1993). In this case, one need not consider the monomeric kinase forms as significant participants in the kinase reaction. Similar experiments performed with CheA indicate a much faster rate of subunit exchange with half-times on the order of minutes (Swanson et al., 1993a). Thus, under physiological conditions, CheA monomers and dimers appear to be in equilibrium. Kinetic studies indicate that the K_d for dissociation of CheA homodimers is approximately 1 μM (Ninfa, 1992). Nucleotide binding, phosphorylation, and CheY binding do not appear to affect CheA dimerization. At concentrations of CheA substantially larger than 1 μM, CheA dimers are the predominant species, and the monomeric forms of the enzyme need not be considered.

Measurements of the kinetics of autophosphorylation of CheA indicate K_ms for MgATP of 0.2 to 0.3 mM, consistent with results obtained from direct measurements of MgATP binding using the method of Hummel and Drier (Tawa and Stewart, 1994b). The time course of CheA phosphorylation fits a simple exponential, with a stoichiometry approaching one phosphoryl group per monomer (Tawa and Stewart, 1994b; Hess et al., 1988b). These results indicate that the two sites within a CheA dimer function independently, with neither positive nor negative cooperativity. Similar results have been obtained with NR_{II} (Ninfa, 1992). From these findings, a simplified kinetic model for the autophosphorylation reaction can be formulated:

$$\text{Kinase} + \text{MgATP} \xrightarrow{K_s} \text{Kinase} \cdot \text{MgATP}$$
$$\xrightarrow{k_{cat}} \text{P}\sim\text{Kinase} + \text{ADP}$$

According to this scheme, the initial rate of autophosphorylation under conditions in which all the kinase is assumed to be in a dimeric state can be expressed in terms of a Michaelis-Menten equation, in which the apparent V_{max} corresponds to k_{cat}[Kinase], with the effective concentration of kinase being the concentration of kinase active sites, which is simply twice the concentration of dimer. The turnover number (k_{cat}) is generally low. The value obtained for CheA is only about 2/min. Under conditions in which there is a significant concentration of monomer, such as occurs at submicromolar concentrations of CheA, the effective kinase concentration ([Kinase]$_{eff}$) is still twice the dimer concentration or

$$[\text{Kinase}]_{eff} = \frac{4[\text{Kinase}] + K_d - \sqrt{(4[\text{Kinase}] + K_d)^2 - 16[\text{Kinase}]^2}}{4}$$

where K_d is the dissociation constant of the kinase dimer. Under these conditions, the apparent V_{max} equals $k_{cat}[\text{Kinase}]_{eff}$.

Histidine protein kinases catalyze an exchange of phosphoryl groups between ATP and ADP (Borkovich and Simon, 1990; Ninfa et al., 1991; Tawa and Stewart, 1994b; Ninfa, 1992). This exchange reaction can be measured by assaying the ATP-dependent formation of [^{3}H]ATP from [^{3}H]ADP:

Kinase + ATP $\leftrightarrow$ Kinase $\cdot$ ATP $\leftrightarrow$
P~Kinase $\cdot$ ADP $\leftrightarrow$ P~Kinase + ADP

P~Kinase + A$\star$DP $\leftrightarrow$ P~Kinase $\cdot$ A$\star$DP $\leftrightarrow$
Kinase $\cdot$ A$\star$TP $\leftrightarrow$ Kinase + A$\star$TP

The reaction has been measured with CheA and NR$_{II}$ as a function of ATP and ADP concentrations (Ninfa, 1992). The results provide independent estimates of k_{cat} as well as nucleotide binding constants for the phosphorylated and dephosphorylated forms of the enzyme. Values obtained for MgATP binding and phosphotransfer to histidine by this method are similar to those derived from measurements of initial rates of autophosphorylation. This approach also provides a measure of MgADP and MgATP binding to phospho-CheA. The values obtained for MgADP are similar to the binding constants of MgATP and MgADP for the dephosphoform of the enzyme. MgATP bound with almost 10-fold less affinity to phospho-CheA, however, as if the phosphorylated histidine interferes with the positioning of the ATP γ-phosphoryl group at the active site. Tawa and Stewart (1994b) obtained a similar value for MgADP binding to dephospho-

CheA using the Hummel-Dreyer method. Measurements of rates of ATP-ADP exchange at submicromolar CheA concentrations are consistent with a K_d for CheA dimer dissociation of approximately 1 μM.

A parallel study of exchange rates with NR$_{II}$ gave results consistent with those obtained from studies of autophosphorylation (Ninfa, 1992). In contrast to CheA, phosphorylated NR$_{II}$ exhibited almost a 10-fold higher affinity for MgADP than dephospho-NR$_{II}$ with K_ds of 0.04 mM versus 0.3 mM. The affinity of MgATP for dephospho-NR$_{II}$ exhibited an intermediate value, 0.1 mM. As in the case of CheA, phospho-NR$_{II}$ had a significantly reduced affinity for MgATP compared with the dephosphorylated kinase. There was no indication of NR$_{II}$ dissociation, even at submicromolar kinase concentrations.

The nucleotide binding properties of the kinases indicate equal or higher affinities for the dephosphorylated enzymes for ADP compared with ATP (Ninfa, 1992; Tawa and Stewart, 1994b). This means that under the saturating nucleotide concentrations one would expect to pertain in vivo, 3 mM ATP and 0.25 mM ADP (Bochner and Ames, 1982), the rate of autophosphorylation (V) should be sensitive to the ratio of ATP to ADP in accord with the relationship

$$V = \frac{V_{max}(\text{ATP})K_{ADP}}{(\text{ATP})K_{ADP} + K_{ADP} \cdot K_{ATP} + (\text{ADP})K_{ATP}}$$

where K_{ATP} and K_{ADP} correspond to the dissociation constants of the corresponding nucleotides for the dephosphorylated kinase. Thus, for CheA, a 50% decrease in ATP coupled to a twofold increase in ADP would be expected to cause more than a 20% decrease in the rate of histidine phosphorylation.

Regulation of Rates of Kinase Autophosphorylation

Rates of kinase autophosphorylation are modulated by sensory inputs through associated regulatory domains. The mechanism of regulation has been most thoroughly investigated in the case of CheA. A domain of CheA at the extreme C terminus of the protein, just distal to the kinase domain, functions to bind an 18,000-molecular-weight monomeric protein, CheW, with a K_d of approximately 15 μM (Gegner and Dahlquist, 1991; Parkinson and Kofoid, 1992; Bourret et al., 1993a). The CheW binding domain can be deleted from CheA without affecting kinase activity assayed with the purified protein (Bourret et al., 1993a), and a kinetic analysis of CheA activity indicates that CheW binding has no effect on nucleotide binding, dimerization, autophosphorylation, or phosphotransfer to the response regulator CheY (Levit and Stock, unpublished data).

CheA together with CheW binds to the signaling domains of transmembrane chemoreceptor proteins to form ternary complexes of CheA, CheW, and receptor (Borkovich et al., 1989; Ninfa et al., 1991; Gegner et al., 1992). CheW appears to have independent binding sites for CheA and receptor and thereby functions to bring these proteins together (Ninfa et al., 1991; Gegner et al., 1992). Like CheA, the receptors are homodimers (Milligan and Koshland, 1988), and the stoichiometry of components within the ternary complex is 2:2:2 (Gegner et al., 1992). Stimulatory ligands bind to the periplasmic sensory domains of the receptors at sites located at the dimer interface (Milburn et al., 1991; Scott et al., 1993; Yeh et al., 1993). Ligand binding causes a conformational change in the structure of this region of the receptor that probably changes the relative disposition of the signaling domains within the cytoplasm (Kim, 1994). This can either inhibit or stimulate kinase activity, presumably by effecting a change in the disposition of CheA monomers with respect to one another (Borkovich et al., 1992; Stock et al., 1991).

Receptor-signaling domain mutants that are locked in a kinase-activating conformation have a decreased tendency to dimerize, and mutations in the signaling domain that lock the kinase in an inactive state tend to facilitate dimerization (Long and Weis, 1992). These results are consistent with a mechanism whereby a principal mode of CheA regulation involves controlling the interaction between kinase and phosphotransfer domains. The activation of CheA autophosphorylation within the ternary complex is one to two orders of magnitude too large to be accounted for by the dimerization mechanism observed with CheA alone, however; and recent results indicate that association with receptor and CheW causes a conformational change in CheA that activates kinase activity independent of dimerization (Wolfe et al., 1994).

Results with other kinases are consistent with mechanisms of regulation that essentially involve control of the monomer to dimer transition. NR_{II} is a relatively tight dimer compared with CheA. It has not been observed to dissociate under physiological conditions, and no subunit exchange has been observed (Ninfa et al., 1993). In NR_{II}, however, no regulation of kinase activity has been detected. An auxiliary protein, P_{II}, forms a complex with NR_{II} to regulate the level of phosphorylation of the cognate response regulator, NR_I, but P_{II} works by increasing an NR_{II}-stimulated phospho-NR_I phosphatase activity rather than by affecting kinase activity (Ninfa, 1986; Kamberov et al., 1994).

The regulation of EnvZ has also been investigated. EnvZ is a transmembrane protein with a periplasmic sensory domain that is linked by a membrane-spanning sequence to a histidine kinase domain at the membrane-cytosol interface (see Forst et al., 1987; and Fig. 1). If the periplasmic and membrane-spanning domains of EnvZ are replaced by comparable regions from the chemotaxis receptors for aspartate, Tar (Yang and Inouye, 1992), or ribose and galactose, Trg (Baumgartner et al., 1994), the chimeric proteins show increased kinase activity in response to aspartate or ribose and galactose, respectively. It is known that these stimulatory ligands bind, either directly or through periplasmic binding proteins, at the receptor sensory domain dimer interface where they would be expected to facilitate dimerization of the EnvZ-receptor hybrids (Milligan and Kosh-

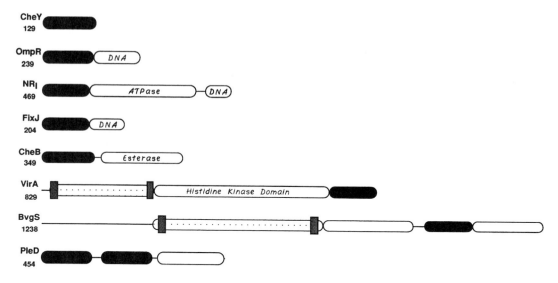

FIGURE 3 Typical domain organizations found in response regulators. For an extensive delineation of response regulator sequences, see Volz (1993).

land, 1993). Another transmembrane sensor kinase, VirA, has been shown to function as a dimer, but the results indicate that it is always a dimer and that stimulatory ligands do not effect a monomer-to-dimer transition (Pan et al., 1993).

RESPONSE REGULATORS

The phosphohistidine produced by a histidine kinase is generally transferred to an aspartate residue in a response regulator (for reviews, see J. B. Stock et al., 1989; Parkinson, 1993; Bourret et al., 1991). Response regulators are characterized by a conserved domain of approximately 125 amino acids (Stock et al., 1990; Volz, 1993), usually attached via a linker sequence to a domain with an effector function (Fig. 3). The effector domains generally have DNA binding activities, and in these cases, response regulator phosphorylation serves to modulate transcription. There are three major classes of DNA binding domains: the OmpR–PhoB subgroup, the NR$_I$ (NtrC) subgroup, and the FixJ subgroup. In each case, there are examples of transcriptional regulators in which the DNA binding domain is not attached to a response

regulator domain and is presumably regulated by a mechanism that does not involve aspartate phosphorylation (J. B. Stock et al., 1989).

There are also several instances in which a different type of effector region is attached to a response regulator. The best characterized of these is a protein, CheB, that functions in chemotaxis to demethylate membrane chemoreceptor proteins. Phosphorylation of the response regulator domain in CheB activates its associated demethylase catalytic domain (Lupas and Stock, 1989).

In some cases, the response regulator domain is produced as an independent unit. The best-characterized instance is the response regulator CheY, which functions in chemotaxis to regulate the flagellar motor. CheY is a 14-kDa monomeric protein that binds to a flagellum-associated protein, FliM, when it is phosphorylated (Welch et al., 1993).

Some histidine kinases have response regulator domains located C-terminal to the kinase domain (see Fig. 1). Examples include VirA and ArcB. Other response regulators usually function with these kinase-regulator fusions to generate a response. In the case of ArcB, the regulator domain associated with the kinase modulates the kinase output (Iuchi, 1993). Thus, the phosphohistidyl group in ArcB is transferred to an aspartyl side chain in the linked regulator domain in preference to the response regulator of the Arc system, ArcA, that functions to produce a transcriptional response. Only under conditions in which the level of phosphorylation of the linked regulator domain of ArcB is high are phosphoryl groups passed to ArcA, however.

To a first approximation, the sequences of response regulator domains do not seem to vary in accord with their attachments to other elements of protein structure. The only exception is the OmpR–PhoB subfamily, whose response regulator sequences tend to be slightly more related to one other than to regulator domains in other contexts (Volz, 1993).

Structure–Function Relationships

The structure of CheY has been determined by X-ray crystallographic (A. M. Stock et al., 1989; Volz and Matsumura, 1991; Stock et al., 1993; Bellsolell et al., 1994) and nuclear magnetic resonance (Bruix et al., 1993; Moy et al., 1994) methods in several laboratories under a variety of conditions. CheY is a doubly wound α/β-protein with five α-helices surrounding a five-stranded parallel β-sheet (Fig. 4). The α- and β-secondary structure elements alternate in the sequence starting with β_1 and ending with α_E. The topology of folding is $\beta_2\beta_1\beta_3\beta_4\beta_5$ with the A and E helices on one side of the sheet and the B, C, and D helices on the other. The sequences of other response regulators clearly conform with this structure, in that all members of the family have hydrophobic residues at positions corresponding to the hydrophobic core of CheY made up of β_1, β_3, and β_4 and the helical surfaces with which they interact (Stock et al., 1990; Volz, 1993).

The site of phosphorylation in CheY is located at Asp-57 in the loop between β_3 and αC (Sanders et al., 1989) (Figs. 4 and 5), and the corresponding aspartates in NR$_I$ (Sanders et al., 1992), PhoB (Makino et al., 1994), and VirG

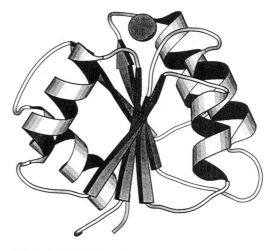

FIGURE 4 Ribbon drawing of CheY with a bound Mg(II) at the active site. From the coordinates of the Mg(II) structure of Stock et al. (1993).

(Jin et al., 1990) have also been shown to be phosphorylated. This residue is completely conserved in all response regulator homologs, and adjacent loops connecting β_1 to αA and β_4 to αD contain residues that are almost as highly conserved (Volz, 1993). There is generally at least one aspartate within the β_1–αA loop at a position corresponding to Asp-13 in CheY. In most response regulators, this residue is flanked by one or more additional acidic residues, most commonly by an aspartate at the position corresponding to Asp-12 in CheY. With only a few exceptions, a serine or threonine is located within the β_4–αD loop at a position corresponding to Thr-87 in CheY. There is only one other residue that is conserved to the same degree as those in the loops adjacent to the phosphorylated aspartate: a lysine residue within the loop connecting β_4 to αD corresponding to Lys-109 in CheY. In CheY, the side chain of this lysine extends over toward Asp-57 so that the lysine ε-amino group and the aspartate β-carboxylate are in close proximity. Thus, all the most highly conserved residues in response regulators cluster together around the site of phosphorylation.

The roles of active site residues have been examined both biochemically and genetically.

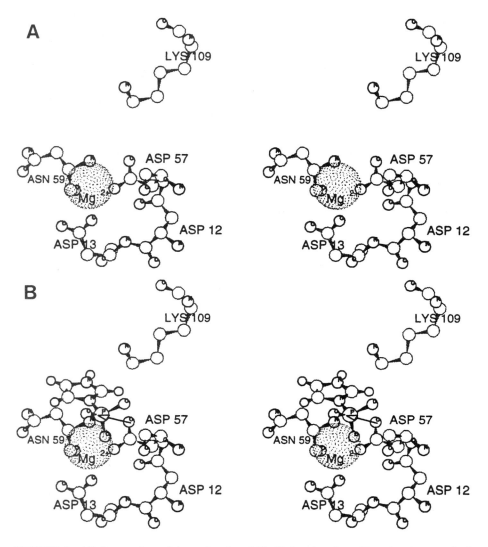

FIGURE 5 (A) Stereo view of the active site of CheY showing the acid pocket composed of Asp-11, Asp-12, and Asp-57 with a bound Mg(II). Asn-59, which coordinates with the metal through a backbone carbonyl oxygen, and Lys-109 are also indicated. (B) Stereo view of the active site of CheY as shown in A, with a phosphoimidazole group positioned to donate its phosphoryl group to the β-carboxyl of Asp-57. From the coordinates of Stock et al. (1993).

The carboxylate side chains in acidic residues in the loop between β_1 and αA (e.g., Asp-12 and Asp-13 in CheY) act together with the β-carboxylate of the phospho-accepting aspartate, Asp-57, to help coordinate a divalent metal ion, generally Mg(II) (Lukat et al., 1990; Lukat et al., 1991; Needham et al., 1993; Stock et al., 1993; Bellsolell et al., 1994). The binding constant of CheY for Mg(II) is approximately 0.5 mM (Lukat et al., 1990; Needham et al., 1993). Metal binding is required for the phosphotransfer reaction (Lukat et al., 1990). Model studies with small-molecule phosphoramidate donors and carboxylate acceptors indicate that

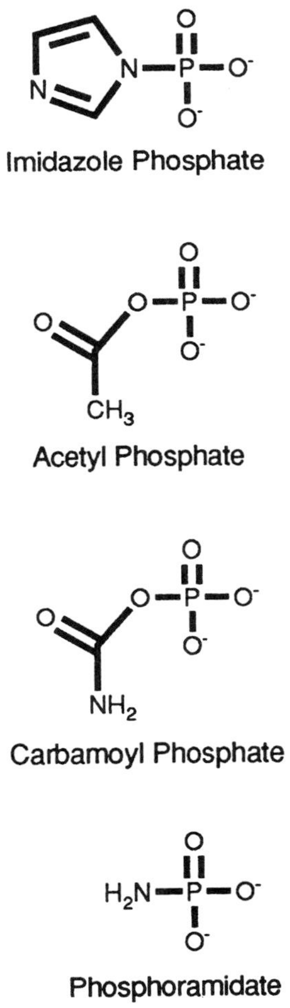

Imidazole Phosphate

Acetyl Phosphate

Carbamoyl Phosphate

Phosphoramidate

FIGURE 6 Small-molecule phosphodonors that can function to phosphorylate response regulators such as CheY.

divalent metals facilitate the formation of acyl phosphates by shielding the charges on the anions and forming a template for the trigonal bipyramide pentavalent phosphate transition state (Herschlag and Jencks, 1990). The geometry of the active site of CheY with Mg(II) bound accommodates such a transition state for phosphotransfer from phosphoimidazole to Asp-57 (Stock et al., 1993) (Fig. 5B). All that is required is a tortional rotation around the β-carbon of Asp-57 to position its β-carboxylate oxygen in proper orientation for nucleophilic

attack of an incoming phosphorus with in-line displacement of the imidazol-leaving group.

Thus, phosphotransfer from the phosphohistidine in a kinase to a response regulator appears to be, in large part, catalyzed by the response regulator rather than the kinase. This supposition is confirmed by the observation that small-molecule phosphodonors such as phosphoramidate, acetyl phosphate, and carbamoyl phosphate can act in place of the phosphorylated histidine kinases to donate phosphoryl groups to response regulator aspartates (Lukat et al., 1992; Feng et al., 1992; McCleary and Stock, 1994; Reyrat et al., 1993; Schroder et al., 1994) (Fig. 6). The affinities of regulators for these donors are low, and phosphotransfer rates appear to be limited by the rate of binding rather than the transfer rate. The reported apparent K_ms correspond to the ratio of the rate constants for phosphoaspartate hydrolysis and phosphotransfer. This is the concentration of phosphodonor required to maintain 50% of the response regulator phosphorylated under steady-state conditions. These values are generally in the millimolar range. The phosphodonor activity of acetyl phosphate seems to be physiologically relevant because its concentrations inside the cell vary from micromolar to millimolar, depending on conditions of growth (McCleary et al., 1993; McCleary and Stock, 1994).

In considering the phosphotransfer chemistry, we have focused on the role of the metal ion in facilitating formation of the transition state. One would expect an additional contribution from a proton donor at the active site because transfer from a 3-phosphohistidine requires protonation at the 1-N atom (Benkovic and Sampson, 1971). The requirement for protonation at this position explains why phosphohistidine groups are relatively stable in base as opposed to acid. A proton donor would also be expected to facilitate the transfer reaction from acyl phosphates through general acid catalysis. The obvious candidate for this function is the ε-amino group of the highly conserved lysine, Lys-109, in CheY. The highly conserved serine or threonine residue in the β_4-αD loop, Thr-87 in CheY, might also be expected to participate in this aspect of the phosphotransfer chemistry. Thr-87 and Lys-109 in CheY are not absolutely required for phosphotransfer (Lukat et al., 1991; Volz, 1993); but at least in the case of Lys-109, the rate is considerably reduced by conservative substitutions at this position (Lukat et al., 1991).

In CheY, Thr-87 and Lys-109 are required for the phosphorylation-induced conformational switch that leads to a response because mutations that introduce conservative substitutions in either of these positions prevent CheY from modulating motility (Lukat et al., 1991; Volz, 1993). Presumably, the mutant proteins cannot undergo a phosphorylation-induced conformational change and therefore cannot interact effectively with the motor. Lys-109 substitutions also block interactions of phospho-CheY with the corresponding phosphatase, CheZ (Lukat et al., 1991; Blat and Eisenbach, 1994). In other response regulators, however, mutations at these positions do not block activation (Stewart et al., 1990). Clearly, any restructuring of residues around the phosphorylated aspartate could lead to a variety of defects in the functioning of the protein.

What is the role of the histidine kinases in the phosphotransfer reaction? Apparently, almost any histidine kinase has at least a limited ability to donate phosphoryl groups to any response regulator (Ninfa et al., 1988; Igo et al., 1989; Olmedo et al., 1990; Amemura et al., 1990). This crosstalk probably reflects the catalytic function of the response regulators more than the kinases. Rates of transfer between noncognate kinase-regulator pairs are at least two orders of magnitude slower than rates of transfer between cognate pairs. The mechanism by which the kinase facilitates the rate of phosphotransfer to a specific regulator does not involve the conserved kinase domain, at least in the case of CheA, where it has been shown that an isolated N-terminal region of approximately 15,000 molecular weight that contains the phosphorylated His-48 transfers phos-phoryl groups to both CheY and CheB at rates approaching those obtained with the intact phosphorylated kinase (Hess et al., 1988a; Swanson et al., 1993b). CheA has a high affinity for the dephosphorylated form of CheY ($K_d \approx 30$ nM), and CheY is released when it is phosphorylated (Schuster et al., 1993). From the effects of mutations and an analysis of patterns of proteolytic cleavage, it has been hypothesized that CheA has two distinct N-terminal domains, one containing the site of histidine phosphorylation and the other a high-affinity CheY binding site (Swanson et al., 1993b; Parkinson and Kofoid, 1992; Morrison and Parkinson, 1994).

Phosphorylation–Induced Conformational Changes

Most response regulators have an effector domain linked to an N-terminal CheY-like domain (see Fig. 3). In several instances, it has been shown that genetic deletion or proteolytic cleavage of the N-terminal domain activates the C terminus. Examples include the esterase activity of CheB (Simms et al., 1985) and transcriptional activation by FixJ (Kahn and Ditta, 1991), PhoB (Makino et al., 1994), and Spo0A (Grimsley et al., 1994). From these results, it is apparent that the N terminus in its dephosphorylated state inhibits the activity of the effector domain(s) and that phosphorylation acts to relieve this inhibitory effect. The simplest mechanism for inhibition would be for the N terminus to sterically interfere with the ability of the C terminus to interact with its macromolecular targets, either DNA or protein. To relieve this inhibition might simply involve a movement of αE that swings the attached C-terminal domain into a different orientation with respect to the regulatory domain. Alternatively, there may be more specific interactions between surfaces on the regulatory and effector domains that are disrupted by phosphorylation-induced conformational changes.

CheY is an unusual response regulator in that it does not have an attached effector domain. Another example of a single-domain response regulator is the SpoOF protein that functions in the initiation of sporulation in *B. subtilis* (Burbulys et al., 1991). Phospho-SpoOF passes its phosphoryl group to a histidine residue in another protein, SpoOB, which is not a member of either the histidine kinase or response regulator superfamilies. The phosphoryl group in SpoOB is then passed to a response regulator, SpoOA, that has an associated DNA binding domain and functions to regulate transcription.

The mode of action of SpoOF raises the possibility that phospho-CheY might function similarly by relaying its phosphoryl group to proteins at the flagellar motor. This is unlikely, however, because a CheY mutant has been identified that causes a motor response but cannot be phosphorylated. This phenotype is produced by substitution of the conserved aspartate in the β_1-αA loop, Asp-13, by a lysine or arginine residue (Bourret et al., 1990; Bourret et al., 1993b). The effects of CheY Asp-13-Lys/Arg on motility are not altered either by deletion of the kinase CheA or by introduction of a second mutation at the phospho-accepting aspartate, Asp-57-Ala or Asp-57-Glu (Bourret et al., 1993b). The structure of CheY Asp-13-Lys has been determined by X-ray crystallographic methods, but no significant conformational changes could be detected (Volz, 1993). Presumably, Asp-13-Lys can achieve an active conformation in the absence of phosphorylation but is not locked in the active conformational state and retains an inactive conformation within the crystals used to determine its structure.

The Asp-13-Lys mutation is of particular interest because a few response regulators have this substitution naturally. These include a *Myxococcus xanthus* regulator, FrzG, which is homologous over its entire length to the *E. coli* CheB protein (McCleary et al., 1990), and a transcriptional regulator, FlbD, from *Caulobacter crescentus,* which belongs to the NR$_1$ family of response regulators (Ramakrishnan and Newton, 1990). These proteins are clearly response regulator homologs, but they are missing all the most highly conserved active site residues except the aspartate corresponding to Asp-57 in CheY. In place of the conserved threonine-serine residue, Thr-87 in CheY, both proteins have a glycine, and in place of the conserved lysine, Lys-109 in CheY, both proteins have a leucine. Despite considerable effort, it has not been possible to phosphorylate FlbD using either small-molecule phosphodonors or histidine kinases such as NR$_{II}$ or CheA (G. Ramakrishnan and A. Newton, personal communication). Moreover, FlbD that has been expressed in *E. coli* and purified has the capacity to activate transcription in the absence of detectable phosphorylation (Benson et al., 1994). These results are consistent with the idea that FlbD and FrzG, like the Asp-13-Lys CheY mutant, can achieve an active conformational state without being phosphorylated.

Phospho-CheY binds at the base of the flagella to a component of the flagellar rotation switch apparatus termed FliM (Welch et al., 1993). Mutants of CheY have been selected as suppressors of Che⁻ mutants in FliM and other components of the switch, and vice versa (Parkinson et al., 1983; Magariyama et al., 1990; Yamaguchi et al., 1986; Sockett et al., 1992; Roman et al., 1992). These mutations often involve substitutions for residues located on the face of CheY that is made by β_5, αD, and αE (Roman et al., 1992; Sockett et al., 1992). From these results, it has been postulated that this is the region of CheY that physically interacts with FliM. One could take a contrary position, however, arguing that the mutations act by causing relatively small changes in CheY structure that perturb its interaction with the kinase or phosphatase to alter rates of phosphorylation or dephosphorylation and thereby rectify the disruptions caused by mutations in flagellar proteins that raise or lower levels of phospho-CheY required to produce a motor response. Mutations in residues that directly participate in binding to the motor might be expected to have more dramatic consequences that would lead to null rather than suppressor phenotypes. This conclusion is supported by the fact that suppressor mutations in the gene that encodes the CheY dephosphorylating enzyme, *cheZ,* are obtained as frequently as suppressors in *cheY* (Parkinson and Parker, 1979; Sockett et al., 1992). The *cheZ* mutants are thought to function by raising levels of phospho-CheY, thereby compensating for mutations in the switch that lower its sensitivity to phospho-CheY. Moreover, a detailed analysis of patterns of suppressor mutagenesis in the switch proteins shows that suppression is not allele-specific (Sockett et al., 1992). In fact, null mutations in CheZ can be suppressed by mutations in motor components.

To further assess the nature of the conformational changes that activate CheY, 4-fluorophenylalanine has been introduced into the protein in place of phenylalanine, and ^{19}F nuclear magnetic resonance has been used to detect which of these residues experiences a change in its environment on phosphorylation (Bourret et al., 1993b; Drake et al., 1993). The results indicate that phenylalanines at the active site, within the hydrophobic core, and near the face of CheY opposite the active site are perturbed by phosphorylation. A phenylalanine residue, Phe-111, in the region that had been thought to interact with the flagellar switch based on suppression mutagenesis studies is relatively unperturbed. Phe-111 contacts Lys-109. From these results, it seems likely that phosphorylation induces a global restructuring of CheY that results in significant perturbations in the opposite side of the molecule. Interestingly, the C terminus extends from this surface, so that this region would be the most likely site of contact with a C-terminal effector domain. CheY variants with activating mutations such as Asp-13-Lys were also examined. The results confirm the finding from X-ray studies that the global structure is unaffected. In these proteins, however, phenylalanines near the active site such as Phe-111 are significantly perturbed. It seems likely that these mutations interfere with the stability of this region of the protein, thereby facilitating the transition to an active conformation.

There has been considerable speculation concerning a salt bridge between Lys-109 and Asp-57 in CheY (Hazelbauer et al., 1993). Phosphorylation at Asp-57 would clearly break this interaction, and it has been posited that this is a trigger that induces the subsequent conformational change leading to regulator activation (Volz and Matsumura, 1991). It has recently been shown, however, that Mg(II) coordinates with Asp-57 and thereby disrupts its interaction with Lys-109 independent of phosphorylation (Stock et al., 1993; Moy et al., 1994). Thus, in vivo, where the response regulators always have an associated metal, the salt bridge probably has no relevance to either the phosphotransfer chemistry or the induction of a conformational transition. It is not even clear that the salt bridge forms in solution under physiological conditions because mutation of Lys-109 does not affect the binding of Mg(II) (Lukat et al., 1990). The X-ray crystal structure of CheY where the salt bridge was observed was determined in the presence of high concentrations of ammonium sulfate (Volz and Matsumura, 1991). In this environment, a sulfate ion is positioned near the active site, and an ammonium ion is positioned within the acid pocket. CheY does not bind Mg(II) under these conditions (Stock et al., 1993). The absence of an Asp-57–Lys-109 interaction under physiological conditions would explain why Mg(II) binding does not cause a significant change in the nuclear magnetic resonance signal from a fluorophenylalanine residue at position 111 (Drake et al., 1993).

Autophosphatase Activities

At physiological temperature and pH, a phosphoaspartate group is relatively unstable compared with other types of phosphorylated residues (Koshland, 1951; DiSabato and Jencks, 1961). Phosphorylated regulators denatured in sodium dodecyl sulfate generally have half-lives of several hours at neutral pH and ambient temperatures (Weiss and Magasanik, 1988; Stock et al., 1988). The half-lives of hydrolysis of phosphoaspartate groups in response regulators under nondenaturing conditions range from only a few seconds (e.g., phospho-CheY and phospho-CheB [Hess et al., 1988b]) to several hours (e.g., phospho-OmpR [Igo et al., 1989]), with most exhibiting intermediate values of several minutes (e.g., phospho-NR$_I$ [Weiss and Magasanik, 1988] and phospho-PhoB [Makino et al., 1989]). Dephosphorylation rates in vivo are generally enhanced by auxiliary phosphatase activities (Ninfa, 1986; Keener and Kustu, 1988; Hess et al., 1988b; Aiba et al., 1989a; Aiba et al., 1989b; Igo et al., 1989; Perego et al., 1994). The intrinsic dephosphorylation rate of a given phosphorylated regulator appears to have been selected in evolution so that it is one to two orders of magnitude slower than the enhanced rate in the presence of the corresponding auxiliary phosphatase. It seems likely that auxiliary phosphatases function by causing a conformational change in the phosphorylated regulator that stimulates its autophosphatase activity rather than by directly participating in the phosphoaspartate hydrolysis reaction.

Mg(II) is required both for dephosphorylation and phosphorylation (Weiss and Magasanik, 1988; Lukat et al., 1990). Mg(II) alone will catalyze the hydrolysis of small-molecular-weight acylphosphates such as acetyl phosphate (Koshland, 1951). Rates of hydrolysis obtained in the presence of molar concentrations of metal are comparable with the autophosphatase rates obtained with phosphorylated response regulators at submillimolar concentrations of metal (Koshland, 1951). The transition state for dephosphorylation would be expected to be essentially the same as the transition state for phosphorylation with water positioned in place of the histidinyl-leaving group. It seems likely that the acidic function that protonates the histidine side chain at the N-1 position to facilitate phosphorylation functions as a base to enhance the nucleophilic attack by water in the dephosphorylation reaction. In CheY, for instance, mutation of Lys-109 to arginine causes a substantial decrease both in rates of phosphorylation and dephosphorylation (Lukat et al., 1991). Auxiliary phosphatases could work by forcing a conformational change that realigns groups at the active site to facilitate this chemistry.

Thermodynamic Considerations

The standard free energy of hydrolysis of an acylphosphate such as acetyl phosphate or phosphoaspartate is −10 to −13 kcal/mol (Jencks, 1968), significantly larger than the standard free energy of hydrolysis of MgATP of about −8 kcal/mol (Guynn et al., 1973). Because the equilibrium for phosphorylation of response regulators by the kinases lies far toward formation of the phosphoaspartate (Ninfa et al., 1988; Keener and Kustu, 1988; Hess et al., 1988a; Wylie et al., 1988; Igo et al., 1989; Aiba et al., 1989a; Makino et al., 1989; Jin et al., 1990), it is clear that the phosphoaspartyl group within these proteins must be considerably more stable. One would expect the conformational change induced by phosphorylation to be a major contributor to this energy of stabilization. It is useful to consider the change in standard free energy associated with response regulator activation in terms of these two processes: the free energy of formation of the acylphosphate bond, $\Delta G^{\circ\prime}_{phospho}$, and the standard free energy of the associated conformational change, $\Delta G^{\circ\prime}_{conform}$:

$$\Delta G^{\circ\prime}_{activation} = \Delta G^{\circ\prime}_{phospho} + \Delta G^{\circ\prime}_{conform}$$

The value of $\Delta G^{\circ\prime}_{phospho}$ for phosphorylation by MgATP will be approximately +3 kcal/mol, assuming standard free energies for hydrolysis of MgATP and acylphosphate of −8 and −11 kcal/mol, respectively. Because the standard free energy of phosphorylation of a response regulator is assumed to correspond to the standard free energy of formation of small-molecule acyl phosphates, the value of $\Delta G^{\circ\prime}_{phospho}$ for phosphorylation by acetyl phosphate will be approximately 0 kcal/mol. The value for $\Delta G^{\circ\prime}_{conform}$ must therefore be relatively large and negative to favor formation of the phosphorylated regulator.

Similar considerations apply for the change in standard free energy associated with inactivation:

$$\Delta G^{\circ\prime}_{\text{inactivation}} = \Delta G^{\circ\prime}_{\text{dephospho}} - \Delta G^{\circ\prime}_{\text{conform}}$$

where $\Delta G^{\circ\prime}_{\text{dephospho}}$ equals the standard free energy of acyl phosphate hydrolysis, approximately -11 kcal/mol.

For most response regulators, it is desirable to have large negative values for $\Delta G_{\text{activation}}$ and $\Delta G_{\text{inactivation}}$ so that these reactions are subject to kinetic control. All the experimental evidence is consistent with this conclusion. Assuming the intracellular ratio of ATP/ADP is 10 (Neuhard and Nygaard, 1987) and the intracellular concentration of P_i is 10 mM (Rao et al., 1993), to have a response regulator work such that were the activation reaction to go to equilibrium there would be at least a 1,000-fold ratio of phospho- to dephosphoregulator and at the same time have the inactivation reaction poised so that if it were to go to equilibrium there would be at least a 1,000-fold ratio of dephospho- to phosphoregulator requires that $\Delta G^{\circ\prime}_{\text{conform}}$ have a value between -5 and -10 kcal/mol. In other words, for the response regulators to serve as kinetically regulated molecular switches, phosphorylation must trigger a large, thermodynamically favorable, conformational transition.

This argument may be turned on its head to understand why the response regulators are phosphorylated at aspartates instead of some other residue such as a serine or threonine. Phosphoaspartates have the highest standard free energy of hydrolysis of any phosphorylated residue. This high energy allows the protein to undergo a large energetically favorable conformational change and still maintain kinetic control of the phosphorylation and dephosphorylation reactions. Phosphoaspartates are hallmarks of proteins that undergo such conformational transitions, the most thoroughly analyzed examples being the ion motive ATPases that pump Na^+ and Ca^{2+} out of eukaryotic cells (Jencks, 1980; Tanford, 1984).

Phosphatase Kinetics

Response regulators generally use small-molecule phosphodonors such as acetyl phosphate (Fig. 6) to phosphorylate themselves and at the same time catalyze the hydrolysis of the phosphoaspartyl group that is formed (Lukat et al., 1992; Feng et al., 1992; McCleary and Stock, 1994). The kinetics of the overall phosphatase reaction is relatively simple and provides a valuable method for analysis of the intrinsic catalytic activities of any regulator in the absence of kinase:

$$\text{Acetyl phosphate} + \text{Regulator} \xrightarrow{k_p} \text{P{\sim}Regulator} + \text{Acetate}$$

$$\text{P{\sim}Regulator} + H_2O \xrightarrow{k_h} \text{Regulator} + P_i$$

The phosphorylation reaction is first order in both acetyl phosphate and regulator and may be characterized in terms of a second-order rate constant for phosphorylation, k_p. The dephosphorylation reaction is first order in phosphoregulator and may be characterized by a pseudo first-order rate constant, k_h. Under steady-state conditions, assuming the concentration of acetyl phosphate does not change, the phosphatase rate, V, follows simple Michaelis-Menten kinetics:

$$V = \frac{V_{\text{max}}(\text{Acetyl phosphate})}{(\text{Acetyl phosphate}) + K_m}$$

where $V_{\text{max}} = k_h(\text{response regulator})$, and $K_m = k_h/k_p$. The enzyme substrate complex in this reaction is not a complex between the regulator and acetyl phosphate but rather the covalent association of a phosphoryl group with the regulator. Even at millimolar concentrations of small-molecule phosphodonors, there is no evidence for the formation of a true enzyme substrate complex. The value of the apparent K_m reflects the relative rates of phosphorylation and dephosphorylation, not the affinity of the regulator for its substrate.

Acetyl phosphate is not a universal phosphodonor. At least one response regulator, CheB, is not phosphorylated by this substrate (Lukat et al., 1992). It seems likely that, because under normal conditions of growth acetyl phosphate levels can reach millimolar concentrations, phosphorylation by this donor has physiological consequences and therefore its phosphodonor activity has been selected for in evolution (McCleary et al., 1993). Phosphoramidate, NH_2PO_3, phosphorylates all regulators that have been examined, and this reaction is probably not physiologically relevant. Rates obtained with this donor may therefore be more informative concerning the general function of a regulator as a phosphotransfer catalyst. Rates of dephosphorylation (i.e., k_h) are independent of the donor used and simply reflect the autophosphatase activity for a given regulator.

SIGNAL TRANSDUCTION

The functionally critical step in two-component signal transduction is the interaction between the histidine kinases and the response regulators. The kinases are the sensors in that their activities are regulated by environmental signals; the response regulators are the receivers in that their activities are controlled by the sensors. The histidine kinases generally function to control both the rate of phosphorylation and the rate of dephosphorylation of the regulators. In view of their associated phosphatase functions, one might be tempted to use the term *kinase-phosphatase,* but the kinase designation refers to the histidine kinase activity of these proteins, not their regulatory interactions with response regulators. Response regulator phosphorylation and dephosphorylation activities are primarily catalyzed by the regulators themselves. Moreover, it seems unlikely that the conserved kinase domain communicates with the regulators directly. This function seems to be served by relatively unconserved portions of the kinase proteins.

Regulation of Phosphotransfer Activity

There are two ways in which the kinases modulate the rate of response regulator phosphorylation. First, the rate of histidine phosphorylation controls the availability of phosphodonor. This aspect of kinase function is an inherent feature of the kinase proteins, independent of the regulators. The second mechanism involves protein-protein contacts between the kinases and their cognate regulators that enhance the rate and determine the specificity of regulator phosphorylation. Phosphotransfer between cognate kinase-regulator pairs is at least two orders of magnitude faster than transfer between noncognate pairs and exhibits a much larger second-order rate constant than is observed for small-molecule phosphodonors. The enhanced rate and specificity both stem from protein-protein interactions that bring the regulator phospho-accepting site into close proximity to the cognate phosphohistidine donor. The reason small-molecule phosphodonors are such poor substrates is because they have such low affinities for the regulators, not because regulators have low intrinsic autophosphorylation catalytic activities for these donors.

Because the interaction of a given regulator with its cognate kinase is highly specific, one would expect the sequences within the kinase and regulator that are involved in this protein-protein interaction not to be conserved. In the CheA protein, an N-terminal 15-kDa domain appears to function as well as the intact protein to donate phosphoryl groups to the two chemotaxis response regulators, CheY and CheB (Hess et al., 1988a). This domain is not homologous to any portion of any other type of histidine kinase. The rate of phosphotransfer from this domain to the regulators is very rapid. CheY does not have a particularly high affinity for this domain. The high-affinity binding site for CheY, K_d of 30 nM, is located in an adjacent domain that is not required for rapid phosphotransfer and may serve an enirely different function (Swanson et al., 1993b). Because the intracellular concentration of CheY is greater than 10 μM (DeFranco and Koshland, 1981), binding in the nanomolar range is not essential to achieve a rapid and specific rate of phosphotransfer. Neither CheY binding to CheA nor the rate of phosphotransfer seems to be influenced by the state of CheA dimerization or the interaction between CheA, CheW, and the receptor. Phosphorylation of CheY does, however, cause its release from the high-affinity binding site (Schuster et al., 1993). This dissociation is probably driven by the phosphorylation-induced conformational change in CheY structure.

Regulation of Phosphatase Activity

Many histidine kinases function to facilitate the rate of dephosphorylation of their cognate response regulators. Genetic studies indicate that the phosphatase functions of the kinases are highly regulated by sensory inputs (J. B. Stock et al., 1989; Parkinson and Kofoid, 1992). Frequently, an auxiliary protein is involved. In the case of NR_{II}, phosphatase activity is dramatically stimulated by a protein called P_{II} in an interaction that is regulated by α-ketoglutarate and ATP binding to P_{II} (Ninfa, 1986; Keener and Kustu, 1988; Kamberov et al., in press).

In other cases, proteins that facilitate the dephosphorylation of the phosphoaspartate group in phosphorylated response regulators function independently of the kinase. For instance, the CheA protein does not appear to function as a phosphatase. In the chemotaxis system, this activity is supplied by CheZ, a protein that binds phospho-CheY and dramatically enhances its rate of dephosphorylation in the absence of any other component (Hess et al., 1988b; Lukat et al., 1992; Blat and Eisenbach, 1994). Moreover, an entire family of phosphorylated response regulator phosphatases that function independently of the corresponding protein histidine kinases has recently been identified in *B. subtilis*.

There is one well-characterized instance of a protein histidine kinase that functions alone to enhance the rate of dephosphorylation of its cognate regulator: EnvZ, which catalyzes the dephosphorylation of phospho-OmpR (Igo et al., 1989; Aiba et al., 1989a; Aiba et al., 1989b). This activity requires nucleotide binding but does not involve hydrolysis of nucleoside phosphates because nonhydrolyzable analogs work as well as ATP. The nucleotide requirement

suggests a role for the kinase domain in the regulation of EnvZ phosphatase activity, but the possibility cannot be excluded that a different nucleotide binding site is involved in EnvZ phosphatase regulation. Mutants in completely conserved residues in the EnvZ kinase domain that abolish kinase activity generally leave EnvZ-mediated phosphatase activity intact (Yang and Inouye, 1993). Similarly, in NR_{II} the conserved kinase domain is not essential because mutations can be introduced in any of the highly conserved residues of this domain with relatively little effect on phosphatase activity. Even a C-terminal deletion through the D, F, and G boxes does not preclude NR_{II}-dependent phosphatase activity (Atkinson and Ninfa, 1993). As expected, most of these mutations completely abolish histidine kinase activity. NR_{II} phosphatase activity is stimulated by nucleotides (Keener and Kustu, 1988). This effect does not depend on the binding of ATP to P_{II} (Kamberov et al., 1994). It is not clear, however, whether a known nucleotide binding site in NR_I (Weiss et al., 1991; Feng et al., 1992) is the locus for the nucleotide effect on NR_{II}-mediated phosphatase activity. It has recently been reported that dephosphorylation of the response regulator NarL is enhanced by its two cognate kinases, NarQ and NarX, in the absence of nucleotide (Schroder et al., 1994).

Histidine kinases must bind the dephosphorylated form of the regulators and release the phosphorylated form. If the free energy of kinase binding to the dephosphorylated conformational form were greater than the free energy of stabilization of the phosphorylated conformation, the bound phosphorylated protein would be forced into its dephospho-conformation, with the phosphoryl group being transferred back to the histidine in the kinase or to water. The latter possibility, which would result in a kinase-mediated phosphatase activity, would be difficult to avoid because the histidine residue that is phosphorylated must be free to move away from the response regulator so that it can interact with MgATP bound to the histidine kinase domain. This movement would be expected to open the site of phosphorylation in the regulator for the entry of water. It is clear, at least in the case of NR_{II}, that the phosphatase mechanism does not require phosphotransfer back to the histidine because mutagenesis of this residue to asparagine causes an enhancement in phosphatase activity (Atkinson and Ninfa, 1993). Most substitutions at this position, however, block both kinase and phosphatase functions.

There is a problem in using tight binding of the dephosphorylated conformation as a mechanism for phosphatase catalysis. Binding of the dephosphorylated regulator in its dephosphoconformation would effectively compete with the binding of phosphoregulator in its dephosphoconformation. If a kinase is to function as a phosphatase, there must be some mechanism by which the kinase avoids binding the dephosphorylated regulator. This could be achieved through unfavorable interactions with the dephosphorylated active site. NR_{II} and EnvZ mutants with altered phosphatase activities tend to cluster around the site of histidine phosphorylation (Atkinson and Ninfa, 1993; Brissette et al., 1991; Waukau and Forst, 1992) or in nonconserved upstream sequences (Atkinson and Ninfa, 1993; Harlocker et al., 1993). The regulator, NR_I, presumably binds to this region of NR_{II} when the histidine is phosphorylated. A phosphoryl group may be an important recognition element for a kinase-regulator interaction.

From these considerations, the most straightforward mechanism for phosphatase regulation in cases such as EnvZ in which no auxiliary protein is involved would entail binding of the dephosphoregulator to the kinase only when the histidine is phosphorylated and binding of phosphoregulator only when the histidine is not phosphorylated. When the rate of histidine phosphorylation is high, the phosphotransfer reaction would predominate; when the rate of histidine phosphorylation is low, the dephosphorylation reaction would be favored. Because the phosphatase reaction can presumably proceed in a monomer, the balance between phosphorylation and dephosphorylation could be controlled by the state of kinase dimerization.

CONCLUSIONS

Two-component systems are complex signal transduction modules composed of many interacting domains. The two conserved domains that give these systems their name have catalytic functions. The conserved histidine kinase domain catalyzes the transfer of a γ-phosphoryl group from MgATP to a histidine residue. The second type of conserved component, the response regulator domains, also function as enzymes. They catalyze the transfer of phosphoryl groups from phosphohistidines in the kinases and from small-molecule phosphodonors such as acetyl phosphate to a highly conserved aspartate residue at their active sites, and they also catalyze the hydrolysis of the resulting phosphoaspartyl group. Considerable progress has been made in understanding the molecular mechanisms that underlie these covalent reactions. In terms of their enzymology, two-component systems are simply a novel class of ATPases, and the general subject of ATPase catalysis is one of the best understood in all of biochemistry.

The most interesting aspect of two-component system activity is the part that is mediated by domains and auxiliary proteins that are not conserved. How is each system regulated, and how does it regulate? These are the central questions that must be addressed to understand function. The answers will depend on understanding reversible protein-protein interactions and the conformational transitions that regulate them. We need to know how protein domains communicate with one another, both in solution and through the membrane. In the future, our understanding of the mechanistic details of the ATPase reaction catalyzed by histidine kinase and regulator domains will provide a basis for beginning to understand the diverse mechanisms of this activity—how membrane receptors function to pass information into the cytoplasm, and how information is processed within the cell to produce appropriate responses to environmental signals.

ACKNOWLEDGMENTS

We thank Alex and Elizabeth Ninfa, Austin Newton, Tom Silhavy, Gudrun Lukat, Ann Stock, Pamela Lane, and Liya Shi for their helpful discussions and Jim Hoch for his insights concerning the kinetics of dephosphorylation.

Work from our laboratory was supported by a grant from the National Institutes of Health (AI20980 to J.B.S.). M.G.S. was supported by a postdoctoral fellowship from the Medical Research Council of Canada.

REFERENCES

Aiba, H., T. Mizuno, and S. Mizushima. 1989a. Transfer of phosphoryl group between two regulatory proteins involved in osmoregulatory expression of the *ompF* and *ompC* genes in *Escherichia coli*. *J. Biol. Chem.* **264:**8563–8567.

Aiba, H., F. Nakasai, S. Mizushima, and T. Mizuno. 1989b. Evidence for physiological importance of the phosphotransfer between the two regulatory components, EnvZ and OmpR, in osmoregulation of *Escherichia coli*. *J. Biol. Chem.* **264:**14090–14094.

Alper, S., L. Duncan, and R. Losick. 1994. An adenosine nucleotide switch controlling the activity of a cell type-specific transcription factor in *B. subtilis*. *Cell* **77:**195–205.

Amemura, M., K. Makino, H. Shinagawa, and A. Nakata. 1990. Cross talk to the phosphate regulon of *Escherichia coli* by PhoM protein: PhoM is a histidine protein kinase and catalyzes phosphorylation of PhoB and PhoM-open reading frame 2. *J. Bacteriol.* **172:**6300–6307.

Atkinson, M. R., and A. J. Ninfa. 1993. Mutational analysis of the bacterial signal-transducing protein kinase/phosphatase nitrogen regulator I (NR$_{II}$ or NtrB). *J. Bacteriol.* **175:**7016–7023.

Baumgartner, J. W., C. Kim, R. E. Brissette, M. Inouye, C. Park, and G. L. Hazelbauer. 1994. Transmembrane signalling by a hybrid protein: communication from the domain of chemoreceptor Trg that recognizes sugar-binding proteins to the kinase/phosphatase domain of osmosensor EnvZ. *J. Bacteriol.* **176:**1157–1163.

Bellsolell, L., J. Prieto, L. Serrano, and M. Coll. 1994. Magnesium binding to the bacterial chemotaxis protein CheY results in large conformational changes involving its functional surface. *J. Mol. Biol.* **238:**489–495.

Benkovic, S. J., and E. J. Sampson. 1971. Structure-reactivity correlation for the hydrolysis of phosphoramidate monoanions. *J. Am. Chem. Soc.* **93:**4009–4016.

Benson, A. K., G. Ramakrishnan, N. Ohta, J. Feng, A. J. Ninfa, and A. Newton. 1994. The *Caulobacter crescentus* FlbD protein acts at *ftr* sequence elements both to activate and to repress transcription of cell cycle-regulated flagellar genes. *Proc. Natl. Acad. Sci. USA* **91:**4989–4993.

Blat, Y., and M. Eisenbach. 1994. Phosphorylation-dependent binding of the chemotaxis signal molecule CheY to its phosphatase, CheZ. *Biochemistry* **33:**902–906.

Bochner, B. R., and B. N. Ames. 1982. Complete analysis of cellular nucleotides by two-dimensional thin layer chromatography. *J. Biol. Chem.* **257:**9759–9769.

Borkovich, K. A., L. A. Alex, and M. I. Simon. 1992. Attenuation of sensory signaling by covalent modification. *Proc. Natl. Acad. Sci. USA* **89:**6756–6760.

Borkovich, K. A., N. Kaplan, J. F. Hess, and M. I. Simon. 1989. Transmembrane signal transduction in bacterial chemotaxis involves ligand-dependent activation of phosphate group transfer. *Proc. Natl. Acad. Sci. USA* **86:**1208–1212.

Borkovich, K. A., and M. I. Simon. 1990. The dynamics of protein phosphorylation in bacterial chemotaxis. *Cell* **63:**1339–1348.

Bourret, R. B., K. A. Borkovich, and M. I. Simon. 1991. Signal transduction pathways involving protein phosphorylation in prokaryotes. *Annu. Rev. Biochem.* **60:**401–441.

Bourret, R. B., J. Davagnino, and M. I. Simon. 1993a. The carboxy-terminal portion of the CheA kinase mediates regulation of autophosphorylation by transducer and CheW. *J. Bacteriol.* **175:**2097–2101.

Bourret, R. B., S. K. Drake, S. A. Chervitz, M. I. Simon, and J. J. Falke. 1993b. Activation of the phosphosignaling protein CheY. Analysis of activated mutants by ^{19}F NMR and protein engineering. *J. Biol. Chem.* **268:**13089–13096.

Bourret, R. B., J. F. Hess, and M. I. Simon. 1990. Conserved aspartate residues and phosphorylation in signal transduction by the chemotaxis protein CheY. *Proc. Natl. Acad. Sci. USA* **87:**41–45.

Brissette, R. E., K. Tsung, and M. Inouye. 1991. Suppression of a mutation in OmpR at the putative phosphorylation center by a mutant EnvZ protein in *Escherichia coli*. *J. Bacteriol.* **173:**601–608.

Bruix, M., J. Pascual, J. Santoro, J. Prieto, L. Serrano, and M. Rico. 1993. ^{1}H- and ^{15}N-NMR assignment and solution structure of the chemotactic *Escherichia coli* CheY protein. *Eur. J. Biochem.* **215:**573–585.

Burbulys, D., K. A. Trach, and J. A. Hoch. 1991. Initiation of sporulation in *B. subtilis* is controlled by a multicomponent phosphorelay. *Cell* **64:**545–552.

Chang, C., S. F. Kwok, A. B. Bleecker, and E. M. Meyerowitz. 1993. *Arabidopsis* ethylene-response gene *ETR1:* similarity of product to two-component regulators. *Science* **262:**539–544.

DeFranco, A. L., and D. E. Koshland, Jr. 1981. Molecular cloning of chemotaxis genes and overproduction of gene products in the bacterial sensing system. *J. Bacteriol.* **147:**390–400.

DiSabato, G., and W. P. Jencks. 1961. Mechanism and catalysis of reactions of acyl phosphates. I. Nucleophilic reactions. *J. Am. Chem. Soc.* **83:**4393–4405.

Drake, S. K., R. B. Bourret, L. A. Luck, M. I. Simon, and J. J. Falke. 1993. Activation of the phosphosignaling protein CheY. *J. Biol. Chem.* **268:**13081–13088.

Dumas, C., I. Lascu, S. Morera, G. Philippe, R. Fourme, V. Wallet, M.-L. Lacombe, M. Veron, and J. Janin. 1992. X-ray structure of nucleoside diphosphate kinase. *EMBO J.* **11:**3202–3208.

Feng, J., M. R. Atkinson, W. McCleary, J. B. Stock, B. L. Wanner, and A. J. Ninfa. 1992. Role of phosphorylated metabolic intermediates in the regulation of glutamine synthetase synthesis in *Escherichia coli. J. Bacteriol.* **174:**6061–6070.

Forst, S., D. Comeau, S. Norioka, and M. Inouye. 1987. Localization and membrane topology of EnvZ, a protein involved in osmoregulation of OmpF and OmpC in *Escherichia coli. J. Biol. Chem.* **262:**16433–16438.

Fukami, Y., and F. Lipmann. 1983. Reversal of Rous sarcoma-specific immunoglobulin phosphorylation on tyrosine (ADP sd phosphate acceptor) catalyzed by the *src* gene kinase. *Proc. Natl. Acad. Sci. USA* **80:**1872–1876.

Gegner, J. A., and F. W. Dahlquist. 1991. Signal transduction in bacteria: CheW forms a reversible complex with the protein kinase CheA. *Proc. Natl. Acad. Sci. USA* **88:**750–754.

Gegner, J. A., D. R. Graham, A. F. Roth, and F. W. Dahlquist. 1992. Assembly of an MCP receptor, CheW, and kinase CheA complex in the bacterial chemotaxis signal transduction pathway. *Cell* **18:**975–982.

Grimsley, J. K., R. B. Tjalkens, M. A. Strauch, T. H. Bird, G. B. Spiegelman, Z. Hostomsky, J. M. Whiteley, and J. A. Hoch. 1994. Subunit composition and domain structure of the Spo0A sporulation transcription factor of *Bacillus subtilis. J. Biol. Chem.* **269:**16977–16982.

Guynn, R., H. L. Gelberg, and R. L. Veech. 1973. Equilibrium constants of the malate dehydrogenase, citrate synthase, citrate lyase, and acetyl coenzyme A hydrolysis reactions under physiological conditions. *J. Biol. Chem.* **248:**6957–6965.

Harlocker, S. L., A. Rampersaud, W.-P. Yang, and M. Inouye. 1993. Phenotypic revertant mutations of a new OmpR2 mutant (V203Q) of *Escherichia*

coli lie in the *envZ* gene, which encodes the OmpR kinase. *J. Bacteriol.* **175:**1956–1960.

Hazelbauer, G. L., H. C. Berg, and P. Matsumura. 1993. Bacterial motility and signal transduction. *Cell* **73:**15–22.

Herschlag, D., and W. P. Jencks. 1990. *J. Am. Chem. Soc.* **112:**1942–1950.

Hess, J. F., R. B. Bourret, and M. I. Simon. 1988a. Histidine phosphorylation and phosphoryl group transfer in bacterial chemotaxis. *Nature* **336:**139–143.

Hess, J. F., K. Oosawa, N. Kaplan, and M. I. Simon. 1988b. Phosphorylation of three proteins in the signaling pathway of bacterial chemotaxis. *Cell* **53:**79–87.

Hubler, L., G. N. Gill, and P. J. Bertics. 1989. Reversibility of the epidermal growth factor receptor self-phosphorylation reaction. Evidence for formation of a high energy phosphotyrosine bond. *J. Biol. Chem.* **264:**1558–1564.

Igo, M. M., A. J. Ninfa, J. B. Stock, and T. J. Silhavy. 1989. Phosphorylation and dephosphorylation of a bacterial transcriptional activator by a transmembrane receptor. *Genes Dev.* **3:**1725–1734.

Iuchi, S. 1993. Phosphorylation/dephosphorylation of the receiver module at the conserved aspartate residue controls transphosphorylation activity of histidine kinase in sensor protein ArcB of *Escherichia coli. J. Biol. Chem.* **268:**23972–23980.

Jencks, W. P. 1968. *Handbook of biochemistry.* CRC, Cleveland, Ohio.

Jencks, W. P. 1980. The utilization of binding energy in coupled vectorial processes. *Adv. Enzymol.* **51:**75–106.

Jin, S., R. K. Prusti, T. Roitsch, R. G. Ankenbauer, and E. W. Nester. 1990. Phosphorylation of the VirG protein of *Agrobacterium tumefaciens* by the autophosphorylated VirA protein: essential role in biological activity of VirG. *J. Bacteriol.* **172:**4945–4950.

Kahn, D., and G. S. Ditta. 1991. Modular structure of FixJ: homology of the transcriptional activator domain with the -35 binding domain of sigma factors. *Mol. Microbiol.* **5:**987–997.

Kamberov, E. S., M. R. Atkinson, P. Chandran, and A. J. Ninfa. 1994. Effect of mutation in *Escherichia coli glnL* (*NtrB*), encoding nitrogen regulator II (NR$_{II}$ or NtrB) on the phosphatase activity involved in bacterial nitrogen regulation. *J. Biol. Chem.* **269:**28294–28299.

Kamberov, E. S., M. R. Atkinson, J. Feng, P. Chandran, and A. J. Ninfa. Sensory components controlling bacterial nitrogen assimilation. *Cell Mol. Biol. Res.,* in press.

Keener, J., and S. Kustu. 1988. Protein kinase and phosphoprotein phosphatase activities of nitrogen regulatory proteins NTRB and NTRC of enteric bacterial: roles of the conserved amino-terminal domain of NTRC. *Proc. Natl. Acad. Sci. USA* **85:**4976–4980.

Kim, S.-H. 1994. "Frozen" dynamic dimer model for transmembrane signaling in bacterial chemotaxis receptors. *Protein Sci.* **3:**159–165.

Kofoid, E. C., and J. S. Parkinson. 1991. Tandem translation starts in the *cheA* locus of *Escherichia coli*. *J. Bacteriol.* **173:**2116–2119.

Koshland, D. E., Jr. 1951. Effect of catalysts on the hydrolysis of acetyl phosphate. Nucleophilic displacement mechanisms in enzymatic reactions. *J. Am. Chem. Soc.* **74:**2286–2292.

Levit, M., and J. B. Stock. Unpublished data.

Long, D. G., and R. M. Weis. 1992. Oligomerization of the cytoplasmic fragment from the aspartate receptor of *Escherichia coli*. *Biochemistry* **31:**9904–9911.

Lukat, G. S., B. H. Lee, J. M. Mottonen, A. M. Stock, and J. B. Stock. 1991. Roles of the highly conserved aspartate and lysine residues in the response regulator of bacterial chemotaxis. *J. Biol. Chem.* **266:**8348–8354.

Lukat, G. S., W. R. McCleary, A. M. Stock, and J. B. Stock. 1992. Phosphorylation of bacterial response regulator proteins by low molecular weight phospho-donors. *Proc. Natl. Acad. Sci. USA* **89:**718–722.

Lukat, G. S., and J. B. Stock. Unpublished data.

Lukat, G. S., A. M. Stock, and J. B. Stock. 1990. Divalent metal ion binding to the CheY protein and its significance to phosphotransfer in bacterial chemotaxis. *Biochemistry* **29:**5436–5442.

Lupas, A., and J. Stock. 1989. Phosphorylation of an N-terminal regulatory domain activates the CheB methylesterase in bacterial chemotaxis. *J. Biol. Chem.* **264:**17337–17342.

Maeda, T., S. M. Wurgler-Murphy, and H. Saito. 1994. A two-component system that regulates an osmosensing MAP kinase cascade in yeast. *Nature* **369:**242–245.

Magariyama, Y., S. Yamaguchi, and S.-I. Aizawa. 1990. Genetic and behavioral analysis of flagellar switch mutants of *Salmonella typhimurium*. *J. Bacteriol.* **172:**4359–4369.

Makino, K., M. Amemura, S.-K. Kim, A. Nakata, and H. Shinagawa. 1994. Mechanism of transcriptional activation of the phosphate regulon in *Escherichia coli*, p. 5–12. *In* A. Torriani-Gorini, E. Yagil, and S. Silver (ed.), *Phosphate in Microorganisms: Cellular and Molecular Biology*. American Society for Microbiology, Washington, D.C.

Makino, K., M. Shinagawa, M. Amemura, T. Kawamoto, M. Yamada, and A. Nakata. 1989. Signal transduction in the phosphate regulon of *Escherichia coli* involves phosphotransfer between

PhoR and PhoB proteins. *J. Mol. Biol.* **210:**551–559.

McCleary, W., M. McBride, and D. Zusman. 1990. Developmental sensory transduction in *Myxococcus xanthus* involves methylation and demethylation of FrzCD. *J. Bacteriol.* **172:**4877–4887.

McCleary, W. R., and J. B. Stock. 1994. Acetyl phosphate and the activation of two-component response regulators. *J. Biol. Chem.* **269:**31567–31572.

McCleary, W. R., J. B. Stock, and A. J. Ninfa. 1993. Is acetyl phosphate a global signal in *Escherichia coli*? *J. Bacteriol.* **175:**2793–2798.

Milburn, M. V., G. G. Prive, D. L. Milligan, W. G. Scott, J. Yeh, J. Jancarik, D. E. J. Koshland, and S.-H. Kim. 1991. Three-dimensional structures of the ligand-binding domain of the bacterial aspartate receptor with and without a ligand. *Science* **254:**1342–1347.

Milligan, D. L., and D. E. Koshland, Jr. 1988. Site-directed cross-linking. Establishment of the dimeric structure of the aspartate receptor of bacterial chemotaxis. *J. Biol. Chem.* **263:**6268–6275.

Milligan, D. L., and D. E. Koshland, Jr. 1993. Purification and characterization of the periplasmic domain of the aspartate receptor. *J. Biol. Chem.* **268:**19991–19997.

Min, K.-T., C. M. Hilditch, B. Diederich, J. Errington, and M. D. Yudkin. 1993. σ^F, the first compartment-specific transcription factor of *B. subtilis*, is regulated by an anti-σ factor that is also a protein kinase. *Cell* **74:**735–742.

Morrison, T. B., and J. S. Parkinson. 1994. Liberation of an interaction domain from the phosphotransfer region of CheA, a signaling kinase of *Escherichia coli*. *Proc. Natl. Acad. Sci. USA* **91:**5485–5489.

Moy, F. J., D. F. Lowry, P. Matsumura, F. W. Dahlquist, J. E. Krywko, and P. J. Domaille. 1994. Assignments, secondary structure, global fold, and dynamics of chemotaxis Y protein using three- and four-dimensional heteronuclear (^{13}C,^{15}N) NMR spectroscopy. *Biochemistry* **33:**10731–10742.

Needham, J. V., T. Y. Chen, and J. J. Falke. 1993. Novel ion specificity of a carboxylate cluster Mg(II) binding site: strong charge selectivity and weak size selectivity. *Biochemistry* **32:**3363–3367.

Neuhard, J., and P. Nygaard. 1987. Purines and pyrimidines, p. 445–473. *In* F. C. Neidhart, J. L. Ingraham, K. B. Low, B. Magasanik, M. Schaechter, and H. E. Umbarger (ed.), *Escherichia coli and Salmonella typhimurium: Cellular and Molecular Biology*. American Society for Microbiology, Washington, D.C.

Ninfa, A. J. 1986. Covalent modification of the *glnG* product, NRI, by the *glnL* product, NRII, regulates the transcription of the *glnALG* operon in *Es-*

cherichia coli. Proc. Natl. Acad. Sci. USA **83**:5909–5913.

Ninfa, A. J., and R. L. Bennett. 1991. Identification of the site of autophosphorylation of the bacterial protein kinase/phosphatase NRII. *J. Biol. Chem.* **266**:6888–6893.

Ninfa, A. J., E. G. Ninfa, A. N. Lupas, A. Stock, B. Magasanik, and J. Stock. 1988. Crosstalk between bacterial chemotaxis signal transduction proteins and the regulators of transcription of the Ntr regulon: evidence that nitrogen assimilation and chemotaxis are controlled by a common phosphotransfer mechanism. *Proc. Natl. Acad. Sci. USA* **85**:5492–5496.

Ninfa, A. J., S. Ueno-Nishio, T. P. Hunt, B. Robustell, and B. Magasanik. 1986. Purification of nitrogen regulator II, the product of the *glnL (ntrB)* gene of *Escherichia coli. J. Bacteriol.* **168**:1002–1004.

Ninfa, E. G. 1992. Ph.D. thesis. Princeton University, Princeton, N.J.

Ninfa, E. G., M. R. Atkinson, E. S. Kamberov, and A. J. Ninfa. 1993. Mechanism of autophosphorylation of *Escherichia coli* nitrogen regulator II (NR$_{II}$ or NtrB): *trans*-phosphorylation between subunits. *J. Bacteriol.* **175**:7024–7032.

Ninfa, E. G., A. Stock, S. Mowbray, and J. Stock. 1991. Reconstitution of the bacterial chemotaxis signal transduction system from purified components. *J. Biol. Chem.* **266**:9764–9770.

Olmedo, G., E. G. Ninfa, J. Stock, and P. Youngman. 1990. Novel mutations that alter the regulation of sporulation in *Bacillus subtilis. J. Mol. Biol.* **215**:359–372.

Ota, I. M., and A. Varshavsky. 1993. A yeast protein similar to bacterial two-component regulators. *Science* **262**:566–570.

Pan, S. Q., T. Charles, S. Jin, Z.-L. Wu, and E. W. Nester. 1993. Preformed dimeric state of the sensor protein VirA is involved in plant-*Agrobacterium* signal transduction. *Proc. Natl. Acad. Sci. USA* **90**:9939–9943.

Parkinson, J. S. 1993. Signal transduction schemes of bacteria. *Cell* **73**:857–871.

Parkinson, J. S., and E. C. Kofoid. 1992. Communication modules in bacterial signaling proteins. *Annu. Rev. Genet.* **26**:71–112.

Parkinson, J. S., and S. R. Parker. 1979. Interaction of the *cheC* and *cheZ* gene products is required for chemotactic behavior in *Escherichia coli. Proc. Natl. Acad. Sci. USA* **76**:2390–2394.

Parkinson, J. S., S. R. Parker, P. B. Talbert, and S. E. Houts. 1983. Interactions between chemotaxis genes and flagellar genes in *Escherichia coli. J. Bacteriol.* **155**:265–274.

Perego, M., C. Hanstein, K. W. Welsh, T. Djavakhishvili, P. Glaser, and J. A. Hoch. 1994. Multiple protein aspartate phosphatases provide a

mechanism for the integration of diverse signals in the control of development in *Bacillus subtilis. Cell* **79**:1047–1055.

Popov, K. M., Y. Zhao, Y. Shimomura, M. J. Kuntz, and R. A. Harris. 1992. Branched-chain α-ketoacid dehydrogenase kinase: molecular cloning, expression, and sequence similarity with histidine protein kinases. *J. Biol. Chem.* **267**:13127–13130.

Ramakrishnan, G., and A. Newton. 1990. FlbD of *Caulobacter crescentus* is a homologue of the NtrC (NR$_I$) protein and activates σ^{54}-dependent flagellar gene promoters. *Proc. Natl. Acad. Sci. USA* **87**:2369–2373.

Ramakrishnan, G., and A. Newton. Personal communication.

Rao, N. N., M. F. Roberts, A. Torriani, and J. Yashphe. 1993. Effect of *glpT* and *glpD* mutations on expression of the *phoA* gene in *Escherichia coli. J. Bacteriol.* **175**:74–79.

Rao, S., and M. G. Rossmann. 1973. Comparison of super-secondary structures in proteins. *J. Mol. Biol.* **76**:241–256.

Reyrat, J. M., M. David, P. Blonski, C. Boiski, P. Boistard, and J. Batut. 1993. Oxygen-regulated transcription of *Rhizobium meliloti nifA* and *fixK* genes. *J. Bacteriol.* **175**:6867–6872.

Roberts, D. L., D. W. Bennett, and S. A. Forst. 1993. Identification of the site of phosphorylation on the osmosensor, EnvZ, of *Escherichia coli. J. Biol. Chem.* **269**:8728–8733.

Roman, S. J., M. Meyers, K. Volz, and P. Matsumura. 1992. A chemotactic signaling surface on CheY defined by suppressors of flagellar switch mutations. *J. Bacteriol.* **174**:6247–6255.

Sanders, D. A., B. L. Gillece-Castro, A. L. Burlingame, and D. E. Koshland, Jr. 1992. Phosphorylation site of NtrC, a protein phosphatase whose covalent intermediate activates transcription. *J. Bacteriol.* **174**:5117–5122.

Sanders, D. A., B. L. Gillece-Castro, A. M. Stock, A. L. Burlingame, and D. E. Koshland, Jr. 1989. Identification of the site of phosphorylation of the chemotaxis response regulator protein, CheY. *J. Biol. Chem.* **264**:21770–21778.

Schneider-Poetsch, H. A. W. 1992. Signal transduction by phytochrome: phytochromes have a module related to the transmitter modules of bacterial sensor proteins. *Photochem. Photobiol.* **56**:839–846.

Schroder, I., C. D. Wolin, R. Cavicchioli, and R. P. Gunsalus. 1994. Phosphorylation and dephosphorylation of the NarQ, NarX, and NarL proteins of the nitrate-dependent two-component regulatory system of *Escherichia coli. J. Bacteriol.* **176**:4985–4992.

Schuster, S. C., R. V. Swanson, L. A. Alex, R. B. Bourret, and M. I. Simon. 1993. Assembly and function of a quaternary signal transduction complex monitored by surface plasmon resonance. *Nature* **365**:343–347.

Scott, W. G., D. L. Milligan, M. V. Milburn, G. G. Prive, J. Yeh, D. E. J. Koshland, and S.-H. Kim. 1993. Refined structures of the ligand-binding domain of the aspartate receptor from *Salmonella typhimurium. J. Mol. Biol.* **232**:555–573.

Shizuta, Y., J. A. Beavo, P. J. Bechtel, F. Hofmann, and E. G. Krebs. 1975. Reversibility of adenosine 3′:5′-monophosphate-dependent protein kinase reactions. *J. Biol. Chem.* **250**:6891–6896.

Simms, S. A., M. G. Keane, and J. Stock. 1985. Multiple forms of the CheB methylesterase in bacterial chemosensing. *J. Biol. Chem.* **260**:10161–10168.

Sockett, H., S. Yamaguchi, M. Kihara, V. M. Irikura, and R. M. Macnab. 1992. Molecular analysis of the flagellar switch protein FliM of *Salmonella typhimurium. J. Bacteriol.* **174**:793–806.

Stewart, R. C., A. F. Roth, and F. W. Dahlquist. 1990. Mutations that affect control of the methylesterase activity of CheB, a component of the chemotaxis adaptation system in *Escherichia coli. J. Bacteriol.* **172**:3388–3399.

Stock, A. M., E. Martinez-Hackert, B. F. Rasmussen, A. H. West, J. B. Stock, D. Ringe, and G. A. Petsko. 1993. Structure of the Mg^{2+}-bound form of CheY and mechanism of phosphoryl transfer in bacterial chemotaxis. *Biochemistry* **32**:13375–13380.

Stock, A. M., J. M. Mottonen, J. B. Stock, and C. E. Schutt. 1989. Three-dimensional structure of CheY, the response regulator of bacterial chemotaxis. *Nature* **337**:745–749.

Stock, A. M., D. C. Wylie, J. M. Mottonen, A. M. Lupas, E. G. Ninfa, A. J. Ninfa, C. E. Schutt, and J. B. Stock. 1988. Phospho-proteins involved in bacterial signal transduction. *Cold Spring Harbor Symp. Quant. Biol.* **53**:49–57.

Stock, J. B., G. S. Lukat, and A. M. Stock. 1991. Bacterial chemotaxis and the molecular logic of intracellular signal transduction networks. *Annu. Rev. Biophys. Biophys. Chem.* **20**:109–136.

Stock, J. B., A. J. Ninfa, and A. M. Stock. 1989. Protein phosphorylation and regulation of adaptive responses in bacteria. *Microbiol. Rev.* **53**:450–490.

Stock, J. B., A. M. Stock, and J. M. Mottonen. 1990. Signal transduction in bacteria. *Nature* **344**:395–400.

Swanson, R. V., R. B. Bourret, and M. I. Simon. 1993a. Intermolecular complementation of the kinase activity of CheA. *Mol. Microbiol.* **8**:435–441.

Swanson, R. V., S. C. Schuster, and M. I. Simon. 1993b. Expression of CheA fragments which define domains encoding kinase, phosphotransfer, and CheY binding activities. *Biochemistry* **32**:7623–7629.

Tanford, C. 1984. Twenty questions concerning the reaction cycle of the sarcoplasmic reticulum calcium pump. *CRC Crit. Rev. Biochem.* **17**:123–151.

Tawa, P., and R. C. Stewart. 1994a. Mutational activation of CheA, the protein kinase in the chemotaxis system of *Escherichia coli. J. Bacteriol.* **176**:4210–4218.

Tawa, P., and R. C. Stewart. 1994b. Kinetics of CheA autophosphorylation and dephosphorylation reactions. *Biochemistry* **33**:7917–7924.

Taylor, S. S., D. R. Knighton, J. Zheng, L. F. Ten Eyck, and J. M. Sowadski. 1992. Structural framework for the protein kinase family. *Annu. Rev. Cell Biol.* **8**:429–462.

Volz, K. 1993. Structural conservation in the CheY superfamily. *Biochemistry* **32**:11741–11753.

Volz, K., and P. Matsumura. 1991. Crystal structure of *Escherichia coli* CheY refined at 1.7-Å resolution. *J. Biol. Chem.* **266**:15511–15519.

Waukau, J., and S. Forst. 1992. Molecular analysis of the signaling pathway between EnvZ and OmpR in *Escherichia coli. J. Bacteriol.* **174**:1522–1527.

Weigel, N., M. A. Kukuruzinska, A. Nakazawa, E. B. Waygood, and S. Roseman. 1982. Sugar transport by the bacterial phosphotransferase system. *J. Biol. Chem.* **257**:14477–14491.

Weiss, D. S., J. Batut, K. E. Klose, J. Keener, and S. Kustu. 1991. The phosphorylated form of the enhancer-binding protein NTRC has an ATPase activity that is essential for activation of transcription. *Cell* **67**:155–167.

Weiss, V., and B. Magasanik. 1988. Phosphorylation of nitrogen regulator I (NRI) of *Escherichia coli. Proc. Natl. Acad. Sci. USA* **85**:8919–8923.

Welch, M., K. Oosawa, S.-I. Aizawa, and M. Eisenbach. 1993. Phosphorylation-dependent binding of a signal molecule to the flagellar switch of bacteria. *Proc. Natl. Acad. Sci. USA* **90**:8787–8793.

Williams, R. L., D. A. Oren, J. Munoz-Dorado, S. Inouye, M. Inouye, and E. Arnold. 1993. Crystal structure of *Myxococcus xanthus* nucleoside diphosphate kinase and its interactions with a nucleotide substrate at 2.0 Å resolution. *J. Mol. Biol.* **234**:1230–1247.

Wolfe, A. J., B. P. McNamara, and R. Stewart. 1994. The short form of CheA couples chemoreception to CheA phosphorylation. *J. Bacteriol.* **176**:4483–4491.

Wolfe, A. J., and R. C. Stewart. 1993. The short form of the CheA protein restores kinase activity and chemotactic ability to kinase-deficient mutants. *Proc. Natl. Acad. Sci. USA* **90**:1518–1522.

Wolodko, W. T., M. E. Fraser, M. N. G. James, and W. A. Bridger. 1994. The crystal structure of succinyl-CoA synthetase from *Escherichia coli* at 2.5-Å resolution. *J. Biol. Chem.* **269**:10883–10890.

Wylie, D., A. Stock, C.-Y. Wong, and J. Stock. 1988. Sensory transduction in bacterial chemotaxis involves phosphotransfer between Che proteins. *Biochem. Biophys. Res. Commun.* **151:**891–896.

Yamaguchi, S., S.-I. Aizawa, M. Kihara, M. Isomura, C. J. Jones, and R. M. Macnab. 1986. Genetic evidence for a switching and energy-transducing complex in the flagellar motor of *Salmonella typhimurium. J. Bacteriol.* **168:**1172–1179.

Yang, Y., and M. Inouye. 1992. Intermolecular complementation between two defective mutant signal-transducing receptors. *Proc. Natl. Acad. Sci. USA* **88:**11057–11061.

Yang, Y., and M. Inouye. 1993. Requirement of both kinase and phosphatase activities of an *Escherichia coli* receptor (Taz1) for ligand-dependent signal transduction. *J. Mol. Biol.* **231:**335–342.

Yeh, J. I., H. P. Biemann, J. Pandit, D. E. Koshland, and S.-H. Kim. 1993. The three-dimensional structure of the ligand-binding domain of a wild-type bacterial chemotaxis receptor. *J. Biol. Chem.* **268:**9787–9792.

Structural and Functional Conservation in Response Regulators

Karl Volz

4

Key control points for bacterial environmental and virulence response systems are signal processing pathways based on an archetypal "two-component" paradigm. The sensing component of such a system is an autophosphorylating protein kinase, whereas the second component—the response regulator—becomes activated after receiving the phosphoryl group from the first. Two-component systems have been identified in more than 20 distinct types of signal transduction pathways, ranging over 30 prokaryotic genera. Operations of the individual systems vary widely, but their primary activation processes are probably all based on the same chemical mechanism of Mg^{2+}-dependent protein phosphorylation. Preservation of a common signaling mechanism in so many different pathways may lie in the survival advantage of finely tuned intersystem crosstalk, enabling a concerted measure of response for optimal environmental adaptation.

Most response regulators are multidomain proteins. They are very diverse in their domain organization and full-length sequences, but because they are all homologous in their regulatory domain, they have been grouped together as a response regulator superfamily. One member of this superfamily is CheY, the single domain regulator of the bacterial chemotaxis system. CheY has become the prototype for investigation of the molecular mechanisms underlying the primary activation events of response regulators. Three-dimensional models are now available for CheY in its apo, Mg^{2+}-bound, and various mutant forms. This review focuses on the details of these structures, with the intent of deriving structure and function principles applicable to the regulatory domains of all two-component systems.

TWO-COMPONENT SYSTEMS AND THEIR RESPONSE REGULATORS

A Very Brief History

The first indication of a family of bacterial response regulators appeared in 1985, when the amino acid sequence of CheY was shown to be related to regulatory proteins of other cellular processes, such as membrane protein synthesis and sporulation (Stock et al., 1985). Soon after, the dual component nature of these systems was described (Nixon et al., 1986), and the details of their molecular commonalities continued to develop. Excellent reviews of the genetics and molecular biology of two-component systems have become available (Stewart and Dahlquist, 1987; Albright et al., 1989; J. B.

Karl Volz, Department of Microbiology and Immunology, University of Illinois at Chicago, Chicago, Illinois 60612.

Two-Component Signal Transduction, Edited by James A. Hoch and Thomas J. Silhavy, © 1995 American Society for Microbiology, Washington, DC 20005

Stock et al., 1989; Bourret et al., 1991; Ninfa, 1991; Stock et al., 1991; Parkinson and Kofoid, 1992; see also relevant chapters in this book). CheY was first crystallized in 1986 (Volz et al., 1986), and the three-dimensional structures for both the *Salmonella typhimurium* (A. M. Stock et al., 1989) and *Escherichia coli* (Volz and Matsumura, 1991) species were solved independently. A comprehensive structure-function analysis of 103 amino acid sequences of response regulators appeared in late 1993 (Volz, 1993); this review is largely based on that report. New two-component systems continue to be discovered; presently, there are more than 150 response regulator sequences with homologous regulatory domains.

Family Relationships

Response regulator sequences have been divided into families based on similarities in their extra domains. Historically, these families have been named after their founder members (i.e., CheY, NtrC, FixJ, and OmpR). The CheY family includes the short single-domain response regulators, whereas all others are multidomain proteins. Most of them are positive transcriptional regulators, containing DNA recognition sequences in their carboxy-terminal domains. The NtrC type includes proteins with full-length homologies to NtrC, typically about 460 ± 20 residues in length, possessing two domains after the response regulator. The last domain of NtrC is homologous to the FIS protein (Johnson et al., 1988; Koch et al., 1988; North et al., 1993), a helix-turn-helix regulatory protein whose three-dimensional structure is known (Kostrewa et al., 1991; Yuan et al., 1991; Kostrewa et al., 1992). The FixJ family contains sequences with one extra FixJ-type domain on their carboxy termini, giving a total length of about 220 residues. The OmpR family includes PhoP, PhoB, VirG, and other full-length OmpR homologs, all two-domain, also about 230 amino acids long.

Amino acid conservation among the regulatory domains is slightly higher within families than across family lines and highest within the OmpR family. Higher intrafamily conservation possibly reflects retention of features specific for interaction with their additional familial domains. However, some sequences show stronger interfamily relationships, such as the FixJ family. After exclusion of scores between highly homologous sequences, the average percent identity score for alignment of the regulatory domains is 23%, with the lowest value of just 6% (Volz, 1993).

Modular Design: The Framework for Functional Variety

It is generally accepted that the regulatory domains of all response regulators share a common structure. It is also believed that most response regulators use the same biochemical mechanism in their primary activation event. This is plausible considering the special conservative relationships among the sequences of the superfamily.

The five standard domain types of multidomain response regulators are found in various combinations, joined together by flexible linkers. This modularity of domain organization illustrates the versatility in functional selection of bacterial signal transduction proteins, where the different accessory domains have been combined with modified versions of the regulatory module to obtain new biochemical activities (Kofoid and Parkinson, 1988; Parkinson and Kofoid, 1992; see also Chapter 2). Functional divergence among these proteins likely arose through selection of the interactions between the unique regions of the CheY-like domain and the family-specific domains. The evolutionary development of this extensive set of regulatory proteins may have occurred through horizontal transmission (Parkinson and Kofoid, 1992).

MOLECULAR STRUCTURE OF CheY: A USER'S GUIDE

Overall Folding and Secondary Structural Features

CheY itself is a small globular protein, made up of 128 amino acid residues. The molecule is compactly folded in a $(\beta/\alpha)_5$ manner, where all five β-strands are in a parallel β-sheet, with the

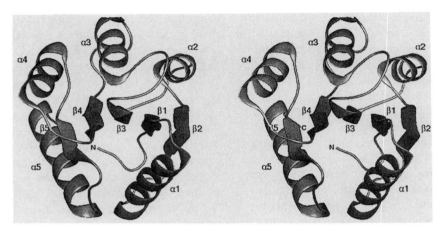

FIGURE 1 Stereo diagram of overall structure of CheY. See discussion in text.

topology $\beta_2\beta_1\beta_3\beta_4\beta_5$ (Fig. 1). CheY is about 40% α-helix and about 20% β-strand, whereas approximately 25% of the molecule is committed to turns connecting the regions of repetitive secondary structure. The remaining approximately 15% of unclassified structure appears in loops of varying sizes. The structural core of CheY is made of the internal β-strands β_1, β_3, and β_4. The five amphiphilic α-helices cluster into two groups lying on opposite sides of the central β-sheet, so that much of the two outermost strands (β_2 and β_5) is solvent-accessible.

Most of the structural features of CheY are discussed in detail in the original report of the high-resolution *E. coli* CheY structure (Volz and Matsumura, 1991; see also entry pdb3chy.ent in the Protein Data Bank [Bernstein et al., 1977]). What follows here is a narrative guide through the CheY molecule, highlighting the interactions that serve as its principal structural determinants. Because of the regularity of the β/α-construction of CheY, it is convenient to review its structure in sections, divided according to five repeating units of the general form "turn-strand-turn-helix."

TURN-β_1-TURN-α_1

The first secondary structure element in CheY appears near the amino terminus, with a type I β-turn at the residues DKEL, positions 3 to 6. K7 is the last solvent-exposed residue before

the polypeptide chain enters the interior of the molecule, where the hydrophobic sequence FLVV (8 to 11) forms the first β-strand. The subsequent β_1-α_1-connection has a β-turn involving residues DDFS, numbered 12 to 15. The aromatic side chain of F14 is entirely solvent-exposed. Helix α_1 begins with the N-capping residue T16 and continues regularly until the tight termination with G29 in an α_L-conformation.

TURN-β_2-TURN-α_2

The polypeptide turns back from α_1 to the β_2-strand, positioned on one solvent-accessible side of the β-sheet, running parallel to β_1. The short connection from α_1 to β_2 involves only residues 30 to 32; they are in extended β-conformation, but with no interstrand hydrogen-bonding support. β_2 is a strand with normal β-configuration, containing a large proportion of hydrophilic amino acid residues. The $\beta_2\leftrightarrow\beta_1$ hydrogen bonds range only from residues 32 to 36 of β_2. The β_2-α_2-link is also short and tight, with the D38 side chain forming the N-capping hydrogen bond on top of α_2. The helix is terminated with a tight turn and hydrogen bond between atoms K45 O and A48 N.

TURN-β_3-TURN-α_3

The loop preceding β_3 is a relatively tenuous link in CheY, where residues 48 to 52 of the sequence AGGYG have some of the highest temperature factors of the molecule. G50 happens to have an α_L-configuration, whereas G52 starts the β_3 strand with a distorted β-bulge. The residues FVIS (53 to 56) of β_3 pass through the center of the molecule, and the strand emerges on the other side after D57. The β_3 to α_3 link from D57 to G65 (DWNMPNMDG) is a complicated and rather rigid loop, supported by many stabilizing hydrogen bonds. Residues DWNM form a common β-turn, followed immediately by a rare classic γ-turn centered on residues PNM. The γ-turn constrains the central residue of N62 to a conformation with a positive ϕ-value. This turn in the loop also serves as a cap for the top of the neighboring helix α_2 through additional hydrogen-bonding interactions. D64, the first residue beyond the γ-turn, is the N-cap side chain for helix α_3. G65 resides in the internal portion of the start of this third helix, where its amide nitrogen has a hydrogen bond back to the carbonyl oxygen of W58. Helix α_3 proceeds normally, until a tight turn and termination by a hydrogen bond between atoms T71 O and A74 N.

TURN-β_4-TURN-α_4

The connection between α_3 and β_4 contains two consecutive turns, with the appearance of being structurally extraneous. Residues 75 to 78 (DGAM) are in a single turn of α-helical conformation, followed immediately by a type I β-turn involving residues 78 to 81 (MSAL). The $\alpha_3 \leftrightarrow \beta_4$-interaction is stabilized a great deal by R73, whose entire guanidinium group is packed internally, forming hydrogen bonds with the backbone carbonyl oxygens of S79, L81, and G102. β_4 is formed by the sequence PVLMVT of residues 82 to 87. The subsequent connection between β_4 and α_4 is through the TAEA sequence of residues 87 to 90, which form another β-turn. Residues 89 to 91 are in extended β-conformation, although no supportive hydrogen-bonding structure exists for them. This long and relatively weak link permits α_4 to be positioned farther away from the body of the CheY molecule compared with

the other four helices, leaving a large interior space, partially occupied by Y106 (see below). The residues of the sequence AAAQAGA (97 to 103) at the end of α_4 are all well ordered; helix 4 ends with a tight turn, having G102 in α_L-conformation.

TURN-β_5-TURN-α_5

The last strand, β_5, begins with a β-bulge distortion at S104, where the Oγ forms the first hydrogen bond to the amide nitrogen of V83. Normal β-conformation for this strand is found only through residues Y106 to V108, whereas the parallel interstrand hydrogen-bonding interactions between β_5 and β_4 extend from G105 to K109. The side chain of Y106 is an unusual two-state rotational conformer, where it is found in two positions of extremely different environments. One position (g) is internal, filling the large cavity between helix α_4 and the strands β_4 and β_5. The other position (t) is fully external and solvent-exposed. At the end of β_5, there is an uncommon type VIb turn involving residues VKPF, numbered 108 to 111. This turn features a *cis* peptide bond between K109 and P110. Residues F111 and T112 continue in the extended backbone conformation for the link between β_5 and α_5, where this last helix begins with the α-conformation of A113. Finally, the termination of α_5 is through a tight helical distortion, with G128 in an α_L-conformation.

Active Site Structure of Apo-CheY

The phosphorylation region of CheY resides atop the carboxy termini of the central β-sheet, and includes five important amino acid side chains from the strands β_1, β_3, β_4, and β_5. At one side of the active site, D12 and D13 are grouped on the same face of the turn after β_1 and project toward the active site. The central strand, β_3, contributes D57, the acidic residue known to be phosphorylated in the activation process of CheY (Sanders et al., 1989). The adjacent strand, β_4, holds T87 in a position such that its γ-hydroxyl group participates in the hydrogen-bonding network of the active site solvent molecules. Lastly, K109, at the end of the outermost β_5-strand, has its side chain in a fully extended configuration, reaching back into the active site, with a firm hydrogen bond to the carboxyl group of D57. The D57–K109 pair is essentially buried, showing a negligible solvent-accessible surface area of only 7.2 Å^2 for the functional groups of both residues. The acidic groups of residues D12 and D13 are also quite inaccessible (6.5 Å^2 total). This is consistent with the anomalous pK_as of the Mg^{2+} binding groups of CheY, estimated to be about 6.0 (Lukat et al., 1990). All potential bonds for the side chains of D57 and K109 are satisfied, and the solvent structure is highly ordered. One SO$_4^{2-}$ molecule is close to the active site, with an ionic bond to the ε-amino group of K109, and another solvent molecule (an NH$_4^+$ ion) resides in the Mg^{2+} binding region near D12 and D13. The presence of the NH$_4^+$ and SO$_4^{2-}$ ions in the structure is a direct result of the crystallization conditions [~2.3 M (NH$_4$)$_2$SO$_4$].

STRUCTURAL AND FUNCTIONAL SIGNIFICANCE OF CONSERVED RESIDUES

Core and Tertiary Structure Conservation

A representation of a multiple-sequence alignment of 79 CheY homologs is shown in the histogram of Fig. 2, where the percent occurrence of each amino acid type is plotted as a function of residue position, with CheY numbering. The most obvious structural feature of the CheY-like domain is the highly hydrophobic nature of the three central β-strands. The hydrophobic core also includes eight highly conserved sites from the surrounding α-helices: sites 21 and 25 from α_1; 46 from α_2; 68, 69, and 72 from α_3; and 116 and 120 from α_5. Twenty-six remaining contributions to the outer shell of hydrophobic residues come from the helices, the intervening loops, and the two outermost β-strands. Three sites—58, 96, and 99—are highly conserved as hydrophobic amino acids, yet in the CheY molecule have a high degree

of solvent accessibility. These positions may be involved in interprotein contacts or inter-domain contacts for multidomain response regulators.

The five helices differ in their degree of variability for structural reasons. For instance, α_3, bordered on either side by α_2 and α_4, is the most conserved, whereas the greatest variability is found in α_2, the most exposed helix. The variability in α_5 might just be due to lack of structural requirements for the flexible linker to the subsequent domain. In general, sections of highest variability in the primary sequence alignment coincide with regions of the CheY molecule most distant from the active site— implying that they are the regions most able to tolerate structural changes without detrimental functional consequences.

Secondary Structure Conservation

All other highly conserved amino acid positions coincide with important structural features of the CheY molecule (Table 1 and Fig. 3). The largest example is the γ-turn loop, prominently located adjacent to the phosphorylation region of the molecule. The apparent "immutability" of this surface loop is the consequence of a combination of five structural elements: the exclusive occurrence of a buried hydrophobic residue at position 60; a conserved proline at position 61; the γ-turn centered at position 62; the N-capping residue of α_3 at position 64; and finally, the internal start of α_3 at the conserved glycine of position 65. The large degree of solvent accessibility and proximity of the γ-turn loop to the phosphorylation region suggests that its conservation may be important for intermolecular recognition. The γ-turn loop might serve as a "trademark" retained by all response regulators for recognition by the kinases, whereas residues in the nearby variable sites might contribute to different systems' specificities. The cognate histidine kinases of these systems might also conservatively retain a complementary recognition surface.

The second uncommon structural feature located in a highly conserved region is the type VIb turn involving the residues KPF, numbered

109 to 111. This turn features a *cis* peptide bond between K109 and P110. K109 is absolutely conserved, whereas the percent identity for proline at 110 is 81%. Conservation of this turn may be related to function, perhaps to confer rigidity to the loop connecting β_5 to α_5. Hypothetically, if repositioning of the K109 backbone atoms occurs during the activation process, the effect would propagate more directly to surrounding parts of the molecule through a rigid link.

Another region in the active site periphery contains the two highly conserved positions 87 and 88. Residue 87 is exclusively a hydroxy-amino acid, likely for functional purposes. The side chain of position 88 appears to be restricted in bulk, perhaps to simply preserve access to the phosphorylation region.

There are also many localized sites of high conservation that serve specialized secondary structural roles in the CheY molecule. For instance, conservation of arginine at position 18 is not for functional reasons but is instead due to the hydrophobic role of the arginine's aliphatic methylene groups in sealing the α_1-β_2-interface. This provides an interesting example in which interchanges between arginine and hydrophobic amino acids (L, A, V, I, and C) are actually conservative substitutions. Another conserved arginine is at position 73 in CheY. The entire guanidinium group of R73 is packed internally, forming multiple hydrogen bonds with three backbone carbonyl oxygens in a strong core-stabilizing manner as seen in many other proteins (Borders et al., 1994). Two more obvious examples of localized high conservation are the helix-terminating glycines

with α_L-configurations: G29 past the end of α_1, and G102 near the end of α_4. The only puzzling conserved site is the alanine at position 103. This site's conservation might be explained by the substitutional covariance it shares with position 106. Residue 106 is predominantly an aromatic, and the side chains for residues 103 and 106 are in van der Waals contact. Substitution by a smaller nonaromatic residue at 106 correlates positively with increased bulk at position 103.

A conservative trend seen in the amino termini of the interior β-strands is the presence of either a charged group or proline residue preceding each strand (CheY positions 7, 52, and 82). For example, position 52 (although a glycine in CheY) is by consensus an aspartate. This type of conservation is apparent in the multiple sequence alignment of other β/α-protein families, as was discussed in a survey of β-breakers in parallel β-sheets (Colloc'h and Cohen, 1991).

One last pattern worth noting is the commonly observed polarized distribution of charged amino acids along the course of the α-helices (Ptitsyn, 1969). The best examples are α_3 and α_5, where acidic residues cluster at the amino termini and basic residues dominate the carboxy termini.

Mutant Sites in Regulatory Domains

Many mutant forms of two-component response regulators have been characterized genetically and biochemically. It is instructive to consider the spatial distribution of point mutations on a generalized response regulator structure. Most mutant sites reside on the surface of

FIGURE 2 (A) Amino acid preference plot for regulatory domains of the response regulator superfamily. A histogram of the percent occurrence of each amino acid type is plotted against CheY numbering. Each bar represents the occurrence of each amino acid type divided by the total number of amino acids per site in the multiple alignment of 79 sequences from Volz (Volz, 1993). Gaps are not counted. Values less than 10% are omitted for clarity, so the sums for most positions do not reach 100%. The color code is acidic (D, E), red; hydroxyl (S, T), rose; hydrophobic (C, I, L, M, V), yellow; aromatic (F, W, Y), green; polar (H, N, Q), light blue; basic (K, R), dark blue; and structural (A, G, P), light gray. The secondary structural elements of the CheY molecule are shown at the top. (B) Stereo diagram of three-dimensional distribution of amino acid preferences on overall structure of CheY. Amino acid residue types are color coded as in A. The orientation is approximately the same as in Fig. 1. Highly conserved positions are labeled. The side chains of D12, D13, D57, T87, and K109 are also shown.

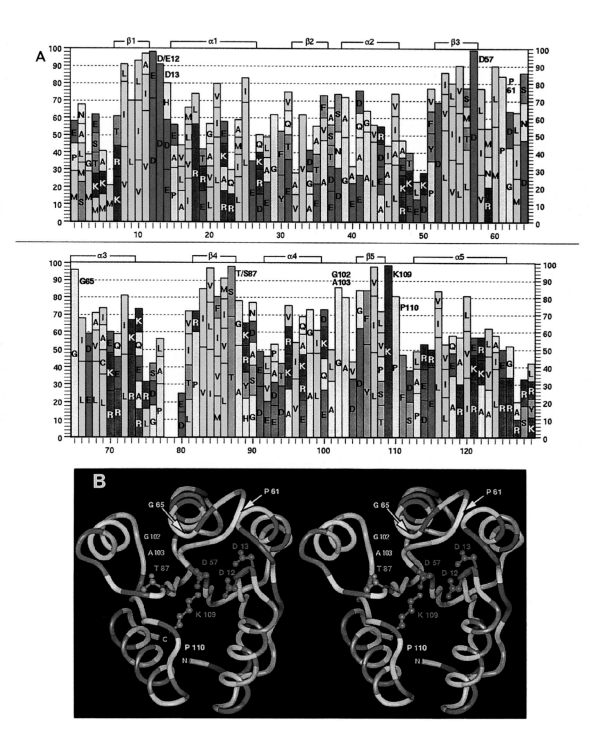

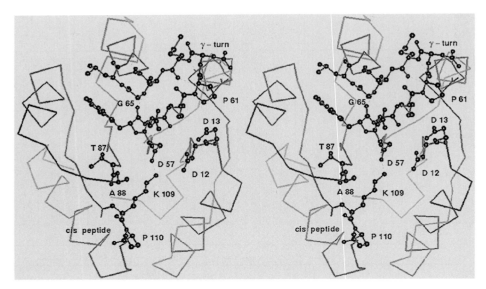

FIGURE 3 Stereo diagram of CheY region containing the most highly conserved residues of the superfamily. The orientation is approximately the same as in Figs. 1 and 2B. See discussion in text.

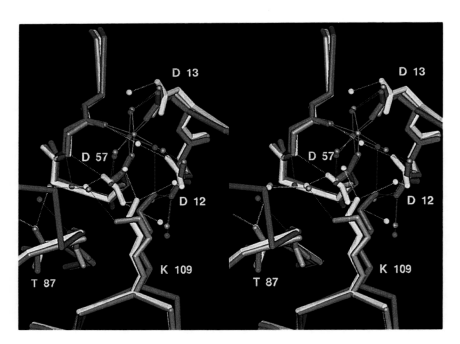

FIGURE 4 Stereo diagram of superposition of apo-CheY and the two CheY:Mg^{2+} complex structures. Details of active sites. The color code is white, wild-type *E. coli* apo-CheY; red, wild-type *E. coli* CheY:Mg^{2+}; blue, wild-type *S. typhimurium* CheY:Mg^{2+}. From the coordinates of Volz and Matsumura (1991); Bellsolell et al. (1994); and Stock et al. (1993). The orientation is rotated about 40° in the plane of the page as compared with Figs. 1, 2B, and 3. See discussion in text.

TABLE 1 Structural roles of "conserved" residues in CheY

Residue	Role
F8	Hydrophobic core sequence of internal β_1 strand
L9	Hydrophobic core sequence of internal β_1 strand
V10	Hydrophobic core sequence of internal β_1 strand
V11	Hydrophobic core sequence of internal β_1 strand
D12	Mg^{2+} binding
D13	Mg^{2+} binding and catalysis
M17	Hydrophobic cover of α_1-α_5-interface
R18	Hydrophobic cover of α_1-β_1-interface
V21	Hydrophobic core
L25	Hydrophobic core
G29	α_L-conformation; C-cap of α_1
F30	Hydrophobic cluster with F8, F53, and F124
V33	Sole hydrophobic residue of exposed β_2-strand; buried
L46	Hydrophobic core
G52	Aspartate by consensus; start of β_3
F53	Hydrophobic cluster with F8, F30, and F124
V54	Hydrophobic core sequence of internal β_3-strand
I55	Hydrophobic core sequence of internal β_3-strand
S56	Hydrophobic core sequence of internal β_3-strand
D57	Invariant; required for phosphorylation-induced activation
W58	Hydrophobic closure to α_3-β_4-α_4-interface
M60	Hydrophobic plug for crevice between β_1, α_2, β_3, and α_3
P61	Rigid group in γ-turn loop
N62	Positive ϕ-conformation; required for γ-turn
D64	N-cap of α_3
G65	Internal start of α_3; severe steric restrictions on β-substituent
L66	Internal residue of α_3
L68	Internal residue of α_3; hydrophobic core
L69	Internal residue of α_3; hydrophobic core
L72	Internal residue of α_3; hydrophobic core
P82	Start of β_4
V83	Hydrophobic core sequence of internal β_4-strand
L84	Hydrophobic core sequence of internal β_4-strand
M85	Hydrophobic core sequence of internal β_4-strand
V86	Hydrophobic core sequence of internal β_4-strand
T87	Invariant hydroxyl; proton donor to active site
A88	Reduced bulk; required for access to D57 active site
I96	Conserved solvent accessible hydrophobic residue
G102	α_L-conformation; C-cap of α_4
A103	Reciprocally covariant with position 106
Y106	Conserved aromatic; reciprocally covariant with 103
V107	Sole hydrophobic residue of exposed β_5-strand; buried
K109	Invariant; required for activation
P110	*cis* peptide following K109; rigidity to β_5-α_5-loop
F111	Outer shell of hydrophobic core
L116	Internal residue of α_5; hydrophobic core
L120	Internal residue of α_5; hydrophobic core
F124	Hydrophobic cluster with F8, F30, and F53

the molecule. There is an obvious clustering of mutations about the functionally important region of the response regulator domain; most are located near the phosphorylation site of the CheY molecule. Other mutations cluster about the two loops following the β_4 and β_5 strands, describing a spatially contiguous surface adjacent to the active site. This same region has been tentatively mapped out as the effector surface of the CheY molecule (Sockett et al., 1992; Roman et al., 1992).

CheY:Mg²⁺ Puzzle

Mg^{2+} is required for the phosphoryl group transfer reactions of CheY (Lukat et al., 1990) and, presumably, all other response regulators. The structural consequences of Mg^{2+} binding to CheY have received much attention, with two independent X-ray structure determinations of CheY:Mg^{2+} complexes (Stock et al., 1993; Bellsolell et al., 1994), as well as several nuclear magnetic resonance (NMR) investigations (Kar et al., 1992; Bruix et al., 1993; Drake et al., 1993; Bourret et al., 1993; Moy et al., 1994). The two X-ray structures agree on the location of the Mg^{2+} ion, placing it only 0.8 Å away from that originally predicted (Volz and Matsumura, 1991). However, the reports differ considerably in the conformational effects of Mg^{2+} binding.

One crystallography group approached the Mg^{2+} substitution problem by back-soaking crystals grown in high ionic strength conditions to low ionic strength, with 10 mM $MgCl_2$ (Stock et al., 1993). The resultant X-ray structure (see entry pdb2che.ent in the Protein Data Bank) shows no large-scale conformational changes due to the bound Mg^{2+} ion when compared with the unliganded form but does reveal significant rearrangements within the active site region. The octahedral coordination sphere of Mg^{2+} comprises three ligands from the protein (D13 Oδ_1, D57 Oδ_2, and N59 O) and three solvent molecules (Fig. 4). These three solvent molecules have hydrogen bonds to the participating amino acid side chains: one to each of the ligating D13 and D57 residues and the other to the Oδ_1 atom of D12. The

terminus of the K109 side chain in this complex is oriented away from the active site, with an apparent high-energy conformation, placing the ϵ-amino group in a 2.8-Å nonbonding contact with A88. The significance of this latter result was not discussed.

The other group solved the problem of Mg^{2+} substitution by cocrystallization in lower ionic strength conditions, with 200 mM magnesium acetate (Bellsolell et al., 1994; see also entry pdb1chn.ent in the Protein Data Bank). This approach has the advantage that it permits large-scale conformational changes to occur that could otherwise be suppressed in a crystal-soaking experiment. And indeed, large-scale changes they found. One turn of the three-turn helix α_4 becomes totally unwound, causing rearrangements of up to 10 Å for the N terminus of the helix. The loop connecting β_4 to α_4 undergoes the most dramatic shift, which, incidentally, results in a 120° rotation of the T87 side chain, pulling its Oγ out of the active site hydrogen-bonding network. The Mg^{2+} ion environment is different from the above-reported structure in one important respect. The three solvent molecules and the three protein ligands in the inner sphere are the same, but the ϵ-amino group of K109 is found *closer* to the active site, establishing hydrogen bonds with both Oδ_2 of D12 and a ligating solvent molecule (Fig. 4). The authors offered the intriguing interpretation that the (unphosphorylated) Mg^{2+}-free state of CheY is the "off" state of the signaling switch, whereas the Mg^{2+}-bound form is the "on" state, and so the phosphorylation event simply stabilizes the Mg^{2+}-bound form. That proposal may be difficult to reconcile if CheY is exclusively in the Mg^{2+}-bound state in vivo. However, the measured K_d of CheY for Mg^{2+} (Lukat et al., 1990; Kar et al., 1992; Needham et al., 1993) is in the same range as estimates for the intracellular Mg^{2+} activity (i.e., effective free concentration [Alatossava et al., 1985; Lusk et al., 1986; Needham et al., 1993]). A related concern is that the Mg^{2+} concentration used in this experiment was much higher than total levels found in the cell.

Even so, the CheY/Mg^{2+} molar ratio in the reported structure is 1:1. For the conformational changes to be artifactual, other Mg^{2+} ions should be detectable.

A variety of NMR investigations has also been applied to the question of the conformational effects of Mg^{2+} binding to CheY, but the issue is still not resolved. As an initial reference point, ^{1}H and ^{15}N-NMR assignments have definitively shown that the metal-free structure of CheY in solution is the same as that in the crystalline state (Bruix et al., 1993). One earlier study showed by nuclear Overhauser effect methods that binding of Mg^{2+} results in a conformational change in CheY, and it argued that the induced conformational change may form part of the phosphorylation-activated switching mechanism (Kar et al., 1992). But a later ^{19}F-NMR investigation gave a contrasting result, concluding that Mg^{2+} binding leads only to local structural changes in the active site, and K109 retains a similar position and conformation in the Mg^{2+}-empty and Mg^{2+}-occupied states (Drake et al., 1993). The authors cautioned, however, that the observed ^{19}F frequency shifts cannot be easily converted to a spatial scale. Other results are consistent with the earlier interpretation, in which a 3D and 4D nuclear Overhauser effect spectroscopy study on ^{13}C- and ^{15}N-enriched CheY provided convincing evidence that the K109-D57 bond is disrupted on Mg^{2+} binding and inferred a separation between the two groups of about 6 Å (Moy et al., 1994).

The exciting variety of results from the above experiments provides several possible models for Mg^{2+} binding to CheY, an essential step in the activation pathway. It is clear that some of these models are not consistent with others. Just as it is difficult to extract concrete structural interpretations from some of the NMR studies, interpretations of the X-ray results must also be tempered by our lack of knowledge of the true molecular conditions in the cell. If one single answer to this puzzle is to be found, it may embody features from several of these different models.

CONCLUSIONS AND PERSPECTIVES

CheY as the Tertiary Template

The scaffold for this superfamily is built on a singular, highly conserved domain. The molecular architecture of CheY provides well-founded explanations for the conservation within the aligned sequences of response regulator homologs. It is inevitable that the overall secondary structure—and even many of the specialized structural features—of CheY will be present in all homologs. These conclusions strongly support a common molecular mechanism of phosphorylation-dependent activation for all members of the superfamily.

Possible Activation Mechanism for CheY

Because of the extremely short half-life of the acylphosphate moiety on CheY (~10 s) (Hess et al., 1988b), a structural determination of phosphorylated CheY is not possible with conventional X-ray diffraction methods. Progress is being made toward solving the solution structure of the phosphorylated, activated structure by NMR methods. Until that goal is achieved, simple model-building experiments with the CheY structures will have to suffice. Several such models have been proposed, most addressing activation-inactivation mechanisms, and some postulating transition state intermediates for phosphorylated CheY:Mg^{2+} (Lukat et al., 1990; Lukat et al., 1991; Volz and Matsumura, 1991; Volz, 1993; Stock et al., 1993; Bellsolell et al., 1994). Details among the activation models vary, but they all invoke a central role for K109 in a phosphate-induced activation event. Most of the models entertain the possibility of a conformational change in the molecule subsequent to phosphorylation. The most direct evidence for this effect is the large-scale chemical shifts observed in an NMR experiment of CheY phosphorylation, summarized as a conformational change propagating from the phosphorylation site to adjacent regions (Lowry et al., 1994).

Phosphorylation is the required primary event in the activation of CheY (Hess et al., 1987; Hess et al., 1988a) and, presumably, all normally functioning response regulators, but phosphorylation and activation can be unlinked. The two most interesting types of phosphorylation mutants are (i) phosphorylatable but not active and (ii) nonphosphorylatable but constitutively active. Biochemical and structural results have been obtained for both types of mutants in CheY.

One constitutively active mutant of CheY has an aspartate-to-lysine substitution at position 13 (Bourret et al., 1990). It is possible that the structure of a constitutively active mutant might represent the activated conformation of the CheY molecule. However, the refined 2.3-Å resolution structure of the D13K mutant reveals no significant differences in backbone conformation when compared with wild-type CheY (Jiang et al., unpublished data). Assuming that the wild-type apo-CheY structure is the inactive form, these results might suggest that no conformational change accompanies the activation process of CheY. But a closer inspection shows that the active site bonding pattern in D13K is extensively rearranged, resulting in a significant overall weakening of the hydrogen-bond interactions. Thus, it is plausible that the D13K crystal structure represents an inactive but metastable state of the molecule, where its transformation to the activated form would proceed dynamically, in solution. This hypothesis has been explored with NMR spectroscopy (Drake et al., 1993; Bourret et al., 1993; Dahlquist et al., unpublished data).

Another window into the activation mechanism of CheY is through structural analysis of phosphorylatable but inactive mutants. The two known CheY mutants of that type are K109R (Lukat et al., 1991) and T87I (Matsumura et al., unpublished data). The T87I mutant structure has been solved and refined at a 2.1-Å resolution (Ganguli et al., submitted). The active site region of the T87I structure is very similar to wild type, as would be expected, but the molecule does show significant conformational differences in the "90s loop," which connects β_4 to α_4. This location partially overlaps the putative signaling surface of the CheY

molecule inferred by conformational suppression analysis (Sockett et al., 1992; Roman et al., 1992) and, interestingly, is the same region where large-scale structural changes were found in one of the CheY:Mg^{2+} complex structures (Bellsolell et al., 1994).

Two–Component Systems as Antimicrobial Targets

The fact that two-component regulatory systems are essential for bacterial survival and virulence makes them attractive candidates for antimicrobial drug design (DiRita and Mekalanos, 1989; Miller et al., 1989; Deretic et al., 1991). Two inhibitors of one signal transduction system have already been characterized (Roychoudhury et al., 1993). Thus, selective inhibition of a bacterial signal transduction pathway is feasible. But because of the widespread use of homologous two-component systems within each bacterial cell, a multitarget approach might be advantageous. If one compound could simultaneously inhibit as many kinase-response regulator pairs as possible, the effects would be cumulative. In other words, maximum inhibition of one critical pathway would not be necessary; even low-level inhibition of many mechanistically similar pathways could cause systemic confusion within the organism, inducing a form of "microbial psychosis" (Dorman, 1994). Also, such a multitarget approach would not be as susceptible to the common chromosomal resistance mechanisms arising from single-point mutations.

REFERENCES

Alatossava, T., H. Jütte, A. Kuhn, and E. Kellenberger. 1985. Manipulation of intracellular magnesium content in polymyxin B nonapeptide-sensitized *Escherichia coli* by ionophore A23187. *J. Bacteriol.* **162**:413–419.

Albright, L. M., E. Huala, and F. M. Ausubel. 1989. Prokaryotic signal transduction mediated by sensor and regulator protein pairs. *Annu. Rev. Genet.* **23**:311–336.

Bellsolell, L., J. Prieto, L. Serrano, and M. Coll. 1994. Magnesium binding to the bacterial chemotaxis protein CheY results in large conformational changes involving its functional surface. *J. Mol. Biol.* **238**:489–495.

Bernstein, F. C., T. F. Koetzle, G. J. B. Williams, E. F. Meyer, Jr., M. D. Brice, J. R. Rodgers, O. Kennard, T. Shimanouchi, and M. Tasumi. 1977. The protein data bank: a computer based archival file for macromolecular structures. *J. Mol. Biol.* **112:**535–542.

Borders, C. L., Jr., J. A. Broadwater, P. A. Bekeny, J. E. Salmon, A. S. Lee, A. M. Eldridge, and V. B. Bett. 1994. A structural role for arginine in proteins: multiple hydrogen bonds to backbone carbonyl oxygens. *Protein Sci.* **3:**541–548.

Bourret, R. B., K. A. Borkovich, and M. I. Simon. 1991. Signal transduction pathways involving protein phosphorylation in prokaryotes. *Annu. Rev. Biochem.* **60:**401–441.

Bourret, R. B., S. K. Drake, S. A. Chervitz, M. I. Simon, and J. J. Falke. 1993. Activation of the phosphosignaling protein CheY. Analysis of activated mutants by ^{19}F NMR and protein engineering. *J. Biol. Chem.* **268:**13089–13096.

Bourret, R. B., J. F. Hess, and M. I. Simon. 1990. Conserved aspartate residues and phosphorylation in signal transduction by the chemotaxis protein CheY. *Proc. Natl. Acad. Sci. USA* **87:**41–45.

Bruix, M., J. Pascual, J. Santoro, J. Prieto, L. Serrano, and M. Rico. 1993. ^{1}H and ^{15}N-NMR assignment and solution structure of the chemotactic *Escherichia coli* CheY protein. *Eur. J. Biochem.* **215:**573–585.

Colloc'h, N., and F. E. Cohen. 1991. β-Breakers: an aperiodic secondary structure. *J. Mol. Biol.* **221:** 603–613.

Dahlquist, F. W., et al. Unpublished data.

Deretic, V., C. D. Mohr, and D. W. Martin. 1991. Mucoid *Pseudomonas aeruginosa* in cystic fibrosis: signal transduction and histone-like elements in the regulation of bacterial virulence. *Mol. Microbiol.* **5:**1577–1583.

DiRita, V. J., and J. J. Mekalanos. 1989. Genetic regulation of bacterial virulence. *Annu. Rev. Genet.* **23:**455–482.

Dorman, C. 1994. *Genetics of Bacterial Virulence.* Blackwell Scientific Publications, Oxford, Great Britain.

Drake, S. K., R. B. Bourret, L. A. Luck, M. I. Simon, and J. J. Falke. 1993. Activation of the phosphosignaling protein CheY. Analysis of the phosphorylated conformation by ^{19}F NMR and protein engineering. *J. Biol. Chem.* **268:**13081–13088.

Ganguli, S., H. Wang, P. Matsumura, and K. Volz. Uncoupling phosphorylation and activation in bacterial chemotaxis: the 2.1 Å structure of a threonine to isoleucine mutant at position 87 of CheY. Submitted for publication.

Hess, J. F., R. B. Bourret, K. Oosawa, P. Matsumura, and M. I. Simon. 1988a. Protein phosphorylation and bacterial chemotaxis. *Cold Spring Harbor Symp. Quant. Biol.* **53:**41–48.

Hess, J. F., K. Oosawa, N. Kaplan, and M. I. Simon. 1988b. Phosphorylation of three proteins in the signaling pathway of bacterial chemotaxis. *Cell* **53:**79–87.

Hess, J. F., K. Oosawa, P. Matsumura, and M. I. Simon. 1987. Protein phosphorylation is involved in bacterial chemotaxis. *Proc. Natl. Acad. Sci. USA* **84:**7609–7613.

Jiang, M., R. B. Bourret, M. I. Simon, and K. Volz. Unpublished data.

Johnson, R. C., C. A. Ball, D. Pfeffer, and M. I. Simon. 1988. Isolation of the gene encoding the Hin recombinational enhancer binding protein. *Proc. Natl. Acad. Sci. USA* **85:**3484–3488.

Kar, L., P. Matsumura, and M. E. Johnson. 1992. Bivalent-metal binding to CheY protein. Effect on protein conformation. *Biochem. J.* **287:**521–531.

Koch, C., J. Vandekerckhove, and R. Kahmann. 1988. *Escherichia coli* host factor for site-specific DNA inversion: cloning and characterization of the *fis* gene. *Proc. Natl. Acad. Sci. USA* **85:**4237–4241.

Kofoid, E. C., and J. S. Parkinson. 1988. Transmitter and receiver modules in bacterial signaling proteins. *Proc. Natl. Acad. Sci. USA* **85:**4981–4985.

Kostrewa, D., J. Granzin, C. Koch, H.-W. Choe, S. Raghunathan, W. Wolf, J. Labahn, R. Kahmann, and W. Saenger. 1991. Three-dimensional structure of the *E. coli* DNA binding protein FIS. *Nature* **349:**178–180.

Kostrewa, D., J. Granzin, D. Stock, H.-W. Choe, J. Labahn, and W. Saenger. 1992. Crystal structure of the factor for inversion stimulation FIS at 2.0 Å resolution. *J. Mol. Biol.* **226:**209–226.

Lowry, D. F., A. F. Roth, P. B. Rupert, F. W. Dahlquist, F. J. Moy, P. J. Domaille, and P. Matsumura. 1994. Signal transduction in chemotaxis: a propagating conformational change upon phosphorylation of CheY. *J. Biol. Chem.* **269:** 26358–26362.

Lukat, G. S., B. H. Lee, J. M. Mottonen, A. M. Stock, and J. B. Stock. 1991. Roles of the highly conserved aspartate and lysine residues in the response regulator of bacterial chemotaxis. *J. Biol. Chem.* **266:**8348–8354.

Lukat, G. S., A. M. Stock, and J. B. Stock. 1990. Divalent metal ion binding to the CheY protein and its significance to phosphotransfer in bacterial chemotaxis. *Biochemistry* **29:**5436–5442.

Lusk, J. E., R. J. P. Williams, and E. P. Kennedy. 1986. Magnesium and the growth of *Escherichia coli*. *J. Biol. Chem.* **243:**2618–2624.

Matsumura, P., et al. Unpublished data.

Miller, J. F., J. J. Mekalanos, and S. Falkow. 1989. Coordinate regulation and sensory transduction in the control of bacterial virulence. *Science* **243:**916–922.

Moy, F. J., D. F. Lowry, P. Matsumura, F. W. Dahlquist, J. E. Krywko, and P. Domaille. 1994. Assignments, secondary structure, global fold, and dynamics of chemotaxis Y protein using three- and four-dimensional heteronuclear (^{13}C, ^{15}N) NMR spectroscopy. *Biochemistry* **33**:10731–10742.

Needham, J. V., T. Y. Chen, and J. J. Falke. 1993. Novel ion specificity of a carboxylate cluster Mg(II) binding site: strong charge selectivity and weak size selectivity. *Biochemistry* **32**:3363–3367.

Ninfa, A. J. 1991. Protein phosphorylation and the regulation of cellular processes by the homologous two-component regulatory systems of bacteria. *Genet. Eng.* **13**:39–72.

Nixon, B. T., C. W. Ronson, and F. M. Ausubel. 1986. Two-component regulatory systems responsive to environmental stimuli share strongly conserved domains with the nitrogen assimilation regulatory genes *ntrB* and *ntrC*. *Proc. Natl. Acad. Sci. USA* **83**:7850–7854.

North, A. K., K. E. Klose, K. M. Stedman, and S. Kustu. 1993. Prokaryotic enhancer-binding proteins reflect eukaryote-like modularity: the puzzle of nitrogen regulatory protein C. *J. Bacteriol.* **175**:4267–4273.

Parkinson, J. S., and E. C. Kofoid. 1992. Communication modules in bacterial signaling proteins. *Annu. Rev. Genet.* **26**:71–112.

Ptitsyn, O. B. 1969. Statistical analysis of the distribution of amino acid residues among helical and nonhelical regions in globular proteins. *J. Mol. Biol.* **42**:501–510.

Roman, S. J., M. Meyers, K. Volz, and P. Matsumura. 1992. A chemotactic signaling surface on CheY defined by supressors of flagellar switch mutations. *J. Bacteriol.* **174**:6247–6255.

Roychoudhury, S., N. A. Zeilinski, A. J. Ninfa, N. E. Allen, L. N. Jungheim, T. I. Nicas, and A. M. Chakrabarty. 1993. Inhibitors of two-component signal transduction systems: inhibition of alginate gene activation in *Pseudomonas aeruginosa*. *Proc. Natl. Acad. Sci. USA* **90**:965–969.

Sanders, D. A., B. L. Gillece-Castro, A. M. Stock, A. L. Burlingame, and D. E. Koshland, Jr. 1989. Identification of the site of phosphorylation of the chemotaxis response regulator protein, CheY. *J. Biol. Chem.* **264**:21770–21778.

Sockett, H., S. Yamaguchi, M. Kihara, V. M. Irikura, and R. M. Macnab. 1992. Molecular analysis of the flagellar switch protein FliM of *Salmonella typhimurium*. *J. Bacteriol.* **174**:793–806.

Stewart, R. C., and F. W. Dahlquist. 1987. Molecular components of bacterial chemotaxis. *Chem. Rev.* **87**:997–1025.

Stock, A. M., D. E. Koshland, Jr., and J. B. Stock. 1985. Homologies between the *Salmonella typhimurium* CheY protein and proteins involved in the regulation of chemotaxis, membrane protein synthesis, and sporulation. *Proc. Natl. Acad. Sci. USA* **82**:7989–7993.

Stock, A. M., E. Martinez-Hackert, B. F. Rasmussen, A. H. West, J. B. Stock, D. Ringe, and G. A. Petsko. 1993. Structure of the Mg^{2+}-bound form of CheY and mechanism of phosphoryl transfer in bacterial chemotaxis. *Biochemistry* **32**:13375–13380.

Stock, A. M., J. M. Mottonen, J. B. Stock, and C. E. Schutt. 1989. Three dimensional structure of CheY, the response regulator of bacterial chemotaxis. *Nature* **337**:745–749.

Stock, J. B., G. S. Lukat, and A. M. Stock. 1991. Bacterial chemotaxis and the molecular logic of intracellular signal transduction networks. *Annu. Rev. Biophys. Biophys. Chem.* **20**:109–136.

Stock, J. B., A. J. Ninfa, and A. M. Stock. 1989. Protein phosphorylation and regulation of adaptive responses in bacteria. *Microbiol. Rev.* **53**:450–490.

Volz, K. 1993. Structural conservation in the CheY superfamily. *Biochemistry* **32**:11741–11753.

Volz, K., J. Beman, and P. Matsumura. 1986. Crystallization and preliminary characterization of CheY, a chemotaxis control protein from *Escherichia coli*. *J. Biol. Chem.* **261**:4723–4725.

Volz, K., and P. Matsumura. 1991. Crystal structure of *Escherichia coli* CheY refined at 1.7 Å resolution. *J. Biol. Chem.* **266**:15511–15519.

Yuan, H. S., S. E. Finkel, J.-A. Feng, M. Kaczor-Grzeskowiak, R. C. Johnson, and R. E. Dickerson. 1991. The molecular structure of wild-type and a mutant Fis protein: relationship between mutational changes and recombinational enhancer function or DNA binding. *Proc. Natl. Acad. Sci. USA* **88**:9558–9562.

PARADIGMS

Control of Nitrogen Assimilation by the NR$_I$-NR$_{II}$ Two-Component System of Enteric Bacteria

Alexander J. Ninfa, Mariette R. Atkinson,
Emmanuel S. Kamberov, Junli Feng, and Elizabeth G. Ninfa

5

Enteric bacteria such as *Escherichia coli, Salmonella typhimurium,* and their relatives regulate the expression of glutamine synthetase (GS) and other enzymes important in nitrogen assimilation in response to changes in the availability of nitrogen. The preferred source of nitrogen, permitting the fastest growth rate, is ammonia. When ammonia is present at a sufficient concentration, only a low level of GS is synthesized. In medium-lacking ammonia or a concentration of ammonia that limits growth, the synthesis of GS is elevated (e.g., see Pahel et al., 1982; Reitzer and Magasanik, 1985). Furthermore, under such conditions of nitrogen limitation, enzymes that catalyze the uptake and catabolism of certain alternative nitrogen sources are synthesized if the substrates of these enzymes are present (Neidhardt and Magasanik, 1957; Kustu et al., 1979; Prival and Magasanik, 1971). For example, arginine can be used as a nitrogen source if ammonia is unavailable or limiting.

The set of genes and operons that are regulated by the availability of nitrogen is known as the Ntr regulon. This regulon is controlled by a two-component system consisting of the products of the *glnL* (*ntrB*) and *glnG* (*ntrC*) genes, referred to as nitrogen regulator II (NR$_{II}$ [NtrB]) and I (NR$_I$ [NtrC]), respectively. NR$_{II}$ is the signal-transducing kinase-phosphatase, and NR$_I$ is the response regulator of this two-component system. In this review, the current state of knowledge about the mechanisms of signal transduction by NR$_I$ and NR$_{II}$ is summarized briefly. In particular, the mechanism of the autophosphorylation of NR$_{II}$, the mechanisms of dephosphorylation of NR$_I$~P, the sensing mechanisms by which the intracellular nitrogen status controls the phosphorylation state of NR$_I$~P, and the mechanism by which nitrogen regulation occurs in cells that lack NR$_{II}$ are discussed. A more detailed review of the role of NR$_I$~P in transcriptional activation is presented in Chapter 9.

INTRODUCTION

Escherichia coli and related bacteria precisely regulate the level of GS activity by three distinct mechanisms. First, the intracellular concentration of the enzyme is regulated in response to the intracellular nitrogen status. This regulation is due to the control of the initiation of transcription of the *glnA* gene, encoding GS, from the nitrogen-regulated *glnA* promoter, *glnAp2* (Reitzer and Magasanik,

Alexander J. Ninfa, Mariette R. Atkinson, Emmanuel S. Kamberov, Junli Feng, and Elizabeth G. Ninfa, Department of Biological Chemistry, University of Michigan Medical School, Ann Arbor, Michigan 48109-0606.

Two-Component Signal Transduction, Edited by James A. Hoch and Thomas J. Silhavy,
© 1995 American Society for Microbiology, Washington, DC 20005

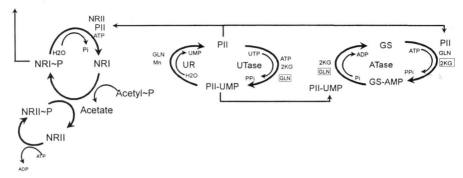

FIGURE 1 Signal transduction system controlling activity of glutamine synthetase and transcription of the Ntr regulon. Abbreviations: GLN, glutamine; 2KG, 2-ketoglutarate. The small molecules activating reactions are shown without boxes, and the small molecules inhibiting reactions are shown in boxes. The figure is similar to Fig. 1 of Atkinson et al. (1994) and Kamberov et al. (1994a, submitted).

1985). Second, the activity of the enzyme is regulated by reversible covalent modification (reviewed in Stadtman et al., 1980; Rhee et al., 1985). Under conditions of nitrogen excess, the dodecameric enzyme is converted to a less active form by adenylylation of a tyrosyl moiety on each subunit. A single sensing mechanism is used to control both the biosynthesis of GS and the covalent modification of GS, as discussed below. Finally, the activity of GS is allosterically controlled by cumulative feedback inhibition by eight small molecules: tryptophan, histidine, carbamyl phosphate, glucosamine-6-phosphate, CTP, AMP, alanine, and glycine. The amide nitrogen of glutamine is directly involved in the biosynthesis of each of these compounds with the exceptions of alanine and glycine. Bacterial GS may be one of the most precisely controlled enzymes in existence.

The covalent modification of GS involves the reversible adenylylation of the enzyme, catalyzed by the enzyme adenylyltransferase (ATase, *glnE* product) (reviewed in Stadtman et al., 1980; Rhee et al., 1985). GS is a dodecamer, and the activity of each subunit is independently controlled by adenylylation. The adenylylated form of the enzyme is less active in the biosynthetic reaction that produces gluta-

mine from glutamate and ammonia. GS also catalyzes the transfer of glutamine to hydroxylamine, producing glutamyl-hydroxamate; this is known as the transferase reaction. Adenylylated and unadenylylated GS are equally active in the transferase activity in the presence of Mn^{2+} (Bender et al., 1977). The existence of a convenient assay for total enzyme (transferase assay in the presence of Mn^{2+}) and nonadenylylated enzyme (transferase assay in the presence of Mg^{2+} or biosynthetic reaction assay) has facilitated the study of the effects of mutations in the signal transduction components that influence the activity of ATase. The adenylylation of GS greatly increases the sensitivity of the enzyme to cumulative feedback inhibition.

In wild-type cells, ATase, which catalyzes both the adenylylation and deadenylylation reactions, is allosterically regulated by glutamine and 2-ketoglutarate and receives additional information on the availability of nitrogen by means of a signal transduction system consisting of two proteins (Engleman and Francis, 1978; Francis and Engleman, 1978; reviewed in Stadtman et al., 1980; Rhee et al., 1985). The first of these is one sensor of the system, a uridylyltransferase/uridylyl-removing enzyme (UTase/UR) encoded by *glnD* (Adler et al.,

1975; Francis and Engleman, 1978; Garcia and Rhee, 1983; Kamberov et al., 1994a). The UTase/UR has one known substrate, the signal transduction protein known as "protein II" or P_{II} (product of *glnB*) (Brown et al., 1971; Adler et al., 1975). P_{II} communicates information from the UTase/UR protein to ATase (Fig. 1). Under conditions of nitrogen limitation, the UTase/UR protein converts P_{II} to a uridylylated form (Brown et al., 1971). P_{II}–UMP then stimulates the deadenylylation of adenylylated GS by ATase. Under conditions of nitrogen excess, the UTase/UR protein has the opposite effect; it catalyzes the removal of uridylyl groups from P_{II}–UMP. The resulting unmodified P_{II} then stimulates the adenylylation of unadenylylated GS by ATase. Thus, the UTase/UR protein controls the activities of ATase indirectly via P_{II} (reviewed in Stadtman et al., 1980; Rhee et al., 1985). The model as discussed so far suggests that P_{II} is a classical second messenger. However, as discussed later, P_{II} is also a sensory component responsible for sensing the intracellular carbon status. Specifically, it seems that the UTase/UR protein is responsible for the sensation of glutamine, and P_{II} is responsible for the sensation of 2-ketoglutarate.

Evidence with purified proteins suggests that the UTase/UR protein and the ATase protein are involved in sensing of the intracellular concentration of glutamine. In vitro, 2-ketoglutarate stimulates and glutamine inhibits the UTase activity, and glutamine stimulates the UR activity (Engleman and Francis, 1978; Francis and Engleman, 1978; Garcia and Rhee, 1983; Kamberov et al., 1994a). Changes in the nitrogen status of the cell are best reflected by changes in the intracellular concentration of these two metabolic intermediates. Glutamine and 2-ketoglutarate also exert allosteric effects on the ATase enzyme (Engleman and Francis, 1978). Glutamine stimulates the adenylylation activity of ATase in the absence of P_{II}, and the combination of glutamine and P_{II} synergistically stimulate the enzyme, whereas 2-ketoglutarate acts to prevent the stimulation of the

ATase adenylylation activity by P_{II}. The deadenylylation of GS-AMP by purified ATase requires P_{II}–UMP, ATP, and 2-ketoglutarate (Engleman and Francis, 1978). Glutamine is a potent inhibitor of the deadenylylation activity, and P_{II} is a weak inhibitor. The combination of P_{II} and glutamine synergistically inhibit the deadenylylation activity (Engleman and Francis, 1978).

Curiously, in cells that lack P_{II}, the ATase enzyme still brings about the modification of GS in the appropriate fashion in response to changes in nitrogen availability. This is unexpected, because the in vitro studies cited above indicated that the deadenylylation reaction required P_{II}–UMP. Apparently, the deadenylylation reaction can occur in vivo in the absence of P_{II}–UMP. A suggestion as to the mechanism of this reaction in cells lacking P_{II} is made later.

The regulation of transcription of the *glnA* gene is also indirectly controlled by the UTase/UR protein and P_{II} (Atkinson et al., submitted; Atkinson and Ninfa, 1992; Bueno et al., 1985; Foor et al., 1980; Kamberov et al., 1994a; Ninfa and Magasanik, 1986; Rhee et al., 1985). The *glnA* gene has two promoters (Reitzer and Magasanik, 1985). A minor promoter, *glnAp1*, is used by the typical *E. coli* RNA polymerase holoenzyme containing σ^{70} and provides for a low level of expression of the gene. The major *glnA* promoter, *glnAp2*, is used by RNA polymerase containing the minor sigma subunit σ^{54} (Hirschman et al., 1985; Hunt and Magasanik, 1985; Ninfa and Magasanik, 1986). Expression of *glnA* from the *glnAp2* promoter is controlled by nitrogen availability, as discussed below.

Like all promoters that are used by σ^{54}-containing RNA polymerase, transcription from *glnAp2* is entirely dependent on the presence of an activator protein (reviewed in Kustu et al., 1989). For *glnAp2* and other nitrogen-regulated promoters, the activator is the phosphorylated form of the *glnG* (*ntrC*) product, referred to as NR_I (also referred to as NtrC) (Ninfa and Magasanik, 1986). NR_I in its unmodified form is a dimer that binds to DNA with little cooperativity and cannot activate transcription. As is the case with other σ^{54}-dependent promoters, the sites at which the activator binds are located relatively far upstream (Hirschman et al., 1985; Ninfa et al., 1987; Reitzer and Magasanik, 1986; Reitzer et al., 1989). NR_I binds with high affinity to two sites centered 110 and 140 base pairs upstream from the site of transcription initiation at *glnAp2* (Hirschman et al., 1985; Ninfa et al., 1987). These sites overlap the -35 and $+1$ sites of the *glnAp1* promoter, and NR_I represses transcription from this promoter (Reitzer and Magasanik, 1985). Phosphorylation of NR_I has several effects. The typically dimeric protein binds to adjacent sites on the DNA with high cooperativity (Porter et al., 1993; Weiss et al., 1992). Furthermore, the protein becomes capable of cleaving ATP (ATPase activity) (Austin and Dixon, 1992; Feng et al., 1992; Weiss et al., 1991). Finally, the protein becomes capable of activating transcription in concert with σ^{54}-RNA polymerase (Hunt and Magasanik, 1985; Ninfa and Magasanik, 1986). Various lines of evidence suggest that on phosphorylation NR_I becomes a tetramer: the ATPase activity, which is necessary for transcriptional activation (Weiss et al., 1991), is highly dependent on the concentration of $NR_I{\sim}P$ (Austin and Dixon, 1992; Feng et al., 1992; Weiss et al., 1991), and transcriptional activation is greatly facilitated by the presence of a pair of adjacent sites with the appropriate stereochemical orientation (Reitzer and Magasanik, 1986; Reitzer et al., 1989).

The regulation of transcription from *glnAp2* is probably achieved by the control of the intracellular concentration of $NR_I{\sim}P$. There are two known (physiologically relevant) sources of phosphoryl groups for the formation of $NR_I{\sim}P$. First, phosphoryl groups can be directly transferred to NR_I from the phosphorylated metabolic intermediate acetyl phosphate (Feng et al., 1992). This reaction requires no additional proteins, and thus NR_I is capable of autophosphorylation. The second source of phosphoryl groups for the phosphorylation of NR_I, and the major source in wild-type cells, is

the autophosphorylated form of the *glnL* (*ntrB*) product, known as NR_{II} (Weiss and Magasanik, 1988). NR_I can be thought of as a phosphatase of $NR_{II}{\sim}P$ (Sanders et al., 1992). NR_{II} binds ATP and becomes autophosphorylated on a histidine residue (His 139) (Ninfa and Bennett, 1991; Ninfa et al., 1993; Weiss and Magasanik, 1988; Keener and Kustu, 1988). On incubation with NR_I, this phosphoryl group is transferred to an aspartate residue, Asp-54, within the N-terminal domain of NR_I (Keener and Kustu, 1988; Sanders et al., 1992; Weiss and Magasanik, 1988).

There are two known reactions by which $NR_I{\sim}P$ becomes dephosphorylated. The acyl-phosphate moiety of $NR_I{\sim}P$ has a half-life of about 4 to 8 min at neutral pH (Keener and Kustu, 1988; Weiss and Magasanik, 1988). Because this half-life is greatly increased (to several hours) on either removal of Mg^{2+} or denaturation of the protein with sodium dodecyl sulfate, this activity has been denoted the "autophosphatase activity" (Keener and Kustu, 1988). In addition to the autophosphatase activity, the combination of NR_{II} and the unmodified form of P_{II} results in the very rapid dephosphorylation of $NR_I{\sim}P$ (Ninfa and Magasanik, 1986). This activity has been denoted the "regulated phosphatase" activity (Keener and Kustu, 1988). The regulated phosphatase activity is stimulated by ATP (Keener and Kustu, 1988). Because the regulated phosphatase activity requires the presence of unmodified P_{II}, the activity of the UTase/UR protein indirectly regulates this activity.

To summarize the points that have been made so far, the UTase/UR protein and P_{II} are part of two bicyclic cascades that regulate the expression of *glnA* and the activity of GS, respectively (Fig. 1). In each case, the unmodified form of P_{II} is an intracellular signal of nitrogen excess, in one case influencing the ATase to adenylylate (and inactivate) GS and in the other case acting in concert with NR_{II} to dephosphorylate (and inactivate) the transcriptional activator, $NR_I{\sim}P$, necessary for *glnA* transcription. On the imposition of nitrogen limitation, conversion of P_{II} to P_{II}-UMP by the UTase/UR influences the ATase to activate inactivated GS and permits the accumulation of the transcriptional activator of *glnAp2*, $NR_I \sim P$.

The *glnG* (*ntrC*) and *glnL* (*ntrB*) genes, encoding NR_I and NR_{II}, are themselves part of the *glnALG* operon, and the activation of expression from *glnAp2* under conditions of nitrogen limitation results in an approximately 10-fold increase in the intracellular concentrations of NR_I and NR_{II} (Atkinson and Ninfa, 1993; Pahel et al., 1982; Reitzer and Magasanik, 1983). Although the *glnAp2* promoter is normally regulated in cells unable to increase the concentration of NR_I (e.g., by genetic manipulations resulting in a low level of constitutive *glnG* expression), other genes that are a part of the Ntr operon are not (Pahel et al., 1982). For example, growth on arginine as the sole nitrogen source requires the capacity to amplify the intracellular concentration of NR_I. One possible explanation for this observation is that the sites to which NR_I binds at the relevant promoters have a lower affinity for $NR_I \sim P$ than do the sites located upstream from *glnAp2*. Another physiologically relevant result of the amplification of NR_I on nitrogen starvation is that the transcription from *glnAp2* is decreased about two- to threefold (Shiau et al., 1992). This may result from $NR_I \sim P$ binding to low-affinity sites located between *glnAp2* and the pair of high-affinity sites. Thus, the data suggest that there is likely to be a burst of transcription from the *glnAp2* promoter on the imposition of nitrogen starvation, with a somewhat lower steady-state level of expression from this promoter during prolonged growth in nitrogen-limiting conditions.

In some cases, the genes expressed in cells containing a high intracellular concentration of $NR_I \sim P$ encode positive regulators of the transcription of additional operons that encode enzymes for the use of poor nitrogen sources. Although no such "intermediate regulators" are known in *E. coli* or *S. typhimurium,* there are two cases of this phenomenon from related

organisms (Wong et al., 1987; Bender et al., 1983). The *nac* gene of *Klebsiella aerogenes* encodes a transcription factor that activates the expression of the *hut* genes, required for histidine use, and several other Ntr operons (Bender et al., 1983; reviewed in Bender, 1991). *E. coli* is likely to contain an analogous system. In *Klebsiella pneumoniae,* the *nifLA* operon encodes the regulators of *nif* genes, which encode the nitrogenase and associated proteins necessary for the fixation of atmospheric N_2 (Drummond et al., 1983). Nitrogen regulation in these organisms can be thought of as a cascade in which $NR_I \sim P$ activates operons at the top (*glnAp2*) and middle levels (*nac, nifLA, aut, glnHp2*), and additional regulators encoded by the middle level of the cascade control operons at the lowest level of the cascade (*hut, put, nif*). A composite figure, showing the organization of such a cascade using examples from several different organisms, is shown in Fig. 2.

NITROGEN REGULATION IN THE ABSENCE OF NR_{II}

If $NR_{II} \sim P$ were the only source of phosphoryl groups for the phosphorylation of NR_I, one would expect that mutations eliminating NR_{II} would result in low levels of *glnA* expression regardless of the nitrogen source. However, it has been known for some time that in glucose-grown cells, approximately normal nitrogen regulation of GS occurs in strains lacking functional NR_{II} (e.g., see Bueno et al., 1985). Experiments in which the *glnAp2* mRNA from whole cells was measured after nitrogen shift indicated that NR_{II} was necessary for the rapid regulation of *glnAp2* (Reitzer and Magasanik, 1985). In the absence of NR_{II}, a slow but nevertheless fairly accurate mechanism for nitrogen regulation of *glnAp2* activity was present. This mechanism of regulation certainly requires NR_I, as mutations eliminating NR_I eliminate all functioning of the *glnAp2* promoter (Reitzer and Magasanik, 1985).

Additional physiology experiments clarified the matter (Feng et al., 1992; Atkinson, unpub-

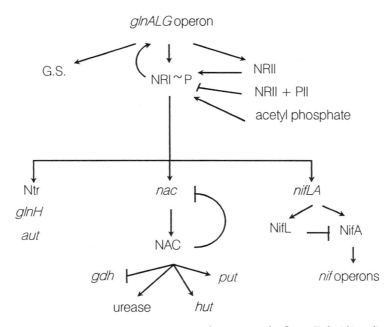

FIGURE 2 Composite nitrogen regulatory cascade from *Escherichia coli*, *Klebsiella aerogenes*, and *Klebsiella pneumoniae*. In this figure, the activation of nitrogen-regulated genes is depicted as a developmental pathway. Only a few examples of nitrogen-regulated genes and operons are shown. At the top of the cascade is the *glnALG* operon (*glnA ntrBC* operon), which is activated by a low intracellular concentration of $NR_I\sim P$. NR_{II} and/or acetyl phosphate can give rise to $NR_I\sim P$ by phosphotransfer to NR_I. $NR_{II} + P_{II}$ result in the destruction of $NR_I\sim P$ (regulated phosphatase activity). One product of the *glnALG* operon is GS. Another result of the activation of the *glnALG* operon is an increase in the intracellular concentration of NR_I. An elevated intracellular concentration of $NR_I\sim P$ results in the activation of genes and operons at the second level of the cascade. For example, the *glnH* (glutamine transport) and *aut* (arginine use) genes of *E. coli* are among the Ntr operons so controlled. In other cases, the elevated intracellular concentration of NR_I results in the activation of genes encoding transcription factors. The *Klebsiella aerogenes nac* gene product, NAC, activates *put* (proline use), *hut* (histidine use), and genes encoding urease and represses its own expression and the expression of *gdh* (glutamate dehydrogenase). The *nifLA* operon of *K. pneumoniae* encodes the activator of *nif* gene expression, NifA, and a regulator of NifA activity, NifL. NifA activates transcription of the *nif* genes, which encode nitrogenase and associated proteins required for the assimilation of N_2.

lished data). It was observed that in cells lacking NR_{II}, only a relatively low intracellular concentration of $NR_I\sim P$ could be formed. This conclusion was deduced by the phenotype of such cells on poor nitrogen sources such as arginine, a measure of the Ntr phenotype. Although cells lacking NR_{II} can grow on glutamine as the sole nitrogen source and activate *glnA* expression under these starvation conditions, they cannot grow on arginine. Furthermore, it was observed that there was a growth-phase component to the nitrogen regulation: at very high culture densities or when grown on solid media, the strain lacking NR_{II} has elevated GS, even in the presence of ammonia (Atkinson and Ninfa, unpublished data).

Thus, the nitrogen regulation observed in strains lacking NR_{II} is only observed at the mid-log conditions that have historically been used to quantitate GS expression. Further experiments indicated that glucose, glycerol, or pyruvate could permit the elevated expression of GS at high culture densities in the strain lacking NR_{II}, but succinate could not (Feng et al., 1992). Therefore, a metabolic intermediate between pyruvate and succinate was apparently the key compound.

At this point, our work was influenced by findings in studies involving two other two-component systems. First, Stock and colleagues discovered that acetyl phosphate, carbamyl phosphate, and phosphoramidate could directly phosphorylate one of the response regulator proteins of the Che system, controlling chemotaxis (Lukat et al., 1992). Those experiments clearly established that the response regulator proteins catalyze their own phosphorylation. An effect of acetate on the chemotactic behavior of cells had long been known (Wolfe et al., 1988); thus, acetyl phosphate was implicated as the cause of this acetate effect. Second, two studies on the regulation of the Pho regulon implicated acetyl phosphate as a regulatory component in that system. In one case, it was observed that overproduction of acetate kinase, owing to the presence of the *ackA* gene on a multicopy plasmid, resulted in the elevation of Pho regulon expression in the absence of a cognate histidine kinase (Lee et al., 1990). This phenomenon was examined with greater rigor by Wanner and colleagues, who demonstrated that the intracellular accumulation of acetyl phosphate could trigger the expression of the Pho regulon in cells that contain the Pho response regulator but lack any kinase known to transfer phosphoryl groups to it (Wanner and Wilmes-Riesenberg, 1992). These findings, along with the results of our physiology experiments, led us to directly examine the role of acetyl phosphate in the Ntr system. Our studies were facilitated by the fact that we had at our disposal not only a fairly defined system for genetic and physiological analysis but also the reconstituted transcription system consisting of purified components.

Pathways for the formation and breakdown of acetyl phosphate are shown in Fig. 3. In cells grown on glucose, glycerol, or pyruvate, acetyl phosphate is formed from acetyl coenzyme A (CoA) by the reversible action of the phosphotransacetylase enzyme, product of *pta*. As indicated in Fig. 3, mutations in *pta* do not prevent the formation of acetyl phosphate in cells grown on acetate, and in fact, growth of such a mutant on acetate leads to the accumulation of acetyl phosphate. The formation of acetyl phosphate from acetate requires the action of the acetate kinase, the product of *ackA*. Mutants lacking both the *pta* and *ackA* enzymes are unable to synthesize acetyl phosphate (Fig. 3, reviewed in McCleary et al., 1993).

In the light of the above, we constructed strains containing various combinations of appropriate mutations and measured GS activity after growth on various carbon and nitrogen sources. In cells containing NR_{II}, there is little effect resulting from mutations affecting the synthesis of acetyl phosphate. By contrast, in cells lacking NR_{II}, acetyl phosphate is required for expression from *glnAp2*. Strains that lack both the capacity to produce acetyl phosphate and NR_{II} do not elevate the expression of GS under any of the growth conditions tested. Conversely, conditions that result in elevated intracellular concentrations of acetyl phosphate, such as growth of a *glnL pta* double mutant on acetate, resulted in elevated expression of GS regardless of the availability of ammonia (Feng et al., 1992).

To further study the mechanism by which acetyl phosphate exerts its effect on GS synthesis, we examined the effect of acetyl phosphate in the in vitro transcription system consisting of purified components: promoter DNA, NR_I, NR_{II}, σ^{54}, and core RNA polymerase. It had been clearly shown that the activation of the *glnAp2* promoter in this system absolutely requires NR_I~P (Ninfa and Magasanik, 1986). There is essentially no transcription from

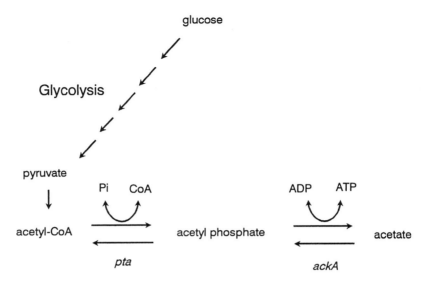

FIGURE 3 Pathways for the formation and breakdown of acetyl phosphate. The *pta* gene encodes the enzyme phosphotransacetylase, and the *ackA* gene encodes the enzyme acetate kinase.

glnAp2 in reaction mixtures containing no source of phosphoryl groups for the phosphorylation of NR_I. As previously demonstrated, the presence of NR_{II} and ATP results in the phosphorylation of NR_I and the activation of *glnAp2* (Ninfa and Magasanik, 1986). In the absence of NR_{II}, acetyl phosphate, carbamyl phosphate, and phosphoramidate could each bring about the NR_I-dependent activation of the *glnAp2* promoter, whereas phosphoenolpyruvate and ATP could not (Feng et al., 1992).

Further studies were performed to characterize the mechanism of action of acetyl phosphate, carbamyl phosphate, and phosphoramidate. Because it is known that phosphorylation of NR_I stimulates the ATPase activity of NR_I, the ability of these compounds to stimulate the ATPase of NR_I was determined. All three compounds stimulated the ATPase activity of NR_I. Previously, the ATPase activity of NR_I~P could only be measured in the presence of NR_{II} and ATP. However, the concerted action of NR_I and NR_{II} itself constitutes an ATPase,

complicating the measurement of the ATPase activity of NR_I~P (Weiss et al., 1991). Having discovered a way to phosphorylate NR_I without recourse to the NR_{II} + ATP reaction permitted us to make the first accurate measurements of the ATPase activity of NR_I~P in an uncomplicated coupled assay system consisting of pyruvate kinase and lactate dehydrogenase (Norby, 1988). Measurements of the concentration dependence of NR_I~P for the ATPase activity confirmed and extended the previous observation that the ATPase activity is highly dependent on the concentration of NR_I~P (Hill coefficients > 2) (Austin and Dixon, 1992; Feng et al., 1992; Weiss et al., 1991). This finding strengthens the conclusion that the activating species for transcription from *glnAp2* consists of a tetramer or higher form of NR_I~P.

Finally, the direct transfer of phosphoryl groups from acetyl-^{32}P to NR_I was observed (Feng et al., 1992). This observation demonstrated conclusively that NR_I is capable of catalyzing its own phosphorylation.

In the light of the findings discussed above, it is interesting to speculate on the nature of the apparent nitrogen regulation observed under certain conditions (i.e., glucose as carbon source, mid-log-phase cultures) in cells lacking NR$_{II}$. This regulation must be due to changes in the intracellular concentration of acetyl phosphate in the presence or absence of ammonia. Presumably, the synthesis of acetyl phosphate must be strongly dependent on the concentration of acetyl-CoA, because the *pta* and *ackA* enzymes are essentially expressed constitutively (reviewed in McCleary et al., 1993). In some way, the presence of ammonia must lower the intracellular concentration of acetyl-CoA. Perhaps this is due to the reactions catalyzed by glutamate dehydrogenase and the concerted action of GS and glutamate synthase. In both of these cases, 2-ketoglutarate is withdrawn from the tricarboxylic acid cycle to form glutamate in an ammonia-dependent fashion. This may, in turn, increase the demand on acetyl-CoA by the reactions drawing it into the tricarboxylic acid cycle. If so, then one would expect that mutations eliminating glutamate dehydrogenase and/or glutamate synthase would have the effect of eliminating the apparent nitrogen regulation in cells lacking NR$_{II}$ and grown under the appropriate conditions (mid-log phase, glucose as carbon source).

In a practical sense, the discovery of the acetyl phosphate-mediated activation of *glnAp2* has clarified the task of measuring, using intact cells, the positive and negative regulatory functions of NR$_{II}$. For example, the positive regulatory function of NR$_{II}$ can most clearly be observed in cells lacking the capacity to form acetyl phosphate, because in such cells NR$_{II}$ is the only source of phosphoryl groups for the phosphorylation of NR$_{I}$ leading to activation of *glnAp2*. The negative regulatory function of NR$_{II}$ can most clearly be observed in cells containing the phosphotransacetylase grown on pyruvate in the presence of ammonia or grown to stationary phase in glucose plus ammonia, because such cells have exceptionally high levels of GS unless NR$_{II}$ is present.

STRUCTURE-FUNCTION ANALYSIS OF NR$_{II}$

Site and Overall Mechanism of Autophosphorylation of NR$_{II}$

NR$_{II}$ becomes autophosphorylated on incubation with ATP. To elucidate the site(s) at which phosphorylation of the protein occurs, NR$_{II}$ was autophosphorylated with labeled ATP, and the stability of the phosphoryl group was measured in the presence of hydroxylamine, base, acid, or pyridine. Most of the phosphoryl groups attached to NR$_{II}$ were acid-labile and base-stable and were slowly hydrolyzed by hydroxylamine, suggesting that the phosphorylated residue was a phosphohistidine (Ninfa et al., 1988; Weiss and Magasanik, 1988). After complete base-catalyzed hydrolysis of the phosphorylated protein, phosphohistidine was directly recovered as the phosphorylated amino acid (Weiss and Magasanik, 1988). The site of the phosphorylated residue in NR$_{II}$~P was determined by sequential cleavage of the autophosphorylated protein with the proteases LysC and trypsin; this analysis revealed that the site of autophosphorylation was in a small peptide (GLAHEIK) containing a single histidine residue, His139 (Ninfa and Bennett, 1991).

The autophosphorylation of the homodimeric NR$_{II}$ protein occurs by the (intramolecular) *trans*-phosphorylation of the subunits within a dimer (Ninfa et al., 1993). That is, NR$_{II}$ subunits bind ATP and phosphorylate the other subunit within the dimer. The NR$_{II}$ homodimer is quite stable, with a dissociation constant (K_d) on the order of 10^{-9}M, as deduced from studies on the dependence of the rate of isotopic exchange (between ATP and ADP) catalyzed by NR$_{II}$ on the enzyme concentration (Ninfa et al., unpublished data). A method was developed, based on mild treatment with urea, that permitted the dissociation of the homodimer and the reassociation of dimers containing dissimilar subunits (Ninfa et al., 1993). Using this method, mixed dimers containing two different mutant subunits, or containing a mutant and wild-type subunit, could be formed. To aid in the visualization of distinct types of subunits, it was observed that subunits could be "tagged" by fusion to the maltose binding protein without affecting the autophosphorylation activity of NR_{II}. Mixed dimers containing particular combinations of mutants were constructed. Homodimers that lack the site of autophosphorylation, His139, are not autophosphorylated. Similarly, when the capacity to bind ATP is greatly diminished (e.g., by alteration of glycine residue at position 313 to alanine [G313A]), the ability of NR_{II} to become autophosphorylated is greatly diminished (Fig. 4). However, when H139N and G313A subunits are combined in mixed dimers, these dimers are able to become autophosphorylated. Similarly, when the G313A subunit is combined with a wild-type subunit in mixed dimers, such dimers are capable of autophosphorylation, which occurs almost exclusively on the G313A subunit. Such results provide strong support for the *trans*-intramolecular mechanism of autophosphorylation (Ninfa et al., 1993).

As discussed above, P_{II} interacts with NR_{II} and $NR_I{\sim}P$ to stimulate the rate of $NR_I{\sim}P$ dephosphorylation (regulated phosphatase activity). One possible explanation for the activity of P_{II} is that P_{II} regulates the interconversion of NR_{II} between two alternative conformations, one of which is associated with the autophosphorylation activity and the other of which is associated with the regulated phosphatase activity. We investigated whether P_{II} affected in any way the autophosphorylation of NR_{II}. Because this reaction is very rapid, the rate and extent of NR_{II} autophosphorylation were measured at 4°C. The results of these studies were that P_{II} had no apparent effect on the rate or extent of NR_{II} autophosphorylation (Kamberov et al., 1994a).

Mutations in the Highly Conserved Residues of the C-Terminal Histidine Kinase Domain

To get a clearer understanding of the roles of the highly conserved residues shared by NR_{II} and the other members of the histidine kinase family of proteins, we systematically altered the most highly conserved of these residues by site-specific mutagenesis (Atkinson and Ninfa, 1993). The position and nature of the changes made are depicted in Fig. 4B. We then proceeded to characterize the effects of these mutations on the positive and negative regulatory functions of NR_{II} in intact cells and, in some cases, in cell-free experiments using purified components. None of the mutations altering highly conserved residues in the C-terminal domain of NR_{II} adversely affected the ability of NR_{II} to negatively regulate *glnA* expression in intact cells, suggesting that these mutations do not affect the regulated phosphatase activity (Atkinson and Ninfa, 1993). However, in experiments with purified components, three of these same mutations were found to decrease the P_{II}-dependent regulated phosphatase activity of NR_{II} (D287N, G289A, G291A) (Kamberov et al., 1994b). This discrepancy suggests that in intact cells, the capacity of the regulated phosphatase activity may exceed that required to maintain physiological regulation.

Mutations in the highly conserved residues of the kinase domain affected the positive regulatory function (i.e., the kinase activity) to various extents (Atkinson and Ninfa, 1993). In some cases (N248D, D287N, G289A, G291A), the amplitude of the defect was negligible when *glnA* expression was measured, but the mutations resulted in the Ntr$^-$ phenotype, indicating that high intracellular concentrations of $NR_I{\sim}P$ could not be formed. In other cases (H139N, H139V, E140Q, E140A, G313A, G315A), a significant defect in *glnA* expression was observed (Atkinson and Ninfa, 1993).

One noteworthy result from the site-specific mutagenesis studies is that certain mutations result in altered NR_{II} proteins that are negative regulators of *glnA* even under nitrogen-limiting conditions and in the absence of P_{II}. For example, the H139N mutation altering the site of autophosphorylation and the adjacent E140Q and E140A mutations result in this phenotype. A termination codon at position 291 in *glnL,* encoding a truncated NR_{II}

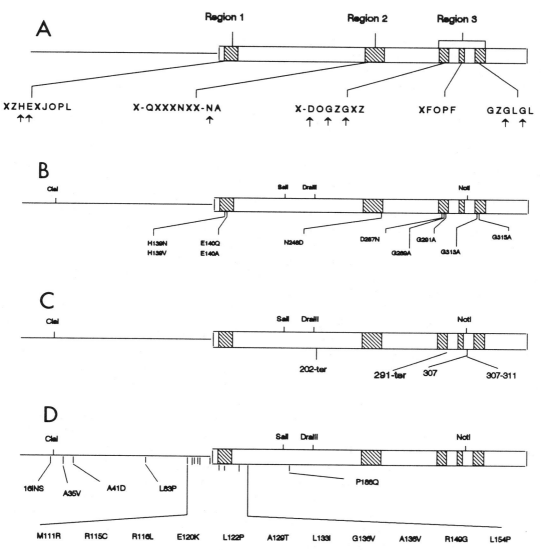

FIGURE 4 Mutations in *glnL (ntrB)*, encoding NR$_{II}$. (A) Schematic depiction of the NR$_{II}$ protein. The nonconserved N-terminal domain is shown as a thin line, and the conserved histidine kinase-phosphatase domain is shown as a thick line. Within this conserved domain, three highly conserved regions are found, depicted with crosshatching and referred to here as region 1, region 2, and region 3. The consensus sequence for the highly conserved regions are shown. The standard single letter amino acid code is used with the following exceptions: X refers to positions where at least 50% of the family have a nonpolar amino acid (I, L, M, or V), Z refers to positions where at least 50% of the family have a polar amino acid (A, G, P, S, or T), J refers to positions where at least 50% of the family have a basic amino acid (H, K, or R), and O refers to positions where at least 50% of the family have an acidic or amidic amino acid (D, E, N, or Q). Positions with less than 50% conservation among the kinase family are shown by dashes. Positions at which mutations were introduced by site-specific mutagenesis are indicated by arrowheads. For a more complete description of the conservation in the kinase family, see Parkinson and Kofoid (1992). Adapted from Fig. 5 of Parkinson and Kofoid (1992) with permission and has also appeared in Atkinson and Ninfa (1993). (B) Mutational analysis of the highly conserved regions of NR$_{II}$. The identities of the alterations made are shown. The phenotypes resulting from these changes are summarized in the text and in Atkinson and Ninfa (1993). (C) Alterations in NR$_{II}$ resulting from the introduction of nonsense codons or small deletions. ter refers to termination codons. 307 refers to the deletion of codon 307, while 307–311 refers to deletion of codons 307 to 311. (D) Mutations selected as suppressors of the Ntr⁻ phenotype resulting from the *glnD99*::Tn*10* mutation. B, C, and D are reproduced from Fig. 2 of Atkinson and Ninfa (1993).

(ter291) that lacks the highly conserved region III of the NR_{II} protein, is also a negative regulator of *glnA* under all conditions.

We examined whether these alterations of the highly conserved residues within the kinase domain affected the regulated phosphatase activity using purified components (Kamberov et al., 1994b). The H139N protein is always a phosphatase, even in the absence of P_{II}. The rate of $NR_I{\sim}P$ dephosphorylation by this protein is further stimulated by P_{II} and is also stimulated by ATP. If this phosphatase activity demonstrated by the H139N protein is the same as the regulated phosphatase activity, as seems likely, then the conserved histidine that is the site of autophosphorylation is not required for the regulated phosphatase activity. Furthermore, the role of P_{II} in this reaction is likely to be entirely regulatory, because the H139N protein can bring about the dephosphorylation of $NR_I{\sim}P$ even in the absence of P_{II}. The purified G313A protein is also an $NR_I{\sim}P$ phosphatase in the absence of P_{II}, and the phosphatase activity of this protein, like the H139N protein, is further stimulated by P_{II} and ATP. However, the stimulation of the regulated phosphatase activity of the G313A protein by ATP was not as dramatic as was the stimulation of this activity of the H139N protein by ATP, which agrees with the observation that G313A is defective in ATP binding (Kamberov et al., 1994b). The ter291 protein is a weak phosphatase in the absence of P_{II}, and for this protein, the rate of $NR_I{\sim}P$ dephosphorylation was not significantly stimulated by P_{II} (Kamberov et al., 1994b).

It seems likely that the nonconserved N-terminal domain of NR_{II} is the P_{II} receptor. Only this portion of NR_{II} is not related to the other members of the histidine kinase family. We constructed a fusion protein in which residues 1 to 110, constituting most of the N-terminal domain of NR_{II}, were removed and replaced by the maltose binding protein. As expected, this fusion protein, MBP-c-NR_{II}, is able to autophosphorylate and transfer phosphoryl groups to NR_I but cannot productively interact with P_{II} to bring about the dephosphorylation of $NR_I{\sim}P$ (Kamberov et al., 1994b).

Mutations in *glnL* That Alter the Interaction of NR_{II} and P_{II}

The *glnD99*::Tn*10* mutation results in the inability to use arginine as the sole nitrogen source (Ntr⁻ phenotype). This mutation results in a defect in the uridylylation of P_{II} under conditions of nitrogen limitation. Consequently, the unmodified P_{II} continuously present in such cells coconstitutes the NR_{II}-P_{II} regulated phosphatase, and the cells are unable to accumulate the high intracellular concentrations of $NR_I{\sim}P$ necessary for the activation of the Ntr regulon (see Figs. 1 and 2). The effect of this mutation on the expression of *glnA* is less severe, diminishing to about 25% the expression of *glnA* on nitrogen starvation. Thus, the mutation is leaky.

We obtained by direct selection a collection of spontaneous suppressor mutations that restore the ability of the *glnD99*::Tn*10* to use arginine as the sole nitrogen source (Atkinson and Ninfa, 1992). Many of these mapped to *glnL* (Fig. 4D), as observed previously (MacNeil et al., 1982; Pahel et al., 1978). When the strains with *glnD99*::Tn*10* and suppressors mapping in *glnL* were analyzed to determine the regulation of *glnA* by nitrogen availability, it was observed that a wide range of regulatory capacity was present. In some strains, *glnA* expression was constitutive and at a high level, whereas in other strains *glnA* expression was observed to be regulated quite well, although not as well as in the wild type. Nucleotide sequencing of the mutant alleles revealed that most of the suppressor mutations mapped to the junction of the nonconserved N-terminal domain of NR_{II} and the conserved C-terminal kinase-phosphatase domain (Fig. 4D). Mutations causing different phenotypes were not clustered but rather distributed with no obvious pattern throughout this region of NR_{II} (Atkinson and Ninfa, 1992).

The source of information causing nitrogen regulation in those suppressor strains that responded well to changes in nitrogen availability was investigated (Atkinson and Ninfa, 1992). It was observed that regulation in such strains depended on both P_{II} (glnB+) and the mutant glnD99::Tn10 alleles. Amplification of the intracellular concentration of P_{II} due to the introduction of a multicopy plasmid containing glnB resulted in the restoration of the Ntr⁻ phenotype, whereas deletion of glnB resulted in the characteristic phenotype associated with the absence of P_{II}. Furthermore, replacement of the glnD99::Tn10 allele with the wild-type allele resulted in the conversion of the regulated phenotype to a constitutive phenotype. Thus, in these regulated suppressor strains, the regulation is due to P_{II}. This indicates that P_{II} in such strains is modified to some extent in response to nitrogen limitation (Atkinson and Ninfa, 1992).

A hypothesis to explain the different phenotypes resulting from the suppressor mutations in glnL is as follows: the suppressors resulting in the constitutive expression of glnA are likely to have either eliminated the capacity of NR_{II} to interact with P_{II} or rendered this interaction unproductive in bringing about the regulated phosphatase activity. For example, a catalytic site involved in the hydrolysis of $NR_I{\sim}P$ to NR_I and P_i may have been destroyed. The suppressors resulting in the regulated phenotype probably have more subtle defects. For example, these mutations may have created altered NR_{II} proteins that interact with P_{II} less well or have a reduced probability, on interacting with P_{II}, of adopting the conformation necessary for the phosphatase activity. Essentially, these mutations have the effect of fine tuning the formation of the NR_{II}-P_{II} phosphatase to the altered modification of P_{II} observed in the glnD99::Tn10 background.

PURIFICATION AND CRYSTALLIZATION OF P_{II} AND RECONSTITUTION OF THE NR_{II}-P_{II} REGULATED PHOSPHATASE

To more fully characterize the nature of the P_{II}-NR_{II} phosphatase activity (regulated phosphatase activity), we reconstituted this activity using purified components. An overexpression vector that dramatically overproduces the P_{II} protein was constructed, and about 100 mg of P_{II} was obtained from 20 g of cells (Kamberov et al., 1994a). This preparation of the protein is essentially homogeneous and crystallizes readily (Kamberov et al., 1994a). Diffraction studies with these crystals are under way (deMel et al., 1994).

As noted above, highly purified P_{II} interacts with NR_{II} to bring about the dephosphorylation of $NR_I{\sim}P$ (Kamberov et al., 1994a; Kamberov et al., 1994b; Keener and Kustu, 1988; Ninfa and Magasanik, 1986). It is clear that P_{II} is not itself a phosphatase but rather works in concert with NR_{II}. For example, the mutant form of NR_{II} containing the A129T alteration (NR_{II}2302) is able to autophosphorylate and transfer phosphoryl groups to NR_I but cannot productively interact with P_{II} to dephosphorylate $NR_I{\sim}P$ (Kamberov et al., 1994a; Kamberov et al., 1994b; Ninfa and Magasanik, 1986). The same property is observed using the fusion protein MBP-c-NR_{II} (Kamberov et al., 1994b).

The regulated phosphatase activity (NR_{II} + P_{II}-mediated phosphatase activity) is highly sensitive to the ratio of $NR_I{\sim}P$ to NR_I (Kamberov et al., in press a). When this ratio is low, P_{II} is much less effective in eliciting the regulated phosphatase activity, which implies that a ternary complex is formed between P_{II}, NR_{II}, and $NR_I{\sim}P$, and that unphosphorylated NR_I competes with the formation of the complex. This probably explains the prior observation that P_{II} was fairly ineffective in bringing about the dephosphorylation of $NR_I{\sim}P$ (Keener and Kustu, 1988); those experiments may have been performed under nonoptimal conditions. At optimal conditions, concentrations of P_{II} that are equivalent to the concentration of NR_{II} result in the rapid and complete dephosphorylation of $NR_I{\sim}P$. This indicates that P_{II} is necessary and sufficient for the elicitation of the regulated phosphatase activity in concert with NR_{II} (Kamberov et al., 1994a).

P_{II} seems to have no effect on either the autophosphorylation of NR_{II} or the transfer of phosphoryl groups between $NR_{II}\sim P$ and NR_I (Kamberov et al., 1994a). Furthermore, P_{II} does not affect the reverse transfer of phosphoryl groups from $NR_I\sim P$ to NR_{II} (Kamberov, unpublished data). Thus, it appears that the regulated phosphatase activity represents a true catalysis of the hydrolysis of $NR_I\sim P$ to unmodified NR_I. One remaining question is whether P_{II} is able to interact at all with NR_{II} in the absence of $NR_I\sim P$. Our results on this issue are presented below.

An unusual feature of the P_{II}-mediated dephosphorylation reaction is that it is greatly stimulated by ATP (Keener and Kustu, 1988). In early experiments, wild-type NR_{II} was used to observe this effect (Keener and Kustu, 1988). This made the results difficult to interpret, because in the presence of ATP, NR_{II} becomes autophosphorylated and transfers phosphoryl groups to NR_I. Our observation that the phosphatase activity of the H139N protein is greatly stimulated by ATP argued that ATP is an allosteric effector, because the H139N protein is not autophosphorylated. Whether the target for ATP binding is NR_{II} or $NR_I\sim P$ is as yet unknown, but ATP does not greatly affect the autophosphatase activity of $NR_I\sim P$ (Kamberov et al., 1994b; Kamberov, unpublished data).

The site of phosphorylation of NR_I is within its N-terminal domain (Keener and Kustu, 1988; Sanders et al., 1992). This domain has been purified after proteolytic digestion of the protein and shown to act as a substrate for phosphorylation by $NR_{II}\sim P$ (Keener and Kustu, 1988). We found that this isolated domain, once phosphorylated, was a substrate for the regulated phosphatase activity (Kamberov et al., 1994a). This has permitted us to assess the regulated phosphatase activity in a coupled assay system in which the hydrolysis of ATP is coupled to the oxidation of NADH to NAD (by pyruvate kinase and lactate dehydrogenase) (Norby, 1988). Because intact $NR_I\sim P$ is an ATPase, this assay method does not work with the intact protein. The coupled assay system can be used to measure the regulated phosphatase activity by adjusting protein concentra-

tions such that the N-terminal domain of NR_I is limiting. The rate of dephosphorylation of this domain, and its subsequent rephosphorylation, then determines the rate at which ADP is produced and, consequently, the rate at which NADH is oxidized. Using this system, we have been able to assess the effectiveness of P_{II} at different concentrations of $NR_I\sim P$ (N-terminal domain). The results of these studies indicated that P_{II} is effective in eliciting the regulated phosphatase activity at very low concentrations of $NR_I\sim P$. This supports the conclusion that the ratio of $NR_I\sim P$ to NR_I, and not the $NR_I\sim P$ concentration itself, determines the effectiveness of P_{II} in eliciting the regulated phosphatase activity (Kamberov et al., 1994a), as noted above.

RECONSTITUTION OF THE SENSORY APPARATUS

To more fully understand the signal transduction mechanisms involved in the control of the phosphorylation and dephosphorylation of NR_I, we have purified the UTase/UR protein and have attempted to reconstitute the signaling by the small molecules glutamine and 2-ketoglutarate. As shown in Fig. 1, we expect that in a reconstituted system containing NR_I, NR_{II}, P_{II}, and the UTase/UR proteins, 2-ketoglutarate should permit the phosphorylation of NR_I to proceed, whereas glutamine should result in the rapid dephosphorylation of $NR_I\sim P$. The key questions to be addressed using the reconstituted system are, (i) do glutamine and 2-ketoglutarate act on different targets (sites) or compete at the same target (site)? (ii) is the modification of P_{II} by UTase/UR cooperative or not (i.e., does the UTase/UR portion of the switch resemble a rheostat or a toggle switch?), (iii) is fully modified P_{II}-UMP innocuous, as depicted in Fig. 1? (iv) is partially modified P_{II} (P_{II} trimers containing less than 4 UMP moieties) active in eliciting the regulated phosphatase activity (i.e., does uridylylation of a subunit of a P_{II} trimer result in a conformational change that alters the activity of the entire trimer)? Each of these questions is addressed below.

Purification of the UTase/UR Protein

Previous experiments that have tried to reconstitute the sensory apparatus had indicated that the purified UTase/UR protein (glnD product) was unstable (Adler et al., 1975; Engleman and Francis, 1978; Garcia and Rhee, 1983). Specifically, the protein lost activity within hours unless kept in buffers containing high concentrations of KCl. We constructed a plasmid that causes the vast overproduction of the UTase/ UR protein and devised a new purification strategy. Using our methods, the UTase/UR activity is quite stable and can be assayed directly in the crude extract after storage of this extract at 4°C for up to 4 days (Kamberov et al., 1994a; Atkinson, unpublished data). Furthermore, the purified protein can be stored in 45% glycerol at −80°C without any loss of activity for more than 6 months (Kamberov et al., 1994a). Our purified UTase/UR protein catalyzes the uridylylation of P_{II} in a reaction that is dependent on the presence of 2-ketoglutarate and greatly stimulated by ATP (Kamberov et al., 1994a), as previously described (Engleman and Francis, 1978; Francis and Engleman, 1978; Garcia and Rhee, 1983). Glutamine completely inhibits the uridylylation reaction (Kamberov et al., 1994a), as expected (Engleman and Francis, 1978; Francis and Engleman, 1978; Garcia and Rhee, 1983). To examine the UR activity, we purified P_{II}–UMP (Kamberov et al., 1994a). Glutamine and Mn^{2+} were required for the UR activity, as described previously (Engleman and Francis, 1978; Francis and Engleman, 1978; Garcia and Rhee, 1983). The presence or absence of ATP had no effect on the UR activity, nor did the presence of 2-ketoglutarate, even at high concentrations, prevent this activity (Kamberov et al., 1994a). These results support the idea that ATP and 2-ketoglutarate, which stimulate the UTase activity, do not compete with glutamine, which stimulates the UR activity.

Flipping the Switch

We manipulated the reaction conditions so that both the UTase and UR activities could occur, depending on the addition of glutamine and 2-ketoglutarate (Atkinson et al., 1994). At these conditions, the uridylylation of P_{II} could be initiated by the addition of 2-ketoglutarate, and subsequent addition of glutamine resulted in the removal of uridylyl groups from P_{II} (Atkinson et al., 1994). We attempted to manipulate the system to allow the switch to be flipped several times. However, once glutamine was added to get the UR activity in the first switch, subsequent additions of 2-ketoglutarate, even at high concentration, never resulted in the complete uridylylation of P_{II}. Instead, the uridylylation of P_{II} proceeded only about halfway. These results again suggested that 2-ketoglutarate and glutamine are affecting different targets, such that no amount of 2-ketoglutarate could effectively offset the presence of glutamine (Atkinson et al., 1994).

Uridylylation and Deuridylylation of P_{II} by the UTase/UR Protein Are Not Cooperative

To determine whether the UTase and UR reactions were cooperative, we needed a way to determine the extent of uridylylation of individual P_{II} trimers. That is, during the course of a reaction, how many singly modified, doubly modified, and completely modified species were present? We observed that native acrylamide gel electrophoresis could be used to separate these species (Atkinson et al., 1994). By examining the time course of the uridylylation and UR reactions on native gels, we could clearly see that neither the uridylylation nor deuridylylation of P_{II} was cooperative (Atkinson et al., 1994). This is as one would expect if P_{II} serves as a rheostat as opposed to a toggle switch.

Fully Modified P_{II} Is Innocuous with Respect to the Elicitation of the Regulated Phosphatase Activity, and Partially Modified P_{II} Is Partially Active in Eliciting the Regulated Phosphatase Activity

Although both P_{II} and P_{II}–UMP have a role in the regulation of the ATase activity, only P_{II} seemed to have a role in the regulation of glnA transcription; the uridylylated form of P_{II} appeared to be innocuous (see Fig. 1). This conclusion is based largely on the genetic evidence of Bueno et al. (1985), who demonstrated that a strain deleted for *glnB*, and thus lacking P_{II}, could fully express *glnA* but lacked the ability to negatively regulate this expression in nitrogen-excess conditions. By contrast, another report (Holtel and Merrick, 1989) suggested that P_{II}-UMP could counter the effects of P_{II} on *glnA* transcription. Specifically, it was observed that a mutant form of P_{II} resulted in the negative regulation of *glnA* under all conditions. From this it was surmised that the mutant P_{II} could not be uridylylated. When this mutant P_{II} was present in the cell along with wild-type P_{II}, codominance was observed (Holtel and Merrick, 1989). That finding suggested that the deleterious effect of the mutant P_{II} (continuous presence of the regulated phosphatase activity) could be offset by the presence of the wild-type P_{II} (i.e., by P_{II}-UMP when the cells are nitrogen-starved). Thus, it was important to see if P_{II}-UMP was really innocuous with regard to the elicitation of the regulated phosphatase activity and whether P_{II}-UMP could counteract the activity of unmodified P_{II}.

We purified P_{II}-UMP after uridylylation of P_{II} by the UTase/UR protein and studied the effect of P_{II}-UMP on the regulated phosphatase activity. Our results indicated that fully modified P_{II}-UMP was innocuous and that the ability to elicit the regulated phosphatase activity could be restored on removal of the uridylyl groups by the UR activity (Atkinson et al., 1994). P_{II}-UMP did not affect the phosphorylation of NR_I by NR_{II} and ATP, nor did P_{II}-UMP counteract the ability of P_{II} to elicit the regulated phosphatase activity in vitro (Atkinson et al., 1994). We also isolated P_{II} trimers that contained various proportions of modified and unmodified subunits. For example, by performing a time course of uridylylation and recovering P_{II} from samples removed at various times, it was possible to isolate populations of P_{II} that contained few completely unmodified trimers and in which about 50% of the total number of monomers are modified. We examined the ability of partially modified P_{II} to elicit the regulated phosphatase activity and found that it was partially active (Atkinson et al., 1994). This suggests that each monomer of P_{II} acts independently and that uridylylation of a subunit does not bring about a concerted change of the trimer to the innocuous form. Thus, the P_{II}-NR_{II} interaction also resembles a rheostat as opposed to a toggle switch.

Reconstitution of the Regulation of NR_I Phosphorylation State by Small Molecules

We attempted to reconstitute the regulation of NR_I phosphorylation by the small molecules 2-ketoglutarate and glutamine. As stated above, we successfully reconstituted the response to glutamine. In those experiments, P_{II}-UMP, UTase/UR, NR_{II}, NR_I, and ATP were combined. This led to the phosphorylation of NR_I, because P_{II}-UMP is innocuous. On addition of glutamine, the level of $NR_I{\sim}P$ decreased due to the appearance of the regulated phosphatase activity (Atkinson et al., 1994). Many control experiments indicated that UTase/UR and P_{II} were necessary for the glutamine signal to have any effect (Atkinson et al., 1994). Curiously, even in the absence of glutamine, UTase/UR brought about a decrease in the steady-state level of $NR_I{\sim}P$ as long as P_{II}-UMP was present. We interpreted this as resulting from basal UR activity of the UTase/UR protein that occurs in the absence of glutamine (Atkinson et al., 1994).

We also tried to reconstitute the response to 2-ketoglutarate, with very surprising results. For these experiments, UTP, P_{II}, NR_{II}, NR_I, and the UTase/UR protein were combined, such that the level of P_{II} is not quite saturating. In the presence of ATP, phosphorylation of NR_I is observed but not to the extent that would have been observed in the absence of P_{II}. We expected that on subsequent addition of 2-ketoglutarate, the level of NR_I-P would increase, owing to the conversion of P_{II} to the innocuous P_{II}-UMP. Instead, what we observed was that addition of 2-ketoglutarate resulted in the immediate and drastic reduction of the

NR_I~P to essentially the baseline value (Atkinson, unpublished data). Subsequent to this, the level of NR_I~P gradually rose, and its level could be again reduced to the baseline level by the addition of glutamine. The initial reduction in NR_I~P seen on addition of 2-ketoglutarate to the system was completely unexpected. We repeated this experiment several times, and each time the addition of 2-ketoglutarate resulted in the unexpected response, namely, the drastic and immediate decrease in the level of NR_I~P. To determine the target being affected by 2-ketoglutarate in these reconstitution experiments, we conducted a similar experiment in which the UTase/UR protein was omitted. Much to our surprise, addition of 2-ketoglutarate to reaction mixtures containing NR_I~P, NR_I, and NR_{II} and limiting P_{II} resulted in a great decrease in NR_I~P; that is, the regulated phosphatase activity seemed to be stimulated directly by 2-ketoglutarate in the absence of the UTase/UR protein (Kamberov et al., submitted). This effect of 2-ketoglutarate required the presence of P_{II} (Kamberov et al., submitted).

We examined whether this 2-ketoglutarate effect was due to a mass action (Kamberov et al., submitted). Different preparations of 2-ketoglutarate were obtained from various suppliers, and highly buffered solutions were checked for the activity, which was uniformly present. Titration of the activity revealed that a concentration of 2-ketoglutarate as low as 0.5 μM resulted in the great stimulation of the regulated phosphatase activity and that this was not due to mass action effects. A mass action effect was also rendered unlikely by the observation that many similar compounds failed to stimulate the regulated phosphatase activity. These included aspartate, asparagine, succinate, citrate, tartrate, glutarate, 2-ketopentanoic acid, 2-ketobutyric acid, and glutamine. Glutamate and pyruvate could weakly stimulate the regulated phosphatase activity (about 3 orders of magnitude less effective than 2-ketoglutarate), whereas 3-ketoglutarate and oxaloacetate were slightly more effective (about 2 orders of magnitude less effective than 2-ketoglutarate). Thus,

it seemed that 2-ketoglutarate was specifically interacting with one or more components of the system to stimulate the regulated phosphatase activity in the absence of the UTase/UR protein (Kamberov et al., submitted).

P_{II} Protein Is the Sensor of Carbon Sufficiency

We conducted binding studies to determine which component (P_{II}, NR_{II}, NR_I, NR_I~P) was interacting with 2-ketoglutarate. The result of these studies was that P_{II} binds 2-ketoglutarate and ATP (Kamberov et al., submitted). The binding of each ligand is strongly dependent on the presence of the other ligand. In the presence of saturating (2 mM) ATP, the K_d of P_{II} for 2-ketoglutarate is about 5.6 μM; in the presence of saturating 2-ketoglutarate (2 mM), the K_d of P_{II} for ATP is about 0.25 μM. Competition experiments revealed that those compounds that weakly affected the regulated phosphatase activity in vitro (i.e., 3-ketoglutarate and oxaloacetate) bound to P_{II} with K_d about 100-fold higher. Similarly, ADP could compete with ATP, but the K_d for ADP was about 100-fold higher than that for ATP (Kamberov et al., submitted).

Because the uridylylation of P_{II} by the UTase/UR protein required both ATP and 2-ketoglutarate, we investigated whether the compounds that weakly stimulated the regulated phosphatase activity would also stimulate the uridylylation of P_{II}. We observed that 3-ketoglutarate and oxaloacetate could indeed stimulate the uridylylation of P_{II} by the UTase/UR protein but only when present at high concentrations (Kamberov et al., submitted). Thus, it seems that the target for 2-ketoglutarate in the uridylylation reaction, and in the stimulation of the regulated phosphatase activity, is P_{II}. This explains why 2-ketoglutarate could not compete with glutamine in reversing the UR reaction, as presented above. Apparently, glutamine controls the activity of the UTase/UR protein, whereas 2-ketoglutarate and ATP regulate the suitability of P_{II} to serve as the substrate in the uridylylation reaction and the ability of P_{II} to elicit the regulated phosphatase activity.

At first glance, it seems to make no sense that the ability of P_{II} to stimulate the regulated phosphatase activity should be enhanced by 2-ketoglutarate, because this activity is associated with nitrogen excess conditions and 2-ketoglutarate is thought to be a signal of nitrogen limitation (carbon excess). Perhaps the reconciliation of these observations will lie in the relative amounts of the UTase/UR and NR_{II} in the cell. If the UTase/UR activity is much greater than the NR_{II} activity, as seems likely, then the stimulation of the uridylylation of P_{II} seen on binding of 2-ketoglutarate would offset the stimulation of the regulated phosphatase activity, such that an increase in 2-ketoglutarate could lead to an increase in $NR_I{\sim}P$. The defect in our reconstituted system apparently was in the ratio of NR_{II} to UTase/UR proteins. This hypothesis is currently being investigated.

We hypothesize that the binding of 2-ketoglutarate and ATP to P_{II} greatly facilitates the interaction of P_{II} with both the UTase/UR protein and NR_{II}. This predicts that there is a conformational alteration of P_{II} on binding of these ligands. Interestingly, P_{II} and 2-ketoglutarate work in opposition to each other in the control of the adenylylation reaction catalyzed by ATase (Engleman and Francis, 1978) (see Fig. 1). Thus, we would predict that the unliganded form of P_{II} interacts with the ATase.

With regard to the NR_{II}-P_{II} interaction, we have recently observed that although NR_{II} does not itself bind 2-ketoglutarate, NR_{II} stimulates the binding of 2-ketoglutarate by P_{II}. The dissociation constant of P_{II} for 2-ketoglutarate is lowered from 5.6 μM to 0.4 μM in the presence of NR_{II} (Kamberov et al., submitted). This result suggested that NR_{II} can interact with P_{II} in the absence of $NR_I{\sim}P$. We examined the ability of immobilized NR_{II} to retain P_{II}, using a column chromatography method (Kamberov et al., submitted). Those experiments revealed that the retention of P_{II} by immobilized NR_{II} required 2-ketoglutarate, Mg^{2+}, and ATP. Thus, P_{II} can bind to NR_{II} in the absence of $NR_I{\sim}P$. Furthermore, these results suggest that 2-ketoglutarate stimulates the

regulated phosphatase activity by stimulating the binding of P_{II} to NR_{II} (Kamberov et al., submitted).

EVIDENCE THAT THE UTase/UR AND P_{II} ARE REDUNDANT

Several lines of evidence indicate that both the UTase/UR protein and P_{II} are likely to be redundant. The case of the UTase/UR protein will be discussed first. It has been noted that P_{II} purified from a strain containing the glnD99::Tn10 mutation and grown under nitrogen-limiting conditions is partially modified (Son and Rhee, 1987). As noted above, some of the suppressors of the glnD99::Tn10 mutation that mapped in glnL resulted in altered proteins that appeared to be less sensitive to P_{II}, while retaining the capacity to interact with P_{II} and coconstitute the NR_{II}-P_{II} phosphatase under conditions of nitrogen excess. This also implies that P_{II} is modified in the appropriate fashion in the glnD99::Tn10 background, albeit a much less efficient modification. Two possibilities are evident: either the glnD99::Tn10 allele produces a product with residual activity, or there is another (minor) UTase/UR protein that can react (cross-react?) with P_{II}. To investigate the first possibility, we selected many independent spontaneous deletion mutations that remove the Tn10-encoded tetracycline resistance determinant from the glnD99::Tn10 strain. One would expect that if the leakiness of this strain is due to leakiness of the glnD99::Tn10 allele (as opposed to there being another UTase/UR), then at least some of the deletions removing tetracycline resistance should now demonstrate a "tight" phenotype. Yet, none of 62 such deletions have a tighter phenotype than does the original strain (Ninfa, unpublished data). Obviously, the DNA alterations in these putative glnD deletion strains will have to be checked, but by their existence, these strains suggest that the leakiness of the glnD99::Tn10 strain was not due to leakiness of the glnD99::Tn10 allele. Furthermore, the Tn10 insertion in the glnD99::Tn10 strain is not at the extreme C terminus of the gene but rather is about two-thirds of the way into the gene (Contreras et al., 1991).

The redundancy of P_{II} is also uncertain but has been surmised for some time based on the phenotypes of strains deleted for *glnB*. Such strains, although having elevated levels of GS under all conditions, are not actually constitutive; they respond weakly to conditions of excess ammonia (Atkinson and Ninfa, 1992; Bueno et al., 1985). More important is the phenotype of double mutants deleted for *glnB* and containing the *glnD99*::Tn*10* mutation. If the *glnB* deletion (which entirely lacks the *glnB* coding sequence) removed all P_{II} activity, then one would expect the *glnD99*::Tn*10 glnB* double mutant and the *glnB* single mutant to have similar levels of *glnA* expression. However, this is not the case; GS is lower in the double mutant (Bueno et al., 1985). Furthermore, although the *glnB* deletion strain is Ntr$^+$, as exemplified by its ability to grow on arginine as the sole nitrogen source, the double mutant is Ntr$^-$ (glucose-arginine negative) (Ninfa and Atkinson, unpublished data). This phenotype is due to the absence of the UTase/UR, as opposed to a possible polarity effect of the Tn*10* on a downstream gene, as shown by the restoration of the Ntr$^+$ phenotype on transformation of the double mutant with a plasmid containing only the *glnD* gene and vector sequences (Atkinson, unpublished data). Thus, the UTase/UR protein apparently has another target in addition to P_{II} (i.e., there is likely to be another protein with P_{II}-like activity). Perhaps this "other P_{II}" is responsible for the ability of the ATase to catalyze the deadenylylation of GS in strains deleted for *glnB*.

The Ntr$^-$ phenotype of the double mutant (*glnB glnD99*::Tn*10*) provides a means to select suppressors (Ninfa and Bai, unpublished data). Suppressor strains in which the downstream components (i.e., NR$_I$, NR$_{II}$, *glnAp2*) of the signal transduction system are normal may have mutations affecting the synthesis or activity of the "other P_{II}." Mapping experiments to identify the site of these mutations are now under way.

OUTSTANDING QUESTIONS

The main issue as yet unresolved concerns the regulated phosphatase activity seen only in the presence of P_{II} and NR$_{II}$. Does this activity represent a stimulation of the autophosphatase activity, or is it a separate activity? There is no convincing evidence supporting or refuting either of these possibilities. One way to answer the question would be to isolate mutations in *glnG* specifically affecting the autophosphatase activity of NR$_I$ and to then examine the regulated phosphatase activity using these altered NR$_I$ proteins. Another way would be to identify an inhibitor that specifically affected only one of these two activities. Efforts in these directions are currently under way, but no answers are yet available.

The other vexing question is, what does P_{II} do to NR$_{II}$ to elicit the regulated phosphatase activity? Because P_{II} does not affect the rate or extent of NR$_{II}$ autophosphorylation or the transfer of phosphoryl groups from NR$_{II}$~P to NR$_I$, either there are not two alternative conformations of NR$_{II}$ ("kinase" and "phosphatase") or the autophosphorylation of NR$_{II}$ can proceed in either conformation. Because the autophosphorylation of NR$_{II}$ is not affected by P_{II} and this activity requires the intact NR$_{II}$ dimer, it seems unlikely that the action of P_{II} will involve an effect on the quaternary structure of NR$_{II}$. Various results suggest that a ternary complex is formed between NR$_{II}$, P_{II}, and NR$_I$~P (e.g., the effect of the NR$_I$/NR$_I$~P ratio on the regulated phosphatase activity). Thus, elucidation of the role of P_{II} in eliciting the regulated phosphatase activity might require more detailed knowledge than is currently available on the interaction between NR$_{II}$ and NR$_I$.

ACKNOWLEDGMENTS

Work in this laboratory is supported by grants from the NIH (GM47460) and NSF (MCB 9318792).

REFERENCES

Adler, S. P., D. Purich, and E. R. Stadtman. 1975. Cascade control of *Escherichia coli* glutamine synthetase: properties of the PII regulatory protein and the uridylyltransferase-uridylyl-removing enzyme. *J. Biol. Chem.* **250**:6264–6272.

Atkinson, M. R. Unpublished data.

Atkinson, M. R., E. S. Kamberov, R. L. Weiss, and A. J. Ninfa. 1994. Reversible uridylylation of the *Escherichia coli* PII signal transduction protein regulates its ability to stimulate the dephosphorylation of the transcription factor nitrogen regulator I (NRI or NtrC). *J. Biol. Chem.* **269**:28288–28293.

Atkinson, M. R., and A. J. Ninfa. 1992. Characterization of *Escherichia coli glnL* mutations affecting nitrogen regulation. *J. Bacteriol.* **174**:4538–4548.

Atkinson, M. R., and A. J. Ninfa. 1993. Mutational analysis of the bacterial signal-transducing protein kinase/phosphatase nitrogen regulator II (NRII or NtrB). *J. Bacteriol.* **175**:7016–7023.

Atkinson, M. R., and A. J. Ninfa. Unpublished data.

Austin, S., and R. Dixon. 1992. The prokaryotic enhancer-binding protein NTRC has an ATPase activity which is phosphorylation and DNA dependent. *EMBO J.* **11**:2219–2228.

Bender, R. A. 1991. The role of the NAC protein in the nitrogen regulation of *Klebsiella aerogenes. Mol. Microbiol.* **5**:2575–2580.

Bender, R. A., K. A. Janssen, A. D. Resnick, M. Blumenberg, F. Foor, and B. Magasanik. 1977. Biochemical parameters of glutamine synthetase from *Klebsiella aerogenes. J. Bacteriol.* **129**:1001–1009.

Bender, R. A., P. M. Snyder, R. Bueno, M. Quinto, and B. Magasanik. 1983. Nitrogen regulation system of *Klebsiella aerogenes*: the *nac* gene. *J. Bacteriol.* **156**:444–446.

Brown, M. S., A. Segal, and E. R. Stadtman. 1971. Modulation of glutamine synthetase adenylylation and deadenylylation is mediated by the metabolic transformation of the PII-regulatory protein. *Proc. Natl. Acad. Sci. USA* **68**:2949–2953.

Bueno, R., G. Pahel, and B. Magasanik. 1985. Role of the *glnB* and *glnD* gene products in regulation of the *glnALG* operon of *Escherichia coli. J. Bacteriol.* **164**:816–822.

Contreras, A., M. Drummond, A. Bali, G. Blanco, E. Garcia, G. Bush, C. Kennedy, and M. Merrick. 1991. The product of nitrogen fixation regulatory gene *nfrX* of *Azotobacter vinelandii* is functionally and structurally homologous to the uridylyltransferase encoded by *glnD* in enteric bacteria. *J. Bacteriol.* **173**:7741–7749.

deMel, V. S. J., E. S. Kamberov, P. D. Martin, J. Zhang, A. J. Ninfa, and B. F. P. Edwards. 1994. Preliminary X-ray diffraction analysis of crystals of the PII protein from *Escherichia coli*. (Crystallization Note). *J. Mol. Biol.* **243**:796–798.

Drummond, M., J. Clements, M. Merrick, and R. Dixon. 1983. Positive control and autogenous regulation of the *nifLA* promoter in *Klebsiella pneumoniae. Nature* (London) **301**:302–313.

Engleman, E. G., and S. H. Francis. 1978. Cascade control of glutamine synthetase. II. Metabolic regulation of the enzymes in the cascade. *Arch. Biochem. Biophys.* **191**:602–612.

Feng, J., M. R. Atkinson, W. McCleary, J. B. Stock, B. L. Wanner, and A. J. Ninfa. 1992. Role of phosphorylated metabolic intermediates in the regulation of glutamine synthetase synthesis in *Escherichia coli. J. Bacteriol.* **174**:6061–6070.

Foor, F., Z. Reuveny, and B. Magasanik. 1980. Regulation of the synthesis of glutamine synthetase by the PII protein in *Klebsiella aerogenes. Proc. Natl. Acad. Sci. USA* **77**:2636–2640.

Francis, S. H., and E. G. Engleman. 1978. Cascade control of glutamine synthetase. I. Studies on the uridylyltransferase and uridylyl removing enzyme(s) from *E. coli. Arch. Biochem. Biophys.* **191**: 590–601.

Garcia, E., and S. G. Rhee. 1983. Cascade control of *Escherichia coli* glutamine synthetase. Purification and properties of PII uridylyltransferase and uridylyl-removing enzyme. *J. Biol. Chem.* **258**: 2246–2253.

Hirschman, J., P.-K. Wong, K. Sei, J. Keener, and S. Kustu. 1985. Products of nitrogen regulatory genes *ntrA* and *ntrC* of enteric bacteria activate *glnA* transcription in vitro: evidence that the *ntrA* product is a sigma factor. *Proc. Natl. Acad. Sci. USA* **82**:7525–7529.

Holtel, A., and M. J. Merrick. 1989. The *Klebsiella pneumoniae* PII protein (*glnB* gene product) is not absolutely required for nitrogen regulation and is not involved in NifL-mediated *nif* gene regulation. *Mol. Gen. Genet.* **217**:474–480.

Hunt, T. P., and B. Magasanik. 1985. Transcription of *glnA* by purified bacterial components: core RNA polymerase and the products of *glnF, glnG,* and *glnL. Proc. Natl. Acad. Sci. USA* **82**:8453–8457.

Kamberov, E. S. Unpublished data.

Kamberov, E. S., M. R. Atkinson, J. Feng, and A. J. Ninfa. 1994a. Sensory components controlling bacterial nitrogen assimilation. *Mol. Cell. Biol. Res.* **40**:175–191.

Kamberov, E. S., M. R. Atkinson, and A. J. Ninfa. 1994b. Effect of mutations in *Escherichia coli glnL* (*ntrB*), encoding nitrogen regulator II (NRII or NtrB), on the regulated phosphatase activity involved in bacterial nitrogen regulation. *J. Biol. Chem.* **269**:28294–28299.

Kamberov, E. S., M. R. Atkinson, and A. J. Ninfa. Complex mechanism of nitrogen sensation in *Escherichia coli*. I. The PII signal transduction protein is the sensor of 2-ketoglutarate and ATP. Submitted for publication.

Keener, J., and S. Kustu. 1988. Protein kinase and phosphoprotein phosphatase activities of nitrogen regulatory proteins NTRB and NTRC of enteric bacteria: roles of conserved amino terminal domain of NTRC. *Proc. Natl. Acad. Sci. USA* **82:**8453–8457.

Kustu, S. G., N. C. MacFarland, S. P. Hui, B. Esmon, and G. F.-L. Ames. 1979. Nitrogen control in *Salmonella typhimurium*: co-regulation of synthesis of glutamine synthetase and amino acid transport systems. *J. Bacteriol.* **138:**218–234.

Kustu, S., E. Santero, J. Keener, D. Popham, and D. Weiss. 1989. Expression of sigma 54 (*ntrA*) dependent genes is probably united by a common mechanism. *Microbiol. Rev.* **53:**367–376.

Lee, T.-Y., K. Makino, H. Shinegawa, and A. Nakata. 1990. Overproduction of acetate kinase activates the phosphate regulon in the absence of the *phoR* and *phoM* functions in *Escherichia coli*. *J. Bacteriol.* **172:**2245–2249.

Lukat, G. S., W. R. McCleary, A. M. Stock, and J. B. Stock. 1992. Phosphorylation of bacterial response regulator proteins by low molecular weight phospho-donors. *Proc. Natl. Acad. Sci. USA* **89:**718–722.

MacNeil, T., G. P. Roberts, D. MacNeil, and B. Tyler. 1982. The products of *glnL* and *glnG* are bifunctional regulatory proteins. *Mol. Gen. Genet.* **188:**325–333.

McCleary, W. R., J. B. Stock, and A. J. Ninfa. 1993. Is acetyl phosphate a global signal in *Escherichia coli*? *J. Bacteriol.* **175:**2793–2798.

Neidhardt, F. C., and B. Magasanik. 1957. Reversal of the glucose inhibition of histidase biosynthesis in *Aerobacter aerogenes*. *J. Bacteriol.* **73:**253–259.

Ninfa, A. J. Unpublished data.

Ninfa, A. J., and M. R. Atkinson. Unpublished data.

Ninfa, A. J., and U. Bai. Unpublished data.

Ninfa, A. J., and R. L. Bennett. 1991. Identification of the site of autophosphorylation of the bacterial protein kinase/phosphatase NRII. *J. Biol. Chem.* **266:**6888–6893.

Ninfa, A. J., and B. Magasanik. 1986. Covalent modification of the *glnG* product, NRI, by the *glnL* product, NRII, regulates the transcription of the *glnALG* operon in *Escherichia coli*. *Proc. Natl. Acad. Sci. USA* **83:**5909–5913.

Ninfa, A. J., E. G. Ninfa, A. Lupas, A. Stock, B. Magasanik, and J. Stock. 1988. Crosstalk between bacterial chemotaxis signal transduction proteins and the regulator of transcription of the Ntr regulon: evidence that nitrogen assimilation and chemotaxis are controlled by a common phosphotransfer mechanism. *Proc. Natl. Acad. Sci. USA* **85:**5492–5496.

Ninfa, A. J., L. J. Reitzer, and B. Magasanik. 1987. Initiation of transcription at the bacterial *glnAp2* promoter by purified *E. coli* components is facilitated by enhancers. *Cell* **50:**1039–1046.

Ninfa, E. G., M. R. Atkinson, E. S. Kamberov, and A. J. Ninfa. 1993. Mechanism of autophosphorylation of *Escherichia coli* nitrogen regulator II (NRII or NtrB): trans-phosphorylation between subunits. *J. Bacteriol.* **175:**7024–7031.

Ninfa, E. G., J. B. Stock, and A. J. Ninfa. Unpublished data.

Norby, J. G. 1988. Coupled assay of Na$^+$, K$^+$-ATPase activity. *Methods Enzymol.* **156:**116–119.

Pahel, G., D. M. Rothstein, and B. Magasanik. 1982. Complex *glnA-glnL-glnG* operon of *Escherichia coli*. *J. Bacteriol.* **150:**202–213.

Pahel, G., D. Zelentz, and B. M. Tyler. 1978. *gltB* gene and the regulation of nitrogen metabolism by glutamine synthetase in *Escherichia coli*. *J. Bacteriol.* **133:**139–148.

Parkinson, J. S. D., and E. C. Kofoid. 1992. Communication modules in bacterial signaling proteins. *Annu. Rev. Genet.* **26:**71–112.

Porter, S. C., A. K. North, A. B. Wedel, and S. Kustu. 1993. Oligomerization of NTRC at the *glnA* enhancer is required for transcriptional activation. *Genes Dev.* **7:**2258–2273.

Prival, M. J., and B. Magasanik. 1971. Resistance to catabolite repression of histidase and proline oxidase during nitrogen-limited growth of *Klebsiella aerogenes*. *J. Biol. Chem.* **246:**6288–6296.

Reitzer, L. J., and B. Magasanik. 1983. Isolation of the nitrogen assimilation regulator, NRI, the product of the *glnG* gene of *Escherichia coli*. *Proc. Natl. Acad. Sci. USA* **80:**5554–5558.

Reitzer, L. J., and B. Magasanik. 1985. Expression of *glnA* in *Escherichia coli* is regulated at tandem promoters. *Proc. Natl. Acad. Sci. USA* **82:**1979–1983.

Reitzer, L. J., and B. Magasanik. 1986. Transcription of *glnA* in *E. coli* is stimulated by activator bound to sites far from the promoter. *Cell* **45:**785–792.

Reitzer, L. J., B. Movsas, and B. Magasanik. 1989. Activation of *glnA* transcription by nitrogen regulator I (NRI)-phosphate in *Escherichia coli*: evidence for a long-range interaction between NRI-phosphate and RNA polymerase. *J. Bacteriol.* **171:**5512–5522.

Rhee, S. G., S. C. Park, and J. H. Koo. 1985. The role of adenylyltransferase and uridylyltransferase in the regulation of glutamine synthetase in *Escherichia coli*. *Curr. Top. Cell. Regul.* **27:**221–232.

Sanders, D. A., B. L. Gillece-Castro, A. L. Burlingame, and D. E. Koshland, Jr. 1992. Phosphorylation siter of NtrC, a protein phosphatase whose covalent intermediate activates transcription. *J. Bacteriol.* **174:**5117–5122.

Shiau, S.-P., B. L. Schneider, W. Gu, and L. J. Reitzer. 1992. Role of nitrogen regulator I (NtrC), the transcriptional activator of *glnA* in enteric bacteria, in reducing expression of *glnA* during nitrogen-limited growth. *J. Bacteriol.* **174**:179–185.

Son, H. S., and S. G. Rhee. 1987. Cascade control of *Escherichia coli* glutamine synthetase. Purification and properties of PII protein and nucleotide sequence of its structural gene. *J. Biol. Chem.* **262**:8690–8695.

Stadtman, E. R., E. Mura, P. B. Chock, and S. G. Rhee. 1980. The interconvertable enzyme cascade that regulates glutamine synthetase activity, p. 41–59. *In* J. Mora and R. Palacios (ed.), *Glutamine: Metabolism, Enzymology, and Regulation.* Academic Press, New York.

Wanner, B. L., and M. R. Wilmes-Riesenberg. 1992. Involvement of phosphotransacetylase, acetate kinase, and acetyl phosphate synthesis in control of the phosphate regulon in *Escherichia coli. J. Bacteriol.* **174**:2124–2130.

Weiss, D. S., J. Batut, K. E. Klose, J. Keener, and S. Kustu. 1991. The phosphorylated form of the enhancer-binding protein NTRC has an ATPase activity that is essential for activation of transcription. *Cell* **67**:155–167.

Weiss, V., F. Claverie-Martin, and B. Magasanik. 1992. Phosphorylation of nitrogen regulator I (NRI) of *Escherichia coli* induces the strong cooperative binding to DNA essential for the activation of transcription. *Proc. Natl. Acad. Sci. USA* **89**:5088–5092.

Weiss, V., and B. Magasanik. 1988. Phosphorylation of nitrogen regulator I (NRI) of *Escherichia coli. Proc. Natl Acad. Sci. USA* **85**:8919–8923.

Wolfe, A. J., M. P. Conley, and H. C. Berg. 1988. Acetyladenylate plays a role in controlling the direction of flagellar rotation. *Proc. Natl. Acad. Sci. USA* **85**:6711–6715.

Wong, P.-K., D. Popham, J. Keener, and S. Kustu. 1987. In vitro transcription of the nitrogen fixation regulatory operon *nifLA* of *Klebsiella pneumoniae. J. Bacteriol.* **169**:2876–2880.

Chemotactic Signal Transduction in
Escherichia coli and *Salmonella typhimurium*

Charles D. Amsler and Philip Matsumura

6

Bacterial chemotaxis is an important model for two-component regulatory systems. The flow of information through the chemotactic signal transduction pathway and the proteins responsible for it have been characterized in great detail, as discussed in this chapter. This has been used as a basis for understanding the flow of information through a wide variety of homologous systems (e.g., J. B. Stock et al., 1989; Bourret et al., 1991; Parkinson and Kofoid, 1992; Stock and Lukat, 1991; Parkinson, 1993; Volz, 1993).

In nature, chemotaxis functions as a response to stressful microenvironments (cf. Amsler et al., 1993), and chemotactic behavior has been observed in a wide variety of bacteria. However, our most detailed knowledge of the chemotactic signal transduction pathway comes from studies of the closely related enteric bacteria *Escherichia coli* and *Salmonella typhimurium*. The information pathway in these species serves as the model to which all other species are usually compared (e.g., Fuhrer and Ordal, 1991; Shaw, 1991; Bischoff and Ordal, 1992; McBride et al., 1992; Zhulin and Armit-

age, 1993). Consequently, this chapter focuses exclusively on chemotaxis in *E. coli* and *S. typhimurium*. The components of the signal pathway in these two species are virtually interchangeable, and everything in this chapter is assumed to be equally applicable to both unless otherwise noted. Furthermore, the discussion is limited to chemotaxis in response to amino acids as well as some sugars and other molecules that involve methylation of the chemotactic receptors. Non-methylation-dependent chemotaxis pathways to other sugars and to oxygen are reviewed by Armitage (1992), Taylor (1983a), Shioi et al. (1987, 1988), and Titgemeyer (1993).

MOTILITY AND EXPRESSION OF THE FLAGELLAR REGULON

Bacteria swim through liquid media using flagella. Bacterial flagella are composed of basal body structures that span the inner and outer membranes and extracellular helical protein strands called *flagellar filaments* (Fig. 1). The flagellar filaments are composed of a single protein called *flagellin* and are attached to the basal bodies by a flexible hook region (Macnab, 1990, 1992). Motor proteins that span the inner membrane rotate part of the basal bodies using proton motive force as the energy source

Charles D. Amsler, Department of Biology, University of Alabama at Birmingham, Birmingham, Alabama 35294–1170. *Philip Matsumura,* Department of Microbiology and Immunology (M/C 790), University of Illinois at Chicago, Chicago, Illinois 60612–7344.

Two-Component Signal Transduction, Edited by James A. Hoch and Thomas J. Silhavy,
© 1995 American Society for Microbiology, Washington, DC 20005

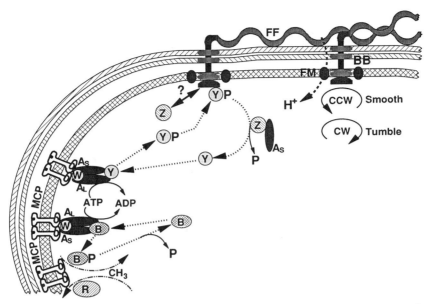

FIGURE 1 Overall model of chemotactic signal transduction. Abbreviations: A_L, CheA$_L$; A_S, CheA$_S$; B, CheB; BB, flagellar basal body; CCW, counterclockwise flagellar rotation; CW, clockwise flagellar rotation; FF, flagellar filament; FM, flagellar motor; MCP, methyl-accepting chemotaxis protein; P, phosphate; R, CheR; W, CheW; Y, CheY; Z, CheZ.

(rather than ATP) (reviewed by Berg et al., 1982; Taylor, 1983b; and Macnab and Aizawa, 1984). This rotation is transmitted through the hook to the filaments, which then propel the cells through the media. Unlike eukaryotic flagella, which are flexible and are moved in an undulating or whiplike pattern, these rigid bacterial flagellar filaments are more analogous to corkscrew-shaped propellers pushing the cell. For more indepth reviews of flagellar structure and function, see Macnab and Aizawa (1984) and Macnab (1987a, 1990, 1992).

There are more than 40 genes in at least 14 operons that code for proteins necessary for chemotaxis. These include the structural components of the flagella, the flagellar motor, the transmembrane receptors, and the chemotactic signal pathway (reviewed by Macnab 1990, 1992). The operons are expressed in a three-tiered hierarchical order. The first level includes only the *flhD* operon, and relief from catabolite repression is necessary for its expression (Silverman and Simon, 1974; Komeda, 1982). The operon contains only the *flhC* and *flhD* genes, and their expression is necessary for expression of all other genes in the regulon (Komeda, 1982). FlhC and FlhD proteins have recently been shown to be transcriptional activators and to function as heterotetramers of two FlhD and two FlhC molecules (Liu and Matsumura, 1994). The second-level operons include most of the basal body structural genes as well as *fliA,* an alternate sigma factor necessary for expression of all the third-level operons. The last level three operons expressed include *fliC, motA,* and *tar.* The *fliC* operon codes for the flagellar filament. The *motA* and *tar* operons code for all the chemotaxis signal transduction proteins, the flagellar motor proteins, and the transmembrane receptor proteins. For a more detailed review of flagellar regulon regulation, see Helmann (1991) and Macnab (1992).

Chemotaxis is a mechanism that bacteria use in response to environmental stress. Although the specific environmental signals that activate flagellar regulon expression are unknown, cells grown in fresh culture media are nonmotile during lag and early exponential growth phases (Adler and Templeton, 1967; Amsler et al., 1993). Presumably, there is a significant metabolic cost to the production of flagella and chemotaxis proteins (Macnab, 1992), which would be of no advantage to a cell in a rich microenvironment. As a culture progresses into late exponential and post-exponential phases, the environment becomes decreasingly suitable for maximal growth and reproduction. During this time, there is an increasing likelihood that motility and chemotaxis would be adaptively advantageous for cells in analogous microenvironments in nature. Correspondingly, expression from the flagellar regulon and cell motility increase throughout this period (Adler and Templeton, 1967; Komeda and Iino, 1979; Amsler et al., 1993). Maximum motility is observed during postexponential growth because both flagellar length and density (number per unit biomass) double relative to early exponential growth-phase cells. As cells enter stationary phase, expression from the flagellar regulon drops somewhat, but flagellar length and density remain nearly constant. Decreased motility in stationary phase is correlated with a threefold drop in torque produced by the flagellar motors, presumably because of decreased proton motive force (Amsler et al., 1993).

CHEMOTACTIC BEHAVIOR

There are two distinct phases of swimming motion in peritrichously flagellated bacteria such as *E. coli* and *S. typhimurium*. These result from the two different directions of flagellar rotation, either clockwise (CW) or counter-clockwise (CCW). When flagella rotate CCW, they generate an inward force toward the body of the bacterium. As the cell begins to move in one direction, the flagella form a thermodynamically stable flagellar bundle behind the cells (Macnab, 1977) (Fig. 2). The resulting linear movement of the cells is called *smooth swimming*. When the flagella switch to CW rotation, however, the bundle is no longer stable and the

SMOOTH SWIMMING
Counterclockwise Flagellar Rotation

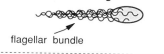

flagellar bundle

TUMBLE
Clockwise Flagellar Rotation

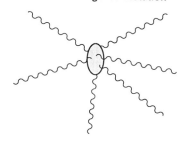

FIGURE 2 Flagellar patterns during smooth swimming and tumble behaviors.

flagella separate. Forward motion ceases immediately, and the forces of the separating flagella cause the cell to rotate to a new, randomly determined orientation in what is referred to as a *tumble* (Fig. 2). Reversals from CCW to CW rotation are not coordinated within a cell (Ichihara et al., 1983; Macnab and Han, 1983), and it is not known whether all the flagella must reverse to CW rotation for a tumble to occur. However, it seems likely that a tumble will occur when some critical number, but not necessarily all, of the flagellar motors rotate CW and that smooth swimming will continue when some number of flagella less than that critical number rotate CW (the "voting hypothesis") (Ichihara et al., 1983; Macnab and Han, 1983).

In the absence of a chemical gradient, *E. coli* and *S. typhimurium* tumble at random intervals, once every 2 to 4 seconds. Therefore, they wander randomly about their environment in what is described mathematically as a "random walk" (Fig. 3). Movement along a chemical gradient is achieved by varying tumble frequency. If the concentration of an attractant is increasing or of a repellent is decreasing, cells do not tumble as frequently as they would in an

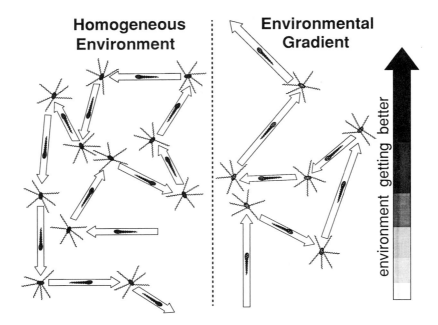

FIGURE 3 Swimming behavior in homogeneous versus heterogeneous environments. Smooth swimming and tumbles as in Fig. 2, with length of arrows indicating the duration of smooth swimming. Bacteria swim smoothly longer when their environment is getting better than when it is unchanging. There is no difference in smooth swimming duration between homogeneous environments and gradients when the environment is getting worse.

unchanging environment. Consequently, cells swim in the direction in which their environment is improving longer than they would have otherwise, and this biases the random walk such that there is net movement up the gradient (Fig. 3). Cells do not respond to decreasing concentrations of attractants or increasing concentrations of repellents; tumble frequency is identical to that in a homogeneous environment until a tumble results in movement in a beneficial direction (Berg and Brown, 1972, 1974; Berg and Tedesco, 1975; Dahlquist et al., 1976). This mechanism is not a true tactic response in the formal sense because tumble frequency rather than orientation of movement is effected but is called *chemotaxis* for historical reasons. It is only about 50% efficient relative to a "true" chemotactic response but presumably requires a much smaller investment in sensory and motor machinery (Dahlquist et al., 1976; Koshland, 1980).

CHEMOTACTIC SIGNAL PATHWAY

For a cell to chemotactically respond to its environment, it must be able to detect specific compounds and changes in their concentration. It must then be able to transmit this information to the flagellar switch that controls rotational direction and, therefore, tumble frequency. The transmembrane receptors discussed below detect compounds in the periplasm and transmit this information to the cytoplasm, but separate transport proteins are required for uptake. Chemical concentrations are "recorded" via reversible methylation of the receptors. Cells swim on the order of 10 body lengths per second (Amsler et al., 1993), and gradients are detected temporally as they move through the environment. Separate groups of receptors not discussed are not methylated and are involved in aerotaxis and in both chemotactic detection and uptake of carbohydrates as part of the phosphoenolpyruvate-sugar phosphotransferase system (PTS) (reviewed by Armitage, 1992; Taylor, 1983a; Titgemeyer, 1993). All types transmit information to the flagellar switch via at least part of a phosphorylation cascade involving six soluble proteins: CheA, CheW, CheY, CheZ, CheR, and CheB. Cells lacking all these proteins do not tumble (e.g., Wolfe et al., 1987). The following discussion is not intended to be an exhaustive review of the literature on the signaling pathway. Rather, our intent is to provide an overview of the function of each component and of the pathway as a whole while calling attention to some recent advances. For more detailed reviews of all or parts of the chemotactic signal pathway, see Ordal (1985), Bartlett and Matsumura (1986), Macnab (1987b), Stewart and Dahlquist (1987), J. B. Stock et al. (1989), Matsumura et al. (1990), Hazelbauer et al. (1990, 1993), Bourret et al. (1991), Stock and Lukat (1991), Armitage (1992), and Parkinson (1993), as well as Chapters 2, 3, 4, and 11 in this book.

Membrane Receptors

Transmembrane receptors transmit chemotactic information from the periplasm into the cytoplasm. Four homologous receptor-transducer proteins called *methyl-accepting chemotaxis proteins* (MCPs) span the inner membrane and generate the original chemotactic signal: Tsr, Tar, Trg, and Tap. They act as receptors by binding directly to some amino acids and repellents and as transducers by interacting with periplasmic binding proteins that are attached to some sugars or dipeptides. These binding proteins can also function as part of transport systems (Oxender, 1972). Attractants that are perceived by MCPs include serine, alanine, and glycine (by Tsr); aspartate, glutamate, and maltose (by Tar); galactose and ribose (by Trg); and various dipeptides (by Tap) (see references in Bartlett and Matsumura [1986] and Stewart and Dahlquist [1987]). Repellents include acetate, leucine, benzoate, and indole (mediated by Tsr) and cobalt and nickel (mediated by Tar) (see references in Bartlett and Matsumura [1986] and Stewart and Dahlquist [1987]). Reversible methylation of the MCPs enables the cells to detect temporal changes in attractants and repellents, as discussed below.

There is a great deal of amino acid sequence homology and presumably structural similarity between the four MCP classes (Stewart and Dahlquist, 1987). A variety of evidence (reviewed by Stewart and Dahlquist [1987] and Ames and Parkinson [1988]) suggests that the N-terminal half of each constitutes the periplasmic chemoreceptor domain flanked by two membrane-spanning regions and that the C-terminal half includes the cytoplasmic signaling region and methylation sites. The three-dimensional crystal structure of the periplasmic domain of Tar has been solved with and without bound aspartate. It contains two long and two short α-helices that form a cylinder 20 Å (2nm) in diameter and more than 70 Å long (Milburn et al., 1991; Yeh et al., 1993). Only small differences are observed between receptors with and without bound aspartate (Milburn et al., 1991; Yeh et al., 1993). The transmembrane region is thought to consist of two α-helices (Milburn et al., 1991; Kim et al., 1992; Pakula and Simon, 1992a).

MCPs exist as stable dimers within the inner membrane (Milligan and Koshland, 1988). Genetic and structural studies have indicated that signaling can occur within a single subunit and that changes in the relative orientation of the subunits are not necessary for signaling (Ames and Parkinson, 1988; Lynch and Koshland, 1991; Milligan and Koshland, 1991). However, it has been argued that such changes can occur (Kim et al., 1992). Two competing models for signaling have been proposed. In the "piston" model, movement between α-helices within a monomer generates the transmembrane signal, and in the "pivot" model, small movements between monomers in the periplasmic domain cause larger relative displacements of the signaling domains (Milburn et al., 1991; Pakula and Simon, 1992b; Kim et al., 1992).

MCPs are not randomly distributed throughout the inner cell membrane as has long been believed. Recently, Maddock and Shapiro (1993) demonstrated that Tar, Tsr, and Trg are arrayed in clusters in the inner membrane and that these MCP aggregates are most commonly observed at a cell pole. CheW is necessary for aggregate formation. The functional significance of these aggregates is unclear because it is believed that chemoreception would be most sensitive if the MCPs were distributed uniformly throughout the cell (Berg and Purcell, 1977). Parkinson and Blair (1993) have speculated that this distribution may be the consequence of a functional trade-off between chemical sensing and uptake.

CheA

CheA is the first link in the sensory chain between the receptors and the flagellar switch. The *cheA* gene codes for two different proteins with different translational start sites and a common carboxy terminus (Matsumura et al., 1977; Smith and Parkinson, 1980). The longer form, $CheA_L$, is a kinase that interacts with the MCPs, autophosphorylates, and in turn, phosphorylates CheY and CheB (see Fig. 1). CheA has been shown to consist of distinct functional domains (Oosawa et al., 1988; Parkinson and Kofoid, 1992; Bourret et al., 1993a; Swanson et al., 1993a; Swanson et al., 1993b). $CheA_L$ is phosphorylated at histidine 48 in the amino-terminal portion of the protein not translated in the short form, $CheA_S$ (Hess et al., 1988b). However, phosphorylation requires a central domain present in both CheA forms (Hess et al., 1988b). CheA phosphorylation can occur via *trans*-phosphorylation between these two regions within a dimer of either $CheA_L$:$CheA_S$ or $CheA_L$:$CheA_L$ (Swanson et al., 1993a; Wolfe and Stewart, 1993). The amino-terminal region containing histidine 48 is capable of phosphotransfer to CheY and CheB (Hess et al., 1988b; Hess et al., 1988c) and has been designated the P_1 region (Parkinson and Kofoid, 1992). A region between P_1 and the central domain known as P_2 appears to be important

in the interaction between CheA and CheY (Parkinson and Kofoid, 1992; Swanson et al., 1993b; Shukla and Matsumura, unpublished data). The carboxy-terminal end of CheA, called the C region, appears to control autokinase activity and probably interacts with CheW, whereas the M region, which is between the central domain and C, is the site of interaction with the MCPs (Parkinson and Kofoid, 1992).

CheW and the $CheA_L$-$CheA_S$-CheW Complex

CheW plays critical roles in CheA phosphorylation and in the interaction between CheA and the MCPs. It is required for CW flagellar rotation in vivo (Conley et al., 1989; Liu and Parkinson, 1989). Stable complexes of $CheA_L$, $CheA_S$, and CheW in a 1:1:1 molar ratio have been isolated from freshly lysed cells (Matsumura et al., 1990; McNally and Matsumura, 1991). The presence of CheW in these complexes enhances CheA autophosphorylation by increasing its affinity for ATP by 68-fold (McNally and Matsumura, 1991). Stable complexes cannot be formed from purified components nor detected if cells have been frozen after lysis (McNally and Matsumura, 1991). However, less stable dimers of purified $CheA_L$ and complexes of purified $CheA_L$ and CheW can be detected in vitro by perturbation chromatography (Gegner and Dahlquist, 1991).

Chemotactic sensory information is passed from the MCPs into the cytoplasmic transduction pathway through MCP control of the phosphorylation rate of $CheA_L$. An attractant signal from the receptors decreases CheA autophosphorylation (Borkovich and Simon, 1990; McNally and Matsumura, 1991; Ninfa et al., 1991), which results in decreased CheY phosphorylation (Borkovich et al., 1989). CheW is necessary for this coupling of MCPs and CheA (Borkovich et al., 1989). CheW is also necessary for MCP aggregation, as mentioned above, and interactions between CheW and MCPs have been detected genetically (Liu and Parkinson, 1991). MCPs are able to form detectable complexes with CheW and CheA$_L$-CheW but not with CheA$_L$ alone (Gegner et al., 1992). CheA and CheW have been shown to colocalize with MCP aggregations in electron micrographs (Maddock and Shapiro, 1993). Therefore, MCPs and the CheA$_L$-CheA$_S$-CheW complex appear to form a quaternary functional complex. However, the level of CheA$_L$-CheA$_S$-CheW complex appears to be constant and not affected by the signaling state of the MCPs or by their absence (McNally and Matsumura, 1991).

CheY

CheY is phosphorylated by CheA, and CheY-phosphate (CheY-P) interacts with the flagellar switch to increase the likelihood of tumbles. The high-resolution, three-dimensional structure of CheY has been solved in both *E. coli* (Volz and Matsumura, 1991) and *S. typhimurium* (A. M. Stock et al., 1989). It is a small globular protein comprising five β-sheets interspersed with five α-helices and short linking regions. Phosphorylation occurs at aspartate 57 (Sanders et al., 1989), which is contained within an acidic pocket on one face of the molecule.

Unphosphorylated CheY is able to bind to the CheA$_L$-CheA$_S$-CheW complex (McNally and Matsumura, 1991). CheY also binds to CheA$_L$-CheA$_S$ alone (McNally and Matsumura, 1991; Shukla and Matsumura, unpublished data) and to CheA$_L$ alone (McNally and Matsumura, 1991; Schuster et al., 1993) but apparently with much lower affinity than to the CheA-CheW complex (McNally and Matsumura, 1991). Ternary complexes of MCPs, CheA, and CheW bind to immobilized CheY when it is unphosphorylated but are released under conditions favoring the formation of CheY-P (Schuster et al., 1993). CheY does not colocalize in electron micrographs with the patches of MCPs, CheA, and CheW described above (Maddock and Shapiro, 1993).

cheY or *cheA* mutations, which do not permit formation of CheY-P, have smooth swimming phenotypes (Oosawa et al., 1988; Bourret et al., 1990). Consequently, interactions be-

tween CheY-P and the flagellar switch are thought to increase the chances of a reversal from CCW to CW rotation. In experiments with partially lysed cells, CheY-P was calculated to be more than 100-fold more active than unphosphorylated CheY in generating CW rotation (Barak and Eisenbach, 1992a). Unphosphorylated CheY could also generate a CW signal, but the signal was weaker and the protein was present at unnaturally high concentrations in these experiments (Barak and Eisenbach, 1992a). When purified MCP and Che proteins are mixed in vitro, addition of attractants dramatically decreases the rate of CheY phosphorylation (Ninfa et al., 1991). This is consistent with the hypothesis that CheY-P is the tumble signal.

The flagellar switch includes the proteins FliG, FliM, and FliN (Yamaguchi et al., 1986a; Yamaguchi et al., 1986b). CheY-P has been shown to bind to FliM, but unphosphorylated CheY does not (Welch et al., 1993). A class of *cheY* mutations has been isolated in *E. coli*, which dominantly suppress point mutations in *fliG*, and also suppresses *fliM* mutations (Roman et al., 1992). These *cheY* mutations do not effect CheY phosphorylation and map to a specific face of CheY that does not overlap with the acidic phosphorylation pocket. A number of *cheY* mutations that suppress mutations in *fliM* have been isolated in *S. typhimurium* (Sockett et al., 1992), and most also map to this face of the molecule. Therefore, these suppressor mutations appear to define the face of the CheY molecule, which interacts with the flagellar switch (Roman et al., 1992). This is consistent with multidimensional nuclear magnetic resonance studies, which have shown substational chemical shift on phosphorylation in two regions that overlap this genetically defined face (in addition to changes in the acidic phosphorylation pocket) (Lowry et al., 1994). Complementary ^{19}F-nuclear magnetic resonance data of fluorine-labeled side chains also detected changes on phosphorylation (Bourret et al., 1993b; Drake et al., 1993), and other nuclear magnetic resonance studies have shown that binding of bivalent cations in the acidic phosphorylation pocket results in significant conformational changes (Kar et al., 1992a; Kar et al., 1992b). These data are consistent with a substantial conformational change triggered by phosphorylation.

CheZ

CheZ counteracts the tumble signal by increasing the conversion of CheY-P to CheY. CheZ has been shown to interact directly with CheY (McNally and Matsumura, 1991; Blat and Eisenbach, 1994). The amount of CheZ that binds to CheY columns is four-fold higher than observed with CheA or CheW (McNally and Matsumura, 1991). In vivo, expression of CheZ antagonizes tumbly behavior that is attributable to CheY-P (Kuo and Koshland, 1987; Wolfe et al., 1987). In vitro, CheY-P autodephosphorylates, but this dephosphorylation is accelerated by the addition of CheZ (Hess et al., 1988a; Hess et al., 1988c). It is not known whether CheZ acts as a phosphatase itself, catalyzes CheY-P autodephosphorylation, or both. CheZ is able to form complexes with CheA$_S$ but not CheA$_L$ in molar ratios of 1 CheA$_S$ to between 5 and 6 CheZ (Wang et al., unpublished data). CheZ-CheA$_S$ complex dephosphorylates CheY-P more rapidly than CheZ alone (Wang et al., unpublished data). This discovery introduces the possibility of an additional level of control in the sensory transduction system and, perhaps, even a second signal via a heretofore unrecognized pathway that modulates CheY-P dephosphorylation.

CheZ may also interact directly with the flagellar switch. Several *cheZ* mutants have been isolated as suppressors of flagellar switch mutations (Parkinson and Parker, 1979; Parkinson et al., 1983; Yamaguchi et al., 1986a). Two of these with unusual phenotypes have been shown to act via increased ability to dephosphorylate CheY-P (Huang and Stewart, 1993), but most show allele-specificity, which suggests direct interaction between CheZ and the flagellar switch proteins (Parkinson and Parker, 1979; Parkinson et al., 1983; Yamaguchi et al., 1986a).

Adaptation: CheR, CheB, and MCP Methylation

The cellular "memory" necessary for bacteria to sense concentration gradients over time is the result of methylation of the MCPs by CheR and demethylation by CheB-phosphate (CheB-P). MCPs can be methylated at four to six sites, depending on receptor class (reviewed by Kehry et al., 1985; Stewart and Dahlquist, 1987). The methyltransferase, CheR, transfers methyl groups from S-adenosylmethionine to specific glutamate residues (Springer and Koshland, 1977). In the absence of chemotactic stimuli, MCPs are continuously methylated and demethylated so that a basal level is maintained (Kort et al., 1975; Goy et al., 1977). The binding of attractants or repellents to MCPs appears to induce conformational changes that make these sites more or less accessible, respectively, to methylation (Stock and Koshland, 1981; Springer et al., 1982). Addition of the attractant aspartate to mixtures of purified Tar and CheR increase the rate of methylation (Ninfa et al., 1991). Adaptation is believed to occur because increasing methylation progressively decreases the attractant signal from the receptor so that its signaling state returns to prestimulus levels (e.g., Macnab, 1987b). However, the mechanism by which this occurs is not known.

CheB-P is a methylesterase that adjusts the signaling state of the MCPs by removing methyl groups to form methanol (Stock and Koshland, 1978; Toews and Adler, 1979). Although methylation by CheR only occurs in the MCP class that has bound a particular attractant (Goy et al., 1977; Silverman and Simon, 1977; Springer et al., 1977), demethylation by CheB-P appears to affect all MCP classes simultaneously (Kehry et al., 1985). CheB is phosphorylated by CheA (Hess et al., 1988c), and the phosphorylated form is thought to be responsible for increased methylesterase activity because mutations in either CheA or CheW decrease the response to chemotactic signals (Springer and Zanolari, 1984; Stewart and Dahlquist, 1988). It has been demonstrated in vitro that CheA phosphorylates CheB and that the influences of CheW and MCPs are similar to those observed on phosphorylation of CheY (Ninfa et al., 1991). Phosphorylation of CheB has been shown to dramatically increase its ability to demethylate MCPs in vitro (Lupas and Stock, 1989). Attractants cause transient decreases in methylesterase activity, whereas repellents transiently increase it (Toews et al., 1979; Kehry et al., 1984, 1985; Springer and Zanolari, 1984). Functionally, CheB is divided into an N-terminal regulatory domain and a C-terminal catalytic domain (Simms et al., 1985). The site of phosphorylation is in the N-terminal domain (Lupas and Stock, 1989), which has strong homology to CheY (Volz, 1993). Genetic evidence suggests that the region surrounding the phosphorylation site interacts both with the C-terminal domain and with CheA (Stewart, 1993).

Summary of Chemotactic Signal Transduction

Our view of the chemotactic signal pathway is summarized in Fig. 1. Homodimeric MCPs spanning the inner membrane are to a large extent clustered at the cell poles. MCPs are activated when they bind to periplasmic attractants, repellents, or specialized binding proteins that have bound attractants. This results in conformational changes in the receptors' cytoplasmic domain. These changes could result from physical displacement either within or between monomers. Protein complexes of $CheA_L$, $CheA_S$, and CheW are closely associated with the MCPs. Conformational signals from the MCPs modulate the rates of phosphorylation within associated CheA dimers. This modulation requires the participation of CheW. If MCPs have bound attractants, phosphorylation rates are decreased. The phosphate signal is then passed from CheA-phosphate in one of two directions, either to CheY as part of the excitation pathway or to CheB as part of adaptation.

Unphosphorylated CheY binds to the CheA-CheW complex; it is released on phos-

phorylation by CheA-phosphate and diffuses from the MCP-CheA-CheW patches to the flagellar basal bodies situated throughout the cell. CheY-P interacts with the flagellar switch proteins, in particular FliM but also FliG. This interaction increases the chances that flagellar rotation will switch from CCW to CW, which is necessary for a tumble to occur. Dephosphorylation of CheY-P is accelerated by CheZ; this occurs more rapidly when CheZ is complexed with $CheA_S$ than when it is not. Unphosphorylated CheY may then bind again to CheA-CheW complex at the MCPs. CheZ may also interact directly with the flagellar switch.

The conformational signaling state of the MCPs makes them more or less accessible to methylation by CheR, depending on whether they have bound attractants or repellents, respectively. Increasing methylation progressively decreases the attractant signal back to prestimulus levels via an unknown mechanism. MCP signaling also influences the rate of demethylation via CheB. MCP signals modulate CheA phosphorylation, as described above, and CheA phosphorylates CheB. CheB-P removes methyl groups from MCPs, thereby antagonizing methylation by CheR. Attractants decrease CheA phosphorylation rates and, therefore, the concentration of CheA-phosphate available to phosphorylate CheB. Overall, attractants increase the rate of addition of methyl groups and decrease the rate of their removal. Conversely, repellents decrease the rate of addition and accelerate removal.

Other Factors Influencing Chemotaxis

Motility and chemotaxis are part of several global stress response networks. Expression of the *flhD* operon, and therefore all flagellar genes, is dependent on the presence of cyclic AMP-catabolite gene activator protein (Silverman and Simon, 1974; Komeda, 1982). This is the regulatory element of the catabolite repression network that also regulates proteins involved with responses to starvation, is an inducer of many catabolic enzymes, and in some species is involved with cell division and a variety of other functions (Botsford and Harman, 1992). Expression of the *flhD* operon is also dependent on the presence of the heat-shock response network proteins DnaK, DnaJ, and GrpE (Shi et al., 1992). Also, CheA, CheY, and CheB are members of the two-component family of homologous signal transduction mechanisms that regulate responses to a wide variety of environmental stresses (J. B. Stock et al., 1989; Parkinson and Kofoid, 1992). Crosstalk or signaling from one system to another occurs in vitro and in some systems under physiologically relevant conditions in vivo (J. B. Stock et al., 1989; Parkinson and Kofoid, 1992). Therefore, signals generated in these other systems have the potential to affect chemotactic behavior. Schuster et al. (1993) speculated that the close physical interactions between MCPs, CheA, CheW, and unphosphorylated CheY may decrease the potential effects of crosstalk. There are also several small molecules that have been shown to effect chemotaxis, but the significance of these effects is unclear. Sudden increases in intracellular calcium ions cause cells to tumble. This effect requires CheA, CheW, and CheY but not MCPs (Tisa and Adler, 1992). Exogenous fumarate affects switching of flagellar rotation in cytoplasm-free cell envelopes to which only it and CheY have been added (Barak and Eisenbach, 1992b). It is speculated that this may be part of a link between the metabolic state of the cell and chemotaxis (Barak and Eisenbach, 1992b). Similarly, addition of acetate to cells deleted for all MCP and chemotaxis genes except *cheY* causes cells to reverse flagellar rotation (Wolfe et al., 1988). This does not appear to be an essential part of chemotactic signal transduction in wild-type cells (Dailey and Berg, 1993), but acetate may be involved in temperature regulation of flagellar expression (Prüss and Wolfe, 1994).

Gaps in Our Knowledge

Although chemotaxis is probably the most thoroughly understood of all signal transduction pathways, significant gaps remain in our understanding of it. In particular, we do not fully understand the signaling steps at the beginning and ends of the pathway. How do conformational changes in MCP signaling domains lead to modulation of CheA phosphorylation rates? How does the interaction between CheY-P and the flagellar switch proteins influence the direction of flagellar rotation? What conformational changes in CheY are responsible for the signal? The roles of CheZ, particularly its interactions with the switch proteins, are not well understood. Also, the exact mechanism by which MCP methylation causes adaptation by returning the signaling state to prestimulus levels is not known. At the center of the pathway, knowledge of the mechanisms by which the functional domains of CheA interact within the CheA-CheW complex and with CheY and CheB are necessary to understand the dynamics and specificity of the phosphorylation cascade. All these questions are important topics of present and future research.

The study of chemotactic signal transduction began with very simple investigations of whole-cell behavior by Adler (e.g., Adler and Dahl, 1967; Adler and Templeton, 1967; Adler, 1973). Most present investigations make use of elaborate genetic and biochemical techniques but combine these with behavioral assays that are either very subjective or very indirect measures of chemotactic behavior. Computer-based quantitative assays of chemotactic behavior are becoming available (e.g., Poole et al., 1988; Sager et al., 1988; Marwan and Osterhelt, 1990; Amsler et al., 1993; Khan et al., 1993) and will facilitate quantitative analysis of chemotaxis in whole free-swimming cells. This will allow investigations to begin at the behavioral level and work toward its biochemical and genetic basis. Such studies would complement most present work, which begins with biochemistry and genetics and extrapolates to their behavioral consequences.

ACKNOWLEDGMENTS

Preparation of this review was supported by Public Health Service grants AI 08547 (C.D.A.) and AI 18985 (P.M.) from the National Institutes of Health.

REFERENCES

Adler, J. 1973. A method for measuring chemotaxis and use of the method to determine optimal conditions for chemotaxis by *Escherichia coli*. *J. Gen. Microbiol.* **74**:77–91.

Adler, J., and M. M. Dahl. 1967. A method for measuring the motility of bacteria and for comparing random and non-random motility. *J. Gen. Microbiol.* **46**:161–173.

Adler, J., and B. Templeton. 1967. The effect of environmental conditions on the motility of *Escherichia coli*. *J. Gen. Microbiol.* **46**:175–184.

Ames, P., and J. S. Parkinson. 1988. Transmembrane signaling by bacterial chemoreceptors: E. *coli* transducers with locked signal output. *Cell* **55**:817–826.

Amsler, C. D., M. Cho, and P. Matsumura. 1993. Multiple factors underlying the maximum motility of *Escherichia coli* as cultures enter post-exponential growth. *J. Bacteriol.* **175**:6238–6244.

Armitage, J. P. 1992. Behavioral responses in bacteria. *Annu. Rev. Physiol.* **54**:683–714.

Barak, R., and M. Eisenbach. 1992a. Correlation between phosphorylation of the chemotaxis protein-CheY and its activity at the flagellar motor. *Biochemistry* **31**:1821–1826.

Barak, R., and M. Eisenbach. 1992b. Fumarate or a fumarate metabolite restores switching ability to rotating flagella of bacterial envelopes. *J. Bacteriol.* **174**:643–645.

Bartlett, D. H., and P. Matsumura. 1986. Behavorial responses to chemical cues by bacteria. *J. Chem. Ecol.* **12**:1071–1089.

Berg, H. C., and D. A. Brown. 1972. Chemotaxis in *Escherichia coli* analysed by three-dimensional tracking. *Nature* (London) **239**:500–504.

Berg, H. C., and D. A. Brown. 1974. Chemotaxis in *Escherichia coli* analysed by three-dimensional tracking. *Antibiot. Chemother.* **19**:55–78.

Berg, H. C., M. D. Manson, and M. P. Conley. 1982. Dynamics and energetics of flagellar rotation in bacteria. *Symp. Soc. Exp. Biol.* **35**:1–31.

Berg, H. C., and E. M. Purcell. 1977. Physics of chemoreception. *Biophys. J.* **20**:193–219.

Berg, H. C., and P. M. Tedesco. 1975. Transient response to chemotactic stimuli in *Escherichia coli*. *Proc. Natl. Acad. Sci. USA* **72**:3235–3239.

Bischoff, D. S., and G. W. Ordal. 1992. *Bacillis subtilis* chemotaxis—a deviation from the paradigm. *Mol. Microbiol.* **6**:23–28.

Blat, Y., and M. Eisenbach. 1994. Phosphorylation-dependent binding of the chemotaxis signal molecule CheY to its phosphatase, CheZ. *Biochemistry* **33**:902–906.

Borkovich, K. A., N. Kaplan, J. F. Hess, and M. I. Simon. 1989. Transmembrane signal transduction in bacterial chemotaxis involves ligand-dependent activation of phosphate group transfer. *Proc. Natl. Acad. Sci. USA* **86**:1208–1212.

Borkovich, K. A., and M. I. Simon. 1990. The dynamics of signal transduction in bacterial chemotaxis. *Cell* **63**:1339–1348.

Botsford, J. L., and J. G. Harman. 1992. Cyclic AMP in prokaryotes. *Microbiol. Rev.* **56**:100–122.

Bourret, R. B., K. A. Borkovich, and M. I. Simon. 1991. Signal transduction pathways involving protein phosphorylation in prokaryotes. *Annu. Rev. Biochem.* **60**:401–441.

Bourret, R. B., J. Davagnino, and M. I. Simon. 1993a. The carboxy-terminal portion of CheA kinase mediates regulation of autophosphorylation by transducer and CheW. *J. Bacteriol.* **175**:2097–2101.

Bourret, R. B., S. K. Drake, S. A. Chervitz, M. I. Simon, and J. J. Falke. 1993b. Activation of the phosphosignaling protein CheY. 2. Analysis of activated mutants by F^{19} NMR and protein engineering. *J. Biol. Chem.* **268**:13089–13096.

Bourret, R. B., J. F. Hess, and M. I. Simon. 1990. Conserved aspartate residues and phosphorylation in signal transduction by the chemotaxis protein CheY. *Proc. Natl. Acad. Sci. USA* **87**:41–45.

Conley, M. P., A. J. Wolfe, D. F. Blair, and H. C. Berg. 1989. Both CheA and CheW are required for reconstitution of signaling in bacterial chemotaxis. *J. Bacteriol.* **171**:5190–5193.

Dahlquist, F. W., R. A. Elwell, and P. S. Lovely. 1976. Studies of bacterial chemotaxis in defined concentration gradients. A model for chemotaxis toward L-serine. *J. Supramol. Struct.* **4**:329–342.

Dailey, F. E., and H. C. Berg. 1993. Change in direction of flagellar rotation in *Escherichia coli* mediated by acetate kinase. *J. Bacteriol.* **175**:3236–3239.

Drake, S. K., R. B. Bourret, L. A. Luck, M. I. Simon, and J. J. Falke. 1993. Activation of the phosphosignaling protein CheY. 1. Analysis of the phosphorylated conformation by F^{19} NMR and protein engineering. *J. Biol. Chem.* **268**:13081–13088.

Fuhrer, D. K., and G. W. Ordal. 1991. *Bacillis subtilis* CheN, a homolog of α, the central regulator of chemotaxis in *Escherichia coli*. *J. Bacteriol.* **173**:7443–7448.

Gegner, J. A., and F. W. Dahlquist. 1991. Signal transduction in bacteria: CheW forms a reversible complex with the protein kinase CheA. *Proc. Natl. Acad. Sci. USA* **88**:750–754.

Gegner, J. A., D. R. Graham, A. F. Roth, and F. W. Dahlquist. 1992. Assembly of an MCP receptor, CheW, and kinase CheA complex in the bacterial chemotaxis signal pathway. *Cell* **70**:975–982.

Goy, M. F., M. S. Springer, and J. Adler. 1977. Sensory transduction in *Escherichia coli*: role of a protein methylation reaction in sensory adaptation. *Proc. Natl. Acad. Sci. USA* **74**:4964–4968.

Hazelbauer, G. L., H. C. Berg, and P. Matsumura. 1993. Bacterial motility and signal transduction. *Cell* **73:**15–22.

Hazelbauer, G. L., R. Yaghmai, G. G. Burrows, J. W. Baumgartner, D. P. Dutton, and D. G. Morgan. 1990. Transducers: transmembrane receptor proteins involved in bacterial chemotaxis. *Soc. Gen. Microbiol. Symp.* **46:**107–134.

Helmann, J. D. 1991. Alternative sigma factors and the regulation of flagellar gene expression. *Mol. Microbiol.* **12:**2875–2882.

Hess, J. F., R. B. Bourret, K. Oosawa, P. Matsumura, and M. I. Simon. 1988a. Protein phosphorylation and bacterial chemotaxis. *Cold Spring Harbor Symp. Quant. Biol.* **53:**41–48.

Hess, J. F., R. B. Bourret, and M. I. Simon. 1988b. Histidine phosphorylation and phosphoryl group transfer in bacterial chemotaxis. *Nature* (London) **336:**139–143.

Hess, J. F., K. Oosawa, N. Kaplan, and M. I. Simon. 1988c. Phosphorylation of three proteins in the signaling pathway of bacterial chemotaxis. *Cell* **53:**79–87.

Huang, C., and R. C. Stewart. 1993. CheZ mutants with enhanced ability to dephosphorylate CheY, the response regulator in bacterial chemotaxis. *Biochim. Biophys. Acta* **1202:**297–304.

Ichihara, A., J. E. Segal, S. M. Block, and H. C. Berg. 1983. Coordination of flagella on filamentous cells of *Escherichia coli. J. Bacteriol.* **155:**228–237.

Kar, L., P. Z. Decroos, S. J. Roman, P. Matsumura, and M. E. Johnson. 1992a. Specificity and affinity of binding of phosphate-containing compounds to CheY protein. *Biochem. J.* **287:**533–543.

Kar, L., P. Matsumura, and M. E. Johnson. 1992b. Bivalent-metal binding to CheY protein—effect on protein conformation. *Biochem. J.* **287:**521–531.

Kehry, M. R., T. G. Doak, and F. W. Dahlquist. 1984. Stimulus-induced changes in methylesterase activity during chemotaxis in *Escherichia coli. J. Biol. Chem.* **259:**11828–11835.

Kehry, M. R., T. G. Doak, and F. W. Dahlquist. 1985. Sensory adaptation in bacterial chemotaxis: regulation of methylesterase activity. *J. Bacteriol.* **163:**983–990.

Khan, S., F. Castellano, J. L. Spudich, J. A. McCray, R. S. Goody, G. P. Reid, and D. R. Trentham. 1993. Excitatory signaling in bacteria probed by chaged chemoeffectors. *Biophys. J.* **65:**2368–2382.

Kim, S.-H., G. G. Privé, J. Yeh, W. G. Scott, and M. V. Milburn. 1992. A model for transmembrane signalling in a bacterial chemotaxis model receptor. *Cold Spring Harbor Symp. Quant. Biol.* **57:**17–24.

Komeda, Y. 1982. Fusions of flagellar operons to lactose genes on a Mu *lac* bacteriophage. *J. Bacteriol.* **150:**16–26.

Komeda, Y., and T. Iino. 1979. Regulation of the flagellin gene (*hag*) in *Escherichia coli* K-12: analysis of *hag-lac* gene fusions. *J. Bacteriol.* **139:**721–729.

Kort, E. N., M. F. Goy, S. H. Larden, and J. Adler. 1975. Methylation of a membrane protein involved in bacterial chemotaxis. *Proc. Natl. Acad. Sci. USA* **72:**3939–3943.

Koshland, D. E. 1980. *Bacterial Chemotaxis as a Model Behavioral System.* Distinguished Lecture Series of the Society of General Physiologists. Vol. 2. Raven Press, New York.

Kuo, S. C., and D. E. Koshland, Jr. 1987. Roles of *cheY* and *cheZ* gene products in controlling flagellar rotation in bacterial chemotaxis of *Escherichia coli. J. Bacteriol.* **169:**1307–1314.

Liu, J., and J. S. Parkinson. 1989. Role of CheW protein in coupling membrane receptors to the intracellular signalling system of bacterial chemotaxis. *Proc. Natl. Acad. Sci. USA* **86:**8703–8707.

Liu, J., and J. S. Parkinson. 1991. Genetic evidence for the interaction between CheW and Tsr proteins during chemoreceptor signaling by *Escherichia coli. J. Bacteriol.* **173:**4941–4951.

Liu, X. Y., and P. Matsumura. 1994. The FlhD/FlhC complex, a transcriptional activator of the *Escherichia coli* flagellar class II operons. *J. Bacteriol.* **176:**7345–7351.

Lowry, D. F., A. Roth, P. Rupert, F. W. Dahlquist, F. Moy, P. Domaile, and P. Matsumura. 1994. Signal transduction in chemotaxis; a propagating conformation change upon phosphorylation of CheY. *J. Biol. Chem.* **269:**26358–26362.

Lupas, A., and J. Stock. 1989. Phosphorylation of an N-terminal regulatory domain activates CheB methylesterase in bacterial chemotaxis. *J. Biol. Chem.* **264:**17337–17342.

Lynch, B. A., and D. E. Koshland, Jr. 1991. Disulfide cross-linking studies of the transmembrane regions of the aspartate sensory receptor of *Escherichia coli. Proc. Natl. Acad. Sci. USA* **88:**10402–10406.

Macnab, R. M. 1977. Bacterial flagella rotating in bundles: a study in helical geometry. *Proc. Natl. Acad. Sci. USA* **74:**221–225.

Macnab, R. M. 1987a. Flagella, p. 70–83. *In* F. C. Neidhardt, J. L. Ingraham, K. B. Low, B. Magasanik, M. Schaechter, and H. E. Umbarger (ed.), *Escherichia coli and Salmonella typhimurium: Cellular and Molecular Biology.* American Society for Microbiology, Washington, D.C.

Macnab, R. M. 1987b. Motility and chemotaxis, p. 732–759. *In* F. C. Neidhardt, J. L. Ingraham, K. B. Low, B. Magasanik, M. Schaechter, and H. E. Umbarger (ed.), *Escherichia coli and Salmonella typhimurium: Cellular and Molecular Biology.* American Society for Microbiology, Washington, D.C.

Macnab, R. M. 1990. Genetics, structure, and assembly of the bacterial flagellum. *Symp. Soc. Gen. Microbiol.* **46**:77–106.

Macnab, R. M. 1992. Genetics and biogenesis of bacterial flagella. *Annu. Rev. Genet.* **26**:131–158.

Macnab, R. M., and S.-I. Aizawa. 1984. Bacterial motility and the bacterial flagellar motor. *Annu. Rev. Biophys. Bioeng.* **13**:51–83.

Macnab, R. M., and D. P. Han. 1983. Asynchronous switching of flagellar motors on a single bacterial cell. *Cell* **32**:109–117.

Maddock, J. R., and L. Shapiro. 1993. Polar localization of the chemoreceptor complex in the *Escherichia coli* cell. *Science* **259**:1717–1723.

Marwan, W., and D. Osterhelt. 1990. Quantitation of photochromism of sensory rhodopsin-I by computerized tracking of *Halobacterium halobium* cells. *J. Biol. Chem.* **215**:277–285.

Matsumura, P., S. Roman, K. Volz, and D. McNally. 1990. Signalling complexes in bacterial chemotaxis. *Soc. Gen. Microbiol. Symp.* **46**:135–154.

Matsumura, P., M. Silverman, and M. Simon. 1977. Synthesis of *mot* and *che* gene products of *Escherichia coli* programmed by hybrid Col El plasmids in minicells. *Bacteriology* **132**:996–1002.

McBride, M. J., T. Köhler, and D. R. Zusman. 1992. Methylation of FrzCD, a methyl-accepting taxis protein of *Myxococcus xanthus,* is correlated with factors affecting cell behavior. *J. Bacteriol.* **174**:4246–4257.

McNally, D. F., and P. Matsumura. 1991. Bacterial chemotaxis signaling complexes: formation of a CheA/CheW complex enhances autophosphorylation and affinity for CheY. *Proc. Natl. Acad. Sci. USA* **88**:6229–6273.

Milburn, M. V., G. G. Privé, D. L. Milligan, W. G. Scott, J. Yeh, J. Jancarik, D. E. Koshland, Jr., and S.-H. Kim. 1991. Three-dimensional structures of the ligand-binding domain of the bacterial aspart receptor with and without a ligand. *Science* **254**:1342–1347.

Milligan, D. L., and D. E. Koshland, Jr. 1988. Site-directed cross-linking establishing the dimeric structure of the aspartate receptor of bacterial chemotaxis. *J. Biol. Chem.* **263**:6268–6275.

Milligan, D. L., and D. E. Koshland, Jr. 1991. Intrasubunit signal transduction by the aspartate chemoreceptor. *Science* **254**:1651–1654.

Ninfa, E. G., A. Stock, S. Mowbray, and J. Stock. 1991. Reconstitution of the bacterial chemotaxis signal transduction system from purified components. *J. Biol. Chem.* **266**:9764–9770.

Oosawa, K., J. F. Hess, and M. I. Simon. 1988. Mutants defective in bacterial chemotaxis show modified protein phosphorylation. *Cell* **53**:89–96.

Ordal, G. W. 1985. Bacterial chemotaxis: biochemistry of behavior in a single cell. *Crit. Rev. Microbiol.* **12**:95–130.

Oxender, D. L. 1972. Membrane transport. *Annu. Rev. Biochem.* **41**:777–814.

Pakula, A. A., and M. I. Simon. 1992a. Determination of transmembrane protein structure by disulfide cross-linking: the *Escherichia coli* Tar receptor. *Proc. Natl. Acad. Sci. USA* **89**:4144–4148.

Pakula, A. A., and M. I. Simon. 1992b. Pivits or pistons? *Nature* (London) **355**:496–497.

Parkinson, J. S. 1993. Signal transduction schemes of bacteria. *Cell* **73**:857–871.

Parkinson, J. S., and D. F. Blair. 1993. Does *E. coli* have a nose? *Science* **259**:1701–1702.

Parkinson, J. S., and E. C. Kofoid. 1992. Communication modules in bacterial signaling proteins. *Annu. Rev. Genet.* **26**:71–112.

Parkinson, J. S., and S. R. Parker. 1979. Identification of the *cheC* and *cheZ* gene products is required for chemotactic behavior in *Escherichia coli. Proc. Natl. Acad. Sci. USA* **76**:2390–2394.

Parkinson, J. S., S. R. Parker, P. B. Talbert, and S. E. Houts. 1983. Interactions between chemotaxis genes and flagellar genes in *Escherichia coli. J. Bacteriol.* **155**:265–274.

Poole, P. S., D. R. Sinclair, and J. P. Armitage. 1988. Real-time computer tracking of free-swimming and tethered rotating cells. *Anal. Biochem.* **175**:52–58.

Prüss, B., and A. J. Wolfe. 1994. Regulation of acetyl phosphate synthesis and degradation and the control of flagellar expression in *Escherichia coli. Mol. Microbiol.* **12**:973–984.

Roman, S. J., M. Meyers, K. Volz, and P. Matsumura. 1992. A chemotactic signaling surface on CheY defined by suppressors of flagellar switch mutations. *J. Bacteriol.* **174**:6247–6255.

Sager, B. M., J. J. Sekelsky, P. Matsumura, and J. Adler. 1988. Use of a computer to assay motility in bacteria. *Anal. Biochem.* **173**:271–277.

Sanders, D. A., B. L. Gillece-Castro, and A. M. Stock. 1989. Identification of the site of phosphorylation of the chemotaxis response regulator protein, CheY. *J. Biol. Chem.* **264**:21770–21778.

Schuster, S. C., R. V. Swanson, L. A. Alex, R. B. Bourret, and M. I. Simon. 1993. Assembly and function of a quaternary signal transduction complex monitored by surface plasmon resonance. *Nature* (London) **365**:343–347.

Shaw, C. H. 1991. Swimming against the tide: chemotaxis in *Agrobacterium. Bioessays* **13**:25–29.

Shi, W., Y. Zhou, J. Wild, J. Adler, and C. A. Gross. 1992. DnaK, DnaJ, and GrpE are required for flagellum synthesis in *Escherichia coli. J. Bacteriol.* **174**:6256–6263.

Shioi, J., C. V. Dang, and B. L. Taylor. 1987. Oxygen as attractant and repellant in bacterial chemotaxis. *J. Bacteriol.* **169:**3118–3123.

Shioi, J., R. C. Tribhuwan, S. T. Berg, and B. L. Taylor. 1988. Signal transduction in chemotaxis to oxygen in *Escherichia coli* and *Salmonella typhimurium. J. Bacteriol.* **170:**5507–5511.

Shukla, D., and P. Matsumura. Unpublished data.

Silverman, M., and M. Simon. 1974. Characterization of *Escherichia coli* mutants that are insensitive to catabolite repression. *J. Bacteriol.* **120:**1196–1203.

Silverman, M., and M. Simon. 1977. Chemotaxis in *Escherichia coli*: methylation of *che* gene products. *Proc. Natl. Acad. Sci. USA* **74:**3317–3321.

Simms, S. A., M. G. Keane, and J. Stock. 1985. Multiple forms of the CheB methyltransferase in bacterial chemosensing. *J. Biol. Chem.* **260:**10161–10168.

Smith, R. A., and J. S. Parkinson. 1980. Overlapping genes at the *cheA* locus of *Escherichia coli*. *Proc. Natl. Acad. Sci. USA* **77:**5370–5374.

Sockett, H., S. Yamaguchi, M. Kihara, V. M. Irikura, and R. M. Macnab. 1992. Molecular analysis of the flagellar switch protein FliM of *Salmonella typhimurium. J. Bacteriol.* **174:**793–806.

Springer, M. S., M. F. Goy, and J. Adler. 1977. Sensory transduction in *Escherichia coli*: two complementary pathways of information processing that involve methylated proteins. *Proc. Natl. Acad. Sci. USA* **74:**3312–3316.

Springer, M. S., and B. Zanolari. 1984. Sensory transduction in *Escherichia coli*: regulation of the demethylation rate by CheA protein. *Proc. Natl. Acad. Sci. USA* **81:**5061–5065.

Springer, M. S., B. Zanolari, and P. A. Pierzchala. 1982. Ordered methylation of the methyl-accepting chemotaxis proteins of *Escherichia coli*. *J. Biol. Chem.* **257:**6861–6866.

Springer, W. R., and D. E. Koshland, Jr. 1977. Identification of a protein methyltransferase as the *cheR* gene product in the bacterial sensing system. *Proc. Natl. Acad. Sci. USA* **74:**533–537.

Stewart, R. C. 1993. Activating and inhibitory mutations in the regulatory domain of CheB, the methylesterase in bacterial chemotaxis. *J. Biol. Chem.* **268:**1921–1930.

Stewart, R. C., and F. W. Dahlquist. 1987. Molecular components of bacterial chemotaxis. *Chem. Rev.* **87:**997–1025.

Stewart, R. C., and F. W. Dahlquist. 1988. N-terminal half of CheB is involved in methylesterase response to negative chemotactic stimuli in *Escherichia coli*. *J. Bacteriol.* **170:**5728–5737.

Stock, A. M., J. M. Mottonen, J. B. Stock, and C. E. Schutt. 1989. Three-dimensional structure of CheY, the response regulator of bacterial chemotaxis. *Nature* (London) **337:**745–749.

Stock, J. B., and D. E. Koshland, Jr. 1978. A protein methylesterase involved in bacterial sensing. *Proc. Natl. Acad. Sci. USA* **75:**3659–3663.

Stock, J. B., and D. E. Koshland, Jr. 1981. Changing reactivity of receptor carboxyl groups during bacterial sensing. *J. Biol. Chem.* **256:**10826–10833.

Stock, J. B., and G. S. Lukat. 1991. Bacterial chemotaxis and the molecular logic of intracellular signal transduction networks. *Annu. Rev. Biophys. Biophys. Chem.* **20:**109–136.

Stock, J. B., A. J. Ninfa, and A. M. Stock. 1989. Protein phosphorylation and regulation of adaptive responses in bacteria. *Microbiol. Rev.* **53:**450–490.

Swanson, R. V., R. B. Bourret, and M. I. Simon. 1993a. Intramolecular compartmentalization of the kinase activity of CheA. *Mol. Microbiol.* **8:**435–441.

Swanson, R. V., S. C. Schyster, and M. I. Simon. 1993b. Expression of CheA fragments which define domains encoding kinase, phosphotransfer, and CheY binding activities. *Biochemistry* **32:**7623–7629.

Taylor, B. L. 1983a. How do bacteria find the optimal concentration of oxygen? *Trends Biochem. Sci.* **8:**438–441.

Taylor, B. L. 1983b. Role of proton motive force in sensory transduction in bacteria. *Annu. Rev. Microbiol.* **37:**551–573.

Tisa, L. S., and J. Adler. 1992. Calcium ions are involved in *Escherichia coli* chemotaxis. *Proc. Natl. Acad. Sci. USA.* **89:**11804–11808.

Titgemeyer, F. 1993. Signal transduction in chemotaxis mediated by the bacterial phosphotransferase system. *J. Cell. Biochem.* **51:**69–74.

Toews, M. L., and J. Adler. 1979. Methanol formation in vivo from methylated chemotaxis proteins in *Escherichia coli*. *J. Biol. Chem.* **254:**1761–1764.

Toews, M. L., M. F. Goy, M. S. Springer, and J. Adler. 1979. Attractants and repellents control demethylation of methylated chemotaxis proteins in *Escherichia coli*. *Proc. Natl. Acad. Sci. USA* **76:**5544–5548.

Volz, K. 1993. Structural conservation in the CheY superfamily. *Biochemistry* **32:**11741–11753.

Volz, K., and P. Matsumura. 1991. Crystal structure of *Escherichia coli* CheY refined at 1.7 Å. *J. Biol. Chem.* **266:**15511–15519.

Wang, H., D. F. McNally, and P. Matsumura. Unpublished data.

Welch, M., K. Oosawa, S.-I. Aizawa, and M. Eisenbach. 1993. Phosphorylation-dependent binding of a signal molecule to the flagellar switch of bacteria. *Proc. Natl. Acad. Sci. USA* **90:**8787–8791.

Wolfe, A. J., P. Conley, and H. C. Berg. 1988. Acetyladenylate plays a role in controlling the direction of flagellar rotation. *Proc. Natl. Acad. Sci. USA* **85:**6711–6715.

Wolfe, A. J., P. Conley, T. J. Cramer, and H. C. Berg. 1987. Reconstitution of signaling in bacterial chemotaxis. *J. Bacteriol.* **169**:1878–1885.

Wolfe, A. J., and R. C. Stewart. 1993. The short form of CheA protein restores kinase activity and chemotactic ability to kinase-deficient mutants. *Proc. Natl. Acad. Sci. USA* **90**:1518–1522.

Yamaguchi, S., S.-I. Aizawa, M. Kihara, M. Isomura, C. J. Jones, and R. M. Macnab. 1986a. Genetic evidence for a switching and energy-transducing complex in the flagellar motor of *Salmonella typhimurium*. *J. Bacteriol.* **168**:1172–1179.

Yamaguchi, S., H. Fujita, A. Ishihara, S.-I. Aizawa, and R. M. Macnab. 1986b. Subdivision of flagellar genes of *Salmonella typhimurium* into regions responsible for assembly, rotation, and switching. *J. Bacteriol.* **166**:187–193.

Yeh, J. I., H.-P. Biemann, J. Pandit, D. E. Koshland, and S.-H. Kim. 1993. The three-dimensional structure of the ligand-binding domain of a wild-type bacterial chemotaxis receptor. *J. Biol. Chem.* **268**:9787–9792.

Zhulin, I. B., and J. P. Armitage. 1993. Motility, chemotaxis, and methylation-independent chemotaxis in *Azospirillum brasilense*. *J. Bacteriol.* **175**:952–958.

Porin Regulon of *Escherichia coli*

Leslie A. Pratt and Thomas J. Silhavy

7

In *Escherichia coli,* the two porin proteins OmpF and OmpC form pores in the outer membrane that allow for the passive diffusion of small hydrophilic molecules across this hydrophobic barrier. The OmpF and OmpC proteins share extensive homology at the amino acid level (Mizuno et al., 1983), and both function as trimers. The main difference between the two porins is the diameter of the pores formed by each: the pore formed by the OmpF porin is slightly larger (1.16 nm) than that formed by the OmpC porin (1.08 nm). This difference can result in rates of diffusion through an OmpF porin that are significantly faster than through an OmpC porin (Nikaido and Vaara, 1987).

In response to the surrounding osmolarity, the total amount of OmpF and OmpC protein remains constant, whereas the relative levels of the two proteins fluctuate in a reciprocal manner. Specifically, under conditions of low osmolarity, expression of OmpF is favored, whereas in high osmolarity, OmpC predominates (Van Alphen and Lugtenberg, 1977). Early studies identified three genetic loci involved in this reciprocal regulation: *ompF,* located at 21 min on the *E. coli* chromosome;

ompC, at 48 min; and *ompB,* at 74 min (Sarma and Reeves, 1977; Sato and Tura, 1979; Verhoef et al., 1977). The *ompF* and *ompC* genes are structural, encoding for the OmpF and OmpC proteins, respectively, and the *ompB* locus plays a regulatory role. The *ompB* operon actually encodes two genes, *envZ* and *ompR,* both of which are important for porin regulation (Hall and Silhavy, 1979, 1981a,b).

Studies using gene and operon fusions to both *ompF* and *ompC* revealed that regulation of porin expression occurs at the transcriptional level (Hall and Silhavy, 1979, 1981a; Inokuchi et al., 1984). This work, combined with additional genetic analysis, led to the proposal of an early model to explain porin regulation. In this model, EnvZ was pictured as an envelope protein responsible for monitoring the osmolarity, and OmpR as a cytoplasmic protein responsible for transcriptionally regulating the porin genes in response to information received from EnvZ (Hall and Silhavy, 1981b). Although modifications and further detail have been added, the basic tenets of this model have withstood the test of time.

EnvZ and OmpR are members of a large family of homologous proteins that are collectively referred to as two-component regulatory systems. These two-component systems enable bacteria to sense and respond to a wide array of

Leslie A. Pratt and Thomas J. Silhavy, Department of Molecular Biology, Princeton University, Princeton, New Jersey 08544.

Two-Component Signal Transduction, Edited by James A. Hoch and Thomas J. Silhavy,
© 1995 American Society for Microbiology, Washington, DC 20005

environmental parameters. In the osmosystem, EnvZ and OmpR form a connected signal transduction pathway that enables cells to sense the external osmolarity and respond to this stimulus by regulating the transcription of the genes encoding the OmpF and OmpC proteins. This signal transduction pathway extends from the periplasm of *E. coli* to the *ompF* and *ompC* promoters and the transcriptional apparatus.

Ultimately, we wish to understand this information flow in molecular terms. Here, our present understanding of how environmental information travels from the periplasm, through the inner membrane, and eventually to the porin promoters and transcriptional machinery, is reviewed. We start with the nature of the stimulus and follow the pathway through each step to reach the final output, the regulation of porin gene transcription.

THE STIMULUS

Most of the well-characterized two-component regulatory systems sense and mediate the response to a particular stimulus or set of stimuli, the molecular nature of which is well understood. For example, the nitrogen system monitors the ratio between two metabolites, glutamine and α-ketoglutarate, and, in this way, determines the availability of nitrogen (see Chapter 5). The chemotaxis system monitors the level of several small-molecule attractants and repellents, the molecular identities of which are also clear (Adler, 1966) (see Chapter 6). By contrast, EnvZ and OmpR mediate the response to osmolarity, a property of the environment altered by varying the concentration of several small molecules, independent of the identity of these molecules (Van Alphen and Lugtenberg, 1977). In other words, osmolarity is a colligative property of the medium that changes with differing levels of a wide array of salts, sugars, or other small molecules. This leaves us with the intriguing question of what constitutes the actual stimulus to which EnvZ responds.

To date, this question remains unanswered. No single molecule has been identified that functions as a direct and specific stimulus of the osmosystem. Indeed, it is possible that EnvZ is able to sense this colligative property of the medium directly; a single molecule that acts as a stimulus simply may not exist. However, dramatic changes (structural and otherwise) occur within the cell in response to changing osmolarity (Csonka, 1989; Stock et al., 1977). Perhaps EnvZ can somehow sense one or more of these alterations.

We do not know the answer to the above question. This is *not* to say, however, that we know nothing concerning the nature of the stimulus. Indeed, studies with a truncated form of EnvZ (EnvZ115) indicate that the stimulus to which EnvZ responds is located within the periplasm. EnvZ is normally localized to the inner membrane with its N-terminal domain in the periplasm (Fig. 1 and see below). Studies using alkaline phosphatase fusions to both EnvZ and EnvZ115 indicate that EnvZ115 is not properly localized and remains in the cytoplasm (Russo, 1992). Interestingly, this truncated form of EnvZ partially suppresses an *envZ* null strain, resulting in an intermediate level of OmpF and OmpC expression. Importantly, these levels do not vary in response to the external osmolarity. EnvZ115 has thus retained the ability to communicate with OmpR but, presumably because of its improper locale, has lost the ability to sense and respond to its stimulus (Igo and Silhavy, 1988). These results imply that the stimulus resides in the periplasm.

Additional studies have revealed, perhaps counterintuitively, that the stimulus sensed by EnvZ is probably present in low osmolarity rather than in high osmolarity. A series of small deletions and insertions in the presumed sensing domain of EnvZ (the N-terminal, periplasmic domain) was created and shown to confer a constitutive high-osmolarity phenotype (Tokishita et al., 1991; Russo, 1992). This observation indicated that when the ability of EnvZ to sense is disrupted, its "default" state is one of high osmolarity. Thus, the stimulus appears to exist in low osmolarity, and when EnvZ senses this stimulus, it shifts from the high-osmolarity default state to that state found in low osmolarity.

It has been proposed that the stimulus sensed by EnvZ may be membrane-derived oligosaccharides (MDO). This proposition was based on observations made concerning porin expression in a strain deficient in the synthesis of MDO (Fiedler and Rotering, 1988). However, later studies revealed that this effect was apparent only in medium of very low ionic strength. Because MDO is highly charged, it may simply be that ionic strength is important for proper EnvZ sensory function. In any case, it is clear that MDO is not the stimulus sensed by EnvZ (Geiger et al., 1992).

Thus, we can say much about what the stimulus is not. It is not found in the cytoplasm, it is not found in high osmolarity, and it is not MDO. Yet, we remain unsure whether EnvZ is somehow able to directly monitor a colligative property of the medium or if a specific molecular signal exists. If a molecular signal does exist, it is not known whether this signal is chemical (some small molecule) or mechanical (e.g., interaction with the cell wall).

THE SENSOR

Whatever the nature of the stimulus, its presence or absence is monitored by the sensor EnvZ. EnvZ is a 450-amino-acid protein that is cotranscribed with its cognate response regulator OmpR from the *ompB* operon (Comeau et al., 1985; Hall and Silhavy, 1979). It is present at approximately 10 copies per cell and is localized to the inner membrane. EnvZ possesses several enzymatic activities, including autophosphorylation, OmpR-kinase, and OmpR-phosphate-phosphatase. All these attributes are critical to the integrity of the many functions EnvZ must perform. EnvZ is responsible for monitoring the osmolarity, transducing this information across the inner membrane, and then relaying the information to the response regulator OmpR to control OmpR's activity. Recent studies focused on understanding how EnvZ accomplishes these many and varied functions.

Sensing Osmolarity

EnvZ must first be able to sense the external osmolarity. It is appropriately localized for this sensory function to the inner membrane (Liljestrom, 1986; Forst et al., 1987). Sequence analysis and the use of β-lactamase fusions indicate that EnvZ possesses two membrane-spanning domains, placing its N-terminal domain in the periplasm and its C-terminal domain in the cytoplasm (Forst et al., 1987) (Fig. 1). The periplasmically located N-terminal region of the protein is perfectly situated to monitor the surrounding osmolarity. Also, the C-terminal domain of EnvZ shows extensive homology with sensors from other two-component systems, whereas the N-terminal domain is not conserved. The homologous sequences in the C termini reflect a conserved mechanism of communication between sensors and their cognate response regulators (Kofoid and Parkinson, 1988; Stock et al., 1990). The diverging sequences in the N termini of the sensors likely reflect the wide array of stimuli monitored by the many two-component systems (Fig. 1).

The results alluded to earlier that suggest the stimulus is periplasmically located are consistent with this idea. Recall that EnvZ115, which is not directed to the inner membrane and therefore has a cytoplasmically located N-terminal domain, is not able to respond to changes in osmolarity (Igo and Silhavy, 1988). Furthermore, the series of small deletions and insertions in the periplasmic domain that result in a constitutive OmpF⁻-OmpC⁺ (F⁻C⁺) phenotype are consistent with this region of the protein sensing something present only in low osmolarity. When EnvZ's ability to sense low osmolarity is disrupted by alterations in its sensing domain, it becomes locked in the high-osmolarity state (Tokishita et. al., 1991; Russo, 1992). Additional studies used a chimeric protein in which the N-terminal domain of EnvZ (consisting of the periplasmic domain, both transmembrane domains, and a small cytoplasmic segment) is replaced with the corresponding region of the chemotaxis receptor Tar. The resulting hybrid protein, Taz1, no longer re-

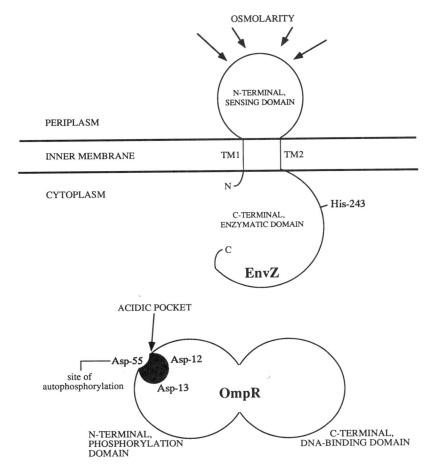

FIGURE 1 Domain structure of OmpR and EnvZ. The sensor, EnvZ, is located in the inner membrane with its N-terminal domain in the periplasm and its C-terminal domain in the cytoplasm. It possesses two transmembrane segments extending from 16 to 46 (TM1) and from 163 to 179 (TM2). The N-terminal domain of EnvZ monitors the osmolarity, and this information is transduced across the inner membrane to the C-terminal cytoplasmic domain. The C-terminal portion of EnvZ relays this information to the response regulator, OmpR, via kinase and phosphatase activities. OmpR is a cytoplasmic protein directly responsible for regulating transcription of the *ompF* and *ompC* genes. The N-terminal half of OmpR is the phosphorylation domain, containing the acidic pocket and the presumed site of phosphorylation (Asp-55), and the C-teminal portion of OmpR is the DNA binding domain.

sponds to varying osmolarity but rather responds (as does Tar) to the presence or absence of aspartate. In response to aspartate, Taz1 mediates an increase in *ompC* expression (Utsumi et al., 1989). Similar results have recently been obtained with an analogous fusion protein (Trz1) between the chemoreceptor Trg and

EnvZ (Baumgartner et al., 1994). These results strongly support the notion that it is the N termini of the transmembrane sensors that possess the sensory function and that the N-terminal periplasmic domain of EnvZ functions directly to monitor the surrounding osmolarity.

Signaling States of EnvZ

Once the information regarding the osmolarity is collected by the N-terminal domain of EnvZ, it must be transduced across the membrane to the C-terminal cytoplasmic domain. To address the problem of the transduction of information across the membrane, and the consequent changes in the activity of EnvZ, it is important to gain an understanding of what constitutes the various signaling states that EnvZ assumes under differing osmotic conditions. One simple model is that EnvZ exists in either an "on" or an "off" state. For example, EnvZ might actively signal OmpR in low osmolarity and not in any way signal OmpR in high osmolarity. Early studies ruled out such a simple model by revealing that whatever the mechanism through which EnvZ communicates osmolarity to OmpR, both low and high osmolarity require an active signal from EnvZ. *envZ* null alleles were isolated and shown to confer an F$^{+/-}$C$^-$ phenotype, expressing *ompF* only to a very low level (Garrett et al., 1983). Because this phenotype corresponds to neither low nor high osmolarity, EnvZ must assume an active signaling state under both conditions. The critical questions then become, what are the active signaling states of EnvZ and how are these signaling states assumed in response to external osmolarity?

As mentioned previously, EnvZ possesses extensive homology in its C-terminal domain to sensors from other two-component systems. Similarly, OmpR possesses extensive homology in its N-terminal domain with response regulators from other two-component systems (Stock et al., 1990). This sequence conservation is suggestive of a conserved mechanism of communication between sensors and their cognate response regulators. In several systems, this mechanism of communication was shown to involve phosphorylation events (Hess et al., 1988; Ninfa and Magasanik, 1986; Wylie et al., 1988). Indeed, a clue concerning the nature of the signaling states of EnvZ came from biochemical analysis of truncated molecules such as EnvZ115 (Aiba et al., 1989a,b; Forst et al.,

1989; Igo and Silhavy, 1988; Igo et al., 1989a,b). First, EnvZ undergoes autophosphorylation in the presence of ATP (Aiba et al., 1989a; Forst et al., 1989; Igo and Silhavy, 1988). The pH stability profile of the phosphorylated EnvZ indicates the existence of a histidyl phosphate. Moreover, EnvZ possesses a histidine residue at position 243, and this histidine is invariant among all known sensors in two-component systems. A substitution at this position (H243V) renders EnvZ refractory to autophosphorylation (Forst et al., 1989; Igo and Silhavy, 1988). Finally, recent in vitro studies have demonstrated directly that the conserved His-243 is the site of EnvZ autophosphorylation (Roberts et al., 1994) (Fig. 1). Second, phosphorylated EnvZ is able to transfer its phosphate group to OmpR, creating OmpR-phosphate (OmpR-P) (Aiba et al., 1989a,b; Igo et al., 1989a,b; Forst et al., 1989). Finally, EnvZ is also capable of promoting the dephosphorylation of OmpR-P. Although ATP is required for EnvZ-mediated dephosphorylation of OmpR-P, nonhydrolyzable analogs function as well. ATP is apparently needed as a cofactor, perhaps helping EnvZ to assume the proper conformation (Aiba et al., 1989a; Igo et al., 1989b). Thus, EnvZ has the ability to covalently modify OmpR through both kinase and phosphatase activities. These results suggest a possible model to explain the signaling states of EnvZ: under differing osmotic conditions, EnvZ might alter its signaling state by varying the relative level of these enzymatic activities.

Consistent with this idea, various *envZ* and *ompR* alleles that confer constitutive aberrant patterns of porin expression in vivo have been shown to display altered phosphorylation properties in vitro. For example, *envZ11* and *envZ473* both confer an F$^-$C$^+$ phenotype irrespective of osmolarity. Studies with purified EnvZ11 revealed that although the mutant protein is still able to phosphorylate OmpR in vitro, it is defective in promoting the dephosphorylation of OmpR-P. Similarly, an *ompR3* allele, which also confers a constitutive F$^-$C$^+$ phenotype, is dephosphorylated less efficiently than is wild-type OmpR (Aiba et al., 1989b). Later studies resulted in the identification of two additional *envZ* alleles, *envZ247* and *envZ250*, that were shown to be defective in their OmpR-kinase activity. These alleles confer an F⁻C⁻ phenotype irrespective of external osmolarity. Importantly, strains merodiploid for *envZ473* and either *envZ247* or *envZ250* exhibit intermediate levels of porin expression that are comparable with that normally seen in low osmolarity. In other words, these alleles are codominant. The *envZ473* and *envZ11* alleles and the *envZ247* and *envZ250* alleles appear to define opposite signaling states of EnvZ. These signaling states do not correspond to the states found in low and high osmolarity; rather, these signaling states appear to be extreme, corresponding to a kinase-dominant and a phosphatase-dominant state of EnvZ (Russo and Silhavy, 1991). The in vitro properties of the various *ompR* and *envZ* alleles described above indicate that the kinase-dominant state is favored in high osmolarity, whereas the relative propensity for the phosphatase-dominant state increases with decreasing osmolarity (Russo and Silhavy, 1991) (Fig. 2). The stimulus, therefore, modulates the conversion between a kinase-dominant and a phosphatase-dominant state of EnvZ, the relative propensity for the latter being increased in its presence.

Regulating the Ratio

How, then, are the enzymatic activities of EnvZ (autophosphorylation, kinase, phosphatase) regulated to ultimately result in the altered ratio of kinase to phosphatase activity? It is possible to envision a variety of models. One possibility is that only one of the two enzymatic activities is altered in response to osmolarity, and this results in changing the ratio of the two activities. Indeed, two models based on this idea have been proposed. In one model, the phosphatase activity is proposed to vary while the kinase activity remains constant (Jin and Inouye, 1993). In the other, it is hypothesized that the kinase activity may fluctuate while the

phosphatase activity remains constant (Russo and Silhavy, 1991). One of these models is based on in vitro patterns of kinase and phosphatase activities obtained with the chimeric sensor Taz1 (Jin and Inouye, 1993). The other was based on mathematical modeling of the porin regulon (Russo and Silhavy, 1991). Although informative, neither approach conclusively distinguishes between the two possibilities. In fact, it is also conceivable that both enzymatic activities vary in response to osmolarity. Further studies should reveal how the enzymatic activities are changed to ultimately result in an altered ratio of kinase-to-phosphatase activity.

A related issue concerns how, in mechanistic terms, EnvZ is altered to change its propensity toward its kinase and/or phosphatase activities. For example, it may be that EnvZ is altered in a way that changes its relative affinities for OmpR and/or OmpR-P; when OmpR is bound, it is phosphorylated by EnvZ, and when OmpR-P is bound, it is dephosphorylated. In this model, altering EnvZ's affinity for either or both substrates is the causal agent in changing the propensity for the kinase and/or phosphatase activity. The basis for an alternative model arises from observations concerning the kinetics of phosphorylation events between components of the chemotaxis regulatory system. In this system, autophosphorylation of CheA (EnvZ homolog), rather than phosphotransfer to CheY (OmpR homolog), was shown to be the rate-limiting step in the phosphorylation of CheY (Ninfa et al., 1991). In the light of these observations, a model could be envisioned in which EnvZ's kinase and phosphatase activities are modulated by regulating its ability to autophosphorylate (Fig. 2). In this model, phosphorylated EnvZ would act as a kinase and unphosphorylated EnvZ as a phosphatase. Controlling the tendency of EnvZ to autophosphorylate would function to regulate the propensity for these opposing enzymatic reactions. The ability of EnvZ to autophosphorylate could be controlled through conformational changes that

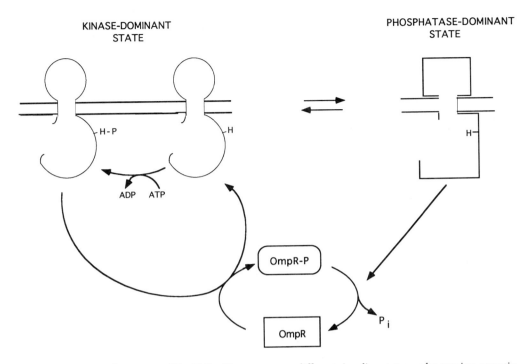

FIGURE 2 Signaling states of EnvZ. EnvZ must assume different signaling states under varying osmotic conditions. This is accomplished by regulating the ratio of kinase to phosphatase activities such that a phosphatase-dominant state is present in low osmolarity whereas a kinase-dominant state predominates in high osmolarity. The ratio of kinase to phosphatase activities could be regulated by altering either or both enzymatic activities. One possibility is that the ratio is regulated by controlling the relative exposure of the autophosphorylation site, thereby regulating the rate of autophosphorylation (see text for further details).

either hide the site of autophosphorylation (increasing the propensity for the phosphatase activity) or expose the site of autophosphorylation (increasing the propensity for kinase activity). Clearly, many possibilities exist to explain how EnvZ might modulate its kinase and/or phosphatase activities, but there is presently no compelling evidence to distinguish among them.

Signal Transduction across the Membrane

Whatever the mechanism involved in internally regulating the relative enzymatic activities of EnvZ, it must be ultimately controlled by information that is transduced from the other side of the membrane. This returns us to the issue raised earlier concerning how EnvZ propa-

gates information across the cytoplasmic membrane. EnvZ possesses only two transmembrane domains (Fig. 1), and therefore one or both of these domains must function to propagate a conformational change across the membrane that is sufficient to significantly alter activity. Are both transmembrane domains involved? Does the mechanism involve rotating, tilting, and/or lateral movements of the transmembrane segments? Or do these transmembrane segments actually move in and out of the lipid bilayer? Is dimerization important, or does EnvZ function as a monomer?

Studies involving complementation analyses with various Taz1 and EnvZ mutant proteins suggest that dimerization is likely to be an important aspect of EnvZ function. When the conserved histidine residue (the site of autophosphorylation) is replaced with a valine, the resulting Taz1-His-277-Val is rendered both unable to mediate the transcriptional activation of *ompC* in response to aspartate in vivo and unable to autophosphorylate and phosphotransfer in vitro. Similarly, when the C-terminal 146 residues of Taz1 are deleted, the resulting mutant protein, Taz1d1A, is unable to mediate transcriptional activation of *ompC* and unable to autophosphorylate and phosphotransfer in vitro. However, if these two mutant proteins are coexpressed, they are together able to confer constitutive transcriptional activation of *ompC*. When the two mutant proteins are mixed in vitro, Taz1-His-277-Val is able to phosphorylate Taz1d1A, and Taz1d1A is in turn able to transfer this phosphate to OmpR. Also, consistent with the constitutive transcriptional activation of *ompC,* the complementation does not restore OmpR-P phosphatase activity (Yang and Inouye, 1991). These studies not only indicate that dimer formation is likely important for EnvZ kinase activity but also further support the notion that both EnvZ kinase and phosphatase activities are important for normal regulation (Yang and Inouye, 1991, 1993).

In addition to the studies described above, knowledge concerning how EnvZ transduces information across the membrane can be gained through comparisons with the more extensively characterized chemoreceptors. EnvZ and the chemoreceptors share similar topologies, and results obtained with the Taz1 and Trz1 hybrid proteins indicate conserved mechanisms of signal transduction (Baumgartner et al., 1994; Utsumi et al., 1989). The chemoreceptors function as dimers that are believed to assume varying signaling states. Based on both genetic and biochemical studies, the first transmembrane domain (TM1) is thought to be largely responsible for dimerization. In one signaling state, the TM1 interaction is thought to be very tight, whereas the cytoplasmic domains are quite flexible. By contrast, the other signaling state is thought to have a looser interaction between the TM1 domains

and a tighter interaction between the cytoplasmic domains (Parkinson, 1993). Consistent with TM1 playing a crucial role in signal transduction for EnvZ as well, *envZ* alleles have been isolated with alterations in TM1 that lock the protein into either a constitutive low- or high-osmolarity state (Tokishita et al., 1992).

Thus, although dimerization appears integral to EnvZ function and TM1 has been implicated as important, we still understand very little of how these factors contribute to the transduction of information across the membrane. Does EnvZ alternate between existing as a monomer and a dimer, each form representing an alternative signaling state of EnvZ? Or does EnvZ persist continuously as a dimer, the conformation of the dimer somehow changing to assume the various signaling states? If so, how does the conformation of the EnvZ dimer, especially TM1, alter to propagate the information across the membrane? The answers to these questions are unknown, and consequently, transmembrane signal transduction remains essentially a mystery.

RESPONSE REGULATOR

Ultimately, the signal must be relayed from EnvZ to OmpR. OmpR, in turn, is directly responsible for eliciting the internal response. Specifically, OmpR regulates the transcription of the *ompF* and *ompC* genes (Fig. 3). OmpR is a 239-amino-acid cytoplasmic protein (Wurtzel et al., 1982) that is present at approximately 1,000 copies per cell. As the response regulator in this system, OmpR performs several important functions. First, OmpR must gather information concerning the osmolarity from EnvZ and then respond appropriately to this information. Depending on the signal it receives from EnvZ regarding the external osmolarity, OmpR favors expression of either *ompF* or *ompC*. It follows, therefore, that OmpR must be capable of assuming at least two states—the state found in low osmolarity and the state found in high osmolarity (Fig. 3, top). In addition to assuming these various forms, OmpR must also be able to bind DNA, activate tran-

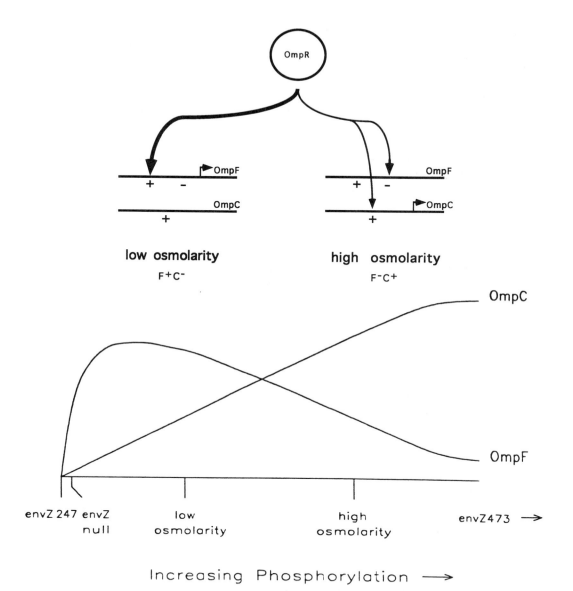

FIGURE 3 Nature of the internal signal. (Top) OmpR assumes different roles under varying osmotic conditions. In low osmolarity (thick line), OmpR functions as a transcriptional activator of *ompF*, whereas in high osmolarity (thin line), OmpR functions to repress *ompF* transcription and activate *ompC* transcription. (Bottom) The distinction between the OmpR in low versus high osmolarity is thought to be a difference in the concentration of OmpR–phosphate. Low concentrations of OmpR favor transcriptional activation of *ompF*, whereas higher concentrations of OmpR–phosphate repress *ompF* transcription and activate *ompC* transcription.

scription, and repress transcription (see below). How OmpR performs these functions has been the focus of much study.

Nature of the Internal Signal

An understanding of the features that distinguish the OmpR found in low versus high osmolarity is critical to comprehending OmpR function. To this end, two genetically distinct states of OmpR were identified in a screen for mutations in *ompR* that confer either a constitutive F^+C^- phenotype (the R2 class) or a constitutive F^-C^+ phenotype (the R3 class). The R2 class genetically defined the low-osmolarity state of OmpR; the R3 class defined the high-osmolarity state. Through diploid analysis, the critical observation was made that any allele (*ompR* or *envZ*) that confers a constitutive high-osmolarity phenotype (F^-C^c) is dominant to any other allele. For example, in an R2/R3 diploid, the phenotype is F^-C^c; although the R2 allele is capable of activating transcription of *ompF,* the R3 allele prevents this activation. This observation was the first indication that OmpR-mediated repression of *ompF* is involved in porin regulation and also revealed that repression of *ompF* is dominant to activation. Importantly, these results led to a clearer understanding of the distinguishing characteristics of OmpR in low versus high osmolarity. It is not simply that OmpR activates transcription of *ompF* in low osmolarity while favoring activation of *ompC* in high osmolarity. Rather, the OmpR in low osmolarity functions as a transcriptional activator of *ompF,* whereas the OmpR in high osmolarity functions as both a transcriptional repressor of *ompF* and a transcriptional activator of *ompC* (Slauch and Silhavy, 1989) (Fig. 3, top).

What is different about the OmpR in low versus high osmolarity that results in such dramatic alterations in OmpR's regulatory activities? In the light of the in vitro evidence illustrating EnvZ-mediated phosphorylation and dephosphorylation of OmpR, both of these activities presumably play a role in modulating OmpR's activity. A simple model might

propose a type of on-off switch in which unphosphorylated OmpR favors expression of one porin gene while phosphorylated OmpR favors expression of the other gene. However, the phenotypes of *envZ* null alleles ($F^{-/+}/C^-$) and the *envZ* kinase⁻ alleles (F^-C^-) reveal that unphosphorylated OmpR activates transcription from neither porin promoter (Garrett et al., 1983; Russo and Silhavy, 1991). Therefore, this simple model cannot account for the reciprocal regulation of the porin genes. The difference between the OmpR found in low versus high osmolarity is not merely the presence or absence of a phosphate group.

Obviously, then, the difference between the low- and high-osmolarity states of OmpR is more subtle. Interestingly, the level of phosphorylated OmpR in vivo has been shown to vary in response to osmolarity, the level of phosphorylated OmpR increasing with increasing osmolarity (Forst et al., 1990). In a similar study, strains containing *envZ* alleles that confer a constitutive high-osmolarity phenotype were shown to possess higher levels of phosphorylated OmpR than the wild-type parent strains (Waukau and Forst, 1992). This is consistent with the in vitro data described earlier illustrating that *envZ* and *ompR* alleles that confer a constitutive high-osmolarity phenotype exhibit defects in the dephosphorylation of OmpR (Aiba et al., 1989b). Additional studies revealed that when increasing levels of phosphorylated OmpR are added to an in vitro transcription assay in which both porin promoters are present, transcription of *ompF* predominates at low concentrations, whereas *ompC* transcription increases as the concentration of phosphorylated OmpR increases (Aiba and Mizuno, 1990). These observations, although providing critical insight into the importance of phosphorylation in osmoregulation, still do not reveal the nature of the molecular difference between the low- and high-osmolarity states of OmpR.

Because three genetically distinguishable states of OmpR have been described (the inactive unphosphorylated OmpR, the low-osmolarity form of OmpR, and the high-osmolarity form of OmpR), EnvZ must be required both to activate OmpR and to regulate the transition between the low- and high-osmolarity forms. Assuming that EnvZ communicates with OmpR solely through previously identified means (i.e., kinase and phosphatase activity), then two possibilities exist. The first possibility is that there are two phosphorylation events, one that activates OmpR and a second that modulates the transition between the low- and high-osmolarity forms. The second possibility is that there is only one phosphorylation site, and the difference is one of quantity; a low concentration of phosphorylated OmpR (OmpR-P) corresponding to the low-osmolarity state and a higher concentration of the same species of OmpR-P corresponding to the high-osmolarity state (Russo and Silhavy, 1991) (Fig. 3, bottom).

Several observations make the second possibility seem most likely. First, based on homology to other response regulators, OmpR is presumed to be phosphorylated on a conserved aspartic acid residue, Asp-55 (Fig. 1). Based on an analogy to CheY (for which the crystal structure has been solved), the conserved phosphorylated aspartic acid residue is present in an acidic pocket, and the residues in this acidic pocket are also strongly conserved. Molecular modeling indicates that there simply is not room for a second phosphate group in this pocket (Stock et al., 1989; Volz and Matsumura, 1991). This observation does not rule out the possibility of a second phosphorylation event; however, it does make it seem unlikely. Importantly, mass spectroscopy analysis shows that phosphorylated OmpR possesses only one phosphate group (Kenney and Silhavy, unpublished data). Furthermore, it is possible to mathematically model the reciprocal regulation of the porin genes by varying the concentration of a single molecular species of OmpR-P. Consequently, there is no need to invoke the existence of a second phosphorylation event for which there is no evidence (Russo and Silhavy, 1991). For these reasons, it is generally

believed that a varying concentration of a single species of OmpR-P creates the necessary internal signal responsible for the reciprocal regulation of the porin genes (Fig. 3, bottom).

DNA Binding

OmpR can activate transcription of both the *ompF* and *ompC* genes and can also repress transcription of *ompF*. This implies the existence of positive DNA binding sites at both promoters and negative DNA binding site(s) at the *ompF* promoter. How OmpR interacts with these sites and how phosphorylation affects these interactions are questions central to understanding the porin regulon.

Both in vivo and in vitro DNA protection studies reveal that OmpR binds to large regions of DNA at both the *ompF* and *ompC* promoters (Norioka et al., 1986; Mizuno et al., 1988; Maeda and Mizuno, 1988; Tsung et al., 1989) (Fig. 4). Despite this information, there is no commonly accepted consensus sequence, and the DNA sequence(s) that OmpR recognizes remains a mystery. There are three homologous 10-bp tandemly repeated sequences at the *ompC* promoter, suggesting that OmpR does not bind to dyad symmetrical sites (Fig. 4). Nonetheless, the large regions of DNA protected by OmpR (more than 60 bp at each promoter) are suggestive of cooperative binding. Although such sequence gazing is suggestive, no evidence has been reported that either supports or refutes these ideas.

Furthermore, very little is known concerning how OmpR recognizes its target DNA sites. OmpR possesses no recognizable helix-turn-helix or other DNA binding motif (Nara et al., 1986). Importantly, OmpR belongs to a subclass of response regulators that, in addition to the N-terminal homology shared by all response regulators, shares extensive homology in their C-terminal domains (Kofoid and Parkinson, 1988; Stock et al., 1990). It is believed that the C termini of these proteins possess the DNA binding domain (see below and Fig. 1), and as with OmpR, it is not known how any of these proteins bind DNA. OmpR is thus a

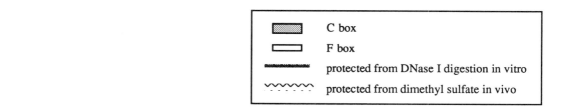

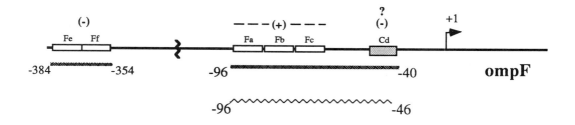

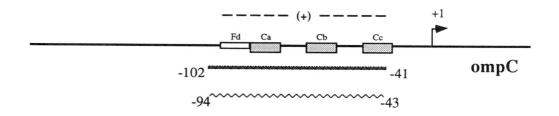

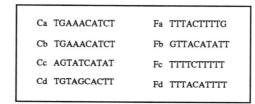

FIGURE 4 Porin promoters. OmpR–mediated transcriptional regulation of the porin genes involves extensive regions of promoter DNA. The regions required for positive and negative regulation are indicated above the promoters by (+) and (–), respectively. The regions protected by OmpR in vivo and in vitro are also indicated (see key).

member of a novel class of DNA binding proteins, and this makes the issue of how OmpR binds DNA one of special importance.

As previously discussed, OmpR assumes very different regulatory roles in low osmolarity versus high osmolarity. This suggests that OmpR might bind to different sites at the porin promoters under these different conditions. Phosphorylation of OmpR could serve to alter the affinity for these various sites, resulting in the observed reciprocal regulation in response to varying osmolarity. Indeed, the mathematical model alluded to previously requires the existence of two types of DNA binding sites, with one type possessing at least a 20-fold higher affinity for OmpR-P (Russo and Silhavy, 1991). Consistent with this notion, the important observation was made that phosphorylation of OmpR does indeed alter its affinity for DNA. Increasing levels of phosphorylated OmpR increases OmpR-DNA binding in vitro (Aiba et al., 1989c). Also, in vitro DNA binding studies with purified, mutant OmpR proteins revealed altered patterns of DNA protection. Specifically, an *ompR2* allele (which confers a constitutive low-osmolarity phenotype) was capable of protecting only a fraction of the sequences normally recognized at the *ompF* promoter and was unable to protect any of the sequences normally recognized at *ompC*. Because this *ompR2* allele confers an F^cC^- phenotype, the sequences unprotected by this mutant protein may reflect the negative and positive sites at the *ompF* and *ompC* promoters, respectively (Mizuno et al., 1988).

These results imply the existence of different recognition sites that are occupied or not occupied, depending on the concentration of OmpR-P. Indeed, a recent study revealed that OmpR binds to sites at the *ompF* promoter in a hierarchical fashion. Specifically, at low concentrations of OmpR-P, the DNA sequences required for transcriptional activation of *ompF* are occupied, whereas at higher concentrations of OmpR-P, negative-acting sequences are occupied (Rampersaud et al., 1994) (Fig. 4 and see below). Presently, the mechanism through

which this differential occupation of sites is accomplished is not well understood. It is not known whether these sites are intrinsically different, recognizing different regions of OmpR-P, or if instead the sites are similar, recognizing the same region of OmpR-P but differing in the strength of this recognition. Also, the molecular consequences of phosphorylation on the interactions between OmpR and the various DNA sites remain unclear.

There have been, however, significant insights gained with regard to the issue of OmpR-DNA binding. An *ompR2* allele, *ompR472*, that was shown to be defective in DNA binding possesses a V203M alteration, consistent with the DNA binding domain corresponding to the C-terminal domain of the protein. Furthermore, the C-terminal portion of OmpR was shown to bind DNA both in vitro (Tate et al., 1988) and in vivo (Tsung et al., 1989). Thus, the amino acid residues required for OmpR-DNA interaction reside in the C-terminal half of the protein (Fig. 1). Encouragingly, the C-terminal domain has recently been crystallized (Kondo et al., 1994), and the forthcoming structure, combined with further genetic analyses, will hopefully prove informative.

OmpR-Mediated Transcriptional Activation

OmpR possesses the ability to activate transcription at both the *ompF* and *ompC* promoters. How OmpR, and indeed transcriptional regulators in general, are able to increase the rate of transcription initiation from the promoters they regulate has been the focus of much attention. OmpR might activate transcription by altering the DNA topology of the promoter to facilitate either the binding or the transcriptional initiation of RNA polymerase (RNAP). Alternatively, OmpR-mediated transcriptional activation may involve direct protein-protein interaction between OmpR and the transcriptional apparatus. If so, this protein interaction could either facilitate the binding of RNAP to the promoter sequences and/or affect a step subsequent to RNAP binding (e.g., closed to open complex formation).

Significant insights into the mechanism of OmpR-mediated transcriptional activation have been gained from a body of evidence that implicates the C-terminal portion of the α-subunit of RNAP as critical for transcriptional activation by OmpR. The first hint of the functional importance of α in the regulation of the porin genes came from results obtained in a suppressor analysis of the *envZ473* allele. *envZ473* confers an F^-C^+ phenotype, irrespective of the osmolarity. Using fusions to *lacZ*, suppressors of this allele were isolated and subsequently shown to map to the α-operon (Garrett and Silhavy, 1987). A later genetic study resulted in the identification of an *rpoA* allele (*rpoA* encodes the α-subunit of RNAP) that interferes with the suppression of *envZ11* by *ompR77* (*envZ11* confers a constitutive F^-C^+ phenotype and *ompR77* alleviates this aberrant pattern of expression in an otherwise wild-type background) (Matsuyama and Mizushima, 1987). Further studies have led to the identification of additional *rpoA* alleles that cause specific decreases in porin gene transcription (Slauch et al., 1991; Sharif and Igo, 1993). Sequence analysis of these *rpoA* alleles reveals most of the mutations map to the C-terminal portion of the protein, implicating this region of α as playing an important role in OmpR-mediated transcriptional activation (Russo and Silhavy, 1992; Slauch et al., 1991; Sharif and Igo, 1993).

Consistent with the in vivo evidence described above, critical in vitro experiments also indicate that OmpR-mediated transcriptional activation requires the integrity of the C-terminal domain of α. In these studies, OmpR-mediated transcriptional activation from the *ompC* promoter was examined in the presence of reconstituted RNAPs containing C-terminally truncated α-subunits. Although these mutant RNAPs were capable of transcription from activator-independent promoters, they were unresponsive to OmpR-mediated activation at the *ompC* promoter (Igarashi et al., 1991). Indeed,

these mutant RNAPs were unresponsive to several transcriptional activators (Igarashi and Ishihama, 1991; Igarashi et al., 1991).

The implication of the above genetic and biochemical results is that there is direct protein-protein communication between OmpR and the C-terminal domain of α and that this interaction is a critical component of OmpR-mediated transcriptional activation. However, little is known concerning how OmpR may contact α and how this proposed interaction functions to stimulate transcription. To address these issues, a recent genetic study focused on identifying the region(s) in OmpR that is both specific and critical for transcriptional activation (Pratt and Silhavy, 1994). Most amino acid residues identified as essential for transcriptional activation cluster within a 20-amino-acid stretch in the C-terminal DNA binding domain of OmpR. This region of OmpR is protease-sensitive and is thus believed to exist as a surface-exposed loop (Kenney and Silhavy, unpublished data). Such a surface locale is certainly consistent with the idea that this region functions as an interactive domain. Accordingly, it has been proposed that this 20-amino-acid stretch may contact α to stimulate transcription. Future work should focus both on testing this hypothesis and on elucidating the mechanistic consequences of this interaction.

Transcriptional Repression

Under conditions of high osmolarity, OmpR represses transcription from the *ompF* promoter (Slauch and Silhavy, 1989). A large body of circumstantial evidence implicates the formation of a repressive loop as important for OmpR-mediated repression of the *ompF* promoter (see the section on the *ompF* promoter for further details). However, the mechanistic implications of this loop are not well understood. It is unclear how, in molecular terms, OmpR mediates the formation of this loop and whether multimerization and/or cooperative binding is involved. It is also unknown how this loop results in a decrease in transcription from the *ompF* promoter. One possibility is that the loop prevents RNAP from binding the promoter. Alternatively, OmpR may interact with RNAP when bound at the promoter and, through this interaction, somehow prevent transcription initiation.

The formation of a DNA loop implies that OmpR may oligomerize in the formation of this loop. Conflicting results have been reported with regard to this issue. It was initially proposed that OmpR may form multimers and that this might be an important distinguishing feature between the OmpR in low versus high osmolarity (Hall and Silhavy, 1981b). However, when initially purified to homogeneity, OmpR was reported to behave as a monomer throughout the purification procedure. Once purified, molecular weight determination and cross-linking studies were also consistent with the native protein existing as a monomer (Jo et al., 1986; Norioka et al., 1986). It is possible that OmpR exists as a monomer under these conditions because it is largely in the unphosphorylated form. Indeed, it has since been reported that OmpR can form higher-order oligomers upon phosphorylation (Nakashima et al., 1991). However, attempts to detect higher-order structure in the absence of cross-linking agents have thus far proved unsuccessful, even upon very high levels of phosphorylation (Kenney and Silhavy, unpublished data). It may be that interaction with DNA is an important component of oligomerization. If so, a better understanding of how both phosphorylation and DNA binding affect OmpR should prove helpful in addressing this problem.

Another central issue concerning OmpR-mediated repression is whether this repression involves communication with the transcriptional machinery or simply the prevention of RNAP from binding to the promoter. Mutations have been isolated in *rpoA* that dramatically decrease transcription from the *ompF* promoter, while causing only subtle decreases in *ompC* transcription (Slauch et al., 1991). There are several possible explanations for this differential effect. For example, it may be that

OmpR-mediated transcriptional activation involves different regions of α at the different promoters. Alternatively, the *rpoA* alleles may affect both transcriptional activation and transcriptional repression at the *ompF* promoter (Garrett and Silhavy, 1987; Slauch et al., 1991). Recently, a novel *rpoA* allele was isolated that affects both OmpR-mediated transcriptional activation and repression. The allele increases OmpR's ability to both activate and repress transcription from the *ompF* promoter, yet does not alter transcriptional activation of *ompC* (Sharif and Igo, 1993). One interpretation of these results is that both OmpR-mediated transcriptional activation and repression involve communication with the transcriptional machinery and that this mutant α-subunit is altered in both of these interactions. However, alternative explanations exist, and we simply do not yet know the answer to this central mechanistic question.

One approach to better understanding OmpR-mediated repression is to identify mutations in *ompR* that render the protein defective for this function. To this end, a recent study focused on identifying the primary defect(s) present in a collection of *ompR* alleles. This work resulted in the classification of two *ompR* alleles as primarily defective in transcriptional repression. The amino acids altered in these mutant proteins reside in what is believed to be a linker region between the N-terminal phosphorylation domain and the C-terminal DNA binding domain of the OmpR protein. Therefore, it was hypothesized that these mutations may impede the transmission of a conformational change that is important for repression (Russo et al., 1993). It remains to be determined whether such mutants fail to interact properly with each other to form the repressive loop or whether they are defective in interacting properly with the polymerase to prevent transcription initiation. The isolation of additional alleles defective for this function and detailed analyses of the underlying causes of the repression defects may prove useful.

PORIN PROMOTERS

To regulate expression of the porin genes, OmpR must interact with regulatory regions of the *ompF* and *ompC* promoters in a manner that results in the activation and/or repression of transcription. The structure and function of the promoters are therefore critical components of the porin regulon. Much study has been focused on elucidating the following: the regions of DNA to which OmpR binds and the nucleotide sequence of these regions; how OmpR recognizes these regions; and finally, how these OmpR-promoter interactions affect DNA topology, functional interactions with the polymerase, and ultimately transcriptional activation and/or repression.

ompF Promoter

Transcriptional regulation of *ompF* involves both OmpR-mediated activation and repression, and the structure of the *ompF* promoter accurately reflects this complexity (Fig. 4). Genetic studies identified poorly conserved −10 and −35 recognition sites (Taylor et al., 1985). A series of upstream and downstream deletions revealed that sequences extending from +17 through −90 relative to the transcriptional start site are essential for OmpR-mediated transcriptional activation. Also, sequences from +17 through −112 were shown to be required for activation by a mutant OmpR protein (Inokuchi et al., 1984). Further studies revealed that, although the above sequences are sufficient for activation of *ompF*, sequences upstream of −240 are required for normal fluctuation in response to varying osmolarity (Ostrow et al., 1986).

The regions described above as important for in vivo OmpR-mediated transcriptional regulation correlate well with both in vitro and in vivo DNA protection studies (Fig. 4). DNase I protection studies revealed that OmpR binds to a region between −40 and −96 relative to the transcriptional start site (Jo et al., 1986; Mizuno et al., 1988). Consistent with these in vitro data, in vivo dimethyl sulfate experiments indicated that OmpR protects guanine residues from −96

through −46 (Tsung et al., 1989). Also, OmpR has been shown to bind to an upstream region, between approximately −350 and −380 from the start site of transcription. Inactivation of this upstream site by insertion of a 22-bp fragment abolishes OmpR-mediated transcriptional repression of *ompF* (Huang et al., 1994). These results confirm and extend the observation that sequences upstream of −240 are required for proper regulation of *ompF* (Ostrow et al., 1986).

All these studies have been crucial both in revealing the regions of DNA important for proper regulation of *ompF* and in further illustrating the complex nature of *ompF* regulation. However, sequence analysis of the identified regions has not proved to be especially informative. There does appear to be a 10-bp sequence centered at −46 that exhibits homology to sequences found at the *ompC* promoter. This sequence has been dubbed the C box (see below) (Tsung et al., 1989). The remaining sequences have the distinguishing characteristic of being very AT-rich, and 10-bp sequences within these AT-rich regions have been called F boxes (Fig. 4). The F and C boxes do not bear any compelling degree of homology with each other (Tsung et al., 1989). Thus, although the regions at the *ompF* promoter to which OmpR binds have been identified, the actual bases recognized by OmpR have remained elusive.

Several important observations indicate that DNA topology is an essential component of transcriptional regulation at the *ompF* promoter. First, *ompF* promoter DNA between −57 and −111 has been shown to contain an intrinsic bend that results in anomalous electrophoretic mobility. Two dA-dT tracts centered at −75 and at −95 are believed responsible for this bend, and indeed perturbation of the periodical spacing of these tracts was shown to greatly diminish the bend (Mizuno, 1987). Importantly, mutations in the *ompF* promoter that alter the dA-dT tracts result in OmpR-dependent constitutive expression of *ompF* (these mutations decrease the intrinsic bend in the DNA as indicated by less anomalous electrophoretic mobility). Given that the sequences altered by these mutations are required for transcriptional activation, the effects cannot be explained by hypothesizing that this region is a negative acting site. Rather the constitutive expression appears to result from the decrease of the intrinsic bend normally present at the promoter. Without this bend in the DNA, OmpR is rendered unable to repress transcription (Slauch and Silhavy, 1991). Finally, integration host factor (IHF) has been shown to be essential for normal *ompF* regulation. IHF is a histone-like protein known to both bind and bend DNA (for review, see Friedman, 1988). Strains deficient for IHF exhibit OmpR-dependent constitutive expression of *ompF* (i.e., the absence of IHF renders OmpR unable to repress *ompF* transcription) (Tsui et al., 1988). Importantly, IHF has been shown to bind the *ompF* promoter DNA in vitro, and this binding strongly increases the intrinsic bend normally present in the *ompF* regulatory DNA (Ramani et al., 1992).

These observations reveal the importance of DNA topology in the regulation of *ompF* transcription and, taken together, strongly suggest the formation of a DNA loop as integral to *ompF* repression. The existence of a distal upstream site required for negative regulation is in itself quite suggestive. Also, the *ompF* promoter is intrinsically bent, and this bend is enhanced by the binding of IHF. Conditions that either diminish this intrinsic bend or interfere with the ability of IHF to enhance this bend abolish OmpR-mediated repression of *ompF*. Thus, the current model for *ompF* transcriptional regulation posits that in low osmolarity OmpR binds to sequences from −60 through −100 and activates transcription. In high osmolarity, OmpR binds to the distal upstream site and to the site centered at −46 and forms a repressive loop (the −46 site is thought to function as a negative site largely because of circumstantial evidence; recall that the OmpR472 protein, which confers an F^{c}C$^-$ phenotype, does not bind this region in vitro). Finally, IHF is required to bend the DNA to form the repressive loop.

Clearly, much information concerning the structure and transcriptional regulation of the *ompF* promoter has been gained. Yet, several important aspects of this model need to be tested, and intriguing questions remain. Rampersaud et al. (1994) have presented evidence that the site at −46 is not absolutely required for repression. However, a more systematic approach is required to confirm this result and establish the mechanistic role of this site in *ompF* regulation. It is also possible that the region from −60 through −100 is involved in negative as well as positive regulation. Furthermore, the effects of small insertions and deletions (corresponding to integral or nonintegral turns of the DNA helix) placed between the two proposed negative sites still need to be examined. Finally, the mechanistic implications of this presumed repressive loop remain unknown. How does OmpR mediate the formation of this DNA loop? And once formed, how does this loop function in the repression of *ompF* transcription?

ompC Promoter

The structure of the *ompC* promoter appears relatively straightforward (Fig. 4). Studies examining the effects of upstream and downstream deletions revealed that sequences extending from +1 through −94 are critical for OmpR-dependent transcriptional activation (Mizuno and Mizushima, 1986). Consistent with these regulatory data, DNase I protection studies indicated that OmpR protects sequences from −78 through −102 (Norioka et al., 1986; Mizuno et al., 1988). With increased concentrations of OmpR, additional protection is observed, extending from −41 through −102 (Maeda and Mizuno, 1988). Also, in vivo dimethyl sulfate protection experiments reveal that OmpR protects guanine residues from −43 through −94 (Tsung et al., 1989).

Within the region extending from −41 through −94, the existence of three 10-bp homologous sequences present in tandem repeat was noted (Mizuno and Mizushima, 1986).

These repeats have since been referred to as C boxes (Tsung et al., 1989). This upstream region was shown to support OmpR-mediated transcriptional activation when the distance between the repeats and the canonical promoter was increased, provided the repeats were present on the correct face of the helix (Maeda and Mizuno, 1988). The repeats were also shown to function in either orientation to support OmpR-mediated transcriptional activation. For these experiments, the sequences from −35 to −107 were placed in either orientation, and both constructs were equally functional (Maeda and Mizuno, 1988).

Finally, the sequences from −72 through −405 were shown to support a significant degree of OmpR-mediated transcriptional activation. Because this region contains only one of the three 10-bp repeats normally present at the *ompC* promoter, it was thought that OmpR recognized this single site (Maeda and Mizuno, 1990). However, OmpR has been shown to protect DNA farther upstream of this repeat (Norioka et al., 1986; Mizuno et al., 1988; Maeda and Mizuno, 1988). This protected DNA is AT-rich, reminiscent of the AT-rich regions found at the *ompF* promoter. Indeed, this region has been dubbed an F box (Tsung et al., 1989). It is thus not presently clear whether OmpR recognizes just the C box or if the AT-rich F box just upstream of the C box is also important.

A great deal has been learned from these studies. The three tandemly repeated C boxes appear to play an important role in OmpR-DNA recognition and in supporting transcriptional activation of *ompC*. Yet many questions remain. How does OmpR recognize these sequences? Does OmpR bind cooperatively to these repeats, and what role do the AT-rich regions play in OmpR-DNA recognition? Does OmpR recognize these AT regions in a similar manner, or are these interactions completely different, perhaps even involving different regions of the protein? Finally, how do these interactions result in transcriptional activation of *ompC*?

OTHER FACTORS AFFECTING PORIN EXPRESSION

Environmental parameters in addition to osmolarity influence expression of the porins. Indeed, the levels of OmpF fluctuate in response to changes in pH, temperature, osmotic stress, and growth phase. Although the mechanisms through which the cell responds to these environmental parameters have not been characterized as extensively as the responses to osmolarity, some key players have been identified.

At high temperatures, a decrease in OmpF expression is mediated through the antisense RNA, *micF* (mRNA-interfering complementary RNA). *micF* was first identified as a 300-bp element upstream of the *ompC* gene that inhibits OmpF expression when present on a multicopy plasmid (Mizuno et al., 1984). This 300-bp region encodes a 93-nucleotide RNA with extensive complementarity to the 5′ end of the *ompF* message. *micF* is believed to inhibit *ompF* translation by binding to the 5′ end of the *ompF* mRNA (Anderson et al., 1989; Mizuno et al., 1984).

It was at first thought that *micF* might be important for regulating porin expression in response to osmolarity (Mizuno et al., 1984). However, further analysis has shown that this antisense RNA does not play a significant role in the negative regulation of OmpF in high osmolarity (Matsuyama and Mizushima, 1985; Ramani et al., 1994). *micF* does appear to be intimately involved in the thermal regulation of OmpF (Anderson et al., 1989). Levels of *micF* RNA increase at high temperatures with a concomitant decrease in *ompF* mRNA over time. In a strain deleted for *micF*, this decrease in *ompF* mRNA is not observed (Anderson et al., 1989).

It has also been shown that *micF* is transcriptionally activated by OmpR, and this transcriptional activation requires the same *cis*-acting elements required for OmpR-mediated transcriptional activation of *ompC* (Takayanagi et al., 1991). However, there appear to be additional *cis*-acting elements that somehow negatively affect *micF* transcription. The manner in which *micF* transcription is regulated is not well understood (Takayanagi et al., 1991). Thus, the present model posits that at high temperatures, transcription of *micF* increases through an unidentified mechanism, and this increase results in the observed post-transcriptional negative regulation of OmpF expression.

OmpF expression is also repressed under conditions of oxidative stress, correlating with an increased resistance to antibiotics (Greenberg et al., 1990). This repression was shown to occur post-transcriptionally and to correlate with an increase in the levels of *micF* RNA. Also, this repression is dependent on both the *micF* gene and the *soxRS* locus (SoxRS regulates the expression of approximately 10 genes in response to oxidative stress). Thus, in response to oxidative stress, SoxRS mediates the repression of OmpF by increasing the transcription of *micF* through an unknown mechanism (Chou et al., 1993).

Finally, OmpF levels decrease on entry into the stationary phase. The stationary phase sigma factor (σ^s), encoded by the *rpoS* gene (Lange and Hengge-Aronis, 1991) has been shown to be essential for this decrease (McCann et al., 1991). It was proposed that σ^s may mediate the decrease in OmpF expression by increasing transcription of *micF*, thus decreasing *ompF* translation (Demple, 1991). However, studies using operon fusions to *lacZ* have shown that transcription is regulated by σ^s (Pratt and Silhavy, unpublished data). The mechanism by which *rpoS* modulates the decrease in *ompF* transcription on entry into the stationary phase is not presently known.

WHY GO TO ALL THE TROUBLE?

Regulation of the porins is clearly both extensive and complex. This complexity implies that porin regulation is normally important to the well-being of the cell. However, neither *ompF* nor *ompC* is essential for cell growth. Furthermore, OmpF and OmpC are extremely similar, the only significant distinction between the two being a small difference in the diameters of the channels formed by each (1.08 nm versus

1.16 nm). Why, then, are such great lengths taken to regulate the relative expression of these two seemingly similar proteins? What is the physiological relevance of the porin regulon?

Studies elucidating the permeability characteristics of the OmpF and OmpC porins have proved informative. Because of the small channels, the exclusion limit of the porins is approximately 600 D. Importantly, the small channel size results in large differences in rates of diffusion for solute molecules of varying sizes (even among molecules that fall below the exclusion limit). Larger molecules diffuse at significantly lower rates than do smaller molecules. Also, diffusion rates have been shown to be negatively affected by both hydrophobicity and by negative charge. And finally, because the OmpF and OmpC channels are so small, the slight difference in their diameters can result in dramatically different rates of diffusion through the two porins. The rates through OmpF are significantly faster than through OmpC. This final point is suggestive, given the differential regulation of the two very similar proteins (Nikaido and Vaara, 1987).

The small channel size of the porins appears to be characteristic of enteric bacteria; other bacteria produce porins with significantly larger channels. With larger channels, the rates of diffusion are not as strongly influenced by the size of the solute molecules (Nikaido and Vaara, 1987). Why might small-channeled porins be unique to enteric bacteria, and why are such elaborate regulatory measures taken to regulate these porins?

Enteric bacteria live in vastly differing environments and must therefore be able to both distinguish and appropriately adjust to these alternative surroundings. When in the intestinal tract of an animal, a bacterium is in an environment in which the temperature is relatively high, nutrients are abundant, bile salts are highly concentrated, and the osmolarity is high. By contrast, when in the external environment, the cell faces lower temperatures, a scarcity of nutrients, and lower osmolarity. It may be that both temperature and osmolarity serve as crucial indicators that enable a bacterium to determine whether it is "in" or "out" of the intestine (Nikaido and Vaara, 1987).

Indeed, both high temperature and high osmolarity result in a net reduction in the level of OmpF expression. Importantly, these conditions mimic those found within the bodies of animals. The change from the larger OmpF channel to the smaller OmpC channel only slightly reduces the permeability of the cell for small nutrients. However, the permeability for larger, more hydrophobic, and/or more negatively charged molecules (e.g., the bile salts) is dramatically decreased. Therefore, switching to the porin with the narrower channel has clear advantages in an environment with an abundance of inhibitory substances (e.g., bile salts). However, both low temperatures and low osmolarity result in an increase in expression of OmpF relative to OmpC. These conditions mimic those found in the external environment. In these surroundings, nutrients are scarce, and the slightly wider OmpF channel enables their more efficient uptake. The smaller, more protective OmpC porin is no longer required in this very dilute environment (Nikaido and Vaara, 1987).

REFERENCES

Adler, H. 1966. Chemotaxis in bacteria. *Science* **153:** 708–716.

Aiba, H., and T. Mizuno. 1990. Phosphorylation of a bacterial activator protein, OmpR, by a protein kinase, EnvZ, stimulates the transcription of the *ompF* and *ompC* genes in *Escherichia coli*. *FEBS Lett.* **261:**19–22.

Aiba, H., T. Mizuno, and S. Mizushima. 1989a. Transfer of phosphoryl group between two regulatory proteins involved in osmoregulatory expression of the *ompF* and *ompC* genes in *Escherichia coli*. *J. Biol. Chem.* **264:**8563–8567.

Aiba, H., F. Nakasai, S. Mizushima, and T. Mizuno. 1989b. Evidence for the physiological importance of the phosphotransfer between the two regulatory components, EnvZ and OmpR, in osmoregulation in *Escherichia coli*. *J. Biol. Chem.* **264:**14090–14094.

Aiba, H., F. Nakasai, S. Mizushima, and T. Mizuno. 1989c. Phosphorylation of a bacterial activator protein, OmpR, by a protein kinase, EnvZ,

results in a stimulation of its DNA-binding ability. *J. Biochem.* **106:**5–7.

Anderson, J., S. A. Forst, K. Zhao, M. Inouye, and N. Delihas. 1989. The function of *micF* RNA. *J. Biol. Chem.* **264:**17961–17970.

Baumgartner, J. W., C. Kim, R. E. Brissette, M. Inouye, C. Park, and G. L. Hazelbauer. 1994. Transmembrane signalling by a hybrid protein: communication from the domain of chemoreceptor Trg that recognizes sugar-binding proteins to the kinase/phosphatase domain of osmosensor EnvZ. *J. Bacteriol.* **176:**1157–1163.

Chou, J. H., J. T. Greenberg, and B. Demple. 1993. Posttranscriptional repression of *Escherichia coli* OmpF protein in response to redox stress: positive control of the *micF* antisense RNA by the *soxRS* locus. *J. Bacteriol.* **175:**1026–1031.

Comeau, D. E., K. Ikenaka, K. Tsung, and M. Inouye. 1985. Primary characterization of the protein products of the *Escherichia coli ompB* locus: structure and regulation of synthesis of the OmpR and EnvZ proteins. *J. Bacteriol.* **164:**578–584.

Csonka, L. N. 1989. Physiological and genetic responses of bacteria to osmotic stress. *Microbiol. Rev.* **53:**121–147.

Demple, B. 1991. Regulation of bacterial oxidative stress genes. *Annu. Rev. Genet.* **25:**315–337.

Fiedler, W., and H. Rotering. 1988. Properties of *Escherichia coli* mutants lacking membrane-derived oligosaccharides. *J. Biol. Chem.* **263:**14684–14689.

Forst, S., D. Comeau, S. Norioka, and M. Inouye. 1987. Localization and membrane topology of EnvZ, a protein involved in osmoregulation of OmpF and OmpC in *Escherichia coli*. *J. Biol. Chem.* **262:**16433–16438.

Forst, S., J. Delgado, and M. Inouye. 1989. Phosphorylation of OmpR by the osmosensor EnvZ modulates expression of the *ompF* and *ompC* genes in *Escherichia coli*. *Proc. Natl. Acad. Sci. USA* **86:**6052–6056.

Forst, S., J. Delgado, A. Rampersaud, and M. Inouye. 1990. In vivo phosphorylation of OmpR, the transcription activator of the *ompF* and *ompC* genes in *Escherichia coli*. *J. Bacteriol.* **172:**3473–3477.

Friedman, D. I. 1988. Integration host factor: a protein for all reasons. *Cell* **55:**545–554.

Garrett, S., and T. J. Silhavy. 1987. Isolation of mutations in the α operon of *Escherichia coli* that suppress the transcriptional defect conferred by a mutation in the porin regulatory gene *envZ*. *J. Bacteriol.* **169:**1379–1385.

Garrett, S., R. Taylor, and T. J. Silhavy. 1983. Isolation and characterization of chain-terminating nonsense mutations in a porin regulator gene, *envZ*. *J. Bacteriol.* **156:**62–69.

Geiger, O., F. D. Russo, T. J. Silhavy, and E. P. Kennedy. 1992. Membrane-derived oligosaccharides affect porin osmoregulation only in media of low ionic strength. *J. Bacteriol.* **174**:1410–1413.

Greenberg, J. T., P. Monach, J. H. Chou, P. D. Josephy, and B. Demple. 1990. Positive control of a global antioxidant defense regulon activated by superoxide-generating agents in *Escherichia coli*. *Proc. Natl. Acad. Sci. USA* **87**:6181–6185.

Hall, M. N., and T. J. Silhavy. 1979. Transcriptional regulation of *Escherichia coli* K-12 major outer membrane protein 1b. *J. Bacteriol.* **140**:342–350.

Hall, M. N., and T. J. Silhavy. 1981a. The *ompB* locus and the regulation of the major outer membrane porin proteins of *Escherichia coli* K-12. *J. Mol. Biol.* **146**:23–43.

Hall, M. N., and T. J. Silhavy. 1981b. Genetic analysis of the *ompB* locus in *Escherichia coli* K-12. *J. Mol. Biol.* **151**:1–15.

Hess, J. F., R. B. Bourret, and M. I. Simon. 1988. Histidine phosphorylation and phosphoryl group transfer in bacterial chemotaxis. *Nature* (London) **336**:139–143.

Huang, K., J. L. Schieberl, and M. M. Igo. 1994. A distant upstream site involved in the negative regulation of the *Escherichia coli ompF* gene. *J. Bacteriol.* **176**:1309–1315.

Igarashi, K., A. Hanamura, K. Makino, H. Aiba, H. Aiba, T. Mizuno, A. Nakata, and A. Ishihama. 1991. Functional map of the α subunit of *Escherichia coli* RNA polymerase: two modes of transcription activation by positive factors. *Proc. Natl. Acad. Sci. USA* **88**:8958–8962.

Igarashi, K., and A. Ishihama. 1991. Bipartite functional map of the *E. coli* RNA polymerase α subunit: involvement of the C-terminal region in transcription activation by cAMP-CRP. *Cell* **65**:1015–1022.

Igo, M. M., A. J. Ninfa, and T. J. Silhavy. 1989a. A bacterial environmental sensor that functions as protein kinase and stimulates transcriptional activation. *Genes Dev.* **3**:598–605.

Igo, M. M., A. J. Ninfa, J. B. Stock, and T. J. Silhavy. 1989b. Phosphorylation and dephosphorylation of a bacterial transcriptional activator by a transmembrane receptor. *Genes Dev.* **3**:1725–1734.

Igo, M. M., and T. J. Silhavy. 1988. EnvZ, a transmembrane environmental sensor of *Escherichia coli* K-12, is phosphorylated in vitro. *J. Bacteriol.* **170**:5971–5973.

Inokuchi, K., H. Furukawa, K. Nakamura, and S. Mizushima. 1984. Characterization by deletion mutagenesis in vitro of the promoter region of *ompF*, a positively regulated gene of *Escherichia coli*. *J. Mol. Biol.* **178**:653–668.

Jin, T., and M. Inouye. 1993. Ligand binding to the receptor domain regulates the ratio of kinase to phosphatase activities of the signalling domain of hybrid *Escherichia coli* transmembrane receptor, Taz1. *J. Mol. Biol.* **232**:484–492.

Jo, Y., F. Nara, S. Ichihara, T. Mizuno, and S. Mizushima. 1986. Purification and characterization of the OmpR protein, a positive regulator involved in osmoregulatory expression of the *ompF* and *ompC* genes in *Escherichia coli*. *J. Biol. Chem.* **261**:15252–15256.

Keeney, L., and T. J. Silhavy. Unpublished data.

Kofoid, E. C., and J. S. Parkinson. 1988. Transmitter and receiver modules in bacterial signaling proteins. *Proc. Natl. Acad. Sci. USA* **85**:4981–4985.

Kondo, H., T. Miyaji, M. Suzuki, S. Tate, T. Mizuno, Y. Nishimura, and I. Tanaka. 1994. Crystallization and X-ray studies of the DNA-binding domain of OmpR protein, a positive regulator involved in activation of osmoregulatory genes in *Escherichia coli*. *J. Mol. Biol.* **235**:780–782.

Lange, R., and R. Hengge-Aronis. 1991. Identification of a central regulator of stationary phase gene expression in *E. coli*. *Mol. Microbiol.* **5**:49–59.

Liljestrom, P. 1986. The EnvZ protein of *Salmonella typhimurium* LT-2 and *Escherichia coli* K-12 is located in the cytoplasmic membrane. *FEMS Microbiol. Lett.* **36**:145–150.

Maeda, S., and T. Mizuno. 1988. Activation of the *ompC* gene by the OmpR protein in *Escherichia coli*. *J. Biol. Chem.* **263**:14629–14633.

Maeda, S., and T. Mizuno. 1990. Evidence for multiple OmpR-binding sites in the upstream activation sequence of the *ompC* promoter in *Escherichia coli*: a single OmpR-binding site is capable of activating the promoter. *J. Bacteriol.* **172**:501–503.

Matsuyama, S., and S. Mizushima. 1985. Construction and characterization of a deletion mutant lacking *micF*, a proposed regulatory gene for OmpF synthesis in *Escherichia coli*. *J. Bacteriol.* **162**:1196–1202.

Matsuyama, S., and S. Mizushima. 1987. Novel *rpoA* mutation that interferes with the function of OmpR and EnvZ, positive regulators of the *ompF* and *ompC* genes that code for outer-membrane proteins in *Escherichia coli* K12. *J. Mol. Biol.* **195**:847–853.

McCann, M. P., J. P. Kidwell, and A. Matin. 1991. The putative σ factor KatF has a central role in development of starvation-mediated general resistance in *Escherichia coli*. *J. Bacteriol.* **173**:4188–4194.

Mizuno, T. 1987. Static bend of DNA helix at the activator recognition site of the *ompF* promoter in *Escherichia coli*. *Gene* **54**:57–64.

Mizuno, T., M. Chou, and M. Inouye. 1983. A comparative study on the genes for three porins of the *Escherichia coli* outer membrane. *J. Biol. Chem.* **258**:6932–6940.

Mizuno, T., M. Chou, and M. Inouye. 1984. A unique mechanism regulating gene expression:

translational inhibition by a complementary RNA transcript (micRNA). *Proc. Natl. Acad. Sci. USA* **81:**1966–1970.

Mizuno, T., M. Kato, Y. Jo, and S. Mizushima. 1988. Interaction of OmpR, a positive regulator, with the osmoregulated *ompC* and *ompF* genes of *Escherichia coli. J. Biol. Chem.* **263:**1008–1012.

Mizuno, T., and S. Mizushima. 1986. Characterization by deletion and localized mutagenesis in vitro of the promoter region of the *Escherichia coli ompC* gene and importance of the upstream DNA domain in positive regulation by the OmpR protein. *J. Bacteriol.* **168:**86–95.

Nakashima, K., K. Kanamuru, H. Aiba, and T. Mizuno. 1991. Signal transduction and osmoregulation in *Escherichia coli. J. Biol. Chem.* **266:**10775–10780.

Nara, F., S. Matsuyama, T. Mizuno, and S. Mizushima. 1986. Molecular analysis of mutant *ompR* genes exhibiting different phenotypes as to osmoregulation of the *ompF* and *ompC* genes of *Escherichia coli. Mol. Gen. Genet.* **202:**194–199.

Nikaido, H., and M. Vaara. 1987. Outer membrane, p. 7–22. *In* F. C. Neidhardt (ed.), *Escherichia coli and Salmonella typhimurium: Cellular and Molecular Biology.* American Society for Microbiology, Washington, D.C.

Ninfa, A. J., and B. Magasanik. 1986. Covalent modification of the *glnG* product, NR$_I$, by the *glnALG* product, NR$_{II}$, regulates the transcription of the *glnALG* operon in *Escherichia coli. Proc. Natl. Acad. Sci. USA* **83:**5909–5913.

Ninfa, G. E., A. Stock, S. Mowbray, and J. Stock. 1991. Reconstitution of the bacterial chemotaxis signal transduction system from purified components. *J. Biol. Chem.* **266:**9764–9770.

Norioka, S., G. Ramakrishnan, K. Ikenaka, and M. Inouye. 1986. Interaction of a transcriptional activator, OmpR, with reciprocally osmoregulated genes, *ompF* and *ompC*, of *Escherichia coli. J. Biol. Chem.* **261:**17113–17119.

Ostrow, K. S., T. J. Silhavy, and S. Garrett. 1986. *cis*-Acting sites required for osmoregulation of *ompF* expression in *Escherichia coli* K-12. *J. Bacteriol.* **168:**1165–1171.

Parkinson, J. S. 1993. Signal transduction schemes in bacteria. *Cell* **73:**857–871.

Pratt, L., and T. J. Silhavy. Unpublished data.

Pratt, L. P., and T. J. Silhavy. 1994. OmpR mutants specifically defective for transcriptional activation. *J. Mol. Biol.* **243:**579–594.

Ramani, N., M. Hedeshian, and M. Freundlich. 1994. *micF* antisense RNA has a major role in osmoregulation of OmpF in *Escherichia coli. J. Bacteriol.* **76:**5005–5010.

Ramani, N., L. Huang, and M. Freundlich. 1992. In vitro interactions of integration host factor with

the *ompF* promoter-regulatory region of *Escherichia coli. Mol. Gen. Genet.* **231:**248–255.

Rampersaud, A., S. L. Harlocker, and M. Inouye. 1994. The OmpR protein of *Escherichia coli* binds to sites in the *ompF* promoter region in a hierarchical manner determined by its degree of phosphorylation. *J. Biol. Chem.* **269:**12559–12566.

Roberts, D. L., D. W. Bennet, and S. A. Forst. 1994. Identification of the site of phosphorylation on the osmosensor, EnvZ, of *Escherichia coli. J. Biol. Chem.* **269:**8728–8733.

Russo, F. D. 1992. Ph.D. thesis. Princeton University, Princeton, N.J.

Russo, F., and T. J. Silhavy. 1991. EnvZ controls the concentration of phosphorylated OmpR to mediate osmoregulation of the porin genes. *J. Mol. Biol.* **222:**567–580.

Russo, F., and T. J. Silhavy. 1992. Alpha: the Cinderella subunit of RNA polymerase. *J. Biol. Chem.* **267:**14515–14518.

Russo, F., J. M. Slauch, and T. J. Silhavy. 1993. Mutations that affect separate functions of OmpR the phosphorylated regulator of porin transcription in *Escherichia coli. J. Mol. Biol.* **231:**261–273.

Sarma, V., and P. Reeves. 1977. Genetic locus (*ompB*) affecting a major outer-membrane protein in *Escherichia coli* K12. *J. Bacteriol.* **132:**23–27.

Sato, T., and T. Tura. 1979. Chromosomal location and expression of the structural gene for major outer membrane protein Ia of *Escherichia coli* K-12 and of the homologous gene of *Salmonella typhimurium. J. Bacteriol.* **139:**468–477.

Sharif, R. T., and M. M. Igo. 1993. Mutations in the alpha subunit of RNA polymerase that affect the regulation of porin gene transcription in *Escherichia coli* K-12. *J. Bacteriol.* **175:**5460–5468.

Slauch, J. M., and T. J. Silhavy. 1989. Genetic analysis of the switch that controls porin gene expression in *Escherichia coli. J. Mol. Biol.* **210:**281–292.

Slauch, J. M., and T. J. Silhavy. 1991. *cis*-Acting *ompF* mutations that result in OmpR-dependent constitutive expression. *J. Bacteriol.* **173:**4039–4048.

Slauch, J. M., F. D. Russo, and T. J. Silhavy. 1991. Suppressor mutations in *rpoA* suggest that OmpR controls transcription by direct interaction with the α subunit of RNA polymerase. *J. Bacteriol.* **173:**7501–7510.

Stock, A. M., J. M. Mottonen, J. B. Stock, and C. E. Schutt. 1989. Three-dimensional structure of CheY, the response regulator of bacterial chemotaxis. *Nature* (London) **337:**745–749.

Stock, J. B., B. Rauch, and S. Roseman. 1977. Periplasmic space in *Salmonella typhimurium* and *Escherichia coli. J. Biol. Chem.* **252:**7850–7861.

Stock, J. B., A. M. Stock, and J. M. Mottonen. 1990. Signal transduction in bacteria. *Nature* (London) **344:**395–400.

Takayanagi, K., S. Maeda, and T. Mizuno. 1991. Expression of *micF* involved in porin synthesis in *Escherichia coli:* two distinct *cis*-acting elements respectively regulate *micF* expression positively and negatively. *FEMS Microbiol. Lett.* **83**:39–44.

Tate, S., M. Kato, Y. Nishimura, Y. Arata, and T. Mizuno. 1988. Location of DNA-binding segment of a positive regulator, OmpR, involved in activation of the *ompF* and *ompC* genes of *Escherichia coli. FEBS Lett.* **242**:27–30.

Taylor, R. T., S. Garrett, E. Sodergren, and T. J. Silhavy. 1985. Mutations that define the promoter of *ompF,* a gene specifying a major outer membrane porin protein. *J. Bacteriol.* **162**:1054–1060.

Tokishita, S., A. Kojima, H. Aiba, and T. Mizuno. 1991. Transmembrane signal transduction and osmoregulation in *Escherichia coli. J. Biol. Chem.* **266**:6780–6785.

Tokishita, S., A. Kojima, and T. Mizuno. 1992. Transmembrane signal transduction and osmoregulation in Escherichia coli: functional importance of the transmembrane regions of membrane-located protein kinase, EnvZ. *J. Biochem.* **111**:707–713.

Tsui, P., V. Helu, and M. Freundlich. 1988. Altered osmoregulation of *ompF* in integration host factor mutants of *Escherichia coli. J. Bacteriol.* **170**:4950–4953.

Tsung, K., R. E. Brissette, and M. Inouye. 1989. Identification of the DNA-binding domain of the OmpR protein required for transcriptional activation of the *ompF* and *ompC* genes of *Escherichia coli* by in vivo DNA footprinting. *J. Biol. Chem.* **264**:10104–10109.

Utsumi, R., R. E. Brissette, A. Rampersaud, S. A. Forst, K. Oosawa, and M. Inouye. 1989. Activation of bacterial porin gene expression by a chimeric signal transducer in response to aspartate. *Science* **245**:1246–1249.

Van Alphen, W., and B. Lugtenberg. 1977. Influence of osmolarity of the growth medium on the outer membrane protein pattern of *Escherichia coli. J. Bacteriol.* **131**:623–630.

Verhoef, C., P. J. de Graaff, and E. J. J. Lugtenberg. 1977. Mapping of a gene for a major outer membrane protein of *Escherichia coli* K12 with the aid of a newly isolated bacteriophage. *Mol. Gen. Genet.* **150**:103–105.

Volz, K., and P. Matsumura. 1991. Crystal structure of *Escherichia coli* CheY refined at 1.7-Å resolution. *J. Biol. Chem.* **266**:15511–15519.

Waukau, J., and S. Forst. 1992. Molecular analysis of the signaling pathway between EnvZ and OmpR in *Escherichia coli. J. Bacteriol.* **174**:1522–1527.

Wurtzel, E. T., M. Y. Chou, and M. Inouye. 1982. Osmoregulation of gene expression I. DNA sequence of the *ompR* gene of the *ompB* operon of *Escherichia coli* and characterization of its gene product. *J. Biol. Chem.* **257**:13685–13691.

Wylie, D., A. Stock, C. Wong, and J. Stock. 1988. Sensory transduction in bacterial chemotaxis involves phosphotransfer between Che proteins. *Biochem. Biophys. Res. Commun.* **151**:891–896.

Yang, Y., and M. Inouye. 1991. Intermolecular complementation between two defective mutant signal-transduction receptors of *Escherichia coli. Proc. Natl. Acad. Sci. USA* **88**:11057–11061.

Yang, Y., and M. Inouye. 1993. Requirement of both kinase and phosphatase activities of an *Escherichia coli* receptor (Taz1) for ligand-dependent signal transduction. *J. Mol. Biol.* **231**:335–342.

Control of Cellular Development in Sporulating Bacteria by the Phosphorelay Two-Component Signal Transduction System

James A. Hoch

8

The initiation of sporulation in bacteria is a cellular response to deteriorating conditions for growth and division. It occurs at a time when readily metabolized carbon sources are near exhaustion and the cell must prepare for the use of less favorable substrates. Cells at this time are bombarded through a myriad of receptors and regulatory molecules with environmental information that they must evaluate, along with metabolic, cell cycle, and cell density information, to decide whether to grow or sporulate. To accomplish this, a mechanism must exist to integrate this information from diverse sources and compute an appropriate response. A cell that decides to undergo another round of division is irreversibly committed to that pathway and can only participate in the decision process again during a window early in the next cell cycle (Mandelstam and Higgs, 1974). Sporulation may be coupled to the cell cycle to suppress division and permit the orderly synthesis of spore membrane structural components in concert with chromosome replication. The nature of the integration mechanism for all this information is now be-

coming apparent, and its various regulatory features will comprise this chapter.

The two-component paradigm is at the heart of the signal transduction system, regulating the initiation of sporulation in sporulating bacteria. This system, the phosphorelay, differs from other two-component signal transduction systems by the mechanism of phosphate flow and types of accessory proteins that control phosphate flow in the system (Burbulys et al., 1991). There are two sensory kinases, KinA and KinB, that input phosphate into the system and two response regulator proteins, Spo0F and Spo0A (Fig. 1). The Spo0F protein consists only of the conserved amino-terminal domain of response regulators such as CheY of the chemotaxis system (J. B. Stock et al., 1989). In this context, Spo0F serves to accept phosphate from either KinA or KinB and has no other apparent function than to serve as a phosphodonor for production of Spo0A~P. This latter reaction is catalyzed by Spo0B, an enzyme unique to the phosphorelay, which acts as a phosphotransferase. The ultimate goal of the phosphorelay is to produce Spo0A~P, the activated form of this transcription factor that recognizes the 7-bp 0A box in sporulation promoters (Strauch et al., 1990). Spo0A~P may serve as a repressor or an activator of transcription, depending on the particular promoter

James A. Hoch, Division of Cellular Biology, Department of Molecular and Experimental Medicine, The Scripps Research Institute, La Jolla, California 92037.

Two-Component Signal Transduction, Edited by James A. Hoch and Thomas J. Silhavy,
© 1995 American Society for Microbiology, Washington, DC 20005

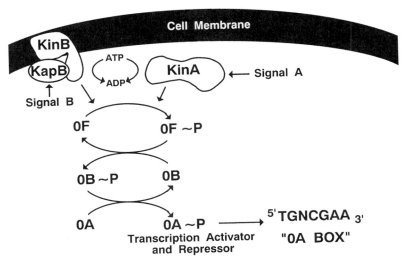

FIGURE 1 Phosphorelay signal transduction pathway of *Bacillus subtilis.* The source of activated phosphate is either KinA or KinB, responding to signal A or signal B, respectively. The nature of the signals is unknown in both cases and, in the case of KinB, is thought to be transmitted through the cytoplasmic protein KapB. The flow of phosphate through the phosphorelay has the goal of producing Spo0A~P, which is a transcription activator or repressor, recognizing the 7-bp sequence shown.

affected. Spo0A~P activates transcription of several genes, including the spoIIA and spoIIG operons coding for σ^F and σ^E, respectively, which are responsible for compartmentalized gene expression in the forespore and mother cell (Trach et al., 1991; Satola et al., 1992).

The conversion of Spo0A to Spo0A~P is believed to be the central factor determining whether a cell will divide or sporulate. That means that any environmental, metabolic, or cell cycle signals, be they positive or negative to sporulation, may impinge directly on the cellular level of Spo0A~P. Because sporulation is a cellular event requiring the precise coordination of many cellular processes, all subordinate to the environmental supply of nutrients and other factors, it should not be surprising that phosphate flow through the phosphorelay is subject to a variety of controls. These controls occur at two levels and take the form of transcriptional regulation of the level of phosphorelay components followed by regulation of the activity of the specific enzymatic processes of the phosphorelay.

TRANSCRIPTIONAL REGULATION OF THE PHOSPHORELAY COMPONENTS

To comprehend the overall regulation of initiation of sporulation and to assess the central role of the phosphorelay, it is necessary to understand the role of transcription regulation in this pathway (Fig. 2). The genes for the key response regulator components of the pathway, *spo0F* and *spo0A,* are the primary subjects of complex transcriptional controls. Both of these genes possess tandem promoters: one a σ^A vegetative promoter and the other a σ^H stationary-phase promoter. The vegetative promoter functions to produce both gene products at a low level during vegetative growth, whereas the stationary-phase promoter is required for higher-level synthesis when the phosphorelay is needed for sporulation. The stationary-phase promoters depend on σ^H and Spo0A~P to function (Lewandoski et al., 1986; Yamashita et al., 1989; Trach et al., 1988; Bai et al., 1990). Thus, the product of the phosphorelay,

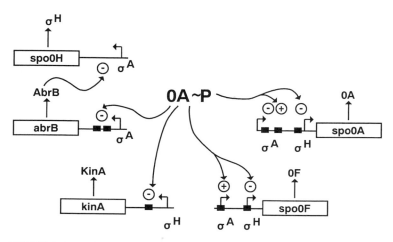

FIGURE 2 Role of Spo0A~P in transcription regulation of phosphorelay. The central goal of Spo0A~P is depicted as either activation or repression of promoters. The small black boxes within the promoter show the location of 0A boxes. The + and − signs refer to the effect of Spo0A~P on the 0A box. Promoters with more than one 0A box with opposite signs means Spo0A~P is both an activator and a repressor of the promoter. Briefly, Spo0A~P is an activator of σ^H promoters on the *spo0F* and *spo0A* genes and a repressor of the vegetative σ^A promoter on the *spo0A* gene and the *abrB* gene. AbrB is a repressor of the transcription of the *spo0H* gene coding for σ^H. Spo0A~P appears to play a role in the repression of the *kinA* gene.

Spo0A~P, acts through a positive feedback loop to stimulate its own synthesis by a promoter switching mechanism (Strauch et al., 1992b) (Fig. 3). Also, Spo0A~P may act as a repressor of both *spo0A* and *spo0F* transcription from the σ^H promoter (Strauch et al., 1992a). The advantage of this arrangement is the ability to rapidly produce both Spo0A and Spo0F and to make the cellular level of both subject to the Spo0A~P level. Expression of σ^H is crucial for this synthesis, and expression of σ^H is, itself, carefully regulated. Transcription of σ^H is controlled by AbrB (Weir et al., 1991) whose level is controlled by Spo0A~P (Strauch et al., 1990), and also, the translation of the σ^H messenger RNA is regulated by an as-yet-undescribed mechanism (Healy et al., 1991). Synthesis of σ^H allows transcription of *kinA*, *spo0F*, and *spo0A* (Predich et al., 1992).

Under conditions of vegetative growth when there is little need or desire to sporulate, the phosphorelay is present at a maintenance level dictated by the constitutive low level of transcription from the σ^A promoters. The spor-

```
                          "-35"                        "-10"
ATCCCTCTTCACTTCTCAGAATACATACGGTAAAATATAC
TAGGGAGAAGTGAAGAGTCTTATGTATGCCATTTTATATG
         ┌─►Pᵥ
AAAAGAAGATTTTTCGACAAATTCACGTTTCCTTGTTTGT
TTTTCTTCTAAAAGCTGTTTAAGTGCAAAGGAACAAACA

CAAATTTCATTTTTAGTCGAAAAACAGAGAAAAACATAGA
GTTTAAAGTAAAAATCAGCTTTTTGTCTCTTTTTGTATCT
                    SspI
ATAACAAAGATATGCCACTAATATTGGTGATTATGATTTT
TATTGTTTCTATACGGTGATTATAACCACTAATACTAAAA

                              ┌─►Pₛ
TTTAGAGGGTATATAGCGGTTTTGTCGAATGTAAACATGT
AAATCTCCCATATATCGCCAAACAGCTTACATTTGTACA
```

FIGURE 3 Promoter switching at Spo0A promoter. The locations of the vegetative promoter P_v and the sporulation promoter P_s transcribed by σ^A and σ^H, respectively, are shown. The boxed nucleotides correspond to the 0A boxes, and the lines demarking both strands are the extent of coverage of each strand by the footprint analysis of Spo0A on this promoter. Note that the Spo0A boxes and positions of Spo0A binding occur between the P_v and the P_s promoters. Spo0A~P is believed to bind preferentially at the top two 0A boxes, simultaneously accomplishing the repression of the P_v promoter and the activation of the P_s promoter. The Spo0A box located at about the −10 of the P_s promoter is believed to be a repression point where the P_s promoter is repressed when Spo0A~P concentrations reach sufficiently high levels.

ulation, σ^H, promoters do not function because σ^H synthesis is shut down and Spo0A~P levels are low. Once the cells reach stationary phase, Spo0A~P levels increase and σ^H is synthesized, allowing production of Spo0F, Spo0A, and KinA from σ^H promoters.

This promoter switching from σ^A to σ^H, at least at the *spo0A* promoter, is further regulated by catabolite repression (Chibazakura et al., 1991). Because the nature of catabolite repression of sporulation is unclear and complex in *Bacillus subtilis* and may result from many factors, its effects have been ignored in this chapter. This has been done to avoid confusion but with a certain level of trepidation because catabolite repression could be affecting regulatory processes in the initiation of sporulation.

Transcription regulation of the genes of the phosphorelay mainly occurs during late exponential growth, and by the beginning of stationary phase, the phosphorelay proteins are maximally expressed. With a full cellular complement of components, the activity or flow of phosphate through the phosphorelay is controlled by intrinsic and extrinsic factors acting at various points to modulate phosphate flow to Spo0A.

SIGNAL INPUT INTO THE PHOSPHORELAY

The two sensor kinases, KinA and KinB, provide signal input in the phosphorelay but function under different environmental and cultural conditions. KinB is expressed during exponential growth from a vegetative σ^A promoter, and its transcription is turned down as the culture approaches the stationary phase (Dartois and Hoch, unpublished data). KinB is likely to be responsible for the low level of sporulation observed during exponential growth. KinB mutants are delayed in transcription of *spoIIA* and *spoIIG* operons by about 1 h, suggesting that Spo0A~P accumulation at the end of exponential growth and the beginning of the stationary phase may be attributable to KinB activity (Dartois and Hoch, unpublished data). KinA appears to be a more powerful

kinase, exerting its influence after the onset of the stationary phase, consistent with its transcriptional pattern and the requirement for σ^H for its transcription (Antoniewski et al., 1990; Predich et al., 1992). Thus, there may be a progression of kinase activity from mainly KinB to mainly KinA.

KinB is a 47,774 M_r protein with a carboxyl domain highly homologous to that of KinA and a hydrophobic amino-terminal domain with six recognizable membrane spanning regions (Trach and Hoch, 1993). KinB is in an operon with another gene, *kapB*, coding for a 14,668 M_r protein. This presumably cytoplasmic protein is required for KinB activity in vivo and could be involved in either its expression or its enzymatic activity. It is assumed that KinB directly phosphorylates Spo0F, but this notion has not been proved by in vitro experiments.

KinA is a 69,170 M_r cytoplasmic protein. Purified KinA phosphorylates Spo0F and, to a slight extent, can directly phosphorylate Spo0A (Perego et al., 1989). However, KinA shows a higher affinity for Spo0F (K_m, 2.0 μM) than for Spo0A (K_m, 23 μM). Furthermore, the k_{cat} with Spo0F is 0.09 s^{-1}, whereas with Spo0A, it is 1.5×10^{-5} s^{-1} (Hoch, unpublished data). Thus, KinA shows a 50,000-fold preference for Spo0F over Spo0A. Nevertheless, if KinA is overproduced in a cell, it can suppress deletion mutations in both *spo0F* and *spo0B* genes, suggesting that KinA can directly phosphorylate Spo0A. Mutations in *spo0F* and *spo0B* genes produce completely sporulation-defective strains, indicating that both KinA and KinB use the phosphorelay and no kinase functions directly on Spo0A in wild-type cells.

What controls the activity of these kinases? That is the central question for which we do not know the answer. Several mutations in genes that appear to be involved in the activation of the *kinB* gene or its product have been isolated (Dartois and Hoch, unpublished data). One of these is defective in both sporulation and competence, but suppressors can be found that revert the sporulation phenotype alone.

This suggests there is some commonality in the signal transduction pathway for KinB, with that of the competence pathway. This conclusion is consistent with earlier results showing that double mutants with mutations in *kinA* and the *comA* gene coding for the response regulator that initiates competence lead to a partially sporulation-defective phenotype (Weinrauch et al., 1990). Mutations in other genes have been isolated that disturb the KinB pathway with no effect on competence. It seems likely that the activation of KinB may be subject to complex controls. Similar studies on KinA were not successful in identifying upstream genes, which may indicate there are no genes upstream, that the genes upstream are lethal if mutated, or that more than one pathway exists to activate KinA. Other scenarios are possible. The regulation of these kinases and the nature of their effector molecules remain a high priority for understanding how the environment influences sporulation.

ENZYMATIC ACTIVITIES OF PHOSPHORELAY PROTEINS

The KinA-Spo0F reaction has been studied in detail, and many of the kinetic parameters have been defined. It is clear that KinA has many properties in common with other sensory kinases, and some differences. The overall reaction from ATP to Spo0F~P has a k_{cat} of 5.4 min^{-1}, which compares favorably with the CheA catalyzed ATP to CheY~P reaction, k_{cat} of 2 min^{-1}. Similarly, the K_m for ATP (100 μM) and K_i for ADP (40 μM) resemble those for CheA, 370 μM and 70 μM, respectively. KinA is also an effective phosphatase of Spo0F~P in vitro. What remains a mystery is the nature of the molecular effector for KinA and whether it stimulates the phosphorylation reactions or activates the dephosphorylation of Spo0F~P by KinA.

The Spo0B phosphotransferase is unique to the phosphorelay. It uses Spo0F~P as a substrate that it dephosphorylates, producing a phosphoamidate enzyme-bound intermediate (Burbulys et al., 1991). It is likely that this is a histidine phosphate, but the actual phosphorylated residue has not been experimentally determined. Spo0A is the acceptor of this phosphate, generating an acyl phosphate. The phosphotransferase reaction is freely reversible, but the equilibrium lies toward Spo0A~P. Under conditions in which Spo0B~P has been preformed and purified, the phosphate group will distribute between equimolar concentrations of Spo0A, Spo0B, and Spo0F at a ratio of 60:30:10 of Spo0A~P/Spo0B~P/Spo0F~P. The phosphotransferase reaction is very fast, and this fact, along with the difficulty in isolating and quantitating phosphorylated Spo0F and Spo0A, has precluded a detailed kinetic study and determination of binding constants. It is of some interest how this enzyme can recognize two different response regulators that are not particularly closely related and maintain specificity for these two. Mutants deleted for Spo0F are completely sporulation-defective, indicating that Spo0B is not subject to promiscuous crosstalk from other phosphorylated response regulators.

Phosphorylation of Spo0A by Spo0B requires the D10, D11, and D56 aspartates of Spo0A. Conversion of D10 or D56 to asparagine results in a completely inactive protein, whereas a protein with the D11N mutation showed diminished activity in vitro. The same relative activity was found for direct phosphorylation of Spo0A by KinA (Burbulys et al., 1991). All three mutations gave rise to sporulation-defective strains, however (Trach et al., 1991).

The acyl phosphates formed on either Spo0F or Spo0A are very stable compared with CheY~P. Spo0F~P has a 40-h half-life at room temperature and greater than 100 h at 4°C (Zapf, unpublished data). These approximate the half-life of acetyl phosphate, which suggests that Spo0F is devoid of the autophosphatase activity postulated for CheY and other response regulators (Sanders et al., 1992). Spo0F is not phosphorylated by acetyl phosphate in vitro, suggesting it is not subject to the putative acetyl phosphate global control mechanism

(McCleary et al., 1993). The stability of Spo0F~P is consistent with its postulated function as an accumulation point for phosphate groups (Burbulys et al., 1991).

QUATERNARY STRUCTURE OF PHOSPHORELAY COMPONENTS

Structural studies of phosphorelay components have focused on Spo0F to obtain a second structure of a response regulator for comparative purposes. Studies of Spo0F by high-resolution multidimensional nuclear magnetic resonance elucidated the secondary structure and general fold of the protein and allowed a determination of the structural responses to magnesium binding. Also, ^{1}H, ^{15}N, and ^{13}C resonance assignments are nearly complete (Feher, unpublished data). Sequential and medium-range nuclear Overhauser enhancements (NOEs), assigned from analysis of a three-dimensional ^{15}N-NOESY-HSQC, showed that Spo0F is an α/β-protein consisting of alternating α-helices and β-strands, β_1-α_1-β_2-α_2-β_3-α_3-β_4-α_4-β_5-α_5. The core of the protein is defined by five β-strands arranged parallel to each other, resulting in an overall fold, β_2-β_1-β_3-β_4-β_5. This general fold and secondary structure was anticipated from analogy to other response regulators (A. M. Stock et al., 1989; Volz and Matsumura, 1991). However, by way of comparison with the published structure of a homologous response regulator, CheY, there are several subtle but interesting differences (Fig. 4). Helices α_4 and α_5 have different registers when the Spo0F sequence is aligned with CheY; α_4 is shifted three residues farther down the sequence, and helix α_5 is three residues shorter than the corresponding α_5 in CheY. Also, strand 5 in Spo0F has four nuclear magnetic resonance characteristics that are not consistent with a typical parallel β-strand. The sequential and across-strand NOEs are weaker than generally observed for β-strands (including those seen for other strands in Spo0F), the ^{3}JHNα coupling constants are less than 8 Hz, there is no protection of the backbone amide protons from deuterium exchange, and there

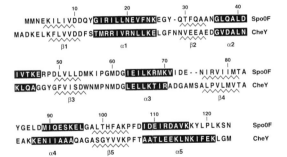

FIGURE 4 Comparison of structural features of Spo0F and CheY. The structural features from the crystal structure of CheY of *Salmonella typhimurium* (A. M. Stock et al., 1989) and the structure of the Spo0F as determined by multidimensional nuclear magnetic resonance analyses (Feher, unpublished data) are presented. Wavy lines indicate regions of β-sheet, and black boxes indicate areas of β-helix.

are strong α/α-NOEs between residues H101 and A103 of β_5 and M81 of β_4. These strong α/α-NOEs are not generally observed between parallel β-strands and suggest that there may be a "pucker" in this region of the β-sheet. In CheY, the regions of the loops between α_4-β_5 and β_5-α_5 have been shown to be important for interactions with the flagellar motor (Sockett et al., 1992; Roman et al., 1992). By analogy, the differences in this region between CheY and Spo0F may reflect the different specificities these response regulators have in their respective signaling pathways.

To characterize the structural consequences of the magnesium-bound form, chemical shift changes of backbone amide proton resonances were followed by ^{1}H-^{15}N HSQC spectra as magnesium concentration was increased. The responses to magnesium fell into three groups: those that did not shift, a group whose shifts reflected a K_d for magnesium of ~20 mM, and a third group whose shifts reflect a K_d for magnesium of 100 mM or greater. Some of the residues showing a 20 mM value for K_d, G36, M55, K56, I57, and M60, are near the conserved residues at the active site. Other residues in this second group are found in other regions of the protein. N20 and V22 are found in helix

α_1. T82, F102, and L119 are residues linked through the hydrogen-bonding network from the conserved site residues through the β-sheet to helix α_5. The residues showing a K_d value of greater than 100 mM are in the loop between β-strand 4 and helix 4. Only one residue reflects both the high- and low-affinity binding sites, E110. Because the low-affinity site would be saturated at magnesium concentrations exceeding 100 mM, it is likely this second site is not physiologically relevant in the resting state of the protein. It is possible, however, that this site becomes important in the phosphorylated form and could play a role in the phosphotransfer mechanism specific to Spo0F and Spo0B.

The Spo0A transcription factor consists of a single polypeptide of 29,691 M_r with two distinguishable domains. The amino-terminal ~125-amino-acid domain is homologous to response regulators. This domain is separated from ~125-amino-acid carboxyl domain by 27 residues of flexible linker. Trypsin cleaves the native protein once within this linker region, whereas both domains are tightly folded and resistant to trypsin attack (Grimsley et al., 1994). Each domain can be isolated free of the other by heparin-agarose chromatography. Footprint analysis of the AbrB promoter by Spo0A, or the isolated carboxyl domain, gives identical patterns showing that the specificity for 0A boxes in promoters is wholly contained within the carboxyl domain of Spo0A. Phosphorylation of Spo0A has been shown to increase its relative affinity for the AbrB promoter by at least 20-fold (Trach et al., 1991). Both the unphosphorylated and the phosphorylated forms of Spo0A are monomers (Grimsley et al., 1994). These results suggest that the amino domain serves to sterically interfere with the DNA binding site on the carboxyl domain, and phosphorylation relieves this inhibition by changing the spatial relationships of the amino and carboxyl domains. Unphosphorylated Spo0A binds specifically to 0A boxes in DNA at higher protein concentrations, indicating that even unphosphorylated protein can isom-

erize to a DNA binding form. Therefore, the primary effect of phosphorylation is to shift the equilibrium in this isomerization toward the DNA binding form.

These properties of Spo0A have regulatory significance, because they differ sharply from those of the NtrC transcription factor, which exists as a dimer and oligomerizes on its target promoters after phosphorylation (Porter et al., 1993). Transcription studies have revealed that phosphorylation has a second important role of promoting the isomerization of the Spo0A~P–RNA polymerase-promoter complex to a state capable of initiating transcription (see Chapter 10).

A study of the *spo0A* genes from several *Bacillus* and *Clostridium* species has revealed striking conservation of amino acid sequence in its homologs (Brown et al., 1994). The carboxyl domain of the proteins is especially well conserved with less conservation in the response regulator domain. The least conserved portion of the molecule is the hinge region, which probably functions only as a flexible linker between the two domains. Interestingly, efforts to find a Spo0A homolog in closely related but non-spore-forming relatives such as *Listeria* or *Staphylococcus* were unsuccessful, suggesting that Spo0A is associated only with endospore formation.

Much less has been done on the structure of the other components of the phosphorelay. Limited gel filtration studies of purified KinA and cross-linking experiments with dimethyl suberimidate show that KinA and KinA~P are both larger than monomers under normal assay conditions (0.1 μM to 1.0 μM KinA). Spo0B was found as a dimer in gel filtration and cross-linking studies.

MODULATION OF PHOSPHATE FLOW BY PHOSPHATASES

One rationale for the complexity of the phosphorelay is the need to provide sites for regulatory action by the variety of effectors that have a potential role in controlling sporulation (Burbulys et al., 1991). Although it is conceivable that the activity of KinA (and therefore all sporulation) might be controlled by the cellular level of an effector molecule whose concentration could be subject to many different regulatory processes, it seems as likely that each component of the phosphorelay could provide a point of entry for regulatory action by various environmental, metabolic, or cellular effectors and processes. One might expect that regulatory molecules involved in preventing phosphate flow in the phosphorelay would be identified genetically as negative regulators. Indeed, one gene identified as such was *spo0E*.

The *spo0E* gene codes for a small protein, 9791 M_r, whose synthesis is controlled by the AbrB transition state regulator (Perego and Hoch, 1987, 1991). Overproduction of Spo0E protein inhibits sporulation, whereas deletion of its gene has no sporulation phenotype. However, deleted mutants segregate sporulation-defective mutants with secondary mutations in genes for components of the phosphorelay. The interpretation of this behavior is that deletion of the *spo0E* gene removes a crucial controlling element, thereby increasing the pressure to sporulate under inappropriate conditions, that selects for secondary mutants unable to sporulate and insensitive to the pressure. This suggested that Spo0E regulates intracellular Spo0A~P concentrations, perhaps by inhibiting or modulating the activity of one of the phosphorelay components. To understand the function of Spo0E, the protein was purified and assayed for its activity on the phosphorelay (Ohlsen et al., 1994). These studies showed that Spo0E was an active phosphatase of the Spo0A~P transcription factor. Spo0E was inactive as a phosphatase on Spo0F~P or on the other phosphorylated components of the phosphorelay. The original Spo0E mutants were known to be chain termination mutations in the carboxyl one-third of the protein. One of these truncated proteins was expressed and found to be a hyperactive phosphatase. It is believed that the carboxyl portion of the Spo0E protein forms a binding site for an

effector molecule, which normally controls the phosphatase activity of Spo0E. Nothing is known of the nature of this putative effector. Spo0E shows no amino acid sequence homology to the CheZ phosphatase of the chemotaxis system (Mutoh and Simon, 1986).

Recently, a new family of phosphatases has been uncovered through the study of an obscure stage 0 mutant, *spo0L* (Perego et al., 1994). Cloning of this locus revealed that it was identical to the previously described *gsiA* locus, which was discovered as a glucose-starvation-inducible gene, dependent for expression on the ComP-ComA two-component system (Mueller and Sonenshein, 1992; Mueller et al., 1992). The phenotype of Spo0L was similar to Spo0E in that overproduction of the Spo0L-GsiA protein prevented sporulation, and deletions of the gene gave sporulation-proficient strains that segregated sporulation-defective secondary mutants. A second related gene, *spo0P*, whose deduced protein is 60% identical to *spo0L*, was uncovered by an independent sequencing study, and the phenotypes of overproduction or deletion of this gene were similar to those of *spo0L*. Purification of both of these proteins allowed a study of their effects on the phosphorelay in which it was discovered that they were phosphatases of Spo0F~P with no activity on Spo0A~P or the other components of the phosphorelay. That these proteins appear to specifically use Spo0F~P as a substrate is consistent with the observation that a suppressor mutant whose sporulation was resistant to a *spo0L* mutation and also to overproduction of Spo0P was found to be a mutation in Spo0F, changing tyrosine residue 13 to a serine. This mutation confers resistance to the phosphatase activity of Spo0L and Spo0P but seems not to influence the activity of the kinases for Spo0F, because back-crosses of these mutants to the wild type did not visibly affect sporulation. Interestingly, the Y13S mutation is close to the aspartate triad active site of this protein.

The original *spo0L* mutations were missense mutations, and purification of one of these mutant proteins revealed a normally active phosphatase. One possibility for the phenotype of SpoOL missense mutations is insensitivity to an effector molecule that may control phosphatase activity in vivo.

The unanswered question in both the case of SpoOE and the SpoOL-SpoOP families is whether these proteins are phosphatases in an enzymatic sense or whether they just bind specifically to either SpoOA or SpoOF response regulator domains and activate an autophosphatase activity. However, there is no evidence for an autophosphatase activity in either SpoOA or SpoOF, because their phosphorylated forms are exceedingly stable in solution.

KINASE-PHOSPHATASE ANTAGONISM

The SpoOL-SpoOP family of proteins are now known as response regulator aspartate phosphatases (or Rap proteins) (Perego et al., 1994). Several more genes in this family have been discovered on the *B. subtilis* chromosome and on plasmids (Perego and Hoch, unpublished data).

The original sequencing studies of the *gsiA* locus (now *rapA*) revealed the gene for a small 44-amino-acid protein in addition to the open major reading frame (Mueller and Sonenshein, 1992). It has now been determined that this small protein controls the activity of RapA and may be an effector recognition subunit (Perego, unpublished data). Not all the members of this family, especially *spoOP* (*rapB*), have a linked gene for a small protein. This does not preclude the existence of an effector molecule, however. The phosphatases of this family are likely to be controlled first by transcriptional activation of their expression and second by molecular effectors regulating their enzymatic activity.

Transcription of the *rapA* gene is regulated by the ComP-ComA two-component regulatory system (Mueller et al., 1992). This system controls the onset of the competent state in *B. subtilis,* which is a semiquiescent stationary-phase physiological state characterized by the activation of a DNA uptake and recombination system (Dubnau, 1991). The ComP-ComA system is induced by a peptide secreted into the culture that may act as a cell density signal (Magnuson et al., 1994). It appears to be important to the competent state to prevent sporulation, and this is accomplished, at least in part, by induction of the *rapA* gene coding for a phosphatase to dephosphorylate SpoOF~P. Thus, even if the kinases are activated to produce SpoOF~P, the phosphate group is prevented from activating SpoOA. Because RapA is likely to be regulated by an effector molecule, it could act to negatively regulate RapA activity, allowing other physiological states or environmental signals to counteract the ComP-ComA system. In fact, mutants in the small protein of the *rapA* locus coding for a putative regulatory subunit are Spo⁻, suggesting this subunit allows a negative regulator to control phosphatase activity (Perego, unpublished data). Transcription of *rapB* is not under ComP-ComA control. Thus, it seems likely that these phosphatases act as sensors for different physiological states and send information of their status to the phosphorelay.

The reactions affecting interconversions of SpoOF~P are shown in Fig. 5. SpoOF~P is depicted as a metastable intermediate that can only transfer its phosphate to SpoOA if the signals activating or inhibiting the phosphatases, RapA and RapB, permit it to accumulate to a level where it can be the substrate for SpoOB. It is unknown how many Rap phosphatases exist for SpoOF~P, but five more members of this family have been found by sequencing studies (Perego, unpublished data). Thus, SpoOF~P may be exposed to a gauntlet of phosphatases, each responsive to different physiological conditions. If SpoOF~P escapes these enzymes, its phosphate can be used to activate SpoOA (Fig. 6). This generates SpoOA~P, which is the substrate for the SpoOE phosphatase. This phosphatase is postulated to have an intracellular effector molecule that may modulate its activity. No other members of the SpoOE family of phosphatases have been found as yet.

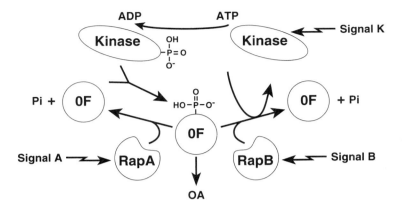

FIGURE 5 Kinase and phosphatase reactions on Spo0F. The kinases for sporulation are depicted as being activated by signal K, which either activates the autophosphorylation reaction of these kinases or converts the kinase to a phosphatase of Spo0F~P. The actual mechanism of signal interpretation for either KinA or KinB on Spo0F has not been determined. Two phosphatases, RapA and RapB, are known to respond to different transcriptional activation signals, and presumably different effector molecules control their activity on Spo0F~P. Sufficient Spo0F~P for the production of Spo0A~P only occurs in the absence of the activity of these (and perhaps other) phosphatases.

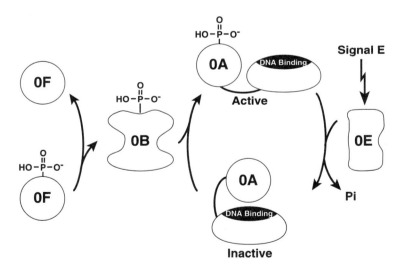

FIGURE 6 Conversion of Spo0F~P to Spo0A~P. Spo0F~P is a substrate for the Spo0B phosphotransferase, which transfers the phosphate group to Spo0A. Spo0A is depicted as inactive in the unphosphorylated state, and this probably occurs because the end-terminal 0A domain prevents access of the DNA binding domain to DNA. Phosphorylation of Spo0A causes a conformational change that exposes the DNA binding site, allowing this transcription factor to bind to 0A boxes in the promoters that it controls. The Spo0E protein is a phosphatase of activated Spo0A~P and is thought to respond to some internal signal, signal E, of unknown nature.

The aspartate phosphatases might be regarded as the procaryotic equivalent of the tyrosine phosphatases of eucaryotes that regulate the state of phosphorylation of proteins critical for cell growth, proliferation, and differentiation (Charbonneau and Tonks, 1992; Walton and Dixon, 1993). Such tyrosine phosphatases are found in many forms, with or without regulatory domains in addition to the catalytic domain. These accessory domains are required for targeting the phosphatase activity to a specific substrate (Hubbard and Cohen, 1993; Mauro and Dixon, 1994). Tyrosine phosphatases are in general rather substrate-nonspecific, which contrasts sharply with the Rap family of aspartate phosphatases or even the Spo0E aspartate phosphatase, both of which seem to be exceedingly specific in the response regulator they will dephosphorylate. If the phosphatase targeting subunit, zip codes, and so on found in tyrosine phosphatases are required for targeting this relatively nonspecific tyrosine phosphatase activity to a specific substrate, then the Rap family with its built-in specificity may not need such extra domains. Also, procaryotes have the advantage of facile transcriptional control, which, along with the possibility of effector molecule control, may provide exquisite regulation of the phosphatase activity. It seems likely that the competition between the opposing activities of the kinases, on one hand, and the phosphatase, on the other, may serve to fine tune the phosphorylation level of Spo0A and provide a basis for the integration of diverse signals that could effect the initiation of sporulation.

This class of phosphatases has implications for two-component mediated signal transduction in general. Kinase-phosphatase antagonism is the basis for specific control of the level of response regulator phosphorylation in many two-component systems, and both activities are inherent in the kinase molecule. Separate phosphatase proteins may only be necessary when two-component signal transduction systems are sensitive to multiple signal input, such as in sporulation, which may exhaust the signal recognition capacity of the kinase.

REPRESSOR-ACTIVATOR ANTAGONISM

The initiation of sporulation depends on mechanisms for recognition of both positive and negative signals, which are integrated by the phosphorelay into the cellular level of Spo0A~P. The activating effects of Spo0A~P on transcription have to counteract several negative regulators acting to prevent sporulation. Many potential negative regulators of sporulation have been described over the years, but in most cases, the mechanism of their action remains murky (Smith, 1993). Two of these, SinR and Hpr (ScoC), however, are better understood.

SinR is a very interesting and important DNA binding protein, with profound effects on sporulation (Gaur et al., 1986; Gaur et al., 1988). SinR inhibits transcription of the protease genes, aprE and nprE, and the sporulation genes, spoIIA, spoIIE, and spoIIG. In the case of aprE, direct binding of Sin to the promoter has been observed (Gaur et al., 1991; Kallio et al., 1991), whereas Sin binds to but will not footprint on the promoters of the spoII genes (Mandic-Mulec et al., 1992). Regardless, overproduction of SinR represses sporulation. The effects of SinR are neutralized by the SinI protein, which binds to it and prevents it from acting as a repressor. During vegetative growth, the sinI gene is repressed by the Hpr and AbrB regulators, allowing SinR to function. In the early stationary phase, the repressive effects of Hpr on the sinI promoter are relieved and transcription of sinI is activated by Spo0A~P (Strauch, personal communication). Presumably, Hpr has to be neutralized before Spo0A~P may activate sinI transcription (Fig. 7).

Hpr is, itself, a negative regulator, controlling both sporulation and protease production (Perego and Hoch, 1988). It is a DNA binding protein that binds directly to the aprE and nprE promoters, as well as to the sinI promoter (Kallio et al., 1991). Deletions of the gene for the regulator results in catabolite-resistant sporulation and the propensity to sporulate under vegetative growth conditions. Whether

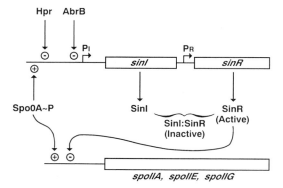

FIGURE 7 Relationship between SinR and Spo0A~P. SinR is a repressor of sporulation capable of binding to the promoters of several protease genes, as well as the promoters for the stage II genes, *spoIIA, spoIIE,* and *spoIIG*. SinI is an inhibitor of the repressive functions of SinR, and this protein is produced at the end of exponential growth to relieve repression of sporulation. Transcription of SinI is activated by the accumulation of Spo0A~P and inhibited by both Hpr and AbrB transition state regulators. Spo0A~P is also a required activator for the transcription of the *spoII* genes.

these sporulation phenotypes are the result of SinI overproduction, resulting in neutralization of *sinR* during vegetative growth as a consequence of the inability to prevent *sinI* transcription, remains to be determined. Experiments need to be performed to ascertain if Hpr's repressive effects are also felt on other sporulation promoters. In the context of this chapter, it is important to remember that cellular Spo0A~P levels are a determining factor in the neutralization of the effects of these regulators.

SinR and Spo0A~P may compete on another level in the transcription of the *spoIIA, spoIIE,* and *spoIIG* genes (Mandic-Mulec et al., 1992). Both SinR and Spo0A~P bind to these promoters but with opposite effects. The repressive effects of SinR binding may be nullified by Spo0A~P competing for similar or overlapping sites on these promoters. Deletion of the *sinR* gene in vivo results in premature and high-level expression of these promoters, suggesting that SinR is, indeed, a repressor whose removal allows lower levels of Spo0A~P to activate transcription of *spoII* genes.

In addition to repression by SinR, the stage II promoters have other barriers to activation during vegetative growth. The *spoIIA* gene uses a σ^H promoter that may be activated only by two stationary-phase regulators, Spo0A~P and σ^H. The *spoIIE* and *spoIIG* promoters, however, use the vegetative σ^A for transcription. Because Spo0A can isomerize to a DNA binding form without phosphorylation, it seems possible that these promoters could be susceptible to vegetative activation. To prevent this from occurring, the promoters have abnormal spacing between the −10 and −35 regions (Kenney et al., 1988; York et al., 1992). Conventional −10 and −35 regions on σ^A promoters are spaced 17 bp apart to place both regions on the same face of the DNA helix. In the *spoIIE* and *spoIIG* promoters, this distance is increased by a few nucleotides to place the −10 and −35 on opposite sides of the DNA helix. Thus, RNA polymerase cannot bind at both regions simultaneously, so the possibility of inadvertent transcription is very low. The binding of both Spo0A and RNA polymerase to such promoters twists the DNA helix, allowing RNA polymerase to contact both regions (see Chapter 10). However, unless Spo0A is phosphorylated, a transcript does not result from this union. This indicates that although the amino domain of Spo0A can isomerize to a form that allows the carboxyl domain to bind to DNA, the unphosphorylated amino domain may be inhibitory to isomerization of the complex to an open productive form. Therefore, these σ^A promoters require both Spo0A accumulation and phosphorylation to function. It seems very important to the cell to prevent spurious activation of stage II promoters.

NATURE OF SIGNALS

In the preceding sections, the pathways for phosphorylation and dephosphorylation of Spo0A were elucidated. It is an article of faith that the kinases and phosphatases are subject to control by effector molecules that act as signals to either increase or depress the probability of sporulation. Unfortunately, the identification of such signaling molecules has seriously lagged behind the characterization of their potential sites of action.

Mutations in the genes for the oligopeptide transport system, *opp,* will render strains sporulation-deficient. Thus, it is believed that a peptide molecule must be transported by the *opp* system to activate some step of the phosphorelay (Rudner et al., 1991; Perego et al., 1991). This peptide may serve as a cell density signal (Grossman and Losick, 1988). Both competence and sporulation are affected by the defects in the *opp* system, but it is clear that the peptides for each differ. Competence is induced by a peptide that comprises the carboxyl amino acids of the *comX* gene (Magnuson et al., 1994). Presumably, the *comX* protein is exported, and its degradation induces competence. The sporulation peptide, however, has not been identified, and its site of action is still unknown.

Sporulation must be coordinated with DNA synthesis, septation, and the cell cycle via one or more signaling pathways (Mandelstam and Higgs, 1974). There is some evidence that the state of DNA synthesis and repair may affect the phosphorelay at the level of Spo0F or Spo0B (Ireton et al., 1993; Ireton and Grossman, 1994). Perhaps this is correlated to the induction of a phosphatase for Spo0F~P. The initiation of DNA synthesis may also be coupled to the activity of the Obg protein, an essential GTP binding protein (Welsh et al., 1994). *obg* is circumstantially connected with *spo0B,* because both of these genes comprise an operon (Trach and Hoch, 1989). The connection between the phosphorelay and the cell cycle is unclear, but it is likely to be more complex than we can imagine at this point.

There is a long-standing observation that lower GTP levels correlate with the onset of sporulation, but the mechanism by which GTP affects sporulation initiation has not been determined (Lopez et al., 1981). There are many other observations of events or processes that occur simultaneously with the onset of sporulation, but a causal relationship is difficult to establish. We now have many potential mechanisms for negative regulation of sporulation through the phosphatases, and some of these observations may soon be understood.

OVERVIEW

The initiation of sporulation has adopted the two-component signal transduction system and made some unique modifications to adapt it to process multiple signal inputs. We are now beginning to understand the proteins and enzymes through which these signals are transmitted, but we have no clue as to the nature and identity of the signals. Clearly, two kinase pathways for the phosphorylation of Spo0F exist, but the molecular effectors for each are obscure and the mechanisms of their actions are unknown. This situation is not unique to sporulation, as the identity of kinase activators for most of the known two-component systems has eluded discovery.

Spo0F was originally postulated to be a second messenger capable of gathering phosphate groups from several kinases and serving as an accumulation point for these groups. The discovery of a family of phosphatases unique for Spo0F~P and controlled by different physiological states suggests that the Spo0F level is continually adjustable and reflective of the sum of the positive and negative signals received from the environment. How many of these phosphatases exist in a cell and the variety of signals controlling either their transcription or their activity are only subjects for speculation at this time. Although the main role for Spo0F~P in sporulation is to serve as a substrate for Spo0A~P production, there is no biochemical or physiological reason why Spo0F~P might not have an independent signaling role in some stationary-phase process. If such a role exists, it is not essential for sporulation.

Spo0A~P is produced by the Spo0B phosphotransferase, an enzyme unique to the phosphorelay. The enzymatic activity is bidirectional with the equilibrium toward Spo0A~P. Does the cell take advantage of this property to allow phosphatases to lower Spo0A~P levels by dephosphorylating Spo0F~P? It certainly happens this way in the reconstituted phosphorelay system in vitro (Perego et al., 1994). Do other proteins or effectors exist that control its activity and its directionality or that short-circuit the Spo0B phosphotransferase, i.e., is Spo0B another point of control on the phosphorelay as originally envisioned?

Spo0A~P production is the goal and raison d'être for the phosphorelay, and its cellular level has assumed a dominant role in the initiation of sporulation. It is a powerful transcription factor, activating at least four sporulation operons and repressing several others. Its influence is magnified by its regulation of other regulators that control a variety of genes (Strauch and Hoch, 1992). Still, it is subject to inactivation by the Spo0E phosphatase and could be dephosphorylated by other phosphatases as yet undiscovered. Some of these phosphatases could provide access for a myriad of environmental conditions, cellular processes, cell cycle constraints, and other conditions that could influence the decision to grow or sporulate. The multiplicity of components of the phosphorelay may have been designed to provide this access. As more and more controls are discovered that act on the phosphorelay, it becomes even more amazing that sporulation occurs at all.

ACKNOWLEDGMENTS

This chapter would never have been written without the cooperation of the people in my laboratory and my many colleagues at The Scripps Research Institute and the University of California, San Diego. I am indebted to Vickie Feher for the nuclear magnetic resonance data and writeup.

Our work is supported in part by grants GM19416 and GM45727 from the National Institute of General Medical Science, National Institutes of Health, United States Public Health Service.

REFERENCES

Antoniewski, C., B. Savelli, and P. Stragier. 1990. The *spoIIJ* gene, which regulates early developmental steps in *Bacillus subtilis*, belongs to a class of environmentally responsive genes. *J. Bacteriol.* **172:** 86–93.

Bai, U., M. Lewandoski, E. Dubnau, and I. Smith. 1990. Temporal regulation of the *Bacillus subtilis* early sporulation gene *spoOF. J. Bacteriol.* **172:**5432–5439.

Brown, D. P., L. Ganova-Raeva, B. D. Green, S. R. Wilkinson, M. Young, and P. Youngman. 1994. Characterization of *spoOA* homologues in diverse *Bacillus* and *Clostridium* species identifies a probable DNA-binding domain. *Mol. Microbiol.* **14:**411–426.

Burbulys, D., K. A. Trach, and J. A. Hoch. 1991. The initiation of sporulation in *Bacillus subtilis* is controlled by a multicomponent phosphorelay. *Cell* **64:**545–552.

Charbonneau, H., and N. K. Tonks. 1992. 1002 protein phosphatases? *Annu. Rev. Cell Biol.* **8:**463–493.

Chibazakura, T., F. Kawamura, and H. Takahashi. 1991. Differential regulation of *spoOA* transcription in *Bacillus subtilis*: glucose represses promoter switching at the initiation of sporulation. *J. Bacteriol.* **173:**2625–2632.

Dartois, V., and J. A. Hoch. Unpublished data.

Dubnau, D. 1991. The regulation of genetic competence in *Bacillus subtilis. Mol. Microbiol.* **5:**11–18.

Feher, V. Unpublished data.

Gaur, N. K., K. Cabane, and I. Smith. 1988. Structure and expression of the *Bacillus subtilis sin* operon. *J. Bacteriol.* **170:**1046–1053.

Gaur, N. K., E. Dubnau, and I. Smith. 1986. Characterization of a cloned *Bacillus subtilis* gene that inhibits sporulation in multiple copies. *J. Bacteriol.* **168:**860–869.

Gaur, N. K., J. Oppenheim, and I. Smith. 1991. The *Bacillus subtilis sin* gene, a regulator of alternate developmental processes, codes for a DNA-binding protein. *J. Bacteriol.* **173:**678–686.

Grimsley, J. K., R. B. Tjalkens, M. A. Strauch, T. H. Bird, G. B. Spiegelman, Z. Hostomsky, J. M. Whiteley, and J. A. Hoch. 1994. Subunit composition and domain structure of the Spo0A sporulation transcription factor of *Bacillus subtilis. J. Biol. Chem.* **269:**16977–16982.

Grossman, A. D., and R. Losick. 1988. Extracellular control of spore formation in *Bacillus subtilis. Proc. Natl. Acad. Sci. USA* **85:**4369–4373.

Healy, J., J. Weir, I. Smith, and R. Losick. 1991. Post-transcriptional control of a sporulation regulatory gene encoding transcription factor sigma H in *Bacillus subtilis. Mol. Microbiol.* **5:**477–487.

Hoch, J. A. Unpublished data.

Hubbard, M. J., and P. Cohen. 1993. On target with a new mechanism for the regulation of protein phosphorylation. *Trends Biochem. Sci.* **18:**172–177.

Ireton, K., and A. D. Grossman. 1994. A developmental checkpoint couples the initiation of sporulation to DNA replication in *Bacillus subtilis. EMBO J.* **13:**1566–1573.

Ireton, K., D. Z. Rudner, K. J. Siranosian, and A. D. Grossman. 1993. Integration of multiple developmental signals in *Bacillus subtilis* through the Spo0A transcription factor. *Genes Dev.* **7**:283–294.

Kallio, P. T., J. E. Fagelson, J. A. Hoch, and M. A. Strauch. 1991. The transition state regulator Hpr of *Bacillus subtilis* is a DNA-binding protein. *J. Biol. Chem.* **266**:13411–13417.

Kenney, T. J., P. A. Kirchman, and C. P. J. Moran. 1988. Gene encoding sigma E is transcribed from a sigma A-like promoter in *Bacillus subtilis*. *J. Bacteriol.* **170**:3058–3064.

Lewandoski, M., E. Dubnau, and I. Smith. 1986. Transcriptional regulation of the *spo0F* gene of *Bacillus subtilis*. *J. Bacteriol.* **168**:870–877.

Lopez, J. M., A. Dromerick, and E. Freese. 1981. Response of guanosine 5'-triphosphate concentration to nutritional changes and its significance for *Bacillus subtilis* sporulation. *J. Bacteriol.* **146**:605–613.

Magnuson, R., J. Solomon, and A. D. Grossman. 1994. Biochemical and genetic characterization of a competence pheromone from *B. subtilis*. *Cell* **77**:207–216.

Mandelstam, J., and S. A. Higgs. 1974. Induction of sporulation during synchronized chromosome replication in *Bacillus subtilis*. *J. Bacteriol.* **120**:38–42.

Mandic-Mulec, I., N. Gaur, U. Bai, and I. Smith. 1992. Sin, a stage-specific repressor of cellular differentiation. *J. Bacteriol.* **174**:3561–3569.

Mauro, L. J., and J. E. Dixon. 1994. "Zip codes" direct intracellular protein tyrosine phosphatases to the correct cellular "address." *Trends Biochem. Sci.* **19**:151–155.

McCleary, W. R., J. B. Stock, and A. J. Ninfa. 1993. Is acetyl phosphate a global signal in *Escherichia coli*? *J. Bacteriol.* **175**:2793–2798.

Mueller, J. P., G. Bukusoglu, and A. L. Sonenshein. 1992. Transcriptional regulation of *Bacillus subtilis* glucose starvation inducible genes: control of *gsiA* by the Comp-ComA signal transduction system. *J. Bacteriol.* **174**:4361–4373.

Mueller, J. P., and A. L. Sonenshein. 1992. Role of the *Bacillus subtilis gsiA* gene in regulation of early sporulation gene expression. *J. Bacteriol.* **174**:4374–4383.

Mutoh, N., and M. I. Simon. 1986. Nucleotide sequence corresponding to five chemotaxis genes in *Escherichia coli*. *J. Bacteriol.* **165**:161–166.

Ohlsen, K. L., J. K. Grimsley, and J. A. Hoch. 1994. Deactivation of the sporulation transcription factor Spo0A by the Spo0E protein phosphatase. *Proc. Natl. Acad. Sci. USA* **91**:1756–1760.

Perego, M. Unpublished data.

Perego, M., S. P. Cole, D. Burbulys, K. Trach, and J. A. Hoch. 1989. Characterization of the gene for a protein kinase which phosphorylates the sporulation-regulatory proteins Spo0A and Spo0F of *Bacillus subtilis*. *J. Bacteriol.* **171**:6187–6196.

Perego, M., C. Hanstein, K. M. Welsh, T. Djavakhishvili, P. Glaser, and J. A. Hoch. 1994. Multiple protein-aspartate phosphatases provide a mechanism for the integration of diverse signals in the control of development in *B. subtilis*. *Cell* **79**:1047–1055.

Perego, M., C. F. Higgins, S. R. Pearce, M. P. Gallagher, and J. A. Hoch. 1991. The oligopeptide transport system of *Bacillus subtilis* plays a role in the initiation of sporulation. *Mol. Microbiol.* **5**:173–185.

Perego, M., and J. A. Hoch. 1987. Isolation and sequence of the *spo0E* gene: its role in initiation of sporulation in *Bacillus subtilis*. *Mol. Microbiol.* **1**:125–132.

Perego, M., and J. A. Hoch. 1988. Sequence analysis and regulation of the *hpr* locus, a regulatory gene for protease production and sporulation in *Bacillus subtilis*. *J. Bacteriol.* **170**:2560–2567.

Perego, M., and J. A. Hoch. 1991. Negative regulation of *Bacillus subtilis* sporulation by the *spo0E* gene product. *J. Bacteriol.* **173**:2514–2520.

Perego, M., and J. A. Hoch. Unpublished data.

Porter, S. C., A. K. North, A. B. Wedel, and S. Kustu. 1993. Oligomerization of NTRC at the *glnA* enhancer is required for transcriptional activation. *Genes Dev.* **7**:2258–2273.

Predich, M., G. Nair, and I. Smith. 1992. *Bacillus subtilis* early sporulation genes *kinA, spo0F,* and *spo0A* are transcribed by the RNA polymerase containing σ^H. *J. Bacteriol.* **174**:2771–2778.

Roman, S. J., M. Meyers, K. Volz, and P. Matsumura. 1992. A chemotactic signaling surface on CheY defined by suppressors of flagellar switch mutations. *J. Bacteriol.* **174**:6247–6255.

Rudner, D. Z., J. R. Ladeaux, K. Breton, and A. D. Grossman. 1991. The *spo0K* locus of *Bacillus subtilis* is homologous to the oligopeptide permease locus and is required for sporulation and competence. *J. Bacteriol.* **173**:1388–1398.

Sanders, D. A., B. L. Gillece-Castro, A. L. Burlingame, and D. E. Koshland, Jr. 1992. Phosphorylation site of NtrC, a protein phosphatase whose covalent intermediate activates transcription. *J. Bacteriol.* **174**:5117–5122.

Satola, S. W., J. M. Baldus, and C. P. Moran, Jr. 1992. Binding of Spo0A stimulates *spoIIG* promoter activity in *Bacillus subtilis*. *J. Bacteriol.* **174**:1448–1453.

Smith, I. 1993. Regulatory proteins that control late growth development, p. 785–800. *In* A. L. Sonenshein, J. A. Hoch, and R. Losick (ed.), *Bacillus subtilis and Other Gram-Positive Bacteria: Biochemistry, Physiology and Molecular Genetics*. American Society for Microbiology, Washington, D.C.

Sockett, H., S. Yamaguchi, M. Kihara, V. M. Irikura, and R. M. Macnab. 1992. Molecular analysis of the flagellar switch protein FliM of *Salmonella typhimurium. J. Bacteriol.* **174:**793–806.

Stock, A. M., J. M. Mottonen, J. B. Stock, and C. E. Schutt. 1989. Three-dimensional structure of CheY, the response regulator of bacterial chemotaxis. *Nature* (London) **337:**745–749.

Stock, J. B., A. J. Ninfa, and A. M. Stock. 1989. Protein phosphorylation and regulation of adaptive response in bacteria. *Microbiol. Rev.* **53:**450–490.

Strauch, M. Personal communication.

Strauch, M., V. Webb, G. Spiegelman, and J. A. Hoch. 1990. The Spo0A protein of *Bacillus subtilis* is a repressor of the *abrB* gene. *Proc. Natl. Acad. Sci. USA* **87:**1801–1805.

Strauch, M. A., and J. A. Hoch. 1992. Transition state regulators: sentinels of *Bacillus subtilis* post-exponential gene expression. *Mol. Microbiol.* **7:**337–342.

Strauch, M. A., K. A. Trach, J. Day, and J. A. Hoch. 1992a. Spo0A activates and represses its own synthesis by binding at its dual promoters. *Biochimie* **74:**619–626.

Strauch, M. A., J.-J. Wu, R. H. Jonas, and J. A. Hoch. 1992b. A positive feedback loop controls transcription of the *spoOF* gene, a component of the sporulation phosphorelay in *Bacillus subtilis. Mol. Microbiol.* **7:**967–974.

Trach, K., D. Burbulys, M. Strauch, J.-J. Wu, N. Dhillon, R. Jonas, C. Hanstein, P. Kallio, M. Perego, T. Bird, G. Spiegelman, C. Fogher, and J. A. Hoch. 1991. Control of the initiation of sporulation in *Bacillus subtilis* by a phosphorelay. *Res. Microbiol.* **142:**815–823.

Trach, K., J. W. Chapman, P. Piggot, D. LeCoq, and J. A. Hoch. 1988. Complete sequence and transcriptional analysis of the *spoOF* region of the *Bacillus subtilis* chromosome. *J. Bacteriol.* **170:**4194–4208.

Trach, K., and J. A. Hoch. 1989. The *Bacillus subtilis spoOB* stage 0 sporulation operon encodes an essential GTP-binding protein. *J. Bacteriol.* **171:**1362–1371.

Trach, K. A., and J. A. Hoch. 1993. Multisensory activation of the phosphorelay initiating sporulation in *Bacillus subtilis:* identification and sequence of the protein kinase of the alternate pathway. *Mol. Microbiol.* **8:**69–79.

Volz, K., and P. Matsumura. 1991. Crystal structure of *Escherichia coli* CheY refined at 1.7-Å resolution. *J. Biol. Chem.* **266:**15511–15519.

Walton, K. M., and J. E. Dixon. 1993. Protein tyrosine phosphatases. *Annu. Rev. Biochem.* **62:**101–120.

Weinrauch, Y., R. Penchev, E. Dubnau, I. Smith, and D. Dubnau. 1990. A *Bacillus subtilis* regulatory gene product for genetic competence and sporulation resembles sensor protein members of the bacterial two-component signal-transduction systems. *Genes Dev.* **4:**860–872.

Weir, J., M. Predich, E. Dubnau, G. Nair, and I. Smith. 1991. Regulation of *spoOH*, a gene coding for the *Bacillus subtilis* σ^H factor. *J. Bacteriol.* **173:**521–529.

Welsh, K. M., K. A. Trach, C. Folger, and J. A. Hoch. 1994. Biochemical characterization of the essential GTP-binding protein Obg of *Bacillus subtilis. J. Bacteriol.* **176:**7161–7168.

Yamashita, S., F. Kawamura, H. Yoshikawa, H. Takahashi, Y. Kobayashi, and H. Saito. 1989. Dissection of the expression signals of the *spoOA* gene of *Bacillus subtilis:* glucose represses sporulation-specific expression. *J. Gen. Microbiol.* **135:**1335–1345.

York, K., T. J. Kenney, S. Satola, C. P. Moran, Jr., H. Poth, and P. Youngman. 1992. Spo0A controls the σ^A-dependent activation of *Bacillus subtilis* sporulation-specific transcription unit *spoIIE. J. Bacteriol.* **174:**2648–2658.

Zapf, J. Unpublished data.

RESPONSE
REGULATOR
FUNCTIONS

Mechanism of Transcriptional Activation by NtrC

Susan C. Porter, Anne K. North, and Sydney Kustu

9

The nitrogen regulatory protein C (NtrC) of enteric bacteria activates transcription by σ^{54}-holoenzyme, an alternative holoenzyme form of RNA polymerase. To do so, NtrC (also called NRI) catalyzes the isomerization of closed complexes between σ^{54}-holoenzyme and a promoter to transcriptionally productive open complexes, a reaction that depends on hydrolysis of ATP. NtrC is a receiver protein of a two-component regulatory system. To catalyze ATP hydrolysis and activate transcription, NtrC must be phosphorylated on an aspartate residue in its amino (N)-terminal receiver domain. Also, activation requires that phosphorylated NtrC dimers form tetramers or higher-order oligomers. Because phosphorylation increases the ability of NtrC dimers to oligomerize, its effects on the activity of NtrC may simply be mediated by effects on oligomerization.

Recent studies indicate that oligomerization determinants of NtrC are located in its central activation domain rather than its N-terminal receiver domain, and hence the effects of the phosphorylated N-terminal domain must

be communicated to the remainder of the protein. The phosphorylated N-terminal domain works positively: removal of this domain does not substitute for phosphorylation.

Preliminary evidence indicates that phosphorylation may increase the activity of members of other subgroups of receiver proteins (i.e., proteins other than activators of σ^{54}-holoenzyme) by facilitating or allowing their oligomerization. By analogy to the case for NtrC, the receiver domains of these other proteins may influence association of their output domains rather than dimerizing themselves.

BACKGROUND

The NtrC protein belongs to a family of activators for the σ^{54}-holoenzyme form of RNA polymerase (Kustu et al., 1989; Thony and Hennecke, 1989; Morett and Segovia, 1993) (Fig. 1). Like NtrC, several but not all these activators are receiver proteins of two-component regulatory systems (Parkinson and Kofoid, 1992). NtrC and other activators of σ^{54}-holoenzyme bind to transcriptional enhancers and contact the polymerase by means of DNA loop formation (Kustu et al., 1991; D. Weiss et al., 1992) (Fig. 2). Binding to an enhancer tethers these activators near the promoters they regulate and thereby increases their frequency of

Susan C. Porter, Anne K. North, and Sydney Kustu, Departments of Plant Biology and Molecular and Cell Biology, University of California, Berkeley, Berkeley, California 94270.

Two-Component Signal Transduction, Edited by James A. Hoch and Thomas J. Silhavy,
© 1995 American Society for Microbiology, Washington, DC 20005

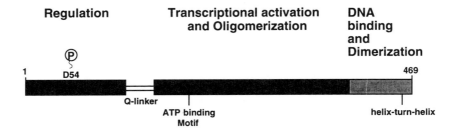

FIGURE 1 Domain structure of NtrC (reviewed by Kustu et al., 1991; D. Weiss et al., 1992; Morett and Segovia, 1993). NtrC is composed of three domains: an amino-terminal receiver domain (regulatory domain), a central activation domain, and a carboxy-terminal DNA binding and dimerization domain. The N-terminal domain (~120 residues) contains the site of phosphorylation, aspartate 54 (D54). It is joined to the remainder of the protein by a flexible glutamine-rich linker (Q-linker). The central activation domain (~240 residues) contains the "phosphate loop" (ATP binding motif), which is known to bind the β-phosphate in several nucleotide binding proteins. Oligomerization determinants lie in this domain (Flashner et al., submitted; see text). The C-terminal region of NtrC (~90 residues) contains a helix-turn-helix DNA binding motif and the major dimerization determinants of the protein. Forms of NtrC lacking the N-terminal domain (ΔN-NtrC and ΔN-NtrC^{S160F}) are missing residues 1 to 133 and begin within the Q-linker. Other activators of σ^{54}-holoenzyme carry a domain homologous to the central domain of NtrC. Some, but not all, of these activators carry an N-terminal receiver domain.

contacts with the polymerase. The activators catalyze the isomerization of closed complexes between σ^{54}-holoenzyme and a promoter to transcriptionally productive open complexes in a reaction that requires hydrolysis of ATP or another nucleoside triphosphate. In the case of NtrC, both nucleotide hydrolysis and transcriptional activation depend on phosphorylation of an aspartate residue in the N-terminal receiver domain of the protein (also called its regulatory domain). DNA binding per se does not depend on phosphorylation, but phosphorylation increases the cooperativity of binding to two sites (see below; V. Weiss et al., 1992; Porter et al., 1993). The phosphorylated receiver domain of NtrC acts positively: removal of this domain does not yield active protein (Drummond et al., 1990; D. Weiss et al., 1992). Interestingly, the receiver domain of the closely related activator dicarboxylate transport protein D (DctD) acts negatively (Huala et al., 1992).

Recent studies indicate that NtrC functions as a simple molecular machine. It couples the energy of ATP hydrolysis to the formation of open complexes by σ^{54}-holoenzyme, a reaction that is otherwise thermodynamically unfavorable (Wedel and Kustu, submitted). It is our working hypothesis that NtrC hydrolyzes ATP to drive a change in conformation of σ^{54}-holoenzyme and that the polymerase itself then denatures the DNA strands around the transcriptional start site. Little is known about the mechanism of energy coupling or about contact between NtrC and σ^{54}-holoenzyme.

In this chapter, we review the evidence that the NtrC protein from enteric bacteria, which is a dimer in solution, must form an appropriate oligomer to hydrolyze nucleotide and activate transcription. Although we focus on experiments from our own laboratory, comparable experiments were reported by V. Weiss et al. (1992) and by Austin and Dixon (1992). Because phosphorylation of the N-terminal receiver domain of NtrC is known to increase oligomerization, effects of phosphorylation on NtrC function may be a consequence of effects on oligomerization (V. Weiss et al., 1992; Porter

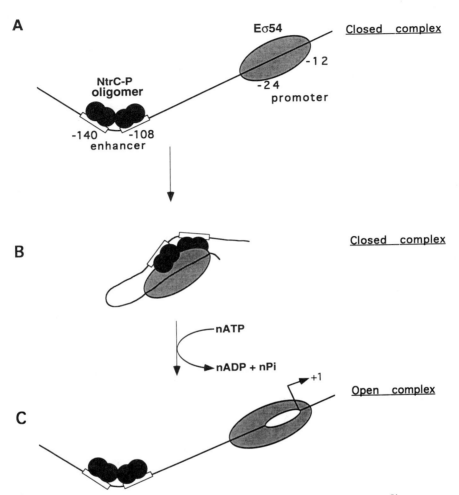

FIGURE 2 Transcriptional activation by NtrC. σ^{54}-holoenzyme by itself ($E\sigma^{54}$) binds to a promoter to form closed complexes, in which the DNA remains double-stranded. NtrC binds to a nearby enhancer, which is composed of two binding sites for dimers (A). Conserved promoter sequences recognized by the polymerase lie at -12 and -24 with respect to the start site of transcription. For the *glnA* gene of *Salmonella typhimurium,* the binding sites that constitute the enhancer lie at -108 and -140. As is the case for eukaryotic transcriptional enhancers, this enhancer functions efficiently at a distance of kilobases from the promoter and downstream as well as upstream of it (Reitzer and Magasanik, 1986; Reitzer et al., 1989). A phosphorylated oligomer of NtrC contacts σ^{54}-holoenzyme by means of DNA loop formation (B). To activate transcription, NtrC catalyzes isomerization of closed complexes between polymerase and the promoter to open complexes, in which there is a region of localized strand denaturation around the transcriptional start site (C). To catalyze open complex formation, NtrC must hydrolyze ATP. In open complexes, the conformation of polymerase is changed such that its DNase I footprint is elongated and extends downstream of the transcriptional start site.

et al., 1993). Finally, recent evidence from our laboratory indicates that oligomerization determinants of NtrC are located in its central activation domain.

EXPERIMENTAL OBSERVATIONS

Mutant Forms of NtrC and Derivatives of the *glnA* Enhancer

Studies of NtrC function have been facilitated by the use of two sorts of tools: mutant forms of NtrC and derivatives of the *glnA* enhancer. Phosphorylated wild-type NtrC has a fairly rapid autophosphatase activity (Keener and Kustu, 1988; Weiss and Magasanik, 1988) and hence is always a mixture of phosphorylated (active) and unphosphorylated (inactive) forms. To obtain a homogeneous population of active molecules, we have used so-called constitutive mutant forms of NtrC, which have some ability to hydrolyze ATP and activate transcription without being phosphorylated (Popham et al., 1989; Weglenski et al., 1989; Austin and Dixon, 1992; D. Weiss et al., 1992). The NtrC^{S160F} constitutive protein, which we have used for many of our studies, has an amino acid substitution in its central activation domain. The activity of unphosphorylated NtrC^{S160F} protein is a few percent that of phosphorylated wild-type NtrC and is greatly increased by phosphorylation.

We have used derivatives of both NtrC^{S160F} and wild-type NtrC that have no detectable ability to bind to DNA (Porter et al., 1993). For most studies, we used forms that contain alanines in place of three hydrophilic residues in the second ("recognition") helix of the DNA binding motif because they are structurally stable. These are designated NtrC^{S160F,3ala} and NtrC^{3ala}.

Finally, we have used derivatives of both wild-type NtrC and NtrC^{S160F} lacking their N-terminal regulatory domains (denoted ΔN-NtrC and ΔN-NtrC^{S160F}, respectively; see Fig. 1). Because ΔN-NtrC^{S160F} retains the ability to activate transcription (see below), failure of ΔN-NtrC to activate is not a trivial consequence of its failure to fold correctly.

The ability of NtrC to activate transcription has been studied intensively at the promoter

for the *glnA* gene, which encodes glutamine synthetase (Kustu et al., 1991; D. Weiss et al., 1992) (Fig. 2). The natural *glnA* enhancer lies about 100 bp upstream of the promoter and is composed of two relatively weak binding sites for NtrC dimers (Fig. 3A and B). (NtrC is known to be a dimer in solution at concentrations greater than 1 μM, the lowest concentration at which this was tested.) The binding sites that constitute the *glnA* enhancer are centered about three turns of the DNA helix apart, and in *Salmonella typhimurium* they lie at −108 and −140 with respect to the transcriptional start site. In addition to the natural *glnA* enhancer, we have used two derivatives of it. One of these, the "strong enhancer," carries two copies of the same high-affinity binding site ("strong site") in place of the two natural sites (Porter et al., 1993) (Fig. 3C). The second derivative, the "weak enhancer," carries one of the natural binding sites and only a half-site at the position of the other (Fig. 3D).

NtrC Dimers Bind Cooperatively to the *glnA* Enhancer and Phosphorylation Increases Cooperativity

In a gel mobility shift assay, two NtrC dimers bind cooperatively to the *glnA* enhancer to yield a single shifted species (Porter et al., 1993). When the two binding sites are separated from each other, NtrC fails to yield a stable shifted species with either binding site alone. Because a single NtrC dimer binds stably to one "strong site" (high-affinity site) in a gel mobility shift assay, we used the "strong enhancer" (Fig. 3C) to study the cooperativity of binding of NtrC quantitatively. Binding studies with the strong enhancer indicated that the presence of one bound dimer of NtrC stimulated binding of a second dimer to the adjacent site by a factor of about 20-fold. Placing the strong binding sites for NtrC on opposite sides of the DNA helix at about two and one-half or three and one-half turns apart essentially eliminated cooperative binding, but cooperativity was (partially) restored when the sites were

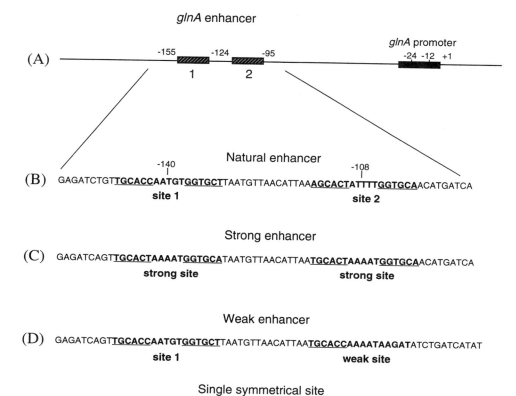

FIGURE 3 *glnA* enhancer and variants of it. (A) *glnA* regulatory region. (B) Sequence of the two dyad-symmetrical sites that compose the *glnA* enhancer. Regions of (imperfect) dyad symmetry are underlined. To produce the "strong enhancer" (C), a single site with increased dyad symmetry was substituted for sites 1 and 2. To produce the "weak enhancer" (D), the dyad symmetry of site 2 in the *glnA* enhancer was destroyed. (E) The "single symmetrical site" has perfect dyad symmetry. (Modified from Fig. 1 of Porter et al., 1993.)

separated by four turns of the helix rather than three. Although the isolated carboxy (C)-terminal DNA binding domain of NtrC (90 residues) had a normal affinity for a single strong binding site, it showed essentially no cooperativity of binding to the two sites that constitute the strong enhancer. This result indicates that determinants of cooperative binding lie outside the C-terminal domain, which carries the major dimerization determinants of the protein (North et al., 1993; Klose et al., 1994) and its helix-turn-helix DNA binding motif (Contreras and Drummond, 1988) (see Fig. 1).

Phosphorylation of NtrC by its cognate transmitter protein nitrogen regulatory protein B (NtrB; also called NR_{II}) increased its cooperativity of binding to the strong enhancer by at least 12-fold (and perhaps considerably more—see V. Weiss et al. [1992] and Porter et al. [1993]). Phosphorylation had no effect on binding of one NtrC dimer to a single site.

Oligomer of NtrC Is Required for Transcriptional Activation

Although it is not yet clear whether the activating species of NtrC is a tetramer or even higher-order oligomer, the following lines of evidence indicate that one dimer is not sufficient: (i) occupancy of a single binding site by a dimer of NtrC does not result in transcriptional activation, whereas occupancy of an enhancer by a tetramer or higher-order oligomer does (V. Weiss et al., 1992; Porter et al., 1993); (ii) activation from a single binding site, which only occurs at much higher concentrations of NtrC than those required for occupancy, can be shown to depend on oligomer formation (Porter et al., 1993); (iii) like cooperativity of binding, synergistic activation from an enhancer is lost if the two sites that constitute it are placed on opposite sides of the DNA helix. These lines of evidence are elaborated below.

A single symmetrical binding site for NtrC (Fig. 3E) can effectively tether a dimer of this protein in the vicinity of σ^{54}-holoenzyme. However, complete occupancy of such a site by the NtrC^{S160F} constitutive protein was not sufficient for transcriptional activation (Fig. 4A and B). By contrast, transcriptional activation from a "weak enhancer" (see Fig. 3D) occurred in parallel to its occupancy by two dimers of this protein (Fig. 4A and B). Activation from the weak enhancer was synergistic with respect to that from a single site (i.e., activation from the enhancer was more than twice that from the single symmetrical site at all concentrations of the NtrC^{S160F} protein of ≤30 nM). For example, at 7 nM NtrC^{S160F}—a concentration at which the symmetrical site was fully occupied by a dimer but the enhancer was not yet fully occupied by a tetramer—activation from the enhancer was nevertheless 24-fold higher than that from the symmetrical site. This experiment separates the role of an enhancer in tethering NtrC near a promoter from its role in facilitating formation of an active oligomer of this protein when it is at low concentration in solution. One possible physiological advantage of the requirement for oligomerization would be to prevent NtrC dimers from activating at σ^{54}-dependent promoters that should be activated by other enhancer binding proteins such as DctD or nitrogen fixation protein A (NifA) (Porter et al., 1993). Although the probability

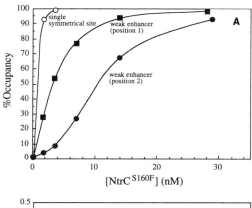

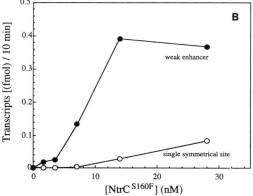

FIGURE 4 Occupancy of the weak enhancer and the single symmetrical binding site by NtrC^{S160F} (A) and transcriptional activation from each (B). Occupancy was assessed by DNase I footprinting at 2 nM DNA, the same concentration used for measuring transcriptional activation (the rate of open complex formation). Conditions used to determine occupancy and transcriptional activation were the same. Templates were linear and carried either the weak enhancer or the single symmetrical site at a distance of ~400 bp upstream of the *glnA* promoter. (Modified from Fig. 5C and D in Porter et al., 1993.)

of there being a single adventitious binding site for NtrC near these other promoters (i.e., within a kilobase) is high, the chances of there being two such sites with appropriate spacing to constitute an enhancer are much lower.

At concentrations of NtrC^{S160F} >7 nM, there was increasing transcriptional activation from the single symmetrical site (Porter et al., 1993) (Fig. 4B). To show that this activation was also a

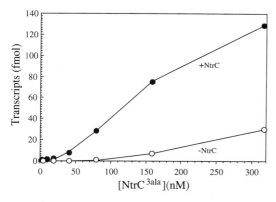

FIGURE 5 Synergy of activation by phosphorylated wild-type NtrC and phosphorylated NtrC^{3ala} on a template carrying a single binding site for NtrC. Stimulation of open complex formation by the NtrC^{3ala} protein (concentrations indicated on the x axis) was assessed in the presence of 5 nM wild-type NtrC (closed circles) or in the absence of wild-type NtrC (open circles). NtrB was present at 100 nM to phosphorylate the NtrC proteins. The supercoiled template, which carries a single symmetrical binding site for NtrC ~400 bp upstream of the *glnA* promoter, was at 20 nM. (From Fig. 6B in Porter et al., 1993.)

consequence of oligomer formation, we demonstrated that the NtrC^{S160F,3ala} protein, which has no detectable ability to bind to DNA, can nevertheless activate transcription synergistically when combined with low concentrations of NtrC^{S160F} itself. Presumably synergistic activation results from the formation of mixed oligomers by means of protein-protein interaction between dimers, with the DNA-bound form (NtrC^{S160F}) serving as tether. We obtained similar but even more striking results when we used phosphorylated wild-type NtrC as the DNA-bound form and NtrC^{3ala} to form oligomers (Fig. 5). Activation by 5 nM phosphorylated wild-type NtrC was only twofold above background, and activation by phosphorylated NtrC^{3ala} was minimal at concentrations less than 75 nM. However, activation by a mixture of phosphorylated wild-type NtrC at 5 nM and phosphorylated NtrC^{3ala} was synergistic, with the maximum degree of synergy reaching at least 40-fold (at 40 nM phosphorylated NtrC^{3ala}).

As expected, if activation from an enhancer depends on protein-protein interaction between bound dimers of NtrC, two binding sites for NtrC yielded little more transcriptional activation than a single site if they were on opposite sides of the DNA helix. Thus, two sites separated by two and one-half or three and one-half turns of the helix, rather than the usual three, stimulated little more than did a single site, even when they were fully occupied (Fig. 6). Stimulation was partially restored if the two sites were separated by four turns of the helix rather than three: activation of transcription, assessed as a rate of open complex formation, was two- to three-fold lower with the sites separated by four turns (Fig. 6). Interestingly, the three best-studied natural enhancers for NtrC are composed of two binding sites with different spacings between them; three turns of the helix at *glnA,* two turns at *nifLA* (the nitrogen fixation regulatory operon), and approximately one turn at *glnH* (the glutamine transport operon) (see discussion in Porter et al., 1993). It is possible that the different enhancers facilitate formation of oligomers with different catalytic efficiencies, another physiological advantage of the requirement for oligomer formation.

Oligomer of NtrC Is Required for ATP Hydrolysis

In solution, phosphorylation increases the ATPase activity of NtrC at least several hundredfold (Weiss et al., 1991; Austin and Dixon, 1992). Initial studies indicated that the rate of ATP hydrolysis showed a steep dependence on the concentration of phosphorylated NtrC, leading to the hypothesis that a phosphorylated oligomer was required. Further, studies of Austin and Dixon (1992) showed that binding to an enhancer (the natural enhancer for the *nifLA* operon) greatly stimulated ATP hydrolysis by the NtrC^{S160F} constitutive protein, but a single site had much less effect. Stimulation was observed for both unphosphorylated and phosphorylated NtrC^{S160F} and for phosphorylated wild-type NtrC.

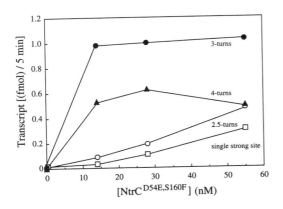

FIGURE 6 Effect of spacing between NtrC binding sites on transcriptional activation (the rate of open complex formation). The effect of the spacing between the two binding sites that constitute the strong enhancer on the rate of open complex formation at the *glnA* promoter was assessed in a single-cycle transcription assay. Rates are expressed in fmol transcript/5 min. The NtrC^{D54E,S160F} constitutive mutant protein, which has higher activity than the NtrC^{S160F} form (Klose et al., 1993), was titrated on templates carrying two strong binding sites separated by three turns of the DNA helix (center-to-center) (Fig. 3C; closed circles), four turns of the helix (closed triangles), or 2.5 turns (open circles). It was also titrated on a template carrying a single strong NtrC binding site (open squares). Templates were fragments of ~700 bp carrying sites ~400 bp upstream of the *glnA* promoter (Porter, 1993; Porter et al., 1993). Specifically, they were (i) the 703-bp KpnI-PstI fragment from pJES534 (three turns); (ii) the 712-bp KpnI-PstI fragment from pJES587 (four turns); (iii) the 698-bp KpnI-PstI fragment from pJES639 (2.5 turns); and (iv) the 635-bp KpnI-PstI fragment from pJES520 (single site). Under the conditions used, transcription from the *glnA* promoter on a template lacking NtrC binding sites was undetectable in the range of concentrations of NtrC^{D54E,S160F} shown.

We extended the studies of Austin and Dixon by showing that an oligonucleotide carrying a "strong enhancer" (derived from the *glnA* enhancer as described above; see Fig. 3C) stimulated the ATPase activity of NtrC^{S160F} at least 50-fold and that maximal stimulation was observed when the number of binding sites was approximately equal to that of NtrC dimers (Fig. 7A). Stimulation was lost as the concentration of oligonucleotide was increased

over that of protein, presumably because NtrC distributed to different DNA molecules as dimers. Thus, it is apparently not DNA binding per se that is stimulatory but rather the formation of oligomers that is facilitated by DNA binding. An oligonucleotide carrying a single binding site for NtrC had some stimulatory effect on the ATPase. This effect also appears to be due to oligomer formation because it, too, was lost as the concentration of oligonucleotide was raised. As expected, an oligonucleotide that lacks binding sites for NtrC did not stimulate the ATPase activity of the NtrC^{S160F} protein.

If the NtrC^{S160F} protein was phosphorylated, stimulatory effects of the strong enhancer on its ATPase activity were readily detected at low concentrations of protein and enhancer (10 nM protein and sites, as opposed to the 250 nM used with unphosphorylated protein) (Fig. 7B). Most striking, stimulatory effects of the enhancer persisted even as the concentration of the oligonucleotide carrying it was raised considerably over that of the protein. The latter observation is presumably accounted for by the fact that phosphorylation increases cooperative interactions between dimers and hence decreases their tendency to distribute to different molecules of DNA. Stimulatory effects of the enhancer on the ATPase activity of phosphorylated wild-type NtrC (Fig. 7C) were very similar to those on the activity of phosphorylated NtrC^{S160F}. We have speculated that the requirement for a phosphorylated oligomer of NtrC to activate transcription may be due to a requirement that this oligomer hydrolyze at least two molecules of ATP simultaneously to catalyze formation of an open complex by σ^{54}-holoenzyme (Porter et al., 1993) (see Fig. 2).

Oligomerization Determinants Lie in the Central Activation Domain of NtrC

We have used a derivative of the NtrC^{S160F} constitutive protein lacking its N-terminal regulatory domain (denoted ΔN-NtrC^{S160F}) to

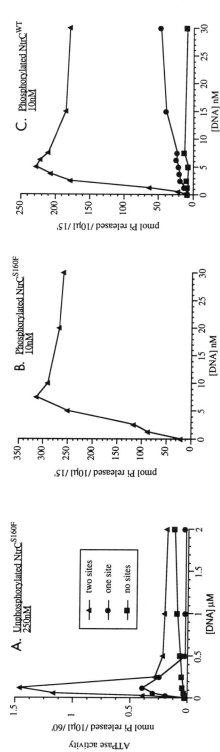

FIGURE 7 Effect of the strong enhancer, a single strong site, or DNA lacking NtrC binding sites on ATPase activity of NtrC. Sites were carried on double-stranded oligonucleotides of 69 bp. The oligonucleotides carried the strong enhancer or derivatives of it in which one or both sites were replaced with random DNA. ATPase assays were performed in acetate buffer, as described (Weiss et al., 1991), and results are expressed in nmol P_i released (A) or pmol P_i released (B and C) during the time indicated. NtrC proteins (NtrC^{S160F} for A and B and wild-type [WT] NtrC for C) were present at the concentrations indicated and were incubated with oligonucleotides for 5 to 10 min at 37°C before the addition of ATP. For the experiments in B and C, NtrB was present at 100 nM to phosphorylate the NtrC proteins, and they were phosphorylated for 10 min at 37°C in the presence of unlabeled ATP before the labeled nucleotide was added.

show that the oligomerization determinants necessary for ATP hydrolysis and transcriptional activation do not lie in the N-terminal receiver domain. Although wild-type NtrC lacking its N-terminal domain is inactive, ΔN-NtrC^{S160F} retained activity, providing a first line of evidence that oligomerization determinants necessary for transcriptional activation lie outside the N-terminal domain. (Conversely, activity of the NtrC^{S160F} protein was not increased by removal of the N-terminal domain, congruent with the view that this domain functions positively.) When ΔN-NtrC^{S160F} was bound to a single site, its ability to activate transcription was synergistic with that of the NtrC^{S160F,3ala} protein, which cannot bind to DNA (Flashner et al., submitted). (As discussed above, the same was true for intact NtrC^{S160F}.) This result confirms that oligomerization determinants required for transcriptional activation lie outside the N-terminal domain. Also, an oligonucleotide carrying the strong enhancer greatly stimulated the ATPase activity of ΔN-NtrC^{S160F}, whereas an oligonucleotide lacking sites had essentially no effect. The latter findings provide an independent line of evidence that oligomerization determinants lie outside the N-terminal domain. Together with previous results indicating that oligomerization determinants of NtrC lie outside its C-terminal DNA binding domain (see above; Porter et al., 1993), results with ΔN-NtrC^{S160F} indicate that these determinants lie in the central activation domain of the protein.

CONCLUSIONS

NtrC must be phosphorylated in its receiver domain to hydrolyze ATP and activate transcription. Phosphorylation stimulates the oligomerization of NtrC, and oligomerization is, in turn, required for ATP hydrolysis and transcriptional activation. Hence, it is a straightforward hypothesis that phosphorylation functions by affecting oligomer formation. Because the phosphorylated receiver domain of NtrC functions positively, it is presumably needed for appropriate oligomerization: removing this domain by proteolysis or genetic engineering does not substitute for phosphorylation. Oligomerization determinants appear to lie in the central activation domain of NtrC (see Fig. 1), and therefore the phosphorylated receiver domain cannot mediate oligomerization directly but must do so through an effect on the remainder of the protein. It is not known how many of the monomeric subunits of NtrC must be phosphorylated to yield an active oligomer or whether there are constraints on how the phosphorylated subunits must be arranged.

One potentially confusing aspect of the requirement for oligomerization is that unphosphorylated NtrC binds cooperatively to the two sites that constitute the glnA enhancer, and hence it can form a tetramer. However, the unphosphorylated protein is essentially incapable of ATP hydrolysis or transcriptional activation, even when bound to the enhancer. It is our working hypothesis that phosphorylation changes and thereby strengthens the interaction between dimers that is detectable in unphosphorylated NtrC and that this altered interaction is essential for transcriptional activation. Because the interaction appears to be a function of the central domain, our proposal is in contrast to the hypothesis that the phosphorylated N-terminal domain itself dimerizes and thereby mediates oligomerization of intact NtrC (Parkinson and Kofoid, 1992; V. Weiss et al., 1992).

Interestingly, effects of phosphorylation on oligomer formation may generalize to subgroups of receiver proteins other than activators of σ^{54}-holoenzyme (e.g., the OmpR and NarL subgroups, which contain activators of σ^{70}-holoenzyme). Phosphorylation increases DNA binding by several members of these subgroups (e.g., Aiba et al., 1989; Makino et al., 1989; Nakashima et al., 1991; Li et al., 1994) and may do so by increasing cooperative interactions between subunits rather than intrinsic DNA binding by a single subunit. By analogy to the case for NtrC, we would postulate that oligomerization of these other proteins is controlled by effects of their receiver domains on their output DNA binding domains and not by dimerization of the phosphorylated receiver domains themselves.

ACKNOWLEDGMENTS

We thank Karl Klose for performing the experiment in Fig. 7A and Franz Narberhaus and Ken Stedman for critical reading of the manuscript.

Work from our laboratory was supported by U.S. Public Health Service grant GM38361 to S.K.

REFERENCES

Aiba, H., F. Nakasai, S. Mizushima, and T. Mizuno. 1989. Phosphorylation of a bacterial activator protein, OmpR, by a protein kinase, EnvZ, results in stimulation of its DNA-binding ability. *J. Biochem.* (Tokyo) **106**:5–7.

Austin, S., and R. Dixon. 1992. The prokaryotic enhancer-binding protein NtrC has an ATPase activity which is phosphorylation and DNA dependent. *EMBO J.* **11**:2219–2228.

Contreras, A., and M. Drummond. 1988. The effect on the function of the transcriptional activator NtrC from *Klebsiella pneumoniae* of mutations in the DNA recognition helix. *Nucl. Acids Res.* **16**:4025–4039.

Drummond, M. H., A. Contreras, and L. A. Mitchenall. 1990. The function of isolated domains and chimaeric proteins constructed from the transcriptional activators NifA and NtrC of *Klebsiella pneumoniae*. *Mol. Microbiol.* **4**:29–37.

Flashner, Y., D. S. Weiss, J. Keener, and S. Kustu. Constitutive forms of the enhancer-binding protein NTRC: evidence that essential oligomerization determinants lie in the central activation domain. Submitted for publication.

Huala, E., J. Stigter, and F. M. Ausubel. 1992. The central domain of *Rhizobium leguminosarum* DctD functions independently to activate transcription. *J. Bacteriol.* **174**:1428–1431.

Keener, J., and S. Kustu. 1988. Protein kinase and phosphoprotein phosphatase activities of nitrogen regulatory proteins NTRB and NTRC of enteric bacteria: roles of the conserved amino-terminal domain of NTRC. *Proc. Natl. Acad. Sci. USA* **85**:4976–4980.

Klose, K. E., A. K. North, K. M. Stedman, and S. Kustu. 1994. The major dimerization determinants of the nitrogen-regulatory protein NTRC from enteric bacteria lie in its carboxy-terminal domain. *J. Mol. Biol.* **241**:233–245.

Klose, K. E., D. S. Weiss, and S. Kustu. 1993. Glutamate at the site of phosphorylation of nitrogen-regulatory protein NTRC mimics aspartyl-phosphate and activates the protein. *J. Mol. Biol.* **232**:67–78.

Kustu, S., A. K. North, and D. S. Weiss. 1991. Prokaryotic transcriptional enhancers and enhancer-binding proteins. *TIBS* **16**:397–402.

Kustu, S., E. Santero, J. Keener, D. Popham, and D. Weiss. 1989. Expression of σ^{54} (*ntrA*)-dependent genes is probably united by a common mechanism. *Microbiol. Rev.* **53**:367–376.

Li, J., S. Kustu, and V. Stewart. 1994. In vitro interaction of nitrate-responsive regulatory protein NarL with DNA target sequences in the *fdnG*, *narG*, *narK*, and *frdA* operon control regions of *Escherichia coli* K-12. *J. Mol. Biol.* **241**:150–165.

Makino, K., H. Shinagawa, M. Amemura, T. Kawamoto, M. Yamada, and A. Makata. 1989. Signal transduction in the phosphate regulon of *Escherichia coli* involves phosphotransfer between PhoR and PhoB proteins. *J. Mol. Biol.* **210**:551–559.

Morett, E., and L. Segovia. 1993. The σ^{54} bacterial enhancer-binding protein family: mechanism of action and phylogenetic relationship of their functional domains. *J. Bacteriol.* **175**:6067–6074.

Nakashima, K., K. Kanamaru, H. Aiba, and T. Mizuno. 1991. Signal transduction and osmoregulation in *Escherichia coli*. *J. Biol. Chem.* **266**:10775–10780.

North, A. K., K. E. Klose, K. M. Stedman, and S. Kustu. 1993. Prokaryotic enhancer-binding proteins reflect eukaryote-like modularity: the puzzle of nitrogen regulatory protein C. *J. Bacteriol.* **175**:4267–4273.

Parkinson, J. S., and E. C. Kofoid. 1992. Communication modules in bacterial signaling proteins. *Annu. Rev. Genet.* **26**:71–112.

Popham, D. L., D. Szeto, J. Keener, and S. Kustu. 1989. Function of a bacterial activator protein that binds to transcriptional enhancers. *Science* **243**:629–635.

Porter, S. C. 1993. The activating complex of the prokaryotic enhancer binding protein, NTRC. Ph.D. thesis, University of California Berkeley.

Porter, S. C., A. K. North, A. B. Wedel, and S. Kustu. 1993. Oligomerization of NTRC at the *glnA* enhancer is required for transcriptional activation. *Genes Dev.* **7**:2258–2273.

Reitzer, L. J., and B. Magasanik. 1986. Transcription of *glnA* in *E. coli* is stimulated by activator bound to sites far from the promoter. *Cell* **45**:785–792.

Reitzer, L. J., B. Movsas, and B. Magasanik. 1989. Activation of *glnA* transcription by nitrogen regulator I (NRI)-phosphate in *Escherichia coli*: evidence for a long range physical interaction between NRI-phosphate and RNA polymerase. *J. Bacteriol.* **171**:5512–5522.

Thony, B., and H. Hennecke. 1989. The −24/−12 promoter comes of age. *FEMS Microbiol. Rev.* **63**:341–358.

Wedel, A. B., and S. Kustu. The bacterial enhancer-binding protein NTRC is a molecular machine: ATP hydrolysis is coupled to transcriptional activation. Submitted for publication.

Weglenski, P., A. J. Ninfa, S. Ueno-Nishio, and B. Magasanik. 1989. Mutations in the *glnG* gene of *Escherichia coli* that result in increased activity of nitrogen regulator I. *J. Bacteriol.* **171:**4479–4485.

Weiss, D. S., J. Batut, K. E. Klose, J. Keener, and S. Kustu. 1991. The phosphorylated form of the enhancer-binding protein NTRC has an ATPase activity that is essential for activation of transcription. *Cell* **67:**155–167.

Weiss, D. S., K. E. Klose, T. R. Hoover, A. K. North, S. C. Porter, A. B. Wedel, and S. Kustu. 1992. Prokaryotic transcriptional enhancers, p. 667–694. *In* S. L. McKnight and K. R. Yamamoto (ed.), *Transcriptional Regulation.* Cold Spring Harbor Laboratory Press, Cold Spring Harbor, N.Y.

Weiss, V., and B. Magasanik. 1988. Phosphorylation of nitrogen regulator I (NRI) of *Escherichia coli. Proc. Natl. Acad. Sci. USA* **85:**8919–8923.

Weiss, V., F. Claverie-Martin, and B. Magasanik. 1992. Phosphorylation of nitrogen regulator I of *Escherichia coli* induces strong cooperative binding to DNA essential for activation of transcription. *Proc. Natl. Acad. Sci. USA* **89:**5088–5092.

Transcription Regulation by the
Bacillus subtilis Response Regulator
Spo0A

George B. Spiegelman, Terry H. Bird, and Valerie Voon

10

Cells integrate multiple signals from the environment to establish their metabolic and gene expression response. These signals indicate conditions such as nutrient status, population density, and physical and chemical parameters such as temperature, water activity, and/or presence of toxic compounds. In the microbiology laboratory, the most common example of this integration occurs when cells reach the stationary stage after growth in an appropriate medium. The ultimate response to the stationary phase varies between species of bacteria, for example, in *Bacillus subtilis* continued starvation leads to the production of a dormant form, the bacterial endospore.

B. subtilis cells ending logarithmic growth enter a "transition state" in which a variety of genes is expressed. The result of this gene activity can be either the return to vegetative growth, if an alternate energy source can be found, or production of a dormant endospore (Strauch and Hoch, 1993). Because the net result of these alternatives is very different, it is not surprising that there would be a complex of feedback-modulated regulatory pathways affecting which alternative dominates. If, in na-

ture, *B. subtilis* cells spend most of their time in a nutrient-depleted state, then teetering on the boundry between return to growth via an alternative energy source and sporulation may be the natural state of the organism.

The wealth of potential regulatory mechanisms, the availability of effective genetic tools, and the intriguing morphology changes that accompany sporulation have led to intense study of the process. Many mutants have been isolated that block sporulation in various stages. Considerable genetic and biochemical evidence is available on the control of gene expression, particularly in the sporulation pathway. This information has been recently reviewed in several places (Errington, 1993; Hoch, 1993; many extensive reviews can be found in Sonenshein et al., 1993).

One of the unanswered questions about sporulation is the nature of the process that commits the transition state cell to enter into the sporulation pathway. The transition state cell must be continually monitoring input signals, and at some summation of signals, the balance is tipped toward sporulation. Central to the summation of the inputs and coordination of the response is the product of the *spo0A* gene. Mutants in *spo0A* were long recognized as having pleiotropic effects (reviewed in Errington, 1993; Losick et al., 1986; Piggot and

George B. Spiegelman, Terry H. Bird, and Valerie Voon, Department of Microbiology and Immunology, University of British Columbia, Vancouver, Canada V6T 1Z3.

Two-Component Signal Transduction, Edited by James A. Hoch and Thomas J. Silhavy,
© 1995 American Society for Microbiology, Washington, DC 20005

Coote, 1976). It is now understood that this is because Spo0A is an "ambiactive" transcription regulator; that is, it both represses and activates transcription. Genes that are regulated by Spo0A include other transcription regulators. For example, Spo0A represses the *abrB* gene (Perego et al., 1988; Strauch et al., 1990). Because AbrB is a negative regulator of many genes expressed during the stationary phase, activation of Spo0A leads to the expression of these genes (reviewed in Strauch and Hoch, 1993). Thus, expression of many of the genes expressed after log growth but not directly required for sporulation is controlled by the same process that controls sporulation initiation.

Genes that are regulated by Spo0A in a direct positive mechanism include alternate σ-factors encoded by the *spoIIA* (Trach et al., 1991; Wu et al., 1989; Wu et al., 1991) and *spoIIG* operons (Bird et al., 1993; Satola et al., 1991; Satola et al., 1992; Baldus et al., 1994) and the *spo0A* operon itself (Strauch et al., 1992). The activity of Spo0A in modulating transcription is affected by its phosphorylation (this process is discussed elsewhere in this book). The purpose of this review is to focus on the mechanism of transcription regulation by Spo0A.

DOMAINS OF Spo0A

The sequencing of the *spo0A* gene identified that it encoded a member of the response regulator family of proteins (Ferrari et al., 1985; Kudoh et al., 1985; Trach et al., 1985). These regulatory proteins contain an N-terminal domain that shows sequence identity to other family members and a C-terminal domain that is unique (reviewed in Parkinson and Kofoid, 1992; Stock et al., 1989). For some response regulators, the N-terminal domain inhibits the activity of the C terminus, and modification of the N terminus by phosphorylation is the signal to release the N-terminal inhibition. This general plan appears true for Spo0A as well (Grimsley et al., 1994). We consider similarities of the two domains of Spo0A to other transcription regulators, beginning with the C terminus.

Spo0A DNA Binding Domain

Grimsley and coworkers (1994) showed that trypsin cleaved Spo0A at a single site, between R142 and S143 (Fig. 1). The C-terminal fragment but not the N-terminal fragment bound DNA. For discussion purposes, we consider the trypsin site the boundary between the two domains and we term the C-terminal fragment the DNA binding domain or Spo0ABD. At the *abrB* promoter (*abrBp*), the DNase I footprint of Spo0ABD was indistinguishable from the footprint of intact Spo0A protein. Spo0A and Spo0A-P (phosphorylated form of Spo0A) also show identical footprints, but the level of protein needed to detect the footprint was at least 20-fold lower with Spo0A-P, showing that phosphorylation increased the binding affinity of Spo0A for DNA (Trach et al., 1991).

The DNA binding domain was able to repress in vitro transcription from *abrBp* and to activate transcription from the promoter for the *spoIIG* operon (*spoIIGp*) in vitro (these data are discussed below). Spo0ABD was less effective at stimulating initiation at *spoIIGp* than was the intact phosphorylated protein but was more effective than the intact nonphosphorylated protein. Thus, as with some other response regulations, the C-terminal region has properties that are repressed by the N terminus, and the repression functions of the N terminus are altered on phosphorylation. Unlike the transcription initiation kinetics with Spo0A-P, there was a pronounced lag in the initiation kinetics with Spo0ABD, suggesting that although the C-terminal fragment was no longer inhibited from transcription activation, it was debilitated in this function relative to Spo0A-P (see Fig. 7). The simple conclusion is that the N-terminal portion of Spo0A stimulates the transcription activation step.

Sporulation-defective mutants of Spo0A have been localized to the C-terminal end of the protein. Replacement of the wild-type gene with one deleted at the C terminus by

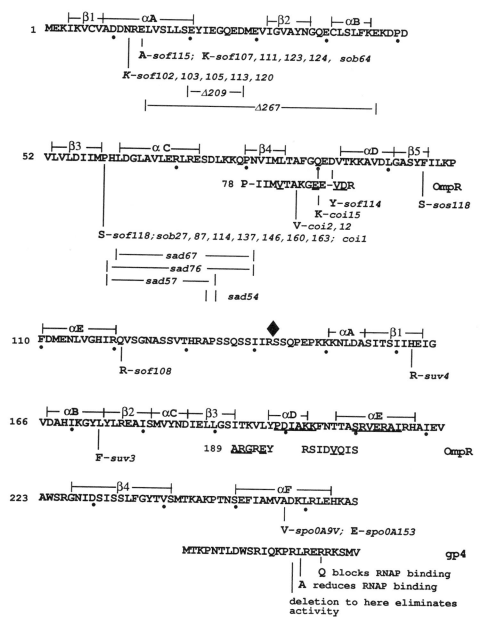

FIGURE 1 Amino acid sequence of Spo0A. The amino acid sequence of Spo0A is taken from Ferrari et al. (1985). Two regions of the amino acid sequence of OmpR are shown where transcription activation mutants (underlined) have been identified (Russo et al., 1993). At the C terminus, the amino acid sequence from ϕ29 gp4 and various mutations of the sequence that affect the activity of the protein are shown. The sequence is from Mizukami et al. (1986) and the mutants are from Mencia et al. (1993) and Rojo et al. (1990). Positions of mutations in Spo0A are shown: *sof* and *sob* mutants are from Spiegelman et al. (1990), *sad* mutants that are deletions of the indicated amino acids are from Ireton et al. (1993), the *coi* mutants are from Olmedo et al. (1990) and Δ209, Δ267 mutants are from Green et al. (1991). The DNA binding motif in Spo0A proposed by Youngman (personal communication) is underlined. Regions marked as helical (α) and extended (β) were calculated using the GGBSM program of PC gene. These regions are not meant to be precise but to allow subdivision of the sequence for discussion. Regions were calculated for the N-terminal and C-terminal domains separately, and so they begin again after the diamond symbol, which indicates the division between the N-terminal and C-terminal domains (as defined in the text). In the C-terminal domain, the positions of the *spo0A9V* and *spo0A153* mutations are indicated (at position 257), and the positions of suppressors of *spo0A9V* are indicated at H162F (*suv4*) and L174F (*suv3*) (Perego et al., 1991).

*Eco*RI cleavage (removing 15 amino acids) prevents sporulation (Ferrari et al., 1985). Perego et al. (1991) investigated the properties of two amino acid substitutions at the same position: *spo0A9V,* an A257V change, and *spo0A153,* an A257E change. Examination of in vivo expression from *abrBp* and *spoIIAp* using *lacZ* fusions showed that Spo0A9V could still repress *abrB,* suggesting that its DNA binding activities were intact, but *spoIIA* was not activated, suggesting that the mutation specifically interfered with promoter activation. Two suppressor mutations of *spo0A9V* were isolated and found to be within the binding domain region: H162R (*suv4*) and L174F (*suv3*). When suppressors of *spo0A153* were sought, wild-type derivatives of this mutant were uniformly found to be revertant rather than suppressor mutations, suggesting that introduction of a negatively charged amino acid in this region is a much more difficult, potentially impossible, change to compensate for.

Recent work on the gp4 protein of the *B. subtilis* phage ϕ29 provides an interesting comparison with Spo0ABD. Gp4 is a 123-amino-acid transcription activator that binds upstream of the ϕ29 major late promoter and stimulates transcription initiation by RNA polymerase containing σ^A (Barthelemy and Salas, 1989). Mutants in gp4 that affect its transcription stimulation properties have been reported. Rojo et al. (1990) made deletions at the C terminus of a cloned copy of the gp4 gene (see Fig. 1 for the amino acid sequence). These deletions replaced the terminal amino acids with those derived by translation of the polylinker of the cloning vector. Mutant proteins were tested for the ability to stimulate in vitro transcription from the ϕ29 late promoter. Replacement of the terminal 10 amino acids eliminated stimulation of in vitro transcription. Gp4 containing the mutation R120Q could still bend DNA but failed to promote RNA polymerase binding as measured by gel retardation and was defective in transcription activation. This introduction of negative charge is reminiscent of the *spo0A153* mutation (A257E)

that prevented sporulation and could not be suppressed intragenically. Gp4 containing the mutation L117A reduced but did not eliminate DNA bending and transcription activation (Mencia et al., 1993).

We compared the overall structure of gp4 and Spo0ABD using predictive secondary structure programs. These programs are not found to be precise but provide a starting point for discussion. Both gp4 and Spo0ABD contain an α-helical C-terminal region (binding domain αF in Fig. 1). The *spo0A9v* and *spo0A153* mutations are found in the helical region, as are the mutations in gp4 affecting transcription activation, suggesting that these regions form homologous domains involved in transcription activation.

Brown et al. (1994) sequenced a variety of Spo0A homologs from *B. subtilis* and related species. They suggested that the underlined sequence in the binding domain in Fig. 1 could form a DNA binding HTH motif. We and others have found that Spo0A is a weak DNA binding protein and does not form complexes that are stable to standard gel retardation conditions or UV laser cross-linking (Bird and Spiegelman, unpublished data), in contrast to gp4, which is highly effective in gel retardation assays (Mencia et al., 1993). DNase I footprinting with Spo0A is readily accomplished, so we presume that the instability of the Spo0A-DNA complex is due to a high dissociation rate.

As discussed in more detail below, Spo0A has significant similarity to OmpR, the response regulator that controls promoters for the *ompC* and *ompF* genes in *Escherichia coli.* Mutations in the C-terminal domain of OmpR that failed to activate transcription of *ompF* and *ompC* fully but did repress *ompF* transcription have been isolated. The fact that *ompF* was repressed suggests that the protein bound DNA, but the low level of activation of *ompF* and *ompC* indicated a defect in transcription activation (Russo et al., 1993). These mutations, which are indicated in Fig. 1, are near the region containing the potential HTH motif in Spo0A suggested by Brown et al. (1994).

N-Terminal Domain

The activation of Spo0A involves phosphorylation of the N terminus of the protein. Mutation of amino acid D56, the site of phosphorylation of Spo0A (Burbulys et al., 1991), demonstrated that it is essential for normal levels of sporulation and for in vitro phosphorylation of the protein by the phosphorelay (Burbulys et al., 1991; Green et al., 1991). Although the mechanism of this activation is not the primary focus of this review, we briefly consider some properties of the N-terminal portion of the molecule.

In vivo studies have shown that the N terminus of Spo0A inhibits the transcription activation properties of the C terminus. Deletion mutants of Spo0A were constructed and placed under the control of an inducible promoter (Ireton et al., 1993). Induction of the promoter led to sporulation initiation, and this induction was not dependent on the phosphorelay components (the positions of these *sad* mutants are indicated in Fig. 1). Remarkably, one *sad* mutant was the deletion of a single amino acid, D75. Analogous results have been obtained with N-terminal deletion mutants of a variety of transcription regulators from gram-negative bacteria (Menon and Lee, 1990; Kahn and Ditta, 1991; Choi and Greenberg, 1991).

Several other changes in the N-terminal domain of Spo0A have been shown to affect its in vivo activity. Mutants that suppressed either deletion of the *spo0F* gene (*sof, rvtA* mutants; Kawamura and Saito, 1983; Sharrock et al., 1984; Hoch et al., 1985; Spiegelman et al., 1990) or inactivation of the *spo0B* gene (*sob* mutants; Shoji et al., 1988; Spiegelman et al., 1990) have been isolated. The suppressor mutants restored sporulation to near-normal levels and had overlapping suppression activities. In all cases, the suppressor mutants were within the N terminus of Spo0A. Several *sof* mutants efficiently suppressed deletion of the two kinases known to be involved in the phosphorylation of Spo0A. These data are compatible with two hypotheses: (i) the *sof* mutations alter the conformation to allow Spo0A to activate

transcription in the absence of phosphorylation, or (ii) *sof* mutations allow another kinase to activate Spo0A. Because the *sof* mutants do not sporulate in log growth, they are not like the N-terminal deletions that are "constitutively active." It is not known if the *sof* forms of Spo0A become phosphorylated in the stationary phase in vivo or if they are more readily phosphorylated by other kinases. *Sad* mutants are active without phosphorylation, because they promote sporulation when the *spo0A* gene also contains a D56Q mutation.

Two other classes of mutations have been mapped and sequenced. Olmedo et al. (1990) isolated mutants that sporulated in the absence of normal nutritional signals (*coi* mutants). These were also found in the N-terminal domain (see Fig. 1). Green et al. (1991) isolated mutations that suppressed either D10Q or D56Q mutations. One suppressor of D10Q was a P60S mutation—this mutation has thus been isolated by four different protocols because it is also a *sof* mutant, a *sob* mutant, and a *coi* mutant. Two mutations suppressing D56Q were found: both were deletions in the N terminus (Δ267 and Δ209: see Fig. 1). The position of these mutations reinforces the conclusion that activation of Spo0A requires conformational change of the N terminus and that this change is accomplished in the wild-type protein by phosphorylation.

There is no simple model for the change induced by phosphorylation of the N-terminal domain. Figure 1 shows predicted helical and extended region for the N-terminal domain. This predicted structure closely matches the structure established for the CheY protein, another member of the response regulator family (reviewed in Stock et al., 1989). The changes induced by phosphorylation are certain to involve the concerted action of multiple regions of the N-terminal domain. As seen in Fig. 1, selection of *sof* mutants recovered mutants in two regions with high frequency: between sequences marked as αA and near the region marked αC. Comparison of the mutants that permit Spo0A activation in the absence of a phosphorelay indicates that the sequence around the αC helix is critical for suppression of the activation function of Spo0A. The deletions reported by Green et al. (1991) and the *sad* mutants both suppress D56Q mutations, suggesting that both regions are involved in suppressing the activity of the C terminus.

The sequence around the N-terminal αC region (Fig. 1) has high homology to OmpR, and mutations that alter the transcription activation properties of OmpR have been described in this region (Tsung et al., 1990; Brissette et al., 1991, 1992; Bowrin et al., 1992; Russo et al., 1993). For example, mutations T83A and G94S suppressed the substitution D55Q. Because D55 is the primary site of phosphorylation of OmpR (but potentially not the only site of phosphorylation [Delgado et al., 1993]), the suppressors allow transcription activation in the absence of phosphorylation. Mutations decreasing the transcription activation properties have also been reported: E87K, V89M, and D90N lead to lower levels of activation of *ompF* and *ompC* and have been interpreted as being deficient in activation (Russo et al., 1993). Similar phenotypes were reported for mutations in the C-terminal region (A189V, R190C, G191S, E193K), suggesting that these regions participate in similar functions (Russo et al., 1993). OmpR containing mutation at either G94D or R182C failed to activate either *ompF* or *ompC,* although they could be phosphorylated in vitro by EnvZ (Brissette et al., 1991). The locations of mutations in OmpR emphasize the complex series of domain interactions involved in activation of transcription.

Two other mutations just N-terminal to the sequences, marked αD and αE in Fig. 1, are interesting. First, *sof-114* is the only *sof* mutation tested that effectively suppressed the *spo0A9V* mutation (Perego et al., 1991). Other suppressors of *spo0A9V* were found in the binding domain: H156R and L163F. This suggests that these regions of the protein may interact in some way, and this interaction could be one aspect of the inhibition of transcription activation. Second, cells containing *sof* or *sob*

mutations were shown to segregate spo⁻ mutants at high frequency (Spiegelman et al., 1990). One such spo⁻ segregant (*sos-118*), which arose from a colony containing the *sof-118* mutation, was found to be a F105S mutation in *spo0A* (this mutant is *sos-118*). This suggests that the changes induced by the original *sof* mutant were overcome by the secondary change in the region marked β5 of the N-terminal domain.

DNA BINDING SITES FOR Spo0A: 0A BOXES

The consensus sequence in the region protected from DNase I digestion by Spo0A, termed the 0A box, was originally identified as 5′TGNCGAA-3′ by Strauch et al. (1990). A recent comparison of 0A boxes revealed an N=T preference (Baldus et al., 1994). A representative collection of 0A boxes and the DNase I protected region surrounding those sites are shown in Fig. 2A. Generally, the 0A box is flanked by an A/T sequence 4 to 8 bp long on the 5′ side. The 0A box appears 5′ to the transcription start site for all promoters identified so far that are stimulated by Spo0A. At promoters repressed by Spo0A, the sequence appears downstream of the transcription start site. The sequence is found in both possible orientations. When the 5′ to 3′ orientation of the 0A box consensus sequence is the same as the direction of transcription, we term this a 0A box; when the opposite orientation is seen, we term this a reverse 0A box. Figure 2B shows the locations and orientations of 0A boxes relative to a variety of promoters that are either repressed or activated by Spo0A. Near several promoters are multiple 0A, or reverse 0A, boxes.

Promoters Activated by Spo0A

Moran and coworkers examined *spoIIGp* by making both deletion and point mutants and examining in vivo and in vitro transcription and DNase I protection by Spo0A (Satola et al., 1991; Satola et al., 1992; Baldus et al., 1994). DNase I protection assays showed two pro-

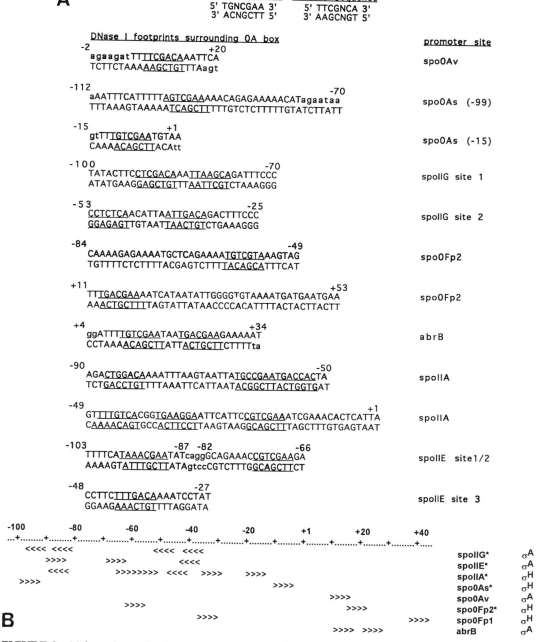

FIGURE 2 0A boxes in regulated promoters. (A) Sequences that have been identified as being protected by Spo0A from digestion by DNase I are indicated in capital letters. In some sequences, the protected regions are not flush on both strands of DNA, and unprotected DNA is indicated by lowercase letters. In examples in which both strands are capitals, comparison of the digestion pattern on the two strands has not been performed. The consensus sequence for the 0A box is underlined. The numbering refers to the +1 site for each promoter. (B) Position of the 0A boxes relative to the start site of transcription for the promoter is indicated. The symbols indicate forward (>>>>) or reverse (<<<<) orientations. Indicated at the side is the gene or operon where the 0A boxes are located and the σ-factor used by the promoter (σ^A or σ^H). Promoters that are activated by Spo0A have an asterisk (★); the other promoters are repressed. The data are taken from *spoIIG* (Satola et al., 1991; Satola et al., 1992; Baldus et al., 1994), *spoIIE* (York et al., 1992), *spoIIA* (Trach et al., 1991; Baldus et al., 1994), *spo0A* (Strauch et al., 1992), *spo0F* (Strauch et al., 1993), and *abrB* (Strauch et al., 1990).

tected regions (the 5' site is site 1, and the 3' site is site 2: see Fig. 2B). Site 2 contains two reverse 0A boxes separated by 4 bp. One of these reverse 0A boxes overlaps the −35 region of the promoter. A nucleotide change in site 2, which increased the consensus of one reverse 0A box (site 2.2 from 5'-ATTGACA to 5'-ATCGACA, change at position −38), resulted in a dramatic increase in promoter activity during the stationary phase. A mutation in the other reverse 0A box (site 2.1 from 5'-CTCAACA to 5'-CTCGACA, change at −47) also stimulated transcription. Mutants in the *spoIIGp* site 1 reverse 0A box that changed the sequence away from the canonical sequence decreased promoter activity. DNase I protection assays suggested that binding to the sites 1 and 2 was not cooperative but did not address the relative binding to the two 0A boxes within site 2.

Three sites near the *spoIIE* operon promoter are protected from DNase I digestion by Spo0A, and each contains a reverse 0A box (York et al., 1992). One of these sites overlaps the −35 region of the promoter, and mutations in this site that decreased consensus of the reverse 0A box within the protected region lowered promoter activity in vivo. Mutational studies of the *spoIIG* and *spoIIE* operons suggested that for promoters recognized by RNA polymerase containing σ^A, Spo0A could interact with the polymerase while bound to a DNA sequence normally in close contact with the σ-subunit of the polymerase. At both of these promoters, the spacing between the −35 and −10 sites is longer (21 or 22 bp) than is optimal for *B. subtilis* promoters (17 to 18 bp: Satola et al., 1991; York et al., 1992). This spacing would place the −35 region on the oppposite side of the DNA helix from its normal location, so one would expect that the polymerase would not naturally contact this site without significant bending of the DNA. For both the *spoIIE* and *spoIIG* promoters, mutations in reverse 0A boxes decrease the level of transcription from *lacZ* fusions in vivo (Satola et al., 1991; York et al., 1992), thus these regions must play some role in promoter activity.

Three other promoters activated by Spo0A have been studied: the promoter for the *spoIIA* operon; the promoter for *spo0A* that is activated during sporulation (*spo0Ap_s*); and one of the promoters associated with the *spo0F* gene (*spo0Fp₂*). All three of these are transcribed by a form of the polymerase containing the σ-factor, σ^H (Yamashita et al., 1986; Wu et al., 1991; Predich et al., 1992). Piggot and coworkers showed that deletion of the 5'-flanking region of *spoIIAp* to −52 eliminated promoter activity in vivo, whereas a construct containing 92 bp upstream had 50% of full activity (Wu et al., 1991). Between these sites are two 0A boxes. Three 0A boxes (one a reverse 0A box) are downstream of the −52 site, and DNase I and methylation protection studies suggested that Spo0A bound to a wide region encompassing −15 to −90 (Trach et al., 1991; Baldus et al., 1994). The deletion results suggested that the 0A boxes important for regulating *spoIIAp* are beyond −50.

There are two promoters for the *spo0F* gene (Lewandowski et al., 1986; Trach et al., 1988). The downstream promoter (*spo0Fp₂*) is activated during sporulation, and deletion studies have shown that a 0A box located 61 bp upstream of *spo0Fp₂*, which bound Spo0A in vitro, was required for the activation (Strauch et al., 1993). The *spo0A* gene also has two promoter sites (Chibazakura et al., 1991). Studies using promoter-*lacZ* fusions indicated that the sporulation-specific promoter lacking the region upstream of −52 (containing a 0A box at −99, which is within a region protected from DNase I by Spo0A) was not stimulated in stationary-phase cultures (Strauch et al., 1992). The current data indicate that 0A boxes that affect transcription by σ^H containing RNA polymerase may be in a different location and orientation than 0A boxes that stimulate σ^A-containing polymerase.

Promoters Repressed by Spo0A

Four promoters repressed by Spo0A have been studied: the two promoters at *abrB*, one promoter at *spo0F*, and the vegetative-specific promoter for the spo0A gene, *spo0Ap*$_v$. For promoters repressed by Spo0A, the binding site for the protein is 3′ of the transcription initation site.

The *abrB* gene appears to have two promoters that are separated by 12 bp (Perego et al., 1988). Downstream of the promoters are tandem 0A boxes separated by 3 bp. The *spo0Ap*$_v$, which is repressed by Spo0A, has a reverse 0A box located 6 bp downstream of the transcription initiation site. Finally, there is a 0A box 37 bp downstream of *spo0Fp*$_1$ (Fig. 2B). Mutation studies examining whether the 0A boxes downstream of the promoter repressed by Spo0A are required for repression have not been carried out. However, in all cases, proper down-regulation of the promoter in vivo was dependent on a functional Spo0A gene (Perego et al., 1988; Chibazakura et al., 1991; Strauch et al., 1992).

One other comment could be made. The length of DNA protected by Spo0A at its binding sites is not uniform. Sixteen base pairs are protected at the site upstream of *spo0Ap*$_s$ (Fig. 2A). If this were the unit length, then the 33 bp protected at the sites 3′ from *abrBp* would be proportional, because there are two 0A boxes. However, some sites such as the *spo0Fp*$_2$, *spo0Ap*$_s$, and *spo0Fp*$_1$ up to 40 bp are protected with only a single obvious consensus sequence. Either the protein can oligomerize at some sites, or the binding of the protein alters the DNase sensitivity, possibly by changing the DNA conformation. Strauch et al. (1993) reported that binding of Spo0A to the site upstream of *spo0Fp*$_2$ blocked binding to a site separated by 60 bp. A mechanism for this effect was not proposed but could be related to a change in DNA conformation.

0A boxes are somewhat reminiscent of the binding sites for OmpR (reviewed in Mizuno and Mizushima, 1990). At *ompF* and *ompC,* there are multiple OmpR binding sites near the −35 region of the promoter. These sites are required for activation but can be inverted, although they cannot be moved to the opposite face of the DNA. Repression of the *ompF*

promoter requires a site distal from the promoter, where it was proposed the OmpR bound and subsequently promoted a DNA loop by interacting with OmpR bound near the promoter (Slauch and Silhavy, 1991; Huang et al., 1994).

IN VITRO TRANSCRIPTION REGULATION BY Spo0A

Transcription Repression

Transcription repression by Spo0A has only been studied at *abrBp* in vitro. Strauch et al. (1990) demonstrated that adding Spo0A to in vitro reactions blocked single round transcripton from the downstream promoter. We recently extended this result to show that both the Spo0A and Spo0ABD showed nonlinear dependence of inhibition on concentration (Fig. 3). The protein concentration yielding 50% inhibition was different for the two proteins; however, the curves are not directly comparable because the specific activity of the preparation could be different. It is possible that inhibition at *abrBp* is a special case because other promoters inhibited by Spo0A (e.g., *spo0Ap*$_v$) do not have tandem Spo0A binding sites. The mechanism of transcription inhibition by Spo0A is unknown at present. If occupancy of 0A boxes by Spo0A or Spo0A-P is part of the mechanism, Spo0A could act by sterically preventing the binding of the polymerase to the promoter or by interacting with the polymerase to tether it to the promoter.

Because Spo0ABD retains the ability to repress transcription in vitro and *spo0A9V* mutants effectively block in vivo transcription from *abrB* (Perego et al., 1988), we presume that neither the N-terminal domain of Spo0A nor the transcription activation domain is required for repression. This supports the idea that repression may simply be due to steric hinderance. A steric hinderance mechanism may also be reflected in the orientation of the 0A boxes near promoters repressed by Spo0A. We assume that the Spo0A protein binds to DNA in a manner determined by the orientation of the 0A box. If Spo0A activates tran-

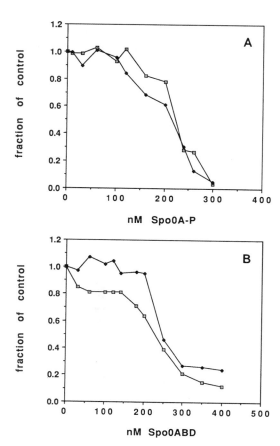

FIGURE 3 Repression of transcription of *abrBp*$_1$ and *abrBp*$_2$ promoters by (A) Spo0A-P and (B) Spo0ABD. An 800-bp *Hin*dIII-*Eco*RI fragment from pJM5134 (Perego et al., 1988) was excised and filled with the Klenow fragment to create blunt ends. This fragment was used as the template (at 5 nM) in transcription assays composed as described in Bird et al. (1993). The transcription products were separated by electrophoresis through polyacrylamide gels containing 7 M urea and localized and quantitated as described in Bird et al. (1993). The results are presented as the fraction of the extent of transcription in the absence of either Spo0A-P or Spo0ABD. Spo0A-P or Spo0ABD were preincubated with the template at 37°C. RNA polymerase was added, and after 3 min, nucleotides allowing transcript elongation were added. After a 3-min initiation-elongation reaction, heparin was added. This is a multiple round initiation assay. Open symbols are transcripts from *abrBp*$_1$; closed symbols are from *abrBp*$_2$.

scription by binding to reverse 0A boxes (e.g., at *spoIIG*), then at *abrBp* the activation region of Spo0A would not be close to the RNA polymerase.

The contribution of phosphorylation to the repression at *abrB*, and most likely *spo0Ap*$_v$, can be explained by the increased binding affinity of the Spo0A-P relative to Spo0A for the 0A boxes. This increase has been documented to be at least 20-fold for the 0A boxes at *abrB* by DNase I footprinting studies (Trach et al., 1991).

Transcription Activation

In vitro studies have been conducted at promoters used by polymerase containing either σ^H or σ^A. The results to date suggest some differences between these cases. Efficient transcription initiation in vitro at the σ^H-dependent *spoIIAp* required a supercoiled DNA template (Bird et al., 1992). The supercoiled form of *spoIIAp* was moderately active, and adding recombinant σ^H to core RNA polymerase allowed production of appreciable levels of RNA in vitro. Spo0A stimulated transcription at this promoter in both phosphorylated and nonphosphorylated forms, but the stimulation by Spo0A-P was far greater. The rates of initiation at this promoter were too fast to permit analysis by standard techniques (Bird and Spiegelman, unpublished data). These data suggest that Spo0A-P is an amplifier of transcription from *spoIIAp* rather than an "on-off" switch. In vivo expression of *spoIIA* is also regulated by Spo0A-P through its repression of *abrB*, because AbrB represses the promoter for the *spo0H* gene that encodes σ^H (Dubnau et al., 1988; Strauch and Hoch, 1993). The direct effect of Spo0A-P stimulation of transcription from *spoIIAp* could be to accelerate the developmental pathway. The 0A boxes at *spoIIAp* and at *spo0Ap*$_s$ are in the forward orientation. Also, they are in position (see Fig. 2B) where they might not interact directly with the polymerase, although this remains to be tested.

One hypothesis for Spo0A stimulation at these promoters is through an effect on DNA structure rather than through direct interaction with the polymerase. As discussed above, DNase I footprinting at *spo0Fp* suggested that Spo0A alters DNA structure. At *spoIIAp*, a wide range of bases was protected by Spo0A in the absence of polymerase (Baldus et al., 1994; Trach et al., 1991). Alternatively, Spo0A could interact differently with polymerases containing different σ-factors. Recent investigations of the α-subunit of *E. coli* RNA polymerase indicate that there may be two different sites of contact on that subunit for transcription activators (reviewed in Ishihama, 1993; Russo and Silhavy, 1992; and see below). Comparison of the effect of mutations in the α-subunit suggested that transcription activators that bind in the −35 region of a promoter contact one site on α and activators that bind to sites farther upstream from the promoter use another site. Thus, it is possible that Spo0A could use two different sites on the polymerase to activate transcription, depending on where it bound.

The interaction of Spo0A-P with the σ^H-containing polymerase may include an effect on the interaction of σ^H with the core enzyme. It is not known how this σ-replacement is carried out. Geiduschek and coworkers studied the σ-replacement during bacteriophage infection in *B. subtilis* and demonstrated that the alternative σ, σ^{28}, could displace σ^A by competition (Chelm et al., 1982). The analogous information is not known for sporulation-specific σ-subunits. We found an apparent synergy between σ^H and Spo0A-P on transcription stimulation of the *spoIIAp* because σ^H was considerably more effective at directing transcription in the presence of Spo0A-P than in its absence. This effect is seen in Fig. 4. These data could imply that the association of σ^H with core polymerase is enhanced in the presence of the *spoIIAp* and the activating factor Spo0A-P, possibly by promoting formation of promoter-polymerase complexes.

The transcription activation by Spo0A at promoters used by RNA polymerase containing σ^A has been most thoroughly studied at *spoIIGp*. Moran and coworkers have shown

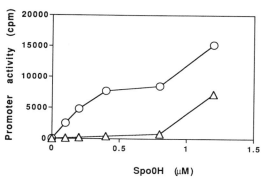

FIGURE 4   Stimulation of transcription from *spoIIAp* by Spo0A-P and σ^H. The template used contained *spoIIAp* cloned adjacent to ribosomal RNA terminators, and the transcription reaction conditions and analysis of product have been described (Bird et al., 1992). Core RNA polymerase was incubated with the DNA template, increasing amounts of recombinant σ^H, either without Spo0A (triangles) or with Spo0A-P (circles). As seen in the presence of Spo0A-P, the stimulation of transcription by low levels of the σ-factor was enhanced. At higher inputs of σ^H, the increase in transcription was the same in the presence and absence of Spo0A-P. Thus, the presence of Spo0A-P increased the stimulation of transcription by the σ-factor.

that phosphorylation of Spo0A increased the binding of the protein to the 0A boxes in the promoter region, and analysis of mutations in these boxes strongly implied that they play a role in transcription initiation (Satola et al., 1991; Satola et al., 1992; Baldus et al., 1994). Also, both Moran's group (Baldus et al., 1994) and our laboratory (Bird et al., unpublished data) have shown that the presence of Spo0A-P enhances the appearance of a DNAse I footprint by the RNA polymerase. However, we found that in the absence of initiating nucleotides, it is not possible to demonstrate a Spo0A-RNA polymerase-DNA complex by gel retardation whether the Spo0A is phosphorylated or not. This result indicates that the effect of Spo0A phosphorylation on the initiation reaction is probably not primarily formation of a stable preinitiation complex.

To examine the mechanism of the activation of *spoIIGp* by Spo0A-P, we measured the kinetics of transcription initiation (Bird et al., unpublished data). The rate of transcription initation at *spoIIGp* was independent of both DNA and RNA polymerase concentration. This finding suggested that RNA polymerase formed a rapid reversible complex with *spoIIGp* regardless of whether Spo0A was present. This complex could not initiate RNA synthesis, and the rate-limiting step for initiation was subsequent to this complex formation. By contrast, both the rate of initiation and the level of complexes formed were stimulated slightly by Spo0A and dramatically by Spo0A-P. These kinetics indicated that Spo0A and Spo0A-P acted on the polymerase-promoter complex and stimulated conversion of this complex to a form that could initiate transcription. The effect of phosphorylation of Spo0A was to increase the rate and extent of the conversion by at least 10-fold.

DNase I footprinting experiments showed that the polymerase was able to bind to the promoter on its own, although the region near the initiation site was poorly protected relative to the protection seen at strong promoters (see below). Addition of Spo0A led to a second form of the footprint, which contained unique sites in the −27 region of the promoter that were hypersensitive to DNase I. When Spo0A-P was added to the footprint assay, a third form of the footprint was detected, which showed considerably more protection of the promoter region and a reduction in the intensity of the DNase I-sensitive sites around position −27.

To examine the effects of Spo0A and Spo0A-P further, we followed the kinetics of formation of the various complexes with the *spoIIGp* by examining a time course of appearance of the DNase I footprint (Fig. 5). RNA polymerase alone rapidly formed a clearly visible complex with the DNA fragment (this is the C_I complex in Fig. 6). The polymerase protected promoter sequences near the −35 consensus sequence but not those near the −10 consensus, indicating that the enzyme was not in a state in which it could initiate transcription. Furthermore, the footprint was not stable for extended incubation and faded by 15 min,

supporting the idea that this association was not stable. The C_I complex is the equivalent of the promoter-polymerase complex predicted from the kinetic data, and as predicted by the kinetic data, its formation was not dependent on Spo0A.

In the presence of Spo0A, an unusual structure formed with conspicuous DNase I-hypersensitive sites near the −27 region of the promoter (the C_{II} complex). In the C_{II} complex, the −10 region was more completely protected than in the C_I complex. A plausible model for this complex is that as a result of interaction with Spo0A, the polymerase contacted the −10 and −35 regions of the promoter in preparation for transcription initiation. Because the −10 and −35 regions are 5 bp farther apart than for the consensus promoter, the simultaneous binding of the two regions was accommodated by helix distortion (Fig. 6). Because Spo0A only weakly stimulated transcription, the Spo0A-polymerase promoter C_{II} complex must not be a form that can initiate rapidly or efficiently. Thus, although Spo0A stimulated the formation of the C_{II} complex, it lacked the ability to stimulate the isomerization of the C_{II} complex to one that was competent for initiation.

When Spo0A-P was added along with polymerase to the DNase I assays, a third type of complex formed very rapidly (the C_{III} complex). In the C_{III} complex, there was extensive protection of the promoter region, and the DNase I-hypersensitive sites seen in the C_{II} complexes were absent. Because Spo0A-P stimulated transcription to high levels, the C_{III} complex must be able to initiate rapidly. We found that at low inputs of Spo0A-P, C_{II} complexes were detected, suggesting that Spo0A-P stimulated first the formation of C_{II} complexes and then the isomerization of these complexes to C_{III} complexes. We propose that the distortion of the DNA was not seen in the kinetic assay in Fig. 5, because in the presence of Spo0A-P, the conversion of C_{II} to C_{III} complexes is very rapid. Thus, phosphorylation of Spo0A enabled it to convert the polymerase-

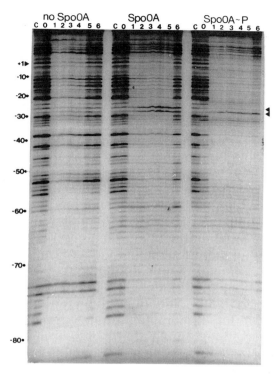

no SpoOA   SpoOA   SpoOA~P
C 0 1 2 3 4 5 6   C 0 1 2 3 4 5 6   C 0 1 2 3 4 5 6

+1▶
-10•
-20•
-30•
-40•
-50•
-60•
-70•
-80•

FIGURE 5   DNase I footprint assay of formation of complexes at *spoIIGp*. A *Hind*III to PvuII DNA fragment containing *spoIIGp* (see Bird et al., 1993) was labeled at the *Hind*III end with [γ-^{32}P]ATP and polynucleotide kinase. The fragment was incubated in transcription reaction buffer either alone or in the presence of either SpoOA or SpoOA-P. All reactions contain 0.4 mM ATP, which allows formation of a dinucleotide at *spoIIGp*. An aliquot of each reaction was removed and treated with 4 μg/ml DNase I for 10 s. The reaction was stopped with 10 mM EDTA, 0.1% sodium dodecyl sulfate. Labeled DNA was recovered by ethanol precipitation, redissolved in formamide gel loading buffer, and electrophoresed on a 6% polyacrylamide gel containing 7 M urea. To examine the rate of complex formation, RNA polymerase (100 nM, final concentration) was added to the template, and after different incubation times, samples were removed and treated with DNase for 10 s as described above. The incubation times (after adding the polymerase) are lane 1, 5 s; lane 2, 30 s; lane 3, 1 min; lane 4, 2 min; lane 5, 5 min, lane 6, 15 min. For each set, the lane marked C shows the pattern obtained when no proteins were added, and the lane marked 0 is the pattern obtained before adding RNA polymerase. The nucleotide positions, relative to the transcription initiation site, are given at the left. Without SpoOA or SpoOA-P, a complex between the polymerase and the DNA was formed very rapidly. From our kinetic data, this complex (C$_I$) cannot initiate RNA synthesis when challenged with nucleotides and heparin and therefore represents a heparin-sensitive complex. When SpoOA was present, RNA polymerase formed a second type of complex (C$_{II}$), as indicated by the DNase I-hypersensitive sites (arrowheads at the right). This complex also cannot initiate RNA synthesis rapidly when challenged with heparin and thus represents a second heparin-sensitive complex. In the presence of SpoOA-P, a complex (C$_{III}$) formed that lacked one of the sites hypersensitive to DNase I and showed increased protection in the −10 region. This complex cannot complete transcripts when challenged with heparin and nucleotides and so represents a stage before the heparin-resistant complex (HR). The kinetic data suggest that the complex formed in the presence of SpoOA-P (C$_{III}$) can initiate RNA synthesis rapidly. We view the C$_{III}$ complex as equivalent to the activated intermediate in the transition to HR complex (Bird et al., unpublished data).

promoter complex to a form that was transcription-competent. The elimination of the hypersensitive sites suggests that the strain in the promoter DNA has been released. Because the polymerase must contact the −10 consensus to be able to initiate rapidly, we propose that the phosphorylated form of SpoOA enables the polymerase to release contacts with the −35 region and still stay in an association with the template that permits efficient transcription initiation.

The transcription kinetics assays we carried out predicted that SpoOA-P stimulated the conversion of an unstable intermediate to one that could initiate RNA synthesis rapidly. The transcription assays used heparin resistance as a measure of transcription initiation. Control experiments showed that heparin resistance of the polymerase at *spoIIGp* required the addition of GTP and ATP, and because the DNase I assays in Fig. 5 contained only ATP, the C$_{III}$ complex must precede the heparin-resistant state.

We examined the kinetics of transcription initiation at *spoIIG* using the SpoOABD. SpoOABD stimulated transcription but to a lower level than did SpoOA-P. Also, the rate of initiation was much slower. This effect is seen in Fig. 7, in which the transcripts versus time of incubation in an initiation reaction are plotted.

FIGURE 6 Reaction mechanism for transcription stimulation at *spoIIGp*. In the cartoon of initiation, the following symbols are used: double lines represent the DNA template; the large oval represents RNA polymerase; the small circle attached to a small oval represents the two-domain structure of Spo0A. The predicted −10 and −35 consensus sequences at the promoter are indicated on the DNA strands. These sequences are separated by 22 bp instead of the expected 17 to 18 bp. The initial binding of the polymerase and DNA, which is independent of Spo0A, protects only the −35 site of the promoter (C_I complex). The binding of Spo0A and RNA polymerase to the DNA creates DNase I-hypersensitive sites in the −27, −28 region of the promoter and increased protection near the −10 region (C_{II} complex). The DNase I-hypersensitive sites are presumed to result from a distortion of the DNA helix, as represented. The binding of two Spo0A proteins is shown because there are two 0A boxes. Phosphorylation of Spo0A (represented by a change in shading pattern) is presumed to cause a change in the shape of the protein. In the presence of Spo0A-P, C_{II} complexes isomerize to C_{III} complexes, which are characterized by increased protection of the −10 region and loss of the hypersensitive sites. The change is represented by release of the −35 contacts between the polymerase and the promoter, which were presumed to cause the helix distortion. The release of the −35 contacts allows the polymerase to contact the −10 region. The contacts with the −10 region permit the polymerase in the C_{III} complex to synthesize a short RNA (represented by the thick wavy line) on addition of the initiating nucleotides ATP and GTP. The initiated complex is designated HR because it is resistant to the polymerase inhibitor heparin.

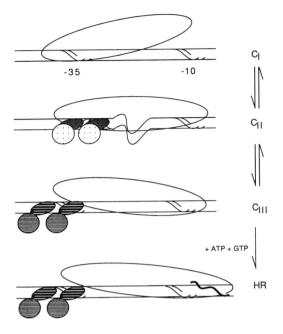

FIGURE 7 Formation of initiation complexes at *spoIIGp*: stimulation by Spo0A, Spo0A-P, and Spo0ABD. Transcription initiation reactions were composed and analyzed as described in Bird et al. (1992). Reactions contained 100 nM Spo0A-P (closed circles), Spo0ABD (triangles), or Spo0A (open circles), GTP, ATP, and a DNA fragment containing *spoIIGp*. At time zero, RNA polymerase was added to the reaction, and after various incubation times, samples were withdrawn and challenged with a mixture containing CTP, UTP, and heparin. The percentage of template containing an initiated complex is plotted as a function of the time of the initiation reaction assay. The reaction containing Spo0A-P initiated very rapidly, whereas those containing the binding domain showed a pronounced lag, although the final level was only slightly lower than that seen with Spo0A-P.

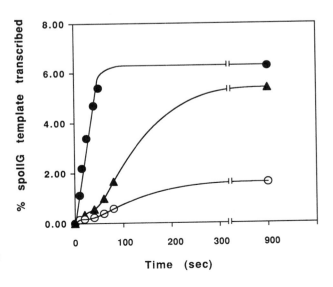

Three conclusions can be drawn from these data. First, the C-terminal domain contains the transcription activation function of Spo0A. Second, the N-terminal domain of Spo0A inhibits transcription activation, so that when it is removed, transcription activation is not dependent on phosphorylation. Third, the region of Spo0A missing from the binding domain contains amino acids that interact with the polymerase to stimulate initiation.

Spo0A AS A MODEL FOR TRANSCRIPTION RESPONSE REGULATORS

Transcription initiation is a multistep process, and potentially each of the steps could be the target for regulation. In general, positive transcription regulators act to overcome a defect in the transcription initiation components for an operon that poses a barrier to initiation (see Adhya and Garges, 1990, for a review). For example, at *spoIIGp*, the defect is likely to be the extended spacing of the −10 and −35 consensus sequences. The role of Spo0A is to aid in a DNA distortion to permit the polymerase to contact both consensus sequences and then to resolve the complex to allow the polymerase to initiate RNA synthesis. Regulation at each promoter should reflect unique characteristics of the combination of the promoter and polymerase. Care should be taken in describing transcription initiation because using the most general description of the process may obscure important differences in mechanisms of action of positive regulators. We briefly review some other positive regulators to compare them with Spo0A.

The transcription regulator for the *mer* operon of Tn*501* has been shown to both repress and activate transcription, and both activities involve binding near the −35 region of the promoter (Frantz and O'Halloran, 1990). In the presence of mercury and the wild-type MerR protein (or in the presence of a mercury-independent mutant form of MerR), the DNA in the promoter is distorted (Parkhill et al., 1993). The distortion has been proposed to align the −10 and −35 regions of the promoter

because the spacing between them is 19 bp instead of the optimal 17 bp for *E. coli* RNA polymerase. This conformation change is similar to the distortion we have seen at *spoIIG*. However, at *spoIIG*, the DNA distortion, indicated by the DNase I-hypersensitive sites, was dependent on the presence of both polymerase and Spo0A, not on Spo0A alone. Furthermore, Spo0A-P was required for the transition between C_{II} and C_{III} complexes, showing that Spo0A-P acts beyond the promoter distortion step. Potentially, the spacing between the −10 and −35 at *spoIIGp* is too great to allow compensation by DNA bending alone, so that a second regulator-dependent step is required.

In contrast to MerR and Spo0A, the *E. coli* transcription factor NtrC acts to overcome a barrier to transcription initiation imposed by the nature of the σ-factor (σ^{54}) used to transcribe the nitrogen regulon. Among the unique characteristics of σ^{54} is its inability to stimulate DNA opening at promoters (Sasse-Dwight and Gralla, 1988; and reviewed in Geiduschek, 1993; Merrick, 1993; Morett and Segovia, 1993). The defect in σ-function may explain the requirement for ATP hydrolysis during NtrC activation (Austin and Dixon, 1992; Weiss et al., 1991), which is not seen with other response regulators (Morett and Segovia, 1993). Also, oligomerization of NtrC appears to be essential to its activity as a transcription activator, and oligomer formation is stimulated by phosphorylation (Porter et al. 1993) (see Chapter 9). This feature separates NtrC from regulators such as Spo0A, which is a monomer in both phosphorylated and unphosphorylated forms (Grimsley et al., 1994) and may also reflect the unusual nature of σ^{54}.

The *E. coli* transcription factor PhoB stimulates transcription from promoters for genes involved with phosphate uptake and metabolism (reviewed in Wanner, 1993). The −35 sequences of the *pho* regulon promoters have little identity to the consensus sequence for a σ^{70}-holoenzyme (Kimura et al., 1989). PhoB binds to a specific sequence near the −10 site of the promoter, and binding is dramatically stimulated by phosphorylation (Makino et al., 1993). Thus, the initiation defect overcome by PhoB is the lack of the −35 consensus in the *pho* regulon promoters. Mutations in the gene for σ^{70}, *rpoD,* have been shown to affect PhoB activation (Makino et al., 1993). For example, point mutations in the first helix of the helix-turn-helix region involved with DNA binding decrease PhoB-directed stimulation, demonstrating that this transcription regulator interacts with the σ^{70}-subunit. By contrast, deletion of the region of the σ-subunit needed for recognition of the −35 consensus sequence does not affect PhoB stimulation of transcription (Kumar et al., 1994). Certain types of promoter mutations (the so-called extended −10 sequences) have been shown to compensate for deletion of −35 sequences, allowing RNA polymerase to open the DNA at these otherwise defective promoters (Keilty and Rosenberg, 1987; Chan and Busby, 1989). PhoB may act as a "−10 extension" by interacting with σ^{70} (Kumar et al., 1994).

The action of PhoB-P at the *pho* regulon promoters is similar to phage λcI protein stimulation of the P_{RM} promoter in that both compensate for lack of consensus at the −35 sequence. There is genetic evidence that cI interacts directly with σ^{70} (Li et al., 1994), although mutations in the α-subunit of the polymerase may also affect cI stimulation (Wegrzyn et al., 1992). CI binds to the −35 region of the P_{RM} and stimulates conversion of closed to open complexes (Meyer and Ptashne, 1980; Hawley and McClure, 1982; Hwang et al., 1988). To a first approximation, cI, PhoB-P, and Spo0A-P appear to act in a similar manner, because all three stimulate conversion of an initial complex to one that can initiate RNA synthesis. There is evidence that Spo0A may interact with the major σ-subunit σ^A, because there are mutants in the gene for σ^A (originally called *crsA* mutants) (Kawamura et al., 1985) that affect regulation of sporulation. However, the precise details of transcription stimulation by PhoB-P, cI, and Spo0A-P are certain to differ in ways that reflect the structures of promoters they activate and their respective binding sites within those promoters.

OmpR, which directly regulates the transcription of genes for the outer membrane porins OmpF and OmpC in *E. coli,* illustrates an alternative mechanism of positive regulation (reviewed in Mizuno and Mizushima, 1990). Multiple OmpR binding sites are needed for transcription stimulation in the −70 to −100 region of the *ompF* and *ompC* promoters, and at *ompF,* sites in the −45 to −70 along with a site between −351 to −384 are needed for OmpR-mediated repression transcription (Tsung et al., 1989; Slauch and Silhavy, 1991; Huang et al., 1994). OmpR stimulation of transcription requires its phosphorylation, and its binding to different sites at the *ompF* promoter is related to the degree of its phosphorylation (Aiba et al., 1989; Rampersaud et al., 1994). The binding of OmpR may be synergistic with RNA polymerase, leading to increased stabilization of closed complexes, particularly at promoters that have low consensus −10 and/or −35 sequences (Tsung et al., 1990). Stabilization of closed complexes could stimulate transcription of the *ompF* and *ompC* genes. This mechanism would distinguish OmpR from NtrC and from Spo0A activity at *spoIIGp* (but perhaps not at *spoIIAp;* see below). A wide variety of mutations of OmpR have been described (Chapter 7). These mutations show that the various transcription regulation functions can be disrupted independently (Russo et al., 1993).

Genetic evidence demonstrates that unlike PhoB, cI, and NtrC, OmpR stimulates transcription through an interaction with the α-subunit of the polymerase (reviewed in Ishihama, 1993; Russo and Silhavy, 1992). Mutations in the α-subunit that affect OmpR stimulation are located in the C terminus and are found within a region of less than 100 amino acids. Mutations can be found that affect not only OmpR but at least six other positive regulators of transcription. α-Mutations may have different effects on the various functions of OmpR in porin gene regulation (e.g., see Slauch et al., 1991; Sharif and Igo, 1993).

Another activator that interacts with the α-subunit is the catabolite repression protein CAP (Ishihama, 1993; Kolb et al., 1993; Russo and Silhavy, 1992; West et al., 1993). Like Spo0A, CAP stimulates transcription from several different promoters and from binding sites located in different places. Furthermore, mutations in α have been used to show that there are two types of binding sites for CAP on the polymerase and that each of these is associated with a DNA binding site in a different place relative to the promoter. Little is known about α-mutations in *B. subtilis,* although a search for mutants that affect sporulation would be very useful.

Further understanding of transcription activation by Spo0A will depend on solving the structure of the Spo0A binding domain alone and the entire Spo0A protein. Further analysis of *spo0A* mutations and their effect on transcription in vitro and in vivo is also required. For example, recent studies of the CAP protein in *E. coli* have identified a region of positive and nonpolar amino acids that are involved in transcription stimulation (Zhang et al., 1992). These residues are in a loop near the DNA binding motif in CAP. The amino acid content of this region is reminiscent of the carboxy terminus of Spo0ABD and gp4, so that the latter should be a prime target for mutagenesis. One of the essential issues is whether the effect of phosphorylation of Spo0A on DNA binding can be separated from an effect on transcription complex isomerization. The kinetic data suggest that there may be two separate effects. Also, the portion of the RNA polymerase that directly contacts Spo0A-P needs to be determined.

PERSPECTIVES

The genetic experiments and the in vitro assays lead to the conclusion that the transcription activation functions of Spo0A lie in the C-terminal domain and that these functions are inhibited by the N-terminal domain. In this regard, Spo0A is like other members of the response regulator family.

The transcription regulation properties of Spo0A are more diverse than have been demonstrated for other response regulators. First, Spo0A both inhibits and stimulates transcription. Given the location of the DNA binding sites that are critical for transcription inhibition, Spo0A may inhibit through a polymerase interaction method rather than an exclusion method. If so, there might be a common thread for inhibition and stimulation. Second, Spo0A appears to be unique in that it stimulates transcription initiation by holoenzymes containing different σ-subunits. The stimulation may result from the interaction of Spo0A with the σ-subunit in some cases (e.g., *spoIIG*) and the α-subunit in other cases (e.g., *spoIIA*). Also, the orientation of Spo0A binding sites differs at different promoters. The different orientations may also reflect interaction with different polymerase subunits. These hypotheses need to be tested directly. Third, the binding sites for Spo0A differ in number and arrangement for different promoters. For example, at *spoIIE,* the 0A box in the region of the −35 consensus is a single box, whereas at *spoIIG,* there are two 0A boxes. The simplest interpretation would be that a single 0A box is sufficient. This would argue that Spo0A-P acts in a monomeric form in support of the studies of its solution structure. However, detailed analysis of the role of the various 0A boxes is required.

Phosphorylation of Spo0A appears to have multiple functions. First, there is evidence that phosphorylation increases the DNA binding affinity and that, at the *spoIIGp*, phosphorylation permits binding of the protein to the site 2 0A boxes, which are necessary for transcription stimulation. Also, the analysis of transcription initiation indicates that phosphorylation confers on Spo0A the ability to drive a conformational change in the ternary complex of Spo0A-P-RNA polymerase-promoter, which must be the inhibited step in the unphosphorylated form. The conformation change appears to be the most significant step in transcription activation of *spoIIGp.* Further structural and mutational analyses are needed to test this model.

ACKNOWLEDGMENTS

We thank A. D. Grossman, C. P. Moran, Jr., and P. Youngman for comments and communicating unpublished results and Loverne Duncan for help in purifying RNA polymerase. We are grateful for the continued enthusiasm, encouragement, and supplies of phosphorelay proteins from J. A. Hoch and the members of his laboratory.

Our work is supported by a grant from the Natural Sciences and Engineering Research Council of Canada.

REFERENCES

Adhya, S., and S. Garges. 1990. Positive Control. *J. Biol. Chem.* **265**:10797–10800.

Aiba, H., F. Kakasai, S. Mizushima, and T. Mizuno. 1989. Phosphorylation of a bacterial activator protein, OmpR, by a protein kinase, EnvZ, results in stimulation of its DNA-binding ability. *J. Biochem.* **106**:5–7.

Austin, S., and R. Dixon. 1992. The procaryotic enhancer binding protein NTRC has an ATPase activity which is phosphorylation and DNA dependent. *EMBO J.* **11**:2219–2228.

Baldus, J. M., B. D. Green, P. Youngman, and C. P. Moran, Jr. 1994. Phosphorylation of *Bacillus subtilis* transcription factor Spo0A stimulates transcription from the *spoIIG* promoter by enhancing binding to weak 0A boxes. *J. Bacteriol.* **176**:296–306.

Barthelemy, I., and M. Salas. 1989. Characterization of a new procaryotic transcription activator and its DNA recognition site. *J. Mol. Biol.* **208**:225–232.

Bird, T., D. Burbulys, J.-J. Wu, M. A. Strauch, J. A. Hoch, and G. B. Spiegelman. 1992. The effect of supercoiling on the in vitro transcription of the *spoIIA* operon from *Bacillus subtilis. Biochemie* **74**:627–634.

Bird, T., J. Grimsley, J. A. Hoch, and G. B. Spiegelman. 1993. Phosphorylation of Spo0A activates its stimulation of in vitro transcription from the *Bacillus subtilis spoIIG* operon. *Mol. Microbiol.* **9**:741–749.

Bird, T., and G. B. Spiegelman. Unpublished data.

Bird, T., J. K. Grimsley, J. A. Hoch, and G. B. Spiegelman. Unpublished data.

Bowrin, V., R. Brissette, and M. Inouye. 1992. Two transcriptionally active OmpR mutants that do not require phosphorylation by EnvZ in an *Escherichia coli* cell free system. *J. Bacteriol.* **174**:6685–6687.

Brissette, R. E., K. Tsung, and M. Inouye. 1991. Intramolecular second site revertants to the phosphorylation site mutation in OmpR, a kinase dependent transcriptional activator in *Escherchia coli. J. Bacteriol.* **173**:3749–3755.

Brissette, R. E., K. Tsung, and M. Inouye. 1992. Mutations in a central highly conserved non-DNA binding region of OmpR, an *Escherichia coli* transcriptional activator, influence its DNA binding ability. *J. Bacteriol.* **174**:4907–4912.

Brown, D. P., L. Ganova-Raeva, B. D. Green, S. R. Wilkinson, M. Young, and P. Youngman. 1994. Characterization of *spo0A* homologues in diverse *Bacillus* and *Clostridium* species identifies a probable DNA-binding domain. *Mol. Microbiol.* **14**:411–436.

Burbulys, D., K. A. Trach, and J. A. Hoch. 1991. Initiation of sporulation in *B. subtilis* is controlled by a multicomponent phosphorelay. *Cell* **64**:545–552.

Chan, B., and S. Busby. 1989. Recognition of nucleotide sequences at the *Escherichia coli* galactose operon P1 promoter by RNA polymerase. *Gene* **84**:227–236.

Chelm, B. K., J. J. Duffy, and E. P. Geiduschek. 1982. Interaction of *Bacillus subtilis* RNA polymerase core with two specificity determining subunits. Competition between σ and the SP01 gene 28 protein. *J. Biol Chem.* **257**:6501–6508.

Chibazakura, T., F. Kawamura, and T. Takahashi. 1991. Differential regulation of *spo0A* transcription in *Bacillus subtilis*: glucose represses promoter switching at the initiation of sporulation. *J. Bacteriol.* **173**:2625–2632.

Choi, S. H., and E. P. Greenberg. 1991. The C-terminal region of the *Vibrio fisheri* LuxR protein contains an inducer independent *lux* gene activation domain. *Proc. Natl. Acad. Sci. USA* **88**:11115–11119.

Delgado, J., S. Forst, S. Harlocker, and M. Inouye. 1993. Identification of a phosphorylation site and functional analysis of conserved aspartic acid residues of OpmR, a transcriptional activator for *ompF* and *ompC* in *Escherichia coli. Mol. Microbiol.* **10**:1037–1047.

Dubnau, E., J. Weir, G. Nair, L. Carter III, C. P. Moran, Jr., and I. Smith. 1988. *Bacillus* sporulation gene *spoOH* codes for σ³⁰ (σ^H). *J. Bacteriol.* **170**:1054–1062.

Errington, J. 1993. *Bacillus subtilis* sporulation: regulation of gene expression and control of morphogenesis. *Microbiol. Rev.* **57**:1–33.

Ferrari, F. A., K. Trach, D. LeCoq, J. Spence, E. Ferrari, and J. A. Hoch. 1985. Characterization of the *spo0A* locus and its deduced product. *Proc. Natl. Acad. Sci. USA* **82**:2647–2651.

Frantz, B., and T. V. O'Halloran. 1990. DNA distortion accompanies transcriptional activation by the metal responsive gene regulatory protein MerR. *Biochemistry* **29**:4747–4751.

Geiduschek, E. P. 1993. Two procaryotic transcriptional enhancer systems. *Prog. Nucleic Acids Res. Mol. Biol.* **43**:109–133.

Green, B. D., M. G. Bramucci, and P. Youngman. 1991. Mutant forms of SpoOA that affect sporulation initiation: a general model for phosphorylation mediated activation of bacterial signal transduction proteins. *Semin. Dev. Biol.* **2**:21–29.

Grimsley, J. K., R. B. Tjalkens, M. A. Strauch, T. H. Bird, G. B. Spiegelman, Z. Hostomsky, J. M. Whiteley, and J. A. Hoch. 1994. Subunit composition and domain structure of the SpoOA sporulation transcription factor of *Bacillus subtilis. J. Biol Chem.* **269**:16977–16982.

Hawley, D. K., and W. R. McClure. 1982. Mechanism of activation of transcription initiation from the λP$_{RM}$ promoter. *J. Mol. Biol.* **157**:493–525 .

Hoch, J. A. 1993. Regulation of the phosphorelay and the initiation of sporulation in *Bacillus subtilis. Annu. Rev. Microbiol.* **74**:441–466.

Hoch, J. A., K. Trach, F. Kawamura, and H. Saito. 1985. Identification of the transcriptional suppressor *sof-1* as an alternation in the SpoOA protein. *J. Bacteriol.* **161**:552–555.

Huang, K.-J., J. L. Schieberl, and M. M. Igo. 1994. A distant upstream site involved in the negative regulation of the *Escherichia coli opmF* gene. *J. Bacteriol.* **176**:1309–1315.

Hwang, J.-J.., S. Brown, and G. N. Gussin. 1988. Characterization of a doubly mutant derivative of the λ P$_{RM}$ promoter. Effects of mutations on activation of P$_{RM}$. *J. Mol. Biol.* **200**:695–708.

Ireton, K., D. Z. Rudner, K. J. Siranosian, and A. D. Grossman. 1993. Integration of multiple developmental signals in *Bacillus subtilis* through the SpoOA transcription factor. *Genes Dev.* **7**:283–294.

Ishihama, A. 1993. Protein-protein communication within the transcription apparatus. *J. Bacteriol.* **175**:2483–2489.

Kahn, D., and G. Ditta. 1991. Molecular structure of *fixJ*: homology of the transcription activator domain with the −35 binding domain of sigma factors. *Mol. Microbiol.* **5**:987–997.

Kawamura, F., and H. Saito. 1983. Isolation and mapping of a new suppressor mutation of an early sporulation gene *spoOF* mutation in *Bacillus subtilis. Mol. Gen. Genet.* **192**:330–334.

Kawamura, F., L. Wang, and R. H. Doi. 1985. Catabolite resistant sporulation (*crsA*) mutations in the *Bacillus subtilis* RNA polymerase σ^{43} gene (*rpoD*) can suppress and be suppressed by mutations in *spoO* genes. *Proc. Natl. Acad. Sci. USA* **83**:8124–8128.

Keilty, S., and M. Rosenberg. 1987. Constitutive function of a positively regulated promoter reveals new sequences essential for activity. *J. Biol. Chem.* **262**:6389–6395.

Kimura, S., K. Makino, H. Shinagawa, M. Amemura, and A. Nakata. 1989. Regulation of the phosphate regulon in *Escherichia coli:* characterization of the promoter of the *pstS* gene. *Mol. Gen. Genet.* **215**:374–380.

Kolb, A., K. Igarashi, A. Ishihama, M. Lavigne, M. Buckle, and H. Buc. 1993. *E. coli* RNA polymerase, deleted in the C-terminal part of its alpha-subunit, interacts differently with the cAMP-CRP complex at the *lac*P1 and at the *gal*P1 promoter. *Nucleic Acids Res.* **21**:319–326.

Kudoh, J., T. Ikeuchi, and K. Kurahashi. 1985. Nucleotide sequence of the sporulation gene *spoOA* and its mutant genes of *Bacillus subtilis. Proc. Natl. Acad. Sci. USA* **82**:2665–2668.

Kumar, A., B. Grimes, N. Fujita, K. Makino, R. A. Malloch, R. S. Hayward, and A. Ishihama. 1994. Role of the sigma^{70} subunit of *Escherichia coli* RNA polymerase in transcription activation. *J. Mol. Biol.* **235**:405–413.

Lewandowski, M., E. Dubnau, and I. Smith. 1986. Transcriptional regulation of the *spoOF* gene of *Bacillus subtilis. J. Bacteriol.* **168**:870–877.

Li, M., M. Moyle, and M. M. Susskind. 1994. Target of transcriptional activation function of phage λ *cI* protein. *Science* **263**:75–77.

Losick, R., P. Youngman, and P. J. Piggot. 1986. Genetics of endospore formation in *Bacillus subtilis. Annu. Rev. Genet.* **20**:625–669.

Makino, K., M. Amemura, S.-K. Kim, A. Nakata, and H. Shinagawa. 1993. Role of the σ^{70} subunit of RNA polymerase in transcriptional activation by activator protein PhoB in *Escherichia coli. Genes Dev.* **7**:149–160.

Mencia, M., M. Salas, and F. Rojo. 1993. Residues of the *Bacillus subtilis* phage φ29 transcriptional activator required both to interact with RNA polymerase and to activate transcription. *J. Mol. Biol.* **233**:695–704.

Menon, K. P., and N. L. Lee. 1990. Activation of *ara* operon by a truncated AraC protein does not require inducer. *Proc. Natl. Acad. Sci. USA* **87**:3708–3712.

Merrick, M. J. 1993. In a class of its own—the RNA polymerase sigma factor σ^{54} (σ^{N}). *Mol. Microbiol.* **10**:903–909.

Meyer, B. J., and M. Ptashne. 1980. Gene regulation at the right operator O$_R$ of bacteriophage λ. III. λ repressor directly activates gene transcription. *J. Mol. Biol.* **139**:195–205.

Mizukami, Y., T. Sekiya, and H. Hirokawa. 1986. Nucleotide sequence of gene F of *Bacillus* phage Nf. *Gene* **43**:231–235.

Mizuno, T., and S. Mizushima. 1990. Signal transduction and gene regulation through the phosphorylation of two regulatory components: the molecular basis for the osmotic regulation of the porin genes. *Mol. Microbiol.* **4:**1077–1082.

Morett, E., and L. Segovia. 1993. The σ^{54} bacterial enhancer binding protein family: mechanism of action and phylogenetic relationship of their functional domains. *J. Bacteriol.* **175:**6067–6074.

Olmedo, G., E. G. Ninfa, J. Stock, and P. Youngman. 1990. Novel mutations that alter regulation of sporulation in *Bacillus subtilis*: evidence that phosphorylation of regulatory protein SpoOA controls the initiation of sporulation. *J. Mol. Biol.* **215:**359–372.

Parkhill, J., A. Z. Ansari, J. G. Wright, N. L. Brown, and T. V. O'Halloran. 1993. Construction and characterization of a mercury-independent MerR activator (MerR^{AC}): transcriptional activation in the absence of Hg(II) is accompanied by DNA distortion. *EMBO J.* **12:**413–421.

Parkinson, J. S., and E. C. Kofoid. 1992. Communication modules in bacterial signaling proteins. *Annu. Rev. Genet.* **26:**71–112.

Perego, M., G. B. Spiegelman, and J. A. Hoch. 1988. Structure of the gene for the transition state regulator, *abrB*: regulator synthesis is controlled by the *spoOA* sporulation gene in *Bacillus subtilis*. *Mol. Microbiol.* **2:**689–699.

Perego, M., J.-J. Wu, G. B. Spiegelman, and J. A. Hoch. 1991. Mutational dissociation of the positive and negative regulatory properties of the SpoOA sporulation transcription factor of *Bacillus subtilis*. *Gene* **100:**207–212.

Piggot, P. J., and J. G. Coote. 1976. Genetic aspects of bacterial endospore formation. *Bacteriol. Rev.* **40:**908–962.

Porter, S. C., A. K. North, A. B. Wedel, and S. Kustu. 1993. Oligomerization of NTRC at the *glnA* enhancer is required for transcriptional activation. *Genes Dev.* **7:**2258–2273.

Predich, M., G. Nair, and I. Smith. 1992. *Bacillus subtilis* early sporulation genes *kinA*, *spoOF* and *spoOA* are transcribed by the RNA polymerase containing σ^H. *J. Bacteriol.* **174:**2771–2778.

Rampersaud, A., S. L. Harlocker, and M. Inouye. 1994. The OmpR protein of *Escherichia coli* binds to sites in the *ompF* promoter region in a hierarchial manner determined by its degree of phosphorylation. *J. Biol. Chem.* **269:**12559–12566.

Rojo, F., A. Zaballos, and M. Salas. 1990. Bend induced by phage ϕ29 transcriptional regulator protein p4 in the viral late promoters is required for activation. *J. Mol. Biol.* **211:**713–725.

Russo, F. D., and T. J. Silhavy. 1992. Alpha the Cinderella subunit of RNA polymerase. *J. Biol Chem.* **267:**14515–14518.

Russo, F. D., J. M. Slauch, and T. J. Silhavy. 1993. Mutations that affect separate functions of OmpR, the phosphorylated regulator of porin transcription in *Escherichia coli*. *J. Mol. Biol.* **231:**261–273.

Sasse-Dwight, S., and J. D. Gralla. 1988. Probing the *E. coli glnALG* upstream activation mechanism *in vivo*. *Proc. Natl. Acad. Sci. USA* **85:**8934–8938.

Satola, S., J. M. Baldus, and C. P. Moran, Jr. 1992. Binding of SpoOA stimulates *spoIIG* promoter activity in *Bacillus subtilis*. *J. Bacteriol.* **174:**1448–1453.

Satola, S., P. A. Kirshman, and C. P. Moran, Jr. 1991. SpoOA binds to a promoter used by σ^A RNA polymerase during sporulation in *Bacillus subtilis*. *Proc. Natl. Acad. Sci. USA* **88:**4533–4537.

Sharif, T. R., and M. M. Igo. 1993. Mutations in the alpha subunit of RNA polymerase that affect regulation of porin gene transcription in *Escherichia coli* K12. *J. Bacteriol.* **175:**5460–5468.

Sharrock, R. A., S. Rubenstein, M. Chan, and T. Leighton. 1984. Intergenic suppression of *spoO* phenotypes by the *Bacillus subtilis* mutation *rvtA*. *Mol. Gen. Genet.* **194:**260–264.

Shoji, K., S. Hiratsuka, F. Kawamura, and Y. Kobayashi. 1988. New suppressor mutation *surOB* of *spoOB* and *spoOF* mutations in *Bacillus subtilis*. *J. Gen. Microbiol.* **134:**3249–3257.

Slauch, J. M., F. D. Russo, and T. J. Silhavy. 1991. Suppressor mutations in *rpoA* suggest that OmpR controls transcription by direct interaction with the alpha subunit of RNA polymerase. *J. Bacteriol.* **173:**7501–7510.

Slauch, J. M., and T. J. Silhavy. 1991. *cis*-acting *ompF* mutations that result in OmpR-dependent constitutive expression. *J. Bacteriol.* **173:**4039–4048.

Sonenshein, A. L., J. A. Hoch, and R. Losick (ed.). 1993. *Bacillus subtilis and Other Gram Positive Bacteria*. ASM Press, Washington, D.C.

Spiegelman, G. B., B. Van Hoy, M. Perego, J. Day, K. Trach, and J. A. Hoch. 1990. Structural alterations in the *Bacillus subtilis* SpoOA regulatory protein which suppress mutations at several *spoO* loci. *J. Bacteriol.* **172:**5011–5019.

Stock, J. B., J. A. Ninfa, and A. M. Stock. 1989. Protein phosphorylation and regulation of adaptive response in bacteria. *Microbiol. Rev.* **53:**450–490.

Strauch, M. A., and J. A. Hoch. 1993. Transition state regulators: sentinels of *Bacillus subtilis* post exponential gene expression. *Mol. Microbiol.* **7:**337–342.

Strauch, M. A., K. Trach, J. Day, and J. A. Hoch. 1992. SpoOA activates and represses its own synthesis by binding at its dual promoters. *Biochemie* **74:**619–626.

Strauch, M. A., V. Webb, G. B. Spiegelman, and J. A. Hoch. 1990. The SpoOA protein of *Bacillus subtilis* is a repressor of the *abrB* gene. *Proc. Natl. Acad. Sci. USA* **87:**1801–1805.

Strauch, M. A., J.-J. Wu, R. H. Jonas, and J. A. Hoch. 1993. A positive feedback loop controls transcription of the *spoOF* gene, a component of the sporulation phosphorelay in *Bacillus subtilis*. *Mol. Microbiol.* **7:**967–974.

Trach, K., D. Burbulys, M. A. Strauch, J.-J. Wu, N. Dhillion, R. Jonas, C. Hanstein, P. Kallio, M. Perego, T. Bird, G. Spiegelman, C. Fogher, and J. A. Hoch. 1991. Control of the initiation of sporulation in *Bacillus subtilis* by a phosphorelay. *Res. Microbiol.* **142:**815–823.

Trach, K., J. W. Chapman, P. Piggot, D. LeCoq, and J. A. Hoch. 1988. Complete sequence and transcriptional analysis of the *spo0F* region of the *Bacillus subtilis* chromosome. *J. Bacteriol.* **170:**4194–4208.

Trach, K., J. W. Chapman, P. J. Piggot, and J. A. Hoch. 1985. Deduced product of the stage 0 sporulation gene *spo0F* shares homology with the Spo0A, OmpR and SfrA proteins. *Proc. Natl. Acad. Sci. USA* **82:**7260–7264.

Tsung, K., R. E. Brissette, and M. Inouye. 1989. Identification of the DNA binding domain of the OmpR protein required for transcriptional activation of the *ompF* and *ompC* genes of *Escherichia coli* by in vivo DNA footprinting. *J. Biol. Chem.* **264:**10104–10109.

Tsung, K., R. E. Brissette, and M. Inouye. 1990. Enhancement of RNA polymerase binding to promoters by a transcriptional activator, OmpR, in *Escherichia coli*: its positive and negative effects on transcription. *Proc. Natl. Acad. Sci. USA* **87:**5940–5944.

Wanner, B. L. 1993. Gene regulation by phosphate in enteric bacteria. *J. Cell. Biochem.* **51:**47–54.

Wegrzyn, G., R. E. Glass, and M. S. Thomas. 1992. Involvement of the *Escherichia coli* RNA polymerase α subunit in transcriptional activation by the bacteriophage lambda CI and CII proteins. *Gene* **122:**1–7.

Weiss, D. S., J. Batut, K. E. Klose, J. Keener, and S. Kustu. 1991. The phosphorylated form of the enhancer binding protein NTRC has an ATPase activity that is essential for activation of transcription. *Cell* **67:**155–167.

West, D., R. Williams, V. Rhodius, A. Bell, N. Sharma, C. Zou, N. Fujita, A. Ishihama, and S. Busby. 1993. Interactions between RNA polymerase cyclic AMP preceptor protein and RNA polymerae at class II promoters. *Mol. Microbiol.* **10:**789–797.

Whipple, F. W., and A. L. Sonenshein. 1992. Mechanism of initiation of transcription by *Bacillus subtilis* RNA polymerase at several promoters. *J. Mol. Biol.* **223:**399–414.

Wu, J.-J., M. G. Howard, and P. J. Piggot. 1989. Regulation of transcription of the *Bacillus subtilis* *spoIIA* locus. *J. Bacteriol.* **171:**692–698.

Wu, J.-J., P. J. Piggot, K. M. Tatti, and C. P. Moran, Jr. 1991. Transcription of the *Bacillus subtilis* *spoIIA* locus. *Gene* **101:**113–116.

Yamashita, S., H. Yoshikawa, F. Kawamura, H. Takahashi, T. Yamamoto, Y. Kobayashi, and H. Saito. 1986. The effect of *spo0* mutations on the expression of *spo0A-* and *spo0F-lacZ* fusions. *Mol. Gen. Genet.* **205:**28–33.

York, K., T. J. Kenney, S. Satola, C. P. Moran, Jr., H. Poth, and P. Youngman. 1992. Spo0A controls the σ^A dependent activation of *Bacillus subtilis* sporulation specific transcription unit *spoIIE*. *J. Bacteriol.* **174:**2648–2658.

Zhang, X., Y. Zhou, Y. W. Ebright, and R. H. Ebright. 1992. Catabolite gene activator protein (CAP) is not an "acidic activating region" transcription activator protein. *J. Biol. Chem.* **276:**8136–8139.

Flagellar Switch

Robert M. Macnab

11

The motility of *Escherichia coli* and *Salmonella typhimurium* cells consists of swimming using a propulsive bundle of several flagella and tumbling by dispersal of the bundle. It involves complex considerations of hydrodynamics, mechanics, and filament structure that go beyond the scope of this chapter (although at the end, I touch on the issue of switch interactions among multiple flagella on a cell). The chemotactic response consists of the modulation of this motility pattern: a cell traveling in a favorable direction senses a temporal gradient that lowers its probability of tumbling compared with one traveling in an unfavorable direction, and so a drift velocity or bias is superposed on an otherwise random trajectory.

At the level of the individual motor, the behavior reduces to two well-defined states—either counterclockwise (CCW) or clockwise (CW) rotation—and their control by information relayed from cell-bound chemoreceptors via the sensory transduction apparatus.

The energy for flagellar rotation is the transmembrane proton motive force. This is a vectorial form of energy, but it is always inwardly directed toward the cytoplasm under physio-

logical conditions and therefore cannot be used to switch the direction of rotation of the motor in the way that reverse air or water flow can reverse a windmill or turbine. One is forced therefore to the conclusion that the motor itself must contain a binary switch.

From the point of view of signal transduction and the motor, the important molecular questions then become (i) what are the components of the switch? (ii) what is the difference between the switch in its two states? and (iii) how does the switch modulator, now known to be CheY-P, modulate the probabilities of the two states? This chapter only covers the switch and its interaction with CheY-P (cf. a related review by Barak and Eisenbach [in press]); the early phases of chemoreception and sensory transduction are covered in Chapter 6.

From the time that it was known that the external portion of the flagellum was a passive structure and that the energy was the transmembrane proton motive force, it was assumed that the motor and switch would be located at the cell surface, at or around a structure called the MS ring (for *m*embrane and *s*upramembrane), which is located in the cytoplasmic membrane (Fig. 1). In earlier papers, it was thought that the MS ring might itself be part of the motor, but it is now evident that it is merely

Robert M. Macnab, Department of Molecular Biophysics and Biochemistry, Yale University, New Haven, Connecticut 06520-8114.

Two-Component Signal Transduction, Edited by James A. Hoch and Thomas J. Silhavy,
© 1995 American Society for Microbiology, Washington, DC 20005

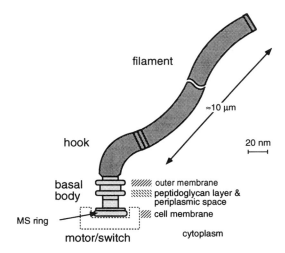

filament

≈10 μm

hook

20 nm

basal
body

outer membrane

peptidoglycan layer &
periplasmic space

cell membrane

MS ring

cytoplasm

motor/switch

FIGURE 1 Bacterial flagellum and its relationship to various elements of cell surface. This represents the state of knowledge until quite recently, with the approximate location of the motor/switch (dotted outline) inferred but not yet demonstrated experimentally.

a mounting plate for the motor. The MS ring is connected to a rod that traverses the periplasmic space; the MS ring, an outer pair of rings, and the rod are collectively called the basal body. External to the cell, the rod connects to a short curved structure called the hook, and the hook connects to the long helical flagellar filament that actually propels the cell (Fig. 1).

Underlying the flagellum is a large genetic system of more than 40 genes. Many of these are structural, whereas others are responsible for control of gene expression and for assembly of the organelle; in this chapter, I focus only on those that are relevant to motor rotation and switching. For more detailed reviews of flagellar structure, function, and genetics, see Blair (1990), Jones and Aizawa (1991), Caplan and Kara-Ivanov (1993), Macnab (1987a,b, 1992, 1995), and Schuster and Khan (1994).

SWITCHING PROCESS

Individual motors switch stochastically, even in the absence of environmental stimuli; in other words, under constant environmental conditions a motor in the CCW state has a constant probability per unit time of switching to the CW state, and a motor in the CW state has a (different) constant probability of switching to the CCW state (Block et al., 1983). Under environmental stimulation by attractants or repellents, those probabilities change.

In experiments in which flagella on a multiflagellate cell are freed from mechanical or hydrodynamic interactions with each other, each motor behaves autonomously, so that switching events for one have no correlation with those for another (Ishihara et al., 1983; Macnab and Han, 1983). This might seem to contradict the observation that when a population of free or tethered cells is given a sudden attractant or repellent stimulus, they all appear to respond simultaneously, but this is simply because the probability of remaining in or entering the CW state (following an attractant stimulus) or the CCW state (repellent stimulus) has suddenly been driven to a very low value. The autonomy of the motors with respect to switching is a fundamentally important point because it eliminates the possibility of a global signal that switches all motors in a temporally coordinated fashion. Knowing now that the signal is almost certainly intracellular CheY-P (Oosawa et al., 1988; Wylie et al., 1988), one assumes that each motor/switch will have a time-averaged occupancy by that ligand and mean probabilities of switching corresponding to that occupancy; when the actual switching events occur, however, remains a matter of chance.

The two directions of rotation appear to be different manifestations of the same process rather than two distinct processes, because time not spent in one direction is spent in the other direction rather than in a stopped state. (Brief stops, or pauses, of the motor are sometimes seen and may represent abortive switching events [Eisenbach et al., 1990].) Generally, it does not seem to be possible to increase the mean time in one state without decreasing the mean time in the other state. A simple model that would conform to this description would be an isomerization in which the absolute stability of one isomeric state of the switch is inversely related to the absolute stability of the other, with the stability of the transition state remaining unchanged (Khan and Macnab, 1980). Steady-state switching frequency does seem to increase with temperature (Welch and Eisenbach, personal communication), indicating that such a thermal isomerization model may be valid.

It is, however, surprising that mutants defective in the switching process have not been observed: there are no reports of switch mutants that spend unusually long mean times in both the CCW and the CW states. There are a few examples of receptor mutants that have this pattern (dubbed slow-switchers), but this is a property of the signaling properties of the receptor rather than the switch itself (Ames and Parkinson, 1988). Block et al. (1983) reported an increase of both CCW and CW intervals of wild-type cells tethered and examined over several hours, but the origin of this effect is unknown. Cell envelopes devoid of cytoplasm, interestingly enough, have essentially no ability to switch: a given envelope remains either CCW or CW indefinitely (Ravid and Eisenbach, 1984; Ravid et al., 1986).

For cells with wild-type motors, the null phenotype with respect to the chemotaxis system (i.e., the phenotype of strains "gutted" for the chemotaxis genes) is the absence of CW rotation (Wolfe et al., 1987). In experiments with cell envelopes, CheY by itself is capable of sustaining CW rotation (Ravid et al., 1986). From these two results, the statement is often made that CCW is the default state of the switch and that CheY is essential for creating the CW state. This approximates the truth, but it is not completely correct, because strains with an extreme CW bias deriving from a switch mutation can still manifest the CW state even in the absence of CheY (Ravid and Eisenbach, 1984). Thus, the motor has an intrinsic capability for both switch states; it is just that for a wild-type motor, the capability is very low.

COMPONENTS OF THE FLAGELLAR SWITCH

What are the molecular components of the switch? Until recently, nothing was known about the switch in terms of its structure, and so its components were defined by genetics: it was presumed that the components of the switch could be identified on the basis of genes giving rise to a mutant phenotype of abnormal switching. Most of the flagellar proteins were eliminated by this criterion, because mutations in them gave rise only to defects in flagellar structure. In fact, only three proteins (FliG, FliM, and FliN) have been found to give rise to defective switching (Parkinson et al., 1983; Yamaguchi et al., 1986b). (Note: before a revision of the flagellar gene nomenclature by Iino et al. [1988], *fliG*, *fliM*, and *fliN* were known in *S. typhimurium* as *flaAII.2*, *flaQII*, and *flaN*, and in *E. coli* as *flaBII*, *flaAII*, and *motD*; the MS-ring gene, which is discussed below, was known as *flaAII.1* in *S. typhimurium* and *flaBI* in *E. coli*.)

It could be argued that if such defects required very specific mutations, these mutations might not have been identified yet for some switch proteins. However, mutant searches have been quite extensive, and for the three switch proteins that have been identified, many examples of switch-defective mutations have been found, as is described in a subsequent section. It therefore seems unlikely that other switch components remain to be discovered. The *fliM* and *fliN* switch genes are adjacent on the same operon (Kutsukake et al., 1988; Kihara et al., 1989; Malakooti et al., 1989), but there is nothing to suggest that the products of the other flagellar genes within the operon play any role in switching; the same is true of the operon containing *fliG*.

Depending on the particular mutations in FliG, FliM, or FliN, cells may either fail to assemble flagella (the most severe phenotype); they may assemble flagella that fail to rotate (the next most severe); or they may assemble flagella that rotate with either an unusually high or an unusually low CCW bias (the least severe) (Yamaguchi et al., 1986b). Thus, these three proteins are important structurally in allowing flagellar assembly to proceed to completion and are important functionally in both generating torque and determining its direction. They may therefore be described as being components of both the motor and the switch; indeed, it is not clear to what extent these two functions are separable. Because each of the three proteins can give rise to both CCW and CW biases, there are not separate CCW subunits and CW subunits within the switch.

As well as the switch proteins, there are two motility-related proteins, MotA and MotB (Dean et al., 1984; Stader et al., 1986); they are thought to be responsible for delivering protons to the motor/switch (Blair and Berg, 1990, 1991) and for anchoring the motor in the cell surface (Chun and Parkinson, 1988), respectively, but because they only give rise to paralyzed mutant phenotype and not switch-defective phenotype, they are not believed to play any role in the interface between the sensory transduction chain and the motor.

Although it is true that mutations in the components of the sensory transduction chain (i.e., the receptors and the various Che proteins) affect switch bias, these components are really direct or indirect inputs to the switch rather than being components of the switch itself. This is most clearly seen by the fact that, with rare exceptions of specially selected gain-of-function alleles (e.g., a *cheZ* mutation causing CCW bias that was isolated as a suppressor of a CW switch mutation [Yamaguchi et al., 1986a]), mutations in a given chemosensory component all cause bias shifts in the same direction as a result of loss of function: for example, *cheY* mutants are CCW-biased, whereas *cheZ* mutants are CW-biased. By contrast, as noted above, mutations in switch components can give rise to either bias.

BIOCHEMICAL PROPERTIES OF THE SWITCH PROTEINS

FliG, FliM, and FliN all have sequences suggesting a location in the cytoplasm or peripheral to the membrane. Because they were found to segregate with cell envelopes (Ravid and Eisenbach, 1984), the latter location seemed most likely and indeed this has since been demonstrated directly (see below). FliG

and FliM are of medium size (37 and 38 kDa, respectively), but FliN is much smaller (15 kDa) (Clegg and Koshland, 1985; Kuo and Koshland, 1986; Homma et al., 1988; Kihara et al., 1989; Malakooti et al., 1989; Roman et al., 1993). For the most part, the amino acid sequences of the switch proteins do not reveal anything unusual. One exception is the clustering of charged residues in FliG, where (in both *S. typhimurium* and *E. coli*) there are six examples of triplets of mixed charge of the form R-(D/E)-(D/E) or (D/E)-(D/E)-R.

LOCATION OF THE SWITCH

In genetic analyses of switch mutations, it was found that some Mot⁻ mutations in a given switch protein could be suppressed extragenically by mutations in the other two proteins and that this suppression was allele-specific (Yamaguchi et al., 1986a). This led to the hypothesis that the switch was a complex made up of subunits of each of the three proteins.

Because the switch is the target for a cytoplasmic protein, CheY, and because it is part of a motor that is imbedded in the cell surface, it was generally assumed that it would be located at the cytoplasmic face of the basal body (e.g., Yamaguchi et al., 1986a) (see Fig. 1). However, when the basal body was analyzed biochemically, it was found that it did not contain the switch proteins (or the Mot proteins) (Aizawa et al., 1985; Jones and Macnab, 1990). This indicated that the basal body did not contain the flagellar motor or switch but instead was a passive structure to which these elements were presumably attached. In this case, the switch must have been a fairly labile structure that dissociated during the isolation process. This led to efforts to moderate the conditions under which the basal body was isolated.

In the meantime, an unexpected result established that one of the switch proteins, FliG, was indeed located at the cytoplasmic face of the MS ring (Francis et al., 1992). A mutation that had been mapped to near the 5′ end of *fliG* turned out to be a small deletion that fused the

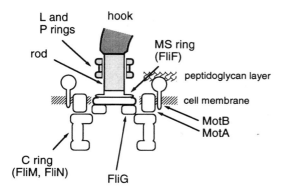

L and
P rings

hook

rod

MS ring
(FliF)

peptidoglycan layer

cell membrane

MotB
MotA

C ring
(FliM, FliN)

FliG

FIGURE 2 Flagellar basal body together with associated structures. The cytoplasmic C ring and the extension of the MS ring are seen in electron micrographs of isolated flagella in vitreous ice or negative stain (Khan et al., 1992; Francis et al., 1994); a related structure has also been seen in freeze-substituted thin sections of cell envelopes (Khan et al., 1991). Several lines of evidence indicate that the three motor/switch proteins are present in these structures, with FliG being located directly below the MS ring (which is made out of FliF subunits), and FliM and FliN forming all or part of the C ring. Protein complexes or "studs" seen in freeze-fracture images of whole cells (Khan et al., 1988) are believed to consist of the Mot proteins, MotA and MotB, located in the membrane around the circumference of the MS ring and perhaps just above the C ring.

upstream gene, which encodes the MS-ring protein FliF, in frame with *fliG*. The mutant was motile and could switch, although with a CW bias compared with wild type. Thus, a fusion protein between the basal-body MS-ring protein and a switch protein sustained almost normal function. Biochemical evidence confirmed the existence of the fusion, and immuno-electron microscopy established that the FliG protein (now stably associated with the basal body via the covalent fusion to the MS-ring protein) was at its predicted location. When the mutant was transformed with a plasmid containing the wild-type *fliF* gene, the cells became nonflagellate, demonstrating that FliG could not function unless it was permitted to reside at the MS ring.

Efforts to refine basal-body isolation methods have since been successful, leading to the discovery of a cylindrical structure (the C ring)

extending from the MS ring into the cytoplasm (Francis et al., 1994) (Fig. 2). Immuno-electron microscopy and image analysis of particles with and without the C ring have established that FliG is located at the junction of the C ring and the MS ring; this is consistent with the location observed in the FliF-FliG fusion strain. These studies have also established that FliM and FliN are located in the C ring itself, although their precise location within that structure is not yet known. It is also not known whether the C ring contains any other proteins besides FliM and FliN. In separate studies (Khan et al., 1991, 1992), an extension of the basal body called the bell (because of its shape) was observed and presumably corresponds to the C ring. This bell was stable in preparations from the wild-type strain but was absent in preparations from temperature-sensitive Mot⁻ mutants whose defect lay in any of the switch genes.

In reconstitution experiments, FliG bound tenaciously to purified MS-ring complexes, FliM bound but could be readily dissociated, and FliN did not bind at all (Oosawa et al., 1994).

LOCATION OF THE Mot PROTEINS

In terms of motor structure, the MotA and MotB proteins, as well as FliG, FliM, and FliN, need to be considered. The Mot proteins are integral to the membrane (Wilson and Macnab, 1988, 1990) and can be incorporated after the rest of the flagellum is complete (Block and Berg, 1984; Blair and Berg, 1988). This suggests that they might be circumferentially located around the MS ring. Consistent with this, freeze-fracture electron microscopy reveals sets of about 12 complexes surrounding structures that, based on their number per cell and their dimensions, are likely to be flagellar basal bodies (Khan et al., 1988). The dimensions of the MS ring (~29-nm outer diameter) and the C ring (~45-nm outer diameter) suggest the Mot complexes may lie in the membrane outside the outer diameter of the MS ring and above the C ring, as is indicated in Fig. 2 (Francis et al., 1994).

STOICHIOMETRIC COMPOSITION OF THE MOTOR/SWITCH

Analysis of the classical basal body (i.e., the structure lacking the switch and Mot proteins) has indicated that the MS ring exists as an annulus of approximately 26 subunits of the FliF protein (Jones et al., 1990). Given that a FliF-FliG fusion protein functions more or less normally, it is likely that the wild-type stoichiometry is also 1:1; the in vitro reconstitution experiments between switch proteins and MS-ring complexes, in fact, yielded a stoichiometry of close to this value. (The wild-type level of FliG has been estimated at about 3,700 copies per cell [Roman et al., 1993]; however, the estimate represents the entire cellular pool, not just the fraction of it that is incorporated into the half-dozen or so flagella on the cell.)

A FliM-FliN fusion protein also sustains rotation and switching (see below), making it likely that the wild-type FliM/FliN stoichiometry is 1:1. In this case, it is not certain what the absolute stoichiometry is. In reconstitution experiments, FliM bound substoichiometrically to MS-ring complexes (FliM/FliF ratio of 1:5), but it was not clear whether this represented the natural stoichiometry or was a consequence of lability (Oosawa et al., 1994). The latter seems likely, given the fact that FliN, which is associated with basal bodies in vivo (Francis et al., 1994), was not detected at all in the reconstitution experiments. If the C ring consisted entirely of equimolar amounts of FliM and FliN, it has been estimated (based on the apparent solid volume of the ring) that there would be about 100 subunits of each. Thus, there is considerable disparity in the available experimental estimates. A direct biochemical estimate may become possible if preparations of basal bodies can be made in which essentially every particle contains an intact C ring.

The finding that FliG (and almost certainly FliM and FliN also) is present in the switch in multiple copies is important, because a switch with multiple binding sites for CheY would be expected to have very different properties (see below) from one with a single site.

Based on electron microscopic evidence, there are likely to be about 12 Mot complexes (Khan et al., 1988), whereas functional analysis suggests a maximum of 8 complexes, with smaller numbers giving rotation with correspondingly reduced torque (Blair and Berg, 1988). It is not clear how these numbers can be reconciled to each other or how they relate to the stoichiometry of the switch. As with FliG, the cellular pools of MotA and MotB are much larger than the amounts that appear to be actually incorporated into the motor structure (Wilson and Macnab, 1988, 1990).

ROTOR VERSUS STATOR OF THE MOTOR/SWITCH

Any rotary motor can be subdivided into its rotating parts (collectively called the rotor) and its stationary parts (the stator). The location of the Mot proteins and especially the postulated function of MotB as an anchor (which has received support from the recent identification of a peptidoglycan binding motif in its sequence [De Mot and Vanderleyden, 1994]) suggest that they should be stator components. The filament and hook are definitely rotor components, and—based on their structure and location—the rod and MS ring are almost certainly part of the rotor also. (An earlier idea that the M ring was a rotor element and the S ring was a stator element became untenable when it was established that both rings represent domains of subunits of the same protein, FliF [Ueno et al., 1992].)

This leaves the question of how to assign the three switch proteins. The FliG protein can be fused to the MS-ring protein and still support motor rotation, indicating it should be assigned to the rotor. The FliM and FliN proteins can be fused to each other and still support motor rotation (Kihara and Macnab, unpublished data), indicating that they must both be rotor elements or they must both be stator elements. There is no clear evidence at this point as to which of the two possibilities is correct; a FliG-FliM fusion engineered to distinguish between them blocked the process of flagellar assembly, preventing the relevant question from being asked (Toker et al., unpublished data).

If all three switch proteins are, in fact, rotor components and the Mot proteins constitute the stator, torque must be developed between the switch proteins and the Mot proteins. In other words, the Mot proteins—most probably MotA—must play a role beyond the mere delivery of protons to the motor.

FUNCTIONAL ANALYSIS OF THE SWITCH

As mentioned above, all three switch proteins are implicated in flagellar assembly, in motor rotation, and in switching. Extensive efforts have been made to determine which parts of these proteins are important for the various aspects of their function by isolating mutants with different phenotypes and establishing the mutations responsible.

Although there is no good selection procedure for single-site switch mutants, it is very easy to select for switch mutations that compensate for a mutation in another component

of the system, either the other switch components or components of the sensory transduction chain. This approach has been successfully applied in both directions, using *cheY* or *cheZ* mutations to select for switch mutations and using switch mutations to select for *cheY* mutations (Parkinson et al., 1983; Yamaguchi et al., 1986a; Magariyama et al., 1990; Roman et al., 1992; Sockett et al., 1992; Irikura et al., 1993).

Figure 3 summarizes the results obtained in *S. typhimurium* for FliG, FliM, and FliN with respect to Mot⁻, Che⁻(CCW), and Che⁻(CW) mutations (Magariyama et al., 1990; Sockett et al., 1992; Irikura et al., 1993). (Not shown are Fla⁻ mutations, which commonly consist of major disruptions such as frameshift mutations.) Most but not all the Che⁻ mutations were derived as CheY or CheZ suppressors; 50 independently isolated suppressors for each of 10 parental CheY and CheZ mutations (i.e., 1,000 mutations in total) were subjected to fine-resolution classical mapping and behavioral analysis, and of those, more than half were identified by DNA sequence analysis. The re-

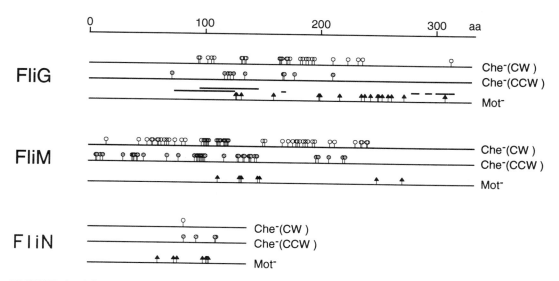

FIGURE 3 Schematic of protein sequences of motor/switch proteins FliG, FliM, and FliN, with sites of mutations giving rise to abnormally high CW bias [Che⁻(CW)], abnormally high CCW bias [Che⁻(CCW)], or paralysis (Mot⁻). Single-site mutations are indicated by flags of open circles, stippled circles, and black triangles, respectively, while in-frame deletions are indicated by bars. FliG and FliM are of essentially the same length, whereas FliN is much shorter. For discussion of the data, see the text. For the actual amino acid changes involved, see Sockett et al. (1992) and Irikura et al. (1993).

mainder of this section addresses the extensive studies from which several conclusions can be drawn:

Suppression is extremely easy to achieve. All mutants tested were capable of being suppressed. This is an unusual situation, because typically suppression involves an "up" mutation that ensures better function with the first one, and such improvement of function is not easy to achieve. The ease in the present case probably reflects the fact that an increase in the time spent in one state (say, CCW) does not have to be achieved by improvement of that state but can be readily accomplished by impairing the other state (CW).

Suppression of a CheY or CheZ mutation by a switch mutation is not allele specific. It turns out that the primary factor in determining how well a given pair of mutations function in combination is a restoration of the unstimulated bias of the motor to close to the wild-type value. Thus, mild bias shifts are best combined with mild bias shifts in the opposite direction; strong bias shifts are best combined with strong bias shifts in the opposite direction. The specifics of the mutations do not seem to be important. Examples of allele-specific suppression may exist, but if so, they are swamped out by the many bias-shift mutations that have essentially the same effect. This absence of evidence for allele-specific suppression, incidentally, prevented the analysis from being used to identify physically interacting components.

The three switch proteins differ greatly in the spectrum of mutations generating a given phenotype. Although all three proteins can give rise to all the various mutant phenotypes, detailed analysis reveals that their propensity to do so differs considerably. Thus, FliM has far more positions in its sequence that affect CCW or CW switching than do FliG or FliN; conversely, it has few positions that affect rotation. The fact that so many positions in FliM contribute to the switching mechanism suggested it might be the target for binding of CheY-P (see below).

FliG has moderate numbers of positions affecting switching and rotation. FliN has only a few positions affecting switching and rotation, although the number of mutants with paralyzed phenotype that have been examined is small.

Positions generating the different phenotypes tend to cluster. As is evident by inspecting Fig. 3, clusters of residues exist that are important for a given mutant phenotype, such as Che⁻(CCW). These presumably represent elements that, in the three-dimensional structure, are important for stabilizing the opposite state. It has been postulated that switching may involve breaking domain interfaces either within a given switch protein or between switch proteins and making alternative interfaces (Sockett et al., 1992). In the absence of structural information (neither the individual switch proteins nor the switch complex have been crystallized), it is not possible to be more specific concerning the significance of these clusters. There are very few examples in which different mutations at the same position can give both the CCW and CW phenotype, arguing that a given position is important to the stability of only one of the two states and that the wild-type amino acid is normally the optimal one in that regard.

Many switch mutations involve charge shifts. In FliM especially, many of the mutations with the most profound destabilizing effect on the CCW state were changes from a basic residue, arginine, to a neutral one, or from a neutral residue to an acidic one, glutamate, indicating that electrostatic interactions may be important in stabilizing the switch states.

Next, the results of this mutant analysis for each of the three proteins are discussed in more detail.

FliG

Only the central portion of FliG seems to be important for switching. The first mutation causing a CCW bias is 22% in from the N terminus and the last is 36% from the C terminus; the corresponding values for mutations causing a CW bias are 29% and (with a single exception) 28%, respectively. Within the central portion, there are fairly localized clusters of residues that are important for stabilizing either the CW state (because mutations cause a CCW bias) or the CCW state (mutations cause a CW bias), with substantial stretches not giving rise to either phenotype. Thus, the switching mechanism appears to be specifically dependent on limited portions of the protein. No obvious theme emerges from the mutations that are observed. A few scattered positions are important for rotation, as well as one cluster of such positions about 25% from the C terminus.

A remarkably large amount of the FliG sequence is nonessential for structure. Thus, a FliF-FliG fusion protein (different from the one referred to earlier) lacking the first 94 acids of FliG still supports flagellation and, to a reduced degree, motility (Francis et al., 1992), and several large internal deletions do so also.

FliM

The C-terminal portion of FliM seems to be unimportant for either rotation or switching, because no mutations causing switch bias or paralyzed phenotype are found within it; it is, however, important for flagellar structure, because frameshift and nonsense mutations cause nonflagellate phenotype. The last of these occurs only 23 residues from the terminus. All the switch-bias mutations in FliM lie in the first 70% of the sequence, with pronounced clusters of positions that are important for stabilizing either the CW state or the CCW state. Of the few positions that cause paralyzed phenotype, almost all are either in or immediately adjacent to the switch-bias clusters, suggesting that the mechanisms of rotation and switching may be linked.

FliN

Initial classical mapping studies proved to be somewhat misleading in terms of the importance of FliN for switching, because of the existence of mutational hotspots; for example, of the 40 independently isolated FliN mutations contributing a CCW bias that were analyzed, 25 were identical. The number of positions established as being involved in switching is actually very small, only one for stabilizing the CCW state and four for the CW state. They all fall within a fairly short region of about 30 residues that also contains positions important for rotation.

The N-terminal half of FliN is fairly divergent between *E. coli* and *S. typhimurium,* whereas the rest of the protein is identical (Kihara et al., 1989; Malakooti et al., 1989). Various frameshift and nonsense mutations in *fliN* gave nonflagellate phenotype, as would be expected. However, several other frameshift mutations early on in the gene gave, surprisingly, paralyzed phenotype; these all gave the same shift of frame and generated a stop codon just before a potential restart, suggesting that N-terminal information might be largely dispensable and that the C terminus was structurally more important. This has since been confirmed (Blair, personal communication): a large in-frame deletion of most of the gene gives the expected nonflagellate null phenotype, but a deletion of the 5′ half including the normal start codon (permitting the putative alternate start codon to be used) permits not only flagellar assembly but also rotation.

CheY, CheZ, AND THE SWITCH

Although all the chemotaxis proteins and the chemoreceptors affect switching, it has been found that extragenic compensation for switch defects derives either from one of the other switch proteins or one of the Che proteins (Yamaguchi et al., 1986a) and that, with rare exceptions (e.g., a single example of suppression of a CW FliG defect by a CheA mutation), the latter category involves either CheY (for CW switch bias) or CheZ (for CCW switch bias). This was the first indication that either or both of these proteins were likely to be the final output from the sensory transduction chain, acting directly on the flagellar switch. It was subsequently demonstrated that CheY was phosphorylated by CheA, which itself had autophosphorylated in a receptor-CheW-mediated reaction, and that CheZ was a phosphatase, accelerating the dephosphorylation of CheY-P (see Chapter 3 for details of the biochemistry of the sensory transduction chain). CheY-P was much more effective in causing CW rotation of cell envelopes than unphosphorylated CheY, although additional cytoplasmic components (not identified at the time) were necessary for the effect to be observed (Barak and Eisenbach, 1992a). The properties of CheY (together with CheA) place it as a member of a large family of two-component regulatory systems (see Chapter 4); it stands out as a special example within the family, in that its effect is on the flagellar motor, whereas the usual regulatory effect is on gene expression.

There is still no direct evidence in whole cells that phosphorylation of CheY is responsible for modulating the switch. However, the evidence from envelopes and extensive biochemical correlations in vitro strongly supports this hypothesis.

CheY, a soluble 14-kDa protein (Matsumura et al., 1984), is present in fairly high copy number in the cell (ca. 7,000 copies per cell, or 8 μM [Kuo and Koshland, 1987]). The fraction that is present as CheY-P in an unstimulated cell is not known, because detection in vivo has proved difficult. The stoichiometry of the switch proteins is much less than that of CheY. FliG, for example, is thought to be present in about 26 copies per flagellum (Francis et al., 1992) or about 150 copies incorporated into flagella per cell.

Recall that the analysis of switch mutants indicated that FliM is the component of the motor most heavily involved in switching (Sockett et al., 1992), suggesting it might be the target for binding of CheY-P. In recent reconstitution experiments, Welch et al. (1993) examined the affinity of each of the purified switch proteins for CheY and CheY-P. They found that FliM bound CheY and that phosphorylation of CheY increased the extent of binding by a factor of about 20. (The binding, however, showed little if any signs of saturating, raising some concern about the specificity of the sites.) FliG bound CheY and CheY-P weakly, and FliN did not bind them at all.

Comparable experiments have not been performed with CheZ, and indeed it is not known whether the latter binds to the switch at all. The fact that CheZ has no effect on switch bias in the absence of CheY (Wolfe et al., 1987) argues that it may not bind directly. Because CheZ is active as a CheY-P phosphatase in the absence of the switch proteins (Hess et al., 1988) and there is probably an appreciable free pool of CheY-P in the cell, there does not seem to be any need to invoke switch-mediated dephosphorylation of CheY-P. It is possible that CheZ can act on CheY-P regardless of whether the latter is free or bound to the switch.

CheY is the only chemotaxis protein (other than the periplasmic domain of the aspartate receptor) to have been crystallized (Stock et al., 1989; Volz and Matsumura, 1991; Stock et al., 1993). Details of the structure together with functional implications are presented in Chapters 3 and 4. It is a compact structure consisting of a sheet of five β-strands flanked by five α-helices, with the phosphorylation site, Asp-57, lying near the end of one of the β-strands in an acidic pocket. The results from two independent CheY-switch intergenic suppression analyses have provided suggestive evidence for the binding site for the switch, because many of the suppressible mutations lay on a surface of the protein adjacent to the active site cleft (Roman et al., 1992; Sockett et al., 1992). It is presumed that a conformation change induced by phosphorylation of the substrate aspartyl residue propagates to this extended surface; phosphorylation per se is not essential, because there are nonphosphorylatable mutant alleles that can affect the switch bias (Bourret et al., 1990; Bourret et al., 1993).

As judged by the behavioral output (percentage time spent in CW rotation), binding of CheY to the switch is a highly cooperative process; Kuo and Koshland (1989), for example, estimated a Hill coefficient of greater than 5, with half-maximal effect at a CheY concentration of about 10 μM. The study was not able to determine the fraction of CheY that was actually phosphorylated under the conditions used.

These authors suggested that a simple two-state model for the motor/switch may be inadequate. They based this on the fact that the mean duration of CW intervals increases with CheY concentration, which would not be expected for the simple model, $M_{CCW} + Y \leftrightarrow M_{CW}Y$. As further evidence for the existence of more than two states, they noted that when the rotational intervals of wild-type cells are measured down to short enough times, they do not conform to a simple Poisson distribution but to a distribution with two decay times, of about 0.06 and 0.7 s, respectively. Interestingly, they found only a single decay time for CheY-induced reversals in strains that had been deleted for the chemotaxis system. In a later section, I argue that there are likely to be many more than two occupancy states for the switch, even though there are only two rotational states.

OTHER FACTORS AFFECTING SWITCHING

Although CheY-P is believed to be the principal agent for modulating the switch, several others appear to play some role.

Fumarate

In experiments with cell envelopes, it was found that CheY alone could cause CW rotation, as could switch mutations, but they could not generate switching events (Ravid and Eisenbach, 1984; Ravid et al., 1986). In other words, a given envelope rotated either CW or CCW but did not switch. When a cytoplasmic fraction was restored to the envelopes, switching was observed. Fumarate alone (which had been found to mediate switching in halobacteria [Marwan et al., 1990]) was subsequently found to be effective in enabling switching in envelopes of *E. coli* or *S. typhimurium* (Barak and Eisenbach, 1992b). The mechanism by which fumarate does this is unknown.

Proton Motive Force

As cells become de-energized, the flagellar motors become increasingly CCW-biased (Khan and Macnab, 1980). At roughly 70% of maximal speed, CW rotation has virtually disappeared. The effect correlates with diminution of proton motive force but not with ATP levels. Thus, it does not appear to be a consequence of interference with CheY phosphorylation, although this has not been tested directly.

Acetylation and Acetyladenylate

Several reports in the literature implicate acetate metabolism in control of switching, starting with the finding that acetate had a profound effect on the CW bias of cells devoid of all known chemotaxis components except CheY. It has variously been suggested that acetyladenylate (Wolfe et al., 1988) or acetyl phosphate (Dailey and Berg, 1993) are the agents responsible. In vitro, acetyladenylate is capable of acetylating CheY and enhancing its ability to generate a CW switch bias in envelopes (Barak et al., 1992). Whether this is a physiologically significant phenomenon is uncertain.

Ca^{2+}

In earlier work, long before the role of phosphorylation cascades in chemotaxis had been established, Ca^{2+} was postulated to be the excitation signal, but this was never demonstrated convincingly. More recently, Tisa and Adler (1992) showed that release of caged Ca^{2+} in a cell causes a CW bias and that release of a caged Ca^{2+} chelator causes a CCW bias. It was suggested that Ca^{2+} might be acting to stabilize

CheY-P. However, a subsequent paper showed that chemotaxis was inhibited by a specific channel blocker, ω-conotoxin (Tisa et al., 1993), suggesting other factors may be involved.

SWITCH MECHANISM

Models for the mechanism of switching are intimately related to models for the mechanism for rotation. These are only described briefly here.

In the model proposed by Läuger (1977, 1988), the protons move down a series of paired half-sites, with one half-site contributed by the stator and one by the rotor. The series of half-sites on the rotor are at a tilt compared with those on the stator, and hence the progressive movement of the protons is accompanied by progressive rotation of the rotor to maintain juxtaposition of the half-sites. Switching is then envisaged as a conformation change that alters the tilt angle from a positive to a negative value.

A model proposed by Berry (1993) has a similar geometry to that of Läuger but invokes arrays of charged sites at a tilt angle to the proton channel that generate torque by electrostatic interaction. Switching of rotation is accomplished by altering the pK values of the charge groups such that they change sign and the electrostatic force is therefore changed from attraction to repulsion.

In the model of Berg and Khan (1983), rotation is envisaged as a series of thermal steps of stator elements with respect to the rotor, with the steps biased in one direction by the presence of a proton at a site that is exposed to the exterior of the cell via a channel in the stator complex and the absence of a proton at a site that is exposed to the interior via another channel in the complex. A step in the "wrong" direction has a prohibitively high activation energy barrier, whereas a step in the "right" direction has a low activation energy barrier and results in transfer of the proton from the exterior to the interior. These steps result in accumulated elastic strain between the stator complexes and a rigid anchor. At steady state, force from the stored elastic energy balances the opposing frictional load of the rotating flagellum. Switching is accomplished by altering the conformation of the channel complex so that the positions of the channels with respect to the proton binding sites on the rotor are inverted.

The model of Oosawa and colleagues (Oosawa and Masai, 1982; Oosawa and Hayashi, 1983) invokes conformational changes (akin to those in actomyosin) loosely coupled to protonation and deprotonation. Switching is accomplished by altering the direction of the conformational changes.

MODEL FOR THE SWITCH

On the basis of what is currently known about the switch, I offer a speculative model with the following properties:

1. It is a large complex built from subcomplexes of the FliG, FliM, and FliN proteins. These subcomplexes are repeated multiple times to form an annulus. For simplicity, assume that the three proteins are present in one copy each within a subcomplex and that there are 26 subcomplexes in the annulus, so that the switch has the composition (FliG-FliM-FliN)$_{26}$.

2. The entire switch complex is a completely cooperative structure that has to be either in the CW or the CCW state (i.e., individual [FliG-FliM-FliN] subcomplexes cannot be in different states). This assumption is based on the fact that the C ring looks like a continuous structure rather than a series of separate complexes and on the fact that major speed fluctuations would be expected if subcomplexes were attempting to drive the motor in opposite directions.

3. Each (FliG-FliM-FliN) subcomplex has a binding site for CheY, located on a face of FliM that is exposed to the cytoplasm.

4. The affinity of CheY for the FliM binding site is enhanced by phosphorylation of CheY.

5. Binding of either CheY or CheY-P increases the stability of the CW state of the switch and decreases the stability of the CCW state of the switch.

6. Because of assumption 5, binding will be highly cooperative.

7. The magnitude of the effects on switch bias for a given extent of binding of CheY or CheY-P may or may not be the same; the simplest assumption would be that they are the same.

8. At any given CheY and CheY-P concentration, there will be a mean extent of occupancy of the sites on the switch, with individual sites undergoing binding and unbinding events on a time scale that is fast compared with the time scale of switching events. Thus, although there are many possible substates, consisting of all combinations of occupancy patterns, they effectively blend into a single state, the time-averaged occupancy.

9. Because of assumption 5, the time-averaged occupancy will not be the same for the two states, the CW state having a higher occupancy than the CCW state.

10. Transitions between the two states at any given set of CheY and CheY-P concentrations can then be expressed by an equilibrium chemical equation that describes the motor in its CCW state with certain amounts of CheY and CheY-P bound, the motor in its CW state with higher amounts of CheY and CheY-P bound, and the differential amounts bound in the two states. The higher the CheY and CheY-P concentrations, the further the equilibrium will lie toward the CW state. (The equation takes the form

$$M_{CCW}CheY_{n_{[CheY,CCW]}}CheY\text{-}P_{n_{[CheY\text{-}P,CCW]}}$$

$$+ (n_{[CheY,CW]} - n_{[CheY,CCW]})CheY$$

$$+ (n_{[CheY\text{-}P,CW]} - n_{[CheY\text{-}P,CCW]})CheY\text{-}P$$

$$\leftrightarrow M_{CW}CheY_{n_{[CheY,CW]}}CheY\text{-}P_{n_{[CheY\text{-}P,CW]}}$$

where M_{CCW} and M_{CW} represent the motor in the CCW and CW states, respectively, $n_{[CheY,CCW]}$ represents the time-averaged number of CheY

molecules bound to the CCW state of the motor, and other coefficients have corresponding meanings. A specific case might be $M_{CCW}CheY_4CheY\text{-}P_6 + 3CheY + 8CheY\text{-}P \leftrightarrow M_{CW}CheY_7CheY\text{-}P_{14}$.)

11. The amounts of CheY and CheY-P taken up or released according to the equation in assumption 10 will not be large enough to appreciably perturb the free pools.

12. For any given pair of extents of occupancy (one for the CCW state and one for the CW state), there will be a defined probability of the motor being in each of its two possible states, CCW and CW.

13. The actual switching events remain stochastic. In other words, there is not a critical extent of occupancy above which the motor is in its CW state and below which it is in its CCW state.

14. FliM is the component undergoing the primary structural change, with FliG and FliN undergoing secondary structural changes as a result of their interactions with FliM.

15. FliG makes the major contribution toward the torque generation mechanism.

16. Either all three motor/switch proteins form the rotor, with the Mot proteins forming the stator, or FliG forms the rotor and FliM and FliN together with the Mot proteins form the stator.

A cartoon illustrating the geometry of the switch in accordance with the above model is given in Fig. 4. It indicates just how complex the structure of the flagellum in the vicinity of the motor/switch is likely to be. However, even the structure shown is probably incomplete, because it does not take into account the components of the apparatus responsible for exporting external flagellar subunits such as flagellin, which are believed to travel down a channel in the nascent structure (see, e.g., Macnab [1992]). This export apparatus, whose molecular composition is not well understood, may be located in the central pore in the MS ring itself, possibly projecting toward the cytoplasm.

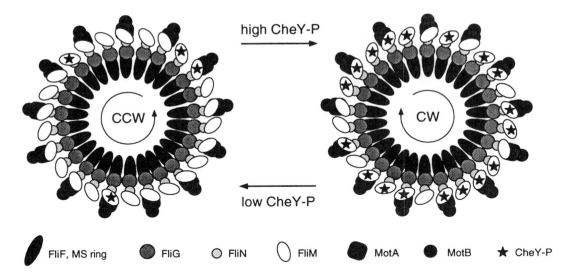

FIGURE 4 Model for flagellar motor and switch, based in part on experimental information and in part on speculation (see text). The view is from the cytoplasm, looking along the basal-body axis toward the exterior. The MS ring, located in the cell membrane, consists of about 26 subunits of FliF (Jones et al., 1990). At its cytoplasmic face and toward the outer radius of the ring are attached subunits of the motor/switch protein FliG, in a 1:1 stoichiometry with FliF subunits (Francis et al., 1992). The other two motor/switch proteins, FliM and FliN, lie at a larger radius than FliG and extend much farther into the cytoplasm, forming the C ring (Francis et al., 1994) (cf. Fig. 2). FliM is shown here to be more cytoplasm-proximal than FliN, on the premise that it must be available for binding CheY-P, but this is not a strong prediction. FliM and FliN are probably in a 1:1 stoichiometry to each other (Kihara and Macnab, unpublished data); here, they are also shown to be present in the same stoichiometry of 26 subunits as FliF and FliG, but their actual stoichiometry is not known. FliG is part of the rotor, but it is not known whether FliM and FliN are part of the rotor or the stator. The Mot proteins are in the cell membrane surrounding the M ring and behind the C ring. Based on electron microscopy, they are thought to be present in about 12 copies (Khan et al., 1988); here, they are arbitrarily shown to be present in 13 copies, and so they are in register with every second FliG-FliM-FliN complex. Depending on the CheY-P concentration (and to a lesser extent the CheY concentration), there will be a time-averaged occupancy of binding sites on FliM, which will determine the probability that the switch will, as a cooperative unit, be in the CCW or CW state (indicated schematically here by the tilt angle of the FliM subunits, although there is no evidence that this is the actual mechanism). It goes back and forth stochastically between the two states at the corresponding probabilities. The arrows indicate the favored state, not the actual switching events (i.e., the figure does not correspond to the switching equilibrium given by the equation in the text).

INTERACTION AMONG SWITCHES ON A MULTIFLAGELLATE CELL

For a cell with multiple flagella that are mechanically and hydrodynamically interacting with each other, interactions also have to be considered among motors and their switch mechanisms. For a cell that is in stable swimming mode, the flagella are in a bundle and are all rotating at the same speed and in the same direction (CCW). This raises the question of whether the switches are all forced to be in the same state. In other words, can external force dictate the state of the switch?

If they cannot, one would expect that any motor that happens to be in the CW state would be acting as a proton pump rather than a proton motor, because CCW rotation should drive protons from inside the cell to outside. (There is disagreement in the literature as to how readily flagella can be driven in the opposite direction from that of the motor torque [Berg and Turner, 1993; Washizu et al., 1993].)

This reverse rotation would be expected to reduce the net torque and thrust to the cell body, and as the number of flagella in the CW state fluctuates, there should be corresponding speed fluctuations. In fact, swimming cells maintain a fairly constant speed until they undergo a tumble.

The simplest hypothesis therefore is that if the CCW probability is fairly high in the absence of interactions, these interactions will make it difficult for a single motor within a CCW rotating bundle to switch to CW or, if in the CW state, to stay that way. How exactly would this effect be achieved? It can be thought about in terms of the effect of torsional stress on the stability of a given state. The switch of a single flagellum rotating CCW as a result of protons flowing down their gradient experiences a stress that will tend to destabilize its CCW state (Fig. 5A). Conversely, a motor that (because it is in the CW state) is being forced by mechanical and hydrodynamic interactions to rotate CCW will experience a stress that will tend to destabilize the CW state (Fig. 5B). Thus, there will be a strong tendency to join the majority state. Only when by chance several motors undergo a switch at about the same time will the bundle become sufficiently destabilized that it disintegrates and a tumble occurs (cf. discussion in Ishihara et al. [1983]).

(As an aside, it is interesting to note that changes in structural state in response to mechanical stress probably occur in other contexts within the flagellum. One is that the rotating hook may back-permute its structural phase, so that the waveform remains stationary [Macnab and Aizawa, 1984]. Another is that when flagella switch to CW rotation, the left-handed helical structure ["normal"] is destabilized and gives way to a completely different right-handed helical structure ["curly"] that contributes to tumbling [Macnab and Ornston, 1977]. Finally, there is the remarkable in vitro experiment of Hotani [1982], in which isolated flagellar filaments attached by one end to a glass surface undergo endless cycles of normal-curly

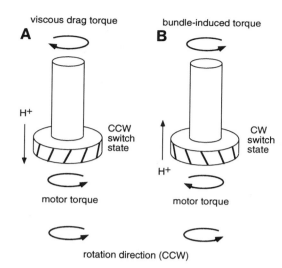

FIGURE 5 Torsional stress in the motor under different conditions. The tilted lines representing switch state (CCW or CW) are similar to the arrays of half-sites on the rotor of the model of Läuger (1988), so that protons moving in an axial direction from outside to inside along pairs of half-sites from the rotor and the stator cause an obligatory rotation of the rotor. (A) A motor in the CCW switch state and rotating CCW as a result of the motor torque generated by the inward proton flux experiences a torsional stress resulting from the opposing drag torque from the medium; this stress will tend to destabilize the CCW state, although presumably not enough to cause a switching event to occur. (B) A motor in the CW switch state is being driven in the CCW direction by torque resulting from mechanical and hydrodynamic interactions within a bundle of flagella whose motors are in the CCW switch state. Here, the motor torque is CW (i.e., it constitutes a drag) because protons are being pumped against their gradient. The resulting torsional stress will destabilize the CW switch state and cause the flagellum to go over to the CCW state like the rest of the flagella in the bundle. Although the effects of torsional stress are most easily visualized in the case of the Läuger model, the principle of stress affecting the stability of the switch states applies to other models as well.

transitions as a result of applied axial flow of medium.)

CONCLUDING COMMENTS

Because of the highly visible nature of its outputs (CCW or CW rotation), the bacterial flagellum represents one of the most striking examples of a biological switch. Starting with the identification of the switch genes roughly a decade ago, our understanding has progressed to the point where the proteins have been located to an actual structure. Much is known about which parts of the proteins are important for which aspects of switch function. There is strong circumstantial evidence for the surface on CheY that is responsible for its binding to the switch and consequent modulation of switch state. There is also biochemical and indirect genetic evidence to suggest that FliM may represent the principal, if not the exclusive, target for CheY binding. Finally, there is a powerful body of indirect evidence that CheY binding is modulated via its phosphorylation in response to sensory stimuli.

There is, however, essentially no information yet about the actual molecular mechanism of switching (or, for that matter, rotation). This will require structural information about the individual components and about the entire motor/switch complex. It is reasonable to hope that the former may be obtained at high resolution by crystallography. Information about the entire complex is more likely to be achieved by continued improvements in image reconstruction of electron micrographic images. Use of mutants that (even in the absence of any of the components of the sensory transduction chain, including CheY) are locked in either of the two switch states offers the prospect of distinguishing them structurally. If the differences are major, they may be discernible by electron microscopy. If they are subtle, high-resolution crystallographic structures will be needed, and it remains to be seen whether the switch states can be recognized at the level of an individual component such as FliM or only at the level of the entire complex.

ACKNOWLEDGMENTS

I am grateful to various colleagues for communicating unpublished data.

Preparation of this chapter has been supported by USPHS grants AI12202 and GM40335.

REFERENCES

Aizawa, S.-I., G. E. Dean, C. J. Jones, R. M. Macnab, and S. Yamaguchi. 1985. Purification and characterization of the flagellar hook-basal body complex of *Salmonella typhimurium. J. Bacteriol.* **161:**836–849.

Ames, P., and J. S. Parkinson. 1988. Transmembrane signaling by bacterial chemoreceptors: *E. coli* transducers with locked signal output. *Cell* **55:**817–826.

Barak, R., and M. Eisenbach. 1992a. Correlation between phosphorylation of the chemotaxis protein CheY and its activity at the flagellar motor. *Biochemistry* **31:**1821–1826.

Barak, R., and M. Eisenbach. 1992b. Fumarate or a fumarate metabolite restores switching ability to rotating flagella of bacterial envelopes. *J. Bacteriol.* **174:**643–645.

Barak, R., and M. Eisenbach. Regulation of the interaction between the signaling protein CheY and the flagellar motor during bacterial chemotaxis. *Curr. Top. Cell. Regul.,* in press.

Barak, R., M. Welch, A. Yanovsky, K. Oosawa, and M. Eisenbach. 1992. Acetyl-adenylate or its derivative acetylates the chemotaxis protein CheY in vitro and increases its activity at the flagellar switch. *Biochemistry* **31:**10099–10107.

Berg, H.C., and S. Khan. 1983. A model for the flagellar rotary motor, p. 486–497. *In* H. Sund and C. Veeger (ed.), *Mobility and Recognition in Cell Biology.* Walter de Gruyter, Berlin.

Berg, H. C., and L. Turner. 1993. Torque generated by the flagellar motor of *Escherichia coli. Biophys. J.* **65:**2201–2216.

Berry, R. M. 1993. Torque and switching in the bacterial flagellar motor. An electrostatic model. *Biophys. J.* **64:**961–973.

Blair, D. F. 1990. The bacterial flagellar motor. *Semin. Cell Biol.* **1:**75–85.

Blair, D. F. Personal communication.

Blair, D. F., and H. C. Berg. 1988. Restoration of torque in defective flagellar motors. *Science* **242:**1678–1681.

Blair, D. F., and H. C. Berg. 1990. The MotA protein of *E. coli* is a proton-conducting component of the flagellar motor. *Cell* **60:**439–449.

Blair, D. F., and H. C. Berg. 1991. Mutations in the MotA protein of *Escherichia coli* reveal domains critical for proton conduction. *J. Mol. Biol.* **221:**1433–1442.

Block, S. M., and H. C. Berg. 1984. Successive incorporation of force-generating units in the bacterial rotary motor. *Nature* (London) **309:**470–472.

Block, S. M., J. E. Segall, and H. C. Berg. 1983. Adaptation kinetics in bacterial chemotaxis. *J. Bacteriol.* **154:**312–323.

Bourret, R. B., S. K. Drake, S. A. Chervitz, M. I. Simon, and J. J. Falke. 1993. Activation of the phosphosignaling protein CheY. II. Analysis of activated mutants by ^{19}F NMR and protein engineering. *J. Biol. Chem.* **268**:13089–13096.

Bourret, R. B., J. F. Hess, and M. I. Simon. 1990. Conserved aspartate residues and phosphorylation in signal transduction by the chemotaxis protein CheY. *Proc. Natl. Acad. Sci. USA* **87**:41–45.

Caplan, S. R., and M. Kara-Ivanov. 1993. The bacterial flagellar motor. *Int. Rev. Cytol.* **147**:97–164.

Chun, S. Y., and J. S. Parkinson. 1988. Bacterial motility: membrane topology of the *Escherichia coli* MotB protein. *Science* **239**:276–278.

Clegg, D. O., and D. E. Koshland, Jr. 1985. Identification of a bacterial sensing protein and effects of its elevated expression. *J. Bacteriol.* **162**:398–405.

Dailey, F. E., and H. C. Berg. 1993. Change in direction of flagellar rotation in *Escherichia coli* mediated by acetate kinase. *J. Bacteriol.* **175**:3236–3239.

Dean, G. E., R. M. Macnab, J. Stader, P. Matsumura, and C. Burks. 1984. Gene sequence and predicted amino acid sequence of the MotA protein, a membrane-associated protein required for flagellar rotation in *Escherichia coli*. *J. Bacteriol.* **159**:991–999.

De Mot, R., and J. Vanderleyden. 1994. The C-terminal sequence conservation between OmpA-related outer membrane proteins and MotB suggests a common function in both gram-positive and gram-negative bacteria, possibly in the interaction of these domains with peptidoglycan. *Mol. Microbiol.* **12**:333–334.

Eisenbach, M., A. Wolf, M. Welch, S. R. Caplan, I. R. Lapidus, R. M. Macnab, H. Aloni, and O. Asher. 1990. Pausing, switching and speed fluctuation of the bacterial flagellar motor and their relation to motility and chemotaxis. *J. Mol. Biol.* **211**:551–563.

Francis, N. R., V. M. Irikura, S. Yamaguchi, D. J. DeRosier, and R. M. Macnab. 1992. Localization of the *Salmonella typhimurium* flagellar switch protein FliG to the cytoplasmic M-ring face of the basal body. *Proc. Natl. Acad. Sci. USA* **89**:6304–6308.

Francis, N. R., G. E. Sosinsky, D. Thomas, and D. J. DeRosier. 1994. Isolation, characterization and structure of bacterial flagellar motors containing the switch complex. *J. Mol. Biol.* **235**:1261–1270.

Hess, J. F., K. Oosawa, N. Kaplan, and M. I. Simon. 1988. Phosphorylation of three proteins in the signaling pathway of bacterial chemotaxis. *Cell* **53**:79–87.

Homma, M., T. Iino, and R. M. Macnab. 1988. Identification and characterization of the products of six region III flagellar genes (*flaAII.3* through *flaQII*) of *Salmonella typhimurium*. *J. Bacteriol.* **170**:2221–2228.

Hotani, H. 1982. Micro-video study of moving bacterial flagellar filaments. III. Cyclic transformation induced by mechanical force. *J. Mol. Biol.* **156**:791–806.

Iino, T., Y. Komeda, K. Kutsukake, R. M. Macnab, P. Matsumura, J. S. Parkinson, M. I. Simon, and S. Yamaguchi. 1988. New unified nomenclature for the flagellar genes of *Escherichia coli* and *Salmonella typhimurium*. *Microbiol. Rev.* **52**:533–535.

Irikura, V. M., M. Kihara, S. Yamaguchi, H. Sockett, and R. M. Macnab. 1993. *Salmonella typhimurium fliG* and *fliN* mutations causing defects in assembly, rotation, and switching of the flagellar motor. *J. Bacteriol.* **175**:802–810.

Ishihara, A., J. E. Segall, S. M. Block, and H. C. Berg. 1983. Coordination of flagella on filamentous cells of *Escherichia coli*. *J. Bacteriol.* **155**:228–237.

Jones, C. J., and S.-I. Aizawa. 1991. The bacterial flagellum and flagellar motor: structure, assembly, and function. *Adv. Microbiol. Physiol.* **32**:109–172.

Jones, C. J., and R. M. Macnab. 1990. Flagellar assembly in *Salmonella typhimurium*: analysis with temperature-sensitive mutants. *J. Bacteriol.* **172**:1327–1339.

Jones, C. J., R. M. Macnab, H. Okino, and S.-I. Aizawa. 1990. Stoichiometric analysis of the flagellar hook- (basal-body) complex of *Salmonella typhimurium*. *J. Mol. Biol.* **212**:377–387.

Khan, I. H., T. S. Reese, and S. Khan. 1992. The cytoplasmic component of the bacterial flagellar motor. *Proc. Natl. Acad. Sci. USA* **89**:5956–5960.

Khan, S., M. Dapice, and T. S. Reese. 1988. Effects of *mot* gene expression on the structure of the flagellar motor. *J. Mol. Biol.* **202**:575–584.

Khan, S., I. H. Khan, and T. S. Reese. 1991. New structural features of the flagellar base in *Salmonella typhimurium* revealed by rapid-freeze electron microscopy. *J. Bacteriol.* **173**:2888–2896.

Khan, S., and R. M. Macnab. 1980. The steady-state counterclockwise/clockwise ratio of bacterial flagellar motors is regulated by protonmotive force. *J. Mol. Biol.* **138**:563–597.

Kihara, M., M. Homma, K. Kutsukake, and R. M. Macnab. 1989. Flagellar switch of *Salmonella typhimurium*: gene sequences and deduced protein sequences. *J. Bacteriol.* **171**:3247–3257.

Kihara, M., and R. Macnab. Unpublished data.

Kuo, S. C., and D. E. Koshland, Jr. 1986. Sequence of the *flaA* (*cheC*) locus of *Escherichia coli* and discovery of a new gene. *J. Bacteriol.* **166**:1007–1012.

Kuo, S. C., and D. E. Koshland, Jr. 1987. Roles of *cheY* and *cheZ* gene products in controlling flagellar rotation in bacterial chemotaxis of *Escherichia coli*. *J. Bacteriol.* **169**:1307–1314.

Kuo, S. C., and D. E. Koshland, Jr. 1989. Multiple kinetic states for the flagellar motor switch. *J. Bacteriol.* **171**:6279–6287.

Kutsukake, K., Y. Ohya, S. Yamaguchi, and T. Iino. 1988. Operon structure of flagellar genes in *Salmonella typhimurium. Mol. Gen. Genet.* **214:**11–15.

Läuger, P. 1977. Ion transport and rotation of bacterial flagella. *Nature* (London) **268:**360–362.

Läuger, P. 1988. Torque and rotation rate of the bacterial flagellar motor. *Biophys. J.* **53:**53–65.

Macnab, R. M. 1987a. Flagella, p. 70–83. *In* F. C. Neidhardt, J. L. Ingraham, K. B. Low, B. Magasanik, M. Schaechter, and H. E. Umbarger (ed.), *Escherichia coli and Salmonella typhimurium: Cellular and Molecular Biology.* American Society for Microbiology, Washington, D.C.

Macnab, R. M. 1987b. Motility and chemotaxis, p. 732–759. *In* F. C. Neidhardt, J. L. Ingraham, K. B. Low, B. Magasanik, M. Schaechter, and H. E. Umbarger (ed.), *Escherichia coli and Salmonella typhimurium: Cellular and Molecular Biology.* American Society for Microbiology, Washington, D.C.

Macnab, R. M. 1992. Genetics and biogenesis of bacterial flagella. *Annu. Rev. Genet.* **26:**129–156.

Macnab, R. M. Flagella and motility. *In* F. C. Neidhardt, R. Curtiss, III, C. A. Gross, J. L. Ingraham, E. C. C. Lin, K. B. Low, Jr., B. Magasanik, W. Reznikoff, M. Riley, M. Schaechter, and H. E. Umbarger (ed.), *Escherichia coli and Salmonella typhimurium: Cellular and Molecular Biology,* 2nd ed. ASM Press, in press.

Macnab, R. M., and S.-I. Aizawa. 1984. Bacterial motility and the bacterial flagellar motor. *Annu. Rev. Biophys. Bioeng.* **13:**51–83.

Macnab, R. M., and D. P. Han. 1983. Asynchronous switching of flagellar motors on a single bacterial cell. *Cell* **32:**109–117.

Macnab, R. M., and M. K. Ornston. 1977. Normal-to-curly flagellar transitions and their role in bacterial tumbling. Stabilization of an alternative quaternary structure by mechanical force. *J. Mol. Biol.* **112:**1–30.

Magariyama, Y., S. Yamaguchi, and S.-I. Aizawa. 1990. Genetic and behavioral analysis of flagellar switch mutants of *Salmonella typhimurium. J. Bacteriol.* **172:**4359–4369.

Malakooti, J., Y. Komeda, and P. Matsumura. 1989. DNA sequence analysis, gene product identification, and localization of flagellar motor components of *Escherichia coli. J. Bacteriol.* **171:**2728–2734.

Marwan, W., W. Schäfer, and D. Oesterhelt. 1990. Signal transduction in *Halobacterium* depends on fumarate. *EMBO J.* **9:**355–362.

Matsumura, P., J. J. Rydel, R. Linzmeier, and D. Vacante. 1984. Overexpression and sequence of the *Escherichia coli cheY* gene and biochemical activities of the CheY protein. *J. Bacteriol.* **160:**36–41.

Oosawa, F., and S. Hayashi. 1983. Coupling between flagellar motor rotation and proton flux in bacteria. *J. Physiol. Soc.* (Japan) **52:**4019–4028.

Oosawa, K., J. F. Hess, and M. I. Simon. 1988. Mutants defective in bacterial chemotaxis show modified protein phosphorylation. *Cell* **53:**89–96.

Oosawa, F., and J. Masai. 1982. Mechanism of flagellar motor rotation in bacteria. *J. Physiol. Soc.* (Japan) **51:**631–641.

Oosawa, K., T. Ueno, and S.-I. Aizawa. 1994. Overproduction of the bacterial flagellar switch proteins and their interactions with the MS ring complex in vitro. *J. Bacteriol.* **176:**3683–3691.

Parkinson, J. S., S. R. Parker, P. B. Talbert, and S. E. Houts. 1983. Interactions between chemotaxis genes and flagellar genes in *Escherichia coli. J. Bacteriol.* **155:**265–274.

Ravid, S., and M. Eisenbach. 1984. Direction of flagellar rotation in bacterial cell envelopes. *J. Bacteriol.* **158:**222–230.

Ravid, S., P. Matsumura, and M. Eisenbach. 1986. Restoration of flagellar clockwise rotation in bacterial envelopes by insertion of the chemotaxis protein CheY. *Proc. Natl. Acad. Sci. USA* **83:**7157–7161.

Roman, S. J., B. B. Frantz, and P. Matsumura. 1993. Gene sequence, overproduction, purification and determination of the wild-type level of the *Escherichia coli* flagellar switch protein FliG. *Gene* **133:**103–108.

Roman, S. J., M. Meyers, K. Volz, and P. Matsumura. 1992. A chemotactic signaling surface on CheY defined by suppressors of flagellar switch mutations. *J. Bacteriol.* **174:**6247–6255.

Schuster, S. C., and S. Khan. 1994. The bacterial flagellar motor. *Annu. Rev. Biophys. Biomol. Struct.* **23:**509–539.

Sockett, H., S. Yamaguchi, M. Kihara, V. M. Irikura, and R. M. Macnab. 1992. Molecular analysis of the flagellar switch protein FliM of *Salmonella typhimurium. J. Bacteriol.* **174:**793–806.

Stader, J., P. Matsumura, D. Vacante, G. E. Dean, and R. M. Macnab. 1986. Nucleotide sequence of the *Escherichia coli motB* gene and site-limited incorporation of its product into the cytoplasmic membrane. *J. Bacteriol.* **166:**244–252.

Stock, A. M., E. Martinez-Hackert, B. F. Rasmussen, A. H. West, J. B. Stock, D. Ringe, and G. A. Petsko. 1993. Structure of the Mg^{2+}-bound form of CheY and mechanism of phosphoryl transfer in bacterial chemotaxis. *Biochemistry* **32:**13375–13380.

Stock, A. M., J. M. Mottonen, J. B. Stock, and C. E. Schutt. 1989. Three-dimensional structure of CheY, the response regulator of bacterial chemotaxis. *Nature* (London) **337:**745–749.

Tisa, L. S., and J. Adler. 1992. Calcium ions are involved in *Escherichia coli* chemotaxis. *Proc. Natl. Acad. Sci. USA* **89:**11804–11808.

Tisa, L. S., B. M. Olivera, and J. Adler. 1993. Inhibition of *Escherichia coli* chemotaxis by ω-conotoxin, a calcium ion channel blocker. *J. Bacteriol.* **175**:1235–1238.

Toker, A., M. Kihara, and R. Macnab. Unpublished data.

Ueno, T., K. Oosawa, and S.-I. Aizawa. 1992. M ring, S ring and proximal rod of the flagellar basal body of *Salmonella typhimurium* are composed of subunits of a single protein, FliF. *J. Mol. Biol.* **227**:672–677.

Volz, K., and P. Matsumura. 1991. Crystal structure of *Escherichia coli* CheY refined at 1.7-Å resolution. *J. Biol. Chem.* **266**:15511–15519.

Washizu, M., Y. Kurahashi, H. Iochi, O. Kurosawa, S.-I. Aizawa, S. Kudo, Y. Magariyama, and H. Hotani. 1993. Dielectrophoretic measurement of bacterial motor characteristics. *IEEE Trans.* **29**:286–294.

Welch, M., and M. Eisenbach. Personal communication.

Welch, M., K. Oosawa, S.-I. Aizawa, and M. Eisenbach. 1993. Phosphorylation-dependent binding of a signal molecule to the flagellar switch of bacteria. *Proc. Natl. Acad. Sci. USA* **90**:8787–8791.

Wilson, M. L., and R. M. Macnab. 1988. Overproduction of the MotA protein of *Escherichia coli* and estimation of its wild-type level. *J. Bacteriol.* **170**:588–597.

Wilson, M. L., and R. M. Macnab. 1990. Co-overproduction and localization of the *Escherichia coli* motility proteins MotA and MotB. *J. Bacteriol.* **172**:3932–3939.

Wolfe, A. J., M. P. Conley, and H. C. Berg. 1988. Acetyladenylate plays a role in controlling the direction of flagellar rotation. *Proc. Natl. Acad. Sci. USA* **85**:6711–6715.

Wolfe, A. J., M. P. Conley, T. J. Kramer, and H. C. Berg. 1987. Reconstitution of signaling in bacterial chemotaxis. *J. Bacteriol.* **169**:1878–1885.

Wylie, D., A. Stock, C.-Y. Wong, and J. Stock. 1988. Sensory transduction in bacterial chemotaxis involves phosphotransfer between Che proteins. *Biochem. Biophys. Res. Commun.* **151**:891–896.

Yamaguchi, S., S.-I. Aizawa, M. Kihara, M. Isomura, C. J. Jones, and R. M. Macnab. 1986a. Genetic evidence for a switching and energy-transducing complex in the flagellar motor of *Salmonella typhimurium*. *J. Bacteriol.* **168**:1172–1179.

Yamaguchi, S., H. Fujita, A. Ishihara, S.-I. Aizawa, and R. M. Macnab. 1986b. Subdivision of flagellar genes of *Salmonella typhimurium* into regions responsible for assembly, rotation, and switching. *J. Bacteriol.* **166**:187–193.

CELLULAR
PHYSIOLOGY

IV

Signal Transduction and Cross Regulation in the *Escherichia coli* Phosphate Regulon by PhoR, CreC, and Acetyl Phosphate

Barry L. Wanner

12

The activation of bacterial alkaline phosphatase (Bap) synthesis by P_i starvation in *Escherichia coli* is a classic example of induced enzyme synthesis in bacteria. This induction is a paradigm of a bacterial signal transduction pathway. It involves a process of transmembrane signaling because it is regulated by the environmental (extracellular) P_i level (Wanner, 1993). Although the P_i-specific transport (Pst) system plays a crucial role in this process, transport per se is not involved. Increased Bap synthesis is due to increased transcription of the Bap structural gene, *phoA*. When the environmental P_i concentration falls to less than about 4 µM (an amount insufficient to saturate fully the Pst system, $K_d \approx 0.8$ µM), the synthesis of mRNA corresponding to the *phoA* gene increases more than 1,000-fold, as judged by examination of a strain carrying a *phoA* promoter fusion to the β-galactosidase structural gene, *lacZ* (Wanner, 1990; and references therein).

The *phoA* gene is a member of the phosphate (Pho) regulon, a system that is comprised of many coregulated genes. These other Pho regulon gene products, like Bap, probably all have roles in the process of phosphorus (P)

assimilation. Regardless of the P source, the overall process of P assimilation has at least two steps in common. First, extracellular P_i (or some form of an alternative P source) is taken up. Subsequently, intracellular P_i (or a phosphoryl group of the alternative P source) is incorporated into ATP, the primary phosphoryl donor in cellular metabolism.

The Pho regulon is regulated by the environmental P_i level. This control is coupled to an early step in P metabolism, P_i uptake. Control by this signal transduction pathway requires an intact Pst system, a protein called PhoU, and PhoR (which appears to act as the P_i sensor). In the absence of this control (due to the absence of PhoR), two additional controls of the Pho regulon are apparent (Wanner and Wilmes-Riesenberg, 1992). Both of these controls are P_i independent and may be examples of "cross regulation" (Wanner, 1992). These controls appear to be coupled to subsequent steps in P_i metabolism, pathways leading to the incorporation of P_i into ATP. One P_i-independent control requires the sensor CreC (formerly called PhoM), which may have a primary role in a signal transduction pathway of a catabolite regulatory system (and which may control an unknown set of target genes). The other P_i-independent control involves acetyl phos-

Barry L. Wanner, Department of Biological Sciences, Purdue University, West Lafayette, Indiana 47907.

Two-Component Signal Transduction, Edited by James A. Hoch and Thomas J. Silhavy,
© 1995 American Society for Microbiology, Washington, DC 20005

phate, an intermediate of the phosphotransacetylase (Pta)-acetate kinase (AckA) pathway in substrate-level phosphorylation. Because the Pta-AckA pathway leads to the incorporation of P_i into ATP, the control of the Pho regulon by acetyl phosphate provides a mechanism for coupling Pho regulon control directly to a late step in the process of P_i assimilation, ATP synthesis.

The transcriptional activation of the Pho regulon in response to P_i limitation requires two proteins, PhoB and PhoR. These proteins share sequence similarities with the response regulator (receiver) and sensor (transmitter) domains, respectively, in common with the large family of two-component regulatory systems in bacteria (Parkinson and Kofoid, 1992; Nixon et al., 1986; Ronson et al., 1987). Accordingly, PhoB is the response regulator, and PhoR is the P_i sensor. PhoB is a DNA binding protein and acts as a transcriptional activator of the Pho regulon. Furthermore, the transcriptional activation of the Pho regulon requires the phosphorylated form of the regulator PhoB (Makino et al., 1989). Under conditions of P_i limitation, the P_i sensor PhoR is thought to be autophosphorylated and to act as a PhoB kinase by transfer of a phosphoryl group to PhoB. Under other conditions, CreC (a PhoR homolog) or acetyl phosphate may act by phosphorylation of PhoB (Wanner and Wilmes-Riesenberg, 1992). Recent studies on the mechanism of transcriptional activation of the Pho regulon showed that activation may involve contact of phospho-PhoB with the σ^{70} subunit of RNA polymerase holoenzyme (Makino et al., 1993).

This chapter focuses on the control of the Pho regulon by the signal transduction pathways involving the P_i sensor PhoR, the catabolite regulatory sensor CreC, and acetyl phosphate. It should be pointed out that the Pho regulon has been studied for a very long time. Yet, there is much more to be learned. In particular, it is still poorly understood how cells detect the environmental P_i level and communicate a corresponding signal across the membrane. It is also poorly understood whether cross regulation by CreC (which acts as an alternative sensor in the absence of PhoR) or by acetyl phosphate (which also acts primarily in the absence of PhoR) has a bona fide role in Pho regulon control under certain conditions in normal cells. Therefore, some speculations are provided about the nature of the P_i signal transduction pathway and about the roles of CreC and acetyl phosphate in Pho regulon control. Admittedly, this was done in part to overcome deficiencies in our understanding of Pho regulon control. Some readers (like one or more recent anonymous reviewers) may even find some of these speculations a bit provocative. These speculations are given not because of compelling evidence favoring them but rather as a means to stimulate new research on this fascinating (and complex) control system. Some aspects of signal transduction and cross regulation of the Pho regulon have been recently described elsewhere (Wanner, 1992, 1993). Also, a historical account of our understanding of these control pathways and of the discovery of acetyl phosphate as an effector of gene expression has been recently published (Wanner, 1994).

SIGNAL TRANSDUCTION PATHWAYS OF THE Pho REGULON

P_i Control by the Pst System, PhoU, and PhoR

The primary control of the Pho regulon is its control by P_i limitation. This control involves two processes: inhibition when P_i is in excess, and activation under conditions of P_i limitation. Activation due to P_i limitation may lead to more than a 1,000-fold induction of an individual Pho regulon gene, depending on the promoter of that gene. P_i repression requires an environmental P_i concentration greater than 4 μM, an intact (but not necessarily functional) Pst system (a cell surface receptor complex for P_i uptake composed of PstS [formerly called PhoS], PstA, PstB, and PstC), the P_i sensor PhoR, and an accessory protein called PhoU. Under conditions of P_i limitation, only PhoR (and none of these other components) and PhoB are required for activation. Because multiple components are required for P_i repression, and many of these (in particular those of the Pst system) are certain to interact, it is likely that the process of P_i repression may involve formation of a "repression complex" and that P_i control may involve protein-protein interactions within such a complex (Wanner, 1990). This complex is likely to be located in the cytoplasmic membrane. The existence of a membrane-associated P_i repression complex containing the Pst system, PhoU, and PhoR is consistent with features of these components. Clearly, the P_i transporter is in the membrane. Also, PhoU is peripherally associated with the cytoplasmic membrane (Surin et al., 1986), even though it does not have features of a membrane protein. PhoU may be membrane-associated due to an interaction with the Pst system or PhoR, or both. Furthermore, PhoR is a cytoplasmic membrane protein. PhoR may be anchored to the membrane via its two N-terminal transmembrane domains (Scholten and Tommassen, 1993). These domains of PhoR appear to flank a short (6 or 7 amino acid) periplasmic domain. PhoR has a large carboxy-terminal (cytoplasmic) domain. Its C-terminal domain is autophosphorylated, and it acts as a phosphoryl transferase and PhoB kinase.

DESCRIPTION OF P_i REPRESSION COMPLEXES

A P_i repression complex may contain all components of the Pst system, PhoU, and PhoR, because all these are required for P_i repression. P_i control involves two forms of PhoR, a repressor form (PhoR^R) that may predominate when P_i is in excess and an activator form (PhoR^A) that may predominate under conditions of P_i limitation (Wanner and Latterell, 1980; Wanner, 1987b). PhoR^R may correspond to a phospho-PhoB phosphatase, and PhoR^A may correspond to autophosphorylated PhoR, which may act as a phosphoryl transferase and PhoB kinase. Two hypothetical P_i repression complexes are depicted in Fig. 1. These repres-

Repression complex I

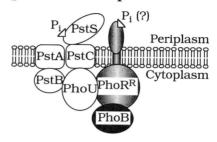

Repression complex II

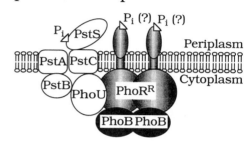

FIGURE 1 Alternative repression complexes for signal transduction by P_i limitation. P_i control may involve P_i binding only to PstS or to both PstS and a regulatory site, which may be on a different Pst component or on PhoR. A hypothetical regulatory site is indicated by a question mark in parentheses. In repression complex I, the PhoR repressor form is shown as a monomer, which is intended to imply any lower oligomeric form. In repression complex II, the PhoR repressor form is shown as a dimer, which is intended to imply any higher oligomeric form. These complexes are compatible with alternative mechanisms for interconversion of the PhoR repressor and activator forms as described in the text.

sion complexes contain PhoR^R. Also, PhoB is shown in contact with PhoR in these complexes, even though PhoB is not considered a component of the repression complex per se. PhoR and PhoB are shown to interact because an interaction is necessary so that PhoR^R may be able to dephosphorylate phospho-PhoB and PhoR^A may be able to phosphorylate PhoB. PhoB is not considered a component of the repression complex because PhoR-PhoB contacts may be transient. The repression complexes shown in Fig. 1 differ only in regard to the oligomeric state of PhoR. These complexes also show two P_i binding sites, one on PstS and a second (regulatory) site on PhoR. Alternative mechanisms for P_i control involving these complexes are considered below. These mechanisms are dependent on different oligomeric forms of PhoR or a (hypothetical) regulatory P_i binding site on a component of the repression complex.

The Pst system is a multicomponent transport system similar to ones for histidine, maltose, ribose, and other bacterial periplasmic transport systems (Ames, 1986a; Adams and Oxender, 1989). It is composed of the P_i binding protein (PstS), two integral membrane proteins (PstA and PstC), and the P_i permease (PstB). These transporters are called "ABC transporters" because each has a permease component containing one or more ATP binding cassette (ABC) motifs. They share similarities at the protein sequence level to a variety of ABC proteins, including those of the mammalian multidrug resistance and cystic fibrosis genes (Gros et al., 1986; Chen et al., 1986; Ames, 1986b; Foote et al., 1990; Hyde et al., 1990). Furthermore, other signal transduction pathways involve regulatory components similar to those of the Pst system. Sporulation and competence in the gram-positive bacterium *Bacillus subtilis* are controlled in part by the *spo0K* locus (Hoch, 1993; Ireton et al., 1993), which encodes an oligopeptide transporter analogous to the Pst system. Also, multicellular fruiting body formation in the gramnegative bacterium *Myxococcus xanthus* requires a wild-type *sasA* locus that encodes another member of the ABC transporter family. The *sasA* locus has a role in negative regulation of signal-dependent genes expressed early after nutrient limitation (Kaplan et al., 1991). One of the three *sasA* open reading frames shares sequence similarity at the protein level to PstB (Kaplan, personal communication). How these diverse processes are controlled by different environmental signals may be fundamentally similar. Our current understanding of P_i control of the Pho regulon by environmental P_i and the Pst system is described below.

Both the P_i binding protein (PstS) and permease (PstB) are important in P_i control. Perhaps the strongest evidence favoring a role for PstS in P_i control is the finding that missense changes that alter P_i binding (Luecke and Quiocho, 1990) abolish P_i repression (Makino et al., 1992). Mutational studies have also implicated PstB in P_i control. PstB has a single ABC motif. Mutational alterations of this motif that abolish P_i uptake simultaneously abolish repression (Cox et al., 1989). Therefore, this motif is essential for both of these processes. Nevertheless, the processes of P_i uptake and repression may be uncoupled. This was shown in mutational studies of an integral membrane component (PstA) of the Pst system. A site-directed missense change in PstA abolishes P_i transport without affecting P_i repression (Cox et al., 1988). However, ABC transporters may contain one or two integral membrane protein species. Those with a single species may contain a dimer of that species. The Pst system contains two species, PstA and PstC. Although both are essential for P_i uptake, both may not be necessary for P_i repression. In this regard, a different *pstA* mutation (the *pstA2* allele) had been previously shown to be unusual (Willsky et al., 1973; Cox et al., 1981). Like most *pst* mutations, the *pstA2* allele abolishes P_i transport. But unlike most *pst* mutations (which lead to full derepression), the *pstA2* mutation leads to only partial (about 10%) derepression of the Pho regulon. Curiously, the *pstA2* allele has a nonsense change near the N terminus of PstA (Cox et al., 1988). The lowered constitutive level observed in the *pstA2* mutant may indicate that PstC can substitute for PstA to form a partially functional repression complex in this mutant, even though PstC apparently cannot substitute for PstA to form a functional transporter. Partially functional substitution by PstC may only result in those mutants in which PstA is totally absent due to an early nonsense mutation.

Also, PhoU is required for P_i repression. How PhoU acts is unknown. It may act in a way similar to P_{II}, the *glnD* gene product of the

Ntr system; P_{II} is required for the regulated dephosphorylation of phospho-NtrC (NR_I) (Magasanik, 1993). Or PhoU may have an entirely different role. In regard to PhoU, until recently all studies on *phoU* function concerned the effect of a single *phoU* mutation (the *phoU35* allele) on P_i control. Effects due to the *phoU35* allele may not indicate effects due to a null allele because this mutation results in a missense (A147E) change (Cox, personal communication). In contrast to mutants carrying the *phoU35* allele, ones deleted of the *phoU* gene show a severe growth defect (Steed and Wanner, 1993). Mutants carrying either the *phoU35* allele or a *phoU* deletion are fully derepressed for the Pho regulon. Also, the severe growth defect of a *phoU* deletion mutant requires a functional Pst system. This suggests that PhoU has a role related to the process of P_i uptake by the Pst system (even though PhoU has no role in transport), in addition to a role as a negative regulator. Because the severe growth defect depends on a functional Pst system, the additional role of PhoU may involve a role in P_i metabolism. Perhaps PhoU acts as an enzyme in intracellular P_i metabolism, providing a new pathway for ATP synthesis when P_i is taken up by the Pst system. A role for PhoU in intracellular P_i metabolism may involve an association between PhoU and the Pst system, whereas a role for PhoU in P_i repression may involve an association between PhoU and PhoR in a repression complex.

Figures 1 and 3 depict PstS in contact with PstC, PstA in contact with PstB, both PstB and PstC in contact with PhoU, and PhoU (and no component of the Pst system) in contact with PhoR. These contacts are shown solely for purpose of illustration. Many alternative contacts are equally feasible. Furthermore, only a single PstB molecule per complex is shown for the Pst system in Figs. 1 and 2. Only one PstB molecule is shown solely for purpose of illustration. It is much more likely that the Pst system may contain two PstB subunits per complex, in agreement with the proposed subunit structure of similar transport systems

(Doige and Ames, 1993). Also, PstB is shown on the cytoplasmic face of the repression complex. PstB may be located within the complex such that PstB is exposed to the periplasm, in which case PstB may contact PstS. Such a location has been inferred for HisP, the permease component of the analogous histidine transporter (Baichwal et al., 1993).

MECHANISMS FOR DETECTION OF ENVIRONMENTAL P_i

The overall process of signal transduction by P_i limitation may be arbitrarily divided into two events: (i) detection of environmental P_i in the periplasm by the P_i sensor and (ii) transduction of a signal for P_i limitation across the cytoplasmic membrane. The former may involve the formation of a repression complex containing the P_i sensor in the membrane; the latter may involve the transmission of a signal for P_i limitation via the regulated interconversion of $PhoR^R$ and $PhoR^A$ within the membrane. Although it is highly likely that these may be concerted events, they are considered separately in the following discussion to allow highlighting of differences among alternative mechanisms. One or more concerted actions are likely due to the involvement of protein-protein interactions and due to the possibility that control is brought about solely by (subtle) conformational changes within a repression complex.

Two mechanisms for detection of environmental P_i that may lead to control of the Pho regulon are illustrated in Fig. 2, a stoichiometric mechanism and a regulatory site mechanism. Either may control the interconversion of $PhoR^R$ and $PhoR^A$. Both mechanisms show PhoR within the repression complex as $PhoR^R$ when P_i is in excess. Under these conditions, all P_i repression complexes (in particular all PstS molecules) are expected to be fully occupied with P_i. Also, full occupancy may be necessary to maintain PhoR as $PhoR^R$. Both mechanisms show PhoR as $PhoR^A$ under conditions of P_i limitation. Under these conditions, decreased P_i occupancy of the repression complex

Stoichiometric mechanism

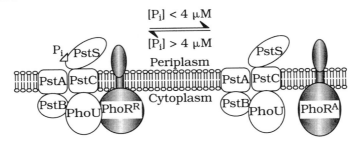

Regulatory site mechanism

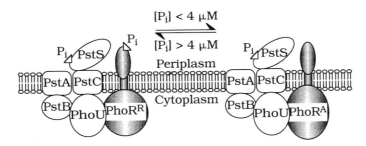

FIGURE 2 Mechanisms of detection of environmental P_i. The stoichiometric and regulatory site mechanisms are illustrated. In accordance with either mechanism, the PhoR repressor form may be associated with a repression complex when P_i is in excess. The stoichiometric mechanism may lead to the release of PhoR from those complexes to which no P_i is bound under conditions of P_i limitation. The regulatory site mechanism may lead to a conformational change in PhoR (without its release from a repression complex) due to absence of P_i occupancy of the regulatory site. PhoR is shown as a monomer solely for the purpose of illustration. No particular oligomeric form of PhoR is implied.

may lead to conversion of $PhoR^R$ to $PhoR^A$. How environmental P_i is detected and how detection of environmental P_i may regulate the interconversion of $PhoR^R$ and $PhoR^A$ are discussed below.

What is the P_i sensor? Several lines of evidence indicate that the Pho regulon is controlled by the level of environmental (extracellular) P_i and not by the level of intracellular P_i and that its control by environmental P_i requires an intact Pst system. The level of P_i necessary for repression (about 4 μM) is an amount sufficient for nearly full saturation of the P_i binding protein PstS ($K_d \approx 0.8$ μM). Inhibition results only when the environmen-

tal P_i level is in excess of this concentration. Activation results at lower concentrations. Under these conditions, intracellular P_i levels (which are usually greater than 10 mM [Shulman et al., 1979]) are maintained at a constant high level (Willsky and Malamy, 1976). Because missense changes in PstS abolish P_i repression, it is tempting to suggest that PstS (and not PhoR) is the P_i sensor. Nevertheless, PhoR is considered as the P_i sensor throughout this discussion. This is because PhoR is a membrane protein and has features in common with several other sensor proteins. Also, the localization of PhoR to the cell surface appears to be essential for P_i control. Furthermore, one proposed mechanism of P_i control may involve P_i binding directly to PhoR (Fig. 2).

The stoichiometric mechanism may (by definition) depend on the ratio of the Pst components, PhoU, and PhoR. When extracellular P_i is greater than 4 µM, the Pst system may be fully saturated with P_i as all PstS molecules may contain bound P_i. Consequently, all PhoR molecules may be associated with the Pst system and PhoU in the repression complex. Under these conditions, PhoR may exist as PhoR^R by virtue of its association with the repression complex. P_i limitation may lead to the formation of some complexes to which no P_i is bound. In the absence of bound P_i, the complex may undergo a conformational change leading to the release of PhoR. On its release, PhoR^R may undergo a conformational change leading to its conversion to PhoR^A. For this mechanism to operate, it may be especially important to maintain a certain stoichiometry between the components of the Pst system, PhoU, and PhoR. A particular stoichiometry may be important under conditions of P_i excess so all PhoR molecules are associated with a repression complex as PhoR^R. A different stoichiometry may be important under conditions of P_i limitation to allow for the operation of a mechanism to reestablish repression after a period of P_i limitation. In regard to a stoichiometric mechanism, it is interesting that synthesis of the Pst system, PhoU, and PhoR is coregulated as the synthesis of each is inhibited when P_i is in excess and activated under conditions of P_i limitation.

The regulatory site mechanism invokes the existence of a second P_i binding site on a component of the repression complex. One P_i binding site exists on the P_i binding protein (PstS) for P_i uptake. The second site may correspond to the regulatory site. This site may be on a different component of the Pst system or it may be on the periplasmic domain of PhoR. According to the regulatory site mechanism, both sites may be occupied only when P_i is in excess, and repression may result only when both sites are occupied. Under conditions of P_i

limitation, the regulatory site may be unoccupied, and activation may result. This may be accomplished by a relatively small difference in the affinity of P_i to these sites. Accordingly, the regulatory site may have a lower affinity. Also, P_i binding to PstS may somehow facilitate P_i binding to the regulatory site. This may explain the effect of PstS mutations on P_i repression. The regulatory site mechanism also provides an explanation for how the processes of P_i repression and P_i uptake may be separable. Under conditions of P_i limitation, P_i may be taken up by the Pst system without causing repression because under these conditions P_i may bind to PstS (as a prerequisite for transport) and the regulatory site may be unoccupied.

MECHANISMS FOR INTERCONVERSION OF PhoR^R AND PhoR^A

Two mechanisms for the regulated interconversion of PhoR^R and PhoR^A are shown in Fig. 3, an association–dissociation mechanism and a conformational change mechanism. An association–dissociation mechanism may result in a change in the oligomeric form of PhoR, and this change may be responsible for interconversion of PhoR^R and PhoR^A. On release of PhoR from the repression complex, PhoR^A may form by association of PhoR subunits. The release of PhoR may be due to a change in the stoichiometry of repression complexes occupied by P_i. Alternatively, the release of PhoR may be due to a conformational change in the repression complex brought about by the absence of bound P_i at a regulatory site. Similarly, a conformational change mechanism may result in interconversion of PhoR^R and PhoR^A. On release of PhoR from the repression complex, a resulting conformational change in PhoR may lead to its conversion to PhoR^A. Alternatively, a conformational change in the repression complex brought about by the absence of bound P_i at a regulatory site may lead to a conformational change in PhoR and its conversion to PhoR^A. At present, there are no data for distinguishing whether the stoichio-

Association-dissociation mechanism

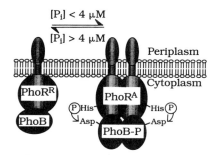

Conformational change mechanism

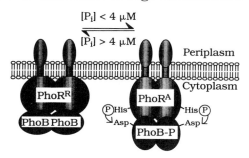

FIGURE 3 Mechanisms for interconversion of PhoR^R and PhoR^A by environmental P$_i$. Association-dissociation and conformational change mechanisms are illustrated. PhoR^R may prevail when the environmental P$_i$ level is in excess (an environmental P$_i$ level greater than about 4 μM); and PhoR^A may prevail under conditions of P$_i$ limitation (an environmental P$_i$ level less than about 4 μM). PhoR^R may be a phospho-PhoB (PhoB-P) phosphatase, and PhoR^A is a phosphoryl transferase and PhoB kinase. Accordingly, PhoR^A may autophosphorylate on a histidine residue and transfer this phosphoryl group to an aspartate residue on PhoB. Although PhoR^R and PhoR^A are depicted in association with PhoB and PhoB-P, respectively, these associations are likely to be transient. No particular oligomeric form of PhoR is implied, except that PhoR^A is probably a dimer or higher oligomer. This is because the mechanism of autophosphorylation for the PhoR homologs EnvZ and CheA involves phosphorylation of one subunit by the other subunit (Yang and Inouye, 1991, 1993; Swanson et al., 1993; Wolfe and Stewart, 1993). The oligomeric structure(s) of PhoR is unknown.

metric mechanism, the regulatory site mechanism, or another mechanism governs P$_i$ control of the Pho regulon.

Catabolite Control of the Pho Regulon by CreC of the Catabolite Regulatory *creABCD* Operon

Two other signal transduction pathways also control the Pho regulon. Although effects due to these signaling pathways are observed (primarily) only in *phoR* mutants (Wanner and Wilmes-Riesenberg, 1992; Wanner, 1993), there is circumstantial evidence suggesting that they may also be important in Pho regulon control in *phoR*$^+$ cells. Both of these controls are P$_i$ independent. They are instead highly regulated by the carbon and energy source (in different ways). How each of these controls affects the Pho regulon in a null-type *phoR* mutant is described below. How they may operate in the presence of PhoR is considered at the end of this chapter.

Both P$_i$-independent controls lead to activation of the response regulator PhoB, probably by phosphorylation of PhoB. One P$_i$-independent control requires the catabolite regulatory sensor CreC (a PhoR homolog). Like PhoR, CreC (formerly called PhoM) is autophosphorylated in an ATP-dependent manner and acts as a phosphoryl transferase and protein kinase of PhoB and of its own partner response regulator CreB (Makino et al., 1989; Amemura et al., 1990). The other P$_i$-independent control requires acetyl phosphate synthesis (Wanner and Wilmes-Riesenberg, 1992). Two ways in which acetyl phosphate may lead to activation of PhoB have been previously suggested (Wanner, 1992). Acetyl phosphate may directly activate PhoB by acting as a chemical phosphorylating agent. Or acetyl phosphate may lead to activation of an unknown sensor (presumably, a PhoR or CreC homolog), which in turn activates PhoB by phosphorylation.

Table 1 shows effects due to P$_i$ limitation and catabolites on the Pho regulon. When P$_i$ is in excess, the Pho regulon is expressed at a very low basal level, as judged by examination of

TABLE 1 Effects of P_i limitation and catabolites on Pho regulon control

Genotype[a]	Growth condition[b]	Response[c]	Sensor, regulator, or effector molecule[d]
Wild-type	Glucose, P_i excess	Inhibition (0.1)	PhoR, PstSCAB, PhoU
Wild-type	Glucose, P_i limitation	Activation (280)	PhoR
phoR	Glucose, P_i-independent	Activation (42.7)	CreC
phoR creC	Glucose, P_i-independent	Basal (0.2)	None
phoR creC	Pyruvate, P_i-independent	Activation (104)	Acetyl phosphate

[a]Nonsense or deletion *phoR* and *creC* mutations behave similarly.

[b]Cells were grown for 24 h on glucose-MOPS (morpholinepropanesulfonic acid) agar with 2 mM P_i, after which colonies were inoculated into glucose- or pyruvate-MOPS broth containing 2 mM P_i (excess) or 0.1 mM P_i (limitation). Cultures were sampled for assay about 16 h later when the culture optical density at 420 nm was near 1.0. P_i-independent indicates that similar results were obtained under conditions of P_i excess and P_i limitation.

[c]Bap-specific activities (nanomole per minute per optical density at 420 nm) from Wanner and Wilmes-Riesenberg (1992) are given in parentheses.

[d]Effects due to acetyl phosphate are inferred by an examination of mutational effects on acetyl phosphate synthesis, as described by Wanner and Wilmes-Riesenberg (1992).

Bap synthesis. Repression requires PhoR, an intact Pst system (composed of PstS, PstC, PstA, and PstB), and PhoU. Under conditions of P_i limitation, the Pho regulon is activated more than 2,000-fold. This activation is dependent on PhoR. Even though P_i control is abolished, the Pho regulon is still highly regulated in the absence of PhoR. In a (null-type) *phoR* mutant, the Pho regulon may be expressed at a high level. Its level of expression is regulated by the carbon and energy source (Wanner et al., 1988). Glucose (as a sole carbon source) leads to greater than 200-fold activation of Bap synthesis. This activation requires CreC. Many other carbon sources also lead to activation via CreC. Yet, certain carbon sources (in particular, pyruvate, as a sole carbon source) lead to greater than 500-fold activation in the absence of both PhoR and CreC. This activation requires acetyl phosphate synthesis. Pyruvate also leads to activation via CreC, in the presence or absence of acetyl phosphate synthesis (Wanner and Wilmes-Riesenberg, 1992).

CreC is encoded by an operon together with three open reading frames of unknown function (Amemura et al., 1986). The *creC* gene was discovered in a search aimed at uncovering mutants altered in localization of Bap to the periplasm (Wanner et al., 1979). A *phoR* mutant was (fortuitously) used in that study to facilitate screening procedures. The *creC* mutations were found to be regulatory, and they defined a new locus (Wanner and Latterell, 1980). These mutations were originally named *phoM* because their mutational effects were masked in a $phoR^+$ strain under conditions of P_i limitation (during growth with glucose as a sole carbon source). DNA sequencing revealed that *creC* was the third gene of an operon (Amemura et al., 1986), the *creABCD* operon, along with CreA, CreB (also called PhoM-Orf2), and CreD. CreA is a periplasmic protein; CreB is a response regulator; and CreD is a membrane protein with a conserved transmitter domain (Table 2). CreB is phosphorylated by CreC (Amemura et al., 1990) and is therefore likely to regulate an unknown set of target genes. Recent studies showed that mutations of *creA*, *creB*, and *creD* also affect Pho regulon control. However, effects due to mutations in these genes may be indirect because these effects are observed only in $creC^+$ *phoR* mutants (Wilmes-Riesenberg and Wanner, unpublished data). CreA, CreB, and CreD may directly or indirectly alter the level of a catabolite, which may be related metabolically to a compound for which CreC acts as a sensor. For example, CreB may regulate a target gene whose product is an enzyme for metabolism of such a catabolite.

What does CreC sense? CreC is a cytoplasmic membrane protein. Its N-terminal half has

TABLE 2 Genes involved in P_i-independent controls of the Pho regulon

Gene[a]	Map location (min)	Protein	Function	Description
creABCD	0	CreA	Unknown	Periplasmic protein
		CreB	Unknown	Response regulator
		CreC	Unknown	Catabolite sensor
		CreD	Unknown	Membrane protein[b]
pta ackA	48	Pta	Mixed-acid fermentation	Phosphotransacetylase
		AckA	Mixed-acid fermentation	Acetate kinase

[a] Two separate P_i-independent controls act on the Pho regulon in the absence of PhoR. One control requires CreC and is affected by mutations in the creABCD operon. The other control requires acetyl phosphate synthesis and is affected by pta and ackA mutations. Also, several other genes (including arcA, cya, crp, icd, mdh, ompR, ops, ppk, ptsHI, pur,, and rpiA) play a role in P_i-independent control because mutations of these genes also alter Pho regulon control in the absence of PhoR. A $creC^+$ phoR or creC phoR mutant was subjected to transposon mutagenesis, and mutants were isolated that showed increased or decreased Bap synthesis (on a variety of growth media). The mutated genes were identified by conventional methods of phenotypic testing and mapping or by DNA sequencing. Genes interrupted by a transposon were often identified by cloning a DNA fragment corresponding to the transposon-chromosomal junction, sequencing the DNA adjacent to the insertion site, and searching GenBank for a match to a previously sequenced gene (Wanner et al., 1988; Wanner and Wilmes-Riesenberg, 1992). Many of these mutations were also tested for effects in $phoR^+$, $creC^+$ phoR pta ackA, and creC phoR pta ackA mutants to assess whether effects involved activation of PhoB by PhoR, CreC, acetyl phosphate, or another signal transduction pathway (Wilmes-Riesenberg and Wanner, unpublished data).

[b] CreD, like CreC, has a conserved transmitter domain common to bacterial signaling proteins (Kofoid and Parkinson, 1988; Parkinson and Kofoid, 1992).

two regions of highly hydrophobic amino acids flanking a hydrophilic region of about 150 amino acids (Amemura et al., 1986). This hydrophilic domain probably corresponds to a periplasmic loop, on the basis of TnphoA fusions to this domain (Wilmes-Riesenberg and Wanner, 1992). Therefore, CreC, like EnvZ (Forst et al., 1987) and many other sensor proteins containing large periplasmic domains, appears to have a membrane topology similar to signal transducers (Tap, Tar, Trg, Tsr) of bacterial chemotaxis (Hazelbauer, 1992). These transducers bind ligands in the periplasm and transmit a signal across the membrane to activate a protein kinase (CheA) in the cytoplasm (Gegner et al., 1992). By analogy, CreC may respond to an extracellular ligand.

Despite its topology, CreC appears to respond to a cellular metabolite. This is because activation by CreC is highly regulated by the carbon and energy source. Yet, CreC does not appear to respond to a particular carbohydrate or growth condition per se. Instead, many different carbohydrates (especially glucose) lead to strong activation. Despite this, mutations of adenylate cyclase (for synthesis of adenosine-3′, 5′-cyclic monophosphate [cyclic AMP]) and of the cyclic AMP activator protein lead to activation by CreC (Wanner, 1987a; Wanner et al., 1988; Wanner and Wilmes-Riesenberg, 1992; Wilmes-Riesenberg and Wanner, unpublished data). These and other results imply that CreC may respond to a catabolite that is a normal metabolic intermediate whose level changes under various growth conditions. Because CreC may respond to an extracellular ligand, such a metabolite may be secreted when produced in large amounts. There is no direct evidence favoring this idea, however.

Catabolite Control of the Pho Regulon by Acetyl Phosphate as an Effector Molecule

Table 1 shows that a phoR creC mutant expresses the Pho regulon at a low basal level during growth on glucose as a sole carbon source. In other words, such a mutant is Bap-negative during growth on glucose. A phoR creC mutant is also Bap-negative on many other compounds as sole carbon sources and on many kinds of complex media as well (Wanner and Latterell, 1980). Yet, a phoR creC mutant synthesizes Bap at a high level on a glucose-enriched complex (an energy-rich) medium (cited in Wanner, 1987a). To understand the basis of this synthesis, Bap-positive

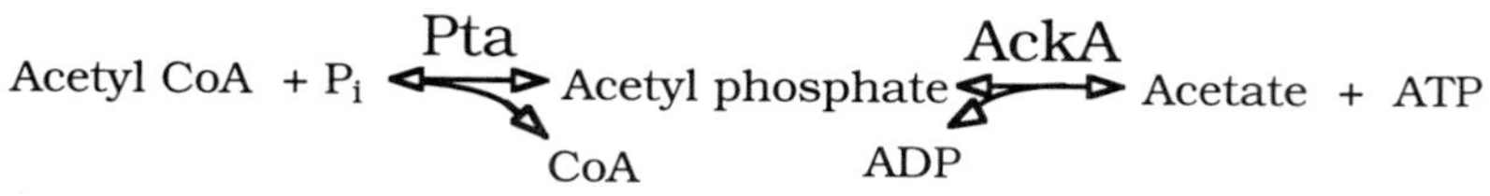

$$\text{Acetyl CoA} + P_i \underset{\text{CoA}}{\overset{\text{Pta}}{\rightleftarrows}} \text{Acetyl phosphate} \underset{\text{ADP}}{\overset{\text{AckA}}{\rightleftarrows}} \text{Acetate} + \text{ATP}$$

FIGURE 4 Pta-AckA pathway for ATP synthesis.

mutants of a *phoR creC* strain were isolated under noninducing conditions. One class of Bap-positive *phoR creC* mutants had lesions in *ackA* (for acetate kinase) (Wanner and Wilmes-Riesenberg, 1992). Other mutants were also isolated. One class of these had lesions in *pta* (for phosphotransacetylase) and abolished Bap synthesis on a glucose-enriched complex medium. By testing effects due to *ackA* and *pta* mutations, it was shown that activation of the Pho regulon in the absence of both PhoR and CreC requires acetyl phosphate synthesis (Wanner and Wilmes-Riesenberg, 1992; Wanner, 1992) (Fig. 4). Effects due to acetyl phosphate (like those due to CreC) are dependent on the response regulator PhoB and are apparent only in the absence of PhoR. Furthermore, activation by CreC or acetyl phosphate appears to be entirely independent of each other. Under certain conditions (e.g., during growth on pyruvate), both CreC and acetyl phosphate may lead to activation. Under these conditions, effects due to these pathways appear to be additive (Wilmes-Riesenberg and Wanner, unpublished data).

Several other genes also affect Pho regulon control in the absence of PhoR, in addition to *creA, creB, creC, creD, pta,* and *ackA*. This was revealed by mutational studies aimed at understanding the genetic basis of P_i-independent Pho regulon control. These other genes include *arcA* (aerobic respiratory control regulator), *cya* (adenylate cyclase), *crp* (catabolite activator protein), *icd* (isocitrate dehydrogenase), *mdh* (malate dehydrogenase), *ompR* (osmoregulator), *ops* (exopolysaccharide production protein), *ppk* (polyphosphate kinase), *pstHI* (phosphohistidinoprotein [HPr] or enzyme I), *pur* (an unidentified gene for guanine biosynthesis), *rpiA* (ribose 5′-phosphate isomerase), an unidenti-

fied *E. coli* open reading frame with sequence similarity to a ribulose 5′-phosphate epimerase gene (from another organism), and others not yet identified (Table 2). Nevertheless, it is reasonable to believe that most (perhaps all) of these gene products lead to activation of PhoB by CreC or acetyl phosphate. This possibility has been tested by construction of the many double or triple mutants (as necessary) to test mutations in these genes for epistasis. In those cases examined so far, regulatory consequences due to lesions in these genes appear to be dependent on CreC or acetyl phosphate synthesis (Wilmes-Riesenberg and Wanner, unpublished data). Also, most of these genes are connected to central metabolism, thus providing circumstantial evidence for linkage of these controls to a central pathway.

CROSS REGULATION OF THE Pho REGULON

Normal Regulation, Cross Regulation, and Crosstalk in Two-Component Regulatory Systems

Environmental sensors and response regulators comprise two large families of evolutionarily related proteins (Ronson et al., 1987). These proteins share sequence similarities at the protein level with other members of the same family. Therefore, sensors are probably structurally and functionally similar to other sensors, and response regulators are probably structurally and functionally similar to other response regulators. Yet, despite these similarities, individual sensors and regulators are able to interact with one another in highly specific ways. Environmental sensors most likely all act as histidine protein kinases that phosphorylate themselves and as phosphotransferases that interact with and phosphorylate response regulators. Phosphorylation, in turn, leads to activation of the response regulator. To guarantee that a particular regulator is activated by a distinct sensor in response to a specific environmental stimulus, it is especially important that a sensor(s) of one two-component regulatory system interacts specifically with a regulator of that system. However, it may also be important that particular response regulators receive input signals from sensors that are also members of a different two-component regulatory system(s). Regulatory interactions of this sort may provide a way of linking different systems in a network in the overall control of a particular cellular process to coordinate cell growth and metabolism.

Normal regulation in a two-component regulatory system usually involves interactions between pairs of environmental sensors and response regulators that act as partner proteins (Stock et al., 1989). Accordingly, normal regulation of the Pho regulon involves phosphorylation by the sensor PhoR of the regulator PhoB under conditions of P_i limitation; normal regulation of a set of other (unknown) target genes (presumably) involves phosphorylation by the sensor CreC of the regulator CreB in response to some catabolite; normal regulation of many other gene systems involves phosphorylation by other sensors of their cognate regulators in response to other environmental stimuli. Normal regulation of a two-component regulatory system may also involve dephosphorylation of the phosphorylated response regulator. Dephosphorylation may involve the cognate sensor or an accessory protein, or both. Furthermore, the control of a particular system may involve regulation of phosphorylation or dephosphorylation, or both.

Cross regulation refers to the control of a response regulator of one two-component response regulator by a different regulatory system (Wanner, 1992). This term was adopted to distinguish those interactions with response regulators that may be especially significant as a biological regulatory process from those interactions (called crosstalk) that may be insignificant in vivo. The most compelling evidence in favor of a role for cross regulation in the control of a two-component regulatory system is seen in the Pho regulon. By definition, cross regulation controls the activity of the response regulator. Also, it is likely that cross regulation always involves the control of a response regulator by phosphorylation (or dephosphorylation). This is because these are the only mechanisms that are known to control the activity of a response regulator. However, cross regulation may involve different mechanisms. Cross regulation may involve the phosphorylation of a response regulator by a nonpartner sensor (or by a chemical phosphorylating agent such as acetyl phosphate); cross regulation may involve a different covalent modification (such as an adenylylation or acetylation); or cross regulation may involve binding of an effector molecule.

Crosstalk refers to those interactions with response regulators that are due to nonspecific interactions or "noise." Accordingly, crosstalk may result from cross-specificities in biochemical reactions in which sensors of similar sequence phosphorylate nonpartner regulators (Wanner, 1992). The phosphorylation of nonpartner regulators by CheA, EnvZ, and NtrB (NR_{II}) is probably due to cross-specificities (Ninfa et al., 1988; Igo et al., 1989). In these cases, much greater amounts of the nonpartner protein are required to show interactions, both in vivo and in vitro. Also, the rates of phosphotransfer between these nonpartner proteins are slower than those between the respective partner proteins. Furthermore, in the absence of other evidence for a regulatory connection among such interacting nonpartner proteins, there is no a priori reason to suppose that such interactions are examples of cross regulation.

Pathways of P Assimilation and P_i-Regulated Genes

The Pho regulon may be subject to both P_i-dependent and P_i-independent controls for the purpose of coordinating the overall process of P assimilation (Wanner, 1990). This process involves two or more metabolic steps, depending

Oxidative phosphorylation

$$ADP + P_i \xrightarrow[\text{PMF}]{\text{AtpIBEFHAGDC}} ATP$$

Glycolysis

$$\text{Glyceraldehyde-3-phosphate} + P_i + NAD \xrightarrow{\text{Gap}} \text{1,3-Diphosphoglycerate} + NADH$$

$$\text{1,3-Diphosphoglycerate} + ADP \xrightarrow{\text{Pgk}} \text{3-Phosphoglycerate} + ATP$$

Tricarboxylic acid cycle

$$\text{Succinyl-CoA} + ADP + P_i \xrightarrow{\text{SucCD}} \text{Succinate} + ATP + CoA$$

Mixed-acid fermentation

$$\text{Acetyl-CoA} + P_i \xrightarrow{\text{Pta}} \text{Acetyl phosphate} + CoA$$

$$\text{Acetyl phosphate} + ADP \xrightarrow{\text{AckA}} \text{Acetate} + ATP$$

FIGURE 5 Pathways for P_i incorporation into ATP. AtpIBEFHAGDC, ATP synthase; Gap, glyceraldehyde-3-phosphate dehydrogenase; Pgk, phosphoglycerate kinase; PMF, proton motive force; SucCD, succinyl coenzyme A synthetase.

on the P source. If the P source is P_i, the first step involves uptake of extracellular P_i, and subsequent steps lead to the incorporation of intracellular P_i into ATP. Pathways by which P_i may be incorporated into ATP are shown in Fig. 5. If the P source is an organophosphate, the first step may involve its extracellular (periplasmic) breakdown with the release of P_i, which is then taken up from the periplasm and metabolized like P_i. Or if the P source is an organophosphate that is taken up intact (such as sn-glycerol-3-phosphate), subsequent steps may involve a series of phosphoryl transfer reactions in which the phosphoryl group eventually enters into ATP (without the formation of P_i as an intermediate). If the P source is a phosphonate (a compound having a direct carbon-phosphorus bond), the first step always involves uptake and subsequent steps (depending on the phosphonate degradative pathway [Lee et al., 1992]) may or may not involve the release of intracellular P_i for ATP synthesis.

Thirty-one (or more) genes are coregulated as members of the Pho regulon in *E. coli* (Wanner, 1993). All these genes are induced by P_i limitation and are regulated by the PhoR-PhoB two-component regulatory system. These genes are arranged in eight transcriptional units, and their gene products probably all have a role in the process of assimilation of different environmental P sources. All sequenced P_i-regulated Pho regulon genes are listed in Table 3. These encode three different binding protein–dependent transport systems. PhnCDE probably comprises a phosphonate transporter; PstSCAB comprises the P_i-specific transporter that is also involved in P_i control; and UgpBAEC comprises a sn-glycerol-3-phosphate transporter. Other *psi* genes encode degradative enzymes (e.g., Bap, a carbon-phosphorus lyase, and a phosphodiesterase), a porin (PhoE), the P_i sensor PhoR, the response regulator PhoB, the negative regulator PhoU (which may also have a role in intracellular P_i metabolism), and three genes of unknown function (*psiE, psiF,* and *phoH* [formerly called *psiH*]). With the possible exception of PhoU, no enzyme for intracellular P_i metabolism appears to be synthesized under Pho regulon control. Whether the CreC-CreB two-component regulatory system or acetyl phosphate control the expression of genes for intracellular P_i metabolism has not been tested.

TABLE 3 P$_i$-regulated genes of the *E. coli* Pho regulon[a]

Name[b]	Map location (min)	Protein	Function[c]	Description[d]
phnCDEFGHIJKLMNOP (*psiD*)	93.3	PhnC	Pn uptake	Pn permease
		PhnD	Pn uptake	Pn binding protein
		PhnE	Pn uptake	Membrane protein
		PhnF	Unknown	GntR family regulator
		PhnGHIJKLM	Degradation	C-P lyase complex
		PhnN, PhnP	Degradation	Lyase accessory proteins
		PhnO	Unknown	Helix-turn-helix regulator
phoA–psiF	8.7	PhoA	Degradation	Bap
		PsiF	Unknown	Periplasmic protein
phoBR	9.0	PhoB	Activator	Response regulator
		PhoR	Kinase	P$_i$ sensor
phoE	5.8	PhoE	Porin	Polyanion entry
phoH (*psiH*)	23.6	PhoH	Unknown	ATPase
psiE	91.5	PsiE	Unknown	Hydrophobic protein
pstSCAB–phoU	84.0	PstS	P$_i$ uptake	P$_i$ binding protein
		PstA, PstC	P$_i$ uptake	Membrane proteins
		PstB	P$_i$ uptake	P$_i$ permease
		PhoU	Regulator and P$_i$ enzyme[d]	Accessory protein
ugpBAECQ (*psiB, psiC*)	75.8	UgpB	G-3-P uptake	G-3-P binding protein
		UgpA, UgpE	G-3-P uptake	Membrane proteins
		UgpC	G-3-P uptake	G-3-P permease
		UgpQ	Degradation	Phosphodiesterase

[a]Only sequenced phosphate starvation inducible (*psi*) genes are listed. Nucleotide sequences were determined by Chen et al. (1990) and Makino et al. (1991) for the functional *E. coli* B and cryptic *E. coli* K-12 *phnC*-to-*phnP* operon, by Chang et al. (1986) and Shuttleworth et al. (1986) for the *phoA* operon (Metcalf et al., 1990), by Overbeeke et al. (1983) for *phoE*,, by Kim et al. (1993) for *phoH,* by Davis and Henderson (1987) for the open reading frame corresponding to *psiE* (Metcalf et al., 1990), by Magota et al. (1984) and Surin et al. (1984) for *pstS* (*phoS*), by Amemura et al. (1985) and Surin et al. (1985) for *pstCAB* and *phoU*, by Overduin et al. (1988) and Kasahara et al. (1991) for *ugpBAEC*, and by Kasahara et al. (1989) and Tommassen et al. (1991) for the downstream *ugpQ* gene.

[b]The *phoA* gene is part of an operon with one or more downstream genes of unknown function (Metcalf et al., 1990). The *phoB* and *phoR* genes lie in an operon because *phoB* mutations may have a polar effect on *phoR* (Wanner and Chang, 1987).

[c]A role for PhoU as an enzyme in P$_i$ metabolism has been suggested to explain the severe growth defect of a $\Delta phoU$ mutant with a functional Pst uptake system (Steed and Wanner, 1993). Pn, phosphonate; G-3-P, *sn*-glycerol-3-phosphate.

[d]The roles of individual gene products are based on biochemical and mutational studies or, in some cases, solely on sequence similarities, as described elsewhere (Wanner, 1993; Metcalf and Wanner, 1993). C-P, carbon-phosphorus bond.

Regulatory Interactions Indicative of Cross Regulation

The Pho regulon appears to be subject to multiple controls (Fig. 6). Each of these is substantial and independent of one another (see Table 1) (Wanner and Latterell, 1980; Wanner, 1987b; Wanner et al., 1988; Wanner and Wilmes-Riesenberg, 1992). Its normal regulation responds to the extracellular P$_i$ level, is subject to control by the PstSCAB transporter and PhoU, and requires the sensor PhoR for activation of PhoB. Both other controls are P$_i$ independent.

These may be examples of cross regulation (Wanner, 1992). One P$_i$-independent control appears to respond to an unknown catabolite (which may be a cellular metabolite excreted to the periplasm), appears to be subject to control by a central pathway of carbon and energy metabolism, and requires the sensor CreC for activation of PhoB. A second P$_i$-independent control may respond to ATP synthesis, is subject to control by the Pta-AckA pathway of substrate-level phosphorylation, and requires acetyl phosphate (which may act as a chemical

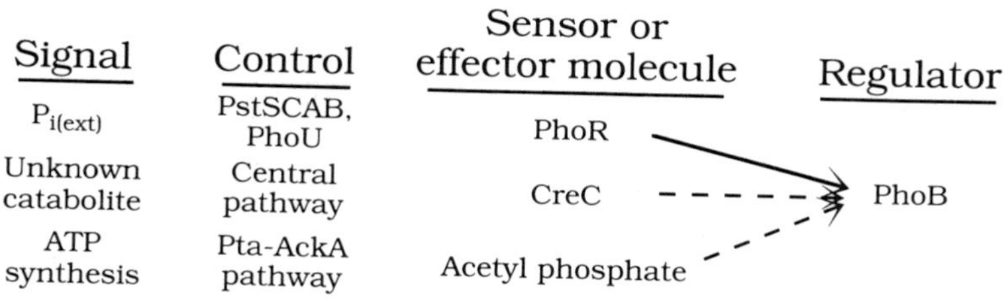

Signal	Control	Sensor or effector molecule	Regulator
$P_{i(ext)}$	PstSCAB, PhoU	PhoR	
Unknown catabolite	Central pathway	CreC	PhoB
ATP synthesis	Pta-AckA pathway	Acetyl phosphate	

FIGURE 6 Multiple controls of PhoB phosphorylation by PhoR, CreC, and acetyl phosphate. A signal for the extracellular (ext) P_i level controls the Pho regulon via the Pst system, PhoU, and the P_i sensor PhoR. A signal for an unknown catabolite (which may be an intermediate of a central pathway of carbon, energy, and P_i metabolism) controls the Pho regulon via a central pathway and the catabolite sensor CreC. A signal for ATP synthesis controls the Pho regulon via the Pta-AckA pathway and acetyl phosphate. PhoR, CreC, and acetyl phosphate, in turn, activate the regulator PhoB by phosphorylation. PhoB has been shown to be directly phosphorylated by acetyl phosphate in vitro (Hiratsu et al., personal communication). Nevertheless, it has not been established whether acetyl phosphate acts directly on PhoB as a phosphoryl donor or indirectly via an unknown sensor in vivo (Wanner and Wilmes-Riesenberg, 1992). The solid arrow symbolizes signal transduction by P_i limitation. The dashed arrows symbolize signal transduction due to (proposed) cross regulation. Adapted from Wanner (1992).

phosphorylating agent) for activation of PhoB. Because large effects due to these P_i-independent controls are seen only in *phoR* mutants, it is uncertain whether they also have a role in Pho regulon control in the presence of PhoR. Also, because both of these controls probably activate PhoB by phosphorylation, a mechanism may be necessary for inactivation of a phospho-PhoB phosphatase (presumably PhoR^R) for these controls to act in the presence of PhoR.

Is there a basis for cross regulation of the Pho regulon? The finding that one P_i-independent control of the Pho regulon is coupled to the Pta-AckA pathway for ATP synthesis provides a teleological basis for cross regulation. This is because regulation involving the Pta-AckA pathway may be coupled to the synthesis of ATP by this pathway. This control involves acetyl phosphate synthesis. Mutations or growth conditions that are expected to lead to increased acetyl phosphate synthesis lead to elevated expression of the Pho regulon. Conversely, mutations or growth conditions that are expected to lead to decreased acetyl phosphate synthesis lead to lowered expression. Because the control of a pathway is

likely to respond to the end product of that pathway, regulation involving the Pta-AckA pathway may detect the ratio of ATP (or another nucleotide) to acetyl phosphate, with a lowered ratio causing activation (Wanner and Wilmes-Riesenberg, 1992; Wanner, 1992).

It is less clear whether there is a basis for cross regulation by CreC. However, it should be pointed out that all effects due to this signaling pathway are observed in mutants with genes in single copy. Under these conditions, regulatory effects are likely to be meaningful. Also, effects due to CreC are highly regulated by the carbon and energy source, suggesting that control by CreC may be coupled to central metabolism. Because many central pathways of carbon and energy metabolism are formally pathways of P_i metabolism as well (see Fig. 5), a teleological basis for cross regulation by CreC may exist. Accordingly, CreC may respond to an intermediate of one of these pathways. Also, the CreC-CreB two-component regulatory system may control the synthesis of enzymes for a central pathway connected with P_i metabolism.

Is cross regulation observed in the presence of PhoR? When P_i is in excess, the Pho regulon is subject to inhibition. This inhibition probably involves a phospho-PhoB phosphatase activity of PhoR^R (or PhoU). Despite this, there are several ways in which cross regulation may act in the presence of PhoR. There may be a control of the PhoR phosphatase activity. In agreement, the PhoR kinase and (presumed) PhoR phosphatase activities appear to be genetically (and, therefore, biochemically) separable. A novel *phoR* mutant has been isolated that displays the phenotype expected of a cell with PhoR kinase activity but lacking PhoR phosphatase activity (Chang and Wanner, unpublished data). There may be growth conditions in which the amount of PhoR is too low to dephosphorylate phospho-PhoB. In this regard, the amounts of PhoB and PhoR are expected to vary as the *phoBR* operon is autogenously regulated (Wanner, 1987b; Wanner and Chang, 1987). Also, this operon may be subject to an antisense RNA control (Chang and Wanner, unpublished data). It may be important to control basal level expression of the Pho regulon; and CreC and acetyl phosphate may play a role in controlling the basal level. That there is genetic control of the basal level of Pho regulon gene expression is not new (Jones, 1969). Alternatively, cross regulation may be important for full activation of only particular Pho regulon promoters (ones other than *phoA*) under conditions of P_i limitation during growth on certain carbon and energy sources. Most studies on Pho regulon control have (for convenience) dealt primarily with *phoA* expression during growth on glucose as a carbon source (Wanner and Wilmes-Riesenberg, 1992). Therefore, effects due to CreC or acetyl phosphate under conditions of P_i limitation may simply have gone unnoticed.

PERSPECTIVES

The primary control of the Pho regulon involves a signal transduction pathway responsive to the extracellular P_i level (Wanner, 1993). In this chapter, a more detailed model for this control has been described. This model invokes the existence of a P_i repression complex, although there is no evidence for such a complex. Is there an alternative model to be considered? Nevertheless, even if one accepts this model, it is still unclear what acts as the P_i sensor. Is the Pst transporter solely responsible for detection of extracellular P_i? If so, how is a signal for P_i limitation communicated to PhoR? Or, is there a P_i regulatory site? Is it on PhoR? In the absence of PhoR, the Pho regulon is subject to two P_i-independent controls (Wanner and Wilmes-Riesenberg, 1992; Wanner, 1992, 1994). One clearly involves the sensor (CreC) of the CreC-CreB two-component regulatory system. However, the target genes of the CreC-CreB system are unknown. If the target genes encode enzymes for central metabolism, this may provide further evidence in favor of a role for cross regulation in Pho regulon control. The other P_i-independent control involves acetyl phosphate synthesis. Based on its effects on Pho regulon control, acetyl phosphate may be an important effector molecule. Does acetyl phosphate have another regulatory role? How does acetyl phosphate act? Even though acetyl phosphate has now been shown to act as a direct phosphorylating agent in several systems in vitro, it remains unclear whether it acts alone or via an (unknown) sensor in vivo. The answers to these and other questions await the outcome of new studies on these signaling pathways of the Pho regulon.

ACKNOWLEDGMENTS

Studies in this laboratory are supported by NIH grant GM35392 and NSF grant MCB 9405929.

REFERENCES

Adams, M. D., and D. L. Oxender. 1989. Bacterial periplasmic binding protein tertiary structures. *J. Biol. Chem.* **264:**15739–15742.

Amemura, M., K. Makino, H. Shinagawa, A. Kobayashi, and A. Nakata. 1985. Nucleotide sequence of the genes involved in phosphate transport and regulation of the phosphate regulon in *Escherichia coli. J. Mol. Biol.* **184:**241–250.

Amemura, M., K. Makino, H. Shinagawa, and A. Nakata. 1986. Nucleotide sequence of the *phoM* region of *Escherichia coli:* four open reading frames may constitute an operon. *J. Bacteriol.* **168:**294–302.

Amemura, M., K. Makino, H. Shinagawa, and A. Nakata. 1990. Cross talk to the phosphate regulon of *Escherichia coli* by PhoM protein: PhoM is a histidine protein kinase and catalyzes phosphorylation of PhoB and PhoM-open reading frame 2. *J. Bacteriol.* **172:**6300–6307.

Ames, G. F.-L. 1986a. Bacterial periplasmic transport systems: structure, mechanism, and evolution. *Annu. Rev. Biochem.* **55:**397–425.

Ames, G. F.-L. 1986b. The basis of multidrug resistance in mammalian cells: homology with bacterial transport. *Cell* **47:**323–324.

Baichwal, V., D. Liu, and G. F.-L. Ames. 1993. The ATP-binding component of a prokaryotic traffic ATPase is exposed to the periplasmic (external) surface. *Proc. Natl. Acad. Sci. USA* **90:**620–624.

Chang, B., and B. L. Wanner. Unpublished data.

Chang, C. N., W.-J. Kuang, and E. Y. Chen. 1986. Nucleotide sequence of the alkaline phosphatase gene of *Escherichia coli. Gene* **44:**121–125.

Chen, C.-J., J. E. Chin, K. Ueda, D. P. Clark, I. Pastan, M. M. Gottesman, and I. B. Roninson. 1986. Internal duplication and homology with bacterial transport proteins in the *mdr1* (P-glycoprotein) gene from multidrug-resistant human cells. *Cell* **47:**381–389.

Chen, C.-M., Q. Ye, Z. Zhu, B. L. Wanner, and C. T. Walsh. 1990. Molecular biology of carbon-phosphorus bond cleavage: cloning and sequencing of the *phn* (*psiD*) genes involved in alkylphosphonate uptake and C-P lyase activity in *Escherichia coli* B. *J. Biol. Chem.* **265:**4461–4471.

Cox, G. Personal communication.

Cox, G. B., H. Rosenberg, J. A. Downie, and S. Silver. 1981. Genetic analysis of mutants affected in the Pst inorganic phosphate transport system. *J. Bacteriol.* **148:**1–9.

Cox, G. B., D. Webb, J. Godovac-Zimmermann, and H. Rosenberg. 1988. Arg-220 of the PstA protein is required for phosphate transport through the phosphate-specific transport system in *Escherichia coli* but not for alkaline phosphatase repression. *J. Bacteriol.* **170:**2283–2286.

Cox, G. B., D. Webb, and H. Rosenberg. 1989. Specific amino acid residues in both the PstB and PstC proteins are required for phosphate transport by the *Escherichia coli* Pst system. *J. Bacteriol.* **171:** 1531–1534.

Davis, E. O., and P. J. F. Henderson. 1987. The cloning and DNA sequence of the gene *xylE* for xylose-proton symport in *Escherichia coli* K12. *J. Biol. Chem.* **262:**13928–13932.

Doige, C. A., and G. F.-L. Ames. 1993. ATP-dependent transport systems in bacteria and humans: rel-evance to cystic fibrosis and multidrug resistance. *Annu. Rev. Microbiol.* **47:**291–319.

Foote, S. J., D. E. Kyle, R. K. Martin, A. M. J. Oduola, K. Forsyth, D. J. Kemp, and A. F. Cowman. 1990. Several alleles of the multidrug-resistance gene are closely linked to chloroquine resistance in *Plasmodium falciparum. Nature* (London) **345:**255–258.

Forst, S., D. Comeau, S. Norioka, and M. Inouye. 1987. Localization and membrane topology of EnvZ, a protein involved in osmoregulation of OmpF and OmpC in *Escherichia coli. J. Biol. Chem.* **262:**16433–16438.

Gegner, J. A., D. R. Graham, A. F. Roth, and F. W. Dahlquist. 1992. Assembly of an MCP receptor, CheW, and kinase CheA complex in the bacterial chemotaxis signal transduction pathway. *Cell* **70:**975–982.

Gros, P., J. Croop, and D. Housman. 1986. Mammalian multidrug resistance gene: complete cDNA sequence indicates strong homology to bacterial transport proteins. *Cell* **47:**371–380.

Hazelbauer, G. L. 1992. Bacterial chemoreceptors. *Curr. Opin. Struct. Biol.* **2:**505–510.

Hiratsu, K., K. Makino, A. Nakata, and H. Shina-gawa. Personal communication.

Hoch, J. A. 1993. Regulation of the phosphorelay and the initiation of sporulation in *Bacillus subtilis. Annu. Rev. Microbiol.* **47:**441–465.

Hyde, S. C., P. Emsley, M. J. Hartshorn, M. M. Mimmack, U. Gileadi, S. R. Pearce, M. P. Gallagher, D. R. Gill, R. E. Hubbard, and C. F. Higgins. 1990. Structural model of ATP-binding proteins associated with cystic fibrosis, multidrug resistance and bacterial transport. *Nature* (London) **346:**362–365.

Igo, M. M., A. J. Ninfa, and T. J. Silhavy. 1989. A bacterial environmental sensor that functions as a protein kinase and stimulates transcriptional activation. *Genes Dev.* **3:**598–605.

Ireton, K., D. Z. Rudner, K. J. Siranosian, and A. D. Grossman. 1993. Integration of multiple developmental signals in *Bacillus subtilis* through the SpoOA transcription factor. *Genes Dev.* **7:**283–294.

Jones, T. C. 1969. Genetic control of basal level of alkaline phosphatase in *Escherichia coli. Mol. Gen. Genet.* **105:**91–100.

Kaplan, H. Personal communication.

Kaplan, H. B., A. Kuspa, and D. Kaiser. 1991. Suppressors that permit A-signal-independent developmental gene expression in *Myxococcus xanthus. J. Bacteriol.* **173:**1460–1470.

Kasahara, M., K. Makino, M. Amemura, and A. Nakata. 1989. Nucleotide sequence of the *ugpQ* gene encoding glycerophosphoryl diester phosphodiesterase of *Escherichia coli* K12. *Nucleic Acids Res.* **17:**28–54.

Kasahara, M., K. Makino, M. Amemura, A. Nakata, and H. Shinagawa. 1991. Dual regulation of the *ugp* operon by phosphate and carbon starvation at two interspaced promoters. *J. Bacteriol.* **173**:549–558.

Kim, S.-K., K. Makino, M. Amemura, H. Shinagawa, and A. Nakata. 1993. Molecular analysis of the *phoH* gene, belonging to the phosphate regulon in *Escherichia coli. J. Bacteriol.* **175**:1316–1324.

Kofoid, E. C., and J. S. Parkinson. 1988. Transmitter and receiver modules in bacterial signaling proteins. *Proc. Natl. Acad. Sci. USA* **85**:4981–4985.

Lee, K.-S., W. W. Metcalf, and B. L. Wanner. 1992. Evidence for two phosphonate degradative pathways in *Enterobacter aerogenes. J. Bacteriol.* **174**:2501–2510.

Luecke, H., and F. A. Quiocho. 1990. High specificity of a phosphate transport protein determined by hydrogen bonds. *Nature* (London) **347**:402–406.

Magasanik, B. 1993. The regulation of nitrogen utilization in enteric bacteria. *J. Cell. Biochem.* **51**:34–40.

Magota, K., N. Otsuji, T. Miki, T. Horiuchi, S. Tsunasawa, J. Kondo, F. Sakiyama, M. Amemura, T. Morita, H. Shinagawa, and A. Nakata. 1984. Nucleotide sequence of the *phoS* gene, the structural gene for the phosphate-binding protein of *Escherichia coli. J. Bacteriol.* **157**:909–917.

Makino, K., M. Amemura, S.-K. Kim, A. Nakata, and H. Shinagawa. 1993. Role of the σ[70] subunit of RNA polymerase in transcriptional activation by activator protein PhoB in *Escherichia coli. Genes Dev.* **7**:149–160.

Makino, K., M. Amemura, S.-K. Kim, H. Shinagawa, and A. Nakata. 1992. Signal transduction of the phosphate regulon in *Escherichia coli* mediated by phosphorylation, p. 191–200. *In* S. Papa, A. Azzi, and J. M. Tager (ed.), *Adenine Nucleotides in Cellular Energy Transfer and Signal Transduction.* Birkhaeuser Verlag, Basel, Switzerland.

Makino, K., S.-K. Kim, H. Shinagawa, M. Amemura, and A. Nakata. 1991. Molecular analysis of the cryptic and functional *phn* operons for phosphonate use in *Escherichia coli* K-12. *J. Bacteriol.* **173**:2665–2672.

Makino, K., H. Shinagawa, M. Amemura, T. Kawamoto, M. Yamada, and A. Nakata. 1989. Signal transduction in the phosphate regulon of *Escherichia coli* involves phosphotransfer between PhoR and PhoB proteins. *J. Mol. Biol.* **210**:551–559.

Metcalf, W. W., P. M. Steed, and B. L. Wanner. 1990. Identification of phosphate-starvation-inducible genes in *Escherichia coli* K-12 by DNA sequence analysis of *psi::lacZ*(Mu d1) transcriptional fusions. *J. Bacteriol.* **172**:3191–3200.

Metcalf, W. W., and B. L. Wanner. 1993. Mutational analysis of an *Escherichia coli* fourteen-gene operon for phosphonate degradation using Tn*phoA'* elements. *J. Bacteriol.* **175**:3430–3442.

Ninfa, A. J., E. G. Ninfa, A. N. Lupas, A. Stock, B. Magasanik, and J. Stock. 1988. Crosstalk between bacterial chemotaxis signal transduction proteins and regulators of transcription of the Ntr regulon: evidence that nitrogen assimilation and chemotaxis are controlled by a common phosphotransfer mechanism. *Proc. Natl. Acad. Sci. USA* **85**:5492–5496.

Nixon, B. T., C. W. Ronson, and F. M. Ausubel. 1986. Two-component regulatory systems responsive to environmental stimuli share strongly conserved domains with the nitrogen assimilation regulatory genes *ntrB* and *ntrC. Proc. Natl. Acad. Sci. USA* **83**:7850–7854.

Overbeeke, N., H. Bergmans, F. V. Mansfeld, and B. Lugtenberg. 1983. Complete nucleotide sequence of *phoE*, the structural gene for the phosphate limitation inducible outer membrane pore protein of *Escherichia coli* K12. *J. Mol. Biol.* **163**:513–532.

Overduin, P., W. Boos, and J. Tommassen. 1988. Nucleotide sequence of the *ugp* genes of *Escherichia coli* K-12: homology to the maltose system. *Mol. Microbiol.* **2**:767–775.

Parkinson, J. S., and E. C. Kofoid. 1992. Communication modules in bacterial signaling proteins. *Annu. Rev. Genet.* **26**:71–112.

Ronson, C. W., B. T. Nixon, and F. M. Ausubel. 1987. Conserved domains in bacterial regulatory proteins that respond to environmental stimuli. *Cell* **49**:579–581.

Scholten, M., and J. Tommassen. 1993. Topology of the PhoR protein of *Escherichia coli* and functional analysis of internal deletion mutants. *Mol. Microbiol.* **8**:269–275.

Shulman, R. G., T. R. Brown, K. Ugurbil, S. Ogawa, S. M. Cohen, and J. A. den Hollander. 1979. Cellular applications of ^{31}P and ^{13}C nuclear magnetic resonance. *Science* **205**:160–166.

Shuttleworth, H., J. Taylor, and N. Minton. 1986. Sequence of the gene for alkaline phosphatase from *Escherichia coli* JM83. *Nucleic Acids Res.* **14**:86–89.

Steed, P. M., and B. L. Wanner. 1993. Use of the *rep* technique for allele replacement to construct mutants with deletions of the *pstSCAB-phoU* operon: evidence of a new role for the PhoU protein in the phosphate regulon. *J. Bacteriol.* **175**:6797–6809.

Stock, J. B., A. J. Ninfa, and A. M. Stock. 1989. Protein phosphorylation and regulation of adaptive responses in bacteria. *Microbiol. Rev.* **53**:450–490.

Surin, B. P., N. E. Dixon, and H. Rosenberg. 1986. Purification of the PhoU protein, a negative regulator of the Pho regulon of *Escherichia coli* K-12. *J. Bacteriol.* **168**:631–635.

Surin, B. P., D. A. Jans, A. L. Fimmel, D. C. Shaw, G. B. Cox, and H. Rosenberg. 1984. Structural gene for the phosphate-repressible phosphate-binding protein of *Escherichia coli* has its own promoter: complete nucleotide sequence of the *phoS* gene. *J. Bacteriol.* **157**:772–778.

Surin, B. P., H. Rosenberg, and G. B. Cox. 1985. Phosphate-specific transport system of *Escherichia coli*: nucleotide sequence and gene-polypeptide relationships. *J. Bacteriol.* **161**:189–198.

Swanson, R. V., R. B. Bourret, and M. I. Simon. 1993. Intermolecular complementation of the kinase activity of CheA. *Mol. Microbiol.* **8**:435–441.

Tommassen, J., K. Eiglmeier, S. T. Cole, P. Overduin, T. J. Larson, and W. Boos. 1991. Characterization of two genes, *glpQ* and *ugpQ*, encoding glycerophosphoryl diester phosphodiesterases of *Escherichia coli*. *Mol. Gen. Genet.* **226**:321–327.

Wanner, B. L. 1987a. Control of *phoR*-dependent bacterial alkaline phosphatase clonal variation by the *phoM* region. *J. Bacteriol.* **169**:900–903.

Wanner, B. L. 1987b. Phosphate regulation of gene expression in *Escherichia coli*, p. 1326–1333. *In* F. C. Neidhardt, J. L. Ingraham, K. B. Low, B. Magasanik, M. Schaechter, and H. E. Umbarger (ed.), *Escherichia coli and Salmonella typhimurium: Cellular and Molecular Biology*, vol. 2. American Society for Microbiology, Washington, D.C.

Wanner, B. L. 1990. Phosphorus assimilation and its control of gene expression in *Escherichia coli*, p. 152–163. *In* G. Hauska and R. Thauer (ed.), *The Molecular Basis of Bacterial Metabolism*. Springer-Verlag, Heidelberg.

Wanner, B. L. 1992. Minireview. Is cross regulation by phosphorylation of two-component response regulator proteins important in bacteria? *J. Bacteriol.* **174**:2053–2058.

Wanner, B. L. 1993. Gene regulation by phosphate in enteric bacteria. *J. Cell. Biochem.* **51**:47–54.

Wanner, B. L. 1994. Multiple controls of the *Escherichia coli* Pho regulon by the P$_i$ sensor PhoR, the catabolite regulatory sensor CreC, and acetyl phosphate, p.13–21. *In* A. Torriani-Gorini, E. Yagil, and S. Silver (ed.), *Phosphate in Microorganisms: Cellular and Molecular Biology*. American Society for Microbiology, Washington, D.C.

Wanner, B. L., and B.-D. Chang. 1987. The *phoBR* operon in *Escherichia coli* K-12. *J. Bacteriol.* **169**:5569–5574.

Wanner, B. L., and P. Latterell. 1980. Mutants affected in alkaline phosphatase expression: evidence for multiple positive regulators of the phosphate regulon in *Escherichia coli*. *Genetics* **96**:242–266.

Wanner, B. L., A. Sarthy, and J. R. Beckwith. 1979. *Escherichia coli* pleiotropic mutant that reduces amounts of several periplasmic and outer membrane proteins. *J. Bacteriol.* **140**:229–239.

Wanner, B. L., M. R. Wilmes, and D. C. Young. 1988. Control of bacterial alkaline phosphatase synthesis and variation in an *Escherichia coli* K-12 *phoR* mutant by adenyl cyclase, the cyclic AMP receptor protein, and the *phoM* operon. *J. Bacteriol.* **170**:1092–1102.

Wanner, B. L., and M. R. Wilmes-Riesenberg. 1992. Involvement of phosphotransacetylase, acetate kinase, and acetyl phosphate synthesis in the control of the phosphate regulon in *Escherichia coli*. *J. Bacteriol.* **174**:2124–2130.

Willsky, G. R., R. L. Bennett, and M. H. Malamy. 1973. Inorganic phosphate transport in *Escherichia coli*: involvement of two genes which play a role in alkaline phosphatase regulation. *J. Bacteriol.* **113**:529–539.

Willsky, G. R., and M. H. Malamy. 1976. Control of the synthesis of alkaline phosphatase and the phosphate-binding protein in *Escherichia coli*. *J. Bacteriol.* **127**:595–609.

Wilmes-Riesenberg, M. R., and B. L. Wanner. 1992. TnphoA and TnphoA' elements for making and switching fusions for study of transcription, translation, and cell surface localization. *J. Bacteriol.* **174**:4558–4575.

Wilmes-Riesenberg, M. R., and B. L. Wanner. Unpublished data.

Wolfe, A. J., and R. C. Stewart. 1993. The short form of the CheA protein restores kinase activity and chemotactic ability to kinase-deficient mutants. *Proc. Natl. Acad. Sci. USA* **90**:1518–1522.

Yang, Y., and M. Inouye. 1991. Intermolecular complementation between two defective mutant signal-transducing receptors of *Escherichia coli*. *Proc. Natl. Acad. Sci. USA* **88**:11057–11061.

Yang, Y., and M. Inouye. 1993. Requirement of both kinase and phosphatase activities of an *Escherichia coli* receptor (Taz1) for ligand-dependent signal transduction. *J. Mol. Biol.* **231**:335–342.

Signal Transduction in the Arc System for Control of Operons Encoding Aerobic Respiratory Enzymes

Shiro Iuchi and E. C. C. Lin

13

Aerobic respiration is the most efficient way to extract energy from carbon and energy sources. It has long been known that aerobically grown *Escherichia coli* contain elevated levels of many enzymes associated with aerobic metabolism. Examples of the enzymes include members of the tricarboxylic acid cycle and the cytochrome *o* complex, the major terminal oxidase (Amarasingham and Davis, 1965; Gray et al., 1966; Smith and Neidhardt, 1983; Kranz and Gennis, 1983). Because the synthesis of so many enzymes apparently depends on a single variable, O_2 tension, there is reason to suspect the existence of a global control mechanism. Supposing further that such a regulation would likely be transcriptional, we decided to search for pleiotropic mutants that highly express operons of aerobic function under anaerobic conditions. To this end, we constructed a chromosomal merodiploid bearing both *sdh*+ (encoding the succinate dehydrogenase complex) and a $\Phi(sdh\text{-}lacZ)$ operon fusion using a Δlac strain and picked a red papilla from each colony on MacConkey lactose agar after anaerobic incubation for 5 days. Retention of *sdh*+ averted perturbation of the network of aerobic

metabolism and allowed us to identify *trans*-regulatory mutations by abnormal activity levels of succinate dehydrogenase. This approach resulted in the collection of many mutants synthesizing high anaerobic levels not only of the enzymes mentioned above but also several primary dehydrogenases and enzymes of the glyoxylate shunt and β-oxidation cycle for fatty acids (Fig. 1). Subsequent genetic characterization of 70 independent mutants identified two regulatory genes: *arcA* at min 0 and *arcB* at min 69.5 (Iuchi and Lin, 1988; Iuchi et al., 1989). We proposed the acronym for *aerobic respiration control* (*arc*) as the genetic symbol for this pair of genes, even though the gene at min 0 had been reported several times and each time given a name according to the investigative interest. The Arc system belongs to a two-component regulatory family. The *arcB* gene encodes the membrane sensor protein, and the *arcA* gene the response regulator (Iuchi et al., 1990b; Drury and Buxton, 1985). Although two-component regulators were initially found to be widely distributed in prokaryotes, these systems have also recently been discovered in yeast and plants and thus seem to be biologically universal (Ota and Varshavsky, 1993; Chang et al., 1993; Ronson et al., 1987; Stock et al., 1989; Parkinson and Kofoid, 1992).

Shiro Iuchi and E. C. C. Lin, Department of Microbiology and Molecular Genetics, Harvard Medical School, Boston, Massachusetts 02115.

Two-Component Signal Transduction, Edited by James A. Hoch and Thomas J. Silhavy,
© 1995 American Society for Microbiology, Washington, DC 20005

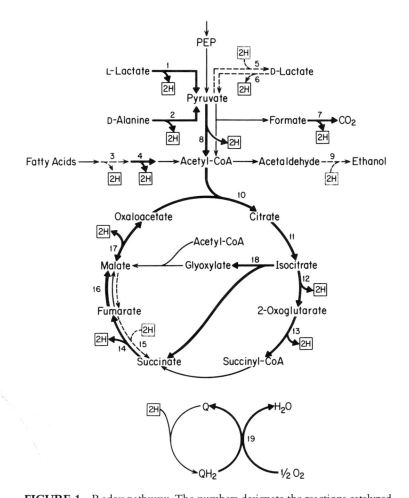

FIGURE 1 Redox pathways. The numbers designate the reactions catalyzed by the enzymes: (1) L-lactate dehydrogenase (flavoprotein); (2) D-amino acid dehydrogenase (flavoprotein); (3) acyl-coenzyme A dehydrogenase (flavoprotein); (4) 3-hydroxyacyl-coenzyme A dehydrogenase (NAD^+-linked); (5) D-lactate:NAD^+ oxidoreductase; (6) D-lactate dehydrogenase (flavoprotein); (7) formate dehydrogenase (the FDH_N enzyme); (8) pyruvate dehydrogenase; (9) ethanol:NAD^+ oxidoreductase; (10) citrate synthase; (11) aconitase; (12) isocitrate dehydrogenase; (13) 2-oxoglutarate dehydrogenase; (14) succinate dehydrogenase; (15) fumarate reductase; (16) fumarase A (Bell et al., 1989); (17) malate dehydrogenase; (18) isocitrate lyase; and (19) ubiquinol-1 oxidase. Reactions catalyzed by enzymes under the *arc* control are represented by thick arrows. Reactions catalyzed by enzymes that may or may not be under the *arc* control are represented by thin uninterrupted arrows. Reactions catalyzed by enzymes not under the *arc* control are represented by interrupted arrows. 2H boxed by solid lines represents reducing equivalents yielded by the reaction, and 2H boxed by dotted lines represents reducing equivalents consumed by the reaction. Modified from Fig. 1 of Iuchi and Lin (1988).

PRIMARY STRUCTURES OF ArcB AND ArcA

Like other typical sensors of two-component regulatory systems (Iuchi et al., 1990b), the deduced amino acid sequence of ArcB predicts two transmembrane portions and a transmitter module containing a canonic conserved histidyl residue (His-292) for autophosphorylation (Fig. 2). There are also, however, several special features of this protein. First, the putative periplasmic portion of ArcB contains only about seven amino acid residues, a brevity that is a departure from other periplasmic domains comprising typically about 100 amino acid residues and serving as a signal detector. Second, a redox sensor protein might be expected to contain an iron-cysteine cluster as in Fnr (Green et al., 1993), but no such structure was found. There are a Cys-180 and a Cys-241, but their redox sensing role was excluded by the almost normal behavior of mutant proteins, ArcB$_{Cys-180-Gly}$ and ArcB$_{Cys-241-Gly}$ (Iuchi and Lin, 1992a). Third, ArcB belongs to the subclass of sensors possessing a receiver module in the C-terminal side. In this module, Asp-576 is expected to be the canonic phosphoryl group receptor. There is, however, no helix-turn-helix motif. Otherwise, the receiver segment is highly homologous to ArcA.

In contrast to ArcB, ArcA is a typical regulator belonging to the OmpR subfamily (Parkinson and Kofoid, 1992). In addition to an expected helix-turn-helix motif in the C-terminal side for binding to DNA, there is a putative acidic pocket in the N-terminal side, consisting of three acidic amino acid residues (Glu-10, Asp-11, and Asp-54).

AUTOPHOSPHORYLATION OF ArcB

As found with previously studied sensors deprived of transmembrane segments (Stock et al., 1989; Aiba et al., 1989; Forst et al., 1989; Igo et al., 1989a; Igo et al., 1989b; Makino et al., 1989; Jin et al., 1990), when purified 'ArcB (genetically truncated from the N-terminal side) was incubated with [γ-^{32}P]ATP, rapid autophosphorylation occurred without any known stimulus. To characterize the phosphorylation reactions further, we prepared everted membrane vesicles from cells hyperproducing ArcB or one of its variants. ArcB in these vesicles rapidly autophosphorylated in the presence of ATP, but the variant ArcB$_{His-292-Gln}$ was inactive. The inertness of this protein provided independent evidence that His-292 is the site of autophosphorylation.

Several lines of evidence point to transphosphorylation from His-292 to Asp-576. First, when incubated at 43°C, pH 12.5, 'ArcB-P exhibited a biphasic loss of the labeled phosphorus. This indicates that both a histidyl and an aspartyl residue are phosphorylated, because it is known that under alkaline conditions Asp-P is more labile than His-P. Second, the phosphorylation level of the protein was severely depressed in the variant ArcB$_{Asp-576-Ala}$. Third, although 'ArcB catalyzed the phosphorylation of ArcB$_{His-292-Gln}$,

100 AA

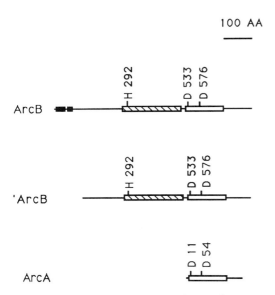

FIGURE 2 Primary structures of ArcB, 'ArcB, and ArcA. Dark boxes indicate putative transmembrane portions. Amino acid sequence between the transmembrane portions is supposed to be in the periplasmic compartment. Hatched boxes indicate the conserved transmitter module. Open boxes indicate the receiver module. Numbers over the diagrams indicate the positions of the highly conserved histidyl (H) and aspartyl (D) residues (Iuchi and Lin, 1992b).

no *substrate* activity was found with ArcB$_{His-292-Gln}$ and ArcB$_{Asp-576-Ala}$ (Iuchi, 1993). It thus appears that the γ-phosphoryl group of ATP is transferred to His-292 and then to Asp-576 and that the liberated His-292 is free to undergo a second round of autophosphorylation.

PHOSPHORYLATION OF ArcA BY ArcB-P

ArcA did not undergo autophosphorylation in the presence of ATP. In the presence of 'ArcB and ATP, a linear initial rate of ArcA phosphorylation occurred for at least 10 min at 25°C. During this period, the concentration of ArcA-P formed was proportional to the concentration of 'ArcB, at least up to 1 nmol/ml (Iuchi and Lin, 1992b).

Rapid Phosphorylation of ArcA Required an Intact ArcB

ArcB variants deprived of the receiver module were significantly diminished in ArcA-phosphorylating activity. By contrast, the variant ArcB$_{Asp-576-Ala}$, possessing the receiver module but lacking its canonic Asp, exhibited no detectable kinase activity (Iuchi, 1993). We believe, therefore, that the receiver module participates in controlling the rate of signal transmission and that its effectiveness depends on the phosphorylation of Asp-576.

COMPOUNDS CONTROLLING THE ACTIVITY OF ArcB

Physiological experiments excluded O_2 as the direct stimulus for the kinase activity of ArcB. Although aerobic growth resulted in the most elevated expression of a target operon, $\Phi(sdh-lac)$, anaerobic growth with NO_3^- or fumarate as the electron acceptor also raised the expression. The degree of stimulation of expression correlated with the oxidizing power (E_0') of the terminal electron acceptor (Iuchi and Lin, 1988).

Genetic experiments also excluded O_2 itself as the signal. Aerobic expression of target operons was muted by deletion of either the *cyo* operon (encoding the terminal oxidase complex with low O_2 affinity) or the *cyd* operon (encoding the

terminal oxidase complex with high O_2 affinity). Deletion of both operons imposed almost full anaerobic repression despite the presence of O_2. It thus appeared that an element of the electron transport chain might signal anaerobiosis (Iuchi et al., 1990a). Although the electron carrier ubiquinone was hypothesized to be a candidate in this control, no such evidence was obtained by study of a *ubiD* mutant blocked in the biosynthesis (Iuchi, unpublished data).

Changing the direction of the search, individual cytosolic metabolites, the concentrations of which were expected to fluctuate with the respiratory mode, were tested with purified 'ArcB or everted vesicles enriched with ArcB. D-Lactate, acetate, pyruvate, and NADH were found to be effective in enhancing the phosphorylation level of both preparations of the sensor protein. (It is apropos to recall that *E. coli* is a mixed-acid fermenter and that D-lactate and acetate are two of the excretion products.) Significant enhancement of phosphorylation occurred at Asp-576 but not at His-292 when any one of the three carboxylic acids was tested at a concentration as low as 0.1 mM. The effect of NADH, however, was clearly observed only at a relatively high concentration of 10 mM. Elevation of the Asp-576-P level seemed to be caused by inhibition of the intrinsic phosphatase activity of the protein (Iuchi, 1993). Our belief that phosphorylation of Asp-576 promotes signal transmission was strongly supported by the observation that the elevation of the Asp-576-P/Asp-576 ratio of ArcB by D-lactate also increased the ArcA kinase activity of the sensor. Because the level of the ArcB$_{Asp-576}$-P should reflect the balance between the rates of intramolecular transphosphorylation and intrinsic dephosphorylation reactions, it is possible that the allosteric effectors control the signal transduction process by influencing the phosphatase activity.

The receiver module is perhaps necessary to provide the binding sites for the cytosolic effectors. A recent study suggested that ArcB also senses proton motive force (Bogachev et al., 1993).

CONTROL OF Φ(sdh-lac) EXPRESSION

To explore the in vivo regulatory properties of ArcB, plasmids encoding wild-type and variant ArcB proteins were introduced into a host bearing a chromosomal Φ(sdh-lac) and an arcB null mutation. As shown in Fig. 3, there was strong anaerobic repression by wild-type ArcB, no repression by ArcB$_{His-292-Gln}$, narrowed range of respiratory control by FN516 and FN517 (versions of ArcB without the receiver module), and strong relief of anaerobic repression by ArcB$_{Asp-533-Ala}$ and ArcB$_{Asp-576-Ala}$ (Iuchi and Lin, 1992a). In vivo regulatory behavior of ArcB variants is therefore consistent with the in vitro rates of auto- and transphosphorylation: the strengths of anaerobic repression correlated with the expected change in ArcA-P levels.

Because certain anaerobic metabolites can enhance the levels of ArcB-P and ArcA-P, we were curious to see whether Fnr, which plays such an extensive role in anaerobic metabolism, would affect the expression of target operons of the Arc system. With Φ(lctD-lac) as the reporter, we found that an fnr mutation relieved anaerobic repression and that this relief was voided by the supplementation of pyruvate. Even in the fnr$^+$ strain, there was a slight but reproducible intensification of anaerobic repression by pyruvate. Similar results were obtained with the Φ(sdh-lac) fusion as the reporter.

STUDIES OF OTHER TRANSCRIPTIONAL CONTROLS

Mutations in arcA or arcB are known to affect the expression of more than 16 operons. Most

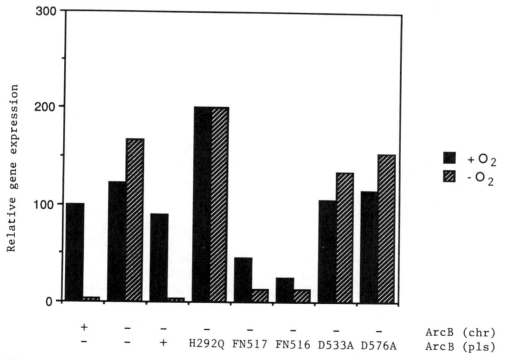

FIGURE 3 Effect of ArcB and its variants on Φ(sdh-lac) expression. Cells were grown either aerobically (solid bar) or anaerobically (hatched bar). pls and chr indicate ArcB encoded by plasmid-borne and chromosome-borne genes, respectively. The + or − sign indicates the presence or absence of wild-type or variant ArcB. Symbols for variant proteins: H292Q (ArcB$_{His-292-Gln}$), FN517 (ArcB possessing 517 amino acid residues from the N-terminal end), FN516 (ArcB possessing 516 amino acid residues from the N-terminal end), D533A (ArcB$_{Asp-533-Ala}$), and D576A (ArcB$_{Asp-576-Ala}$)H292 (Iuchi and Lin, 1992a).

of them are anaerobically repressed by ArcA-P, but two of them are activated. The positive control of the *cydAB* and *pfl* (encoding pyruvate-formate lyase) operons is comprehensible, because they have a vital role to play when there is a lack of O_2 (Iuchi et al., 1990a; Fu et al., 1991; Sawers and Suppmann, 1992; Iuchi, 1993). The *traY* operon on the F-plasmid encoding proteins necessary for F-pili biogenesis is also under positive control of the Arc system (Silverman et al., 1993). The physiological significance of this is less clear.

Negative control by ArcA has been studied in the *sodA* gene (encoding the Mn-dependent superoxide dismutase). Footprinting experiments indicated that the ArcA regulator protects a 60-bp span downstream of −35 and −10 (Tardat and Touati, 1993).

Expression of *arcB* seems to be independent of the environmental redox (Iuchi and Lin, 1992a). By contrast, *arcA* expression is increased up to fourfold anaerobically by Fnr (Compan and Touati, 1994). Significant basal expression of the two *arc* genes should expedite responses to changes of redox environments.

WORKING MODEL FOR SIGNAL TRANSMISSION BY THE Arc SYSTEM

Our current working model for the Arc system is schematized in Fig. 4. Under aerobic conditions, the receiver module of ArcB docks on its transmitter module. This quiescent or "closed" state is periodically activated by phosphorylation of His-292 in the transmitter module by ATP. Because Asp-576 in the receiver module is in close vicinity, intramolecular transphosphorylation readily occurs. The phosphorylation of Asp-576 changes the conformation of the receiver module, allowing it to disengage from the transmitter module and allowing ArcB to assume the "open" state. Subsequently, Asp-576 might be dephosphorylated (mainly by the phosphatase activity), thereby returning ArcB to the closed state. On the other hand, the free His-292 might undergo a second round of autophosphorylation before Asp-576 loses its phosphoryl group. However, the doubly phosphorylated ArcB might also lose its phosphoryl group at Asp-576, resulting in another futile process, thereby demoting ArcB to the singly phosphorylated state. Only rarely would ArcA have a chance to dock on the transmitter module of a doubly phosphorylated ArcB and thus allow transphosphorylation to take place.

Under anaerobic conditions, a different situation exists. One or more anaerobic metabolites can noncovalently associate with the receiver module, and this engagement lengthens the half-life of Asp-576-P. Prolonging the existence of the open form of ArcB increases the chance for ArcA to be phosphorylated. The increase in ArcA-P either represses or activates a target operon (Iuchi, 1993). This model will undoubtedly undergo refinements as more experimental results of the signal transmission machinery become available.

CONCLUDING REMARKS

As a membrane sensor, a special feature of ArcB is its possession of a receiver module, but a surprising feature is the cytosolic origin of at least some of the signals. It is tempting to suggest that other sensors with a receiver module (Parkinson and Kofoid, 1992) can also detect changes within the cell, including those in yeast and plants (Ota and Varshavsky, 1993; Chang et al., 1993). These sophisticated sensors might have arisen from a simpler two-component system by some mechanism involving gene duplication.

Like other global regulators, ArcA-P positively regulates some promoters and negatively regulates others. Furthermore, it has become progressively more evident that most of the targets of one global regulator overlap with those of other global regulators (Iuchi and Lin, 1993). Such interlocking control not only allows the target operon to respond in a complex manner to environmental changes but also allows cellular processes to respond in concert.

Under aerobic conditions

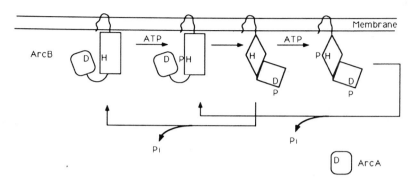

Under anaerobic conditions

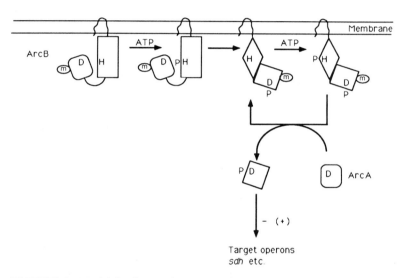

FIGURE 4 Model for the signaling process by ArcB under aerobic and anaerobic conditions. H, D, P, and m, respectively, indicate conserved His-292, conserved Asp-576, phosphoryl group, and cellular metabolites such as D-lactate and NADH. Arrows indicate reaction steps. The signs + and − indicate a positive and negative effect, respectively. ArcB could be a dimer or oligomer, and the phosphoryl group could be transferred between the subunits (Iuchi, 1993).

ACKNOWLEDGMENT

Work in this laboratory was supported by grants 5 R01 GM40993 and 5 R01 11983 of the National Institutes of Health.

REFERENCES

Aiba, H., T. Mizuno, and S. Mizushima. 1989. Transfer of phosphoryl group between two regulatory proteins involved in osmoregulatory expression of the *ompF* and *ompC* genes in *Escherichia coli*. *J. Biol. Chem.* **264**:8563–8567.

Amarasingham, C. R., and B. J. Davis. 1965. Regulation of α-ketoglutarate dehydrogenase formation in *Escherichia coli*. *J. Biol. Chem.* **240**:3664–3668.

Bell, P. J., S. C. Andrews, M. N. Sivak, and J. R. Guest. 1989. Nucleotide sequence of the FNR-regulated fumarase gene (*fumB*) of *Escherichia coli* K-12. *J. Bacteriol.* **171**:3494–3503.

Bogachev, A. V., R. A. Murtazina, and V. P. Skulachev. 1993. Cytochrome *d* induction in *Escherichia coli* growing under unfavorable conditions. *FEBS Lett.* **336**:75–78.

Chang, C., S. F. Kwok, A. B. Bleeker, and E. M. Meyerowitz. 1993. *Arabidopsis* ethylene-response gene ETR1: similarity of product to two-component regulators. *Science* **262**: 539–544.

Compan, I., and D. Touati. 1994. Anaerobic activation of *arcA* transcription in *Escherichia coli*: roles of Fnr and ArcA. *Mol. Microbiol.* **11**:955–964.

Drury, L. S., and R. S. Buxton. 1985. DNA sequence analysis of the *dye* gene of *Escherichia coli* reveals amino acid homology between the Dye and OmpR proteins. *J. Biol. Chem.* **260**: 4236–4242.

Forst, S., J. Delgado, and M. Inouye. 1989. Phosphorylation of OmpR by the osmosensor EnvZ modulates expression of the *ompF* amd *ompC* genes in *Escherichia coli*. *Proc. Natl. Acad. Sci. USA* **86**: 6052–6056.

Fu, H.-A., S. Iuchi, and E. C. C. Lin. 1991. The requirement of ArcA and Fnr for peak expression of the *cyd* operon in *Escherichia coli* under microaerobic conditions. *Mol. Gen. Genet.* **226**: 209–213.

Gray, C. T., J. W. T. Wimpenny, D. E. Hughes, and M. R. Mossman. 1966. Regulation of metabolism in facultative bacteria. 1. Structural and functional changes in *Escherichia coli* associated with shifts between the aerobic and anaerobic states. *Biochim. Biophys. Acta* **117**:22–32.

Green, J., A. D. Sharrocks, B. Green, M. Geisow, and J. R. Guest. 1993. Properties of FNR proteins substituted at each of the five cysteine residues. *Mol. Microbiol.* **8**:61–68.

Igo, M. M., A. J. Ninfa, and T. J. Silhavy. 1989a. A bacterial environmental sensor that functions as a protein kinase and stimulates transcriptional activation. *Genes Dev.* **3**:598–605.

Igo, M. M., A. J. Ninfa, J. B. Stock, and T. J. Silhavy. 1989b. Phosphorylation and dephosphorylation of a bacterial transcriptional activator by a transmembrane receptor. *Genes Dev.* **3**:1725–1734.

Iuchi, S. 1993. Phosphorylation/dephosphorylation of the receiver module at the conserved aspartate residue controls transphosphorylation activity of histidine kinase in sensor protein ArcB of *Escherichia coli*. *J. Biol. Chem.* **263**:23972–23980.

Iuchi, S. Unpublished data.

Iuchi, S., D. C. Cameron, and E. C. C. Lin. 1989. A second global regulator gene (*arcB*) mediating

repression of enzymes in aerobic pathways of *Escherichia coli*. *J. Bacteriol.* **171**:868–873.

Iuchi, S., V. Chepuri, H.-A. Fu, R. B. Gennis, and E. C. C. Lin. 1990a. Requirement for terminal cytochromes in generation of the aerobic signal for the *arc* regulatory system in *Escherichia coli*: study utilizing deletions and *lac* fusions of *cyo* and *cyd*. *J. Bacteriol.* **172**:6020–6025.

Iuchi, S., and E. C. C. Lin. 1988. *arcA* (*dye*), a global regulatory gene in *Escherichia coli* mediating repression of enzymes in aerobic pathways. *Proc. Natl. Acad. Sci. USA* **85**:1888–1892.

Iuchi, S., and E. C. C. Lin. 1992a. Mutational analysis of signal transduction by ArcB: a membrane sensor protein for anaerobic expression of operons involved in the central aerobic pathways in *Escherichia coli*. *J. Bacteriol.* **174**:3972–3980.

Iuchi, S., and E. C. C. Lin. 1992b. Purification and phosphorylation of the Arc regulatory components of *Escherichia coli*. *J. Bacteriol.* **174**:5617–5623.

Iuchi, S., and E. C. C. Lin. 1993. Adaptation of *Escherichia coli* to redox environments by gene expression. *Mol. Microbiol.* **9**:9–15.

Iuchi, S., Z. Matsuda, T. Fujiwara, and E. C. C. Lin. 1990b. The *arcB* gene of *Escherichia coli* encodes a sensor-regulator protein for anaerobic repression of the *arc* modulon. *Mol. Microbiol.* **4**:715–727.

Jin, S., T. Roitsch, R. G. Ankenbauer, M. P. Gordon, and E. W. Nester. 1990. The VirA protein of *Agrobacterium tumefaciens* is autophosphorylated and is essential for *vir* gene regulation. *J. Bacteriol.* **172**: 525–530.

Kranz, R. G., and R. B. Gennis. 1983. Immunological characterization of the cytochrome *o* terminal oxidase from *Escherichia coli*. *J. Biol. Chem.* **258**:10614–10621.

Makino, K., H. Shinagawa, M. Amemura, T. Kawamoto, M. Yamada, and A. Nakata. 1989. Signal transduction in the phosphate regulon of *Escherichia coli* involves phosphotransfer between PhoR and PhoB proteins. *J. Mol. Biol.* **210**:551–559.

Ota, I. M., and A. Varshavsky. 1993. A yeast protein similar to bacterial two-component regulators. *Science* **262**:566–569.

Parkinson, J. S., and E. C. Kofoid. 1992. Communication modules in bacterial signaling proteins. *Annu. Rev. Genet.* **26**:71–112.

Ronson, C. W., B. T. Nixon, and F. M. Ausubel. 1987. Conserved domains in bacterial regulatory proteins that respond to environmental stimuli. *Cell* **49**:579–581.

Sawers, G., and B. Suppmann. 1992. Anaerobic induction of pyruvate formate-lyase gene expression is mediated by the ArcA and FNR proteins. *J. Bacteriol.* **174**:3474–3478.

Silverman, P. M., L. Tran, R. Harris, and H. M. Gaudin. 1993. Accumulation of the F plasmid TraJ protein in *cpx* mutants of *Escherichia coli*. *J. Bacteriol.* **175:**921–925.

Smith, M. W., and F. C. Neidhardt. 1983. Proteins induced by aerobiosis in *Escherichia coli*. *J. Bacteriol.* **154:**344–350.

Stock, J., A. J. Ninfa, and A. M. Stock. 1989. Protein phosphorylation and regulation of adaptive responses in bacteria. *Microbiol. Rev.* **53:**450–490.

Tardat, B., and D. Touati. 1993. Iron and oxygen regulation of *Escherichia coli* MnSOD expression: competition between the global regulators Fur and ArcA for binding to DNA. *Mol. Microbiol.* **9:**53–63.

Dual Sensors and Dual Response Regulators Interact to Control Nitrate- and Nitrite-Responsive Gene Expression in *Escherichia coli*

Valley Stewart and Ross S. Rabin

14

Enterobacteria are facultative aerobes, and several distinct respiratory pathways function to provide energy for anaerobic metabolism. These pathways are built up from dehydrogenases, a common quinone pool, and reductases. Electron donors include formate, hydrogen, NADH, glycerol 3-phosphate, succinate, and lactate, whereas electron acceptors include nitrate, nitrite, dimethyl sulfoxide, trimethylamine N-oxide, and fumarate (reviewed by Stewart, 1988). Synthesis of many of the corresponding dehydrogenases and reductases is subject to two levels of global transcriptional regulation. First, anaerobic respiratory enzymes are synthesized only in the absence of oxygen. This anaerobic control of respiratory enzyme gene expression requires the Fnr (*f*umarate and *n*itrate *r*eductase) protein, a transcriptional regulatory protein that shares sequence similarity with Crp, the cyclic AMP receptor protein. The Fnr protein binds to sequences (consensus TTGAT-N_4-ATCAA) centered at about −40 with respect to the transcriptional start point. It is proposed that redox transitions of a bound iron atom are involved in modulating Fnr activity (reviewed by Spiro and Guest, 1990; Gunsalus, 1992; Iuchi and Lin, 1993).

Anaerobic respiratory enzyme synthesis is subject to a second level of control by nitrate (NO_3^-), which is the preferred anaerobic electron acceptor. Nitrate induces the synthesis of the nitrate respiratory chain components (formate dehydrogenase-N and nitrate reductase, encoded by the *fdnG* and *narG* operons, respectively), while simultaneously repressing the synthesis of other anaerobic respiratory chain components (e.g., fumarate reductase, encoded by the *frdA* operon). The reduction product of nitrate, nitrite (NO_2^-), is also involved in regulating anaerobic respiratory gene expression. Nitrate and nitrite control is mediated by the Nar (*n*itrate *r*eductase) dual interacting two-component regulatory systems, which consist of homologous membrane-bound sensors (the NarX and NarQ proteins) and homologous DNA binding response regulators (the NarL and NarP proteins). Genes that are known to be controlled by the Nar system are listed in Table 1.

We emphasize that the Nar system controls nitrate and nitrite metabolism strictly in response to the needs of anaerobic respiration and has nothing to do with the use of nitrate or nitrite as nitrogen sources for biosynthesis.

Valley Stewart, Sections of Microbiology and Genetics and Development, Cornell University, Ithaca, New York 14853–8101. *Ross S. Rabin,* NeXagen, Inc., Boulder, Colorado 80301.

Two-Component Signal Transduction, Edited by James A. Hoch and Thomas J. Silhavy,
© 1995 American Society for Microbiology, Washington, DC 20005

TABLE 1 Known NarL- and NarP-regulated operons

NarL[a]	NarP	Operon	Function	References
+	Ø	narGHJI	Nitrate reductase	Stewart, 1982; Rabin and Stewart, 1993
+	Ø	narK	Nitrite extrusion	Stewart and Parales, 1988; Rabin and Stewart, unpublished data
+	+	fdnGHI	Formate dehydrogenase-N	Berg and Stewart, 1990; Rabin and Stewart, 1993
+	+	nirBDC	NADH–nitrite reductase	Tyson et al., 1993; Tyson et al., 1994
+	?	modABCD	Molybdate uptake	Miller et al., 1987
+	−	None known		
−	Ø	frdABCD	Fumarate reductase	Iuchi and Lin, 1987; Rabin and Stewart, 1993
−	+	aeg-46.5	Periplasmic nitrate reductase	Choe and Reznikoff, 1991; Rabin and Stewart, 1993
±[b]	+	nrfABCDEFG	Formate-nitrite reductase	Rabin and Stewart, 1993; Tyson et al., 1994
−	−	pfl	Pyruvate-formate lyase	Sawers and Böck, 1988; Sawers, personal communication
−	?	adhE	Alcohol dehydrogenase	Kalman and Gunsalus, 1988; Chen and Lin, 1991; Leonardo et al., 1993
−	?	dmsABC	Dimethyl sulfoxide reductase	Cotter and Gunsalus, 1989
Ø	+	None known		
Ø	−	None known		

[a]+, positive regulation (activation); −, negative regulation (repression); Ø, no effect; ?, not yet examined.

[b]NarL protein represses nrfA operon expression in response to nitrate, and activates nrfA operon expression in response to nitrite.

These latter processes in enterobacteria are conducted by distinct enzyme systems that are under separate genetic control (Goldman et al., 1994; Lin et al., 1994).

DUAL TWO-COMPONENT REGULATORY SYSTEMS

Identification and Characterization of the narL, narX, narQ, and narP Genes

The narL gene was originally identified by an insertion mutation that eliminates nitrate induction of narG (nitrate reductase) operon expression. This narL null allele is recessive to the narL⁺ gene, indicating that the NarL protein is a transcriptional activator (Stewart, 1982). The narL gene is located at 27 min on the genetic map, adjacent to the narGHJI operon. The narL⁺ gene is also essential for nitrate induction of fdnG (formate dehydrogenase-N) operon expression (Berg and Stewart, 1990) and for nitrate repression of frdA (fumarate reductase)

operon expression (Iuchi and Lin, 1987; Kalman and Gunsalus, 1988; Stewart and Berg, 1988). Thus, the NarL protein is both an activator and a repressor of anaerobic respiratory gene expression (Table 1).

The narX gene was originally identified during the physical analysis of the narL region (Stewart and Parales, 1988), and the narX and narL genes form a complex operon (Egan and Stewart, 1990). DNA sequence analysis suggested that the NarX protein serves as a cognate sensor for the NarL protein (Nohno et al., 1989; Stewart et al., 1989). However, insertions in the narX gene cause only a subtle regulatory phenotype, and nonpolar deletions are phenotypically silent (Stewart and Berg, 1988; Stewart and Parales, 1988; Egan and Stewart, 1990). These observations led to the proposal that a second nitrate-responsive sensor can substitute for NarX protein function (Egan and Stewart, 1990). This sensor, encoded by the narQ gene at 53 min on the genetic map, was identified in screens for mutations that eliminate nitrate regulation in a *narX* null strain (Chiang et al., 1992; Rabin and Stewart, 1992). Genetic studies show that either the *narX*[+] or the *narQ*[+] gene is sufficient for essentially normal nitrate regulation, whereas *narX narQ* double null strains phenotypically resemble *narL* null strains (Rabin and Stewart, 1992).

Identification of dual nitrate-responsive sensors led to the notion that dual response regulators may also be involved in the Nar regulatory circuit. A means for identifying the second response regulator was provided by Choe and Reznikoff's (1991) discovery of the *aeg-46.5* locus (*anaerobically expressed gene at 46.5 min on the genetic map*). Genes in the *aeg-46.5* region are provisionally designated as *nap* (periplasmic *nitrate* reductase) and as *cyc* (*cytochrome c* biogenesis; see below). Nitrate is an efficient inducer of *aeg-46.5* operon expression in a *narL* null strain (Choe and Reznikoff, 1991), and introduction of both *narX* and *narQ* null mutations eliminates this nitrate induction (Rabin and Stewart, 1993). Taken together, these results suggested that the activity of a second response regulator (the NarP protein) is also modulated by the NarX and NarQ sensors. A search for mutations that eliminate nitrate induction of *aeg-46.5* operon expression led to the identification of the *narP* gene (Rabin and Stewart, 1993), which is located immediately adjacent to the *aeg-46.5* operon (Richterich et al., 1993). Thus, unlike the *narXL* complex operon, the *narQ* and *narP* genes are unlinked.

Structures of the NarL, NarP, NarX, and NarQ Proteins

The DNA sequences of the *narL* and *narP* genes reveal that the encoded proteins are homologous response regulators of approximately 24 kDa. The amino-terminal segments (approximately 120 residues) are homologous to the phosphorylated domains of response regulators and contain the conserved Asp and Lys residues involved in the transphosphorylation reactions. The carboxyl-terminal segments (approximately 95 residues) share similarity with

the MalT-LuxR-FixJ-RcsA family of DNA binding proteins (Kahn and Ditta, 1991; Stout et al., 1991), many but not all of which are response regulators. This class of proteins appears to contain a "probe helix" type of DNA binding motif (Suzuki, 1993).

The DNA sequences of the *narX* and *narQ* genes reveal that the encoded proteins are homologous sensors of approximately 65 kDa. Both proteins apparently share structural similarity with the methyl-accepting chemotaxis proteins (MCPs), with periplasmic amino-terminal and cytoplasmic carboxyl-terminal regions. Gene fusion analysis using Tn*phoA* (reviewed by Manoil et al., 1990) supports this topological model for NarX protein architecture (Williams and Stewart, unpublished data). The amino-terminal periplasmic regions of the NarX and NarQ proteins (approximately 115 residues) are largely dissimilar except for a stretch immediately following the first transmembrane region in which 15 of 17 consecutive residues are identical in the two proteins (Chiang et al., 1992). The carboxyl-terminal regions (approximately 400 residues) share similarity with histidine protein "kinase" sensors, with the exception that the most distal conserved region (block G2) is absent (Stock et al., 1989; Parkinson and Kofoid, 1992). Finally, the NarX and NarQ sequences share a stretch of 11 identical residues in the area between blocks H and N (Chiang et al., 1992; Rabin and Stewart, 1992). This general region of sensor proteins may be important in conferring specificity on sensor-response regulator interactions (Parkinson and Kofoid, 1992).

Effects of *narL* and *narP* Null Alleles on Nitrate and Nitrite Regulation

Analysis of target operon expression in *narL* and *narP* null mutants has revealed a diversity of regulatory patterns. An attempt to summarize these patterns is presented in Fig. 1. The NarL and NarP proteins can both act either as activators or as repressors of gene expression, in response to either nitrate or nitrite, depending on the target operon examined. Different combi-

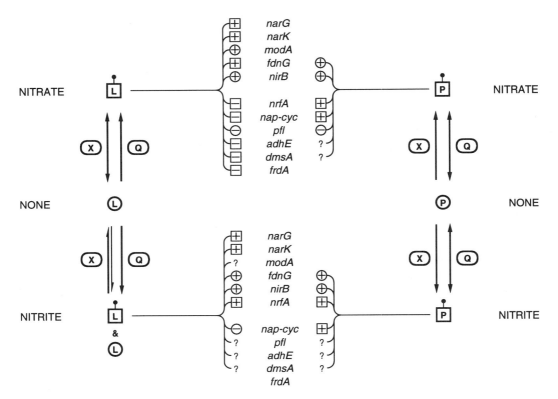

FIGURE 1 Model for nitrate- and nitrite-regulated gene expression. Activation and repression of target operon expression are indicated by + and −, respectively; question marks denote situations that have not yet been tested. Strong interactions (10-fold or greater regulation) are boxed, whereas weaker interactions (less than 10-fold regulation) are circled. The NarL and NarP proteins are represented by circled L and circled P, respectively; the phosphorylated forms are represented by the same letters boxed. The NarX and NarQ proteins are represented by the letters X and Q, respectively. In the presence of nitrite, a relatively small proportion of NarL molecules is phosphorylated. Target operon designations are listed in Table 1. The *aeg-46.5* operon is denoted as "*nap-cyc*" in this figure. Modified from Rabin and Stewart (1993).

nations of these activities (e.g., NarL-mediated repression and NarP-mediated activation) allow adaptability in controlling the expression of different target operons. This flexibility undoubtedly allows for finely tuned synthesis of anaerobic respiratory components in response to changing concentrations of nitrate and nitrite, but the overall physiological rationale for these regulatory circuits remains to be detailed.

Nitrate induction of the *narG* and *narK* (nitrite efflux) (DeMoss and Hsu, 1991; Rowe et al., 1994) operons and nitrate repression of the *frdA* operon is fully dependent on the *narL⁺* gene, and expression is virtually unaffected by a *narP* null allele (Rabin and Stewart, 1993;

Rabin and Stewart, unpublished data). Thus, these operons appear to be solely regulated by the NarL protein. Nitrate induction of the *fdnG* operon is similar, except that a residual three- to fourfold nitrate induction remains in *narL* null strains (Berg and Stewart, 1990); this residual induction is eliminated on introduction of a *narP* null allele (Rabin and Stewart, 1993). In *narX⁺ narQ⁺* strains, nitrite has moderate (*narG, narK*), weak (*fdnG*), or undetectable (*frdA*) effects on expression of these operons (Rabin and Stewart, 1993; Rabin and Stewart, unpublished data).

The *aeg-46.5* (*nap-cyc*) region (Choe and Reznikoff, 1991) apparently encodes a cytochrome *c*-linked periplasmic nitrate reductase homologous to the NapAB enzyme of *Alcaligenes eutrophus* (Richterich et al., 1993; Siddiqui et al., 1993). Other genes in this region apparently encode cytochrome *c* biogenesis functions (Richterich et al., 1993). These genes are likely arranged in an operon. Nitrate and nitrite induction of *aeg-46.5* operon expression is eliminated in a *narP* null strain, indicating that the NarP protein activates *aeg-46.5* operon transcription (Rabin and Stewart, 1993). Conversely, nitrate and nitrite induction of *aeg-46.5* operon expression is enhanced in a *narL* null *narP+* strain, indicating that the NarL protein antagonizes NarP protein activation of *aeg-46.5* operon transcription (Choe and Reznikoff, 1991; Rabin and Stewart, 1993).

The *nirB* operon encodes a cytoplasmic NADH-linked, ammonium-forming nitrite reductase that is involved in regenerating NAD$^+$ and also in detoxifying nitrite (Page et al., 1990). *nirB* operon expression is weakly induced by either nitrate or nitrite. Studies with *narL* null (Page et al., 1990; Tyson et al., 1993) and *narP* null (Tyson et al., 1994) strains show that both the NarL and the NarP proteins activate *nirB* operon expression in response to nitrate but that only the NarL protein is involved in induction by nitrite.

The *nrfA* operon encodes respiratory nitrite reductase (periplasmic cytochrome c_{552}) and associated components, and probably corresponds to the *aeg-93* operon (Choe and Reznikoff, 1991; Darwin et al., 1993a). In wild-type strains, *nrfA* operon expression is induced by nitrite but not by nitrate. Introduction of a *narL* null allele leads to nitrate induction as well, suggesting that the NarP protein activates *nrfA* operon expression in response to either nitrate or nitrite and that the NarL protein represses *nrfA* operon expression only in response to nitrate. *narL+* *narP* null strains are similar to the wild type, exhibiting nitrite but not nitrate induction of *nrfA* operon expression, suggesting that the NarL protein activates *nrfA* operon expression in response to nitrite (Rabin and Stewart, 1993; Tyson et al.,

1994). Thus, the NarP protein is a nitrate- and nitrite-responsive activator of *nrfA* operon expression, whereas the NarL protein is a nitrate-responsive repressor but a nitrite-responsive activator.

The *pfl* gene encodes pyruvate-formate lyase, the enzyme responsible for anaerobic conversion of pyruvate to formate and acetyl-coenzyme A. *pfl* gene expression is repressed by nitrate, and this repression is fully alleviated only in *narL narP* or in *narX narQ* double null strains (Sawers and Böck, 1988; Sawers, personal communication). This indicates that the NarL and NarP proteins both act as repressors of *pfl* gene expression. The effects of nitrite on *pfl* gene expression have not been reported.

The *adhE* gene encodes a trifunctional enzyme (alcohol dehydrogenase/acetaldehyde dehydrogenase/pyruvate-formate lyase deactivase) that plays a central role in the mixed-acid fermentation. Nitrate repression of *adhE* expression is largely but incompletely relieved by a *narL* null allele (Chen and Lin, 1991; Leonardo et al., 1993). It is argued that the residual nitrate repression in *narL* null strains is due to redox control by nitrate (Chen and Lin, 1991). The effects of a *narP* null allele on this residual nitrate repression have not been reported.

The *dmsA* operon encodes dimethylsulfoxide reductase. Nitrate repression of *dmsA* operon expression is largely but incompletely relieved by an uncharacterized *narL* allele (Cotter and Gunsalus, 1989). The effects of a *narP* null allele on this residual nitrate repression have not been reported.

The *modA* (*chlD*) operon encodes components of a molybdate uptake system, and molybdenum cofactor is an essential component of formate dehydrogenase, nitrate reductase, and dimethylsulfoxide reductase. Nitrate induces *modA* operon transcription by approximately twofold, and this induction is abolished in a *narL* null strain (Miller et al., 1987), suggesting that the NarL protein is the sole nitrate-responsive regulator of *modA* operon expression.

Thus, the Nar system can adopt a diversity of regulatory patterns (Table 1; Fig. 1). Much work remains to be done to characterize the interactions of these proteins with their respective DNA targets and to understand the physiological rationale for the observed regulatory patterns.

Effects of *narX* and *narQ* Null Alleles on Nitrate and Nitrite Regulation

In *narL*⁺ *narP*⁺ strains, either the *narX*⁺ or the *narQ*⁺ gene is sufficient for essentially normal NarL- and NarP-dependent nitrate regulation, although a subtle exception is provided by the *aeg-46.5* (*nap-cyc*) operon (Rabin and Stewart, 1992, 1993). The situation with respect to nitrite regulation is quite different. Several NarL-regulated operons, including *narG, fdnG,* and *frdA,* are strongly regulated by nitrate but only weakly regulated by nitrite. Introduction of a *narQ* null allele has little effect on this pattern. However, introduction of a nonpolar *narX* null allele leads to strong nitrite induction of *narG* and *fdnG* operon expression and strong nitrite repression of *frdA* operon expression. These nitrite effects require the *narQ*⁺ gene (Rabin and Stewart, 1993). Thus, it seems that the NarQ protein can activate (phosphorylate) the NarL protein equally well in response to either nitrate or nitrite. However, the NarX protein appears to activate the NarL protein only in response to nitrate, and the presence of a functional NarX protein seems to block NarQ-dependent NarL activation in response to nitrite.

Our interpretation of these observations is that the NarX protein is both a positive ("protein kinase") and a negative ("phosphoprotein phosphatase") regulator of NarL protein function, whereas the NarQ protein is predominately a positive regulator of NarL protein function. In the presence of nitrate, both the NarX and NarQ proteins act to phosphorylate the NarL protein. In the presence of nitrite, the NarQ protein acts to phosphorylate the NarL protein, but the NarX protein counters by acting as a phospho-NarL phosphatase. This would result in a low but significant level of phospho-NarL, leading to detectable nitrite induction of *narG* and *fdnG* operon expression in the wild type. Additional genetic and biochemical observations support the hypothesis that the NarX protein is both a positive and a negative regulator of NarL function (see below).

INTERACTIONS BETWEEN THE NarX AND NarL PROTEINS

The *narQ* and *narP* genes have only recently been recognized, whereas the *narX* and *narL* genes have been studied for many years. Likewise, the regulatory role of nitrite has only recently become generally appreciated (Page et al., 1990; Rabin and Stewart, 1993). Therefore, we now focus our attention on aspects of *narX* and *narL* gene function in response to nitrate, studies of which have mainly been performed in *narQ*⁺ *narP*⁺ strain backgrounds. Among the most studied targets of NarL protein action are the NarL-activated *narG* and *fdnG* operons and the NarL-repressed *frdA* operon.

Phosphoryl Group Transfer between the NarX and NarL Proteins

Sequence comparisons strongly suggest that the NarX/NarQ and NarL/NarP proteins function in phosphoryl transfer reactions. Studies with purified proteins have examined some of these interactions in vitro (Schröder et al., 1994; Walker and DeMoss, 1993). The NarX protein (in both full-length and amino-terminal truncated versions) autophosphorylates when incubated with ATP, and phospho-NarX serves as an efficient substrate for NarL protein phosphorylation (kinase activity). The resultant phospho-NarL is relatively long-lived, but addition of NarX protein accelerates phospho-NarL dephosphorylation (Schröder et al., 1994; Walker and DeMoss, 1993). This supports previous genetic experiments suggesting that the NarX protein additionally has a phospho-NarL phosphatase activity (Stewart, 1993). An amino-terminal truncated version of the NarQ protein has little effect on phospho-NarL dephosphorylation (Schröder et al., 1994), again consistent with conclusions from genetic analysis (Rabin and Stewart, 1993). The stability of phospho-NarL and phospho-NarX is affected by acidic and alkaline conditions in a manner consistent with the formation of phosphoaspartyl and phosphohistidyl residues, respectively (Walker and DeMoss, 1993). Finally, inclusion of nitrate has no influence on the rates of full-length NarX protein autophosphorylation, phosphoryl transfer to the NarL protein, or phospho-NarL dephosphorylation in the absence or presence of NarX protein, suggesting that the solubilized NarX protein has lost the ability for ligand-dependent signaling (Walker and DeMoss, 1993).

narL Mutations

The Asp-59 residue in the NarL protein corresponds to the site of phosphorylation in all response regulators studied (Stock et al., 1989; Parkinson and Kofoid, 1992). A site-specific change of Asp-59 to Asn (D59N) results in a recessive null phenotype, consistent with the notion that the phosphorylated form of the NarL protein both activates and represses transcription (Egan and Stewart, 1991).

Mutational analysis of the narL gene used a λ ΔnarX narL+ specialized transducing phage. Constitutive alleles of the narL gene were isolated by mutagenizing the transducing phage and selecting for lysogens that expressed a Φ(fdnG-lacZ) operon fusion in the absence of nitrate. One allele in particular, a change of Val-88 to Ala (V88A), confers a robust constitutive phenotype (Egan and Stewart, 1991). The narL(V88A) allele is dominant to the narL+ gene, indicating that the NarL(V88A) protein has gained a function. In a narX+ strain cultured in the absence of nitrate, the narL(V88A) allele confers strong narG operon expression, weaker fdnG operon expression, and negligible frdA operon repression (Egan and Stewart, 1991).

The constitutive phenotype conferred by the narL(V88A) allele remains in narX narQ double null strains (Rabin and Stewart, 1992).

Indeed, introduction of a narX null allele exacerbates the constitutive phenotype. This is most easily seen for the fdnG operon, which is expressed at about 10% of the fully induced level in a narX+ strain but at about 100% of the fully induced level in a narX null strain (Egan and Stewart, 1991).

Thus, the NarX protein acts as a negative regulator of NarL(V88A) protein function, presumably through phospho-NarL phosphatase activity. This implies that phosphorylation plays a role in activating the NarL(V88A) protein. One possibility is that the NarL(V88A) protein may have a relaxed specificity for phosphorylation by noncognate sensors. Whether nonphosphorylated NarL(V88A) protein retains some activity remains to be determined.

A Φ(narG-lacZ) operon fusion is expressed at about 40% of the fully induced level in a narL(V88A) strain cultured in the absence of nitrate (Rabin and Stewart, 1992). Introduction of a narQ null allele has no effect on this constitutive phenotype, whereas introduction of a narX null allele increases nitrate-independent expression to about 130% of the fully induced level, irrespective of the presence of the narQ+ gene (Rabin and Stewart, 1992). This observation reinforces the idea that the NarX protein, but not the NarQ protein, acts as a negative regulator of NarL protein function (Rabin and Stewart, 1993).

narX Null Mutations

Insertions in the narX gene result in subtle nitrate-regulatory phenotypes (Stewart and Parales, 1988; Stewart and Berg, 1988), apparently because of weak polarity on narL gene expression (Egan and Stewart, 1990). Two relatively large in-frame deletions of the narX gene were constructed, and in narQ+ strains, both are phenotypically silent (Egan and Stewart, 1990). However, the phenotypes of ΔnarX narQ::Tn10 strains resemble those of narL null strains, indicating that these ΔnarX deletions are true null alleles (Rabin and Stewart, 1992). Our reference to narX and narQ null alleles refers to

TABLE 2 Effects of *narX* mutations on nitrate and nitrite regulation[a]

| | β-Galactosidase sp act[c] | | | | | |
| | *narQ*[+] | | | *narQ*::Tn*10* | | |
Genotype[b]	None[d]	+ NO$_3^-$	+ NO$_2^-$	None	+ NO$_3^-$	+ NO$_2^-$
Φ(*narG-lacZ*)						
narX[+]	30	3,240	340	26	3,130	91
Δ*narX242*	47	3,290	3,610	29	27	26
narX512(H399Q)	29	3,400	840	31	29	27
narX★*511*(A224V)	410	3,430	2,850	960	3,300	2,490
narX★*32*(E208K)	940	3,130	2,530	2,040	2,070	2,080
Φ(*fdnG-lacZ*)						
narX[+]	8	880	29	8	930	11
Δ*narX242*	10	950	630	8	7	6
narX512(H399Q)	7	990	57	7	8	6
narX★*511*(A224V)	24	910	390	38	900	340
narX★*32*(E208K)	53	730	280	80	170	130
Φ(*frdA-lacZ*)						
narX[+]	110	12	140	110	16	140
Δ*narX242*	120	11	14	120	130	130
narX512(H399Q)	120	9	120	120	130	120
narX★*511*(A224V)	110	8	71	110	10	81
narX★*32*(E208K)	110	43	89	99	92	100

[a]See Rabin and Stewart (1993) for details of strain constructions, enzyme assay, and culture conditions.

[b]All strains are derived from strain VJS676 (Δ*lac*) and carry gene fusions on single-copy λ specialized transducing phage.

[c]β-Galactosidase activity is expressed in Miller units.

[d]Anaerobic growth in defined mininal morpholinepropanesulfonic acid medium with no acceptor, nitrate (40 mM), or nitrite (5 mM), as indicated.

in-frame *narX* deletion and *narQ*::Tn*10* insertion alleles, respectively.

narX(H399Q) Mutation

The His-399 residue in the NarX protein corresponds to the site of phosphorylation in all sensors studied (with the exception of CheA) (Stock et al., 1989; Parkinson and Kofoid, 1992). A site-specific change of His-399 to Gln (H399Q) was constructed to examine the role of this residue in NarX protein function (Rabin and Stewart, 1992).

Either the NarX or the NarQ protein is sufficient for essentially wild-type nitrate induction of *narG* and *fdnG* operon expression and nitrate repression of *frdA* operon expression in *narL*[+] strains. Nitrate regulation is abolished both in *narX narQ* double null strains and in *narX*(H399Q) *narQ* null strains (Rabin and

Stewart, 1992) (Table 2), indicating that the NarX(H399Q) protein has lost its positive regulatory function.

In *narX* null *narQ*[+] strains, induction of *narG* and *fdnG* operon expression and repression of *frdA* operon expression is effected equally well by nitrite and nitrate (Rabin and Stewart, 1993) (Table 2). In striking contrast, expression of these operons was nearly insensitive to nitrite in *narX*(H399Q) *narQ*[+] strains, mimicking the wild-type response to nitrite (Table 2). Thus, the *narX*(H399Q) allele resembles the *narX* null allele only in that its product lacks the positive function required for nitrate regulation. Conversely, the *narX*(H399Q) and the *narX*[+] allele resemble each other because their products retain the negative function involved in response to nitrite. (Note that the *narQ*[+] gene must be present to detect the negative function of the NarX(H399Q) protein [Table 2].) The simplest explanation for this result is that the NarX(H399Q) protein cannot autophosphorylate and therefore lacks kinase activity but that it retains near–wild-type phosphoprotein phosphatase activity. This result is consistent with studies of the NtrB sensor, in which changes to conserved histidine protein kinase residues eliminate positive regulation but leave negative regulation relatively intact (Atkinson and Ninfa, 1993).

narX★ Mutations

Kalman and Gunsalus (1990) isolated constitutive alleles of the *narX* gene (*narX*★ mutations) by screening for constitutive *frdA* operon repression in the absence of nitrate. Three alleles carry different single amino acid substitutions in the region immediately following the second transmembrane region of the NarX protein (Kalman and Gunsalus, 1990). Screens for constitutive *fdnG* operon expression, using mutagenized λ *narX*⁺ Δ*narL* specialized transducing phage, yielded two additional *narX*★ alleles (Collins et al., 1992).

The *narX*★ mutations lie in a region of shared similarity with MCP proteins, termed the linker (Collins et al., 1992), and mutational alterations in this region of the Tsr protein (the serine chemoreceptor) lead to altered chemotactic signaling (Ames and Parkinson, 1988). It is proposed that the linker region is involved in signal transduction from the periplasmic receptor domain to the cytoplasmic output domain of MCPs (Ames and Parkinson, 1988). Two of the *narX*★ mutational changes (E208K and M219I) are identical to two of those in *tsr* (E248K and M259I) and affect conserved residues (Ames and Parkinson, 1988; Collins et al., 1992) (Fig. 2). A third *narX*★ change (G205R)

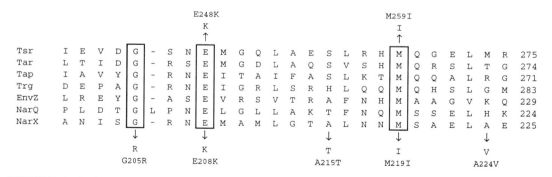

FIGURE 2 Linker regions of MCPs (Tsr, serine chemoreceptor; Tar, aspartate chemoreceptor; Tap, peptide chemoreceptor; Trg, ribose chemoreceptor) and sensors (EnvZ, osmolarity sensor; NarX and NarQ, nitrate/nitrite sensors). The segments immediately following the second transmembrane region are shown. Residues that are identical in all seven sequences are boxed. Mutational changes in *tsr* (Ames and Parkinson, 1988) and *narX* (Kalman and Gunsalus, 1990; Collins et al., 1992) are indicated.

alters another conserved residue (Fig. 2). Recent analysis reveals that the sensor EnvZ also contains a homologous linker region (Baumgartner et al., 1994) (Fig. 2). However, similar linker regions are not apparent in many other sensors examined, including the PhoR, UhpB, and FixL proteins.

Complementation analysis in partial diploid strains reveals that the four narX* alleles studied are recessive to the narX+ gene, suggesting that these mutations result in the loss of a function (Collins et al., 1992). The impaired function is suggested to be phosphoprotein phosphatase activity. This relative deficiency in NarX* negative function might be due to either a lesion that directly affects phosphatase activity or to an equilibrium shift alteration that creates an inappropriate bias toward the kinase form (see below). We favor this second idea because the NarX* lesions are located in the linker region, which is probably involved in intramolecular signal transduction (Ames and Parkinson, 1988). We consider it unlikely that phosphatase activity is directly impaired by lesions in the linker region. If we are correct, this implies that the NarX* proteins are deficient in adopting the phosphatase conformation (Collins et al., 1992).

In both narQ+ and narQ null backgrounds, the narX*(A224V) and narX*(E208K) strains exhibited a substantial increase in nitrite induction of narG and fdnG operon expression in comparison with the narX+ strains (Table 2). This nitrite effect mimics that of the narX null narQ+ strains in which the NarQ protein's positive function strongly induces narG and fdnG operon expression in the absence of the inhibiting influence of NarX protein negative function (Rabin and Stewart, 1993) (Table 2). Interestingly, nitrite induction in the narX*-(A224V) strains was essentially independent of NarQ protein function, whereas nitrite induction in the narX*(E208K) strains was augmented by NarQ protein positive function. Therefore, we believe that in addition to their respective defects in negative function, the NarX*(A224V) protein retains substantial positive function, whereas that of the NarX*(E208K) protein is somewhat deficient, consistent with previous interpretations (Kalman and Gunsalus, 1990; Rabin and Stewart, 1992).

Hierarchy of NarL–Mediated Regulation

Why do the narX* alleles exert only subtle effects on nitrite repression of frdA operon expression, whereas they yield intermediate nitrite induction of fdnG operon expression and strong nitrite induction of narG operon expression (Table 2)? This pattern (narG > fdnG > frdA) is also observed in narL(V88A) strains cultured in the absence of nitrate (Egan and Stewart, 1991; Rabin and Stewart, 1992) and in narX insertion strains cultured in the presence of nitrate (Stewart and Berg, 1988). These narX insertion strains apparently synthesize less NarL protein because of partial polarity (Egan and Stewart, 1990). Together, these results suggest that the narG operon is more sensitive to NarL regulation and retains substantial expression when NarL protein activity is submaximal (either through decreased synthesis or decreased phosphorylation). By contrast, the frdA operon seems to be less sensitive to NarL regulation and exhibits significant repression only when the NarL protein is fully active (Rabin and Stewart, 1993). This apparent hierarchy may be relevant to the wild-type organism in physiological conditions, in which only low concentrations of nitrate are available. Synthesis of alternate respiratory pathways in this situation could allow maximum flexibility in respiration.

Therefore, we presume that the narX* alleles have a weak effect on frdA operon expression and a strong effect on narG operon expression because only a relatively low amount of phospho-NarL is produced by interaction with the NarX* proteins in strains grown either with no acceptor or in the presence of nitrite (Collins et al., 1992) (Table 2).

Nitrate versus Nitrite: Equilibrium Model for NarX and NarQ Protein Functions

The response to nitrite presents an enigma. In narQ narL double null strains, the narX+ gene is sufficient for full aeg-46.5 (nap-cyc) and nrfA

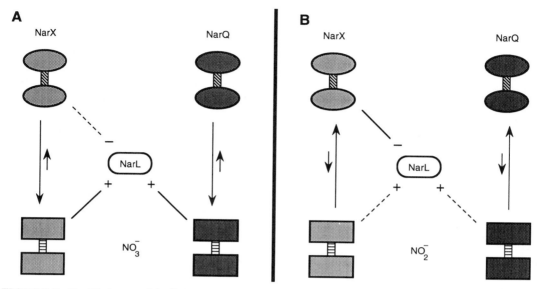

FIGURE 3 Equilibrium model of NarX and NarQ protein functions in the presence of (A) nitrate or (B) nitrite. See text for details. The NarX and NarQ sensor proteins are shown in equilibria between their negative (rounded) and positive (rectangular) forms. The relative sizes of the equilibrium arrows indicate qualitatively how much of a given sensor population is found in each of the two forms. Positive interactions with the NarL response regulator are designated + (i.e., NarL kinase), and negative interactions are designated − (i.e., phospho-NarL phosphatase). A solid or dashed line (at 45° angle to the sensor protein) emphasizes whether a large or small proportion, respectively, of a given sensor population is responsible for a specific influence on the NarL protein.

(*aeg-93*) operon expression in response to nitrite, suggesting that nitrite-stimulated NarX protein efficiently phosphorylates the NarP protein. However, only in *narX* null mutants does nitrite become an efficient regulator of *narG, fdnG,* and *frdA* operon expression, suggesting that nitrite-stimulated NarX protein acts to dephosphorylate the NarL protein (Rabin and Stewart, 1993). The same studies lead to the conclusion that nitrate-stimulated NarX protein acts to phosphorylate both the NarL and the NarP proteins. Mechanistic models to explain this conundrum, involving different types of sensor-response regulator interactions in response to different signal ligands, are plausible but awkward. We currently entertain a simpler model based on the equilibrium between kinase and phosphatase forms of the NarX sensor.

Russo et al. (1993) described the overall function of the EnvZ sensor as a ratio of kinase to phosphatase activities (K:P). In their model, the level of phospho-OmpR protein is controlled by both activities of the EnvZ protein. Thus, the K:P ratio provides a direct indication of the signaling state of the EnvZ protein and its resulting control over the phosphorylation state of the OmpR protein. This model implicitly represents a given signaling state as an equilibrium between kinase and phosphatase conformations.

We similarly view NarX and NarQ protein molecules as populations in equilibria between their positive (NarL kinase) and negative (phospho-NarL phosphatase) functional forms (Fig. 3). In this model, nitrate and nitrite elicit different regulatory responses only because of the degree to which they affect sensor equilibria and not because the two stimuli affect intrinsic sensor function differently. A specific equilibrium state produces a regulatory outcome based on the net effect of the positive and negative sensor activities, which, in turn, controls the phosphorylation state of the NarL response regulator. Nitrate stimulates both sensors to shift strongly toward the positive form and away from the negative form. Nitrate regulation results from the positive activities of both sensors; the contribution of NarX negative function is negligible. We imagine that nitrite elicits a weaker equilibrium shift for both sensors, presumably only slightly different from the equilibrium state found with no added inducer. Nitrite regulation in the wild type would thus result primarily from a combination of NarQ positive function and NarX negative function, with only a minor contribution from NarX positive function (Fig. 3).

In this model, wild-type NarL-dependent nitrite regulation is relatively weak because the rate of phospho-NarL dephosphorylation exceeds that of NarL phosphorylation (Fig. 3). We believe that the relatively low level of kinase function provided by either sensor in response to nitrite would be sufficient to saturate the population of NarL protein molecules with phosphoryl groups were phosphatase function inconsequential. Such a situation probably occurs in nitrite-grown narX null narQ$^+$ strains: the amount of NarQ protein in kinase form is fairly low, but the phospho-NarL produced will persist if the NarQ protein lacks significant phospho-NarL phosphatase activity. Biochemical experiments reveal that phospho-NarL is relatively long-lived, at least in vitro (Schröder et al., 1994; Walker and DeMoss, 1993). Laboratory cultures are routinely grown with excess nitrate or nitrite, and our model assumes that excess inducer is more than sufficient to sustain the sensor equilibria as indicated in Fig. 3.

A remaining conundrum concerns the second nitrate-nitrite response regulator protein, NarP. We previously reported that the NarX protein positively regulates the NarP protein in response to nitrite (Rabin and Stewart, 1993). Yet, the equilibrium model shows that nitrite-stimulated NarX protein is shifted toward its negative form (Fig. 3). We reconcile this apparent contradiction by postulating that the NarX protein has a negligible phospho-NarP phosphatase function and that its NarP kinase function works efficiently (Rabin and Stewart, 1993). As stated above for the NarL protein, we presume that the sensor kinase function provided in response to nitrite is sufficient to saturate the population of NarP response regulators with phosphoryl groups, assuming that sensor phosphatase functions are negligible. Clearly, all these ideas await direct experimental tests.

Negative Regulation by the NarX (and NarQ?) Sensors

Several lines of evidence, described more fully above, suggest that the NarX protein negatively regulates NarL protein function by acting as a phospho-NarL phosphatase (Stewart, 1993). First, nitrite-dependent NarL activation of narG and fdnG operon expression is greatly enhanced in narX null mutants (Rabin and Stewart, 1993). Second, the constitutive phenotype conferred by the narL(V88A) alteration is also greatly enhanced in narX null mutants (Egan and Stewart, 1991; Rabin and Stewart, 1992). Finally, purified NarX protein acts as a phospho-NarL phosphatase in vitro (Schröder et al., 1994; Walker and DeMoss, 1993). Indeed, the phenotypes conferred by both narX* and the narX(H399Q) alleles are rationally interpreted as selective alterations in negative and positive functions, respectively (Collins et al., 1992) (Table 2). The presumed role of this negative regulation is to dephosphorylate phospho-NarL when nitrate (or nitrite) is exhausted from the growth medium (see Fig. 1).

How is phospho-NarP dephosphorylated when nitrate and nitrite are exhausted? In a narL narQ double null strain, expression of the nrfA (aeg-93) operon (in a nrf insertion mutant) is elevated during growth in the absence of nitrate or nitrite, whereas expression in a narL narX double null strain resembles that of the narX$^+$ narQ$^+$ and narX narQ double null strains (Rabin and Stewart, 1993). This observation suggests that the NarQ protein (but not the NarX protein) acts to negatively regulate NarP protein function, and we have depicted this hypothetical interaction in Fig. 1. We emphasize that this idea is currently tenuous and requires substantial additional experimentation.

TARGET OPERON CONTROL REGIONS

The control regions of several nitrate- and nitrite-regulated operons have now been studied in some detail. All operons known to be regulated by the Nar system are also subject to Fnr protein control in response to anaerobiosis, and most such operons studied contain apparent Fnr protein binding sites centered in the region between −40 to −50 with respect to their transcription initiation sites.

NarL protein binding sites apparently conform to a consensus sequence termed the NarL heptamer (consensus TACYNMT, where Y = C or T, M = A or C, and N = any nucleotide) (Tyson et al., 1993). The NarL heptamers in different control regions exhibit great diversity with respect to number, location, orientation, and spacing (Stewart, 1993) (Fig. 4). For several control regions, NarL heptamers have been identified by mutational analysis (Dong et al., 1992; Li and Stewart, 1992; Tyson et al., 1993; Tyson et al., 1994) and/or by DNase I footprinting experiments using purified NarL protein (Li et al., 1994; Walker and DeMoss, 1994). In other cases, heptamers have been identified solely through sequence inspection. Finally, DNase I footprinting experiments have revealed several NarL heptamers that were not apparent through sequence inspection alone (Li et al., 1994; Walker and DeMoss, 1994).

This section summarizes current information on the architecture of several NarL- and NarP-sensitive operon control regions. NarL and NarP heptamers are designated by their position with respect to the transcription initiation site. For example, the heptamer spanning positions −106 through −112 in the *fdnG* operon control region is designated as "heptamer −109."

narG Operon Control Region

The *narG* operon control region is perhaps the best-studied target of NarL protein activation. Mutational analysis has identified the Fnr pro-

tein binding site centered at −41.5 (Walker and DeMoss, 1991) (Fig. 4), and further studies have shown that the requirement for Fnr protein action can be circumvented by changing the −10 region to the consensus sequence (Walker and DeMoss, 1992). This −10 consensus promoter retains NarL-dependent nitrate stimulation even in an *fnr* null strain background, indicating that the Fnr protein is not essential for NarL protein action at this control region (Walker and DeMoss, 1992).

Early deletion analysis demonstrated that the region around −200 is essential for nitrate induction (Li and DeMoss, 1988), and mutational characterization identified heptamer −195 (Dong et al., 1992). Footprint studies identified three heptamers in the −200 region and at least three additional heptamers in the −104 to −54 region (Li et al., 1994; Walker and DeMoss, 1994) (Fig. 4). To date, further mutational studies have confirmed that at least the heptamer at position −89 is critical for full nitrate induction of *narG* operon expression (Li et al., 1994; Walker and DeMoss, 1994). However, deletion of the −205 heptamer confers no observable nitrate regulatory phenotype (Li and DeMoss, 1988; Rabin and Stewart, unpublished data).

The spacing between the −200 region and the transcription initiation site is critical for nitrate induction (Li and DeMoss, 1988), as is the geometry of the intervening region (Dong et al., 1992). One explanation for these constraints came from the discovery that the integration host factor (IHF), a sequence-specific DNA bending protein, is required for nitrate induction of *narG* operon expression. Indeed, IHF protein specifically binds to the −106 to −144 region of the *narG* operon control region (Rabin et al., 1992; Schröder et al., 1993; Walker and DeMoss, 1994) (Fig. 4). By analogy with certain σ^{54}-dependent control regions (Hoover et al., 1990), it is postulated that the IHF protein mediates the formation of a looped DNA structure that brings the −200 region into closer proximity to the transcription initiation site. The nature of the resultant

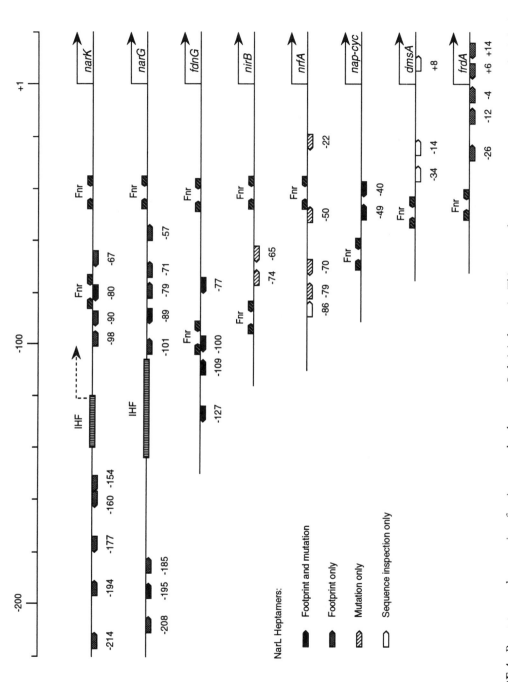

FIGURE 4 Promoter-regulatory regions for nitrate-regulated operons. Scale is in base pairs. Thin arrows denote transcription initiation sites; the dashed arrow for the *narK* control region denotes a minor constitutive transcript. Fnr protein binding sites are shown as dark hatched inverted arrows. NarL heptamer sequences are indicated by their position with respect to the transcription initiation site. Heptamers denoted by black arrows have been identified by both mutational analysis and by DNase I footprinting. Heptamers denoted by gray arrows have been identified by DNase I footprinting only. Heptamers denoted by light hatched arrows have been identified by mutational analysis only. Heptamers denoted by white arrows have been identified by sequence inspection only. The IHF protein binding sites in the *narK* and the *narG* control regions are denoted by striped boxes. The *aeg-46.5* operon is denoted as "*nap-cyc*" in this figure. Modified from Stewart (1993) and Li et al. (1994).

protein-protein and protein-DNA interactions and their respective roles in transcriptional activation remains to be determined (Walker and DeMoss, 1994).

Thus, the *narG* operon exhibits a complexity that is unusual for prokaryotic transcriptional control regions (Fig. 4). The NarL protein binds to three tandem heptamers in the −200 region and to at least three heptamers in both tandem and inverted orientation in the −80 region. These two regions are separated by an IHF protein binding site, and the Fnr protein binding site is adjacent to the promoter.

narK Gene Control Region

The NarK protein functions in nitrite extrusion (DeMoss and Hsu, 1991; Rowe et al., 1994). The *narK* gene is transcribed divergently from the *narXL* operon, but the *narX* transcriptional initiation site has not yet been reported. Discussion of this control region is therefore from the perspective of *narK* gene expression, although it is not known which features are involved in *narK* gene transcription, in *narXL* operon transcription, or both (Bonnefoy and DeMoss, 1992; Kolesnikow et al., 1992).

The *narK* control region superficially resembles that of the *narG* operon (Fig. 4). Both upstream and downstream NarL heptamers have been identified by footprint analysis, and IHF protein binds (albeit weakly) to the intervening region (Li et al., 1994) (Fig. 4). Indeed, the requirement for IHF protein in *narK* gene transcription is much less stringent than that for the *narG* operon (Kolesnikow et al., 1992; Rabin et al., 1992; Rabin and Stewart, unpublished data). This control region also contains dual Fnr protein binding sites, one centered at −41.5 and a second centered at −79.5; however, the role of this second Fnr protein site is not clear (Bonnefoy and DeMoss, 1992; but see also Li et al., 1994).

fdnG Operon Control Region

The patterns of *narG* and *fdnG* operon expression are indistinguishable in wild-type strains (Berg and Stewart, 1990), but the architecture of their respective control regions differs considerably (Fig. 4). An Fnr protein binding site centered at −42.5 is essential for anaerobic induction, and a second site centered at −97.5 seems to be involved in depressing *fdnG* operon expression in the absence of nitrate (Li and Stewart, 1992). Combined deletion, mutational, and footprint analyses have identified four NarL heptamers spanning the region from −130 to −74 (Li and Stewart, 1992; Li et al., 1994). Each of these heptamers is required for full nitrate induction of *fdnG* operon expression. Finally, the IHF protein plays no role in *fdnG* operon expression, and it does not bind to the control region (Rabin et al., 1992).

nirB Operon Control Region

The *nirB* operon is activated by the NarL protein in response to nitrate and nitrite and by the NarP protein in response to nitrate only (Page et al., 1990; Tyson et al., 1993; Tyson et al., 1994). An Fnr protein binding site, defined by extensive mutational analysis, is centered at position −41.5, whereas a second potential Fnr protein binding site at position −89.5 does not play an obvious role (Jayaraman et al., 1988; Jayaraman et al., 1989; Bell et al., 1990). Mutational analysis has identified an inverted pair of heptamers centered at positions −65 and −74 (Tyson et al., 1993) (Fig. 4). NarL-dependent nitrite induction and NarP-dependent nitrate induction are both abolished by mutational alteration of either the −65 or the −74 heptamers, whereas NarL-dependent nitrate induction is abolished only by alterations in both (Tyson et al., 1993; Tyson et al., 1994). This indicates that the NarL and NarP proteins both bind to the same sites (Tyson et al., 1993; Tyson et al., 1994), although footprint analysis of these heptamers has not been reported.

Taken together, these results suggest that (i) the NarL and NarP proteins both bind to the same sites; (ii) phospho-NarL has a higher affinity for these sites than does phospho-NarP; and (iii) NarL and NarP protein phosphorylation is stimulated more by nitrate than by nitrite (Tyson et al., 1993; Tyson et al., 1994). This last conclusion for NarP differs from that reached by study of other target operons (Rabin and Stewart, 1993) but may reflect different binding affinities or requirements for transcription activation at different control regions.

nrfA Operon Control Region

The nrfA (aeg-93) operon exhibits NarL-dependent nitrate repression and NarL- or NarP-dependent nitrite activation (Rabin and Stewart, 1993; Tyson et al., 1994). Mutational analysis has identified two inverted heptamers, centered at positions −79 and at −70, that are required for NarL- and NarP-dependent nitrite activation and NarP-dependent nitrate activation (Tyson et al., 1994) (Fig. 4). Again, this suggests that both the NarL and the NarP proteins bind to the same sites for transcriptional activation. NarL-dependent nitrate repression requires two other heptamers, centered at positions −50 and −22 (Tyson et al., 1994).

The control of nrfA operon expression is thus imagined to depend on the relative affinities of various protein-DNA interactions, which have yet to be directly measured (Tyson et al., 1994). Nitrite presumably results in the formation of relatively low amounts of phospho-NarL, which is hypothesized to have a strong affinity for binding to the −70 and −79 heptamers. Formation of additional phospho-NarL in response to nitrate would result also in occupancy of the −50 and −22 heptamers, thereby leading to nrfA operon repression. Conversely, phospho-NarP is suggested to significantly bind only to the −70 and −79 heptamers, leading to NarP-dependent activation in response to nitrite (and nitrate, in narL null mutants).

aeg-46.5 Operon Control Region

Activation of aeg-46.5 (nap-cyc) operon expression in response to nitrate or nitrite is wholly dependent on the NarP protein, whereas the NarL protein antagonizes nitrate and nitrite induction to greater and lesser extents, respectively (Choe and Reznikoff, 1991; Rabin and Stewart, 1993). DNA sequence analysis has identified two inverted heptamers centered at positions −40 and −49 (Choe and Reznikoff, 1993) (Fig. 4), and mutational analysis supports the role of these heptamers in NarP-dependent activation. Additional mutational analysis has identified an apparent Fnr protein binding site centered at position −64.5 (Darwin and Stewart, unpublished data). Preliminary footprinting experiments indicate that phospho-NarP and phospho-NarL both protect heptamers −40 and −49 from DNase I attack (Darwin and Stewart, unpublished data). The manner by which the NarL protein inhibits aeg-46.5 operon expression remains unknown.

frdA Operon Control Region

The frdA operon is repressed by the NarL protein in response to nitrate. An Fnr protein binding site is centered at position −46 (Eiglmeier et al., 1989). Footprint analysis has identified at least five NarL heptamers located in the region around the transcriptional initiation site, consistent with the role of the NarL protein as a repressor of frdA operon transcription (Li et al., 1994) (Fig. 4). Mutational analysis of this region has not been reported.

dmsA Operon Control Region

Expression of the dmsA operon is repressed by nitrate (Cotter and Gunsalus, 1989). Potential NarL heptamer sequences, identified by sequence inspection, are present in this control region (Tyson et al., 1993) (Fig. 4), but mutational and footprint analyses of their functions remain to be reported.

AFTERWORD

Nitrate and nitrite regulation provide a rich and unexpected complexity that will challenge experimentalists for years to come. Dual homologous sensors respond to nitrate and nitrite in similar but distinct ways, and each communicates with dual homologous response regulators in a likewise manner. The DNA targets of response regulator action present unmatched diversity in number, organization, and spacing.

Why has such a complicated regulatory circuit evolved for response to two small inorganic molecules? In our view, at least two physiological considerations justify a complex regulatory network.

First, use of a favored electron acceptor (nitrate) leads to the production of a less-favored, albeit relatively efficient, electron acceptor (nitrite). Enterobacteria growing anaerobically in the presence of nitrate thus experience a dynamic equilibrium between nitrate and nitrite concentrations. As nitrite is formed at the expense of nitrate, the organsim must respond by synthesizing appropriate levels of the relevant enzymes. Respiratory nitrite reductase (*nrfA* operon product) is used only when nitrate concentrations are low and nitrite concentrations are high, whereas NADH-nitrite reductase (*nirB* operon product) may serve to detoxify some of the nitrite produced by nitrate respiration (Page et al., 1990). Formate dehydrogenase-N (*fdnG* operon product) donates electrons to both respiratory nitrate and nitrite reductases (Darwin et al., 1993b). Thus, diverse targets of nitrate and nitrite induction have different requirements for expression at different ratios of the two substrates/effectors.

Second, alternate respiratory enzymes, such as fumarate reductase (*frdA* operon product), may play a significant physiological role in environments that contain relatively low amounts of nitrate. We believe that this is reflected in the apparent hierarchy of nitrate regulation (*narG* > *fdnG* > *frdA*) as revealed by analysis of various mutants growing with relatively high nitrate concentrations (see above).

Most studies have used batch cultures grown with excess nitrate or nitrite. A full physiological appreciation for the complexity of Nar-mediated regulation will require more subtle experimental analysis.

ACKNOWLEDGMENTS

We are grateful to Sandy Parkinson and Tom Silhavy for helpful discussions and to our colleagues in the laboratory for their continued interest and support. We appreciate the open exchange of ideas and information with Steve Busby, Jeff Cole, Jack DeMoss, and Gary Sawers. Nada Bsat and Andrew Darwin provided many thoughtful comments on the manuscript.

Original research reported in Table 3 was supported by Public Health Service grant GM36877 from the National Institute of General Medical Sciences (V.S.) and by a National Science Foundation graduate fellowship (R.S.R.).

REFERENCES

Ames, P., and J. S. Parkinson. 1988. Transmembrane signaling by bacterial chemoreceptors: *E. coli* transducers with locked signal output. *Cell* **55:**817–826.

Atkinson, M. R., and A. J. Ninfa. 1993. Mutational analysis of the bacterial signal transducing protein kinase/phosphatase nitrogen regulator II (NR$_{II}$ or NtrB). *J. Bacteriol.* **175:**7016–7023.

Baumgartner, J. W., C. Kim, R. E. Brissette, M. Inouye, C. Park, and G. L. Hazelbauer. 1994. Transmembrane signalling by a hybrid protein: communication from the domain of chemoreceptor Trg that recognizes sugar-binding proteins to the kinase/phosphatase domain of osmosensor EnvZ. *J. Bacteriol.* **176:**1157–1163.

Bell, A. I., J. A. Cole, and S. J. W. Busby. 1990. Molecular genetic analysis of an Fnr-dependent anaerobically inducible *Escherichia coli* promoter. *Mol. Microbiol.* **4:**1753–1763.

Berg, B. L., and V. Stewart. 1990. Structural genes for nitrate-inducible formate dehydrogenase in *Escherichia coli* K-12. *Genetics* **125:**691–702.

Bonnefoy, V., and J. A. DeMoss. 1992. Identification of functional *cis*-acting sequences involved in regulation of *narK* gene expression in *Escherichia coli*. *Mol. Microbiol.* **6:**3595–3602.

Chen, Y.-M., and E. C. C. Lin. 1991. Regulation of the *adhE* gene, which encodes ethanol dehydrogenase in *Escherichia coli*. *J. Bacteriol.* **173:**8009–8013.

Chiang, R. C., R. Cavicchioli, and R. P. Gunsalus. 1992. Identification and characterization of *narQ*, a second nitrate sensor for nitrate-dependent gene regulation in *Escherichia coli*. *Mol. Microbiol.* **6:**1913–1923.

Choe, M., and W. S. Reznikoff. 1991. Anaerobically expressed *Escherichia coli* genes identified by operon fusion techniques. *J. Bacteriol.* **173:**6139–6146.

Choe, M., and W. S. Reznikoff. 1993. Identification of the regulatory sequence of anaerobically expressed locus *aeg-46.5. J. Bacteriol.* **175:**1165–1172.

Collins, L. A., S. M. Egan, and V. Stewart. 1992. Mutational analysis reveals functional similarity between NARX, a nitrate sensor in *Escherichia coli* K-12, and the methyl-accepting chemotaxis proteins. *J. Bacteriol.* **174:**3667–3675.

Cotter, P. A., and R. P. Gunsalus. 1989. Oxygen, nitrate, and molybdenum regulation of *dmsABC* gene expression in *Escherichia coli*. *J. Bacteriol.* **171**:3817–3823.

Darwin, A., H. Hussain, L. Griffiths, J. Grove, Y. Sambongi, S. Busby, and J. Cole. 1993a. Regulation and sequence of the structural gene for cytochrome c_{552} from *Escherichia coli*: not a hexahaem but a 50 kDa tetrahaem nitrite reductase. *Mol. Microbiol.* **9**:1255–1265.

Darwin, A., and V. Stewart. Unpublished data.

Darwin, A., P. Tormay, L. Page, L. Griffiths, and J. Cole. 1993b. Identification of the formate dehydrogenases and genetic determinants of formate-dependent nitrite reduction by *Escherichia coli* K12. *J. Gen. Microbiol.* **139**:1829–1840.

DeMoss, J. A., and P.-Y. Hsu. 1991. NarK enhances nitrate uptake and nitrite excretion in *Escherichia coli*. *J. Bacteriol.* **173**:3303–3310.

Dong, X. R., S. F. Li, and J. A. DeMoss. 1992. Upstream sequence elements required for NarL-mediated activation of transcription from the *narGHJI* promoter of *Escherichia coli*. *J. Biol. Chem.* **267**:14122–14128.

Egan, S. M., and V. Stewart. 1990. Nitrate regulation of anaerobic respiratory gene expression in *narX* deletion mutants of *Escherichia coli* K-12. *J. Bacteriol.* **172**:5020–5029.

Egan, S. M., and V. Stewart. 1991. Mutational analysis of nitrate regulatory gene *narL* in *Escherichia coli* K-12. *J. Bacteriol.* **173**:4424–4432.

Eiglmeier, K., N. Honoré, S. Iuchi, E. C. C. Lin, and S. T. Cole. 1989. Molecular genetic analysis of FNR-dependent promoters. *Mol. Microbiol.* **3**:869–878.

Goldman, B. S., J. T. Lin, and V. Stewart. 1994. Identification and structure of the *nasR* gene encoding a nitrate- and nitrite-responsive positive regulator of *nasFEDCBA* (nitrate assimilation) operon expression in *Klebsiella pneumoniae* M5al. *J. Bacteriol.* **176**:5077–5085.

Gunsalus, R. P. 1992. Control of electron flow in *Escherichia coli*: coordinated transcription of respiratory pathway genes. *J. Bacteriol.* **174**:7069–7074.

Hoover, T. R., E. Santero, S. Porter, and S. Kustu. 1990. The integration host factor stimulates interaction of RNA polymerase with NIFA, the transcriptional activator for nitrogen fixation operons. *Cell* **63**:11–22.

Iuchi, S., and E. C. C. Lin. 1987. The *narL* gene product activates the nitrate reductase operon and represses the fumarate reductase and trimethylamine N-oxide reductase operons in *Escherichia coli*. *Proc. Natl. Acad. Sci. USA* **84**:3901–3905.

Iuchi, S., and E. C. C. Lin. 1993. Adaptation of *Escherichia coli* to redox environments by gene expression. *Mol. Microbiol.* **9**:9–15.

Jayaraman, P.-S., J. A. Cole, and S. J. W. Busby. 1989. Mutational analysis of the nucleotide sequence at the FNR-dependent *nirB* promoter in *Escherichia coli*. *Nucleic Acids Res.* **17**:135–146.

Jayaraman, P.-S., K. L. Gaston, J. A. Cole, and S. J. W. Busby. 1988. The *nirB* promoter of *Escherichia coli*: location of nucleotide sequences essential for regulation by oxygen, the FNR protein, and nitrite. *Mol. Microbiol.* **2**:527–530.

Kahn, D., and G. Ditta. 1991. Modular structure of FixJ: homology of the transcriptional activator domain with the −35 binding domain of sigma factors. *Mol. Microbiol.* **5**:987–997.

Kalman, L. V., and R. P. Gunsalus. 1988. The *frdR* gene of *Escherichia coli* globally regulates several operons involved in anaerobic growth in response to nitrate. *J. Bacteriol.* **170**:623–629.

Kalman, L. V., and R. P. Gunsalus. 1990. Nitrate-independent and molybdenum-independent signal transduction mutations in *narX* that alter regulation of anaerobic respiratory genes in *Escherichia coli*. *J. Bacteriol.* **172**:7049–7056.

Kolesnikow, T., I. Schröder, and R. P. Gunsalus. 1992. Regulation of *narK* gene expression in *Escherichia coli* in response to anaerobiosis, nitrate, iron, and molybdenum. *J. Bacteriol.* **174**:7104–7111.

Leonardo, M. R., P. R. Cunningham, and D. P. Clark. 1993. Anaerobic regulation of the *adhE* gene, encoding the fermentative alcohol dehydrogenase of *Escherichia coli*. *J. Bacteriol.* **175**:870–878.

Li, J., S. Kustu, and V. Stewart. 1994. In vitro interaction of nitrate-responsive regulatory protein NarL with DNA target sequences in the *fdnG*, *narG*, *narK*, and *frdA* operon control regions of *Escherichia coli* K-12. *J. Mol. Biol.* **241**:150–165.

Li, J., and V. Stewart. 1992. Localization of upstream sequences required for nitrate and anaerobic induction of formate dehydrogenase-N operon expression in *Escherichia coli* K-12. *J. Bacteriol.* **174**:4935–4942.

Li, S.-F., and J. A. DeMoss. 1988. Location of sequences in the *nar* promoter of *Escherichia coli* required for regulation by Fnr and NarL. *J. Biol. Chem.* **263**:13700–13705.

Lin, J. T., B. S. Goldman, and V. Stewart. 1994. The *nasFEDCBA* operon for nitrate and nitrite assimilation in *Klebsiella pneumoniae* M5al. *J. Bacteriol.* **176**:2551–2559.

Manoil, C., J. J. Mekalanos, and J. Beckwith. 1990. Alkaline phosphatase fusions: sensors of subcellular location. *J. Bacteriol.* **172**:515–518.

Miller, J. B., D. J. Scott, and N. K. Amy. 1987. Molybdenum-sensitive transcriptional regulation of the *chlD* locus of *Escherichia coli*. *J. Bacteriol.* **169**:1853–1860.

Nohno, T., S. Noji, S. Taniguchi, and T. Saito. 1989. The *narX* and *narL* genes encoding the nitrate-sensing regulators of *Escherichia coli* are homologous to a family of prokaryotic two-component regulatory genes. *Nucleic Acids Res.* **17**:2947–2957.

Page, L., L. Griffiths, and J. A. Cole. 1990. Different physiological roles of two independent pathways for nitrite reduction to ammonia in enteric bacteria. *Arch. Microbiol.* **154**:349–354.

Parkinson, J. S., and E. C. Kofoid. 1992. Communication modules in bacterial signaling proteins. *Annu. Rev. Genet.* **26**:71–112.

Rabin, R. S., L. A. Collins, and V. Stewart. 1992. In vivo requirement of integration host factor for nitrate reductase (*nar*) operon expression in *Escherichia coli* K-12. *Proc. Natl. Acad. Sci. USA* **89**:8701–8705.

Rabin, R. S., and V. Stewart. 1992. Either of two functionally redundant sensor proteins, NarX and NarQ, is sufficient for nitrate regulation in *Escherichia coli* K-12. *Proc. Natl. Acad. Sci. USA* **89**: 8419–8423.

Rabin, R. S., and V. Stewart. 1993. Dual response regulators (NarL and NarP) interact with dual sensors (NarX and NarQ) to control nitrate- and nitrite-regulated gene expression in *Escherichia coli* K-12. *J. Bacteriol.* **175**:3259–3268.

Rabin, R. S., and V. Stewart. Unpublished data.

Richterich, P., N. Lakey, G. Gryan, L. Jaehn, L. Mintz, K. Robison, and G. M. Church. 1993. Unpublished DNA sequence. GenBank accession number U00008.

Rowe, J. J., T. Ubbink-Kok, D. Molenaar, W. N. Konings, and A. J. M. Driessen. 1994. NarK is a nitrite-extrusion system involved in anaerobic nitrate respiration by *Escherichia coli. Mol. Microbiol.* **12**:579–586.

Russo, F. D., J. M. Slauch, and T. J. Silhavy. 1993. Mutations that affect separate functions of OmpR the phosphorylated regulator of porin transcription in *Escherichia coli. J. Mol. Biol.* **231**:261–273.

Sawers, G. Personal communication.

Sawers, G., and A. Böck. 1988. Anaerobic regulation of pyruvate-formate lyase from *Escherichia coli* K-12. *J. Bacteriol.* **170**:5330–5336.

Schröder, I., S. Darie, and R. P. Gunsalus. 1993. Activation of the *Escherichia coli* nitrate reductase (*narGHJI*) operon by NarL and Fnr requires integration host factor. *J. Biol. Chem.* **268**:771–774.

Schröder, I., C. Wolin, R. Cavicchioli, and R. P. Gunsalus. 1994. Phosphorylation and dephosphorylation of the NarQ, NarX, and NarL proteins of the nitrate-dependent two-component regulatory system of *Escherichia coli. J. Bacteriol.* **176**:4985–4992.

Siddiqui, R. A., U. Warnecke-Eberz, A. Hengsberger, B. Schneider, S. Kostka, and B. Friedrich. 1993. Structure and function of a periplasmic nitrate reductase in *Alcaligines eutrophus* H16. *J. Bacteriol.* **175**:5867–5876.

Spiro, S., and J. R. Guest. 1990. FNR and its role in oxygen-regulated gene expression in *Escherichia coli. FEMS Microbiol. Rev.* **75**:399–428.

Stewart, V. 1982. Requirement of Fnr and NarL functions for nitrate reductase expression in *Escherichia coli* K-12. *J. Bacteriol.* **151**:1320–1325.

Stewart, V. 1988. Nitrate respiration in relation to facultative metabolism in enterobacteria. *Microbiol. Rev.* **52**:190–232.

Stewart, V. 1993. Nitrate regulation of anaerobic respiratory gene expression in *Escherichia coli. Mol. Microbiol.* **9**:425–434.

Stewart, V., and B. L. Berg. 1988. Influence of *nar* (nitrate reductase) genes on nitrate inhibition of formate-hydrogen lyase and fumarate reductase synthesis in *Escherichia coli* K-12. *J. Bacteriol.* **170**:4437–4444.

Stewart, V., and J. Parales. 1988. Identification and expression of genes *narL* and *narX* of the *nar* (nitrate reductase) locus in *Escherichia coli* K-12. *J. Bacteriol.* **170**:1589–1597.

Stewart, V., J. Parales, and S. M. Merkel. 1989. Structure of genes *narL* and *narX* of the *nar* (nitrate reductase) locus in *Escherichia coli* K-12. *J. Bacteriol.* **171**:2229–2234.

Stock, J. B., A. J. Ninfa, and A. M. Stock. 1989. Protein phosphorylation and regulation of adaptive responses in bacteria. *Microbiol. Rev.* **53**:450–490.

Stout, V., A. Torres-Cabassa, M. R. Maurizi, D. Gutnick, and S. Gottesman. 1991. RcsA, an unstable positive regulator of capsular polysaccharide biosynthesis. *J. Bacteriol.* **173**:1738–1747.

Suzuki, M. 1993. Common features in DNA recognition helices of eukaryotic transcription factors. *EMBO J.* **12**:3221–3226.

Tyson, K. L., A. I. Bell, J. A. Cole, and S. J. W. Busby. 1993. Definition of nitrite and nitrate response elements at the anaerobically inducible *Escherichia coli nirB* promoter: interactions between FNR and NarL. *Mol. Microbiol.* **7**:151–157.

Tyson, K. L., J. A. Cole, and S. J. W. Busby. 1994. Nitrite and nitrate regulation at the promoters of two *Escherichia coli* operons encoding nitrite reductase: identification of common target heptamers for both NarP- and NarL-dependent regulation. *Mol. Microbiol.* **13**:1045–1055.

Walker, M. S., and J. A. DeMoss. 1991. Promoter sequence requirements for Fnr-dependent activation of transcription of the *narGHJI* operon. *Mol. Microbiol.* **5**:353–360.

Walker, M. S., and J. A. DeMoss. 1992. Role of alternative promoter elements in transcription from the *nar* promoter of *Escherichia coli. J. Bacteriol.* **174:** 1119–1123.

Walker, M. S., and J. A. DeMoss. 1993. Phosphorylation and dephosphorylation catalyzed in vitro by purified components of the nitrate sensing system, NarX and NarL. *J. Biol. Chem.* **268:**8391–8393.

Walker, M. S., and J. A. DeMoss. 1994. NarL-phosphate must bind to multiple upstream sites to activate transcription from the *narG* promoter of *Escherichia coli. Mol. Microbiol.* **14:**633–641.

Williams, S. B., and V. Stewart. Unpublished data.

Regulation of Capsule Synthesis: Modification of the Two-Component Paradigm by an Accessory Unstable Regulator

Susan Gottesman

15

The most obvious phenotype of *Escherichia coli* mutant in the ATP-dependent Lon protease is the appearance of mucoid colonies, due to overproduction of colanic acid capsular polysaccharide. Activation of transcription from genes encoding proteins of the capsule synthesis pathway depends on a two-component regulator, encoded by *rcsC* and *rcsB,* acting in concert with the product of the *rcsA* gene, an unstable and limiting regulatory protein that is a substrate for the Lon protease (reviewed in Gottesman and Stout, 1991). Thus, when Lon is absent, the increased accumulation of RcsA leads to overproduction of the capsule. The role of RcsA as an accessory factor for RcsB, described below, provides a unique variation on the two-component paradigm.

GENERAL DESCRIPTION OF THE Rcs REGULATORY NETWORK

Figure 1 shows a schematic for the components of the capsule regulatory network. Capsule synthesis in *E. coli* K-12 under laboratory growth conditions is generally low, and transcription of the *cps* genes, necessary for capsule synthesis, is correspondingly low (*cps-lac* fusions express 1 to 5 U of β-galactosidase). Dramatic increases in *cps* expression correlate with increased capsule synthesis and generally can be ascribed to either of two control points for the regulatory system. Thus, some increases in capsule synthesis in mutants or under particular growth conditions are consistent with an increase in activation of RcsB, the regulator of the RcsC/RcsB two-component system (see below; Fig. 1 and Table 1). A second set of mutants appears to increase the level (or in some cases, the activity) of RcsA (see below; Fig. 1 and Table 1). Capsule synthesis under all these conditions is completely dependent on RcsB; RcsA appears to act as an accessory factor that allows modulation of RcsB activity (Brill et al., 1988). Environmental signals that turn on capsule synthesis to the same extent as the mutations described here have not yet been found.

END PRODUCT: FUNCTION OF COLANIC ACID CAPSULE

The capsular polysaccharide produced when the RcsA/B/C system is active in *E. coli* K-12 is colanic acid, a polymer of repeating units of glucose, galactose, fucose, and glucuronic acid (Markovitz, 1977). The primary cluster of *cps* genes maps near the *rfb* (encoding proteins for O antigen LPS synthesis) and *his* genes, at 41 min (Trisler and Gottesman, 1984). Colanic

Susan Gottesman, Laboratory of Molecular Biology, National Cancer Institute, Bethesda, Maryland 20892-4255.

Two-Component Signal Transduction, Edited by James A. Hoch and Thomas J. Silhavy,
© 1995 American Society for Microbiology, Washington, DC 20005

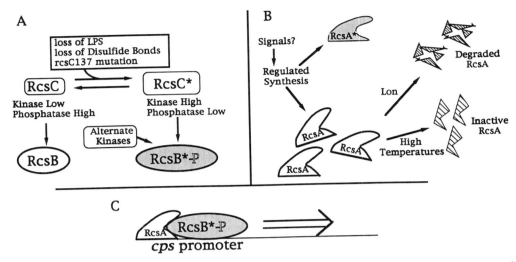

FIGURE 1 Points of control for capsular polysaccharide synthesis. (A) Control pathway I: increased RcsB phosphorylation. In response to environmental signals or to changes in the cell surface as indicated in the box, the membrane protein sensor RcsC is activated, probably by a shift from phosphatase activity to kinase activity. The target of RcsC, the response regulator RcsB, is believed to act as a transcriptional activator when phosphorylated (shaded oval). High levels of activated RcsB allow *cps* transcription even when RcsA levels are low. Alternate kinase pathways (possibly involving RcsF) may also lead to RcsB activation. (See text for references and more detail.) (B) Control pathway II: increased accumulation of active RcsA. Increased capsule synthesis occurs when the amounts of active RcsA are increased. This can occur either when synthesis is increased, when degradation by Lon is blocked, or when RcsA is mutated (RcsA★) to improve its interaction with RcsB (shaded figure). RcsA activity is blocked at high temperatures. (C) Activation of *cps* gene transcription. In the presence of activated RcsB and RcsA, *cps* transcription occurs at high levels. High RcsB-P or high RcsA is sufficient to give high-level transcription. RcsA appears to be an auxiliary factor that stimulates RcsB activity. In this possible model for its action, we imagine cobinding of RcsA and RcsB to the promoter.

TABLE 1 Genes affecting capsular polysaccharide synthesis in *E. coli*

Gene	Location	Probable function	Mutant phenotypes
rcsA	43.5	Positive regulator; accessory for RcsB; unstable; target of Lon protease	*rcsA* null: low *cps* expression; *rcsA★*: high *cps* expression
rcsB	49.8	Positive regulator (target for phosphorylation)	*rcsB* null: low *cps* expression
rcsC	49.9	Sensor kinase/phosphatase; may sense status of surface	*rcsC* null: no phenotype alone; blocks *cps* production in *rfa* mutants, on drying; *rcsC137* and others: high, temp.-independent *cps* expression
rcsF	4.9	? Alternative RcsB phosphorylation pathway	Multicopy: high *cps* expression; Mutant: lower *cps* expression
lon	10.0	Protease that degrades RcsA	*lon* null: high *cps* expression
rfaPG	81	Modulation of RcsC kinase activity	High *cps* expression
cpsA-E	45	Structural genes for capsule synthesis; necessary for high-level capsule synthesis	Necessary for capsule synthesis; no regulatory effects seen
cpsF	90		

acid can also be synthesized by several other enteric bacteria and some, but not all, clinical isolates of *E. coli*. Although some of these bacteria primarily synthesize other capsular polysaccharides, overproduction of RcsA or RcsB can lead to increased synthesis of colanic acid (Jayaratne et al., 1993; Russo and Singh, 1993).

Also, some bacterial strains not capable of synthesizing colanic acid instead synthesize one of a variety of related polysaccharides, also under control of the RcsA/RcsB system (Keenleyside et al., 1992; Keenleyside et al., 1993). C. Whitfield and coworkers recently demonstrated that all tested capsules of the IA group, composed of repeating units lacking amino sugars, are subject to RcsA/RcsB regulation (Jayaratne et al., 1993). In the cases that have been examined, group IA capsules are encoded by genes mapping near the *his* region (which are presumably allelic with *cps*) and are expressed preferentially at low temperatures. In cells expressing group IB capsules (containing amino sugars), overproduction of the RcsA and RcsB regulators leads to synthesis of colanic acid (Jayaratne et al., 1993), suggesting that these strains contain genes for the synthesis of group IB capsules elsewhere, in addition to the colanic acid synthesis *cps* genes at min 41. Class II capsules, which contain sialic acid and are frequently found in *E. coli* pathogens, are also encoded elsewhere and are expressed at higher temperatures (37°C); expression of these capsules is apparently not dependent on RcsA and B (Russo and Singh, 1993; Keenleyside et al., 1992).

The regulation of type I capsule synthesis by RcsA and RcsB extends to other gram-negative organisms as well. In plant pathogens *Erwinia stewartii* and *Erwinia amylovora,* synthesis of capsules necessary for pathogenesis is also under RcsA/B/C regulation (Torres-Cabassa et al., 1987; Bernhard et al., 1990; Chatterjee et al., 1990; Coleman et al., 1990; Coplin et al., 1990). RcsA or RcsB from *Erwinia* species is able to function in *E. coli* and vice versa, suggesting that the targets as well as the regulatory proteins have been well conserved. In *Klebsiella*

species, a type I capsule with a role in pathogenesis is under RcsA/B regulation (McCallum and Whitfield, 1991; Wacharotayankun et al., 1992), and the Vi antigen of *Citrobacter freundii* appears to be as well (Houng et al., 1992). In at least some cases (*E. coli, Erwinia*), the temperature regulation of capsule synthesis (expressed only at temperatures less than 37°C) suggests these capsules have roles in non-mammalian environments. However, for *Klebsiella* species, synthesis is significant even at 37°C, suggesting such systems may also be important for mammalian hosts.

The precise function of the colanic acid capsule is not known. Recent data indicate a role for such capsules in protecting cells from dehydration (Ophir and Gutnick, 1994), and the modest induction of capsule synthesis on dehydration supports a functional and regulatory link. The absence of other known environmental signals that lead to significant induction levels leaves us without further clues to possible function.

RcsB EFFECTOR IS ESSENTIAL FOR CAPSULE SYNTHESIS

RcsB was genetically defined by the isolation of mutations that block expression of *cps* genes in *lon* mutant hosts (Gottesman et al., 1985). Null mutations in this locus cannot be overcome by overproducing RcsA, the second positive regulator of the system, whereas *rcsA* null mutations can be overcome by RcsB overproduction, suggesting that RcsB is the primary positive regulator of *cps* gene transcription and RcsA aids RcsB activity (Brill et al., 1988; see below).

The sequence of RcsB shows reasonable homology in the N terminus to other effectors of two-component systems. The C terminus shows similarity to a family of proteins including LuxR, FixJ, and MalT and may include a helix-turn-helix motif; recent experiments with these systems suggest that residues in the region of homology at the C terminus encode DNA binding and positive activation (Choi and Greenberg, 1991; Kahn and Ditta, 1991; Vidal-Ingigliardi et al., 1993) (Fig. 2).

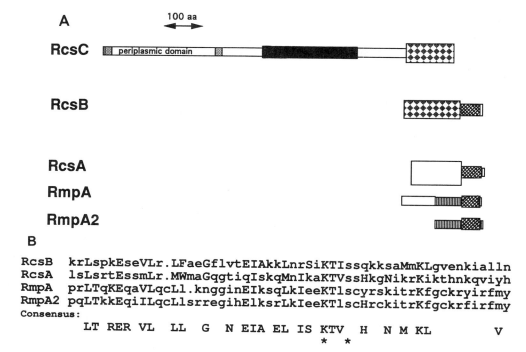

FIGURE 2 Conserved sequence elements in capsule regulatory proteins. (A) Domains of capsule regulators. The domains are shown to scale. The shaded regions bordering the periplasmic domain in RcsC are the hydrophobic regions believed to span the inner membrane. The kinase domain (solid) and effector domains (checkered) are defined as in Stout and Gottesman (1990), and the DNA binding/activation domain (cross-hatched) as in Stout et al. (1991). The RmpA domain (hatched) is the region of high homology between RmpA and RmpA2, in addition to the homology seen in the DNA binding/activation domain (Nassif et al., 1989; Wacharotayankun et al., 1993). (B) Alignment of DNA binding/activation domains from RcsB, RcsA, and RmpA proteins. The alignment between RcsA and RcsB is as in Stout et al. (1991), as is the consensus sequence. RmpA and RmpA2 were aligned with members of the conserved family using Blast (Altschul et al., 1990). The two positions marked with an asterisk are those expected to encode DNA binding site recognition residues, as predicted from mutagenesis studies on similar proteins and the position within the DNA binding helix of the putative helix-turn-helix (Choi and Greenberg, 1991; Kahn and Ditta, 1991; Vidal-Ingigliardi et al., 1993).

Site-directed mutation of the conserved aspartate (at the position that is phosphorylated in other two-component regulators) to asparagine leads to an inactive protein, as expected if phosphorylation of this site is necessary for activity (Gottesman et al., 1994). The primary candidate for a kinase donating phosphate to RcsB is RcsC, a membrane protein with homology to sensors of two-component systems, encoded by a gene right next to *rcsB* but separately transcribed (Brill et al., 1988; Stout and Gottesman, 1990) (see below). Also, alternative sensors can activate capsule synthesis when overproduced (Gervais and Drapeau, 1992), presumably by phosphorylating RcsB, and a gene, *rcsF*, has been identified, which also appears to play a role in activation (Gervais and Drapeau, 1992).

C-terminal truncated forms of RcsB, overexpressed from plasmids, turn off capsule synthesis, and this can be overcome by overproducing RcsA, suggesting that RcsB interacts directly with RcsA (Brill et al., 1988; Stout et al., 1991). The ability of RcsB to partially stabilize RcsA also supports this model (Stout et al., 1991).

Recent experiments by Drapeau and co-workers (Gervais and Drapeau, 1992; Gervais et al., 1992) suggest that RcsB has a role in regulation of *ftsZ*, an essential cell division gene encoding a critical septum protein. They identified multicopy clones carrying *rcsB* among those that suppressed the inability of *ftsZ* mutants to grow on medium without salt; this effect is independent of RcsA. Increasing the amount of RcsB led to an increase in expression of an *ftsAZ-lacZ* fusion (Gervais and Drapeau, 1992). It is not clear how directly this effect occurs, because the site of RcsB action has not been determined.

PATHWAY I FOR CAPSULE ACTIVATION: INCREASED PHOSPHORYLATION OF RcsB

When high-level expression of *cps-lac* fusions is selected at elevated temperatures, the primary location of the resulting mutations (which increase capsule synthesis and *cps* transcription at all temperatures) is in *rcsC*, the gene immediately clockwise to *rcsB* (Gottesman et al., 1985). These recessive mutations, exemplified by *rcsC137*, appear to increase the activation of RcsB, presumably by increasing phosphorylation.

RcsC is a 104,000-D protein with a small periplasmic domain and a large C-terminal cytoplasmic domain, as judged both by sequence analysis and Tn*phoA* fusions (Stout and Gottesman, 1990). The cytoplasmic domain shows a region of homology to protein kinases of two-component regulators and a C-terminal additional domain with homology to the receptor domain of such regulators (Fig. 2). This puts RcsC in a class similar to several other sensors with a modulating receptor domain (Parkinson and Kofoid, 1992).

RcsC appears to act both positively and negatively to regulate RcsB activity. Although null mutants in *rcsC* have little or no effect on the expression of *cps* genes in wild-type *lon*+ or *lon* mutant cells (Brill et al., 1988; Gervais et al., 1992), recent evidence of secondary conditions that lead to *cps* activation in an *rcsC*-dependent fashion supports a positive role for the kinase domain of *rcsC*. Mutants in *lps* synthesis or in disulfide bond formation in the periplasm (*dsbA* or *dsbB* mutants) increase capsule synthesis; this increase is abolished in *rcsC* null mutants (Parker et al., 1992; Bardwell et al., 1991). Mutations in *rcsD*, previously identified as a negative regulator of *cps* expression, have proved to be in *dsbB* and to be dependent on RcsC for *cps* stimulation (Klopotowski, personal communication). Also, an increase in capsule synthesis on dehydration observed in *lon* mutants is mostly abolished in an *rcsC* mutant (Ophir and Gutnick, 1994). These observations strongly support a positive action of RcsC in the stimulation of RcsB activity under particular conditions and are inconsistent with a solely negative role for RcsC.

There is also good genetic evidence to support a negative activity of RcsC. Mutations in *rcsC* that increase capsule synthesis in *lon*+ cells, such as *rcsC137*, are recessive to wild type, suggesting the loss of phosphatase activity rather than the acquisition of a new (increased kinase) activity, although we cannot rule out possible effects of multimerization of the mutant and wild-type protein on expression of a possible increased kinase activity. The apparent high level of RcsB activation in *rcsC137* mutants, compared with the absence of activation seen when *rcsC* is inactivated by transposon insertion, suggests that either the wild-type RcsC protein is responsible for a high level of phosphatase activity or the *rcsC137* mutant both abolishes phosphatase and increases kinase activity simultaneously, both of which are recessive to wild-type protein. Overproduction of RcsC decreases capsule synthesis modestly under some conditions (Brill et al., 1988; Gervais and Drapeau, 1992). Finally, mutations in *rcsC* do not tolerate transformation with *rcsB*+ plasmids in *cps* mutant hosts (Brill et al., 1988), presumably because these cells lack the cognate phosphatase activity necessary to dephosphorylate overproduced RcsB, which has been subject to activation by crosstalk; accumulation of high levels of intermediates in *cps* mutants is apparently lethal.

The C-terminal receptor domain in RcsC carries a conserved aspartate, the putative site for phosphorylation, but no DNA binding domain. Little has been done to investigate the function of this domain in RcsC; C-terminal truncations of this region impair but do not totally abolish the ability of the overproduced protein to complement an *rcsC137* mutant (Brill et al., 1988), consistent with a primarily negative role for this domain.

RcsF has been identified as a second function contributing to RcsB activation (Gervais and Drapeau, 1992). Overproduction of RcsF increases capsule synthesis twofold, and mutations in *rcsF* decrease capsule synthesis two- to threefold. The RcsF gene does not itself show homology to other protein kinases and may act indirectly to activate RcsB.

PATHWAY II FOR STIMULATION OF RcsB ACTIVITY: RcsA ACCUMULATION

cps genes can be expressed at high levels under conditions that do not appear to require activation of a protein kinase; in these cases, a second pathway for stimulation of RcsB activity seems to operate via increased levels of RcsA and the accessory activity of RcsA. It seems likely that the basal level of phosphorylation of RcsB is sufficient for activation under these conditions.

RcsA is apparently normally limiting for capsule synthesis, because increasing its accumulation (by eliminating Lon, the protease that normally rapidly degrades it, or by overproduction from a plasmid) or improving its ability to act (by mutation to a form that appears to interact more tightly with RcsB, RcsA★) increases *cps* expression in a manner that is absolutely dependent on RcsB (Brill et al., 1988; Torres-Cabassa and Gottesman, 1987) (see Fig. 1). In the absence of *rcsA*, *cps* expression can be seen, although the levels of expression are about 10% of those seen in the presence of RcsA (Brill et al., 1988).

Under wild-type growth conditions, very little RcsA accumulates in cells, partially because synthesis is limited and to a large extent

because RcsA is rapidly degraded (half-life < 5 min) by the Lon protease (Torres-Cabassa and Gottesman, 1987). Any situation that causes increased RcsA synthesis (increased copy number or mutations in genes that affect RcsA synthesis; see below) increases capsule synthesis significantly. Silverman and Simon (1973), mapping genes of flagella synthesis that map near *rcsA*, noticed that cells carrying an F′ for the region could be scored by their increased mucoidy, presumably due to overproduction of RcsA due to presence of an extra copy of *rcsA* on the F′.

RcsA, like RcsB, was identified by mutations that abolished *cps* gene expression in a *lon* mutant (Gottesman et al., 1985). The *rcsA* gene encodes a 23,500-molecular weight protein; the N terminus shows little similarity to other proteins, but like RcsB, the C terminus is a member of the LuxR/FixJ family and contains an apparent helix-turn-helix (Stout et al., 1991) (Fig. 2). Because the requirement for RcsA can be partially bypassed by overproducing RcsB (Brill et al., 1988) and varies from extremely low (in a *lon*⁺ cell in which RcsB is activated by mutations in RcsC) to relatively high (in a *lon* mutant in which no other pathway of RcsB activation is present), RcsA appears to act as an accessory factor capable of stimulating RcsB activity. The similarity of the C-terminal domains of RcsA and RcsB, which include the helix-turn-helix, as well as evidence that RcsA and RcsB directly interact, leads to a model in which RcsA and RcsB bind together to sites in the *cps* promoter to stimulate transcription (Gottesman and Stout, 1991) (see Fig. 1).

Theoretically, RcsA might act in one or more of the following ways: (i) RcsA may allow RcsB to activate transcription in the absence of phosphorylation; (ii) RcsA might block dephosphorylation or directly stimulate RcsB phosphorylation; (iii) RcsA may increase the activity of activated (phosphorylated) RcsB, by increasing binding or activation of transcription for bound RcsB.

The lack of activity of RcsB mutants in which the conserved aspartate is changed to asparagine makes the first of these mechanisms less likely, although we cannot rule out other roles for this aspartate residue for RcsB activity (Gottesman et al., 1994). If phosphorylation is essential for RcsB activity (mechanisms ii or iii), the basal level of RcsB must be derived from sources other than RcsC, because *rcsC* null mutations still show good activity when RcsA is abundant. Also, one would expect that if mechanism ii were true, eliminating the presumed cognate phosphatase, RcsC, would mimic the effect of RcsA by decreasing dephosphorylation. Because *rcsC* null mutants have little or no effect on capsule synthesis levels either when RcsA is limiting (*lon*+ cells) or when it is abundant (*lon*− cells), I do not favor mechanism ii. Mechanism iii (see Fig. 1) fits our current understanding of this system best.

The activity of RcsA is also regulated. Although there is no evidence (none has been sought) for phosphorylation of RcsA, the temperature sensitivity of capsule synthesis (capsule synthesis is very low at temperatures of 37°C and above) is apparently best explained by temperature sensitivity of RcsA activity. Conditions that require only low or no RcsA (abundance of activated RcsB) escape the temperature regulation. Also, the chaperone proteins DnaJ and DnaK are necessary for RcsA activity even at low temperatures (Jubete et al., unpublished data). RcsA protein may not be capable of maintaining itself in an active and soluble form on its own; it requires interaction with chaperones to maintain activity, which is best done at low temperatures. At higher temperatures, chaperones may preferentially interact with other cellular proteins, interfering with capsule activation. Both the *dnaJ*/*dnaK* requirement and the temperature sensitivity of capsule synthesis are bypassed by mutations such as *rcsC137*, which are believed to increase RcsB phosphorylation (Jubete et al., unpublished data). Because under these conditions there is still a requirement for *rcsA*+ for maximal capsule synthesis, either a small amount of still active RcsA will suffice or interaction with

the activated form of RcsB is sufficient to maintain RcsA in an active form even at high temperatures.

RmpA, an Alternative Activator of Capsule Synthesis

In *Klebsiella pneumoniae,* mucoidy is dependent on the *rmpA* gene carried by a virulence plasmid (Nassif et al., 1989). When a plasmid carrying this gene is introduced into *lon* mutants of *E. coli* K-12, colanic acid is overproduced. This overproduction is independent of *rcsA* but requires *rcsB* (Nassif et al., 1989). RmpA and a larger protein with similar activity, also found on *Klebsiella* virulence plasmids, RmpA2, bear limited resemblance to RcsA and RcsB, primarily in the conserved C terminus that includes the putative helix-turn-helix (Vasselon et al., 1991; Wacharotayankun et al., 1993) (Fig. 2). Given the dependence of the increase in capsule synthesis on a *lon* mutant, the data are consistent with RmpA acting as an inefficient analog of RcsA in *E. coli,* both as a target of Lon and as an activator of RcsB. In *Klebsiella* species, RmpA and RmpA2 may provide additional pathways for capsule activation, possibly which respond to different signals or are more temperature-resistant.

ADDITIONAL LEVELS OF CONTROL: REGULATION OF REGULATOR SYNTHESIS

The RcsB protein is relatively abundant in cells, and it is unclear whether there is any significant regulation of its activity during various growth conditions. Gervais et al. (1992) found a modest positive (twofold) effect of RcsB on expression of a *rcsB-lac* fusion; they did not investigate whether this stimulation was dependent on activated RcsB or RcsA. A σ^{54} promoter for *rcsB* has been identified by sequence comparisons, and certain *rcsB-lacZ* fusions are not expressed in σ^{54} mutants (Stout and Gottesman, 1990). However, in experiments using other *rcsB-lacZ* constructs, no effect of a σ^{54} mutation was found (Gervais et al., 1992), and there is no clear evidence that σ^{54}

mutations either block capsule synthesis (an expectation if RcsB was drastically reduced) or reduce the amount of RcsB protein (Stout, personal communication). No investigation of RcsC synthesis regulation has been carried out.

RcsA is normally synthesized in very low amounts, and the protein is difficult to detect in wild-type cells, due to both low levels of synthesis and rapid degradation. Among the mechanisms keeping RcsA synthesis low, *rcsA* is one of several loci that are "silenced" by the histonelike protein HNS, so that transcription from an appropriate *rcsA-lac* fusion is extremely low unless HNS is mutant (Sledjeski and Gottesman, in press). The HNS silencing can be overcome by overproduction of DsrA, an 85-nucleotide RNA encoded by an independently transcribed gene just downstream from *rcsA* and also found in *K. pneumoniae*. Because this effect of DsrA is only seen when the *dsrA* gene is present on a multicopy plasmid, it is unclear what its significance will be for the normal regulation of *rcsA*. However, the existence of a closely linked and conserved gene such as *dsrA* suggests that increased synthesis of DsrA may play a role in increasing RcsA synthesis and therefore capsule synthesis under specific conditions. Because of the instability of RcsA, increasing RcsA synthesis provides a mechanism for the rapid but transient stimulation of capsule synthesis when it is needed.

OVERVIEW

Capsular polysaccharide synthesis is apparently a strong drain on the cell's resources. Because capsule synthesis can be increased either by increasing RcsB activation or by increasing RcsA amounts, it would appear that the use of an auxiliary factor such as RcsA provides a strategy for feeding multiple signals into the system for a single output (increased capsule synthesis). Also, because activated RcsB overcomes the temperature sensitivity of RcsA activity, positive signals from one pathway (RcsB activation) are epistatic to negative signals from the second (RcsA) pathway. Currently missing from our understanding of this complex system are the molecular details of the interactions and the nature of the environmental signals that modulate the activation pathways. Both are likely to prove interesting.

ACKNOWLEDGMENTS

I thank T. Klopotowski for communicating unpublished results, Darren Sledjeski for comments on the manuscript, and the members of my laboratory for discussions.

REFERENCES

Altschul, S. F., W. Gish, W. Miller, E. W. Myers, and D. Lipman. 1990. A basic local alignment search tool. *J. Mol. Biol.* **215:**403–410.

Bardwell, J. C. A., K. McGovern, and J. Beckwith. 1991. Identification of a protein required for disulfide bond formation in vivo. *Cell* **67:**581–589.

Bernhard, F., K. Poetter, K. Geider, and D. L. Coplin. 1990. The *rcsA* gene from *Erwinia amylovora:* identification, nucleotide sequence analysis, and regulation of exopolysaccharide biosynthesis. *Mol. Plant-Microbe Interaction* **3:**429–437.

Brill, J. A., C. Quinlan-Walshe, and S. Gottesman. 1988. Fine-structure mapping and identification of two regulators of capsule synthesis in *Escherichia coli* K-12. *J. Bacteriol.* **170:**2599–2611.

Chatterjee, A., W. A. Chun, and A. K. Chatterjee. 1990. Isolation and characterization of an *rcsA*-like gene of *Erwinia amylovora* that activates extracellular polysaccharide production in *Erwinia* species, *Escherichia coli*, and *Salmonella typhimurium. Mol. Plant-Microbe Interaction* **3:**144–148.

Choi, S. H., and E. P. Greenberg. 1991. The C-terminal region of the *Vibrio fischeri* LuxR protein contains an inducer-independent *lux* gene activating domain. *Proc. Natl. Acad. Sci. USA* **88:**11115–11119.

Coleman, M., R. Pearce, E. Hitchin, F. Busfield, J. W. Mansfield, and I. S. Roberts. 1990. Molecular cloning, expression and nucleotide sequence of the *rcsA* gene of *Erwinia amylovora*, encoding a positive regulator of capsule expression: evidence for a family of related capsule activator proteins. *J. Gen. Microbiol.* **136:**1799–1806.

Coplin, D. L., K. Poetter, and D. R. Majerczak. 1990. Regulation of exopolysaccharide synthesis in *Erwinia stewartii* by *rcsB* and *rcsC. Phytopathology* **81:**1220.

Gervais, F. G., and G. R. Drapeau. 1992. Identification, cloning, and characterization of *rcsF,* a new regulator gene for exopolysaccharide synthesis that suppresses the division mutation *ftsZ84* in *Escherichia coli* K-12. *J. Bacteriol.* **174:**8016–8022.

Gervais, F. G., P. Phoenix, and G. R. Drapeau. 1992. The *rcsB* gene, a positive regulator of colanic acid biosynthesis in *Escherichia coli,* is also an activator of *ftsZ* expression. *J. Bacteriol.* **174:**3964–3971.

Gottesman, S., W. Clark, and V. Stout. 1994. Unpublished data.

Gottesman, S., and V. Stout. 1991. Regulation of capsular polysaccharide synthesis in *Escherichia coli* K12. *Mol. Microbiol.* **5:**1599–1606.

Gottesman, S., P. Trisler, and A. S. Torres-Cabassa. 1985. Regulation of capsular polysaccharide synthesis in *Escherichia coli* K12: characterization of three regulatory genes. *J. Bacteriol.* **162:**1111–1119.

Houng, H.-S. H., K. F. Noon, J. T. Ou, and L. S. Baron. 1992. Expression of Vi antigen in *Escherichia coli* K-12: characterization of *viaB* from *Citrobacter freundii* and identity with *rcsB. J. Bacteriol.* **174:**5910–5915.

Jayaratne, P., W. J. Keenleyside, P. R. MacLachlan, C. Dodgson, and C. Whitfield. 1993. Characterization of *rcsB* and *rcsC* from *Escherichia coli* 09:K30:H12 and examination of the role of the *rcs* regulatory system in expression of group I capsular polysaccharides. *J. Bacteriol.* **175:**5384–5394.

Jubete, Y., M. Maurizi, and S. Gottesman. Unpublished data.

Kahn, D., and G. Ditta. 1991. Modular structure of FixJ: homology of the transcriptional activator domain with the −35 binding domain of sigma factors. *Mol. Microbiol.* **5:**987–997.

Keenleyside, W. J., D. Bronner, K. Jann, B. Jann, and C. Whitfield. 1993. Coexpression of colanic acid and serotype-specific capsular polysaccharides in *Escherichia coli* strains with group II K antigens. *J. Bacteriol.* **175:**6725–6730.

Keenleyside, W. J., P. Jayaratne, P. R. MacLachlan, and C. Whitfield. 1992. The *rcsA* gene of *Escherichia coli* 09:K30:H12 is involved in the expression of the serotype-specific group I K (capsular) antigen. *J. Bacteriol.* **174:**8–16.

Klopotowski, T. 1994. Personal communication.

Markovitz, A. 1977. Genetics and regulation of bacterial capsular polysaccharide biosynthesis and radiation sensitivity, p. 415–462. *In* I. Sutherland (ed.), *Surface Carbohydrates of the Prokaryotic Cell.* Academic Press, London.

McCallum, K. L., and C. Whitfield. 1991. The *rcsA* gene of *Klebsiella pneumoniae* 01:K20 is involved in expression of the serotype-specific K (capsular) antigen. *Infect. Immun.* **59:**494–502.

Nassif, X., J.-M. Fournier, J. Arondel, and P. J. Sansonetti. 1989. Mucoid phenotype of *Klebsiella pneumoniae* is a plasmid-encoded virulence factor. *Infect. Immun.* **57:**546–552.

Nassif, X., N. Honoré, T. Vasselon, S. T. Cole, and P. J. Sansonetti. 1989. Positive control of colanic acid synthesis in *Escherichia coli* by *rmpA* and *rmpB,* two virulence-plasmid genes of *Klebsiella pneumoniae. Mol. Microbiol.* **3:**1349–1359.

Ophir, T., and D. L. Gutnick. 1994. A role for exopolysaccharides in the protection of microorganisms from dessication. *Appl. Environ. Microbiol.* **60:**740–745.

Parker, C. T., A. W. Kloser, C. A. Schnaitman, M. A. Stein, S. Gottesman, and B. W. Gibson. 1992. Role of the *rfaG* and *rfaP* genes in determining the lipopolysaccharide core structure and cell surface properties of *Escherichia coli* K-12. *J. Bacteriol.* **174:**2525–2538.

Parkinson, J. S., and E. C. Kofoid. 1992. Communication modules in bacterial signaling proteins. *Annu. Rev. Genet.* **26:**71–112.

Russo, T. A., and G. Singh. 1993. An extraintestinal, pathogenic isolate of *Escherichia coli* (04/K54/H5) can produce a group 1 capsule which is divergently regulated from its constitutively produced group 2, K54 capsular polysaccharide. *J. Bacteriol.* **175:**7617–7623.

Silverman, M., and M. Simon. 1973. Genetic analysis of flagellar mutants in *Escherichia coli. J. Bacteriol.* **113:**105–113.

Sledjeski, D., and S. Gottesman. A small RNA acts as an antisilencer of the HNS-silenced *rcsA* gene of *Escherichia coli. Proc. Natl. Acad. Sci. USA* **92,** in press.

Stout, V. Personal communication.

Stout, V., and S. Gottesman. 1990. RcsB and RcsC: a two-component regulator of capsule synthesis in *Escherichia coli. J. Bacteriol.* **172:**659–669.

Stout, V., A. Torres-Cabassa, M. R. Maurizi, D. Gutnick, and S. Gottesman. 1991. RcsA, an unstable positive regulator of capsular polysaccharide synthesis. *J. Bacteriol.* **173:**1738–1747.

Torres-Cabassa, A., S. Gottesman, R. D. Frederick, P. J. Dolph, and D. L. Coplin. 1987. Control of extracellular polysaccharide synthesis in *Erwinia stewartii* and *Escherichia coli* K-12: a common regulatory function. *J. Bacteriol.* **169:**4525–4531.

Torres-Cabassa, A. S., and S. Gottesman. 1987. Capsule synthesis in *Escherichia coli* K-12 is regulated by proteolysis. *J. Bacteriol.* **169:**981–989.

Trisler, P., and S. Gottesman. 1984. *lon* transcriptional regulation of genes necessary for capsular polysaccharide synthesis in *Escherichia coli* K-12. *J. Bacteriol.* **160:**184–191.

Vasselon, T., P. J. Sansonetti, and X. Nassif. 1991. Nucleotide sequence of *rmpB,* a *Klebsiella pneumoniae* gene that positively controls colanic biosynthesis in *Escherichia coli. Res. Microbiol.* **142:**47–54.

Vidal-Ingigliardi, D., E. Richet, O. Danot, and O. Raibaud. 1993. A small C-terminal region of the *Escherichia coli* MalT protein contains the DNA-binding domain. *J. Biol. Chem.* **268:**24527–24530.

Wacharotayankun, R., Y. Arakawa, M. Ohta, T. Hasegawa, M. Mori, T. Horii, and N. Kato. 1992. Involvement of *rcsB* in *Klebsiella* K2 capsule synthesis in *Escherichia coli* K-12. *J. Bacteriol.* **174:** 1063–1067.

Wacharotayankun, R., Y. Arakawa, M. Ohta, K. Tanaka, T. Akashi, M. Mori, and N. Kato. 1993. Enhancement of extracapsular polysaccharide synthesis in *Klebsiella pneumoniae* by RmpA2, which shows homology to NtrC and FixJ. *Infect. Immun.* **61:**3164–3174.

Expression of the Uhp Sugar-Phosphate Transport System of *Escherichia coli*

Robert J. Kadner

16

This book documents the excitement generated by the recent discovery of the widespread distribution of two-component regulatory systems throughout the eubacterial kingdom and even in eukaryotic systems as well. These systems control a wide variety of metabolic processes, usually by regulating specific gene transcription in response to environmental conditions. Many of these systems respond to external signals by sensing the presence of a particular physical state or of a chemical effector in the environment. The term *two-component* refers to the involvement of two proteins, called the sensor kinase and the response regulator, which contain several strongly conserved regions: the kinase or transmitter module and the phosphorylation or receiver module (Parkinson and Kofoid, 1992). In addition to these conserved segments, many of the two-component proteins carry segments involved in signal input and response output. The rationale for the term *two-component* is now seen to be inaccurate, because many systems use additional proteins as necessary parts of the signaling process. Furthermore, some proteins contain both transmitter and receiver modules, although the consequences of their presence in a single

polypeptide is not clear. Whatever the arrangement of the regulatory components, the key feature of all two-component systems is their action through protein phosphorylation and the regulation of the phosphate transfer steps. It is likely that covalent phosphorylation plays as important a role in metabolic coordination and response to changes in the environment for bacteria as it does for eukaryotic cells.

Uhp SYSTEM

This chapter reviews the regulation of the Uhp system, which is a sugar-phosphate transport protein whose expression is induced by external glucose 6-phosphate (Glu6P). This transmembrane regulatory system displays several unusual features, primarily centered around the two membrane-associated signaling proteins. One of these proteins is similar in sequence and topology to the transporter protein itself, whereas the other membrane protein has a very different transmembrane topology than other sensor kinase proteins. Aspects of this regulatory system have been reviewed recently (Kadner et al., 1994) and are diagrammatically illustrated in Fig. 1.

Many strains of bacteria possess transport systems for uptake in an unaltered form of various organophosphates, usually sugar phosphates or related compounds. Such transport

Robert J. Kadner, Department of Microbiology, School of Medicine, University of Virginia, Charlottesville, Virginia 22908.

Two-Component Signal Transduction, Edited by James A. Hoch and Thomas J. Silhavy,
© 1995 American Society for Microbiology, Washington, DC 20005

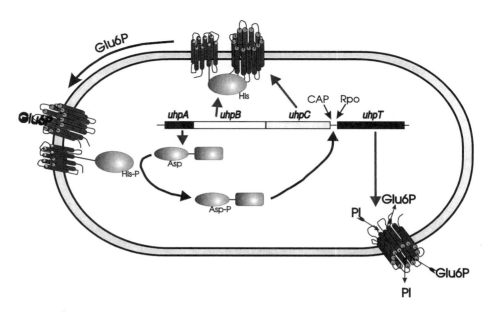

FIGURE 1 Schematic model of Uhp components. The membrane-associated proteins are indicated with their predicted transmembrane topology. My results suggest that UhpB and UhpC act as a signaling complex, in which the binding of Glu6P to UhpC affects the accessibility or activity of the kinase portion in the C-terminal half of UhpB. We presume that phosphate transfer occurs from UhpB to Asp-54 in UhpA and that this phosphorylation allows UhpA to activate transcription at the *uhpT* promoter, in conjunction with RNA polymerase holoenzyme and the cyclic AMP-CAP complex. From Kadner et al. (1994), with permission.

systems provide greater metabolic efficiency than the more common process in which an external sugar phosphate must be dephosphorylated by an external phosphatase, transported as the free sugar, and rephosphorylated in a phosphoenolpyruvate- or ATP-dependent reaction. The presence of organophosphate transport systems in many gram-positive and gram-negative bacteria suggests that their organophosphate substrates are widely available. Organophosphates are generated by phospholipid and nucleic acid turnover, by release of metabolic intermediates on cell lysis, and possibly by phosphorylase action on polysaccharides. However, a potential drawback to the possession of organophosphate transport systems is that they could provide a route for loss of internal organophosphates, including many key metabolic intermediates. It might thus be expected that expression or activity of organophosphate transport systems is regulated to prevent their expression when suitable substrates are not available and to reduce the efflux of internal substrates.

Several organophosphate transport systems have been analyzed in *Escherichia coli* or *Salmonella typhimurium*. The Ugp system allows uptake of glycerol phosphate mono- and diesters for use as phosphate source but not as carbon source (Brzoska et al., 1994). The Ugp transport system is a typical member of the family of ATP-driven periplasmic permeases, and its expression is regulated as part of the *pho* regulon by protein phosphate transfer reactions, as described by Wanner in Chapter 12. Three structurally similar transporters carry out anion exchange, whereby the uptake of their organophosphate substrates is coupled to the downhill movement of inorganic phosphate down its concentration gradient out of the cell (Sonna et al., 1988; Maloney et al., 1990). These three transporters are UhpT, which has a broad range of substrates; GlpT, which carries glycerol 3-phosphate (Gly3P); and PgtP, a *S. typhimurium* transporter of phosphoglycerates and phosphoenolpyruvate. Expression of GlpT is regulated by the GlpR repressor in response to internal Gly3P (Schweizer et al., 1985). Expression of UhpT and PgtP is induced in response to external Glu6P and phosphoglycerates, respectively, but not in response to the invariably present internal pools of these compounds (Winkler, 1970; Dietz, 1976). The Uhp and Pgt systems are regulated by means of three regulatory proteins, two of which belong to the group of two-component regulatory proteins. However, the individual features of these two systems are quite different (Yang et al., 1988; Yu and Hong, 1986).

The Uhp system (named for the *u*ptake of *h*exose *p*hosphates) allows bacteria to transport and use as a carbon source a wide range of compounds, including phosphate esters of glucose, fructose, mannose, galactose, glucosamine, mannitol, sedoheptulose, arabinose, and ribose (Pogell et al., 1966). Because most of these compounds do not induce expression of the Uhp system, they do not support growth of wild-type *E. coli* but only of variants in which the Uhp system is expressed constitutively. Also, the UhpT and GlpT systems confer susceptibility to fosfomycin, an antibiotic inhibitor of muramic acid synthesis, and this Glu6P-induced susceptibility to fosfomycin provides a strong selection for Uhp⁻ mutants. Uhp⁺ revertants are selected by their growth on Glu6P or fructose 6-phosphate, whereas Uhp-constitutive mutants can be selected for their growth on glucosamine 6-phosphate, glucose 1-phosphate, arabinose 5-phosphate, and so on, under conditions that prevent the action of periplasmic acid or alkaline phosphatases (Pradel and Boquet, 1989).

uhp GENES

The nucleotide sequence of the *uhp* locus (at 82.8 min on the *E. coli* genetic map) and genetic analyses revealed the presence of four *uhp* genes, which are transcribed in the same direction in the order *uhpABCT* (Friedrich and Kadner, 1987; Weston and Kadner, 1988). The *uhpABCT* genes are quite closely packed, with minimal intercistronic distances. The *uhpA* and *uhpT* promoter regions overlap the upstream genes. Comparison of the deduced Uhp polypeptides from *E. coli* and *S. typhimurium* showed a high degree of similarity, with 90 to 95% amino acid identity (Island et al., 1992). The *uhpT* gene is solely responsible for transport activity, and its expression from a heterologous promoter in the absence of other *uhp* products provides full transport activity (Ambudkar et al., 1990). The *uhpABC* products are required for *uhpT* transcription, as shown by the lack of expression of a *uhpT-lacZ* reporter gene in the presence of nonpolar deletion mutations in any one of the regulatory genes (Island et al., 1992).

Sequence comparison shows that the membrane-associated UhpC regulatory protein and the UhpT transporter protein are similar in size and hydropathy pattern and contain about 30% identical amino acid residue positions. Both proteins show a similar degree of sequence relatedness to the GlpT and PgtP transporters (Eiglmeier et al., 1987; Goldrick et al., 1988). The apparent recruitment of a transport protein to serve a regulatory role is one of the unusual features of the Uhp system. The UhpA and UhpB regulatory proteins contain the sequence motifs characteristic of receiver and transmitter modules of two-component regulatory proteins, respectively. The main difference from other two-component systems is that the amino-terminal half of UhpB is highly nonpolar, in contrast to most sensor kinases, which possess none or two transmembrane segments. More details of the organization of these proteins are presented below.

SIGNAL AND RESPONSE

An experimental advantage of the Uhp system is that its effector signal and its response are simple and clearly defined. Although the UhpT transport system has a very broad substrate specificity, its expression is induced only in response to extracellular Glu6P or 2-deoxy-glucose-6-phosphate. As far as I know, the only event induced by exposure to extracellular Glu6P is production of the UhpT transporter. As shown by expression of *uhpT-lacZ* reporters and by the level of *uhp*-specific mRNA, regulation of UhpT expression occurs at the level of transcription. Entry of inducer or detectable levels of UhpT transport activity is not needed for induction (Shattuck-Eidens and Kadner, 1981). In the absence of inducer or UhpA function, expression of *uhpT* is not detectable, showing that Uhp is regulated by an on-off switch, in which a single chemical inducer in a defined cellular location enables expression of a gene that otherwise is silent. Expression of UhpT also occurs as part of the OxyR peroxide response system (Storz and Altuvia, personal communication). Also, the maximal level of induced UhpT expression is modulated by catabolite repression mediated through the catabolite gene activator protein (CAP) system.

GENETIC APPROACH

Although two-component systems are constructed from modules of similar nature, there is likely to be considerable variety in different systems in the details of their actions, such as the mechanism of communication between modules; the mechanism of coupling between the kinase, phosphatase, and phosphate transfer reactions; and the mechanism of transcriptional activation. We are taking two genetic approaches to investigate the functions and interactions of the Uhp regulatory proteins. The first approach uses linker insertions to place defined mutations at easily identified sites. The resulting phenotype can indicate whether the affected portion of the protein is necessary for the normal regulatory process. These linker insertions introduce four amino acids of primarily charged or polar character and are thus more likely to cause a substantial local disruption of the structure of the protein than might single amino acid substitutions. Linker insertions also provide a convenient way to generate internal in-frame deletion mutations and fusions to topological (PhoA) and regulatory

(LacZ) reporters. The complementary second genetic approach is to select mutants with a particular behavior, usually those that restore function to another *uhp* mutant. This approach can potentially identify the location and nature of all the single base changes that confer a defined phenotype, such as phosphorylation-independent or kinase-constitutive behavior.

The role of the *uhp* regulatory genes was examined in several ways. In-frame deletions of individual or multiple *uhp* genes were constructed using the collection of linker insertions (Island et al., 1992). The effect of these nonpolar null mutations on expression of a *uhpT-lacZ* reporter was determined with all participants present in single copy in the chromosome. As expected, loss of *uhpA* resulted in complete loss of detectable *uhpT* expression. The absence of UhpB or UhpC resulted in complete loss of induction by Glu6P but an elevated basal level of expression relative to the uninduced wild-type strain. One interpretation of this behavior is that UhpB and UhpC are both required for activation of sensor kinase activity in the presence of inducer and that they both contribute to a co-phosphatase activity that removes from UhpA any phosphate residues that might have been inadvertently acquired by crosstalk from other sensor kinases.

SIGNAL TRANSDUCERS: UhpBC

The 500-residue UhpB protein has a strongly bipartite structure. Its amino-terminal half (residues 1 to 273) is highly nonpolar, bears no obvious homology to other proteins, and is predicted from hydropathy and PhoA fusion analyses to cross the membrane eight times. The carboxyl-terminal half of UhpB faces the cytoplasm and contains the sequence motifs characteristic of sensor kinase proteins. Many of the other sensor kinase proteins possess two transmembrane segments and present a large periplasmic domain that contains the binding site for the appropriate ligand in the periplasmic space. Why does UhpB not follow that paradigm and use a periplasmic ligand binding domain but instead place so much of its mass in

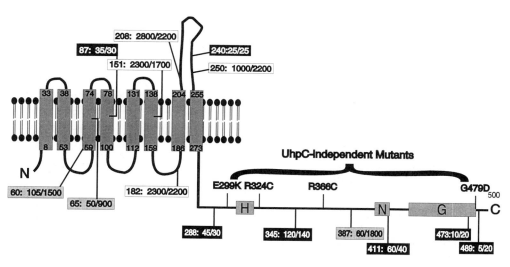

FIGURE 2 Schematic model of the location and phenotype of mutants in UhpB. The location of 12-bp linker insertions is indicated in rectangles, relative to the predicted transmembrane topology of the protein. Insertions conferred a range of regulatory phenotypes, including fully constitutive (open boxes), inducible behavior with elevated uninduced level (shaded boxes), or lack of expression (filled boxes). The location and sequence change of single base substitutions that allow Uhp expression in the absence of UhpC function are indicated in the C-terminal portion of UhpB.

the membrane? Why is another large membrane protein, UhpC, required for regulation? The similarity of UhpC protein to the UhpT/PgtP/GlpT family of transport proteins suggests that UhpC contains an organophosphate binding site, and it is a challenge to identify the structural features that allow UhpC to participate in signaling and UhpT to act in transport.

Whereas most of the 12-bp linker insertions in *uhpA* and *uhpT* obliterated their respective function, about half of the insertions in either *uhpB* or *uhpC* were still Uhp+, although many of these showed altered regulation of *uhpT* expression. For example, all the Uhp+ insertions in the membrane-embedded amino-terminal half of UhpB displayed partially or fully constitutive behavior, suggesting that the transmembrane portion of UhpB serves not just as a membrane anchor but plays a direct role in the transmembrane signaling process (Fig. 2). It is not known yet whether the Uhp− insertions in *uhpB* cause defective incorporation or stability of the variant protein in the

membrane, although this is likely. Most of the Uhp+ linker insertions in *uhpC* retained normal inducible behavior. This result contrasts with the loss of transport function due to insertions in *uhpT* and suggests that the function of UhpC in regulation is much more tolerant of the local structural disruptions caused by the insertion of four amino acids than is the structurally similar UhpT transporter protein. Particularly noteworthy is the finding that three insertions in the large central cytoplasmic loop obliterated transport by UhpT, whereas three insertions in the same region of UhpC did not interfere with signaling. One insertion of four amino acids at residue 91 conferred high-level constitutive Uhp expression, although addition of four more amino acids at that site resulted in a Uhp-negative phenotype. This dominant constitutive insertion in UhpC suggests that the cytoplasmic loop containing residue 91 might be involved in transmembrane signaling because its alteration can transmit a signal indicative of inducer binding.

SIGNAL TRANSDUCTION COMPLEX

Because both UhpB and UhpC appear to play positive roles in the response to inducer, do they act separately and sequentially or cooperatively in signaling? Analysis of the dominance and epistasis properties of the *uhp* linker insertions could provide an answer. If a pathway contains independent signaling elements, elements acting upstream in a pathway might not affect the phenotype conferred by mutants acting downstream in that pathway. To test these relationships, linker insertions in *uhpA* or *uhpB* were combined with wild-type, constitutive, or negative alleles of *uhpC* by in vitro techniques and were transferred to the chromosome in single copy (Island and Kadner, 1993). As expected, UhpA and UhpB functions were absolutely required for *uhpT* expression, even in the presence of a constitutive form of UhpC (Table 1). When partially or fully constitutive forms of UhpB were combined with a constitutive form of UhpC, high-level constitutive behavior was seen, suggesting that these altered forms of UhpB can still respond to signals from UhpC. The surprising result was that when most of the constitutive forms of UhpB were combined with any Uhp⁻ form of UhpC, *uhpT* expression was lost. These results show that the activity of UhpB mutants altered in their transmembrane segment required the presence of an active form of UhpC. One insertion in the transmembrane portion of UhpB (at residue 151) conferred high-level constitutive expression in the presence of either active or inactive forms of UhpC, suggesting that at least this form of UhpB is stable and functional in the membrane in the absence of UhpC function. This observation that many constitutive forms of UhpB still require a functional form of UhpC for activity suggests that UhpB and UhpC function together as a complex (i.e., that they may act sequentially but not necessarily separately).

Tests of dominant properties of the linker insertion mutations were made by placing a *uhp* cluster containing a linker insertion as a tandem duplication with either a deletion of the *uhpABC* region or the wild-type *uhp*⁺ region. Insertions in all three *uhp* regulatory genes showed a fairly simple pattern, at least in qualitative terms. Constitutive behavior was dominant over inducible behavior, and inducible behavior was at least partially dominant over the negative phenotype. The insertion mutants in *uhpA* seemed to interfere somewhat with the action of wild-type *uhpA*, as is more extensively described below.

Although UhpB and UhpC function are required for normal inducible expression of *uhpT*, several conditions bypass this requirement. Expression of Uhp in the absence of UhpBC function results either from overexpression of UhpA on multicopy plasmids or from mutations that were selected by Uhp⁺ behavior in the absence of UhpB or UhpC. Mutations that bypass the requirement for UhpA or UhpT are not obtained (our use of high-phosphate growth medium presumably prevents growth on Glu6P of strains that constitutively express alkaline or acid phosphatases). Starting with a strain deleted for *uhpBC*, Uhp⁺ variants were obtained, all of which retained the *uhpBC* deletion and expressed *uhpT* in a constitutive manner. Transduction analysis showed that the trait of UhpBC independence was genetically linked to the *uhp* region. The *uhpA* gene from four independent isolates was isolated by polymerase chain reaction amplification and subjected to nucleotide sequence analysis, which revealed single nucleotide substitutions in each. In three isolates, Phe-17 was changed to Val, and in the fourth isolate, His-170 was changed to Tyr (Wei, unpublished data). It thus appears that amino acid substitutions at a limited number of sites in either the phosphorylation module or the output module can overcome the requirement for phosphorylation by UhpB kinase.

Uhp⁺ revertants were similarly isolated in a strain deleted for *uhpC*. These revertants also conferred constitutive expression of *uhpT* as a result of secondary mutations in the *uhp* region. However, no sequence changes were found in the *uhpA* gene from several isolates. Instead, the

TABLE 1 Uhp regulation in double linker insertion mutants[a]

| | β-Galactosidase activity from *uhpT-lacZ* gene fusion in cells grown in the absence/presence of Glu6P | | | | | | | | | | |
| | *uhpA* or *uhpB* alleles | | | | | | | | | | |
uhpC alleles	A+B+	A15	A169 A189	B151	B182	B208	B250	B60	B65	B387	B87, B240 B288, B345 B411
C+	2/1550										
C91::4	3950/3350	1/1	2/2	2260/1700	2330/2150	2770/2210	1040/2150	105/1540	45/860	60/1750	55/55
C224	7/300	1/1	2/2	2530/1620	3020/2250		3100/2130	2700/2140	55/35	3020/2020	100/75
C91::8	10/10	3/1	2/2	2420/1960	400/1150	2780/2420	1130/1460				
C41, C141	9/9	2/1		2510/1940	15/15	5/4	12/11	25/25	185/140	75/50	125/100
C361, C399		1/1		2310/1780	8/8	4/3	12/14				
C404, delC											

[a] All alleles are designated by the site in the amino acid sequence after which the additional amino acids are added by the linker insertion mutation.

uhpB genes from 12 independent isolates were found to encode variants with one or, in one case, two amino acid substitutions, as diagrammed in Fig. 2. Many isolates carried the same change. These results identified four sites in the carboxyl half of UhpB, at which an amino acid substitution resulted in UhpC independence. All these amino acid substitutions resulted in a change of residue charge. Three of these sites were near conserved motifs that characterize the sensor kinase proteins.

Thus, two separate types of alterations in UhpB can result in constitutive behavior. Mutations affecting many sites throughout the transmembrane portion of UhpB conferred constitutive expression of *uhpT*, but most of these were only active in the presence of an active form of UhpC. However, amino acid substitutions at many sites in the carboxyl half of UhpB resulted in constitutive expression that was now independent of UhpC. Almost all the randomly inserted linker substitutions in this half of UhpB resulted in loss of function, showing that only particular alterations can release the normal regulatory control while retaining (putative) protein kinase and phosphate transfer activity. These results suggested a model in which UhpB kinase activity, which is needed for phosphorylation and activation of UhpA, is normally held in an inactive state by an interaction with cytoplasmically exposed segments of the transmembrane part of UhpB or of a UhpB-UhpC complex.

COMPLEXITY OF SIGNAL TRANSDUCTION

Many two-component systems that respond to an external signal follow a simple paradigm, in which the transmembrane signal transducing protein serves both as receptor for the signal, often via the periplasmic ligand binding domain, and as the kinase whose autophosphorylation or phosphate transfer activity is regulated by ligand occupancy. The sensor kinase may also cooperate in removal of phosphate from the response regulator protein. It is becoming apparent that many of these regulatory systems

are substantially more complex than this paradigm and use additional proteins. In the chemotaxis system, the CheA kinase is separate from the membrane signal transducers and is coupled to them by action of CheW. In the *pho* regulon, the sensor kinase PhoR protein is anchored in the membrane by two transmembrane segments but with no significant periplasmic domain (Scholten and Tommassen, 1993). This arrangement allows the kinase to be held in proximity to PhoU or other components of the phosphate-specific transport system, which appear to provide the repressive signal of phosphate supply (Wanner, 1993). The regulation of NtrB kinase-phosphatase activity is coupled to the state of covalent modification of the cytoplasmic protein P_{II}, whose degree of uridylylation is controlled in response to the amount of fixed nitrogen available to the cell (Atkinson and Ninfa, 1993).

INTRACELLULAR SIGNAL: UhpA

Multiple types of output modules are connected to phosphorylation modules and discriminate between the NtrC, OmpR, and UhpA families of transcription activators, which suggests that they might activate transcription in different ways. Sequence comparisons show that the 196-residue UhpA protein can be divided into a 119-residue phosphorylation-receiver module characteristic of all response regulator proteins, a variable 18-residue linker region, and a 59-residue output module. This output module is predicted to contain a DNA binding helix-turn-helix motif that is well conserved in many transcription regulatory proteins, including members of the UhpA family of two-component response regulator proteins as well as others that are not regulated by protein phosphorylation, such as MalT and LuxR (Henikoff et al., 1990). The response regulators have been divided into several subfamilies based both on the phylogenetic comparisons within the phosphorylation-transmitter module and on the structures of the output modules (Volz, 1993; Parkinson and Kofoid, 1992). UhpA belongs to the subfamily that includes NarL, NarP, RcsB, and UvrC2 of *E. coli;* DegU and ComA of *Bacillus subtilis;* rhizobial FixJ; MoxX of *Paracoccus denitrificans;* and an open reading frame from *Streptomyces griseus.*

Overexpression of UhpA results in the constitutive expression of uhpT that no longer requires the action of UhpB or UhpC (Shattuck-Eidens and Kadner, 1983). This effect could result from inadvertent phosphorylation of excess UhpA by other sensor kinases, or it could result from a phosphorylation-independent increase in the concentration of a transcriptionally active form of UhpA in equilibrium with the inactive species. One hint to this question comes from construction of a variant form of UhpA in which the putative site of phosphorylation, Asp-54, is converted to Asn. This UhpA-D54N mutant is inactive when present in single copy but almost fully as active as the wild-type allele when present on a multicopy plasmid, suggesting that the normal site of phosphorylation is not required when UhpA is overproduced (Webber and Merkel, unpublished data).

The role of the carboxyl-terminal output module in UhpA action was investigated by preparation of a series of gene truncations that removed increasing lengths of UhpA from the carboxyl end. The ability of these truncated genes on multicopy plasmids to complement for UhpA function in a Δ*uhpA* host or to interfere with Uhp signaling in a *uhp⁺* host was tested. Removal of seven amino acids completely eliminated UhpA activation of *uhpT-lac* transcription; removal of even one amino acid caused partial loss of transcription activation. Most deletions that removed part of the output module but left intact the phosphorylation module (residues 1 to 120) exhibited a dominant negative phenotype in a wild-type strain (Webber and Kadner, in press). Further deletion into the phosphorylation module resulted in loss of any detectable consequence, probably owing to the instability of the truncated fragment. The degree of impairment of Uhp function by carboxyl-terminal truncations of

UhpA was related to the relative gene copy number of the variant and wild-type forms of *uhpA* (i.e., inhibition of Uhp function was lesser when plasmid vectors of lower copy number were used; inhibition was minimal when the truncated genes were present at single copy).

The dominant negative effect of overproduction of the phosphorylation module of UhpA could result from competition between the variant and wild-type forms for interaction with some component of the signaling pathway, such as the *uhpT* promoter, the UhpB kinase, or other molecules of UhpA, if oligomerization of UhpA is an important step in its action. Another possibility is that UhpA regulates its own synthesis. The dominant negative behavior is explained if the truncated variants retain repression of *uhpA* but lack the ability to activate transcription of *uhpT,* analogous to the situation in LuxR (Choi and Greenberg, 1992). Several genetic experiments have been performed to initiate dissection of the mechanism of this interference. First, truncated forms carrying the D54N change, which blocks the normal phosphorylation site, lost most or all their inhibitory effect on expression of the normal Uhp system. This result shows that interference by the liberated receiver module requires that it be able to be phosphorylated. Second, overexpression of the truncated forms of UhpA, which interfere strongly with a chromosomal copy of wild-type UhpA, poorly interferes with the action of the phosphorylation-independent F17V form of UhpA. We interpret these results to show that the dominant negative effect resulting from overexpression of the receiver module results primarily from its ability to remove phosphate residues from UhpB rather than through the formation of mixed oligomers.

TARGET OF UhpA

The putative phosphorylation of UhpA results in its conversion to a form that allows transcription initiation at the *uhpT* promoter. Deletion analysis of the *uhpT* promoter region revealed that all the sequence elements required for normal expression and regulation are located within 120 bp upstream of the transcription start site. This region contains several elements that serve for recognition of transcription factors (Merkel et al., 1992). A −10 element typical of σ^{70}-dependent promoters is absolutely required for transcription. The −35 region bears no resemblance to the canonical sequence for this region. Because base substitutions in this region eliminated promoter function, it is probably recognized in a specific manner by the transcription system. A 40-bp region of hyphenated dyad symmetry is centered at −64. The behavior of the *uhpT* promoter in gel mobility shift assays is complex, and direct evidence that this region serves as the target for UhpA binding and action has not been obtained yet. Genetic support for the role of the −64 element included the fact that deletions entering this region from upstream completely eliminate promoter activity. Linker substitution mutations in which the native sequences in the −64 region are converted to an *Nco*I restriction site reduced or eliminated promoter function, depending on the location of the substitution and the number of base pair residues that were changed. The presence of this −64 region on a multicopy plasmid rendered an otherwise *uhp*+ host strain unable to grow on sugar phosphates, apparently as the result of its titration of the limiting amount of activated UhpA in the cell. Deletions or linker substitutions in this element abrogated this competitive effect, as did overproduction of UhpA. This region may contain multiple binding sites for UhpA, because the phenotypes of even symmetrically comparable mutations in this region were not identical and because single base changes within the region had only minor effects on activity (Merkel and Kadner, unpublished data). It was shown that spacing between the −64 element and the remainder of the promoter was crucial for promoter function (Merkel et al., 1992). The UhpA protein has been purified from expression systems (Wei, unpublished data), and preliminary studies show that it binds specifically to four regions within the −64 element.

Upstream of the −64 element is a region centered at −103.5 that has a close match to the consensus for a DNA binding site for CAP. This region was shown to bind purified CAP in vitro and its deletion resulted in a 7- to 15-fold decrease in induced *uhpT* promoter activity (Merkel et al., 1992; Merkel et al., 1995). Absence of this CAP binding site or mutational elimination of CAP activity affected the maximal level of the *uhpT* promoter activity but not its inducibility by external Glu6P. As is often the case with CAP-activated transcription, the exact placement of the CAP site is not crucial, and it can function at various distances from the remainder of the promoter, as long as it lies on the same face of the helix (reviewed in Kolb et al., 1993). Promoter activity was present but progressively decreased when the spacing of the CAP site from the other elements in the *uhpT* promoter was increased in multiples of the DNA helical repeat. When the CAP site was moved a nonintegral number of helical turns, promoter activity was as low as when the CAP protein was missing (Merkel et al., 1995). Several surface-exposed regions of the CAP protein have been found to be necessary for transcription activation at CAP-dependent promoters that do not require the action of other transcription activator proteins. Mutants of CAP that lack positive control function at these other promoters were still fully functional at the *uhpT* promoter. I am studying whether a different surface of the CAP protein serves for activation of the *uhpT* promoter or whether perhaps CAP activation is mediated by changes in DNA conformation, such as DNA bending, rather than direct protein-protein contact.

Whatever the mechanism for CAP activation, I think that the purpose of catabolite repression of the Uhp system is to limit the flux of substrate entry and hence the flux of the glycolytic or other catabolic pathways. Catabolite repressibility of other systems appears to coordinate the cellular consumption of various available carbon sources, providing precedence to the catabolism of glucose over other sources.

Because the Uhp substrate is the intracellular form of glucose, it does not seem that glucose would be a preferred nutrient over Glu6P. The need for a governor of Uhp function is apparent from the behavior of cells that overexpress UhpT, owing either to amplification of the *uhpT* gene or supplementation with cyclic AMP. These cells are sensitive to the presence of Glu6P or other substrates (Ackerman et al., 1974; Kadner et al., 1992). Addition of Glu6P and other immediate substrates for glycolysis results in cell killing, coincident with the production of methylglyoxal, a lethal side product of glycolysis. Sugar phosphates that are not directly metabolized by glycolysis also inhibit growth of those cells but do not elicit cell killing or methylglyoxal production. Thus, catabolite repressibility of Uhp appears to be very important to maintain rates of carbohydrate metabolism within appropriate limits.

THE FUTURE

In conclusion, the results obtained so far indicate that the Uhp system has the properties of a simple switch composed of complex components. The simplicity of the switch is reflected in the simple and well-defined nature of the inducing signal and the on-off nature of the output, in which UhpT expression is apparently regulated only by the presence of external Glu6P, as modulated by catabolite repression. The *uhpT* promoter appears to be a relatively simple example of a promoter controlled by two transcriptional regulatory proteins, although it is not yet known whether other proteins besides RNA polymerase holoenzyme might also participate. The unusual complexity of the system is based on the apparent involvement of UhpB and UhpC in recognition of inducer and transmembrane signal transduction. Why is this system so much more complicated than most other two-component systems that use a transmembrane sensor kinase with a periplasmic signal binding domain? There is no obvious answer, except the consideration that this more complex system may be needed to provide a very high level of inducer

specificity. Perhaps the presence of two signaling proteins provides an absolute vectorial accuracy, so that the system can respond solely to external inducer and not at all to the same molecule in the cytoplasm.

ACKNOWLEDGMENT

Work from my laboratory was supported by NIH research grant GM38681.

REFERENCES

Ackerman, R. S., N. R. Cozzarelli, and W. Epstein. 1974. Accumulation of toxic concentrations of methylglyoxal by wild-type *Escherichia coli* K-12. *J. Bacteriol.* **119:**357–362.

Ambudkar, S. V., V. Anantharam, and P. C. Maloney. 1990. UhpT, the sugar phosphate antiporter on *Escherichia coli,* functions as a monomer. *J. Biol. Chem.* **265:**12287–12292.

Atkinson, M. R., and A. J. Ninfa. 1993. Mutational analysis of the bacterial signal-transducing protein kinase/phosphatase nitrogen regulator II (NR$_{II}$ or NtrB). *J. Bacteriol.* **175:**7016–7023.

Brzoska, P., M. Rimmele, K. Brzostek, and W. Boos. 1994. The *pho* regulon-dependent Ugp uptake system for glycerol-3-phosphate in *Escherichia coli* is *trans* inhibited by P$_i$. *J. Bacteriol.* **176:**15–20.

Choi, S. H., and E. P. Greenberg. 1992. Genetic dissection of DNA binding and luminescence gene activation by the *Vibrio fischeri* LuxR protein. *J. Bacteriol.* **174:**4064–4069.

Dietz, G. W. 1976. The hexose phosphate transport system of *Escherichia coli. Adv. Enzymol.* **44:**237–259.

Eiglmeier, K., W. Boos, and S. Cole. 1987. Nucleotide sequence and transcriptional startpoint of the *glpT* gene of *Escherichia coli:* extensive sequence homology of the glycerol-3-phosphate transport protein with components of the hexose-6-phosphate transport system. *Mol. Microbiol.* **1:**251–258.

Friedrich, M. J., and R. J. Kadner. 1987. Nucleotide sequence of the *uhp* region of *Escherichia coli. J. Bacteriol.* **169:**3556–3563.

Goldrick, D., G.-Q. Yu, S.-Q. Jiang, and J.-S. Hong. 1988. Nucleotide sequence and transcription start point of the phosphoglycerate transporter gene of *Salmonella typhimurium. J. Bacteriol.* **170:** 3421–3426.

Henikoff, S., J. C. Wallace, and J. P. Brown. 1990. Finding protein similarities with nucleotide sequence databases. *Methods Enzymol.* **183:**111–132.

Island, M. D., and R. J. Kadner. 1993. Interplay between the membrane-associated UhpB and UhpC regulatory proteins. *J. Bacteriol.* **174:**5028–5034.

Island, M. D., B.-Y. Wei, and R. J. Kadner. 1992. Structure and function of the *uhp* genes for the sugar phosphate transport system in *Escherichia coli* and *Salmonella typhimurium*. *J. Bacteriol.* **174:**2754–2762.

Kadner, R. J., M. D. Island, T. J. Merkel, and C. A. Webber. 1994. Transmembrane control of the Uhp sugar-phosphate transport system: the sensation of Glu6P, p. 78–84. *In* A. Torriani-Gorini, E. Yagil, and S. Silver (ed.), *Phosphate in Microorganisms: Cellular and Molecular Biology*. American Society for Microbiology, Washington, D.C.

Kadner, R. J., G. P. Murphy, and C. M. Stephens. 1992. Two mechanisms for growth inhibition by elevated transport of sugar phosphates in *Escherichia coli*. *J. Gen. Microbiol.* **138:**2007–2014.

Kolb, A., S. Busby, H. Buc, S. Garges, and S. Adhya. 1993. Transcriptional regulation by cAMP and its receptor protein. *Annu. Rev. Biochem.* **62:**749–795.

Maloney, P. C., S. V. Ambudkar, V. Anantharam, L. A. Sonna, and A. Varadhachary. 1990. Anion-exchange mechanisms in bacteria. *Microbiol. Rev.* **54:**1–17.

Merkel, T. J., J. L. Dahl, R. H. Ebright, and R. J. Kadner. 1995. Transcription activation at the *Escherichia coli uhpT* promoter by the catabolite gene activator protein. *J. Bacteriol.* **177:**1712–1718.

Merkel, T. J., and R. J. Kadner. Unpublished data.

Merkel, T. J., D. M. Nelson, C. L. Brauer, and R. J. Kadner. 1992. Promoter elements required for positive control of transcription of the *Escherichia coli uhpT* gene. *J. Bacteriol.* **174:**2763–2770.

Parkinson, J. S., and E. C. Kofoid. 1992. Communication modules in bacterial signaling proteins. *Annu. Rev. Genet.* **26:**71–112.

Pogell, B. M., B. R. Maity, S. Frumkin, and S. Shapiro. 1966. Induction of an active transport system for glucose-6-phosphate in *Escherichia coli*. *Arch. Biochem. Biophys.* **116:**406–415.

Pradel, E., and P. L. Boquet. 1989. Mapping of the *Escherichia coli* acid glucose-1-phosphatase gene *agp* and analysis of its expression in vivo by use of an *agp-phoA* protein fusion. *J. Bacteriol.* **171:**3511–3517.

Scholten, M., and J. Tommassen. 1993. Topology of the PhoR protein of *Escherichia coli* and functional analysis of internal deletion mutants. *Mol. Microbiol.* **8:**269–275.

Schweizer, H., W. Boos, and T. J. Larson. 1985. Repressor for the *sn*-glycerol-3-phosphate regulon of *Escherichia coli* K-12: cloning of the *glpR* gene and identification of its product. *J. Bacteriol.* **161:**563–566.

Shattuck-Eidens, D. M., and R. J. Kadner. 1981. Exogenous induction of the *Escherichia coli* hexose phosphate transport system defined by *uhp-lac* operon fusions. *J. Bacteriol.* **148:**203–209.

Shattuck-Eidens, D. M., and R. J. Kadner. 1983. Molecular cloning of the *uhp* region and evidence for a positive activator for expression of the hexose phosphate transport system of *Escherichia coli*. *J. Bacteriol.* **155:**1060–1070.

Sonna, L. A., S. V. Ambudkar, and P. C. Maloney. 1988. The mechanism of glucose-6-phosphate transport by *Escherichia coli*. *J. Biol. Chem.* **263:**6625–6630.

Storz, G., and S. Altuvia. Personal communication.

Volz, K. 1993. Structural conservation in the CheY superfamily. *Biochemistry* **32:**11741–11753.

Wanner, B. L. 1993. Gene regulation by phosphate in enteric bacteria. *J. Cell. Biochem.* **51:**47–54.

Webber, C. A., and R. J. Kadner. Action of receiver and activator modules of UhpA in transcriptional control of the *Escherichia coli* sugar phosphate transport system. *Mol. Microbiol.* **15,** in press.

Webber, C. A., and T. J. Merkel. Unpublished data.

Wei, B.-Y. Unpublished data.

Weston, L. A., and R. J. Kadner. 1988. Role of *uhp* genes in expression of the *Escherichia coli* sugar-phosphate transport system. *J. Bacteriol.* **170:**3375–3383.

Winkler, H. H. 1970. Compartmentation in the induction of the hexose-6-phosphate transport system of *Escherichia coli*. *J. Bacteriol.* **101:**470–475.

Yang, Y.-L., D. Goldrick, and J.-S. Hong. 1988. Identification of the products and nucleotide sequence of two regulatory genes involved in exogenous induction of phosphoglycerate transport in *Salmonella typhimurium*. *J. Bacteriol.* **170:**4299–4303.

Yu, G.-Q., and J.-S. Hong. 1986. Identification and nucleotide sequence of the activator gene of the externally induced phosphoglycerate transport system of *Salmonella typhimurium*. *Gene* **45:**51–57.

Symbiotic Expression of *Rhizobium meliloti* Nitrogen Fixation Genes Is Regulated by Oxygen

Peter G. Agron and Donald R. Helinski

17

A common nutritional challenge faced by plants is nitrogen starvation. Many plants confront this challenge by participating in symbiotic relationships with soil microbes that provide nitrogen to the plant. Plant partners include mycorrhizal fungi, which possess very efficient uptake mechanisms for soil nutrients including nitrogen sources, and nitrogen-fixing bacteria that are able to reduce atmospheric dinitrogen to ammonia. An example of the latter class is the gram-negative bacterium *Rhizobium meliloti,* which, in response to nitrogen starvation of a potential host such as alfalfa, can establish an intricate partnership with its plant host (Denarie and Cullimore, 1993; Fisher and Long, 1992; Long, 1989).

R. meliloti elicits the development of specialized organs on the plant roots called nodules. As a nodule develops, the endosymbiont invades and colonizes the interior of the structure and ultimately resides inside plant cells surrounded by a plant-derived membrane (Fisher and Long, 1992; Long, 1989; Denarie and Cullimore, 1993). The final stage of the developmental program is bacterial nitrogen fixation, which can allow host plants to thrive even when there are no alternative nitrogen sources.

The nodule provides a distinct environment that is highly specialized to facilitate nitrogen fixation. Within the nodule, the bacteria are amply supplied with fixed carbon from host photosynthesis; this is critical because nitrogen fixation requires a great deal of energy. Estimates of the number of moles of ATP required to reduce 1 mol of dinitrogen range from about 30 to 50 (Haaker and Klugkist, 1987). Molecular oxygen is also required to support the actively respiring bacteria, which are obligate aerobes. Oxygen, however, severely inhibits nitrogenase, the enzyme that catalyzes nitrogen fixation (Hill, 1988). This conundrum is solved in large part by the plant-derived oxygen-carrier protein leghemoglobin. While leghemoglobin supplies oxygen to the bacteria, it also simultaneously reduces the concentration of free oxygen, thus protecting nitrogenase (Appleby, 1984). Leghemoglobin is aided in this task by the hard outer cortical layer of the nodule, which functions as a gas diffusion barrier. Therefore, the nodule interior is a microaerobic environment.

A key step in the differentiation of the soil bacteria into intracellular nitrogen-fixing en-

Peter G. Agron and Donald R. Helinski, Department of Biology and Center for Molecular Genetics, University of California, San Diego, La Jolla, California 92093-0634. *Present address for Peter G. Agron:* Division of Infectious Disease, University of California, San Francisco, San Francisco, California 94143-0654.

Two-Component Signal Transduction, Edited by James A. Hoch and Thomas J. Silhavy, © 1995 American Society for Microbiology, Washington, DC 20005

dosymbionts is the coordinated expression of the nitrogen fixation genes in the interior of the mature nodule. How these genes are coordinately expressed specifically within the nodule is an important question in understanding the plant-microbe symbiosis. In the asymbiotic free-living nitrogen-fixing bacterium *Klebsiella pneumoniae,* the expression of the nitrogen fixation (*nif*) genes is controlled by the *ntrB/ntrC* two-component system, which induces expression in response to a low supply of alternative nitrogen sources (Gussin et al., 1986). Homologs of these genes have been identified in *R. meliloti,* but inactivation of these genes by transposon mutagenesis does not affect symbiosis (Szeto et al., 1984). These observations suggested that *R. meliloti* possesses a different mechanism for controlling nitrogen fixation gene expression. A seminal development in understanding how these genes are specifically expressed in the nodule interior was the finding that low oxygen tension induces nitrogen fixation gene expression in cultures grown ex planta (Ditta et al., 1987). It was, therefore, proposed that microaerobiosis is the signal for nitrogen fixation gene expression in root nodules, a hypothesis supported by the fact that the interior of the nodule is microaerobic (Appleby, 1984). Low oxygen tension is an appropriate signal, because this cue couples gene expression to a crucial physiological requirement for nitrogen fixation. Gene expression, therefore, coincides with the arrival of the bacteria into an environment in which the gene products can function.

REGULATION OF SYMBIOTIC NITROGEN FIXATION GENES IN *R. MELILOTI*

Twenty-three genes specifically required for nitrogen fixation have been identified in *R. meliloti.* They can be divided into two groups. The first group contains several homologs of nitrogen fixation genes in *K. pneumoniae* (*nif*), as well as several genes that do not have enteric counterparts (*fix*). Among these genes is *nifA,* a regulatory gene encoding a transcriptional activator responsible for the expression of the other members of this group, for example the genes encoding the three polypeptides of nitrogenase (*nifHDK*) (Fig. 1). It was the expression of *nifA* that was originally shown to be oxygen-regulated (Ditta et al., 1987). Except for *nifA* itself, a defining feature of these genes is *nifA*-dependent expression.

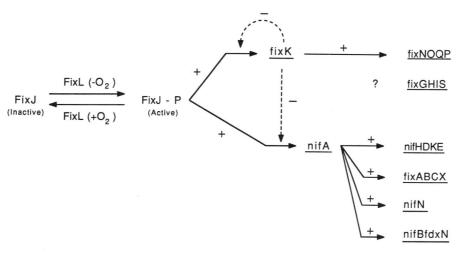

FIGURE 1 Regulation of symbiotic nitrogen fixation genes of *R. meliloti.* A plus (+) indicates transcriptional activation of the target gene or operon; a minus (–) indicates repression of transcription.

Using DNA-RNA hybridization techniques, another group of genes was identified in a search for new megaplasmid genes that are specifically expressed during symbiosis (David et al., 1987; Renalier et al., 1987). None of the members of this group has known homologs in *K. pneumoniae*. Portions of the *R. meliloti* megaplasmid were used as probes against RNA isolated from either nodules or vegetative bacteria. The expression of these genes was established to be *nifA*-independent. One of these genes, designated *fixK*, is highly expressed during symbiosis and encodes a transcriptional activator that causes high expression of the *fixNOQP* operon (Batut et al., 1989). The expression of these genes is also induced by microaerobiosis ex planta. This suggested that a common regulatory mechanism was likely to be involved in the transcriptional control of both groups of genes. Closely linked to *fixK* and *fixNOQP* are two *fix* operons that are not highly expressed in nodules, *fixGHIS* (Kahn et al., 1989) and *fixLJ* (David et al., 1988). The transcriptional regulation of these operons, if any, is currently unclear. Sequence analysis suggested that *fixL* and *fixJ* encode a two-component regulatory system (David et al., 1988). Genetic experiments showed that expression of both *nifA* (David et al., 1988) and *fixK* (Batut et al., 1989) was dependent on *fixLJ*, both during symbiosis and in microaerobic cultures.

The *fixLJ* operon was also isolated by complementation of mutants defective for microaerobic induction of *nifA*, further substantiating the role of these regulatory genes (Virts et al., 1988). Isolation of these oxygen regulation mutants (*oxr*) was based on the observation that when cells containing a *nifA::lacZ* reporter fusion are grown on solid medium containing the chromogenic indicator X-Gal, a blue coloration appears in the center of the colonies. This was presumed to be caused by microaerobiosis in the colony centers. Mutants that did not produce blue-center colonies were isolated and were subsequently found to be defective for microaerobic *nifA* induction ex planta and also defective in symbiosis with alfalfa. A region of DNA able to complement these mutants was found to contain *fixLJ* (Virts et al., 1988).

Based on these and other observations, the following mechanism of nitrogen fixation gene regulation in *R. meliloti* was proposed (David et al., 1988; Batut et al., 1989). FixL, a two-component sensor and membrane protein, responds to microaerobiosis by phosphorylating FixJ, a two-component response regulator, which then activates transcription of *nifA* and *fixK*. The products of these regulatory genes, in turn, induce the expression of at least 15 of the remaining *nif/fix* genes (Fig. 1). In contrast to FixJ, NifA and FixK do not appear to be regulated by phosphorylation. Aside from the regulatory genes (*fixLJ*, *nifA*, *fixK*) and the genes encoding the subunits of nitrogenase (*nifHDK*), the remaining genes appear to be required for either synthesis of the molybdenum-iron cofactor of nitrogenase or various aspects of electron transport to nitrogenase.

It has also been shown that feedback control plays a role in the *nif/fix* regulon. FixK is homologous to the *Escherichia coli* proteins Fnr and Crp. It negatively regulates its own transcription, and this activity requires an Fnr consensus binding site located about 490 bp upstream of the transcription start site (Waelkens et al., 1992). The mechanism by which FixK acts as both a positive and negative regulator of transcription is presently unclear. It has also been observed that *nifA* expression is elevated in a *fixK* mutant (Batut et al., 1989), suggesting that FixK negatively regulates *nifA* expression. Although in Fig. 1 this negative activity is depicted as a direct interaction between FixK and *nifA*, it is also possible that the effect is indirect, perhaps by affecting *fixLJ* expression. The transcriptional activity of NifA is eliminated by atmospheric oxygen tension (Huala and Ausubel, 1989), showing yet another level at which oxygen regulates symbiotic nitrogen fixation.

FixL, THE SENSOR KINASE

Oxygen Sensor and Hemoprotein
The *fixL* and *fixJ* genes are sufficient to regulate the expression of *nifA* and *fixK* in response to anaerobiosis in *E. coli* (de Philip et al., 1990).

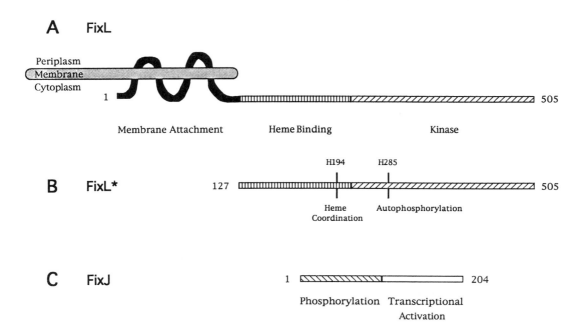

FIGURE 2 Schematic representation of the two-component system controlling nitrogen fixation gene expression in *R. meliloti*. (A) The sensor kinase, FixL. The three principal domains and the predicted membrane topology of FixL are shown. (B) FixL★, a water-soluble truncated derivative of FixL used for in vitro studies. Two critical histidine residues and their corresponding functions are shown. (C) FixJ, the response regulator. The two principal domains of FixJ are indicated. Coordinates are the positions of amino acids relative to the N terminus of the native protein (1). The amino acid numbering of FixL is according to Lois et al. (1993a).

Fully consistent with the prediction that FixJ is a transcriptional activator of *nifA* and *fixK*, overexpression of *fixJ* induces the expression of these genes in the absence of *fixL* in aerobically growing *E. coli* (Hertig et al., 1989). At lower levels of *fixJ* expression, however, both *fixL* and low oxygen concentrations are also required for gene expression, showing that FixL is an oxygen sensor and mediates the low oxygen signal through FixJ (de Philip et al., 1990). These results also indicated that any auxiliary factors that may be required for sensing must be conserved between *E. coli* and *R. meliloti*.

Evidence that FixL senses oxygen directly came from purification of a water-soluble truncated derivative of FixL, designated FixL★. This protein lacks the N-terminal membrane-spanning regions (compare Fig. 2A and 2B). A striking feature of this protein is the presence of an oxygen-binding heme moiety; its visible spectrum is nearly identical to that of hemoglobin (Gilles-Gonzalez et al., 1991). Deoxygenation of FixL★ resulted in spectral shifts similar to deoxyhemoglobin, and this effect was reversible by exposure to oxygen. FixL★ autophosphorylates using the γ-phosphate of ATP and phosphorylates purified FixJ (Gilles-Gonzalez et al., 1991). This led to the proposal that FixL senses oxygen through its heme moiety and transduces the signal by controlling FixJ phosphorylation. Anaerobiosis increases the rate of FixL★ autophosphorylation, leading to increased phosphorylation of the response regulator FixJ (Monson et al., 1992; Lois et al., 1993b; Gilles-Gonzalez and Gonzalez, 1993). These results confirmed that the enzyme activity of FixL★ can be modulated directly in response to oxygen concentration.

Three Principal Domains of FixL

Deletion analyses of FixL have shown that separate domains are responsible for the activities of this protein. A purified deletion derivative of FixL, corresponding to the C-terminal region that is conserved among two-component sensors, functions as a kinase. This fragment is capable of both autophosphorylation and phosphorylation of FixJ in vitro, albeit with less activity than FixL★ (Monson et al., 1992). In contrast to FixL★, these activities are not oxygen regulated. A deletion derivative corresponding to the central nonconserved region of the protein exhibits a visible spectrum that is identical to FixL★, indicating that this region of the protein represents a heme binding domain (Monson et al., 1992). Because the absence of this heme binding domain results in the loss of oxygen regulation of the kinase domain, it has been inferred that the heme binding domain is responsible for the regulation of enzyme activity in response to oxygen. In vivo experiments in *E. coli* are consistent with these results. In contrast to full-length FixL, which induces gene expression only under anaerobic conditions, a FixL derivative lacking the central nonconserved region of the protein activates gene expression even under aerobic growth conditions (de Philip et al., 1992). It was, therefore, proposed that the central nonconserved region represses the activity of the C-terminal kinase domain and that derepression occurs when oxygen tension is decreased.

Based on hydropathy profiles and Tn*phoA* mutagenesis, it has been proposed that FixL, a membrane protein, has four membrane-spanning regions with the topology shown in Fig. 2A (Lois et al., 1993a). The fact that FixL★ retains oxygen-regulated activity in vitro demonstrates that the amino-terminal membrane attachment domain is not essential for FixL function. The activity of FixL derivatives lacking the membrane attachment domain have also been tested in vivo. Although oxygen regulation is still observed in both *E. coli* (de Philip et al., 1992) and *R. meliloti* (Lois et al., 1993a), activity is reduced relative to the full-length protein. It is possible that the membrane environment is necessary for FixL to function optimally. Alternatively, the truncated proteins may be slightly misfolded or may be subject to proteolysis. Additional evidence that the membrane-spanning regions comprise a separate domain is suggested by the finding that the FixL homolog of *Bradyrhizobium japonicum*, a nitrogen-fixing endosymbiont of soybean, lacks this region (Anthamatten and Hennecke, 1991).

Residues Critical for FixL Function

In vitro analysis of site-directed mutants of FixL★ has allowed the identification of two critical residues. Sequence comparisons suggested that histidine 285, located in the kinase domain, is the site of autophosphorylation. Consistent with this assignment, when His-285 is changed to glutamine in FixL★, the protein cannot autophosphorylate or phosphorylate FixJ (Monson et al., in press). It has long been known that histidines are critical for the coordination of heme in proteins such as hemoglobin and myoglobin. The heme of FixL★ is noncovalently bound because detergent denaturation results in loss of this moiety. Therefore, mutation of critical histidine residues can be expected to prevent heme binding. When His-194 is changed to an asparagine, the FixL★ derivative no longer contains heme, implying that this residue is functionally analogous to coordinating histidines in hemoglobin and other heme proteins (Monson et al., in press). Further biochemical and structural studies are necessary to determine the precise role of this residue in coordination and oxygen binding.

Oxygen Regulation of FixL Activity

The rate of FixL★ autophosphorylation (reaction 1_F, Fig. 3) increases dramatically when oxygen is removed (Monson et al., 1992; Lois et al., 1993b; Gilles-Gonzalez and Gonzalez, 1993). This rate is highly dependent on the divalent cation present, with manganese exhibiting the greatest stimulation (Gilles-Gonzalez

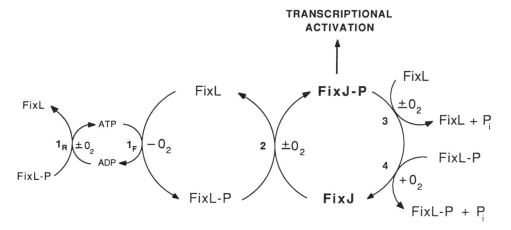

FIGURE 3 Current model for modulation of FixJ-phosphate levels by FixL in *R. meliloti*. $-O_2$ indicates that the designated reaction is stimulated by micro- or anaerobiosis. $+O_2$ indicates that the reaction is stimulated by atmospheric oxygen tension and/or repressed by low oxygen tension. $\pm O_2$ indicates that the reaction rate is not affected by oxygen concentration. The numbered reactions are described in the text; 1_F and 1_R refer to forward and reverse reactions, respectively.

and Gonzalez, 1993). Because manganese concentrations are quite high in nodules, this finding may be physiologically relevant. This reaction is also greatly affected by the ATP-to-ADP ratio, because the autophosphorylation reaction is reversible (Gilles-Gonzalez and Gonzalez, 1993) (reaction 1_R, Fig. 3). In contrast to the forward reaction, the reverse reaction is not oxygen regulated. This reverse reaction explains why it has not been possible to prepare pure FixL*-phosphate, despite the fact that this form of the protein is extremely stable (Lois et al., 1993b; Gilles-Gonzalez and Gonzalez, 1993). It has been proposed that this equilibrium may play a regulatory role because a high cellular energy charge is required for nitrogen fixation (Gilles-Gonzalez and Gonzalez, 1993). Therefore, if the bacteria encounter microaerobic conditions outside the root nodule, where nitrogen fixation does not occur, inappropriate *nif/fix* gene expression may be prevented by low levels of carbon sources, even though a low oxygen signal is present.

Because of the relatively high stability of phospho-FixL*, radiolabeled preparations can be separated from ATP/ADP and combined with FixJ to monitor FixJ phosphorylation in the absence of kinase autophosphorylation. In contrast to the autophosphorylation reaction, phosphotransfer from FixL*-phosphate to FixJ is not affected by oxygen (reaction 2, Fig. 3). FixJ phosphorylation is, therefore, regulated at the level of sensor autophosphorylation (Lois et al., 1993b; Gilles-Gonzalez and Gonzalez, 1993).

FixL* in Dephosphorylation of Phospho-FixJ

Phospho-FixJ is quite stable relative to other response regulators (Weinstein et al., 1993), thus allowing the separation of radiolabeled protein from phosphodonors. If phospho-FixJ is prepared by phosphotransfer from FixL*, the kinase can be removed by immunoaffinity chromatography using an anti-FixL* antiserum (Weinstein et al., 1993). Alternatively, as has been shown for other response regulators, FixJ can be phosphorylated by low-molecular-weight compounds such as ammonium hydrogen phosphoramidate or acetyl phosphate (Reyrat et al., 1993). In this case, removal of the phosphodonor can be achieved by size fractionation. Using radiolabeled FixJ, dephos-phorylation can be specifically monitored by measuring the decrease of radiolabeled protein under various conditions. Phospho-FixJ alone is dephosphorylated at a modest rate by hydrolysis, but the rate of phosphate loss is substantially increased in the presence of FixL★ (Lois et al., 1993b). The phosphatase activity of FixL★ is not oxygen regulated (Lois et al., 1993b) (reaction 3, Fig. 3). Unlike several other two-component sensors, this activity is not dependent on nucleotide cofactors.

In contrast to FixL★, Lois et al. (1993b) observed an oxygen-regulated phosphatase activity when an enriched phospho-FixL★ preparation was combined with radiolabeled phospho-FixJ. The aerobic phosphatase activity was the same as with pure FixL★, but this activity was substantially reduced under anaerobic conditions. ATP was present in these reactions to regenerate phospho-FixL★ that is lost by phosphotransfer to FixJ. Therefore, it is not clear whether the phosphatase activity of phospho-FixL★ is ATP-dependent. Lois et al. (1993b) proposed that phospho-FixL phosphatase activity, but not that of FixL, is repressed by low oxygen (reaction 4, Fig. 3). According to this model, it follows that under aerobic conditions, the level of phospho-FixJ is very low due to the reduced rate of FixL autophosphorylation (reaction 1_F) combined with the phosphatase activity of both FixL (reaction 3) and phospho-FixL (reaction 4). This also implies that FixJ-phosphate would be rapidly dephosphorylated after a sudden shift from microaerobiosis to aerobiosis. When oxygen tension is reduced, FixL autophosphorylation increases (reaction 1_F), simultaneously increasing the level of FixJ phosphodonor (FixL★-phosphate) and decreasing the level of unphosphorylated FixL, thus reducing the effect of reaction 3. Also, low oxygen also decreases the effect of reaction 4. The net effect, then, is an increase in the level of phospho-FixJ. Therefore, in the current model, gene expression is reciprocally coordinated by the kinase and phosphatase activities of FixL in response to changes in oxygen tension.

One aspect of the phosphatase experiments that must be noted is that it has not been possible to prepare pure phospho-FixL★, most likely because of reaction 1_R. Lois et al. (1993b) reported a preparation of 85% purity. A rigorous comparison of FixL★ and phospho-FixL★ would be greatly aided by the availability of pure phosphorylated preparations that also can be maintained in the phosphorylated form. This might be accomplished by devising a method that continuously removes ADP from the autophosphorylation reaction, thus shifting the equilibrium of reaction 1 toward FixL★ phosphorylation. This could also be important in measuring the absolute rates of the dephosphorylation reactions so that they may be compared with the rate of autophosphorylation. A comparison of rates is essential for understanding the contribution of the phosphatase activity to the overall regulation of the system. This determination is particularly important in light of the fact that even in air, phospho-FixJ accumulates in the presence of FixL★ and ATP, albeit at a relatively low rate (Monson et al., 1992; Lois et al., 1993b; Gilles-Gonzalez et al., 1991; Gilles-Gonzalez and Gonzalez, 1993). Therefore, under the conditions in which these experiments were performed, the phosphatase activity of FixL★ was insufficient to counteract fully the aerobic phosphorylation of FixJ, suggesting either that the in vitro system at present does not completely mimic the behavior of the system in vivo, or that the phosphatase activity serves to reduce, but not entirely prevent, phospho-FixJ accumulation. FixL mutants affected in phosphatase activity would also be useful for the further examination of these coupled reactions.

Negative and Positive Regulation of *nifA* Expression in *E. coli*

As previously stated, overexpression of *fixJ* in *E. coli* causes adventitious expression of *nifA* and *fixK* independently of oxygen tension in the absence of *fixL* (Hertig et al., 1989). If *fixJ* expression is reduced, target gene expression is induced by *fixL* and low oxygen, mimicking the behavior of these genes in *R. meliloti* (de Philip et al., 1990). It has also been observed, however, that the adventitious expression of *nifA* by the overexpression of *fixJ* is reduced by the coexpression of *fixL* (de Philip et al., 1990; Weinstein et al., 1993). Interestingly, this negative effect is substantially reduced under anaerobic growth conditions. The negative effect is particularly pronounced under conditions established by Weinstein et al. (1993). This has allowed the isolation of both *fixL* (Lois et al., 1993b) and *fixJ* mutants (Weinstein et al., 1993) that prevent the negative effect, using a *nifA::lacZ* fusion and growth on solid medium containing a chromogenic indicator to assay expression. A *fixL* mutant isolated from such a screen exhibits both increased kinase activity and reduced phosphatase activity when tested in vitro in the truncated (FixL★) form (Lois et al., 1993). A *fixJ* mutant that when overexpressed induces *nifA* expression even when *fixL* is also expressed displays a resistance to FixL★ phosphatase activity when the purified mutant FixJ is tested in vitro (Weinstein et al., 1993b). Given the behavior of this system, it is tempting to speculate that the negative effect of *fixL* in *E. coli* is due to a FixL phosphatase activity acting on phospho-FixJ. This is based on the hypothesis that *fixL*-independent *fixJ* activity is the result of fortuitous FixJ phosphorylation, perhaps by crosstalk from other two-component sensors or phosphorylation from acetyl phosphate. Conditions have been established in which the truncated derivative of *fixL* (*fixL★*) also displays a strong negative effect in *E. coli* (Agron, 1993). This result is significant because it suggests a relationship between the negative effect and the observations made using the in vitro system, which uses the truncated derivative of FixL. This negative activity of *fixL* observed in *E. coli* will likely continue to be useful in the study of FixL function.

FixJ, THE RESPONSE REGULATOR

Transcriptional Activator In Vitro

As previously mentioned, the observation that *fixJ* induces *nifA* and *fixK* expression in vivo strongly suggested that FixJ is a positive activator of these genes. This activity is independent of both *fixL* and oxygen tension when *fixJ* is overexpressed in *E. coli* (Hertig et al., 1989), but at lower levels of *fixJ* expression, induction of *nifA* also requires both *fixL* and low oxygen, as is the case in *R. meliloti* (de Philip et al., 1990). These results indicate that FixL controls the activity of FixJ in response to oxygen availability. Furthermore, the observations that low oxygen tension both induces gene expression in vivo (Ditta et al., 1987; Batut et al., 1989) and increases FixJ phosphorylation in vitro (Monson et al., 1992; Lois et al., 1993b; Gilles-Gonzalez and Gonzalez, 1993) implied that phospho-FixJ is the active form of this protein. These properties of FixJ are consistent with its assignment as a response regulator based on amino acid sequence comparisons and in vitro transcription experiments.

High concentrations of purified FixJ will activate transcription of both *nifA* and *fixK* in vitro, using plasmid DNA templates and either *E. coli* (Agron et al., 1993; Reyrat et al., 1993) or *R. meliloti* (Agron et al., 1993) RNA polymerase. These results show that FixJ is indeed a transcriptional activator and is not, for example, a coactivator in a more complex transcription initiation system. Therefore, FixJ and RNA polymerase are necessary and sufficient protein components for the transcription reaction. Furthermore, when *E. coli* RNA polymerase is used, the σ^{70} (RpoD) form of this enzyme is required (Reyrat et al., 1993). Because σ^{70} is the major σ-subunit in *E. coli,* it is likely that the corresponding major σ-subunit is also required for *nifA* and *fixK* expression in *R. meliloti*. Therefore, it is unlikely that expression of these genes in *R. meliloti* requires an alternative σ-factor for promoter recognition. This is in contrast to many of the other *nif/fix* genes. The *nifA*-dependent genes (see Fig. 1) are transcribed by the σ^{54} (RpoN) form of RNA polymerase holoenzyme (Gussin et al., 1986).

When FixJ concentrations are decreased, transcription can be greatly stimulated by preincubation with both FixL★ and ATP (Agron et al., 1993; Reyrat et al., 1993). Because FixL★ phosphorylates FixJ in an ATP-dependent manner, this suggests that phosphorylation stimulates FixJ activity. Furthermore, FixJ activity is also greatly stimulated in the presence of either ammonium hydrogen phosphoramidate (Agron et al., 1993) or acetyl phosphate (Reyrat et al., 1993), two low-molecular-weight phosphodonors that have been shown to phosphorylate the response regulators CheB and CheY at the appropriate aspartate residues (Lukat et al., 1992). A direct correlation between FixJ phosphorylation by these alternative phosphodonors and FixJ activity has been demonstrated using radiolabeled acetyl phosphate and showing both FixJ phosphorylation and transcriptional activation by this compound (Reyrat et al., 1993). Transcriptional stimulation by alternative phosphodonors is consistent with the view that FixL does not directly participate in transcription but rather acts indirectly through FixJ phosphorylation-dephosphorylation.

Although data are not presented, Reyrat et al. (1993) reported that under aerobic conditions, a FixL derivative similar to FixL★ exerts a negative effect on in vitro transcription in the presence of *E. coli* RNA polymerase ATP and FixJ that had been previously phosphorylated with acetyl phosphate. The effect of anaerobiosis on this reaction was not reported. This is reminiscent of the in vivo negative effect in *E. coli* discussed earlier. This observation is important because this effect may be the result of catalyzed dephosphorylation of FixJ-phosphate. Agron et al. (1993), however, using FixJ that had not been prephosphorylated, observed an ATP-dependent stimulatory effect of FixL★ under aerobic conditions with RNA polymerase from *E. coli,* but not with RNA polymerase from *R. meliloti.* The reasons for the different results obtained with *E. coli* RNA polymerase are unclear, but it may be that tighter regulation is obtained with the RNA polymerase from the native host, which would not be unexpected. More work is necessary to understand the down-regulation by FixL derivatives in the in vitro transcription system.

Conditions in which anaerobiosis greatly stimulates transcription have also been established. This can be shown with either *E. coli* RNA polymerase (Agron et al., 1993; Reyrat et al., 1993) or polymerase purified from *R. meliloti,* the native host (Agron et al., 1993). Thus, with the FixL/FixJ system, signal transduction has been reconstituted in its entirety from stimulus (low oxygen tension) to response (transcription). This signal transduction in vitro requires only three water-soluble purified protein components: FixL★ (or an equivalent FixL derivative), FixJ, and RNA polymerase. The development of an oxygen-regulated in vitro transcription system is an important step in the study of FixJ action at target promoters. Various biochemical techniques can now be used to examine the structure and behavior of the target promoter complexes.

Two Principal Domains of FixJ

A deletion mutant of *fixJ* that encodes a truncated protein lacking the N-terminal region that is conserved among all two-component response regulators is active as a transcriptional activator in vivo (Kahn and Ditta, 1991). This shows that the N terminus is not required for transcriptional activity. Furthermore, this derivative is active in *R. meliloti* even under aerobic conditions (Kahn and Ditta, 1991). This implies that the N terminus represses the activity of the C terminus and that a low oxygen concentration causes derepression. Because FixJ activity is controlled by phosphorylation, it is likely that protein phosphorylation is responsible for the derepression. Based on sequence similarity to other response regulators, FixJ is thought to be phosphorylated at an aspartate residue in the N terminus. Phosphorylation of this residue may cause a conformational change that releases the activity of the C-terminal domain. Alternatively, phosphorylation could regulate FixJ activity by influencing intermolecular interactions such as dimerization or monomerization. A schematic diagram of the proposed domain structure of FixJ is presented in Fig. 2C.

A helix-turn-helix DNA binding motif is located near the C terminus of FixJ. This region is homologous to the −35 binding region of σ-factors; at present, the significance of this homology with respect to FixJ activity is unclear (Kahn and Ditta, 1991). This motif may be responsible for sequence-specific DNA binding by FixJ that would allow promoter recognition. Consistent with this hypothesis, many mutations in the putative helix-turn-helix result in a loss of *nifA* expression in *E. coli* (Kahn and Ditta, 1991). Despite these observations, sequence-specific DNA binding by FixJ or phospho-FixJ has not yet been reported. Attempts at demonstrating sequence-specific DNA binding to either the *nifA* or *fixK* promoters using purifed proteins or cell extracts have been thus far unsuccessful (Agron, 1993). Consequently, the basic mechanism of FixJ action at target promoters is unknown and is an important question for further study.

Two-component response regulators that function as transcription factors can be divided into three subclasses based on sequence similarity outside the universally conserved N-terminal phosphoryl acceptor domain (Stock et al., 1989). The first two subclasses are represented by NtrC and OmpR. The basics of the mode of action of these response regulators at target promoters are reasonably well understood. The mechanisms are very different between these two subclasses and appear to be very similar within the two subclasses. FixJ, however, is a member of a third subclass that includes UhpA, the response regulator of the hexose-phosphate uptake system of *E. coli*. The basic mechanism of action of any member of this subclass is not understood. Studies of FixJ therefore have important implications for understanding regulation by these proteins. The primary structure of UhpA is highly similar to that of FixJ over its entire length (David et al., 1988). Although an in vitro assay for studying UhpA has not been reported, in vivo titration experiments have shown that this protein binds to an inverted repeat in the upstream region of the *uhpT* promoter (Merkel et al., 1992).

FixJ Target Promoters

Several mutations in both the *nifA* and *fixK* promoters have helped define sequences required for transcription in *R. meliloti*. The 5′ end of the *nifA* promoter (PnifA) lies between positions −57 and −54 bp relative to the start site of transcription, defined as +1 (Agron et al., 1992). Sequences upstream of −67 can be deleted from the *fixK* promoter (PfixK) without decreasing microaerobic induction (Agron et al., 1992). As discussed earlier, transcription from these promoters requires the σ^{70} holoenzyme when *E. coli* RNA polymerase is used (Reyrat et al., 1993), suggesting that sequences around −35 and −10 are required for polymerase binding, as is the case with other σ^{70} promoters. Essential sequences have been found upstream of −35 in each promoter and are, therefore, candidates for FixJ binding sites. PnifA sequences between −54 and −39 are essential for microaerobic induction in *R. meliloti* (Agron et al., 1992). Several single point mutations in this region abolish promoter activity. In PfixK, point mutations in the −60, −45, and −35 regions result in decreased promoter activity (Waelkens et al., 1992). The sequences of PnifA and PfixK do not reveal any conspicuous inverted or direct repeats that might comprise a FixJ binding site(s). A sequence comparison of four FixJ-dependent promoters, however, reveals sequence conservation in the −54 to −33 region (Waelkens et al., 1992). The four promoters are from *nifA*, *fixK*, a second copy of *fixK* found in *R. meliloti*, and a *fixK* homolog from *Azorhizobium caulinodans*, the nitrogen-fixing endosymbiont of *Sesbania rostrata*, a tropical legume. The expression of the *A. caulinodans fixK* is controlled by *fixL/fixJ* homologs (Kaminski et al., 1991). Despite this sequence conservation, PfixK sequences upstream of −57, outside the conserved region, are required for normal promoter function (Waelkens et al., 1992). This is not the case for PnifA, suggesting that although both promoters are FixJ-dependent, the sequence requirements of each promoter are different. One possible explanation is that

PfixK contains an additional FixJ binding site(s). Furthermore, point mutations that change specific bases pairs of PnifA to the corresponding nucleotide residue of PfixK (Agron et al., 1992) and vice versa (Waelkens et al., 1992) unexpectedly lead to a loss of promoter activity.

The mutants generated in these studies will be useful for biochemical studies of transcription initiation at these promoters. For example, a PnifA mutant with nucleotide changes in the −52 region of the promoter is severely defective when assayed in planta, in *E. coli*, in *R. meliloti* after microaerobic induction ex planta (Agron et al., 1992), and for in vitro transcription (Agron et al., 1993), further suggesting that the in vitro system faithfully reflects the in vivo mechanism of promoter activation.

CONCLUSION AND PERSPECTIVES

In addition to the importance of FixL/FixJ in the biology of *R. meliloti* and related bacteria such as *B. japonicum* (Anthamatten and Hennecke, 1991) and *A. caulinodans* (Kaminski and Elmerich, 1991), the relative simplicity of the system makes it an attractive model for the study of two-component signal transduction. Perhaps the most remarkable feature of this system is its autonomy. Signal transduction involving FixL/FixJ does not appear to require auxiliary proteins as in many other systems, but rather FixL senses an environmental stimulus (low oxygen tension) directly. However, it is possible that oxygen levels are not the only environmental stimulus that is sensed by FixL. The response of the sensor to molecular oxygen concentration is clear at a basic level because FixL is a hemoprotein. In contrast with many other systems, the phosphorylated forms of both the sensor kinase (FixL) and the response regulator (FixJ) are very stable, greatly facilitating many biochemical approaches (Lois et al., 1993b; Gilles-Gonzalez and Gonzalez, 1993; Weinstein et al., 1993). *E. coli* assays are available to screen for mutants affected in regulation, and these mutants cannot only be tested in vitro but can also be introduced into *R. meliloti* to correlate biochemical behavior with phenotype (Weinstein et al., 1992). Furthermore, this system conforms well to the overall architecture of the standard two-component system. The sensor consists of an input (heme binding) domain and a transmitter (kinase) domain, as well as a transmembrane domain, whereas the response regulator consists of a receiver (phosphorylation) domain and an output (transcriptional activation) domain. Many other systems are more complex variations on this theme. Finally, signal transduction can be reconstituted in vitro, from stimulus to gene expression, with a minimum of water-soluble components (Agron et al., 1993; Reyrat et al., 1993). Although many of these studies used an *R. meliloti* FixL derivative lacking the membrane attachment domain, the FixL homolog of *B. japonicum* is water soluble in its native form and therefore provides a useful basis of comparison. Given these features, the FixL/FixJ system provides an excellent paradigm for the study of two-component signal transduction.

The study of FixL/FixJ may have important consequences in understanding other oxygen-regulated biological processes. Oxygen is an important regulator of many processes in bacteria. Therefore, FixL/FixJ homologs may play a role in other bacterial processes aside from nitrogen fixation. Oxygen is also an important regulator of cell function in eukaryotes. In mammals, for example, the transcription of the erythropoietin gene, which encodes a stimulator of erythrocyte production, is induced by hypoxia (low oxygen tension). There is evidence that a heme protein is required for the signal transduction (Goldberg et al., 1988). It is possible that the mammalian oxygen sensor may be homologous to FixL, because two-component systems have been found in eukaryotes. Even if this is not the case, mammalian oxygen sensors may be similar to FixL in other ways, because they respond to similar oxygen levels. Therefore, biochemical studies of FixL may have wide-ranging importance.

Intensive research on oxygen-binding hemo-proteins for more than 40 years has provided a formidable battery of experimental approaches that can be used to study FixL. Because FixL is an oxygen sensor and not an oxygen carrier like hemoglobin and myoglobin, its oxygen binding characteristics may prove to be quite different and fundamentally important. How oxygen binding influences protein structure and enzyme activity is also an important question for future work. The structural basis for communication between the various domains within both FixL and FixJ is not yet understood. In addition to the intermolecular interactions between the kinase (transmitter) and phosphoryl acceptor (receiver) domains, other important intermolecular interactions may be important, such as monomer-dimer transitions. The subunit structures of these proteins have not been rigorously determined. Approaches such as X-ray crystallography and raman spectroscopy are also likely to be important in elucidating structure-function relationships. The final critical step of signal transduction, transcription initiation by phospho-FixJ, is poorly understood. The successful development of an in vitro transcription system provides a foothold toward understanding this process. Although many significant strides have been made in understanding the FixL/FixJ system, it is clear that a great deal of exciting work lies ahead.

ACKNOWLEDGMENTS

The studies carried out in our laboratory on the FixL/FixJ system were supported by research grants from the National Science Foundation (DMB-8716673) and the National Institutes of Health (5 RO1 GM 44400/02). P.G.A. was supported by training grants HG0005–01/02 from the National Institutes of Health.

REFERENCES

Agron, P. G. 1993. Transcriptional regulation of *Rhizobium meliloti* nitrogen fixation genes by oxygen. Ph.D. thesis. University of California, San Diego.

Agron, P. G., G. S. Ditta, and D. R. Helinski. 1992. Mutational analysis of the *Rhizobium meliloti nifA* promoter. *J. Bacteriol.* **174**:4120–4129.

Agron, P. G., G. S. Ditta, and D. R. Helinski. 1993. Oxygen regulation of *nifA* transcription in vitro. *Proc. Natl. Acad. Sci. USA* **90**:3506–3510.

Anthamatten, D., and H. Hennecke. 1991. The regulatory status of the *fixL*- and *fixJ*-like genes in *Bradyrhizobium japonicum* may be different from that in *Rhizobium meliloti*. *Mol. Gen. Genet.* **225**:38–48.

Appleby, C. A. 1984. Leghemoglobin and *Rhizobium* respiration. *Annu. Rev. Plant Physiol.* **35**:443–478.

Batut, J., M.-L. Daveran-Mingot, M. David, J. Jacobs, A. M. Garnerone, and D. Kahn. 1989. *fixK*, a gene homologous with *fnr* and *crp* from *Escherichia coli*, regulates nitrogen fixation genes both positively and negatively in *Rhizobium meliloti*. *EMBO J.* **8**:1279–1286.

David, M., M.-L. Daveran, J. Batut, A. Dedieu, O. Domergue, J. Ghai, C. Hertig, P. Boistard, and D. Kahn. 1988. Cascade regulation of *nif* gene expression in *Rhizobium meliloti*. *Cell* **54**:671–683.

David, M., O. Domergue, P. Pognonec, and D. Kahn. 1987. Transcription patterns of *Rhizobium meliloti* symbiotic plasmid pSym: identification of *nifA*-independent *fix* genes. *J. Bacteriol.* **169**:2239–2244.

Denarie, J., and J. Cullimore. 1993. Lipo-oligosaccharide nodulation factors: a new class of signalling molecules mediating recogniton and morphogenesis. *Cell* **74**:951–954.

de Philip, P., J. Batut, and P. Boistard. 1990. *Rhizobium meliloti* FixL is an oxygen sensor and regulates *R. meliloti nifA* and *fixK* genes differently in *Escherichia coli*. *J. Bacteriol.* **172**:4255–4262.

de Philip, P., E. Soupene, J. Batut, and P. Boistard. 1992. Modular structure of the FixL protein of *Rhizobium meliloti*. *Mol. Gen. Genet.* **235**:49–54.

Ditta, G., E. Virts, A. Palomares, and C.-H. Kim. 1987. The *nifA* gene of *Rhizobium meliloti* is oxygen regulated. *J. Bacteriol.* **169**:3217–3223.

Fisher, R. F., and S. R. Long. 1992. *Rhizobium*-plant signal exchange. *Nature* (London) **357**:655–660.

Gilles-Gonzalez, M. A., G. S. Ditta, and D. R. Helinski. 1991. A haemoprotein with kinase activity encoded by the oxygen sensor of *Rhizobium meliloti*. *Nature* **350**:170–172.

Gilles-Gonzalez, M. A., and G. Gonzalez. 1993. Regulation of the kinase activity of heme protein FixL from the two-component system FixL/FixJ of *Rhizobium meliloti*. *J. Biol. Chem.* **268**:16293–16297.

Goldberg, M. A., S. P. Dunning, and H. F. Bunn. 1988. Regulation of the erythropoietin gene: evidence that the oxygen sensor is a heme protein. *Science* **242**:1412–1415.

Gussin, G. N., C. W. Ronson, and F. M. Ausubel. 1986. Regulation of nitrogen fixation genes. *Annu. Rev. Genet.* **20**:567–591.

Haaker, H., and J. Klugkist. 1987. The bioenergetics of electron transport to nitrogenase. *FEMS Microbiol. Rev.* **46**:57–71.

Hertig, C., R. Y. Li, A.-M. Louarn, A.-M. Garnerone, M. David, J. Batut, D. Kahn, and P. Boistard. 1989. *Rhizobium meliloti* regulatory gene *fixJ* activates transcription of *R. meliloti nifA* and *fixK* genes in *Escherichia coli. J. Bacteriol.* **171**:1736–1738.

Hill, S. 1988. How is nitrogenase regulated by oxygen? *FEMS Microbiol. Rev.* **54**:111–130.

Huala, E., and F. M. Ausubel. 1989. The central domain of *Rhizobium meliloti* NifA is sufficient to activate transcription from the *R. meliloti nifH* promoter. *J. Bacteriol.* **170**:3354–3365.

Kahn, D., M. David, O. Domergue, M.-L. Daveran, J. Ghai, P. R. Hirsch, and J. Batut. 1989. *Rhizobium meliloti fixGHI* sequence predicts involvement of a specific cation pump in symbiotic nitrogen fixation. *J. Bacteriol.* **171**:929–939.

Kahn, D., and G. Ditta. 1991. Modular structure of FixJ: homology of the transcriptional activator domain with the −35 binding domain of sigma factors. *Mol. Microbiol.* **5**:987–997.

Kaminski, P. A., and C. Elmerich. 1991. Involvement of *fixLJ* in the regulation of nitrogen fixation in *Azorhizobium caulinodans. Mol. Microbiol.* **5**:665–673.

Kaminski, P. A., K. Mandon, F. Arigoni, N. Desnoues, and C. Elmerich. 1991. Regulation of nitrogen fixation in *Azorhizobium caulinodans:* identification of a *fixK*-like gene, a positive regulator of *nifA. Mol. Microbiol.* **5**:1983–1991.

Lois, A. F., M. Weinstein, G. S. Ditta, and D. R. Helinski. 1993a. The oxygen sensor FixL of *Rhizobium meliloti* is a membrane protein containing four possible transmembrane segments. *J. Bacteriol.* **175**:1103–1109.

Lois, A. F., M. Weinstein, G. S. Ditta, and D. R. Helinski. 1993b. Autophosphorylation and phosphatase activities of the oxygen-sensing protein FixL of *Rhizobium meliloti* are coordinately regulated by oxygen. *J. Biol. Chem.* **268**:4370–4375.

Long, S. R. 1989. Rhizobium-legume nodulation: life together in the underground. *Cell* **56**:203–214.

Lukat, G. S., W. R. McCleary, A. M. Stock, and J. B. Stock. 1992. Phosphorylation of bacterial response regulator proteins by low molecular weight phospho-donors. *Proc. Natl. Acad. Sci. USA* **89**:718–722.

Merkel, T. J., D. M. Nelson, B. L. Brauer, and R. J. Kadner. 1992. Promoter elements required for positive control of transcription of the *Escherichia coli uhpT* gene. *J. Bacteriol.* **174**:2763–2770.

Monson, E. K., G. S. Ditta, and D. R. Helinski. The oxygen sensor protein, FixL, of *Rhizobium meliloti:* role of histidine residues in heme binding, phosphorylation and signal transduction. *J. Biol. Chem.*, in press.

Monson, E. K., M. Weinstein, G. S. Ditta, and D. R. Helinski. 1992. The FixL protein of *Rhizobium meliloti* can be separated into a heme-binding oxygen-sensing domain and a functional C-terminal kinase domain. *Proc. Natl. Acad. Sci. USA* **89**:4280–4284.

Renalier, M.-H., J. Batut, J. Ghai, B. Terzaghi, M. Gherardi, A.-M. Garnerone, J. Vasse, G. Truchet, T. Huguet, and P. Boistard. 1987. A new symbiotic cluster on the pSym megaplasmid of *Rhizobium meliloti* 2011 carries a functional *fix* gene repeat and a *nod* locus. *J. Bacteriol.* **169**:2231–2238.

Reyrat, J.-M., M. David, C. Blonski, P. Boistard, and J. Batut. 1993. Oxygen-regulated in vitro transcription of *Rhizobium meliloti nifA* and *fixK* genes. *J. Bacteriol.* **175**:6867–6872.

Stock, J. B., A. J. Ninfa, and A. M. Stock. 1989. Protein phosphorylation and regulation of adaptive responses in bacteria. *Microbiol. Rev.* **53**:450–490.

Szeto, W. W., J. L. Zimmerman, V. Sundaresan, and F. M. Ausubel. 1984. Identification and characterization of the *Rhizobium meliloti ntrC* gene: *R. meliloti* has separate regulatory pathways for activating nitrogen fixation genes in free-living and symbiotic cells. *J. Bacteriol.* **169**:1423–1432.

Virts, E. L., S. W. Stanfield, D. R. Helinski, and G. S. Ditta. 1988. Common regulatory elements control symbiotic and microaerobic induction of *nifA* in *Rhizobium meliloti. Proc. Natl. Acad. Sci. USA* **85**:3062–3065.

Waelkens, F., A. Foglia, J.-B. Morel, J. Fourment, J. Batut, and P. Boistard. 1992. Molecular genetic analysis of the *Rhizobium meliloti fixK* promoter: identification of sequences involved in positive and negative regulation. *Mol. Microbiol.* **6**:1447–1456.

Weinstein, M., A. F. Lois, G. S. Ditta, and D. R. Helinski. 1993. Mutants of the two-component regulatory protein FixJ of *Rhizobium meliloti* that have increased activity at the *nifA* promoter. *Gene* **134**:145–152.

Weinstein, M., A. F. Lois, E. K. Monson, G. S. Ditta, and D. R. Helinski. 1992. Isolation of phosphorylation-deficient mutants of the *Rhizobium meliloti* two-component regulatory protein, FixJ. *Mol. Microbiol.* **6**:2041–2049.

Complex Phosphate Regulation by Sequential Switches in *Bacillus subtilis*

F. Marion Hulett

18

The formation of two cell types with differing developmental fates, a small forespore and a large mother cell, is the first morphological indication of early sporulation in *Bacillus subtilis*. This apparently simple morphological structure belies the complex network of interconnected regulatory pathways that are activated during late growth in response to nutritional stress and cell cycle-related signals. These interconnected regulatory pathways control which genes are expressed and the changing physiological state of the cell, often culminating in the initiation of sporulation. Such signal transduction pathways in *B. subtilis* cannot be viewed as linear regulatory pathways but rather as a signal transduction network that gathers diverse input from the environment and intracellular signals and processes that information to determine which program of late growth response is most appropriate, thus determining the hierarchy of environmental signal responses.

Depletion of nutrients, including phosphate, is a stress often encountered by a bacterial cell and results in slowed growth, marking the cessation of exponential growth. P_i is the critical limiting nutrient for biological growth in soil, the natural environment of *B. subtilis*. Soil bacteria, including *B. subtilis*, have evolved complex regulatory systems for using this limited nutrient, which is often present at levels 2 to 3 orders of magnitude lower than those of other required ions (Ozanne, 1980).

Genes that are transcriptionally activated by phosphate starvation have been used to examine the signal transduction mechanisms governing gene expression during postexponential growth. The *phoA* and *phoB* loci, encoding alkaline phosphatases APase A and APase B, are two such genes and have been used to identify a network of at least three signal transduction systems that have a role in the phosphate deficiency response of *B. subtilis*. The interconnected pathways involve the PhoP-PhoR system, whose primary role is the phosphate deficiency response, the Spo0 phosphorelay required for the initiation of sporulation, and a new signal transduction system ResD-ResE, which also has a role in respiratory regulation during late growth. The PhoP-PhoR and the ResD-ResE systems positively regulate the Pho response, whereas the Spo0 system represses both activator signal transduction systems, thereby repressing the Pho response. Perhaps the most novel aspect to come from studies of the Pho signal transduction system is

F. Marion Hulett, Laboratory for Molecular Biology, Department of Biological Sciences, University of Illinois at Chicago, Chicago, Illinois 60607.

Two-Component Signal Transduction, Edited by James A. Hoch and Thomas J. Silhavy,
© 1995 American Society for Microbiology, Washington, DC 20005

the discovery that the nature of the communication between two signal transduction systems may differ during late growth responses and during the developmental process of spore formation.

ALKALINE PHOSPHATASE GENE FAMILY AS REPORTERS OF THE Pho REGULON

The study of phosphate metabolism in *Bacillus* species in general and in *B. subtilis* in particular has been complicated by two facts that reflect the potential importance of this process. First, APase, the usual enzyme of choice as a reporter of phosphate starvation-regulated gene expression, is encoded by a multigene family composed of at least five genes (Hulett, 1993; Bookstein et al., 1990; Hulett et al., 1990; Kapp et al., 1990). Second, APases are induced under phosphate starvation conditions, but they are also expressed during sporulation development, independent of phosphate concentration. It was therefore necessary to define a promoter region sufficient for phosphate starvation-induced expression and use it in a reporter construct to assess phosphate-regulated gene transcription uncomplicated by the sporulation-induced transcription.

Two unlinked APase structural genes, *phoA* (formerly *phoAIV*) and *phoB* (formerly *phoAIII*), have been cloned using reverse genetics (Hulett et al., 1991). Characterization of the genes showed that the sequences encoding the mature proteins are 64% identical and that the deduced amino acid sequences are 63% identical. Both genes are expressed during phosphate-limited growth (Bookstein et al., 1990; Kapp et al., 1990) and during the developmental process leading to sporulation (Birkey et al., unpublished data; Chesnut et al., 1991). That APase transcriptional regulation during phosphate-limited stress appears to be controlled by a different signal transduction regime from APase transcription during spore development (Chesnut et al., 1991; Piggot and Taylor, 1977) is discussed below.

PhoP (RESPONSE REGULATOR) AND PhoR (SENSOR KINASE): THE PRIMARY Pho REGULON SIGNAL TRANSDUCTION SYSTEM

Two genes, *phoP* and *phoR*, encoding proteins that show similarity in sequence to procaryotic two-component signal transduction regulators (Lee and Hulett, 1992; Miki et al., 1965; Seki et al., 1987, 1988) are required for phosphate starvation induction of APases (Chesnut et al., 1991; Miki et al., 1965; Piggot and Taylor, 1977). APase A and APase B are the main APase proteins expressed during phosphate limitation, accounting for 98% of that activity. Both members of the signal transduction system that responds to limiting phosphate, PhoP (response regulator) and PhoR (histidine kinase), are equally required for the transcription of either APase gene, *phoA* or *phoB*, during phosphate limitation (Table 1). This fact indicates that no P_i-independent controls activate PhoP in the absence of its cognate sensor, PhoR, at least under the culture conditions used in the reported study (Hulett et al., 1994). In *Escherichia coli*, a mutation in *phoR*, the gene encoding the histidine kinase, results in constitutive expression of the Pho regulon, removing both repression and induction of the regulon. This constitutive expression is believed to involve either cross regulation by CreC (formerly PhoM), a sensor kinase induced by growth on glucose, or acetyl phosphate during growth on pyruvate (Wanner, 1992; Wanner and Wilmes-Riesenberg, 1992).

Comparison of the *B. subtilis phoA* (APase A) and *phoB* (APase B) promoter sequences used during phosphate limitation showed certain similarities but no sequences similar to the *E. coli* "Pho box." Both promoters have −10 regions that are similar (*phoA*) or identical (*phoB*) to the consensus sequence for a σ^A-promoter. However, there is no similarity at the −35 region to any σ-consensus sequence or to each other. Interestingly, both APase promoters have an identical 6-bp sequence, TTAACA, positioned 9 bp 5′ of the −10. The significance of this sequence, if any, is unknown. No se-

TABLE 1   Regulation of Pho regulon genes by Pho, Spo, and Res two-component signal transduction systems

Genotype	Sp act (%)			
	APase[a]	phoA[b]	phoB[c]	phoPR[d]
spo+pho+res+ [e]	100	100	100	100
phoP	<5.0	<2.0	<2.0	<1
phoR	<5.0	<2.0	<2.0	<1
spo0A	>500	>500	>500	≈200
resD	10	ND[f]	ND	<1
resE	30–40	ND	ND	ND
resDE spo0A	≈250	ND	ND	≈110

[a]Total alkaline phosphatase specific activity expressed as percentage of specific activity measured in the parent strain, JH642, grown in a low-phosphate defined medium.

[b]β-Gal from phoA::lacZ fusion. Percentage specific activity and culture conditions as above.

[c]β-Gal from phoB::lacZ fusion. Percentage specific activity and culture conditions as above.

[d]β-Gal from phoPR::lacZ fusion. The difference between specific activity during low-level constitutive expression and maximum induced expression in the parent strain was considered as 100%. Percentage of that specific activity and culture conditions as above.

[e]JH642 trp phe is the parent strain for all lacZ promoter fusions.

[f]ND, value not determined.

quence similar to the Pho box consensus sequence (Makino et al., 1988) for PhoB binding in E. coli (or presumably PhoP in B. subtilis) was found in either APase gene promoter used during phosphate starvation induction. A sequence with similarity to the E. coli Pho box sequence, which was noted previously (Bookstein et al., 1990) in the phoB (APase B) 5′ region, could be deleted without changing the phosphate limitation-inducible (Pv) promoter activity (Chesnut et al., 1991). This suggests that if PhoP binds directly to the APase promoters, the binding sequence is different than for PhoB of E. coli (Chesnut et al., 1991). Preliminary evidence, using partially purified preparations of PhoP for gel retardation studies, suggests that PhoP may bind directly to the phoA promoter region (Liu and Hulett, unpublished data).

The sequence comparison of the Pho regulon two-component regulatory proteins of E. coli and B. subtilis shows extensive similarities but also reveals intriguing differences, especially in the PhoR protein sequences. PhoP is 40% identical to PhoB of E. coli, the transcription activator protein for the Pho regulon of E. coli (Makino et al., 1988; Seki et al., 1987; Yamada et al., 1989). Attempts to complement E. coli phoB mutants with B.

subtilis phoP have been unsuccessful (Liu and Hulett, unpublished data). It has not been determined if the PhoR of E. coli is unable to phosphorylate PhoP or if PhoP~P cannot activate the expected reporter gene transcription, E. coli phoA.

The kinase for PhoB of E. coli, PhoR, and the carboxyl-terminal three-fourths of PhoR from B. subtilis show significant similarity (Seki et al., 1988). However, the B. subtilis PhoR has an amino-terminal extension of 137 amino acids, which is not found in the E. coli PhoR protein (Fig. 1). This amino-terminal extension in the PhoR of B. subtilis includes a hydrophobic domain close to the amino terminus, followed by a long hydrophilic amino acid sequence. Sequence similarity beginning with the N-terminal Met of the E. coli PhoR begins just before the second hydrophobic domain of the B. subtilis PhoR. The predicted membrane topology of the Bacillus PhoR (Fig. 1) includes a large "periplasmic" domain, a domain that is believed to function as the receiver domain in many sensor kinases (Tokishita et al., 1991) but is absent in the E. coli PhoR. The PhoR from both organisms contains an elongated C2 region (Fig. 1). This extended cytoplasmic

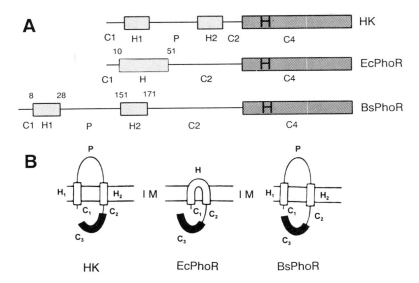

FIGURE 1 Comparison of linear domains and predicted membrane topology of PhoR proteins of *E. coli* and *B. subtilis* and the generalized structure of two-component sensor kinases. Cytoplasmic domains are designated as lines or, in the case of the histidine kinase, a shaded or darkened box. Each cytoplasmic domain is lettered C for cytoplasmic and numbered according to its position from the amino terminus to the carboxy terminus. Hydrophobic amino acid sequences, presumed to be transmembrane domains, are designated as stippled boxes lettered H and numbered as described for cytoplasmic domains. The positions of the amino acids in the primary sequence that comprise the hydrophobic domains are indicated at the top left end and right end of the stippled boxes in PhoR linear domain illustrations. The protein domain located outside the cytoplasmic membrane is lettered P (for periplasmic in gram-negative bacteria). (A) Comparison of linear domains of a generalized sensor-histidine kinase (HK), the PhoR from *E. coli* (EcPhoR), and the PhoR of *Bacillus subtilis* (BsPhoR). (B) Predicted membrane topology of the three sensor kinases shown in A. This figure is an extension of a model presented by Scholten and Tommassen (1993) to include the *Bacillus* PhoR.

region, which is between the second transmembrane domain and the histidine kinase domain, was proposed by Scholten and Tommassen (1993) to function as an internal sensing domain for an internal signal that represses the kinase function of the *E. coli* PhoR protein. Thus, the *Bacillus* PhoR protein contains two possible sensing domains, the extended cytoplasmic C_2 region similar to the proposed sensor domain in *E. coli* PhoR (Scholten and Tommassen, 1993) and a large periplasmic domain that is the assumed or implicated sensing domain of several sensor kinases (Tokishita et al., 1991; Miller et al., 1992). Possible functions for two sensing domains are discussed below.

Pho REGULON IS CONTROLLED BY A REGULATORY NETWORK COMPOSED OF THREE TWO-COMPONENT SIGNAL TRANSDUCTION SYSTEMS

Induction of the *phoPR* Operon Is Autoregulated

The *B. subtilis* Pho regulon regulator genes, *phoP* and *phoR,* and genes they regulate, alkaline phosphatase genes *phoA* and *phoB,* are induced at the end of exponential growth as a culture enters stationary growth (Hulett et al., 1994; Hulett et al., 1994) in response to extracellular phosphate concentrations dropping to less than 0.1 mM (Hulett and Jensen, 1988). Low-level constitutive expression of the *phoPR* operon during cell growth under phosphate-replete conditions is independent of PhoP and PhoR, but the two- to threefold induction of this operon initiated on phosphate limitation requires both PhoP and PhoR (Birkey et al., 1994; Hulett et al., 1994) (Table 1). Thus, autoinduction of the *phoPR* operon accompanies the induction of Pho regulon genes, both presumably dependent on phosphorylated PhoP. In a Spo⁺ strain, the induction of the Pho regulators and Pho regulon genes continues for several hours and then stops, even as conditions of limiting phosphate persist or worsen.

Spo0A~P Represses *phoPR* Induction

Unexpectedly, as *B. subtilis* cultures continue to experience phosphate-limiting stress conditions, a second genetic switch, Spo0A~P (Ferrari et al., 1985; Burbulys et al., 1991) terminates the phosphate response (Jensen et al., 1993). Spo0A is the main regulator of transition-stage regulation in *B. subtilis* and is the subject of Chapter 10. A common role for Spo0A~P in postexponential gene expression is to repress transcription of *abrB* (Perego et al., 1991), a gene encoding the exponential-phase repressor, AbrB (Strauch et al., 1989). Removal of AbrB allows expression of certain genes postexponentially (Marahiel et al., 1987; Robertson et al., 1989) and frees other genes to be activated by their genetic switches, should the correct environmental conditions provide the appropriate signal (Strauch and Hoch, 1993). This is not the role of Spo0A~P in Pho regulon regulation.

In contrast with the situation in Spo⁺ cells, neither transcription of the *phoPR* operon nor that of the genes they regulate, *phoA* and *phoB*, is terminated after 2 to 3 h of induction in a *spo0A* mutant. This results in hyperinduction of Pho regulon genes and the accumulation of the Pho gene products assayed (Table 1), PhoP, APase A, and APase B (Hulett et al., 1994; Hulett et al., 1994). The evidence suggests that the phosphorylated form of Spo0A is required

for repression of Pho regulon genes because mutations in genes whose products are involved in the phosphorylation of Spo0A (Burbulys et al., 1991), *spo0F* and *spo0B*, cause the same hyperinduction phenotype as a *spo0A* mutation (Hulett et al., 1994; Sun and Hulett, unpublished data). The mechanism of repression of the Pho regulon genes by Spo0A~P is not clear. However, the Spo0A~P repression of the *phoPR* operon or of the APase genes is apparently indirect because gel retardation of or footprinting on the *phoB* or the *phoPR* operon promoter by Spo0A was not observed (Strauch et al., unpublished data).

A model of the regulation of APase and other Pho regulon genes during phosphate limitation, based on data described above and on general properties conserved among two-component signal transduction systems, is diagrammed in Fig. 2. When *B. subtilis* experiences phosphate depletion, APases, which are dependent on PhoP and PhoR for expression, are synthesized. Because mutations in PhoR are as deleterious to APase production as mutations in PhoP, we propose that (under the experimental conditions used) PhoR alone is responsible for activation of PhoP, presumably by phosphorylation (Makino et al., 1989). The activated PhoP acts as a transcription activator of Pho regulon genes including *phoA*, *phoB*, and *phoPR*. Induction continues for several hours while cellular phosphate reserves are depleted. If P_i or certain complex phosphate compounds that can serve as a source of P_i are added, exponential growth resumes. By contrast, if no phosphorylated substrates are made available that would allow the phosphate response system to provide the limiting nutrient, P_i, the nutritional stress worsens and the phosphate response fails. The genes encoding the phosphate-sensing two-component system, the *phoPR* operon, are repressed by a second regulator (Burbulys et al., 1991; Ferrari et al., 1985), Spo0A~P, which concomitantly releases other adaptive response systems from repression by AbrB (Strauch et al., 1990) or enables the cell to sporulate (Burbulys et al., 1991). The

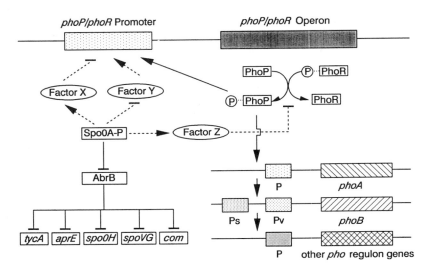

FIGURE 2 Model for the roles of PhoP, PhoR, and Spo0A in Pho regulon regulation. Arrowheads indicate positive regulation. Vertical bars indicate negative regulation. Factor X, factor Y, and factor Z are hypothetical intermediates, indicating (via dashed lines) that Spo0A~P regulation on *phoPR* transcription may be indirect or possibly affect phosphorylation of PhoP (factor Z). Arrows with solid lines identify promoters believed to require PhoP~P for induction. Solid lines with bars identify negative regulation by Spo0A or AbrB. PhoP (response regulator); PhoR (histidine kinase); Spo0A (response regulator); AbrB (repressor of many postexponentially expressed genes). Genes regulated by Spo0A and/or AbrB are not limited to those shown.

Spo0A~P repression of the *phoPR* operon is apparently indirect (Strauch et al., unpublished data). At least two mechanisms of Spo0A~P repression of the induction of the *phoPR* operon seem possible based on the current data. Spo0A~P may regulate genes whose products (Fig. 2, factor X or factor Y) regulate *phoPR* transcription via a pathway antagonistic to or enhancing the role of PhoP~P, or conversely, through a pathway (Fig. 2, factor Z) that affects the phosphorylation state of PhoP, which, in turn, affects *phoPR* transcription. We favor the latter hypothesis.

A Third Signal Transduction System, ResD and ResE, Is Required for Full Induction of the Pho Regulon

Efforts of the European *Bacillus* genome sequencing project identified two new genes with significant similarity to *phoP* and *phoR* (Sorokin et al., 1993; Azevedo et al., 1993); ResD (formerly Orf17) is 44% identical to PhoP and ResE (formerly Orf18) 28% identical to PhoR. Mutations in the gene-encoding response regulator, ResD, or both the response regulator and the sensor kinase, ResE, caused decreased *phoPR* transcription and decreased production of phosphate starvation-induced APases to levels approximately 10% of the parent strain (Table 1). Also, these mutations cause a pleiotropic phenotype affecting sporulation, cytochrome oxidase biosynthesis, lysozyme sensitivity, and carbon use (Hulett et al., unpublished data). Postexponential induction of the operon including *resDE* reaches a maximum 2 to 3 h into the stationary phase and is then repressed. Induction occurs as the culture departs from exponential growth due to limiting phosphate availability or from nutrient exhaustion of complex media. Induction of the *res* operon is not dependent on early sporulation genes such as *spo0A, spo0B,* or *spo0F.*

ResD and ResE also have an activation role in postexponential respiration. The genes encoding the ResD and ResE two-component signal transduction system are at the 3′ end of an operon with genes showing similarity to cytochrome *c* biogenesis proteins (Sorokin et al., 1993; Hulett et al., unpublished data). ResD is required for transcription of its own operon (autoregulation) and is required for *ctaA* transcription (Sharkova and Hulett, unpublished data). CtaA is required for *a*-type cytochrome formation (Mueller and Taber, 1989), and the CtaA protein appears to function specifically in biosynthesis of heme A (Svensson et al., 1993). The *a*-type cytochromes are not produced in a *resDE* mutant (Sun and Hulett, unpublished data). This establishes a function for ResD in regulation of respiration as well as in Pho regulation.

Mutations in *resE* (sensor kinase) are less deleterious than *resD* (response regulator) or *resDE* double mutants. Transcription in a *resE* mutant strain of the *ctaA* or *resA* promoters, which are activated by ResD, is only reduced to levels 30 to 40% of that observed in the *resE*[+] parent strain. The timing of the temporal expression of ResD-dependent operons in a *resE* mutant strain is identical to wild-type strains; only the amplitude of induction is changed. At least two explanations might account for this observation. First, ResD may function as an activator without being phosphorylated, albeit not as well. There is precedence for a regulatory role by an unphosphorylated response regulator. The unphosphorylated form of a response regulator, DegU of *B. subtilis,* is reported to be active but has a specific function different from the phosphorylated function (Msadek et al., 1990). Alternatively, another sensor kinase in addition to ResE may be necessary for full activation of ResD or may substitute for ResE. Precedence for multiple histidine kinases serving a sensor function that indirectly activate the same response regulator comes from the signal transduction pathway necessary for the initiation of *B. subtilis* sporulation, the phosphorelay (Burbulys et al., 1991).

The pathway for Pho regulation can be separated from the respiration regulation pathway in an *resDE spo0A* strain in which Pho regulon function is restored, but neither the transcription of the operon including *resD* nor the *ctaA* gene is restored (Sun and Hulett, unpublished data) (Table 1). Pho regulon gene expression, as measured by APase induction and *phoPR* induction dependent on decreased phosphate concentrations, is restored, and the 6-carbon sugar requirement of the *resDE* mutant is eliminated by a mutation in *spo0A.*

How does a *spo0A* mutation bypass the ResD requirement for Pho induction? Our working hypothesis is that ResD and Spo0A~P are antagonistic in their regulation of an "activation factor" that is required for phosphorylation of PhoP by PhoR, perhaps factor Z (Fig. 3). The existence of other genes essential for Pho regulon induction is supported by recent genetic evidence; mutations that abolish *phoPR* operon induction and Pho regulon gene expression but do not map to the *phoPR* locus or the *resDE* locus have been identified (Hulett et al., unpublished data). These mutations identify a gene(s) that is epistatic to *spo0A.* Also, several classes of suppressor mutants have been isolated in the *resDE* background that suppress the Pho phenotype but vary with respect to other phenotypes of the *resDE* mutants.

DO THE PhoP-PhoR TWO-COMPONENT REGULATORS HAVE A DEVELOPMENTAL SIGNAL TRANSDUCTION ROLE DURING SPORULATION?

phoPR Operon Induction during Spore Development Is Spo0A Dependent

As discussed above, the PhoP-PhoR signal transduction system has a well-documented role in the *B. subtilis* phosphate deficiency response in vegetative cells. We recently showed that the *phoPR* operon is induced postexponentially in sporulating cultures independent of phosphate concentrations in the medium

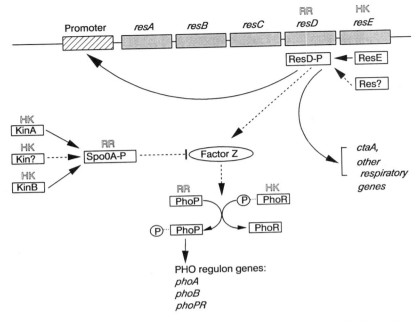

FIGURE 3 Model for ResD/E regulation of genes involved in respiration and in Pho regulon regulation. Arrowheads indicate positive regulation. An arrow with a solid line indicates that the upstream gene product may act directly. Arrows with dashed lines identify proposed regulation. Flat arrowheads indicate negative regulation. ResD (response regulator); ResE (sensor-histidine kinase); SpoOA (response regulator); PhoP (response regulator); PhoR (sensor-histidine kinase); factor Z (hypothetical activation factor involved in transfer of phosphate from PhoR~P to PhoP). Working model suggests that SpoOA and ResD are antagonist in the regulation of factor Z? Additional, proposed but not yet identified, histidine kinases are marked with a question mark (?).

(Birkey et al., 1994). This developmentally regulated induction of *phoPR* operon transcription is dependent on SpoOA~P but not on PhoP or PhoR (Table 2), in direct contrast with the phosphate starvation induction of the *phoPR* operon, which is repressed by SpoOA~P but requires PhoP and PhoR (see Table 1). Consistent with sporulation developmental regulation, the *phoPR* operon was also induced independent of the phosphate concentration by decoyinine (Sun and Hulett, unpublished data), a drug that lowers GTP levels and artificially induces sporulation in nutrient excess.

Primer extension data suggest that the *phoPR* operon is transcribed from multiple promoters and that one of them is responsible for transcription during sporulation induction (Sun and Hulett, unpublished data). Although

genes regulated by PhoP and PhoR during sporulation have not been characterized, candidate genes have been identified by Tn*917-lacZ* transcriptional fusions to promoters dependent on SpoOA, PhoP, and PhoR. It is tempting to hypothesize that a signal recognized by PhoR for a sporulation response would be different than the signal recognized by PhoR for a phosphate deficiency response and that the two potential sensor domains of PhoR (see Fig. 1) may each have sensor function, one for the signal generated during phosphate deprivation and the other for a developmental signal generated during sporulation.

Further analysis is required to determine what role, if any, PhoP and PhoR might have during sporulation.

TABLE 2 *sap* mutations do not bypass a *spoII* or *spo0* block of sporulation APase (P$_s$) expression but result in expression of the *phoPR* operon and a Pho regulon promoter, P$_v$, under phosphate-replete, sporulation induction conditions

Genotype	Promoter activity[a]			
	phoPR	*phoB*[b]	P$_v$	P$_s$
spo+ [c]	+	+	−	+
phoPR	+	+	−	+
spo0	−	−	−	−
spoII	−	−	−	−
spoII sap[d]	+	+	+	−
spoII sap phoP	−	−	−	−
spoII sap phoR	−	−	−	−
spo+ *sap*	ND[e]	+	−	+

[a]Promoter activity was measured as levels of β-galactosidase specific activity from promoter *lacZ* transcriptional fusions when strains were grown in Schaeffer's sporulation medium and was determined by colony color (blue or white) on nutrient agar plates containing X-Gal (Birkey et al., 1994).

[b]Full-length *phoB* promoter contains tandem P$_s$ and P$_v$ promoters.

[c]Parent strain of mutants listed below. MB24 (*trpC2 metC3 rif-2*).

[d]*spoIIA sapA* and *spoIIA sapB* mutant strains gave similar induction results. *spoIIA sapB* strains resulted in higher levels of expression from *phoPR* and P$_v$ promoters than *spoIIA sapA* strains (Birkey et al., 1994).

[e]ND, not determined.

ANALYSIS OF PUTATIVE SPORULATION APase (*sap*) REGULATORY MUTANTS REVEALS NEW INSIGHT INTO REGULATION OF THE *phoPR* OPERON AND ACTIVATION OF PhoP

B. subtilis produces APases under sporulation conditions, independent of P$_i$ concentrations and independent of PhoP and PhoR (Glenn and Mandelstam, 1971; Warren, 1968; Bookstein et al., 1990; Chesnut et al., 1991). *phoB* (formerly *pho*AIII) encodes an APase that is produced under both phosphate-limiting and sporulation conditions from tandem promoters (Bookstein et al., 1990; Chesnut et al., 1991). Reporter fusions to each individual promoter of *phoB* (Fig. 4) have been used to identify the regulatory genes responsible for Pho regulon induction and sporulation induction of APase. Transcriptional fusions of the full-length, sporulation-specific (P$_s$) or Pho-regulated (P$_v$) promoters of *phoB* to a *lacZ* reporter gene confirmed earlier studies that suggested sporulation APase induction requires *spo0* genes and

certain stage II sporulation genes including *spoIIA, spoIIE,* and *spoIIG* (Waites et al., 1970; Piggot and Taylor, 1977; Chesnut et al., 1991; Birkey et al., 1994) (Table 2). These promoter constructs were used to analyze mutations that can bypass this *spoII* block on sporulation APase production but do not restore the sporulation phenotype. These loci were designated *sapA* and *sapB* (sporulation alkaline phosphatase) and mapped to different regions of the *B. subtilis* chromosome (Piggot and Taylor, 1977). Thus, although *spoII* mutant strains do not produce sporulation APase, *spoIIA sapA* and *spoIIA sapB* mutant strains produce APase during sporulation.

sap Mutations Restore *phoPR* Induction and P$_v$ Promoter Function in Early Sporulation Mutants but Do Not Restore Sporulation or Sporulation APase Promoter (P$_s$) Function

The *spoII sap* strains both expressed β-galactosidase from the full-length *phoB* promoter under sporulation conditions. Interestingly, this

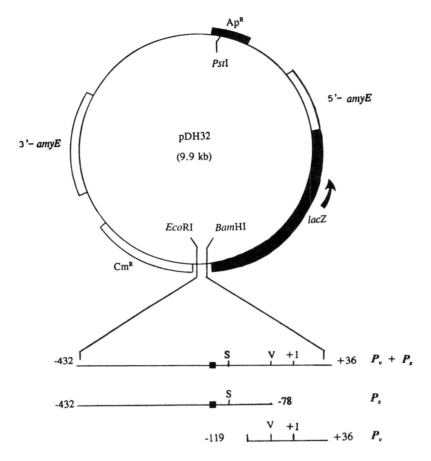

FIGURE 4 Deletion derivatives of *phoB* promoter. The numbering of the promoter fragments is identical to that in Chesnut et al. (1991). S represents the sporulation-specific promoter region (P$_s$), whereas V represents the phosphate starvation-inducible (P$_v$) promoter region. The translation start site is indicated by +1. The full-length *phoB* promoter (P$_s$+P$_v$) and each promoter fragment containing a single promoter, P$_s$ or P$_v$, were cloned into pDH32; the plasmid was linearized and transformed into *B. subtilis* MB24 and MB24 derivative strains used in Table 2. Chloramphenicol-resistant transformants were selected and screened for an *amy⁻* phenotype.

expression of *phoB* in the *spoII sap* strains was from P$_v$, the phosphate starvation-inducible promoter of *phoB*, rather than from P$_s$, the sporulation-specific promoter (Birkey et al., 1994) (Table 2). The induction of phosphate starvation-inducible promoters, including P$_v$, requires PhoP and PhoR for induction during phosphate-limited growth (Miki et al., 1965; Chesnut et al., 1991). Were the *phoPR* products involved in regulating P$_v$ expression under phosphate-replete sporulation conditions in a *spoII* mutant, a mutation known to abolish sporulation induction of *phoPR* transcription (Table 2)? The answer is yes. In these *spoII sap* strains, the *phoPR* genes are induced under sporulation conditions and require *sapA* or *sapB* mutations in the Spo⁻ background for induction, and both PhoP and PhoR are required for expression from P$_v$ in phosphate-replete growth conditions (Birkey et al., 1994).

The *sap* mutations function to allow *phoPR* transcription and alkaline phosphatase expression from P_v only in *spo0* or *spoII* mutant strains.

Because the *sap* mutations are uncharacterized, the mechanism by which they regulate expression of *phoPR* or activate PhoP is unknown. However, we can conclude that the proteins encoded by the *sap* genes do not bypass the *spoII* requirement for sporulation APase production, but rather they are involved in activation of PhoP by PhoR or in transcription of the *phoPR* operon, thus regulating Pho regulon promoters (Birkey et al.,1994). Consistent with this hypothesis is the fact that another mutation, designated *phoS,* was identified that produced APases constitutively under both vegetative and sporulation conditions, and mapping data suggested that *phoS* was an allele of *sapA* (Piggot and Taylor, 1977).

SUMMARY AND PERSPECTIVE

At least three two-component signal transduction systems interact in a network to regulate the Pho regulon in *B. subtilis.* Transcription of the operon containing the genes encoding the primary two-component regulators, PhoP and PhoR, is maintained at low constitutive levels during exponential growth. By definition, the *phoPR* operon is part of the Pho regulon because induction of the operon depends on both PhoP and PhoR. The concentration of PhoP (presumably as PhoP~P) and PhoR is amplified in response to an unknown signal, resulting in induction of other Pho regulon genes whose expression coincides with the induction of their regulators.

A second genetic switch, ResD, is required for full amplification of the Pho-regulated phosphate deficiency response. The molecular components that transmit the signal, a signal dependent on ResD function, to PhoP-PhoR are unknown. What seems clear is that the ResD genetic switch is activated in response to nutrient deprivation and has an essential regulatory role in postexponential and developmental respiration, suggesting that it may be activated in response to the reduced energy state of the cell. In the case of phosphate deprivation, ResD also initiates a signal that is relayed to the PhoP-PhoR system that regulates genes whose products could potentially exploit sources of complex phosphate. Because ResD is activated by deprivation of other nutrients in addition to phosphate and is only partially dependent on ResE for activation, it is plausible that ResD is a global regulator activated by several alternative sensor kinases that could channel information through ResD to activate postexponential respiration systems and send positive signals to the appropriate signal transduction system whose activation might allow adaptation to new situations.

A third signal transduction system represses the Pho regulon. Under culture conditions in which the phosphate response fails to supply P_i for continued growth because suitable phosphate sources are unavailable, Spo0A~P, the response regulator for the sensory transduction system that is essential for the initiation of the sporulation, turns off the failed Pho regulon. Spo0A~P also initiates the changes necessary for the developmental program leading to spore formation while the environmental concentrations of P_i are still adequate to complete the sporulation process. (In this respect, it is interesting to note that the concentration of P_i that results in activation of the Pho regulon in *B. subtilis* is 20 to 25 times higher than in *E. coli*.) The Spo0A, Spo0F, and Spo0B components of the Spo0 phosphorelay are required for repression of the Pho regulon; mutations in any of these components result in hyperinduction of the Pho regulon genes even in a *resDE* mutant. Thus, a *spo0A* mutation is epistatic to a *resD* mutation such that hyperinduction of the Pho regulon in a *spo0A resDE* double mutant is similar to that in a *spo0A* mutant strain. We propose that Spo0A and ResD have roles as antagonists in the regulation of the Pho regulon by modulating the function of a factor involved in activation of PhoP by PhoR.

Another two-component system network controlling postexponential-phase responses that govern the choice between competence gene expression, degradative enzyme production, and sporulation in *B. subtilis* has been extensively studied. DegU is the response regulator for signaling degradative enzyme production. *degU*(Hy) mutations, which lead to accumulation of the phosphorylated DegU response regulator, appear to increase the rate of transcription initiation of their target genes *sacB* (levansucrase structural gene) (Aymerich et al., 1986; Shimotsu and Henner, 1986) and *aprE* (alkaline protease structural gene)(Henner et al., 1988). Preliminary studies showed an increased level of accumulation of another secreted degradative enzyme(s), alkaline phosphatase, in the *degU*(Hy) strain (Jensen and Hulett, unpublished data). It will be of interest to determine the extent of this apparent overlap between the DegS-DegU, ComP-ComA, Spo0A signal transduction network and the ResD-ResE, PhoP-PhoR, Spo0A regulatory network described here.

In *B. subtilis,* the Spo0A/PhoP-PhoR system provides evidence that the relationship of the individual regulatory systems is dependent on the developmental state of the cell. In the vegetative cell type during phosphate-limiting conditions, Spo0A~P represses autoinduction of the operon encoding the Pho regulon regulators, shutting down the Pho response as just described. However, during early sporulation under phosphate-replete conditions, Spo0A~P is essential for induction of the *phoPR* operon. One possibility is that accessory protein factors in the transduction pathway between Spo0A~P and the Pho regulators are different in the two cases. This hypothesis predicts that a specific accessory factor(s) is responsible for the positive nature of the signal, resulting in induction of the *phoPR* operon during sporulation development, and that different accessory factors are responsible for the negative nature of the signal from Spo0A~P, which represses the Pho response. Also, the signal to PhoR and the functional sensor domain of PhoR may be different in the two cases.

REFERENCES

Aymerich, S., G. Gonzy-Treboul, and M. Steinmetz. 1986. 5'-Noncoding region *sacR* is the target of all identified regulation affecting the levansucrase gene in *Bacillus subtilis. J. Bacteriol.* **166:**993–998.

Azevedo, V., A. Sorokin, S. Ehrlich, and P. Serror. 1993. The transcriptional organization of the *Bacillus subtilis* 168 chromosome region between the *spoVAF* and *serA* genetic loci. *Mol. Microbiol.* **10:**397–405.

Birkey, S., L. Shi, and F. M. Hulett. Unpublished data.

Birkey, S., G. Sun, P. Piggot, and F. M. Hulett. 1994. A PHO regulon promoter induced under sporulation conditions. *Gene* **147:**95–100.

Bookstein, C., C. W. Edwards, N. V. Kapp, and F. M. Hulett. 1990. The *Bacillus subtilis* 168 alkaline phosphatase III gene: impact of a *phoAIII* mutation on total alkaline phosphatase synthesis. *J. Bacteriol.* **172:**3730–3737.

Burbulys, D., K. A. Trach, and J. A. Hoch. 1991. Initiation of sporulation in *B. subtilis* is controlled by a multicomponent phosphorelay. *Cell* **64:**545–552.

Chesnut, R. S., C. Bookstein, and F. M. Hulett. 1991. Separate promoters direct expression of *phoAIII*, a member of the *Bacillus subtilis* multigene family, during phosphate starvation and sporulation. *Mol. Microbiol.* **5:**2181–2190.

Ferrari, F. A., K. Trach, D. Le Coq, J. Spence, E. Ferrari, and J. A. Hoch. 1985. Char characterization of the *spo0A* locus and its deduced product. *J. Bacteriol.* **82:**2647–2651.

Glenn, A. R., and J. Mandelstam. 1971. Sporulation in *Bacillus subtilis* 168: comparison of alkaline phosphatase from sporulating and vegetative cells. *J. Biochem.* **123:**129–138.

Henner, D. J., E. Ferrari, M. Perego, and J. A. Hoch. 1988. Location of the targets of the *hpr-97, sacU32* (Hy), and *sacQ36* (Hy) mutations in upstream regions of the subtilisin promoter. *J. Bacteriol.* **170:**296–300.

Hulett, F. M. 1993. Regulation of phosphorus metabolism, p. 229–235. *In* A. Sonenshein, J. Hoch, and R. Losick (ed.), *Bacillus subtilis and Other Gram-Positive Bacteria: Biochemistry, Physiology, and Molecular Genetics.* American Society for Microbiology, Washington, D.C.

Hulett, F. M., C. Bookstein, and K. Jensen. 1990. Evidence for two structural genes for alkaline phosphatase in *Bacillus subtilis. J. Bacteriol.* **172:**735–740.

Hulett, F. M., and K. Jensen. 1988. Critical roles of *spo0A* and *spo0H* in vegetative alkaline phosphatase production in *Bacillus subtilis. J. Bacteriol.* **170:**3765–3768.

Hulett, F. M., E. E. Kim, C. Bookstein, N. V. Kapp, C. W. Edwards, and H. W. Wyckoff. 1991. *Bacillus subtilis* alkaline phosphatases III and IV. Cloning, sequencing, and comparisons of deduced amino acid sequence with *Escherichia coli* alkaline phosphatase three-dimensional structure. *J. Biol. Chem.* **266:**1077–1084.

Hulett, F. M., J. Lee, L. Shi, G. Sun, R. Chesnut, E. Sharkova, M. A. Duggan, and N. Kapp. 1994. Sequential action of two component genetic switches regulates the Pho regulon in *Bacillus subtilis. J. Bacteriol.* **176:**1348–1358.

Hulett, F. M., G. Sun, and W. Liu. 1994. The Pho regulation of *Bacillus subtilis* is regulated by sequential action of two genetic switches, p. 50–54. *In* A. Torriani-Gorini, E. Yagil, and S. Silver (ed.), *Phosphate in Microorganisms: Cellular and Molecular Biology.* American Society for Microbiology, Washington, D.C.

Hulett, F. M., G. Sun, E. Sharkova, R. Chesnut, S. Birkey, M. F. Duggan, D. Erlich, and A. Sorokin. 1994. Unpublished data.

Jensen, K., and F. M. Hulett. Unpublished data.

Jensen, K. K., E. Sharkova, M. F. Duggan, Y. Qi, A. Koide, J. A. Hoch, and F. M. Hulett. 1993. *Bacillus subtilis* transcription regulator, Spo0A, decreases alkaline phosphatase levels induced by phosphate starvation. *J. Bacteriol.* **175:**3749–3756.

Kapp, N. V., C. W. Edwards, R. S. Chesnut, and F. M. Hulett. 1990. The *Bacillus subtilis phoAIV* gene: effects of in vitro inactivation on total alkaline phosphatase production. *Gene* **96:**95–100.

Lee, J., and F. M. Hulett. 1992. Nucleotide sequence of the *phoP* gene encoding PhoP, the response regulator of the phosphate regulon of *Bacillus subtilis. Nucleic Acids Res.* **19:**5848.

Liu, W., and F. M. Hulett.Unpublished data.

Makino, K., H. Shinagawa, M. Amemura, T. Kawamoto, M. Yamada, and A. Nakata. 1989. Signal transduction in the phosphate regulon of *Escherichia coli* involves phosphotransfer between PhoR and PhoB proteins. *J. Mol. Biol.* **210:**551–559.

Makino, K., H. Shinagawa, M. Amemura, K. Kimura, and A. Nakata. 1988. Regulation of the phosphate regulon of *Escherichia coli.* Activation of *pstS* transcription by PhoB protein in vitro. *J. Mol. Biol.* **203:**85–95.

Marahiel, M. A., P. Zuber, G. Czekay, and R. Losick. 1987. Identification of the promoter for a peptide antibiotic biosynthesis gene from *Bacillus brevis* and its regulation in *Bacillus subtilis. J. Bacteriol.* **169:**2215–2222

Miki, T., A. Minami, and Y. Ikeda. 1965. The genetics of alkaline phosphatase formation in *Bacillus subtilis. Genetics* **52:**1093–1100.

Miller, J., S. Johnson, W. Black, D. Beattie, J. Mekalanos, and S. Falkow. 1992. Constitutive sensory transduction mutations in the *Bordetella pertussis bvgS* gene. *J. Bacteriol.* **174:** 970–979.

Msadek, T., F. Kunst, D. Henner, A. Klier, G. Rapoport, and R. Dedonder. 1990. Signal transduction pathway controlling synthesis of a class of degradative enzymes in *Bacillus subtilis:* expression of the regulatory genes and analysis of mutations in *degS* and *degU. J. Bacteriol.* **172:**824–834.

Mueller, J., and H. Taber. 1989. Isolation and sequence of *ctaA,* a gene required for cytochrome aa3 biosynthesis and sporulation in *Bacillus subtilis. J. Bacteriol.* **171:**4967–4978.

Ozanne, P. G. 1980. Phosphate nutrition of plants—a general treatise, p. 559–585. *In* E. Khasswenh (ed.), *The Role of Phosphorus in Agriculture.* American Society of Agronomy, Madison, Wis.

Perego, M., J.-J. Wu, G. B. Spiegelman, and J. A. Hoch. 1991. Mutational dissociation of the positive and negative regulatory properties of the Spo0A sporulation transcription factor of *Bacillus subtilis. Gene* **100:**207–212.

Piggot, P. J., and S. Y. Taylor. 1977. New types of mutations affecting formation of alkaline phosphatase by *Bacillus subtilis* in sporulation conditions. *J. Gen. Microbiol.* **128:**663–669.

Robertson, J. B., M. Gocht, M. A. Marahiel, and P. Zuber. 1989. AbrB, a regulator of gene expression in *Bacillus,* interacts with the transcription initiation regions of a sporulation gene and an antibiotic biosynthesis gene. *Proc. Natl. Acad. Sci. USA* **86:**8457–8461.

Scholten, M., and J. Tommassen. 1993. Topology of the PhoR protein of *Escherichia coli* and the functional analysis of internal deletion mutants. *Mol. Microbiol.* **8:**269–275.

Seki, T., H. Yoshikawa, H. Takahashi, and H. Saito. 1987. Cloning and nucleotide sequence of *phoP,* the regulatory gene for alkaline phosphatase and phosphodiesterase in *Bacillus subtilis. J. Bacteriol.* **169:**2913–2916.

Seki, T., H. Yoshikawa, H. Takahashi, and H. Saito. 1988. Nucleotide sequence of the *Bacillus subtilis phoR* gene. *J. Bacteriol.* **170:**5935–5938.

Sharkova, E., and F. M. Hulett. Unpublished data.

Shimotsu, H., and D. J. Henner. 1986. Modulation of *Bacillus subtilis* levansucrase gene expression by sucrose and regulation of the steady-state mRNA level by *sacU* and *sacQ* genes. *J. Bacteriol.* **168:**380–388.

Sorokin, A., E. Zumstein, V. Azevedo, S. D. Ehrlich, and P. Serror. 1993. The organization of the *Bacillus subtilis* 168 chromosome region between the *spoVA* and *serA* genetic loci, based on sequence data. *Mol. Microbiol.* **10:**385–395.

Strauch, M., and J. A. Hoch. 1993. MicroReview. Transition-state regulators: sentinels of *Bacillus subtilis* post-exponential gene expression. *Mol. Microbiol.* **7:**337–342.

Strauch, M., V. Webb, G. Spiegelman, and J. A. Hoch. 1990. The SpoOA protein of *Bacillus subtilis* is a repressor of the *abrB* gene. *Proc. Natl. Acad. Sci. USA* **87:**1801–1805.

Strauch, M. A., J. A. Hoch, and F. M. Hulett. Unpublished data.

Strauch, M. A., G. B. Spiegelman, M. Perego, W. C. Johnson, D. Burbulys, and J. A. Hoch. 1989. The transition state transcription regulator *abrB* of *Bacillus subtilis* is a DNA binding protein. *EMBO J.* **8:**1615–1621.

Sun, G., and F. M. Hulett. Unpublished data.

Svensson, B., M. Lubben, and L. Hederstedt. 1993. *Bacillus subtilis* CtaA and CtaB function in haem A biosynthesis. *Mol. Microbiol.* **10:**193–201.

Tokishita, S., A. Kojima, H. Aiba, and T. Mizuno. 1991. Transmembrane signal transduction and osmoregulation in *Escherichia coli* K-12. *J. Biol. Chem.* **266:**6780–6785.

Waites, W. M., D. Kay, I. W. Dawes, D. A. Wood, S. C. Warren, and J. Mandelstam. 1970. Sporulation in *Bacillus subtilis*. Correlation of biochemical events with morphological changes in asporogenous mutants. *Biochem. J.* **118:**667–676.

Wanner, B. L. 1992. Is cross regulation by phosphorylation of two-component response regulator proteins important to bacteria? *J. Bacteriol.* **174:**2053–2058.

Wanner, B. L., and M. R. Wilmes-Riesenberg. 1992. Involvement of phosphotransacetylase, acetate kinase, and acetyl phosphate synthesis in control of the phosphate regulon in *Escherichia coli*. *J. Bacteriol.* **174:**2124–2130.

Warren, S. C. 1968. Sporulation in *Bacillus subtilis*. Biochemical changes. *Biochem. J.* **109:**811–818.

Yamada, M., K. Makino, M. Amemura, H. Shinagawa, and A. Nakata. 1989. Regulation of the phosphate regulon of *Escherichia coli:* analysis of mutant *phoB* and *phoR* genes causing different phenotypes. *J. Bacteriol.* **171:**5601–5606.

PATHOGENESIS

V

Two-Component Signal Transduction and Its Role in the Expression of Bacterial Virulence Factors

Michelle Dziejman and John J. Mekalanos

19

In contrast to saprophytic bacteria, pathogenic bacteria live on or within living host tissues. Like all bacteria, their prime objectives are survival and multiplication. Unfortunately, the appearance of disease within the host is often the inadvertent consequence of bacteria successfully achieving these goals.

Pathogenic bacteria coordinate an intricate network of virulence factors, whose expression must be precisely controlled to maximize the chance of establishing a successful infection. Adherence factors, toxins, capsules, and other complex properties (e.g., motility, intracellular invasion) are among the virulence factors used by different pathogens. In addition, many pathogenic microbes display tactics for evading the host immune system. Not all virulence determinants confer a selective advantage to the bacteria at the same stage of infection, and so expression of these factors must be precisely timed. Groups of virulence factors acting at the same stage of the infection are often expressed in a coordinate fashion via common regulatory systems. Presumably, these regulatory systems recognize environmental cues during the transition from environmental reservoir to host and subsequently throughout the infection process.

For example, elevated temperature and low pH might be early signals heralding entry of an enteric pathogen into the host from an environmental source. Chemotactic signals affecting motility behavior might then enable the pathogen to reach its desired local anatomical site. Once there, the pathogen must express new genes to compete with normal host flora for essential and limited nutrients. Establishment of the pathogen at a primary site might then facilitate its passage to new host sites, where further alterations in gene expression might occur. Some pathogens, such as invasive bacteria, would further need to recognize intra- versus extracellular location and even subcellular compartments within host cells (e.g., endosome, lysosome, cytoplasm) before they initiate a program of growth. Thus, the pathogenic "lifestyle" presents the enormous challenge to bacterial agents of monitoring the full range of environmental parameters encountered in a host and then translating that information accurately into changes in the expression of virulence genes or an adaptive response.

Certain common themes have emerged from the study of the control of bacterial virulence gene expression (Mekalanos, 1992).

Michelle Dziejman and John J. Mekalanos, Department of Microbiology and Molecular Genetics, Harvard Medical School, Boston, Massachusetts 02115-5701.

Two-Component Signal Transduction, Edited by James A. Hoch and Thomas J. Silhavy,
© 1995 American Society for Microbiology, Washington, DC 20005

TABLE 1 Two-component regulation of virulence

Organism	Signals	Adaptive response	Regulatory proteins
Agrobacterium tumefaciens	Phenolic compounds, monosaccharides	Crown gall tumor formation	VirA/VirG
Bacteroides fragilis	Tetracycline	Porin expression	RprX/RprY
		Self-transfer of DNA	RtcA/RtcB
Bordetella pertussis	Temperature, SO_4, nicotinic acid	Activation and repression of virulence factors, including HA and toxin	BvgA/BvgS
Citrobacter freudii	Unknown	Vi antigen expression	ViaA/ViaB
Clostridium perfringens	Unknown	Perfringolysin-O, HA, and collagenase expression	VirR
Enterococcus faecium	Vancomycin	Vancomycin resistance	VanR/VanS
Erwinia amylovora	Unknown	Exopolysaccharide synthesis	RcsB/RcsC
		Hypersensitive response	HrpS/?
Klebsiella pneumoniae	N_2	Urease production	NtrA/NtrC
	O_2, N_2	Nitrogen fixation	NifA/NifL
	Unknown	Capsule production	RmpA2/?
Neisseria gonorrhoeae	Stress	Pilin production	PilA/PilB
Pseudomonas aeruginosa	Unknown	Pilin expression	PilR/PilS
	Enterobactin, iron	Ferric enterobactin receptor	PfeR/PfeS
	Osmolarity	Alginate synthesis	AlgR1/AlgR2
Pseudomonas spp.	Aromatic effectors	Phenol catabolism	DmpR/?
Pseudomonas syringae	Copper	Copper resistance	CopR/CopS
Shigella flexneri	Osmolarity	Porin expression	OmpR/EnvZ
Streptococci spp.	CO_2	M-protein expression	Mry
Vibrio cholerae	pH, osmolarity, temperature	Virulence factor expression, including toxin and pili	ToxR, (ToxS)
Vibrio parahemolyticus	pH	Thermostable direct hemolysin production	ToxR, (ToxS)

Among these is the fact that certain environmental parameters are frequently found associated with the control of virulence gene transcription. Low iron concentration and elevated temperature have been viewed as parameters that signal entry of the microbe into host tissue. As examples, *Shigella flexneri, Yersinia* spp., and *Bordetella pertussis* all coordinately regulate a subset of virulence factors in response to temperature, and iron-regulated virulence genes have been identified in *Escherichia coli, Corynebacterium diphtheria, Pseudomonas aeruginosa,* and *Vibrio cholerae.* Other common signals regulating virulence gene expression include ions such as Ca^{2+}, variations in pH or osmolarity, and the levels of gases such as O_2 and CO_2. It is more difficult to assign these parameters to a particular host compartment because these signals may vary considerably, depending not only on the site within the host but also on the resulting pathological changes in tissues.

This volume is dedicated to the discussion of the modulation of gene expression by two-component systems in response to various environmental stimuli. Such systems have certainly been implicated in the control of virulence in several different microorganisms, as shown in Table 1. In the case of most sensor-regulator protein pairs described thus far for pathogenic organisms, the specific host cues that the sensor protein detects and how this sensing is achieved on a molecular level are not very well understood. Moreover, in contrast to virulence regulatory proteins, it is becoming evident that a subset of two-component regulatory systems are conserved among different organisms for the same purpose. As examples, OmpR–EnvZ proteins have been implicated in osmoregulation of porin synthesis not only in *E. coli* but also in *Salmonella* and *Shigella* spp., whereas NtrC homologs are involved in nitrogen regulation in *Klebsiella* and *Salmonella* spp. and *E. coli* (see Chapters 5 and 9). Furthermore, these homologs can often functionally substitute for one another in different organisms. It seems less likely that such conservation is going to become the rule for the genes involved in virulence control. Thus, toxin production by *V. cholerae,* *B. pertussis,* *P. aeruginosa,* and *C. diphtheriae* is controlled by distinct regulatory proteins, but these proteins bear little resemblance to each other. Perhaps this should not be surprising, considering the greater genetic distance between these various pathogenic organisms. However, this lack of similarity may also simply reflect the fact that these organisms grow in distinctly different anatomical sites and that the in vitro environmental conditions that modulate toxin expression in these different organisms have little in common.

Nevertheless, several well-characterized two-component systems controlling virulence gene expression, such as VirA–VirG in *Agrobacterium tumefaciens,* BvgA–BvgS in *Bordetella pertussis,* and PhoP–PhoQ proteins of *Salmo-*

nella do fit the classic paradigms of two-component systems. Investigations of gene regulation in *Pseudomonas* spp. have yielded at least seven different two-component systems, which modulate expression of such diverse factors as the ferric enterobactin receptor (PfeR–PfeS) (Dean and Poole, 1993), alginate biosynthesis (AlgR1–AlgR2), fimbriae expression (PilS–PilR) (Ishimoto and Lory, 1992; Hobbs et al., 1993), copper resistance (CopS–CopR) (Mills et al., 1993), and capsule synthesis (HrpR–HrpS) (Grimm and Panopoulos, 1989).

In addition to the classic two-component motif, a variety of other protein families have been implicated in regulating expression of virulence determinants. A brief discussion of these other mechanisms follows. Also, some of these systems are described in Table 2.

The AraC proteins of *E. coli* and *Salmonella* spp. regulate expression of genes in the arabinose operon. Homologs include the ToxT protein of *V. cholerae,* which acts downstream of ToxR to activate expression of the toxin coregulated pilus and accessory colonization factor genes (Higgins et al., 1992); VirF, which controls plasmid-encoded virulence genes in *Yersinia* spp. (Cornelis et al., 1989); and FapR, which is involved in pilus regulation in *E. coli* (Klaasen and Graaf, 1990). Fur proteins that mediate the coordinate regulation of genes by iron have been found in *E. coli,* *Yersinia pestis,*

TABLE 2 Non-two-component regulation

Organism	Signals	Adaptive response	Regulatory proteins
Agrobacterium tumefaciens	Opines	Ti plasmid transfer	TraR, TraI
Bacillus anthracis	CO_2	Capsule and toxin production	AtxA
Corynebacterium diphtheriae	Iron	Toxin production	DtxR
Escherichia coli	Iron	Shiga-like toxin	Fur
	Stress	Pilus expression	FapR
Listeria monocytogenes	Heat shock	Virulence factors including listerolysin and phospholipase C	Unknown
Pseudomonas aeruginosa	Iron	Exotoxin A	RegA, RegB
	Homoserine lactone	Elastin-specific protease	LasR, LasI
Shigella flexneri	Temperature	Plasmid-encoded cell invasion genes	VirB, VirF, VirR
Vibrio cholerae	Iron	Outer membrane proteins	IrgB
Yersinia spp.	Temperature, Ca^{2+}	Yops	VirF, LcrH

P. aeruginosa, and various *Vibrio* spp. (reviewed in Mekalanos, 1992). As mentioned, other ions can serve as important signals for global regulation of virulence factors in vitro. The low Ca^{2+} response of *Yersinia* spp. is modulated by the plasmid encoded *lcr* locus, whose products coordinate the expression of V antigen and surface proteins (Yops) in response to temperature and Ca^{2+} levels. Specifically, the YopN protein appears to be surface-localized and plays a role in calcium signal transduction by sensing calcium concentration (Forsberg et al., 1991). Similar surface sensing by *Salmonella typhimurium* has recently been postulated as critical for formation of surface appendages (Ginocchio et al., 1994). These events precede the formation of epithelial membrane ruffles, resulting in bacterial uptake (Galan et al., 1992; Pace et al., 1993). Finally, the LuxR-LuxI proteins, originally characterized in the luminescent symbiont *Vibrio fischeri,* have recently been recognized to head a family of homologous systems whose hallmark is regulation based on population density-dependent accumulation of autoinducers (for review, see Fuqua et al., 1994). Both the conjugal transfer of Ti plasmids in *Agrobacterium tumefaciens* and the production of extracellular virulence determinants of *P. aeruginosa* are regulated by similar systems, and LuxR homologs have been identified for other gram-negative bacteria as well.

Chapters following this one encompass detailed descriptions of the two-component systems in the pathogens *A. tumefaciens,* *B. pertussis,* *S. typhimurium, Pseudomonas* spp., and *V. cholerae.* What follows here is a brief overview of two-component regulation in other pathogens.

SHIGELLA FLEXNERI

S. flexneri is an invasive pathogen that causes bacillary dysentery. Bacteria replicate within colonic epithelial cells, and inter- and intracellular spread occurs by actin polymerization and formation of membrane protrusions, similar to that observed in *Listeria monocytogenes* (Bernardini et al., 1989; Goldberg et al., 1993; Makino et al., 1986; Mounier et al., 1990; Tilney and

Portnoy, 1989). *S. flexneri* has adapted the OmpR-EnvZ system, initially described in *E. coli* and also present in *Salmonella* spp., for its own use to regulate expression of virulence factors and the OmpF and OmpC porins (Bernardini et al., 1990). The *ompR-envZ* genes lie in the *ompB* operon, as in *E. coli.* However, there is at least one interesting regulatory modification. OmpF is produced only under conditions of low osmolarity as in *E. coli,* but OmpC is highly expressed whether the bacteria are grown under conditions of either high or low osmolarity. OmpF pores are slightly larger than OmpC pores, and this observation has led to the hypothesis that a smaller pore is more advantageous in the host environment, allowing for reduced diffusion of toxic and potentially harmful molecules elicited by the host. Considering that *S. flexneri* is transmitted by an oral-fecal route and that bacteria do not survive well in the environment, regulation itself may have adapted to provide protection to the pathogen in its most common environment, the host.

In *S. flexneri,* the *ompB* locus is capable of regulating plasmid genes as well as the chromosomal *ompC* and *ompF* genes. Osmoregulation of a *vir::lac* reporter fusion within an invasion gene located on the virulence plasmid of *S. flexneri* has been demonstrated (Bernardini et al., 1990). Regulation of this reporter fusion does not occur in an *ompB* deletion mutant, and osmoregulation could not be restored by providing the *ompB* locus on a multicopy plasmid. This suggests that OmpR-EnvZ protein concentration is a factor affecting regulation, and a similar phenomenon has been observed in *E. coli* for its OmpR-EnvZ system. However, regulation could be restored by transduction of the *E. coli* chromosomal *ompB* locus, which is again indicative of the high degree of conservation of these proteins and their mechanisms of action.

A mutation in *envZ* alone severely reduced virulence as assayed in a guinea pig keratoconjunctivitis model (Sereny test) and by the inability to invade and form plaques on HeLa cell monolayers. Transcription of the *vir::lac* fusion still responded in an osmodependent manner, however, suggesting that invasion is regulated by other factors in addition to *envZ* or that crosstalk from a heterologous sensing component could result in enough active OmpR to activate transcription. Attempts to provide *envZ in trans* to an *ompR* mutant were unsuccessful, as expression of *envZ* at physiological levels could not be achieved (Bernardini et al., 1990).

In vitro studies showed that both *ompB* and *ompC* mutants were defective in two important steps in *S. flexneri* pathogenesis: spreading and host cell killing (Bernardini et al., 1993). An *ompC* mutant was able to form only tiny plaques in a HeLa or Caco-2 cell monolayer, whereas the *ompB* mutant was unable to form plaques at all. Similarly, although the *ompC* mutant showed reduced invasiveness, it was not as deficient as the *ompB* mutant in either the invasion assay or in the Sereny test. Both mutants appeared to move intracellularly (via actin polymerization) but were unable to form extracellular protrusions, suggesting that this defect was directly responsible for the inability to spread from cell to cell. Although these studies indicate that the *ompB* locus regulates additional genes important for pathogenesis, they also show that the OmpC protein itself contributes significantly to the ability to establish infection. One hypothesis for a role for OmpC involves its interaction with cytoskeletal proteins, whose reorganization has recently been correlated with movement and protrusion formation in *S. flexneri* (Prevost et al., 1992; Kadurugamuwa et al., 1991).

STAPHYLOCOCCUS AUREUS

Staphylococcus aureus is the causative agent of a variety of skin ulcerations and has also been implicated in cases of endocarditis, meningitis, and impetigo. *S. aureus* also constitutes the principal cause of nongonococcal bacterial arthritis in humans. Many virulence factors have been identified and include a variety of extracellular toxins: four different hemolysins, toxic shock syndrome toxin 1, and an enterotoxin.

Secreted enzymes include a lipase, nuclease, proteases, and a hyaluronate lyase. Cell wall-associated proteins include protein A, collagen binding protein, and fibronectin binding protein. Also, an impressive array of receptors for mammalian matrix proteins is produced: fibrinogen, laminin, plasminogen, vitronectin, thrombospondin, and bone sialoprotein.

The synthesis of these virulence factors appears to be temporally regulated, as alpha- and beta-hemolysin and toxic shock syndrome toxin 1 are produced only after exponential growth has ceased, and coagulase and protein A are synthesized during exponential growth. Proteins such as enterotoxin B, metalloprotease, and nuclease are produced at low levels during the growth cycle but more actively after exponential growth. Synthesis of these and other exoproteins is regulated in part by the *agr* system, which has a unique mechanism of regulation (Kornblum et al., 1990; Morfeldt et al., 1988; Recsei et al., 1986). The *agr* locus consists of two divergently transcribed operons, one of which contains four genes (Peng et al., 1988). Two of these gene products, AgrA and AgrB, conform well to the standard two-component motif (Kornblum et al., 1990). These proteins are capable of activating transcription from their own promoter, forming an autocatalytic circuit similar to that of the Mry protein of *Streptococcus pyogenes,* which is described later. Regulation of other genes under *agr* control is apparently mediated by an RNA molecule termed RNA III, whose transcription from the divergently transcribed promoter in the locus is directed by the AgrA and AgrB proteins (Janzon et al., 1989). Secreted proteins are up-regulated by RNA III, whereas surface proteins are down-regulated. RNA III encodes the 26-amino-acid δ-hemolysin polypeptide, but studies that have disrupted the open reading frame by site-directed mutagenesis indicate that it is not the protein but rather the RNA molecule itself that is responsible for Agr regulation (Janzon and Arvidson, 1990). RNA III alone is not sufficient to result in transcriptional activation, however, and an unknown temporal signal is postulated to be important for regulation (Vandenesch et al., 1991).

One environmental condition that appears to be important for Agr regulation is pH (Regassa and Betley, 1992). Alkaline pH results in a decrease in RNA III expression and, consequently, in extracellular hemolysin production. Capsular polysaccharide biosynthesis has also been shown to be pH-regulated, and again, alkaline pH results in a decrease of expression (Dassy et al., 1993). This effect could be partially overcome by growth in some media, suggesting that an additional signal(s) or effects on metabolism may contribute to regulation.

An in vivo model for bacterial arthritis has indicated that AgrA may be involved in the direct regulation of additional genes whose expression is not mediated by RNA III (Abdelnour et al., 1993). In a murine model that mimics the human disease, infection by bacterial mutants defective in RNA III produced a phenotype that was intermediate between that seen for an *agr* mutant and wild-type bacteria. The ability of wild-type and mutant bacteria to bind bone sialoglycoprotein was also assayed. Again, the RNA III mutant showed an intermediate ability to bind, suggesting that the receptor gene is regulated both by RNA III and another, as yet unidentified, Agr-regulated component. Bacteria deleted for the *agr* locus were unable to migrate through blood to the joint cavities and could not be recovered from these locations. Unlike wild-type *S. aureus*, mutant bacteria could be recovered from the spleen, and one possible explanation is that the increased expression of cell wall-associated proteins in an *agr* mutant renders it more easily phagocytosed or immobilized in the reticuloendothelial system (Abdelnour et al., 1993). Little is known about how *S. aureus* penetrates blood vessels into joints, but it appears that Agr proteins have an important role in this stage of infection.

It is of interest to note that because two-component proteins share several domains of significant homology, the use of degenerate, sensor-specific oligodeoxyribonucleotide primers to amplify DNA from *S. aureus* resulted in the identification of several proteins that shared similar motifs (Bayles, 1993). An EnvZ homolog was identified, along with unrelated families of transport proteins and metabolic enzymes, whose activity is regulated by histidine residue phosphorylation (Robertson et al., 1988; Deutscher and Sauerwald, 1986). Also, Parkinson and Kofoid (1988), using two-component conserved domains, conducted a computer analysis that searched for homologous proteins. They identified a transporter protein: the PtsG protein of *E. coli* that is specific for glucose transport. As suggested by both groups that independently identified non-two-component homologs sharing a similar phosphochemistry, this type of molecular signal transduction may have been evolutionarily conserved to suit a variety of adaptive mechanisms.

NEISSERIA GONORRHOEAE

Pili are a major virulence factor of *N. gonorrhoeae,* and piliated bacteria are able to attach more successfully to susceptible host tissue and establish an infection in human hosts, whereas nonpiliated bacteria cannot (Mardh and Westrom, 1976; McGee et al., 1981; Swanson, 1973; Boslego et al., 1991). This is likely due to the inability of nonpiliated bacteria to attach successfully to various cells of susceptible tissue. Variation in piliation may be important for the disease process or host-to-host spread (Keevil et al., 1986), and in vitro it can change with culture conditions. Expression of the pilin gene, *pilE,* is controlled by the PilA and PilB proteins. *pilA* and *pilB* are divergently transcribed, with overlapping promoters (Taha et al., 1991). A *pilB* mutant is hyperpiliated, but *pilA* mutants are lethal. However, a *pilA* mutant that is heterodiploid for the wild-type allele exhibits a nonpiliated, pleiotropic phenotype (Taha et al., 1988; Taha et al., 1991).

PilB shows homology to the EnvZ class of sensors. PilA shows homology to OmpR and NtrC but also shows striking homology to the *E. coli* FtsY protein, which is involved in cell division (Taha et al., 1991). Pilin regulation appears to have been conserved among *Neisseria* spp., as the use of *pilA* and *pilB* probes for Southern blots (of DNA from various *Neisseria* spp.) under stringent conditions revealed the existence of homologs in nearly all strains tested (Taha and Marchal, 1990).

The *pilE* upstream region contains a σ^{54}-activated promoter consensus sequence. By itself, PilA can activate the pilin gene promoter, but in conjunction with PilB, it functions to repress pilin expression, although the mechanism of action remains to be elucidated (Taha et al., 1988). PilA and PilB are believed to control the expression of other genes as well, because two-dimensional gels of *pilA* mutants showed significant differences when compared with wild-type cells. One affected protein of approximately 68 kDa could be immunoprecipitated with GroEL antibodies, suggesting that PilA may have a role in initiation of the stress response (Taha et al., 1992).

KLEBSIELLA PNEUMONIAE

K. pneumoniae is frequently associated with lobar pneumonia and urinary tract infection but can also cause sinusitis, meningitis, endocarditis, septicemia, and liver abscess. Under conditions of limited ammonia, synthesis of urease is induced (Friedrich and Magasanik, 1977; Mulrooney et al., 1989; Macaluso et al., 1990). Urease is believed to increase the virulence of pathogens when expressed in an infected urinary tract. Also, the capsule appears to play an important role in virulence, as it has been demonstrated that, in general, highly virulent strains are highly mucoid, whereas avirulent strains are only slightly mucoid (Mizuta et al., 1983). Both these virulence factors appear to be nitrogen-regulated.

Expression of the nitrogen fixation (*nif*) regulon in *K. pneumoniae* is controlled by both the oxygen status and the nitrogen status of the cell. Recently, mutations were isolated that separated the nitrogen- and oxygen-sensing abilities of the sensor component, NifL, although it is still unclear by

what mechanisms this protein is able to distinguish between the two different signals (Sidoti et al., 1993; see Chapter 17).

The NtrA–NtrC nitrogen regulatory system of *K. pneumoniae* is highly homologous to those of *E. coli* and *Salmonella* spp. (see Chapters 5 and 9). Briefly, a multicomponent cascade is responsible for the direct activation of σ^{54}-dependent nitrogen regulated genes and for the indirect regulation of σ^{70}-dependent promoters. This indirect regulation occurs by the NAC protein, which is a LysR-like regulator (Bender, 1991; Macaluso et al., 1990). However, additional loci appear to be regulated by the Ntr system, including urease gene expression (Macaluso et al., 1990; Collins et al., 1993). A series of *lacZYA* transcriptional fusions in the urease operon has revealed a region essential for NAC activation. This activation was also shown to be dependent on NtrA and NtrC as well (Collins et al., 1993).

Also, *K. pneumoniae* regulates synthesis of K antigen extracapsular polysaccharides by a two-component homolog known as RmpA2 (Wacharotayankun et al., 1993). This protein contains several regions of homology to response regulator proteins: its central domain is homologous to those of NtrC proteins in both *K. pneumoniae* and *E. coli,* and both amino- and carboxy-terminal domains show some homology to similar domains of RcsA. RcsA is a transcriptional activator in *E. coli,* and it functions with the two-component proteins, RcsB and RcsC, to regulate colonic acid synthesis (Stout and Gottesman, 1990; Stout et al., 1991; see Chapter 15). These proteins can serve to regulate synthesis of the *Klebsiella* capsule polysaccharides in *E. coli,* suggesting that homologs exist in *K. pneumoniae* (Wacharotayankun et al., 1993). Southern analysis suggests that *rmpA2* exists on the large plasmids of all virulent strains of *Klebsiella* but not on those of avirulent strains.

CLOSTRIDIUM PERFRINGENS

Clostridium perfringens can cause wound infections, including gas gangrene, by eliciting a variety of exotoxins such as α-toxin, τ-toxin (perfringolysin-O), κ-toxin (a collagenase), and λ-toxin (a protease) (Hatheway, 1990; McDonel, 1980). A hemagglutin (HA) is also produced, and although several toxin genes have been cloned, their regulation remains to be elucidated (Okabe et al., 1989; Shimizu et al., 1991; Rood and Cole, 1991).

A recent finding indicates that these virulence factors may, in part, be regulated by a two-component system. The *virR* gene, which shows amino-terminal similarity to AlgR from *P. aeruginosa* and AgrA from *S. aureus,* was identified by its ability to complement a *C. perfringens* mutant strain that did not produce perfringolysin-O, collagenase, or HA (Shimizu et al., 1994). No cognate sensor was identified, although an overlapping, downstream aminoterminal reading frame contained highly hydrophobic residues, suggestive of membrane-spanning regions. Complete cloning of that gene is currently being pursued.

Other regulatory mechanisms have been identified, including regulation of the gene for perfringolysin-O by an adjacent gene, *pfoR* (Shimizu et al., 1991). Also, perfringolysin-O, collagenase, protease, and HA production were shown to be induced by a small diffusible factor known as substance A (Imagawa et al., 1981; Imagawa and Higashi, 1992). At this time, however, it is not clear how, or if, these regulatory systems are integrated with VirR to regulate toxin production.

ERWINIA AMYLOVORA

The phytopathogen *Erwinia amylovora* causes a disease known as fireblight, which affects both fruit-bearing and ornamental plant species. The pathogenesis of the disease involves severe disruption of the host cell membrane, followed by cellular collapse and spreading tissue necrosis (Bachmann, 1913; Nixon, 1927; Youle and Cooper, 1987). Although the disease is well understood at the cellular level, the virulence factors that allow progression of disease have not yet been well characterized. One important observation, that unencapsulated bacteria are avirulent, has led to the identification of an extracellular polysaccharide as important for virulence (reviewed in Leigh and Coplin, 1992). The precise role extracellular polysaccharide plays in causing disease, however, remains to be elucidated.

Interestingly, regulation of extracellular polysaccharide synthesis appears to be controlled by functional homologs of the *E. coli* rcsB and rcsC genes, as described for *K. pneumoniae* (Coplin et al., 1990; Roberts and Coleman, 1991). As discussed in Chapter 15, these two-component proteins regulate colanic acid capsule synthesis in *E. coli* in response to low growth temperature and availability of nitrogen and carbon sources. Conservation of similar regulatory proteins and mechanisms suggests that the regulation of capsule synthesis is important for both human and plant pathogens.

Avirulent mutants of *E. amylovora* were isolated and found to be completely encapsulated and unaltered in any cell-surface polysaccharides, indicating a role for additional factors in virulence. Insertion mutations have identified a 15-kb region that is essential for pathogenicity and for eliciting a necrotic hypersensitive response in nonsusceptible host tissue, and this locus has been designated *hrp* (reviewed in Roberts and Coleman, 1991). Thus far, only one *hrp* gene has been sequenced, and it shows homology to the HrpS protein of *Pseudomonas syringae,* which itself has homology to NifA and NtrC genes in *Rhizobium* and *Klebsiella* spp. (Sneath et al., 1990). All proteins contain a conserved ATP binding site and show homology in their carboxy-terminal domains, suggesting that they belong to the two-component family of response regulator proteins.

GROUP A STREPTOCOCCI

Streptococcus pyogenes can cause a variety of diseases, including impetigo, rheumatic fever, toxic streptococcal syndrome, scarlet fever, and sepsis. The list of proposed virulence factors contains proteins whose diverse functions may serve to evade the immune system as well as to actively cause disease. These include the pyrogenic exotoxins that are important in scarlet fever and toxic streptococcal syndrome; a protease that may contribute to invasive infections; C5a peptidase, which can prevent recognition of chemotactic signals by polymorphonuclear leukocytes; an immunoglobulin G Fc receptor protein, which may bind opsonic antibodies; plasminogen-activated streptokinases, which may aid in the dissemination of bacteria in the host tissue; and a hyaluronic acid capsule and the M protein, which both protect the bacteria from phagocytosis.

Despite this impressive list of potential virulence factors, only M protein has been shown to be essential for survival within the host and the ability to cause disease (reviewed by Fischetti, 1989; Scott, 1990). M protein is a fibrillar surface protein composed of a dimeric coiled-coil structure that is extremely variable both in length and sequence. The primary immune response of the host is against M-protein epitopes, and opsonic antibodies are protective against further infection. However, protection is type-specific, and more than 80 different antigenic types have been defined (Lancefield, 1962). Also, M protein is antigenically similar to heart muscle, and cross-reactivity of anti-M-protein antibodies with cardiac tissue may contribute to the rheumatic sequelae that sometimes follow infection (Dale and Beachey, 1985).

Although homology is very limited, certain highly conserved residues within a protein called Mry have led to the conclusion that the regulation of M protein may occur through a two-component system. The *mry* locus was originally defined as a Tn*916* insertion upstream of the M-protein structural gene (*emm6*) that reduced transcription of that gene, and the gene product was classified as a response regulator protein (Perez-Casal et al., 1991). The Mry protein contains two helix-turn-helix motifs in its amino terminus, and some limited homology to the amino terminus of response regulators is found in its middle and carboxy-terminal regions. The *mry* gene is expressed from two promoters: the proximal promoter is responsible for low-level transcription under

nonpermissive conditions, and a second, distal promoter is used under other appropriate environmental conditions (Okada et al., 1993). This distal promoter is autoregulated (by the Mry protein itself), suggesting that activation of the low-level constitutively produced protein is sufficient to allow transcription from the distal, major promoter to proceed. The resulting higher levels of Mry protein are then able to activate promoters for Mry-regulated genes, including M protein. It is unclear what other virulence determinants are regulated by Mry. Production of the capsule and streptokinase were unaffected in an *mry* mutant, suggesting that Mry does not have a global regulatory role (Perez-Casal et al., 1991).

One of the signals that causes activation of the distal promoter appears to be CO_2, as conditions of 10% CO_2, 5% O_2 resulted in a 25-fold stimulation of expression of an *emm6*::chloramphenicol acetyltransferase reporter fusion over ambient conditions (Caparon et al., 1992). Simply reducing the O_2 level caused only minimal stimulation. Also, mid-logarithmically growing cells showed greater reporter fusion activity than cells from any other growth stage. This temporal regulation is similar to that seen for the Agr system in *S. aureus* discussed previously.

Interestingly, both *Cryptococcus neoformans* and *Bacillus anthracis* have CO_2-regulated capsules, and the protective antigen subunit of anthrax toxin is also CO_2-regulated, although the mechanisms of control are unknown (Granger et al., 1985; Makino et al., 1988; Koehler et al., 1994; Bartkus and Leppla, 1989). A locus, previously identified as *atx,* has been shown to be necessary for *trans* activation of anthrax toxin genes, but its protein product shows no homology to response regulators nor does it contain a known DNA binding motif.

CONCLUSION

It is clear that the wide spectrum of virulence factors encompassed by pathogenic bacteria provides an impressive and complicated defense that the host must overcome to suppress infection. The use of two-component systems to regulate virulence has proved to be an important aspect of pathogenesis. Whereas our understanding of bacterial pathogenesis was initially defined by an understanding of the disease state, we are currently learning to understand the pathogen itself. Enumeration of the factors required for virulence and elucidation of the mechanism of their action and regulation have provided insight into how these pathogens cause the disease state. The ultimate application of this knowledge will be better treatments and better prevention of disease. To outwit the pathogen, we must first discover how it outwits our own mechanisms of protection and defense. Understanding the regulatory mechanisms used by pathogenic microbes may provide a window to this knowledge.

REFERENCES

Abdelnour, A., S. Arvidson, T. Bremell, C. Ryden, and A. Tarkowski. 1993. The accessory gene regulator (*agr*) controls *Staphylococcus aureus* virulence in a murine arthritis model. *Infect. Immun.* **61:**3879–3885.

Bachmann, F. M. 1913. The migration of *Bacillus amylovorus* in the host tissues. *Phytopathology* **3:**3–14.

Bartkus, J. M., and S. H. Leppla. 1989. Transcriptional regulation of the protective antigen gene of *Bacillus anthracis. Infect. Immun.* **57:**2295–2300.

Bayles, K. W. 1993. The use of degenerate, sensor gene-specific, oligodeoxyribonucleotide primers to amplify DNA fragments from *Staphylococcus aureus. Gene* **123:**99–103.

Bender, R. A. 1991. The role of NAC protein in the nitrogen regulation of *Klebsiella aerogenes. Mol. Microbiol.* **5:**2575–2580.

Bernardini, M. L., A. Fontaine, and P. J. Sansonetti. 1990. The two-component regulatory system ompR-envZ controls the virulence of *Shigella flexneri. J. Bacteriol.* **172:**6274–6281.

Bernardini, M. L., J. Mounier, H. d'Hautevile, M. Coquis-Rondon, and P. J. Sansonetti. 1989. Identification of *icsA,* a plasmid locus of *Shigella flexneri* that governs intra- and inter-cellular spread through interaction with F-actin. *Proc. Natl. Acad. Sci. USA* **86:**3867–3871.

Bernardini, M. L., M. G. Sanna, A. Fontaine, and P. Sansonetti. 1993. OmpC is involved in invasion of epithelial cells by *Shigella flexneri. Infect. Immun.* **61:**3625–3635.

Boslego, J. W., E. C. Tramont, R. C. Chung, D. G. McChesney, J. Ciak, J. C. Sadoff, M. V. Piziak,

J. D. Brown, C. J. Brinton, S. W. Wood, and J. R. Bryan. 1991. Efficacy trial of a parenteral gonococcal pilus vaccine in men. *Vaccine* **9:**154–162.

Caparon, M. G., R. T. Geist, J. Perez-Casal, and J. R. Scott. 1992. Environmental regulation of virulence in group A Streptococci: transcription of the gene encoding M protein is stimulated by carbon dioxide. *J. Bacteriol.* **174:**5693–5701.

Collins, C. M., D. M. Gutman, and H. Laman. 1993. Identification of a nitrogen-regulated promoter controlling expression of *Klebsiella pneumoniae* urease genes. *Mol. Microbiol.* **8:**187–198.

Coplin, D. L., K. Poetter, and D. R. Majerczak. 1990. Regulation of exopolysaccharide synthesis in *Erwinia stewartii* by *rcsB* and *rcsC. Phytopathology* **81:**1220.

Cornelis, G., C. Sluiters, C. L. de Rouvroit, and T. Michiels. 1989. Homology between virF, the transcriptional activator of the *Yersinia* virulence regulon, and AraC, the *Escherichia coli* arabinose operon regulator. *J. Bacteriol.* **171:**254–262.

Dale, J. B., and E. H. Beachey. 1985. Multiple, heart-cross-reactive epitopes of streptococcal M proteins. *J. Exp. Med.* **161:**113–122.

Dassy, B., T. Hogan, T. J. Foster, and J. Fournier. 1993. Involvement of the accessory gene regulator (*agr*) in expression of type 5 capsular polysaccharide by *Staphylococcus aureus. J. Gen. Microbiol.* **139:**1301–1306.

Dean, C. R., and K. Poole. 1993. Expression of the ferric enterobactin receptor (PfeA) of *Pseudomonas aeruginosa:* involvement of a two-component regulatory system. *Mol. Microbiol.* **8:**1095–1103.

Deutscher, J., and H. Sauerwald. 1986. Stimulation of dihydroxyacetone and glycerol kinase activity in *Streptococcus faecalis* by phosphoenolpyruvate-dependent phosphorylation catalyzed by enzyme I and HPr of the phosphotransferase system. *J. Bacteriol.* **166:**829–836.

Fischetti, V. A. 1989. Streptococcal M protein: molecular design and biological behavior. *Clin. Microbiol. Rev.* **2:**285–314.

Forsberg, A., A. M. Viitanen, M. Skurnik, and W. H. Wolf. 1991. The surface-located YopN protein is involved in calcium signal transduction in *Yersinia pseudotuberculosis. Mol. Microbiol.* **5:**977–986.

Friedrich, B., and B. Magasanik. 1977. Urease of *Klebsiella aerogenes:* control of its synthesis by glutamine synthetase. *J. Bacteriol.* **131:**446–452.

Fuqua, W. C., S. C. Winans, and E. P. Greenberg. 1994. Quorum sensing in bacteria: the LuxR-LuxI family of cell density-responsive transcriptional regulators. *J. Bacteriol.* **176:**269–275.

Galan, J. E., C. C. Ginocchio, and P. Costeas. 1992. Molecular and functional characterization of the *Salmonella typhimurium* invasion gene *invA:* homology of InvA to members of a new protein family. *J. Bacteriol.* **17:**4338–4349.

Ginocchio, C. C., S. B. Olmsted, C. L. Wells, and J. E. Galan. 1994. Contact with epithelial cells induces the formation of surface appendages on *Salmonella typhimurium. Cell* **76**:717–724.

Goldberg, M. B., O. Barzu, C. Parsot, and P. J. Sansonetti. 1993. Unipolar localization and ATPase activity of IcsA, a *Shigella flexneri* protein involved in intracellular movement. *J. Bacteriol.* **175**:2189–2196.

Granger, D. L., J. R. Perfect, and D. T. Durack. 1985. Virulence of *Cryptococcus neoformans*. Regulation of capsule synthesis by carbon dioxide. *J. Clin. Invest.* **76**:508–516.

Grimm, C., and N. J. Panopoulos. 1989. The predicted protein product of a pathogenicity locus from *Pseudomonas syringae* pv. *phaseolicola* is homologous to a highly conserved domain of several prokaryotic regulatory proteins. *J. Bacteriol.* **171**: 5031–5038.

Hatheway, C. L. 1990. Toxigenic clostridia. *Clin. Microbiol. Rev.* **3**:66–98.

Higgins, D. E., E. Nazareno, and V. J. DiRita. 1992. The virulence gene activator ToxT from *Vibrio cholerae* is a member of the AraC family of transcriptional activators. *J. Bacteriol.* **174**:6874–6980.

Hobbs, M., E. S. R. Collie, P. D. Free, S. P. Livingston, and J. S. Mattick. 1993. PilS and PilR, a two-component transcriptional regulatory system controlling expression of type 4 fimbriae in *Pseudomonas aeruginosa. Mol. Microbiol.* **7**:669–682.

Imagawa, T., and Y. Higashi. 1992. An activity which restores theta toxin activity in some theta toxin-deficient mutants of *Clostridium perfringens. Microbiol. Immunol.* **36**:523–527.

Imagawa, T., T. Tatsuki, T. Higashi, and T. Amano. 1981. Complementation characteristics of newly isolated mutants from two groups of strains of *Clostridium perfringens. Biken J.* **24**:13–21.

Ishimoto, K. S., and S. Lory. 1992. Identification of *pilR*, which encodes a transcriptional activator of the *Pseudomonas aeruginosa* pilin gene. *J. Bacteriol.* **174**:3514–3521.

Janzon, L., and S. Arvidson. 1990. The role of the -lysin gene (*htd*) in the regulation of virulence genes by the accessory gene regulator (*agr*) in *Staphylococcus aureus. EMBO J.* **9**:1391–1399.

Janzon, L., S. Lofdahl, and S. Arvidson. 1989. Identification and nucleotide sequence of the delta-lysin gene, *hld,* adjacent to the accessory gene regulator (*agr*) in *Staphylococcus aureus. Mol. Gen. Genet.* **219**:480–485.

Kadurugamuwa, J. L., M. Rohde, J. Wehland, and K. N. Timmis. 1991. Intercellular spread of *Shigella flexneri* through a monolayer mediated by membranous protrusions and associated with reorganization of the cytoskeletal protein vinculin. *Infect. Immun.* **59**:3463–3471.

Keevil, C. W., N. C. Major, D. B. Davies, and A. Robinson. 1986. Physiology and virulence determinants of *Neisseria gonorrhoeae* grown in glucose-, oxygen- or cysteine-limited continuous culture. *J. Gen. Microbiol.* **132**:3289–3302.

Klaasen, P., and F. K. Graaf. 1990. Characterization of FapR, a positive regulator of expression of the 987P operon in enterotoxigenic *Escherichia coli. Mol. Microbiol.* **4**:1779–1783.

Koehler, T. M., Z. Dai, and M. Kaufman-Yarbray. 1994. Regulation of the *Bacillus anthracis* protective antigen gene: CO_2 and a *trans*-acting element activate transcription from one or two promoters. *J. Bacteriol.* **176**:586–595.

Kofoid, E. C., and J. S. Parkinson. 1988. Transmitter and receiver modules in bacterial signaling proteins. *Proc. Natl. Acad. Sci. USA* **85**:4981–4985.

Kornblum, J., B. N. Kreiswirth, S. J. Projan, H. Ross, and R. P. Novick. 1990. Agr: a polycistronic locus regulating exoprotein synthesis in *Staphylococcus aureus,* p. 373–402. *In* R. Novick and R. Skurray (ed.), *The Molecular Biology of Staphylococci.* VCH Publishers, New York.

Lancefield, R. C. 1962. Current knowledge of type-specific M antigens of group A streptococci. *J. Immunol.* **89**:307–313.

Leigh, J. A., and D. L. Coplin. 1992. Exopolysaccharides in plant-bacterial interactions. *Annu. Rev. Microbiol.* **46**:307–346.

Macaluso, A., E. A. Best, and R. A. Bender. 1990. Role of the *nac* gene product in the nitrogen regulation of some NTR-regulated operons of *Klebsiella aerogenes. J. Bacteriol.* **172**:7249–7255.

Makino, C., C. Sasakawa, K. Kamata, T. Kurata, and M. Yoshikawa. 1986. A genetic determinant required for continuous reinfection of adjacent cells on large plasmid in *Shigella flexneri* 2a. *Cell* **46**:551–555.

Makino, S. I., C. Sasakawa, I. Uchida, M. Terekado, and M. Yoshikawa. 1988. Cloning and CO_2-dependent expression of the genetic region for encapsulation from *Bacillus anthracis. Mol. Microbiol.* **2**:371–376.

Mardh, P. A., and I. Westrom. 1976. Adherence of bacteria to vaginal epithelial cells. *Infect. Immun.* **13**:661–666.

McDonel, J. L. 1980. *Clostridium perfringens* toxins (type A, B, C, D, E). *Pharmacol. Ther.* **10**:617–635.

McGee, Z. A., A. P. Johnson, and D. Taylor-Robinson. 1981. Pathogenic mechanisms of *Neisseria gonorrhoeae*: observation on damage to human fallopian tubes in organ culture by gonococci of colony type 1 or type 4. *J. Infect. Dis.* **143**:413–422.

Mekalanos, J. J. 1992. Environmental signals controlling expression of virulence determinants in bacteria. *J. Bacteriol.* **174**:1–7.

Mills, S. D., C. A. Jasalavich, and D. A. Cooksey. 1993. A two-component regulatory system required for copper-inducible expression of the copper resistance operon of *Pseudomonas syringae. J. Bacteriol.* **175:**1656–1664.

Mizuta, K., M. Ohta, M. Mori, T. Hasagawa, I. Nakashima, and M. Kato. 1983. Virulence for mice of *Klebsiella* strains belonging to the 01 group: relationship to their capsular (K) types. *Infect. Immun.* **40:**56–61.

Morfeldt, E., L. Janzon, S. Arvidson, and L. Lofdahl. 1988. Cloning of a chromosomal locus (*exp*) which regulates the expression of several exoprotein genes. *Mol. Gen. Genet.* **211:**435–440.

Mounier, J., A. Ryter, R. M. Coquis, and P. J. Sansonetti. 1990. Intracellular and cell-to-cell spread of *Listeria monocytogenes* involves interaction with F-actin in the enterocyte like cell line Caco-2. *Infect. Immun.* **58:**1048–1058.

Mulrooney, S. B., H. S. Pankratz, and R. P. Hausinger. 1989. Regulation of gene expression and cellular localization of cloned *Klebsiella aerogenes* (*K. pneumoniae*) urease. *J. Gen. Microbiol.* **135:**1769–1776.

Nixon, E. L. 1927. The migration of *Bacillus amylovorus* in apple tissue and its effects on host cells. *Bull. Pennsylvania Agric. Exp. Station* **212:**3–16.

Okabe, A., T. Shimizu, and H. Hayashi. 1989. Cloning and sequencing of a phospholipase C gene of *Clostridium perfringens. Biochem. Biophys. Res. Commun.* **160:**33–39.

Okada, M., R. T. Geist, and M. G. Caparon. 1993. Positive transcriptional control of *mry* regulates virulence in the group A streptococcus. *Mol. Microbiol.* **7:**893–903.

Pace, J., M. J. Hayman, and J. E. Galan. 1993. Signal transduction and invasion of epithelial cells by *S. typhimurium. Cell* **72:**505–514.

Peng, H. L., R. P. Novick, B. Kreiswirth, J. Kornblum, and P. Schlievert. 1988. Cloning, characterization, and sequencing of an accessory gene regulator (*agr*) in *Staphylococcus aureus. J. Bacteriol.* **170:**4365–4372.

Perez-Casal, J., M. G. Caparon, and J. R. Scott. 1991. Mry, a *trans*-acting positive regulator of the M protein gene of *Streptococcus pyogenes* with similarity to the receptor proteins of two-component regulatory systems. *J. Bacteriol.* **173:**2617–2624.

Prevost, M. C., M. Lesourd, M. Arplin, F. Vernel, J. Mounier, R. Hellio, and P. J. Sansonetti. 1992. Unipolar reorganization of F-actin layer at bacterial division and bundling of actin filaments by plastin correlate with movement of *Shigella flexneri* within HeLa cells. *Infect. Immun.* **60:**4088–4099.

Regassa, L. B., and M. J. Betley. 1992. Alkaline pH decreases expression of the accessory gene regulator (*agr*) in *Staphylococcus aureus. J. Bacteriol.* **174:**5095–5100.

Recsei, P., B. Kreiswirth, M. O'Reilly, P. Schlievert, A. Gruss, and R. Novick. 1986. Regulation of exoprotein gene expression by *agr. Mol. Gen. Genet.* **202:**58–61.

Roberts, I. S., and M. J. Coleman. 1991. The virulence of *Erwinia amylovora:* molecular genetic perspectives. *J. Gen. Microbiol.* **137:**1453–1457.

Robertson, E. S., J. C. Hoyt, and H. C. Reeves. 1988. Evidence of histidine phosphorylation in isocitrate lyase from *Escherichia coli. J. Biol. Chem.* **263:**2477–2482.

Rood, J. I., and S. T. Cole. 1991. Molecular genetics and pathogenesis of *Clostridium perfringens. Microbiol. Rev.* **55:**621–648.

Scott, J. R. 1990. *The M Protein of Group A Streptococcus: Evolution and Regulation.* Academic Press, San Diego, Calif.

Shimizu, T., W. Ba-thein, M. Tamaki, and H. Hayashi. 1994. The *virR* gene, a member of a class of two-component response regulators, regulates the production of perfringolysin O, collagenase, and hemagglutinin in *Clostridium perfringens. J. Bacteriol.* **176:**1616–1623.

Shimizu, T., A. Okabe, J. Minami, and H. Hayashi. 1991. An upstream regulatory sequence stimulates expression of the perfringolysin O gene of *Clostridium perfringens. Infect. Immun.* **59:**137–142.

Sidoti, C., G. Harwood, R. Ackerman, J. Coppard, and M. Merrick. 1993. Characterisation of mutations in the *Klebsiella pneumoniae* nitrogen fixation regulatory gene *nifL* which impair oxygen regulation. *Arch. Microbiol.* **159:**276–281.

Sneath, B. J., X.-H. Jiang, H. M. Howson, and S. V. Beer. 1990. A pathogenicity gene from *Erwinia amylovora* encoded a predicted protein product homologous to a family of prokaryotic response regulators. *Phytopathology* **80:**1038.

Stout, V., and S. Gottesman. 1990. RcsB and RcsC: a two-component regulator of capsule synthesis in *Escherichia coli. J. Bacteriol.* **172:**659–669.

Stout, V., A. Torres-Cabassa, M. R. Maurizi, D. Gutnick, and S. Gottesman. 1991. RcsA, an unstable positive regulator of capsular polysaccharide synthesis. *J. Bacteriol.* **173:**1738–1747.

Swanson, J. 1973. Studies on gonococcal infection. IV. Pili: their role in attachment of gonococci to tissue culture cells. *J. Exp. Med.* **137:**571–589.

Taha, M. K., B. Dupuy, W. Saurin, M. So, and C. Marchal. 1991. Control of pilus expression in *Neisseria gonorrhoeae* as an original system in the family of two-component regulators. *Mol. Microbiol.* **5:**137–148.

Taha, M. K., M. Larribe, B. Dupuy, D. Giorgini, and C. Marchal. 1992. Role of *pilA*, an essential regulatory gene of *Neisseria gonorrhoeae,* in the stress response. *J. Bacteriol.* **174:**5978–5981.

Taha, M. K., M. So, H. S. Seifert, E. Billyard, and C. Marchal. 1988. Pilin expression in *Neisseria gonorrhoeae* is under both positive and negative transcriptional control. *EMBO J.* **7:**4367–4378.

Taha, M. L., and C. Marchal. 1990. Conservation of *Neisseria gonorrhoeae* pilus expression regulatory genes *pilA* and *pilB* in the genus *Neisseria. Infect. Immun.* **58:**4145–4148.

Tilney, L. G., and D. A. Portnoy. 1989. Actin filaments and the growth, movement, and spread of the intracellular bacterial parasite, *Listeria monocytogenes. J. Cell Biol.* **109:**1597–1608.

Vandenesch, F., J. Kornblum, and R. P. Novick. 1991. A temporal signal, independent of *agr,* is required for *hla* but not *spa* transcription in *Staphylococcus aureus. J. Bacteriol.* **173:**6313–6320.

Wacharotayankun, R., Y. Arakawa, M. Ohta, K. Tanaka, T. Akashi, M. Mori, and N. Kato. 1993. Enhancement of extracapsular polysaccharide synthesis in *Klebsiella pneumoniae* by RmpA2, which shows homology to NtrC and FixJ. *Infect. Immun.* **61:**3164–3174.

Youle, D., and R. M. Cooper. 1987. Possible determinants of pathogenicity of *Erwinia amylovora,* evidence for an induced toxin. *Acta Hort.* **217:** 161–167.

Regulation of *Salmonella* Virulence by Two-Component Regulatory Systems

Eduardo A. Groisman and Fred Heffron

20

Microorganisms exhibit a wide variety of adaptive responses to changes in environmental conditions. Specific sensory systems detect modifications in the environment and act to transduce these signals into appropriate cellular activities that result in the adaptive behavior. These activities include changes in locomotion so that the microbe swims toward attractants and away from repellents (see Chapter 6) as well as the synthesis of different sets of proteins that allow the microbe to cope with different physical or nutritional environments. The capacity to adjust to novel conditions is particularly remarkable in microbial pathogens because they encounter variable habitats during the course of infection and must modulate the repertoire of expressed genes to profit from the new environmental conditions (Mekalanos, 1992).

The recent use of classical bacterial genetics for the analysis of microbial pathogenesis has allowed the identification of loci previously not recognized as virulence determinants. This approach compares the behavior of isogenic wild-type and mutant strains for their patho-

genic properties in defined animal or tissue culture model systems, and it has allowed the identification of transcription regulatory factors and of products required for the export and assembly of crucial virulence factors. In *Salmonella typhimurium,* in which successful genetic screenings have defined loci required for intramacrophage survival (Fields et al., 1986), resistance to antimicrobial peptides (Groisman et al., 1992b), and entry into epithelial cells (Stone et al., 1992), four chromosomally encoded transcription factors (Crp, PhoP, OmpR, and FruR) have been identified as absolutely required for mouse virulence in vivo. And in *Vibrio cholerae,* mutations in either the structural or the regulatory genes for cholera toxin diminish its virulence properties (DiRita and Mekalanos, 1989).

BIOLOGY OF *SALMONELLA* PATHOGENESIS

Salmonellae are gram-negative microorganisms that have been isolated from a wide variety of animal species including insects, amphibians, and mammals (Brenner, 1992). In humans, they give rise to a variety of disease conditions collectively known as human salmonellosis that include gastroenteritis, enteric fever, and bacteremia (Groisman et al., 1990). Salmonellae are responsible for more than 12 million annual

Eduardo A. Groisman, Department of Molecular Microbiology, Washington University School of Medicine, St. Louis, Missouri 63110. *Fred Heffron,* Department of Microbiology and Immunology, Oregon Health Sciences University, Portland, Oregon 97201.

cases of typhoid fever worldwide (resulting in >500,000 deaths), and they are one of the three most important causes of infant mortality in many underdeveloped countries. In the United States, salmonellae contribute to the 6.5 million cases of food-borne infections that result in more than 9,000 deaths each year (Todd, 1989). Annual costs associated with salmonellosis have been estimated at about $4 billion in the United States alone, making salmonellosis the food-borne disease with the greatest cost (Todd, 1989). Certain salmonellae have a very narrow host range, whereas others are poorly host-adapted and produce different diseases in different hosts. For example, typhoid fever is caused primarily by the human-adapted *S. typhi,* whereas *S. typhimurium* is the leading cause of gastroenteritis in humans but causes a typhoidlike disease in mice (Groisman and Saier, 1990). In AIDS patients or otherwise immunocompromised individuals, *Salmonella* serotypes normally associated with gastroenteritis give rise to septicemia (Sperber and Scheulpner, 1987). *Salmonella* infections affect several economically important animal species including chickens, cows, and racing horses.

Most *Salmonella* infections result from oral ingestion of contaminated water or foodstuff, passage through the stomach, and engulfment by epithelial or M cells in the small intestine (Finlay and Falkow, 1989). Species causing enteric fever proceed into the deeper tissue before entering the reticuloendothelial system, where they multiply and disseminate to other organs. Bacteria eventually reseed the bloodstream, resulting in septicemia followed soon afterward by death of the animal. However, gastroenteritis-causing species do not usually penetrate deeper than the basement membrane of the epithelium. The ability to survive within professional phagocytic cells makes *Salmonella* a facultative intracellular parasite. Several host genetic factors have been identified that control an animal susceptibility to *Salmonella* (Groisman et al., 1990) and often to other intracellular parasites.

TWO-COMPONENT SYSTEMS AND THE CONTROL OF *SALMONELLA* VIRULENCE

Let us consider the different environments faced by *Salmonella* over the course of infection. The bacteria must resist killing by the acid pH of the stomach, and by bile salts, antimicrobial peptides, and immunoglobulin A in the small intestine. They must also resist clumping by the viscous intestinal mucin. As bacteria pass the stomach into the small intestine, they are subjected to osmotic stress and to an alkaline pH. *Salmonella* must adhere to the proper cells of the intestine, resist killing by inflammatory neutrophils attracted to the adherent bacteria, and enter and transverse the epithelial cell into the underlying tissues. In the lymphatics, *Salmonella* encounters antibody, complement, macrophages, neutrophils, and natural killer cells. Later in the infection process, *Salmonella* faces compounds that constitute the adaptive immune response. Therefore, many different microenvironments are faced by *Salmonella* between the time of oral ingestion and the death of the animal.

How many two-component regulatory systems are required for virulence in *Salmonella*? To our knowledge, no one has carried out a systematic study of this area, although we would expect to find that several of the estimated 50 systems that are present in *S. typhimurium* contribute to its pathogenic potential. These systems could account, in part, for the some 200 genes that are necessary to cause a lethal infection in mice (Tsolis et al., unpublished data). For example, OmpR/EnvZ and PhoP/PhoQ are absolutely required for *Salmonella* virulence, whereas other systems, such as PmrA/PmrB and ArcA/ArcB, could also be involved in virulence, as suggested by certain in vitro phenotypes associated with mutations in these loci.

The PmrA/PmrB system controls modification of the positive charge of the lipopolysaccharide (LPS), affecting resistance of *Salmonella* to positively charged antibiotics such as polymyxin and to cationic microbicidal peptides such as defensins (Vaara, 1981). *pmrA* and *pmrB* encode the response regulator and sensory element, respectively (Roland et al., 1993). The genetic structure of the *pmrAB* locus is peculiar in that the *pmrAB* genes are preceded by *pmrC*, which has no homologs in the data bases (Roland et al., 1993). Certain mutations within *pmrA* confer increased resistance to polymyxin, a positively charged compound containing a short cyclic peptide sequence joined to a fatty acid. The antimicrobial effects of polymyxin on gram-negative bacteria result primarily from permeabilization of the inner membrane, which follows the electrostatic interactions between the positively charged peptide and the negatively charged outer membrane (Vaara, 1981). Strains with missense mutations in *pmrA* that exhibit increased resistance to polymyxin show an increased substitution by both 4-aminoarabinose and phosphorylethanolamine in the lipid A moiety of the LPS (Helander et al., 1994). This makes the outer membrane of the bacterial cell less negatively charged and, consequently, not able to bind as much polymyxin. These mutants are also more resistant to several antimicrobial polypeptides present in human neutrophil granules, perhaps accounting for their increased resistance to killing by neutrophils in culture (Stinavage et al., 1989). To our knowledge, virulence studies have not been reported for mutants harboring either loss or gain of function alleles of *pmrA*.

It is likely that certain mouse tissues present an environment of reduced oxygen tension. In fact, induction of the invasion determinants in *S. typhimurium* requires growth of the organism under low-oxygen conditions (Lee and Falkow, 1990), which could mimic those encountered in the small intestine. With the exception of *orgA* (Jones and Falkow, 1994), the expression of which is oxygen-modulated, the pattern of regulation of the majority of loci required for epithelial cell invasion (Galán et al., 1992; Lee et al., 1992; Stone et al., 1992; Groisman and Ochman, 1993) is yet to be examined. The bacterial response to low-oxygen conditions

could be mediated by ArcA/ArcB, a two-component system that represses the expression of many genes under anaerobic conditions. However, the virulence role and the particular signal that is detected by this system remain unknown.

S. typhimurium strains defective in the EnvZ/OmpR system are attenuated for mouse virulence (Groisman et al., 1992c). This two-component system controls the expression of several products, including the outer membrane porins OmpC and OmpF, in response to changes in osmolarity. Although *ompR* mutants of *S. typhimurium* are attenuated for mouse virulence by both the oral and the intravenous routes (Dorman et al., 1989), strains with single mutations in the OmpR-regulated genes *ompC, ompF,* or *tpp* retain their virulence properties (Chatfield et al., 1991). A strain defective in both *ompC* and *ompF* was attenuated but not to the extent of the *ompR* mutant (Chatfield et al., 1991). This indicates that there must be other OmpR-regulated genes that contribute to the pathogenic properties of virulent *S. typhimurium*.

Most regulatory studies on the EnvZ/OmpR system have been conducted in *Escherichia coli* (Chapter 13), and some of the conclusions may not be applicable to certain *Salmonella* species. For example, expression of *E. coli* OmpC is induced in media of high osmolarity and repressed under low-osmolarity conditions. However, osmolarity changes did not affect the amount of OmpC in *S. typhi* (Puente et al., 1991). Interestingly, expression of a plasmid-borne *S. typhi ompC* was still dependent on an active *ompR* in *E. coli* (Puente et al., 1991). However, OmpF expression was induced in media of low osmolarity and repressed in media with high osmolarity in both *E. coli* and *S. typhi* (Puente et al., 1991). These data should be examined carefully not only due to their potential biological significance but also because outer membrane preparations of *Salmonella* are being evaluated for their ability to elicit protective immunity.

ROLE OF THE PhoP/PhoQ SYSTEM IN SALMONELLA PATHOGENESIS

The response of *S. typhimurium* to environments encountered during the course of infection is partially under the transcriptional regulation of the PhoP/PhoQ two-component system (Groisman et al., 1989; Miller et al., 1989). This system includes the regulator PhoP, the kinase-phosphatase PhoQ, and an assemblage of some 40 PhoP-regulated genes, many of which are essential for various stages of infection (García-Véscovi et al., 1994). PhoP is the best characterized of the eight proteins (RpoS, CRP, OmpR, SpvR, Fur, FlgM, and FruR) identified as controlling virulence gene transcription in *Salmonella* (Groisman and Ochman, 1994) and the only one for which a correlation has been established between attenuation of pathogenicity and susceptibility to particular host defense mechanisms (Fields et al., 1989; Groisman and Saier, 1990).

Discovery

phoP was first identified in a mutant hunt for genes involved in phosphate metabolism in *S. typhimurium*. Kier et al. (1977) characterized the regulation of three phosphatases in *S. typhimurium* in response to various nutrional limitations and found that an acid phosphatase (later shown to be the product of *phoN*) was induced in response to carbon, nitrogen, phosphorus, or sulfur limitation. This phosphatase was not regulated by either cyclicAMP-CRP or the two-component system PhoB/PhoR (Chapter 12) as might have been expected from the nutritional signals that control its expression. Mutations in two loci affected synthesis of this phosphatase: *phoN*, mapping to the 96-min region, was postulated to be the structural gene based on temperature-sensitive alleles that mapped to this region (Kier et al., 1979). And *phoP*, mapping at 25 min, was postulated to be a regulator because of the high level of phosphatase expression that resulted from a constitutive mutant (Kier et al., 1979).

Pathogenicity

That PhoP is required for *Salmonella* virulence was first established when a class of transposon Tn*10*-generated, macrophage-sensitive mutants (Fields et al., 1986) was mapped to the *phoP* locus (Fields et al., 1989). These mutant strains had a very high median lethal dose (LD_{50}) in mice and were among the most sensitive mutants ever tested in macrophages. These macrophage-sensitive mutants were allelic with those isolated by Kier et al. (1979) because (i) they did not produce nonspecific acid phosphatase and (ii) the original *phoP* alleles (Kier et al., 1979) attenuated *S. typhimurium* for mouse virulence to the same extent as the macrophage-sensitive mutants (Fields et al., 1989). Further analysis showed that the transposon Tn*10* had inserted within 500 bp of each other in three analyzed mutants: two within the *phoP* gene and one in the *phoQ* open reading frame. An attempt was made to determine the specific microbicidal mechanism of the macrophage that was responsible for the survival defect of these mutants. No increase in sensitivity was observed to toxic oxidizing compounds such as menedione, paraquat, and hydrogen peroxide (Fields et al., 1989). However, the *phoP* mutants exhibited heightened sensitivity to crude granulocyte extracts from human neutrophils and rabbit macrophages, suggesting sensitivity to nonoxidative killing mechanisms (Fields et al., 1989). Active fractions of these extracts were rich in small microbicidal peptides that could correspond to the arginine- and cysteine-rich antimicrobial peptides known as defensins (Lehrer et al., 1993). When purified rabbit NP-1 was tested, survival of the *phoP* mutants was about 500 times lower than that corresponding to the wild-type parent (Fields et al., 1989). Heightened susceptibility seemed to be fairly specific because the mutants were as resistant as the wild-type strain to other granule components including lysozyme, cathepsin G, and bacterial permeability increasing factor (Fields et al., 1989).

Work from Miller et al. (1989) and Groisman et al. (1989) demonstrated that PhoP/

PhoQ was a member of the family of two-component regulatory systems. The 224-amino-acid PhoP is the regulator and has been placed in the OmpR subgroup of this protein family (Volz, 1993), being most closely related to PhoB (35% identity, 76% similarity). The 487-amino-acid PhoQ is the sensory element with a predicted structure that includes two transmembrane domains and defines a putative periplasmic region of 146 residues that could be involved in sensing environmental changes (Miller et al., 1989). Analysis of transcriptional *lac* fusions with several genes including *psiD, pagA,* and *pagB* showed that PhoP controls expression at the transcriptional level (Groisman et al., 1989; Miller et al., 1989), as predicted for a regulator of the two-component family.

The high LD_{50} of strains with mutations in *phoP* and *phoQ* could perhaps be a reflection that PhoP/PhoQ is required for several crucial steps during infection. The PhoP/PhoQ system is unique in that both null and constitutive alleles attenuate strains for mouse virulence (García-Véscovi et al., 1994). However, these mutant strains differ with respect to the particular virulence phenotypes associated with mutations in the *phoPQ* operon. For example, strains with either null (Fields et al., 1986) or constitutive (Miller and Mekalanos, 1990) alleles of *phoP* are unable to survive within macrophages. However, only *phoP⁻ S. typhimurium* exhibit hypersusceptibility to antimicrobial peptides (Fields et al., 1989; Groisman et al., 1989; Miller et al., 1990) and acid pH (Foster and Hall, 1990). However, constitutive mutants exhibit epithelial cell invasion defects (Behlau and Miller, 1993) and reside in normal rather than the spacious vacuoles detected with wild-type and *phoP⁻ Salmonella* (Alpuche-Aranda et al., 1994). In a *phoP⁻* mutant, virulence determinants such as PagC (Miller et al., 1989) are not made, whereas in a strain with a *phoP*-linked constitutive allele, attenuation is presumably due to the continuous expression of PhoP-activated genes and repression of PhoP-repressed loci (Miller and Mekalanos, 1990).

The original constitutive mutant (allele *pho-24*) was isolated by Kier et al. (1979) as a strain that produced nonspecific acid phosphatase under noninducing conditions. DNA sequence analysis of this mutant showed no differences in the *phoP* coding region (Groisman, unpublished data) but the presence of a base pair substitution that resulted in a single amino acid change in PhoQ (cited in Miller et al. [1993]). Because a *S. typhimurium phoP⁻* strain harboring *phoP⁺ phoQ⁺* on a multicopy number plasmid retains its virulence properties (Fields et al., 1989), higher amounts of PhoP and PhoQ per se do not confer attenuation. The phenotype of *pho-24* mutants could be interpreted in one of two ways: (i) the mutated PhoQ is better at phosphorylating PhoP so that activation-repression of PhoP-modulated genes takes place regardless of the environmental signals; or (ii) the mutated PhoQ lost its specificity while becoming a constitutive kinase, and the avirulence phenotype results from the activation-repression of several genes (not just those normally regulated by PhoP). In other words, one would like to know whether the phenotypic consequences of the *pho-24* mutation is dependent on a functional *phoP* gene. Therefore, a critical evaluation of the various hypotheses will entail testing the phenotype of a constitutive mutant that maps to *phoP* itself.

Virulence Phenotypes

Several virulence defects have been described for strains harboring mutations in *phoP* or *phoQ* (Table 1). These include attenuation for mouse virulence in vivo, intramacrophage survival, susceptibility to defensins and other antimicrobial peptides, susceptibility to acid pH, susceptibility to serum, inability to invade epithelial cells, altered processing and presentation of antigens by macrophages, and a failure to reside within spacious vacuoles. Some of these phenotypes were observed with strains harboring null alleles of *phoP* or *phoQ*, whereas others were detected in the mutant with the constitutive *pho-24* allele. A *phoP⁻ S. typhimurium* is unable to synthesize at least 9 of 34 proteins

TABLE 1 Virulence phenotypes associated with *Salmonella* null and constitutive mutants

Property	Phenotype of *S. typhimurium* mutant[a]		
	phoP+	*phoP*-	*pho-24*
Mouse virulence in vivo	Virulent	Attenuated	Attenuated
Intramacrophage survival	Proficient	Defective	Defective
Resistance to defensins	Resistant	Susceptible	Resistant
Resistance to acid pH	Resistant	Susceptible	n.d.
Invasion of epithelial cells	Proficient	Proficient	Defective
Spacious vacuoles	Proficient	Proficient	Defective
Antigen presentation	Wild type[b]	Increased	Decreased

[a]n.d., not determined.

[b]The levels of antigen presentation exhibited by wild-type *S. typhimurium* are lower than that of *E. coli* K-12.

whose expression is induced on interaction with macrophages (Buchmeier and Heffron, 1990). This could account for its macrophage survival and in vivo mouse virulence defects (Fields et al., 1989; Miller et al., 1989), its hypersensitivity to acid pH (Foster and Hall, 1990) and defensins (Fields et al., 1989; Miller et al., 1990), and its inability to modulate antigen processing and presentation (Wick et al., in press).

S. typhimurium strains with mutations in *phoP* or *phoQ* are hypersusceptible to several antimicrobial peptides including rabbit and human defensins (Fields et al., 1989; Miller et al., 1990; Groisman et al., 1992b), frog-derived magainins, insect and pig cecropins, and mastoparan and melittin, which are components of wasp and honey bee venoms, respectively (Groisman et al., 1992a). Although these peptides do not share primary amino acid sequence identity, they are all small and cationic and can adopt structures that show amphiphilicity. Moreover, they have the same postulated mechanism of action: membrane depolarization provoked by the generation of voltage-gated ion channels. Interestingly, salmonellae have been isolated from a wide variety of animal species known to produce antimicrobial peptides and might have evolved mechanisms of resistance against these structurally different compounds (Groisman, 1994). During infection of humans, *S. typhi* is likely to face defensins in the lumen of the small intestine and

within the phagocytic cells of liver and spleen. Several experiments highlight the importance of *Salmonella* resistance to defensin-mediated killing for human disease: (i) patients with "specific granule deficiency" lack defensins and have frequent bacterial infections (Lehrer et al., 1993); (ii) a *phoP*- derivative of the human-adapted *S. typhi* is hypersusceptible to human defensins (Groisman and Heffron, unpublished data); and (iii) in experimental infections of rabbit neutrophils, *S. typhimurium* was found to reside in vacuoles rich in azurophilic granule components, prominently defensins (Joiner et al., 1989).

Sensitivity to defensins could be due to the inability of PhoP to regulate a resistance gene(s) or result from the absence of PhoQ, which harbors a periplasmic acidic domain that could interact with cationic peptides. However, a direct involvement of PhoQ in defensin resistance has been ruled out in experiments showing that a *phoQ*+ plasmid clone could not rescue the defensin-susceptibility phenotype of a *phoP*- mutant (Miller et al., 1990). Genetic evidence suggests that resistance to antimicrobial peptides is complex and multifactorial (Groisman, 1994). Starting with a peptide susceptible *S. typhimurium phoP*- mutant, the Groisman laboratory isolated (i) second-site suppressors that recover resistance to magainin 2 and cecropin P1 but do not suppress hypersensitivity to mastoparan and melittin, and (ii) multicopy number suppressors that rescued resistance to defensin but not to magainin 2. Furthermore, a PhoP-activated gene was identified that is necessary for resistance to melittin and protamine but has no effect on magainin 2 or mastoparan (Soncini and Groisman, unpublished data).

Wild-type *Salmonella* appears to inhibit phagosome acidification. Normally, when a microbe is engulfed by a macrophage, the phagocytic vesicle reaches a pH of 6 within minutes; after fusion to the lysosomes, it reaches a pH of 4.5. Alpuche-Aranda et al. (1992) showed that *S. typhimurium* inhibits acidification of the vacuole and that this property requires a functional *phoP* gene. Furthermore, they demonstrated that the PhoP-regulated *pagC* gene is strongly induced only within the phagocytic vesicle of the macrophage (see below). In work conducted in vitro, Foster and Hall (1990) showed that there is an acid-tolerance response in *S. typhimurium* that results in an increased ability to survive exposure to an acid pH by previous adaptation to a mildly acidic pH. These investigators went on to show that *phoP⁻* mutants retain the ability to adapt to an acid pH, but their survival is 1,000 times lower than that of the wild-type strain in both adapted and nonadapted conditions (Foster and Hall, 1990). Cumulatively, these data

suggest a correlation between tolerance, adaptation to acid pH, and virulence in *S. typhimurium*.

PhoP/PhoQ Regulon

Based on the differential pattern of proteins visualized in two-dimensional gels, Miller and Mekalanos (1990) originally suggested that about 40 proteins were under control of the PhoP/PhoQ system. Several loci have been identified as being PhoP-regulated based on the differential β-galactosidase and alkaline phosphatase activity of *lac* and *phoA* fusions, respectively, between *phoP⁺* and *phoP⁻* strains. To date, only 11 loci are reported as harboring insertions in PhoP-modulated genes (Table 2), and only two of them—*pagC* (Pulkkinen and Miller, 1991) and *phoN* (Kasahara et al., 1991; Groisman et al., 1992c)—have been cloned and analyzed at the molecular level. Miller et al. (1989) identified *pagC,* encoding an outer membrane protein that is required for macrophage survival and virulence in vivo. PagC is a member of a protein family that includes Ail, a *Yersinia* protein required for resistance to serum complement and invasion of epithelial cells, Rck, a *S. typhimurium* protein that confers complement resistance to certain strains of *E. coli* and is encoded by the *Salmonella* virulence plasmid, Lom, a bacteriophage λ-

TABLE 2 PhoP/PhoQ-modulated genes[a]

Gene	Map position (min)	Virulence phenotype	Gene product
pagA	42	No effect	n.d.
pagB	n.m.	No effect	n.d.
pagC	25	Macrophage survival	Outer membrane protein
pagD	n.m.	Macrophage survival	Envelope protein
pbgG	96	Macrophage survival	Envelope protein
pcgG	n.m.	Peptide resistance	n.d.
phoN	95	No effect	Periplasmic nonspecific acid phosphatase
psiD	93	No effect	n.d.
prgA	n.m.	No effect	Envelope protein
prgB	28–32	No effect	Envelope protein
prgC	n.m.	No effect	Envelope protein
prgE	n.m.	No effect	Envelope protein
prgH	59	Epithelial cell invasion	n.d.

[a]n.m., not mapped; n.d., not determined.

encoded protein of unknown function, and to OmpX of *Enterobacter cloacae* (Pulkkinen and Miller, 1991). Although *pagC* is required for full in vivo virulence and survival within murine macrophages, mutations in *pagC* diminished *Salmonella* virulence but not to the level of *phoP⁻* or *phoQ⁻* mutants (Miller et al., 1989), and its function remains unknown. Despite the similarity of PagC with Ail in their predicted transmembrane domains (Pulkkinen and Miller, 1991) and the phenotype of certain *pagC*::Tn*phoA* mutants (Miller et al., 1992), PagC does not seem to play a role in epithelial cell invasion.

Behlau and Miller (1993) identified a collection of strains with mutations in PhoP-repressed genes as those that were turned off in the *pho-24* constitutive mutant. Although most of these mutant strains did not exhibit virulence phenotypes, one mutant was attenuated for mouse virulence and defective for epithelial cell invasion (Behlau and Miller, 1993). The affected locus, *prgH,* was mapped next to a cluster of genes that have been shown to be required for invasion in the 59-min region (Galán et al., 1992; Lee et al., 1992; Groisman and Ochman, 1993; Jones and Falkow, 1994). The transcriptional orientation of *prgH* is opposite to that of *hil,* a locus identified by Lee and Falkow as one that on overexpression allowed invasion of epithelial cells under nonphysiological conditions (Lee et al., 1992). The *prgH* mutant could be rescued by a DNA fragment located adjacent to *hil.* That the *prgH Salmonella* was attenuated for virulence by both the oral and intraperitoneal routes of inoculation suggests that PrgH plays a role beyond entry into the intestinal epithelial cells.

In an effort to identify all PhoP-regulated genes, the laboratory of one of us (E.A.G.) has conducted a screening of 50,000 mutants looking for strains harboring MudJ insertions whose β-galactosidase activity was modulated by PhoP. Mutants were isolated in a strain harboring a chromosomal *phoP*::Tn*10* and a plasmid with a wild-type *phoP* gene under the trancriptional control of the *lac* promoter. Candidate genes were initially identified as those

expressing different levels of β-galactosidase activity in the presence or absence of a gratuitous inducer of the *lac* promoter. Fifty-four mutants that still exhibited differences in β-galactosidase activity when comparing isogenic *phoP⁺* and *phoP⁻* strains (i.e., in the absence of plasmid) are being currently characterized phenotypically and genotypically. This screening has already yielded new fusions to *psiD,* a previously known PhoP-activated gene (Groisman et al., 1989), and also established that the *phoPQ* operon is autoregulated (Soncini et al., in press). Because the original *phoP* and *phoQ* mutations were isolated during a screening for macrophage-sensitive mutants (Fields et al., 1986), it would be interesting to establish whether some of the other genes identified in that screening are under PhoP/PhoQ control. The recent molecular characterization of 30 such mutants (Bäumler et al., 1994) will make this possible.

Like several other two-component regulatory systems, the *phoPQ* system is autoregulated. Using chromosomal *lac* transcriptional fusions to *phoP* and *phoQ,* it was determined that autoregulation requires both PhoP and PhoQ. When PhoQ levels are low, the transcriptional activity of the *phoPQ* operon increases as the amount of PhoP increases. However, this activity is repressed as the amount of PhoQ increases. This regulation could also be observed at the protein level by Western blot analysis with anti-PhoP and anti-PhoQ antibodies. Finally, we presently know very little about the *cis* DNA sequences recognized by PhoP. Groisman et al. (1992c) identified a stretch of 16 nucleotides upstream of the predicted Pribnow box of the *phoN* gene and suggested that this motif could correspond to a PhoP binding site because 13 of these 16 nucleotides were also present upstream of the postulated Pribnow box of the *phoPQ* operon. However, such motif could not be found in the predicted regulatory region of *pagC* (Pulkkinen and Miller, 1991). The *pagC* transcript is about 1,100 nucleotides, unusually long for a protein of only 188 amino acids (Pulkkinen and Miller, 1991).

Signals

Several environmental cues control expression of different PhoP-regulated genes (Table 3). For example, starvation for phosphate, carbon, nitrogen, and sulfur stimulates production of the nonspecific acid phosphatase (Kier et al., 1977), yet expression of *psiD* is regulated by phosphate but not carbon starvation (Foster and Spector, 1986). A similar situation exists for the PhoB-controlled phosphate regulon of *E. coli*: certain PhoB-regulated genes (i.e., *phoA*) are induced exclusively by phosphate starvation, whereas others (i.e., *ugpAB*) are pleiotropically regulated in response to starvation for other groups of compounds as well (Chapter 12).

Miller and coworkers reported that β-galactosidase activity originating from a plasmid-linked *pagC-lac* transcriptional fusion is stimulated in macrophages with kinetics that followed phagosome acidification (Alpuche-Aranda et al., 1992). Treatments that raised the acid pH of the phagosome abolished *pagC* gene expression, but exposure to an acid pH in vitro was not enough to achieve the induction levels observed within macrophages (Alpuche-Aranda et al., 1992). Furthermore, the *pagC-lac* fusion was not expressed in two different epithelial cells (Caco-2 and MDCK) tested (Alpuche-Aranda et al., 1992). Work by Buchmeier and Heffron (1990) suggested that *S. typhimurium* inhabits two different vesicles within the macrophage—unfused vesicles in which the bacteria replicate, and fused vesicles in which the bacteria replicate poorly. It is not clear which intracellular environment Miller and coworkers were examining, although it is likely that most of the pH-sensitive dye used to make pH measurements was taken up by fluid-phase uptake and concentrated in late endosomes/lysosomes; this would have resulted in pH determinations being made on fused rather than unfused vesicles. If this is true the pH may be even higher in unfused vesicles—a measurement that should be made. Presumably, the induction kinetics observed for *pagC* is also a composite, and it may be expressed in only one intracellular compartment. Work on growth rate measurements performed by Abshire and Neidhardt (1993a,b) indicates the presence of two distinct populations of *Salmonella* within macrophages and appears to corroborate the two intracellular environments hypothesis.

One might expect *prg* genes to be normally repressed under conditions in which *pag* genes are activated, and indeed, Miller proposed that PhoP-activated genes are expressed in macrophages whereas those that are PhoP-repressed are turned on in epithelial cells (Alpuche-Aranda et al., 1992; Behlau and Miller, 1993). This correlation seems to hold for *pagC,* which is induced under starvation conditions, and for *prgH,* which is turned on in rich media (Behlau and Miller, 1993). However, the regulation of many other genes belonging to the PhoP regulon is more complex. For example, *pagC* and *prgH* are similarly regulated by anaerobiosis, which is different from that observed for another PhoP-repressed gene (*prgB*). Furthermore, both *pagC* and *prgB* are induced by acid pH, but *prgH* expression remains unaffected (Behlau and Miller, 1993).

Finally, the wide phylogenetic distribution of *phoP* (see below), which includes both pathogenic and nonpathogenic bacteria (Groisman et al., 1989), implies that PhoQ serves as a sensor of a particular stress condition(s) that must also exist outside animal hosts. It also indicates that the physiological role of PhoP/

TABLE 3 Signals known to modulate expression of PhoP/PhoQ-modulated genes

Inducing condition	Examples
Acid pH	*pagA, pagB, prgB*
Carbon starvation	*phoN*
Nitrogen starvation	*phoN*
Phosphate starvation	*phoN, psiD*
Sulfur starvation	*phoN*
General starvation	*pagC*
Anaerobiosis	*prgB*
Aerobiosis	*prgH*
Phosphate starvation	*phoN*
Intramacrophage	*pagA, pagB, pagC*

PhoQ cannot be limited to the regulation of *Salmonella* pathogenesis and must be responsible for the adaptive response to environments or conditions faced by other gram-negative bacteria.

Phylogenetic Distribution of *phoP* and PhoP-Regulated Genes

Hybridization experiments in which *phoP* was used as a probe showed that it was widely distributed among gram-negative bacteria (Groisman et al., 1989). Interestingly, hybridization was also detected in two different strains of the yeast *Saccharomyces cerevisiae*. Although two-component regulatory systems have recently been identified in both plants and yeasts (Chang and Meyerowitz, 1994), the reported yeast sequences do not appear to correspond to the sequences detected in the Southern hybridization experiments in which *phoP* was used as a probe (Groisman et al., 1989).

phoP homologs have been cloned and sequenced from *E. coli* (Kasahara et al., 1992; Groisman et al., 1992a), *Shigella flexneri* (Grois-

man, unpublished data), and *Brucella abortus* (Halingeserth, personal communication), and partial sequence information is also available for *phoP* homologs amplified by the PCR from the genomes of *Yersinia enterocolitica* and *Yersinia pestis* (Fig. 1). The *E. coli phoP* gene maps to the 25-min region, as in *S. typhimurium,* and its deduced amino acid sequence shows 93% identity with the *Salmonella* counterpart (Kasahara et al., 1992; Groisman et al., 1992a). Kasahara et al. (1992) showed that in *E. coli, phoP* and *phoQ* are also organized as an operon in which the *phoP* is the promoter proximal gene. The amino acid sequence of PhoQ is somewhat less conserved (86% identity) than that of PhoP (Kasahara et al., 1992). That the *E. coli* K-12 and *Shigella* homologs were cloned by complementation of a *S. typhimurium phoP⁻* mutant indicates that these genes are functional (i.e., not pseudogenes). A *phoP⁻* mutant of *E. coli* was constructed by reverse genetics and shown to exhibit increased susceptibility to the frog-derived antimicrobial peptide magainin 2 (Groisman et al., 1992a). This mutant

```
                    1                                                          60
StyPhoP  MMRVLVVEDN ALLRHHLKVQ LQDSGHQVDA AEDAREADYY LNEHLPDIAI VDLGLPDEDG
 EcPhoP  .MRVLVVEDN ALLRHHLKVQ IQDAGHQVDD AEDAKEADYY LNEHIPDIAI VDLGLPDEDG
ShfPhoP  .MRVLVVEDN ALLRHHLKVQ IQDAGYQVDD AEDAKEADYY LNEHIPDIAI VDLGLPDEDG
 YePhoP  .........N ALLRHHLKVQ IQDAGYQVDD AEDAKEADYY LNEHIPDIAI VDLGLPDEDG
 YpPhoP  .........N ALLRHHLNVQ IEDSGYQVDD AEDAKEADYF LQEHGPDIAI IDLGLPGEDG

                    61                                                        120
StyPhoP  LSLIRRWRSS DVSLPVLVLT AREGWQDKVE VLSSGADDYY TKPFHIEEVM ARMQALMRRN
 EcPhoP  LSLIRRWRSN DVSLPILVLT ARESWQDKVE VLSAGADDYV TKPFHIEEVM ARMQALMRRN
ShfPhoP  LSLIRRWRSN DVSLPILVLT ARESWQDKVE VLSAGADDYV TKPFHIEEVM ARMQALMRRN
 YePhoP  LSLIRRWRSN DVSLPILVLT ARESWQDKVE VLSAGA.... .......... ..........
 YpPhoP  LSLIRRWRSH QTNLPILVLT ARESWQDKVA DLEAGA.... .......... ..........

                    121                                                       180
StyPhoP  SGLASQVINI PPFQVDLSRR ELSVNEEVIK LTAFEYTIME TLIRNNGKVV SKDSLMLQLY
 EcPhoP  SGLASQVISL PPFQVDLSRR ELSINDEVIK LTAFEYTIME TLIRNNGKVV SKDSLMLQLY
ShfPhoP  SGLASQVISL PPFQVDLSRR ELSINDEVIK LTAFEYTIME TLIRNNGKVV SKDSLMLQLY
 YePhoP  .......... .......... .......... .......... .......... ..........
 YpPhoP  .......... .......... .......... .......... .......... ..........

                    181
StyPhoP  PDAELRESHT IDVLMGRLRK KIQAQYPHDV ITTVRGQGYL FELR*
 EcPhoP  PDAELRESHT IDVLMGRLRK KIQAQYPQEV ITTVRGQGYL FELR*
ShfPhoP  PDAELRESHT IDVLMGRLRK KIQAQYPQEV ITTVRGQGYL FELR*
 YePhoP  .......... .......... .......... .......... .....
 YpPhoP  .......... .......... .......... .......... .....
```

FIGURE 1 Alignment of the deduced amino acid sequences of PhoP from different gram-negative bacteria. Sty, *S. typhimurium;* Ec, *E. coli;* Shf, *Shigella flexneri;* Ye, *Yersinia enterocolitica;* Yp, *Yersinia pestis.*

should make it possible to evaluate whether the PhoP/PhoQ system is responsible for the transcriptional control of phosphate starvation-inducible genes of *E. coli* that do not respond to the PhoB/PhoR system (Chapter 12).

In contrast to *phoP*, the *phoN* gene shows an extremely narrow phylogenetic distribution and maps to one of the recognized regions of the *S. typhimurium* chromosome known to be absent from the *E. coli* K-12 genome (Groisman et al., 1992c). A molecular genetic analysis of *phoN* indicates that this gene was acquired horizontally by an ancestral *Salmonella* (Groisman et al., 1992c). First, *phoN* has a G+C content of only 45%, much lower than the 52% typical of the *S. typhimurium* genome. Second, its codon usage does not correspond to that of frequently expressed genes in *Salmonella*. Fi-

nally, the *phoN* gene is preceded by what appears to be the leading end of a plasmid origin of transfer (*oriT*) region, suggesting that *phoN* might have been acquired as a consequence of an aborted plasmid transfer event. Likewise, *pagC* appears to be absent from many enteric species (cited in Miller et al. [1993]), and *prgH* maps to the 59-min region (Behlau and Miller, 1993) in what appears to be a "pathogenicity island" that harbors a cluster of invasion determinants including the *spa/inv*, *hil*, and *ogrA* loci. This raises the question of why *pagC*, *prgH*, and *phoN* recruited PhoP/PhoQ to control their expression. It could not have been due to a requirement for growth within animal tissues because *phoN* does not appear to have a role in virulence whereas *pagC* and *prgH* do.

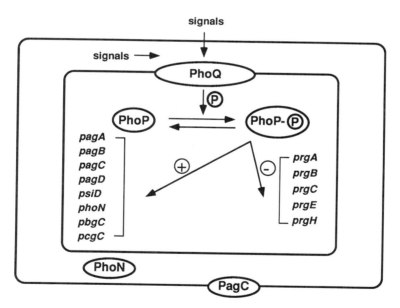

FIGURE 2 Model for transcriptional regulation of PhoP/PhoQ regulon. On sensing particular signals from the environment, PhoQ is predicted to autophosphorylate at a histidine residue and then transfer that phosphoryl group to an aspartic acid in PhoP. The phosphorylated form of PhoP is predicted to activate and repress some 40 loci. PhoP-activated genes are designated as *pag*, *pbg*, and *pcg* and PhoP-repressed genes as *prg* and *psg*. *pagC* encodes an outer membrane protein required for intramacrophage survival. *phoN* encodes periplasmic nonspecific acid phosphatase with no apparent role in virulence.

Applications

The study of bacterial pathogenesis has as one of its goals the identification of all determinants that can be targeted for the prevention or treatment of disease. Analysis of the PhoP/PhoQ regulon has been particularly valuable for several reasons (Fig. 2). First, it provided a hypersusceptible strain that facilitates the identification of novel antimicrobial compounds from host tissues. Second, despite their inability to survive in macrophages, both *phoP⁻* and *pho-24* constitutive mutants retain their immunogenic properties and seem to be good vaccine strains (Galan and Curtiss, 1989; Miller et al., 1989; Miller and Mekalanos, 1990). Moreover, the PhoP-regulated *pagC* gene appears to be preferentially expressed in certain host tissues providing a means of turning on foreign antigens cloned behind its promoter for expression in *Salmonella* vaccine strains (Miller et al., 1993). Finally, PhoP-regulated determinants could serve as tools to manipulate basic cell biological processes such as maturation of vacuolar compartments and presentation of antigens.

CONCLUSIONS AND PERSPECTIVE

A recent compilation of genes implicated in *Salmonella* virulence listed 42 distinct loci scattered around the genome (Groisman and Ochman, 1994). Seven such loci encode products required for the regulation of transcription including two two-component regulatory systems: *phoPQ* and *envZ/ompR*. Because most of the determinants conferring pathogenic properties to *Salmonella* are yet to be discovered, some of them are likely to encode other members of the two-component family that could be responsible for sensing the many environments faced by *Salmonella* during the course of infection.

The wide phylogenetic distribution of *phoP/phoQ* and *envZ/ompR* indicates that they evolved to sense signals present both inside and outside animal hosts. Therefore, the virulence role of these regulators could rely on either the particular subset of genes they recruited to be part of their respective regulons in different bacterial species or in the presence of allelic variants within essentially the same set of regulated genes.

ACKNOWLEDGMENTS

Work in our laboratories is supported by the National Institutes of Health grants AI29554-02 (EAG) and AI22933-08 (FH). E.A.G. is a recipient of a Junior Faculty Research Award from the American Cancer Society.

REFERENCES

Abshire, K. Z., and F. C. Neidhardt. 1993a. Analysis of proteins synthesized by *Salmonella typhimurium* during growth within a host macrophage. *J. Bacteriol.* **175**:3734–3743.

Abshire, K. Z., and F. C. Neidhardt. 1993b. Growth rate paradox of *Salmonella typhimurium* within host macrophages. *J. Bacteriol.* **175**:3744–3748.

Alpuche-Aranda, C. M., E. L. Racoosin, J. A. Swanson, and S. I. Miller. 1994. *Salmonella* stimulates macrophage macropinocytosis and persists within spacious phagosomes. *J. Exp. Med.* **179**:601–608.

Alpuche-Aranda, C. M., J. A. Swanson, W. P. Loomis, and S. I. Miller. 1992. *Salmonella typhimurium* activates virulence gene transcription within acidified macrophage phagosomes. *Proc. Natl. Acad. Sci. USA* **89**:10079–10083.

Bäumler, A. J., J. G. Kusters, I. Stojiljkovic, and F. Heffron. 1994. *Salmonella typhimurium* loci involved in survival within macrophages. *Infect. Immun.* **62**:1623–1630.

Behlau, I., and S. I. Miller. 1993. A PhoP-repressed gene promotes *Salmonella typhimurium* invasion of epithelial cells. *J. Bacteriol.* **175**:4475–4484.

Brenner, D. J. 1992. Introduction to the family *Enterobacteriaceae*, p. 2673–2696. *In* H. Balows, H. G. Trüper, M. Dworkin, W. Harder, and K.-H. Schliefer (ed.), *The Prokaryotes*, 2nd ed., vol. III. Springer-Verlag, New York.

Buchmeier, N. A., and F. Heffron. 1990. Induction of *Salmonella* stress proteins upon infection of macrophages. *Science* **248**:730–732.

Chang, C., and Meyerowitz, E. M. 1994. Eukaryotes have "two-component" signal transducers. *Res. Microbiol.* **145**:481–486.

Chatfield, S. N., C. J. Dorman, C. Hayward, and G. Dougan. 1991. Role of *ompR*-dependent genes in *Salmonella typhimurium* virulence: mutants defective in both OmpC and OmpF are attenuated in vivo. *Infect. Immun.* **59**:449–452.

DiRita, V. J., and J. J. Mekalanos. 1989. Genetic regulation of bacterial virulence. *Annu. Rev. Genet.* **232**:455–482.

Dorman, C. J., S. Chatfield, C. F. Higgins, C. Hayward, and G. Dougan. 1989. Characterization of porin and *ompR* mutants of a virulent strain of *Salmonella typhimurium: ompR* mutants are attenuated in vivo. *Infect. Immun.* **57**:2136–2140.

Fields, P. I., E. A. Groisman, and F. Heffron. 1989. A *Salmonella* locus that controls resistance to microbicidal proteins from phagocytic cells. *Science* **243**:1059–1062.

Fields, P. I., R. V. Swanson, C. G. Haidaris, and F. Heffron. 1986. Mutants of *Salmonella typhimurium* that cannot survive within the macrophage are avirulent. *Proc. Natl. Acad. Sci. USA* **83**:5189–5193.

Finlay, B. B., and S. Falkow. 1989. *Salmonella* as an intracellular parasite. *Mol. Microbiol.* **3**:1833–1841.

Foster, J. W., and H. K. Hall. 1990. Adaptive acidification tolerance response of *Salmonella typhimurium*. *J. Bacteriol.* **172**:771–778.

Foster, J. W., and M. P. Spector. 1986. Phosphate starvation regulon of *Salmonella typhimurium*. *J. Bacteriol.* **166**:666–669.

Galán, J., C. Ginocchio, and P. Costeas. 1992. Molecular and functional characterization of the *Salmonella* invasion gene *invA*: homology of *invA* to members of a new protein family. *J. Bacteriol.* **174**:4338–4349.

Galan, J. E., and R. Curtiss III. 1989. Virulence and vaccine potential of *phoP* mutants of *Salmonella typhimurium*. *Microb. Pathog.* **6**:433–443.

García-Véscovi, E., F. Soncini, and E. A. Groisman. 1994. The role of the PhoP/PhoQ regulon in *Salmonella* virulence. *Res. Microbiol.* **145**:473–480.

Groisman, E. A. 1994. How bacteria resist killing by host-defense peptides. *Trends Microbiol.* **2**:444–449.

Groisman, E. A. Unpublished data.

Groisman, E. A., E. Chiao, C. J. Lipps, and F. Heffron. 1989. *Salmonella typhimurium phoP* virulence gene is a transcriptional regulator. *Proc. Natl. Acad. Sci. USA* **86**:7077–7081.

Groisman, E. A., P. I. Fields, and F. Heffron. 1990. Molecular biology of *Salmonella* pathogenesis, p. 251–272. *In* B. H. Iglewski and V. L. Clark (ed.), *Molecular Basis of Bacterial Pathogenesis*. Academic Press, San Diego, Calif.

Groisman, E. A., and F. Heffron. Unpublished data.

Groisman, E. A., F. Heffron, and F. Solomon. 1992a. Molecular genetic analysis of the *Escherichia coli phoP* locus. *J. Bacteriol.* **174**:486–491.

Groisman, E. A., and H. Ochman. 1993. Cognate gene clusters govern invasion of host epithelial cells by *Salmonella typhimurium* and *Shigella flexneri*. *EMBO J.* **12**:3779–3787.

Groisman, E. A., and H. Ochman. 1994. How to become a pathogen. *Trends Microbiol.* **2**:289–294.

Groisman, E. A., C. A. Parra, M. Salcedo, C. J. Lipps, and F. Heffron. 1992b. Resistance to host antimicrobial peptides is necessary for *Salmonella* virulence. *Proc. Natl. Acad. Sci. USA* **89**:11939–11943.

Groisman, E. A., and M. H. Saier, Jr. 1990. *Salmonella* virulence: new clues to intramacrophage survival. *Trends Biochem. Sci.* **15**:30–33.

Groisman, E. A., M. H. Saier, Jr., and H. Ochman. 1992c. Horizontal transfer of a phosphatase gene as evidence for mosaic structure of the *Salmonella* genome. *EMBO J.* **11**:1309–1316.

Halingeserth, S. Personal communication.

Helander, I. M., I. Kilpeläinen, and M. Vaara. 1994. Increased substitution of phosphate groups in lipopolysaccharides and lipid A of the polymyxin-resistant *pmrA* mutants of *Salmonella typhimurium*: a ^{31}P-NMR study. *Mol. Microbiol.* **11**:481–487.

Joiner, K. A., T. Ganz, J. Albert, and D. Rotrosen. 1989. The opsonizing ligand on *Salmonella typhimurium* influences incorporation of specific, but not azurophil, granule constituent into neutrophil phagosomes. *J. Cell. Biol.* **109**:2771–2782.

Jones, B. D., and S. Falkow. 1994. Identification and characterization of a *Salmonella typhimurium* oxygen-regulated gene required for bacterial internalization. *Infect. Immun.* **62**:3745–3752.

Kasahara, M., A. Nakata, and H. Shinagawa. 1991. Molecular analysis of the *Salmonella typhimurium phoN* gene, which encodes nonspecific acid phosphatase. *J. Bacteriol.* **173**:6760–6765.

Kasahara, M., A. Nakata, and H. Shinagawa. 1992. Molecular analysis of the *Escherichia coli phoP-phoQ* operon. *J. Bacteriol.* **174**:492–498.

Kier, L. D., R. Weppelman, and B. N. Ames. 1977. Regulation of two phosphatases and a cyclic phosphodiesterase of *Salmonella typhimurium*. *J. Bacteriol.* **130**:420–428.

Kier, L. D., R. M. Weppelman, and B. N. Ames. 1979. Regulation of nonspecific acid phosphatase in *Salmonella: phoN* and *phoP* genes. *J. Bacteriol.* **138**:155–161.

Lee, C., and S. Falkow. 1990. The ability of *Salmonella* to enter mammalian cells is affected by bacterial growth state. *Proc. Natl. Acad. Sci. USA* **87**:4304–4308.

Lee, C. A., B. D. Jones, and S. Falkow. 1992. Identification of a *Salmonella typhimurium* invasion locus by selection for hyperinvasive mutants. *Proc. Natl. Acad. Sci. USA* **89**:1847–1851.

Lehrer, R. I., A. K. Lichtenstein, and T. Ganz. 1993. Defensins: antimicrobial and cytotoxic peptides of mammalian cells. *Annu. Rev. Immunol.* **11**:105–128.

Mekalanos, J. J. 1992. Environmental signals controlling expression of virulence determinants in bacteria. *J. Bacteriol.* **174**:1–7.

Miller, S. I., A. M. Kukral, and J. J. Mekalanos. 1989. A two-component regulatory system (*phoP phoQ*) controls *Salmonella typhimurium* virulence. *Proc. Natl. Acad. Sci. USA* **86**:5054–5058.

Miller, S. I., W. P. Loomis, C. Alpuche-Aranda, I. Behlau, and E. Hohmann. 1993. The PhoP virulence regulon and live oral *Salmonella* vaccines. *Vaccine* **11**:122–125.

Miller, S. I., and J. J. Mekalanos. 1990. Constitutive expression of the *phoP* regulon attenuates *Salmonella* virulence and survival within macrophages. *J. Bacteriol.* **172**:2485–2490.

Miller, S. I., W. S. Pulkkinen, M. E. Selsted, and J. J. Mekalanos. 1990. Characterization of defensin resistance phenotypes associated with mutations in the *phoP* virulence regulon of *Salmonella typhimurium*. *Infect. Immun.* **58**:3706–3710.

Miller, V. L., K. B. Beer, W. P. Loomis, J. A. Olson, and S. I. Miller. 1992. An unusual *pagC*::Tn*phoA* mutation leads to an invasion- and virulence-defective phenotype in *Salmonellae*. *Infect. Immun.* **60**:3763–3770.

Puente, J. L., A. Verdugo-Rodríguez, and E. Calva. 1991. Expression of *Salmonella typhi* and *Escherichia coli* OmpC is influenced differently by medium osmolarity; dependence on *Escherichia coli* OmpR. *Mol. Microbiol.* **5**:1205–1210.

Pulkkinen, W. S., and S. I. Miller. 1991. A *Salmonella typhimurium* virulence protein is similar to a *Yersinia enterocolitica* invasion protein and a bacteriophage lambda outer membrane protein. *J. Bacteriol.* **173**:86–93.

Roland, K. L., L. E. Martin, C. R. Esther, and J. K. Spitznagel. 1993. Spontaneous *pmrA* mutants of *Salmonella typhimurium* LT2 define a new two-component regulatory system with a possible role in virulence. *J. Bacteriol.* **175**:4154–4164.

Soncini, F., and E. A. Groisman. Unpublished data.

Soncini, F. C., E. García-Véscovi, and E. A. Groisman. Transcriptional autoregulation of the *Salmonella typhimurium phoPQ* operon. *J. Bacteriol.*, in press.

Sperber, S. J., and C. J. Scheulpner. 1987. Salmonellosis during infection with human immunodeficiency virus. *Rev. Infect. Dis.* **9**:925–934.

Stinavage, P., L. E. Martin, and J. K. Spitznagel. 1989. O-antigen and lipid A phosphoryl groups in resistance of *Salmonella typhimurium* LT-2 to nonoxidative killing in human polymorphonuclear cells. *Infect. Immun.* **57**:3894–3900.

Stone, B. J., C. M. Garcia, J. L. Badger, T. Hassett, R. I. F. Smith, and V. L. Miller. 1992. Identification of novel loci affecting entry of *Salmonella enteritidis* into eukaryotic cells. *J. Bacteriol.* **174**:3945–3952.

Todd, E. C. D. 1989. Preliminary estimates of costs of foodborne disease in the United States. *J. Food Prot.* **52**:595–601.

Tsolis, R., J. Lipps, E. A. Groisman, and F. Heffron. Unpublished data.

Vaara, M. 1981. Increased outer membrane resistance to ethylenediaminetetraacetate and cations in novel lipid A mutants. *J. Bacteriol.* **148**:426–434.

Volz, K. 1993. Structural conservation in the CheY superfamily. *Biophys. J.* **32**:11741–11753.

Wick, M. J., C. V. Harding, N. J. Twesten, S. J. Normark, and J. D. Pfeifer. The *phoP* locus influences processing and presentation of *S. typhimurium* antigens by activated macrophages. *Mol. Microbiol.*, in press.

Bordetella pertussis BvgAS
Virulence Control System

M. Andrew Uhl and Jeff F. Miller

21

It is estimated that more than 50 million people worldwide become infected with *Bordetella pertussis,* and approximately 0.5 million people die annually of the resulting disease (Hewlett, 1990). Pertussis is a severe childhood affliction and is characterized by prolonged episodes of paroxysmal coughing, which led to its designation as whooping cough or "cough of 100 days" in Mandarin Chinese. The virulence potential of this organism is controlled through the action of a two-component regulatory system (Weiss et al., 1983; Weiss and Falkow, 1984; Aricò et al., 1989; Stibitz et al., 1989).

During infection of a suitable host, *B. pertussis* adheres to ciliated epithelial cells and expresses several toxins responsible for both local damage and systemic effects (Weiss and Hewlett, 1986; Pittman, 1984; Goldman et al., 1982). The outer membrane protein pertactin, fimbriae, and filamentous hemagglutinin have been implicated in adherence (Roberts et al., 1991; Leininger et al., 1991; Mooi et al., 1992; Tuomanen and Weiss, 1985; Tuomanen et al., 1988; Relman et al., 1990). Pertussis toxin and adenylate cyclase toxin are among the toxins secreted by *B. pertussis,* both functioning to increase host cell intracellular cyclic AMP levels (Irons and Gorringe, 1988; Ui, 1988; Masure et al., 1987; Hewlett and Gordon, 1988). These toxins have multiple effects on the immune response, including reduction of migratory and phagocytic capability of macrophages and suppression of the chemotactic response of neutrophils. Tracheal cytotoxin, a muramyl peptide secreted from growing cultures, damages ciliated epithelial cells and leads to death and extrusion from the mucosal surface (Goldman et al., 1982). A role for tracheal cytotoxin in transmission of the organism has also been suggested (Luker et al., 1993). Presumably, toxins and adhesins of *B. pertussis* allow colonization of the respiratory tract and maintain a favorable environment for bacterial survival, multiplication, and transmission.

Coordinate regulation of virulence factors in response to environmental signals has been established as a conserved strategy of microbial pathogens (Miller et al., 1989a; DiRita and Mekalanos, 1989; Mekalanos, 1992). Virulence genes of bordetellae, with the exception of tracheal cytotoxin, are not expressed under certain growth conditions. This phenomenon, known as phenotypic modulation, involves

M. Andrew Uhl and Jeff F. Miller, Department of Microbiology and Immunology, School of Medicine, University of California, Los Angeles, Los Angeles, California 90095.

Two-Component Signal Transduction, Edited by James A. Hoch and Thomas J. Silhavy, © 1995 American Society for Microbiology, Washington, DC 20005

lack of expression of virulence genes at low temperature or by addition of $MgSO_4$ or nicotinic acid to the growth medium (Lacey, 1960; Melton and Weiss, 1989; Miller et al., 1992). Expression of virulence genes is controlled through an additional mechanism. Avirulent variants of *B. pertussis* arise from virulent-phase cultures and no longer express most adhesins and toxins. Reversion to a virulent phase has been noted (Leslie and Gardner, 1931; Monack et al., 1989; Weiss and Falkow, 1984; Stibitz et al., 1989). An important advance in understanding virulence gene regulation in bordetellae came when it was realized that phenotypic modulation and phase variation acted through a single locus, *bvg* (Weiss et al., 1983; Weiss and Falkow, 1984; Aricò et al., 1989; Aricò et al., 1991; Stibitz et al., 1988; Stibitz et al., 1989; Miller et al., 1992).

bvg VIRULENCE CONTROL SYSTEM

Alteration between two distinct phases of *B. pertussis* is mediated by the products of the *bvg* locus. Bvg⁺-phase bordetellae express most of the defined toxins and adhesins. When the *bvg* locus is inactivated through modulating signals or by mutation, the organism switches to the Bvg⁻ phase. *bvg* (initially designated and also referred to as *vir*) was originally identified in *B. pertussis* as the site of a transposon insertion that eliminated expression of a collection of virulence factors including pertussis toxin, adenylate cyclase/hemolysin, and filamentous hemagglutinin (Weiss et al., 1983; Weiss and Falkow, 1984). Sequence analysis of *bvg* revealed that the proteins encoded by this locus, BvgA and BvgS, belong to the two-component regulatory system family (Aricò et al., 1989; Stibitz et al., 1989; Stibitz and Yang, 1991). The *bvg* loci from two closely related species, *B. bronchiseptica* and *B. parapertussis,* have also been sequenced (Aricò et al., 1991). The *bvg* loci of these species are 96% identical at the amino acid level, and functional conservation has been demonstrated (Monack et al., 1989). Furthermore, *B. bronchiseptica* and *B. parapertussis* respond to identical modulating signals as

B. pertussis, showing conservation of signal recognition. Sequences that hybridize to *bvgS* have also been detected in *B. avium,* but the *bvg* locus from that species has not yet been characterized (Gentry-Weeks et al., 1991).

DOMAIN ORGANIZATION OF BvgA AND BvgS

BvgA is a typical response regulator with an N-terminal receiver and a C-terminal helix-turn-helix motif (Aricò et al., 1989; Stibitz and Yang, 1991) (Fig. 1). BvgS is a 134-kDa cytoplasmic membrane sensor protein with a large N-terminal region in the periplasm and several cytoplasmic domains that can be defined by primary sequence similarities (Stibitz and Yang, 1991) (Figs. 1 and 2). The BvgS linker, located adjacent to the second transmembrane sequence, connects the periplasmic region of BvgS to cytoplasmic effector domains and is the location of signal-insensitive mutations (Miller et al., 1992). As illustrated in Fig. 1, the transmitter of BvgS is similar to other transmitters, with both the conserved histidine (His-729) and a proposed nucleotide binding region (ATP). Although periplasmic, linker, and transmitter domains are common features of sensor proteins, BvgS is distinguished from many sensor proteins by containing additional sequences. The transmitter of BvgS is followed by a consensus receiver domain and a C-terminal domain. We and others have noted regions of sequence similarity among the C termini of a subset of sensor proteins, including BvgS (Ishige et al., 1994) (Fig. 2). Although small regions of similarity exist throughout the C-terminal domains of these proteins, the most notable is the presence of an absolutely conserved histidine with conservation of flanking sequence. The conserved region in the C terminus is similar to the site of autophosphorylation in CheA (Stock et al., 1989). Between the transmitter, receiver, and C terminus are A/P-rich regions (Fig. 1), which are also found in FrzE of *Myxococcus xanthus,* pyruvate dehydrogenase, and other enzymes. A/P-rich sequences are believed to represent conformationally flex-

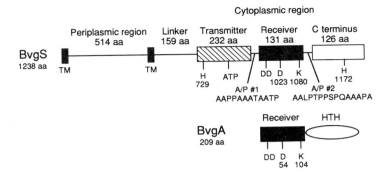

FIGURE 1 Selected features of the BvgA and BvgS proteins (Aricò et al., 1989). Shaded boxes depict regions sharing sequence similarity to two-component systems. The consensus ATP binding motif (ATP) and amino acids conserved among two-component systems are shown for BvgS: His-729 (H729), Asp-1023 (D1023), and Lys-1080 (K1080). Selected amino acids conserved among receivers are also noted for BvgA: Asp-54 (D54) and Lys-104 (K104). Proposed hydrophobic transmembrane sequences (TM) and A/P-rich regions in BvgS and the helix-turn-helix motif in BvgA (HTH) are noted.

ible sites that promote intramolecular interactions (McCleary and Zusman, 1990; Wootton and Drummond, 1989; Radford et al., 1989).

TRANSCRIPTIONAL REGULATION OF THE *bvg* OPERON

The *bvg* locus is linked to the *fha/fim* locus, involved in production and export of filamentous hemagglutinin and fimbriae, and is transcribed in an opposite direction (Stibitz et al., 1988; Willems et al., 1992; Locht et al., 1992; Roy et al., 1990; Scarlato et al., 1990; Willems et al., 1994) (Fig. 3). At least two promoters, *bvg*P$_1$ and *bvg*P$_2$, regulate transcription of *bvgAS* (Fig. 3). In the Bvg$^+$ phase, the *bvg* locus is strongly transcribed from the *bvg*P$_1$ promoter, whereas a basal level of transcription of the *bvg* locus is directed from the *bvg*P$_2$ promoter in the Bvg$^-$ phase. Transcription of the *bvg*P$_1$ and P$_2$ promoters is mutually exclusive, and it has been hypothesized that activation of the *bvg*P$_1$ promoter prevents transcription from the *bvg*P$_2$ promoter. Additional promoters, the *bvg*P$_3$ and P$_4$ promoters, have been identified in Bvg$^+$-phase organisms (Scarlato et al., 1990). *bvg*P$_3$ activates transcription of *bvg* in the Bvg$^+$ phase, whereas *bvg*P$_4$ has been proposed to synthesize

an antisense transcript that may function to down-regulate *bvgAS* mRNA. The observed transcriptional pattern of the *bvg* locus suggests that a basal level of BvgAS is constitutively present, and on proper signal inputs, the signal transduction system is amplified for activation of virulence genes.

GENETIC ANALYSIS OF BvgAS

Signals Sensed by Bvg

In 1960, Lacey reported that altering growth conditions affected the expression of several *B. pertussis* antigens. Certain culture conditions caused a switch to what is now recognized as the Bvg$^-$ phase (Lacey, 1960). Further analysis and extension of this work demonstrated that chlorate and sulfate anions, nicotinic acid derivatives, and low temperature can reversibly down-regulate expression of virulence factors by *bvg* (Schneider and Parker, 1982; Melton and Weiss, 1989, 1993; Miller et al., 1992). A detailed characterization of analogs of nicotinic acid has shown that the carboxyl group is required for modulation (Melton and Weiss, 1993). Both the resonance and planar structure of the ring are also important. In vivo signals

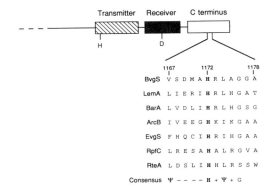

FIGURE 2 Alignment of the conserved histidine region in the C termini of BvgS family members. Alignments of a small portion of the C-terminus domain are shown with consensus amino acids noted for matches in six of the eight amino acids sequences. Consensus abbreviations are Ψ for hydrophobic amino acid and + for positively charged amino acid. Amino acid numbering is noted for BvgS, with the conserved histidine in bold. The location of the consensus sequence in the C terminus is shown, with the transmitter and receiver domains included for comparison. BvgS family members are LemA of *Pseudomonas syringae* pv. *syringae;* BarA, ArcB, and EvgS of *E. coli;* RpfC of *Xanthomonas campestris* pv. *campestris;* and RteA of *Bacteroides thetaiotamicron* (Parkinson and Kofoid, 1992; Utsumi et al., 1994). Sequences were aligned with the PILEUP program from the Genetics Computer Group sequence analysis software package. The consensus motif has also been noticed by others (Ishige et al., 1994).

that may promote phenotypic modulation and the mechanism through which modulators interact with BvgAS to regulate its function are not known.

Signal–Insensitive Mutations

Several mutations that render *B. pertussis* insensitive to modulators have been isolated and characterized (Knapp and Mekalanos, 1988; Miller et al., 1992). Single amino acid substitutions mapping to the BvgS linker region eliminated modulation by sulfate anion, nicotinic acid, and low temperature (Miller et al., 1992). Three of the four identified constitutive mutations are located in the highly positively charged N-terminal portion of the linker

(Fig. 4). Similar mutations have been described in other sensory proteins, NarX and Tsr (Kalman and Gunsalus, 1990; Collins et al., 1992; Ames and Parkinson, 1988). Constitutive BvgS mutations eliminate the need for the BvgS periplasmic domain, which is necessary for activity of wild-type BvgS (Miller et al., 1992) (Fig. 4). Constitutive mutations do not abrogate the requirement for the BvgS transmitter, receiver, or C-terminal domains (Uhl and Miller, unpublished data). These observations suggest the BvgS linker acts in concert with the periplasmic region to receive and transmit signals to the BvgS cytoplasmic domains.

Requirements for BvgS Domains In Vivo

Genetic studies of BvgAS have focused on determining the requirements for various domains of BvgS as well as specific amino acids conserved among two–component family members. These studies have been facilitated by the development of genetic systems in *B. pertussis* and by functional transfer of BvgAS into *Escherichia coli* (Weiss and Falkow, 1982; Stibitz et al., 1986; Stibitz et al., 1988; Miller et al., 1989b). In *E. coli,* BvgAS responds to identical modulating signals as in *B. pertussis* and can activate *fhaB::lacZYA, bvg::lacZYA,* and *ptx::lacZYA* transcriptional fusions (Miller et al., 1989b; Uhl and Miller, submitted; Roy et al., 1990). A listing of mutations in BvgS and their phenotypes in vivo and in vitro is presented in Table 1.

Mutational analysis of the BvgS periplasmic, transmitter, receiver, and C-terminal domains has shown that they are all required for function. As listed in Table 1, several mutations in the periplasmic region eliminate activity. Deletions or small insertion mutations in the BvgS receiver and C terminus or precise deletions of the Bvg receiver and C terminus also abolish activity (Miller et al., 1992; Stibitz, submitted; Uhl and Miller, 1994; Beier and Gross, unpublished data). Amino acids in the BvgS transmitter and receiver that are conserved among

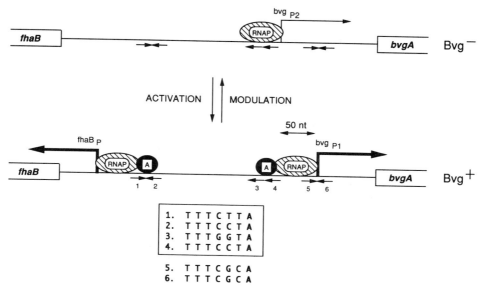

FIGURE 3 Model for activation of the *bvg* and *fhaB* loci (Roy et al., 1990; Scarlato et al., 1990). Transcription of the *bvgAS* operon initiates in the Bvg⁻ phase from the *bvg*P2 promoter. In the Bvg⁺ phase, following activation of BvgAS, BvgA (A) is able to direct RNA polymerase (RNAP) to activate transcription of the *bvg* and *fhaB* loci at the *bvg*P1 and *fhaB* promoters. Relative levels of transcription are represented by arrow thickness. Locations of inverted and direct repeats in the *bvg* and *fhaB* intergenic regions are illustrated by small numbered arrows, and the sequences comprising these repeats are listed below. Two additional Bvg⁺-phase promoters, *bvg*P3 and *bvg*P4, have also been identified and are discussed in the text (Scarlato et al., 1990).

two-component family members are essential for BvgS function, as alteration of these amino acids (His-729, Asp-979, Asp-980, Asp-1023, Lys-1080) render BvgS inactive.

Intramolecular Complementation of BvgS Mutations

Detection of multimeric complexes and intermolecular complementation of mutations has provided evidence that multimerization is a conserved property of sensor proteins (Parkinson and Kofoid, 1992, and references within). Mutations in the BvgS periplasmic region (*bvgS729, bvgS789, bvgS742, bvgS788, bvgS840*) or a mutation in the transmitter (*bvgS910*) can be complemented by *bvgS* alleles with mutations in the C terminus (*bvgS326, bvgS739*; Stibitz, submitted) (Table 1). Individually, these mutations eliminate BvgS activity, but coexpression restores activity to wild-type levels,

consistent with multimerization of BvgS occurring. This also provides evidence that some domains may function as discrete modules. Multimerization of BvgS is likely to be essential for function, as has been hypothesized for other sensor proteins, but whether multimerization is a constitutive feature of BvgS signal transduction or regulates its function is not known.

BvgA Mutational Analysis

BvgS regulates transcription of virulence genes in *B. pertussis* through the action of BvgA. Overproduction of BvgA was sufficient for activation of an *fhaB::lacZYA* promoter fusion in *E. coli* and removed both the requirement for BvgS and the effects of modulating agents (Roy et al., 1989). As shown in Table 1, inframe deletions in *bvgA* abolished expression of virulence determinants in *B. pertussis* (Roy et

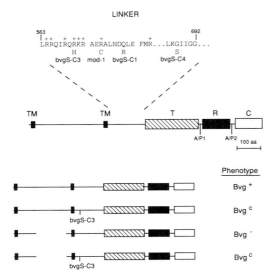

LINKER

563   692
|++ + +++   +   +
LRRQIRQRKR AERALNDQLE FMR...LKGIIGG...
     H   C   R   S
   bvgS-C3 mod-1 bvgS-C1   bvgS-C4

TM   TM   T   R   C

                      A/P1 A/P2 100 aa

Phenotype

Bvg +

bvgS-C3   Bvg c

Bvg -

bvgS-C3   Bvg c

FIGURE 4 Sensory transduction mutations in the BvgS linker region (Miller et al., 1992). The structure of the BvgS protein is indicated. Abbreviations: TM, transmembrane region; T, transmitter; A/P1, AP-rich sequence AAPPAAATAATP; R, receiver; A/P2, AALPTPPSPQAAAPA; C, C-terminal region. The amino acid sequence of the linker region is shown along with substitutions resulting from the constitutive mutations in *bvgS* (Bvg^c, insensitive to modulation). Phenotypes of BvgS wild-type and mutations (*bvgS-C3*; Δ366P; *bvgS-C3* Δ366P) in an *E. coli* reporter system are shown.

al., 1989; Stibitz and Yang, 1991). Functions of BvgA can be separated through deletion mutations (Boucher et al., 1994) (Table 1). DNA binding ability was localized to the C-terminal portion of the molecule and does not require the BvgA receiver, although the receiver was necessary for activation of BvgA. Mutations in the BvgA C terminus have been characterized that eliminate expression of *ptx* and *cya* without much effect on *fha* expression (Stibitz, 1994) (Table 1). These BvgA C-terminal mutations may alter a function of BvgA that is more critical for *ptx* and *cya* expression than for *fhaB*, such as BvgA multimerization or interaction with components of the transcriptional initiation apparatus.

A BvgA constitutive mutation (*bvgA-C1*) has been isolated that activated an *fhaB::lacZYA*

transcriptional fusion greater than 50-fold over background levels (Uhl and Miller, unpublished data). The mutation, Val13-Ile, maps near two of the conserved aspartic acids in the BvgA receiver (Asp-9 and Asp-10) and removes the requirement for BvgS for BvgA-dependent activation of the *fhaB* promoter. Addition of BvgS to *bvgA-C1* increased activation of *fhaB::lacZYA* an additional eightfold, suggesting *bvgA-C1* is capable of being phosphorylated by BvgS. BvgA constitutive mutations and in vitro assays with phosphorylated BvgA should facilitate the study of activation and repression of Bvg-regulated promoters.

BIOCHEMISTRY OF BvgAS SIGNAL TRANSDUCTION

Autophosphorylation of BvgS: Roles of Multiple Domains

As members of the two-component system family, BvgA and BvgS can be anticipated to follow certain parameters of signal transduction. Autophosphorylation of the cytoplasmic portion of BvgS ('BvgS) and phosphotransfer to BvgA have been demonstrated using an in vitro phosphorylation assay (Uhl and Miller, 1994). Like other sensor proteins, 'BvgS is a kinase that preferentially uses the γ-phosphate of ATP as a substrate for phosphorylation (Fig. 5, lane 1). The rapid decay of phosphorylated 'BvgS in the presence of excess ATP, a half-life of 2 min in the absence of BvgA and a half-life of 30 s with BvgA, suggests an intrinsic phosphatase activity, although this has not been examined in detail.

For FrzE of *Myxococcus xanthus*, which has a linked transmitter and receiver domain, and ArcB of *E. coli*, which contains transmitter, receiver, and C-terminal domains, chemical stability analysis of phosphoryl group linkages suggested that these molecules are phosphorylated at both a histidine and the conserved aspartic acid in the receiver (McCleary and Zusman, 1990; Iuchi and Lin, 1992). BvgS has three proposed sites of phosphorylation: His-729 of the transmitter, Asp-1023 of the receiver, and His-1172 of the C terminus. In

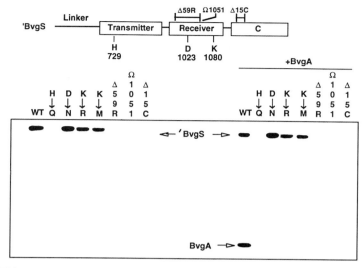

FIGURE 5 In vitro activity of 'BvgS and mutant derivatives. Wild-type 'BvgS and mutant derivatives were phosphorylated in the absence (lanes 1 to 8) or presence (lanes 9 to 16) of BvgA. Lanes 1 and 9, wild-type 'BvgS; lanes 2 and 10, 'BvgS H729Q; lanes 3 and 11, 'BvgS D1023N; lanes 4 and 12, 'BvgS K1080R; lanes 5 and 13, 'BvgS K1080M; lanes 6 and 14, 'BvgS Δ59R; lanes 7 and 15, 'BvgS Ω1051; lanes 8 and 16, 'BvgS Δ15C. Locations of 'BvgS and BvgA are marked with arrows. Adapted from Uhl and Miller, 1994, with permission.

contrast to FrzE and ArcB, chemical stability analysis of 'BvgS indicated that the phosphoprotein contained primarily amidyl phosphate, consistent with BvgS being phosphorylated predominantly at His-729 and/or His-1172 (Uhl and Miller, 1994).

Autophosphorylation for a subset of BvgS mutations are shown in Fig. 5. As expected, mutation of the proposed primary site of autophosphorylation, His-729 in the transmitter, abolishes activity in vivo and autophosphorylation in vitro (Fig. 5, lane 2). An unexpected property of BvgS is that precise deletion of both the receiver and C terminus eliminates autophosphorylation. Linker insertion and deletion mutations of the receiver and C-terminal domains also have the same effect (lanes 4 to 6). Accordingly, autophosphorylation of 'BvgS in vitro has been determined to require the following features: (i) the conserved histidine at position 729 in the BvgS transmitter, (ii) an intact receiver domain, and (iii) an intact

C-terminal domain. In most sensor proteins, the transmitter contains sufficient information for autophosphorylation. 'BvgS is unusual in its requirements for a structurally intact receiver and C terminus for autophosphorylation; the multiple domains necessary for autophosphorylation may relate to the complex organization of BvgS (see Fig. 1). Mutations outside the transmitter domain that affect autophosphorylation have been isolated in CheA, and the BvgS receiver and C terminus may regulate autophosphorylation as has been proposed for the C terminus of CheA (Swanson et al., 1993).

BvgS contains sites of phosphorylation in addition to the transmitter. The BvgS C terminus is able to be phosphorylated by 'BvgS in vitro (Uhl and Miller, unpublished data). The phosphoryl group linkage of the BvgS C terminus is base-stable, and the presence of a conserved histidine in BvgS family members suggests this may be a site of phosphorylation

TABLE 1 BvgA and BvgS mutations and phenotypes in vivo and in vitro[a]

Allele	Region	Location/description	Phenotype	Reference
BvgA				
ΔBvgA	Receiver	Deletion from aa 20–116	Bvg⁻	Roy et al., 1989
bvgAΔ907	Receiver and C terminus	Deletion from aa 78–188	Bvg⁻	Stibitz and Yang, 1991
ΔN-BvgA	Receiver	Deletion from aa 1–120	Binds specific DNA sequences, does not transactivate	Boucher et al., 1994
bvgA972	Receiver	Insertion at aa 27	Bvg⁻	Stibitz, submitted
bvgA1060	C terminus	Glu-198-Lys	Has little effect on *fhaB* activation, decreases *ptx* and *cya* activation	Stibitz, 1994
bvgA1056	C terminus	Asp-201-Asn	Has little effect on *fhaB* activation, decreases *ptx* and *cya* activation	Stibitz, 1994
bvgA-C1	Receiver	Val-13-Ile	BvgA constitutive, activates *fhaB* independent of BvgS	Uhl and Miller, unpublished data
ΔC–BvgA	C terminus	Deletion from aa 189–209	Does not bind DNA or transactivate	Boucher et al., 1994
BvgS				
Ω150	Periplasmic	Insertion at aa 150	Bvg⁻	Miller et al., 1992
bvgS729	Periplasmic	Insertion at aa 174	Bvg⁻	Stibitz, submitted
bvgS789	Periplasmic	Insertion at aa 338	Bvg⁻	Stibitz, submitted
bvgS742	Periplasmic	Insertion at aa 353	Bvg⁻	Stibitz, submitted
bvgS788	Periplasmic	Insertion at aa 430	Bvg⁻	Stibitz, submitted
bvgS840	Periplasmic	Insertion at aa 514	Bvg⁻	Stibitz, submitted
Ω516	Periplasmic	Insertion at aa 516	Bvg⁻	Miller et al., 1992
Δ366P	Periplasmic	Deletion from aa 150–366	Bvg⁻	Miller et al., 1992
bvgS-C3	Linker	Arg-570-His	Insensitive to modulators (Bvg^c)	Miller et al., 1992
mod-1	Linker	Arg-575-Cys	Insensitive to modulators (Bvg^c)	Miller et al., 1992
bvgS-C1	Linker	Gln-580-Arg	Insensitive to modulators (Bvg^c)	Miller et al., 1992
bvgS-C4	Linker	Gly-688-Ser	Insensitive to modulators (Bvg^c)	Miller et al., 1992

Mutation	Location	Description	Phenotype	Reference
His-729-Gln	Transmitter	Conserved histidine among sensor proteins	Bvg⁻, unable to autophosphorylate	Uhl and Miller, 1994
bvgS910	Transmitter	Insertion at aa 827	Bvg⁻	Stibitz, submitted
Pro957-Ter	A/P linker	Results in deletion of BvgS receiver and C terminus	Bvg⁻	Beier and Gross, unpublished data
Δ284RC	Receiver and C terminus	Deletion of BvgS receiver and C terminus	Bvg⁻, unable to autophosphorylate	Uhl and Miller, 1994
Δ59R	Receiver	In-frame 59 aa deletion	Bvg⁻, unable to autophosphorylate	Uhl and Miller, 1994
Ω1051	Receiver	Insertion at aa 1051	Bvg⁻, unable to autophosphorylate	Uhl and Miller, 1994
Asp-980-Gly	Receiver	Conserved aspartic acid in receivers	Bvg⁻	Beier and Gross, unpublished data
Asp-979-Gly/Asp-980-Gly	Receiver	Conserved aspartic acids in receivers	Bvg⁻	Beier and Gross, unpublished data
Asp-1023-Asn	Receiver	Aspartic acid corresponding to Asp-57 of CheY	Bvg⁻, can autophosphorylate, deficient in phosphotransfer to the BvgS C terminus and BvgA	Uhl and Miller, 1994, and unpublished data
Asp-1023-Gly	Receiver	Aspartic acid corresponding to Asp-57 of CheY	Bvg⁻	Uhl and Miller, 1994
Lys-1080-Arg	Receiver	Conserved lysine among receivers	Bvg⁻, can autophosphorylate, deficient in phosphotransfer to BvgA	Uhl and Miller, unpublished data
Lys-1080-Leu	Receiver	Conserved lysine among receivers	Bvg⁻	Beier and Gross, unpublished data
bvgS326	A/P linker	Frameshift at aa 1104	Bvg⁻	Stibitz, submitted
bvgS739	C terminus	Insertion at aa 1151	Bvg⁻	Stibitz, submitted
ΔArg-1146/Thr-1147	C terminus		Bvg⁻	Beier and Gross, unpublished data
Ω1211	C terminus	Insertion at aa 1211	Bvg⁻, unable to autophosphorylate	Uhl and Miller, 1994
Δ15C	C terminus	In-frame 15 aa deletion	Bvg⁻, unable to autophosphorylate	Uhl and Miller, 1994

[a] Insertion mutations in all cases are linker insertions that result in the introduction of three to five amino acids. Comparison of conserved amino acids is taken from Parkinson and Kofoid (1992) and Stock et al. (1989).

(see Fig. 2). Phosphorylation of C-terminal domains has also been suggested for ArcB and BarA (Ishige et al., 1994). 'BvgS Asp-1023-Asn, which eliminates the proposed site of phosphorylation in the BvgS receiver, is deficient for transphosphorylation of the BvgS C terminus (Uhl and Miller, unpublished data). This invokes a model in which phosphorylation of the BvgS receiver is required for subsequent phosphorylation of the BvgS C terminus (see below). Mutational analysis of the BvgS C terminus has shown an absolute requirement for this domain for Bvg function (Table 1); the relationship of C-terminal phosphorylation to signal transduction is only now being determined.

Phosphotransfer to BvgA

Both phosphorylated 'BvgS and acetyl phosphate can serve as phosphodonors to BvgA (Uhl and Miller, 1994; Boucher et al., 1994) (Fig. 5, lane 9). Phosphotransfer to BvgA from BvgS can be uncoupled from BvgS autophosphorylation by mutations in BvgS (Uhl and Miller, 1994; unpublished data). These mutations, Asp-1023-Asn, Lys-1080-Arg, and Lys-1080-Met, alter two of the conserved amino acids in the BvgS receiver and are nonfunctional in vivo but can autophosphorylate to levels indistinguishable from wild-type 'BvgS in vitro (Fig. 5; lane 1: wild type, lane 3: Asp-1023-Asn, lane 4: Lys-1080-Arg, lane 5: Lys-1080-Met). 'BvgS Asp-1023-Asn is completely deficient for phosphotransfer to BvgA (lane 11) in addition to its inability to transphosphorylate the BvgS C terminus. However, 'BvgS Asp-1023-Asn has similar decay kinetics as wild-type 'BvgS and similar chemical stability of the phosphate linkage.

Function of Phosphorylated BvgA

Initial indications that BvgA is responsible for transcriptional activation included studies demonstrating that BvgA is a sequence-specific DNA binding protein recognizing specific targets with the consensus sequence TTTCCTA (Roy and Falkow, 1991). This sequence is present as an inverted repeat upstream of the *fhaB* promoter and as direct repeats upstream of the *bvgAS* promoter (Fig. 3). Although this and other studies demonstrated that BvgA was responsible for transcriptional activation, they left unanswered the question regarding the effect of phosphorylation on BvgA function. The role of phosphorylation of BvgA in transcriptional activation has recently been addressed. Phosphorylated BvgA has an enhanced capacity for binding specific DNA target sites as compared with nonphosphorylated BvgA (Boucher et al., 1994). The C-terminal portion of BvgA, which contains a consensus helix-turn-helix motif, can still bind specific DNA sequences in the absence of a receiver domain and form multimeric complexes when expressed as a fusion protein. The location of phosphorylation of BvgA is presumed to be Asp-54, and the phosphate linkage of BvgA is characteristic of an acyl phosphate, using either 'BvgS or acetyl phosphate as a phosphodonor (Uhl and Miller, 1994; Boucher et al., 1994).

Model of BvgAS Signal Transduction

A proposed model for BvgAS signal transduction is shown in Fig. 6. In the modulated state (presence of sulfate anions, nicotinic acid, or low temperature), BvgS is not phosphorylated, and *bvg*-repressed genes (*vrg*s) are expressed. In the absence of modulating factors, BvgS can autophosphorylate at His-729 in the transmitter. Phosphorylated BvgS is then proposed to intramolecularly transfer the phosphoryl group to Asp-1023 of the BvgS receiver. Mutational analysis of BvgS (Asp-1023-Asn, Lys-1080-Arg) has suggested that phosphorylation-activation of the BvgS receiver precedes phosphorylation of the C terminus. Removal of the proposed site of phosphorylation in the BvgS receiver eliminates both transphosphorylation of the BvgS C terminus and BvgA phosphorylation. 'BvgS donates a phosphoryl group to BvgA in a phosphotransfer reaction (Uhl and Miller, 1994). Once phosphorylated, BvgA appears to exhibit an increased affinity for DNA (Boucher et al., 1994). Heightened ability to form multimeric complexes or the ability to

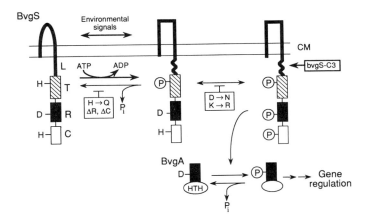

FIGURE 6 Hypothetical model for steps involved in BvgAS signal transduction (Uhl and Miller, 1994). After multimerization of BvgS, signal inputs (low temperature, nicotinic acid, MgSO$_4$) regulate activity of the protein. BvgS in a conformationally "active" state is able to autophosphorylate at His-729 of the transmitter. After autophosphorylation, the BvgS transmitter intramolecularly transfers to Asp-1023 in the receiver. The BvgS C terminus is also phosphorylated subsequent to receiver phosphorylation (Uhl and Miller, unpublished data). BvgA is phosphorylated by BvgS in a phosphotransfer reaction. Phosphorylated BvgA is able to activate transcription of virulence genes, and initial reports indicate this is through increased affinity for DNA (Boucher et al., 1994). Autophosphorylation and phosphotransfer have been demonstrated for BvgA and BvgS. Mutations that alter the proposed site of autophosphorylation (H→Q) or remove the receiver or C terminus (ΔR, ΔC) prevent autophosphorylation. Both Asp-1023 (D→N) and Lys-1080 (K→R) are necessary for phosphotransfer to BvgA, and Asp-1023 is required for transphosphorylation of the BvgS C terminus (Uhl and Miller, 1994, and unpublished data). BvgS signal-insensitive muutations, such as *bvgS-C3,* are proposed to lock the protein into an active conformation (Miller et al., 1992). Multimerization of BvgS has been suggested by intermolecular complementation of BvgS mutations (Stibitz, submitted).

form productive contacts with RNA polymerase could also depend on phosphorylation.

BvgAS AND PATHOGENESIS

Activation of Virulence Factors by BvgAS

Factors positively regulated by the *bvg* locus include pertussis toxin, adenylate cyclase toxin-hemolysin, filamentous hemagglutinin, dermonecrotic toxin, cytochrome *d*-629, fimbriae, pertactin, capsule formation, and several additional uncharacterized genetic loci (Weiss and Hewlett, 1986; Charles et al., 1989; Weiss et al., 1989; Finn et al., 1991). *bvg*-activated promot-

ers have been differentiated into distinct categories based on temporal activation in *B. pertussis,* sensitivity to modulators, and requirements for activation in *E. coli.* Transcription of promoters thought to be directly activated by *bvg,* including *fha* (filamentous hemagglutinin), *fim* (fimbriae), and *bvg* (the operon itself), occurs within 10 min after a shift from modulating to nonmodulating conditions (Scarlato et al., 1991). By contrast, regulation of the pertussis toxin (*ptx*) and adenylate cyclase (*cya*) promoters differs from the promoters immediately activated by *bvgAS.* Activation of the *ptx* and *cya* promoters begins a few hours after a shift from modulating to nonmodulating conditions, and these promoters appear to be more sensitive to modulation than the *bvg* and *fha* promoters (Scarlato et al., 1991; Scarlato and Rappuoli, 1991; Stibitz, submitted). Transcriptional fusions of the *ptx, cya, fha,* and *bvg* promoters to *lacZYA* have been constructed and tested in *E. coli*. The *bvgAS* locus in *trans* is sufficient for activation of the *fha* and *bvg* promoters (Roy et al., 1989, 1990; Miller et al., 1989b). Although it was initially thought that the *bvg* locus was not sufficient for activation of *ptx* in *E. coli* (Miller et al., 1989b), it has since been determined that *ptx* can be expressed in a *bvg*-dependent manner in *E. coli* but only under specific culture conditions (Uhl and Miller, submitted). Although *cya* expression in *E. coli* was not detected in the presence of *bvg* under the conditions tested (Goyard and Ullmann, 1991), the culture conditions required for *ptx* expression may also be necessary for *cya* expression. Factors in addition to BvgAS that may be required for activation of *ptx* and *cya* have been proposed. These include a 23-kDa protein (Huh and Weiss, 1991) and the supercoiled state of the promoter (Scarlato et al., 1993). Inhibition of DNA gyrase by addition of novobiocin increased transcription of *ptx* but did not affect *fhaB* transcription using a plasmid-encoded reporter system. Differential response of *bvg*-activated promoters to gyrase inhibitors has also been noted in *B. pertussis* (Graeff et al., unpublished data). However, conditions that influence supercoiling of the *B. pertussis* chromosome and the manner by which supercoiling relates to *bvg* regulation of virulence genes are not well defined.

Mutations that specifically decrease expression of *ptx* and *cya* have been mapped to the BvgA C terminus and upstream of the *B. pertussis rpoA* locus (Stibitz, 1994; Carbonetti et al., 1993; Carbonetti et al., 1994). The *rpoA* mutations increased expression of the α-subunit of RNA polymerase as determined by Western blot analysis. It has been proposed that overexpression of RpoA may decrease the availability of BvgA, providing evidence that *ptx* expression requires an excess of phosphorylated BvgA as compared with *fhaB* and *bvg* (Scarlato et al., 1990). In support of this hypothesis, overexpression of BvgA in the presence of BvgS or overexpression of a constitutively active BvgA allele is able to activate *ptx* expression in *E. coli* (Uhl and Miller, submitted). Activation of virulence genes appears to depend on both phosphorylation of BvgA and on the amount of phosphorylated BvgA present in the cell. A role for specific regulatory factors in addition to those encoded by *bvg* remains to be proved.

bvg-Repressed Genes

Growing evidence indicates that alteration between the Bvg$^+$ and Bvg$^-$ phases is a biphasic transition with differential expression of genes in both phases, as opposed to an "on-off" switch of virulence factors. In the Bvg$^-$ phase, *vir*-repressed genes (*vrgs*) are activated although *vir*-activated genes are not expressed. *vrgs* were initially identified by Tn*phoA* mutagenesis of *B. pertussis* and screening for PhoA expression under Bvg$^-$ conditions (Knapp and Mekalanos, 1988; Beattie et al., 1993). *vrg-6* has been sequenced, and the predicted product is a small protein with several proline-rich repeats (Beattie et al., 1990).

A class of *vir*-repressed genes has also been identified in *B. bronchiseptica,* a member of the *Bordetella* genus that is closely related to *B. pertussis*. The motility phenotype of *B. bronchiseptica* is regulated by *bvgAS* in a reciprocal manner to *bvg*-activated genes (Akerley et al., 1992; Akerley and Miller, 1993). Siderophore production in some strains of *B. bronchiseptica* is also negatively regulated by *bvg* (Agiato Foster et al., 1994). Two motility loci in *B. bronchiseptica, flaA* (the structural gene for flagellin) and *frlAB* (the motility master regulatory locus, similar to *flhCD* of *E. coli*) have been characterized. Transcription of both *flaA* and *frlAB* is negatively regulated by *bvgAS* (Akerley and Miller, 1993; Akerley et al., in press). *B. pertussis* contains sequences that hybridize to both *flaA* and *frl,* and the *frl* locus of *B. pertussis* is functional when expressed in *B. bronchiseptica*

(Akerley and Miller, unpublished data). The role of *frl* in the nonmotile *B. pertussis* is still under investigation.

Role of *bvgAS* In Vivo

It has been firmly established that the Bvg$^+$ phase of *B. pertussis* and *B. bronchiseptica* is required for colonization and disease in animal models (Weiss and Hewlett, 1986; Weiss and Goodwin, 1989; Gueirard and Guiso, 1993; Cotter and Miller, 1994; Akerley et al., in press). However, a definitive role for the Bvg$^-$ phase or for BvgAS signal transduction in vivo has not been established. Several roles for the Bvg$^-$ phase have been proposed, including establishment of infection, modulation of disease, persistence of the organism by down-regulation of antigenic proteins, transmission, or survival in an undefined environmental reservoir. Bvg$^-$-phase factors of *B. pertussis* have only recently begun to be characterized, but demonstration that a transposon insertion in a Bvg$^-$-phase gene (*vrg6*::Tn*phoA*) is less virulent in a mouse model of infection suggests that Bvg-mediated signal transduction by *B. pertussis* is important in vivo (Beattie et al., 1992). Identification of additional Bvg$^-$-phase genes and their functional roles in *B. pertussis* will further our understanding of virulence gene regulation as it relates to pathogenesis.

A more comprehensive view of the role of *bvgAS*-mediated signal transduction has emerged from studies with *B. bronchiseptica*. Analysis of *B. bronchiseptica* infection of rabbits by Bvg-phase-locked strains (Bvg^c or Bvg$^-$) revealed that the Bvg$^+$ phase was both necessary and sufficient for colonization and short-term infection and that alteration to the Bvg$^-$ phase was not required (Cotter and Miller, 1994). An immunological response to flagella, a marker for Bvg$^-$-phase gene expression in vivo, was not detected. However, when Bvg-phase-locked strains were grown in vitro in nutrient-poor media, the Bvg$^-$ phase was optimal for growth. Bvg$^+$-phase-locked organisms either perished or gained compensatory mutations to allow survival in nutrient-poor media. In *B. bronchiseptica*, *bvgAS* may regulate the switch from nutrient-rich to nutrient-poor conditions in either internal host environments or external reservoirs. Although the Bvg$^-$ phase may be advantageous for nutrient-limiting conditions, expression of Bvg$^-$-phase genes in vivo can be deleterious (Akerley et al., in press). When motility is ectopically expressed in the Bvg$^+$ phase of *B. bronchiseptica,* an immune response to flagella is generated in a rat model of bordetellosis. Motile organisms are either cleared from the trachea or revert to a motility-negative phenotype (Akerley et al., in press). Although the Bvg$^+$ phase of *B. pertussis* and *B. bronchiseptica* appears to play similar roles during infection, the Bvg$^-$ phases of bordetellae may serve distinct functions.

SUMMARY AND CONCLUSIONS

Although BvgA is a typical response regulator, BvgS is a complex sensor protein containing multiple cytoplasmic domains involved in signaling. Mutations that lock BvgS into an active state map to the BvgS linker and eliminate the requirement for the BvgS periplasmic region (Miller et al., 1992). The BvgS periplasmic domain is thought to recognize signals and transmit information to the BvgS cytoplasmic domains through the BvgS linker. The cytoplasmic domains function to control a subsequent phosphorylation cascade. The ability of BvgS to autophosphorylate and transfer a phosphoryl group to BvgA in vitro correlates with its ability to activate virulence gene expression in vivo (Uhl and Miller, 1994).

Sensor proteins are defined by the presence of a transmitter domain (Stock et al., 1989; Parkinson and Kofoid, 1992). Although several sensor proteins contain linked transmitter and receiver domains, BvgS belongs to a smaller subset that contains transmitter, receiver, and additional C-terminal domains. Hybrid kinases include both prokaryotic and eukaryotic sensor proteins (Parkinson and Kofoid, 1992; Ota and Varshavsky, 1993; Chang et al., 1993; Utsumi et al., 1994) (see Fig. 2). Genetic evidence suggests the receiver in some sensor proteins represents an autoinhibitory domain that must be phosphorylated to allow phosphorylation of the cognate response regulator (Iuchi and Lin, 1992; Chang and Winans, 1992; Pazour et al., 1991). Results from in vitro studies of ArcB support this contention (Iuchi, 1993). Phosphotransfer from the ArcB transmitter to the ArcB receiver was demonstrated, and phenotypes of ArcB receiver mutations suggested that phosphorylation of the ArcB receiver is required for phosphotransfer to ArcA, the cognate response regulator. In BvgS, similar receiver mutations block phosphoryl group transfer to BvgA, yet genetic and biochemical evidence contradict an exclusively autoinhibitory model for the BvgS receiver and C terminus (Uhl and Miller, 1994; Beier and Gross, unpublished data). Although the BvgS C terminus is phosphorylated in vitro, it is not yet known if the C terminus is responsible for phosphotransfer to BvgA or if the phosphorylated C terminus regulates another function of BvgS (Uhl and Miller, unpublished data). Further characterization of the biochemistry of BvgS and other sensor proteins with linked transmitter, receiver, and C-terminal domains is necessary to determine whether receiver/C-terminal modules on sensor proteins regulate signal transduction through common mechanisms.

In considering the BvgAS signal transduction system, as with many other two-component systems, we are left with several important questions. What is the nature of the signals sensed by *Bordetella* through *bvg*, and how are these signals transmitted to activation of BvgS and BvgA? How do the BvgS receiver and C-terminal domains regulate phosphorylation-dephosphorylation of BvgA and BvgS? Which regions of BvgA and BvgS govern specificity of signal transduction for *bvg*? These questions directly relate to the ability of *Bordetella* to coordinately regulate genes involved in colonization, multiplication, and localized spread of the bacteria inside the host. Pathogenesis of *Bordetella* begins with *bvg*; characterization of the mechanisms by which BvgAS senses and responds to changing environments is essential for understanding *B. pertussis*.

ACKNOWLEDGMENTS

We are grateful to Nick Carbonetti, Roy Gross, Camille Locht, Fritz Mooi, and Scott Stibitz for making data available before publication. We also thank Nick Carbonetti, Roy Gross, Camille Locht, Olaf Schneewind, and members of our laboratory for helpful comments.

Work in our laboratory is funded by NIH grant AI31548. M.A.U. is supported by NIH predoctoral training grant GM07185. J.F.M. is a Pew scholar in the biomedical sciences.

REFERENCES

Agiato Foster, L. A., P. C. Giardina, M. Wang, B. J. Akerley, J. F. Miller, and D. W. Dyer. 1994. Siderophore biosynthesis in *Bordetella bronchiseptica* is controlled by the *bvg* regulon, abstr. 1558. Abstr. 93rd Gen. Meet. Am. Soc. Microbiol. 1994. American Society for Microbiology, Washington, D.C.

Akerley, B. J., P. A. Cotter, and J. F. Miller. Ectopic expression of the flagellar regulon alters development of the *Bordetella*-host interaction. *Cell,* in press.

Akerley, B. J., and J. F. Miller. 1993. Flagellin transcription in *Bordetella bronchiseptica* is regulated by the BvgAS virulence control system. *J. Bacteriol.* **175:**3468–3479.

Akerley, B. J., and J. F. Miller. Unpublished data.

Akerley, B. J., D. M. Monack, S. Falkow, and J. F. Miller. 1992. The *bvgAS* locus negatively controls motility and synthesis of flagella in *Bordetella bronchiseptica. J. Bacteriol.* **174:**980–990.

Ames, P., and J. S. Parkinson. 1988. Transmembrane signaling by bacterial chemoreceptors: *E. coli* transducers with locked signal input. *Cell* **55:**817–826.

Aricò, B., J. F. Miller, C. Roy, S. Stibitz, D. Monack, S. Falkow, R. Gross, and R. Rappuoli. 1989. Sequences required for expression of *Bordetella pertussis* virulence factors share homology with prokaryotic signal transduction proteins. *Proc. Natl. Acad. Sci. USA* **86:**6671–6675.

Aricò, B., V. Scarlato, D. M. Monack, S. Falkow, and R. Rappuoli. 1991. Structural and genetic analysis of the *bvg* locus in *Bordetella* species. *Mol. Microbiol.* **5:**2481–2491.

Beattie, D. T., S. Knapp, and J. J. Mekalanos. 1990. Evidence that modulation requires sequences downstream of the promoters of two *vir*-repressed genes of *Bordetella pertussis. J. Bacteriol.* **172:**6997–7004.

Beattie, D. T., M. J. Mahan, and J. J. Mekalanos. 1993. Repressor binding to a regulatory site in the DNA coding sequence is sufficient to confer transcriptional regulation of the *vir*-repressed genes (*vrg* genes) in *Bordetella pertussis. J. Bacteriol.* **175:**519–527.

Beattie, D. T., R. Shahin, and J. J. Mekalanos. 1992. A *vir*-repressed gene of *Bordetella pertussis* is required for virulence. *Infect. Immun.* **60:**571–577.

Beier, D., and R. Gross. Unpublished data.

Boucher, P. E., F. D. Menozzi, and C. Locht. 1994. The modular architecture of bacterial response regulators: insights into the activation mechanism of the BvgA transactivator of *Bordetella pertussis. J. Mol. Biol.* **241:**363–377.

Carbonetti, N., T. M. Fuchs, A. A. Patamawenu, T. J. Irish, H. Deppisch, and R. Gross. 1994. Effect of mutations causing overexpression of RNA polymerase α subunit on regulation of virulence factors in *Bordetella pertussis. J. Bacteriol* **176:**7267–7273.

Carbonetti, N. H., N. Khelef, N. Guiso, and R. Gross. 1993. A phase variant of *Bordetella pertussis* with a mutation in a new locus involved in the regulation of pertussis toxin and adenylate cyclase toxin expression. *J. Bacteriol.* **175:**6679–6688.

Chang, C., S. F. Kwok, A. B. Bleecker, and E. M. Meyerowitz. 1993. *Arabidopsis* ethylene-response gene ETR1: similarity of product to two-component regulators. *Science* **262:**539–544.

Chang, C., and S. C. Winans. 1992. Functional roles assigned to the periplasmic, linker, and receiver domains of the *Agrobacterium tumefaciens* VirA protein. *J. Bacteriol.* **174:**7033–7039.

Charles, I. G., G. Dougan, D. Pickard, S. Chatfield, M. Smith, P. Novotny, P. Morrissey, and N. F. Fairweather. 1989. Molecular cloning and characterization of protective outer membrane protein P.69 from *Bordetella pertussis. Proc. Natl. Acad. Sci. USA* **86:**3554–3558.

Collins, L. A., S. M. Egan, and V. Stewart. 1992. Mutational analysis reveals functional similarity between NARX, a nitrate sensor in *Escherichia coli* K-12, and the methyl-accepting chemotaxis proteins. *J. Bacteriol.* **174:**3667–3675.

Cotter, P. A., and J. F. Miller. 1994. The Bvg$^+$ phase is necessary and sufficient for the establishment of respiratory tract colonization by *Bordetella bronchiseptica*: analysis of phase-locked regulatory mutants in a rabbit model. *Infect. Immun.* **62:**3381–3390.

DiRita, V. J., and J. J. Mekalanos. 1989. Genetic regulation of bacterial virulence. *Annu. Rev. Genet.* **23:**455–482.

Finn, T. M., R. Shahin, and J. J. Mekalanos. 1991. Characterization of *vir*-activated TnphoA gene fusions in *Bordetella pertussis. Infect. Immun.* **59:**3272–3279.

Gentry-Weeks, C. R., D. L. Provence, J. M. Keith, and R. Curtiss III. 1991. Isolation and characterization of *Bordetella avium* phase variants. *Infect. Immun.* **59:**4026–4033.

Goldman, W. E., D. G. Klapper, and J. B. Baseman. 1982. Detection, isolation, and analysis of a released *Bordetella pertussis* product toxic to cultured tracheal cells. *Infect. Immun.* **36:**782–794.

Goyard, S., and A. Ullmann. 1991. Analysis of *Bordetella pertussis cya* operon regulation by use of *cya-lac* fusions. *FEMS Microbiol. Lett.* **77:**251–256.

Graeff, H., D. Beier, H. Deppisch, and R. Gross. Unpublished data.

Gueirard, P., and N. Guiso. 1993. Virulence of *Bordetella bronchiseptica*: role of adenylate cyclase-hemolysin. *Infect. Immun.* **61:**4072–4078.

Hewlett, E. L. 1990. *Bordetella* species, p. 1756–1762. *In* G. L. Mandell, R. G. Douglas, Jr., and J. E. Bennett (ed.), *Principles and Practice of Infectious Diseases,* 3rd ed. Churchill Livingstone, New York.

Hewlett, E. L., and V. M. Gordon. 1988. Adenylate cyclase toxin of *Bordetella pertussis,* p. 193–209. *In* A. Wardlaw and R. Parton (ed.), *Pathogenesis and Immunity in Pertussis.* John Wiley & Sons, Inc., New York.

Huh, Y. J., and A. A. Weiss. 1991. A 23-kilodalton protein, distinct from BvgA, expressed by virulent *Bordetella pertussis* binds to the promoter region of *vir*-regulated genes. *Infect. Immun.* **59:**2389–2395.

Irons, L. I., and A. R. Gorringe. 1988. Pertussis toxin: production, purification, molecular structure, and assay, p. 95–120. *In* A. Wardlaw and R. Parton (ed.), *Pathogenesis and Immunity in Pertussis.* John Wiley & Sons, Inc., New York.

Ishige, K., S. Nagasawa, S. Tokishita, and T. Mizuno. 1994. A novel device of bacterial signal transducers. *EMBO J.* **13:**5195–5202.

Iuchi, S. 1993. Phosphorylation/dephosphorylation of the receiver module at the conserved aspartate residue controls transphosphorylation activity of histidine kinase in sensor protein ArcB of *Escherichia coli. J. Biol. Chem.* **268:**23972–23980.

Iuchi, S., and E. C. C. Lin. 1992. Mutational analysis of signal transduction by ArcB, a membrane sensor protein for anaerobic repression of operons involved in the central aerobic pathways in *Escherichia coli. J. Bacteriol.* **174:**3972–3980.

Kalman, L. V., and R. P. Gunsalus. 1990. Nitrate- and molybdenum-independent signal transduction mutations in *narX* that alter regulation of anaerobic respiratory genes in *Escherichia coli. J. Bacteriol.* **172:**7049–7056.

Knapp, S., and J. J. Mekalanos. 1988. Two *trans*-acting regulatory genes (*vir* and *mod*) control antigenic modulation in *Bordetella pertussis. J. Bacteriol.* **170:**5059–5066.

Lacey, B. W. 1960. Antigenic modulation of *Bordetella pertussis. J. Hyg.* **58:**57–93.

Leininger, E., M. Roberts, J. G. Kenimer, I. G. Charles, N. Fairweather, P. Novotny, and M. J. Brennan. 1991. Pertactin, an Arg-Gly-Asp-containing *Bordetella pertussis* surface protein that promotes adherence of mammalian cells. *Proc. Natl. Acad. Sci. USA* **88**:345–349.

Leslie, P. H., and A. D. Gardner. 1931. The phases of *Haemophilus pertussis. J. Hyg.* **31**:423–434.

Locht, C., M. C. Geoffroy, and G. Renauld. 1992. Common accessory genes for the *Bordetella pertussis* filamentous hemagglutinin and fimbriae share sequence similarities with the *papC* and *papD* gene families. *EMBO J.* **11**:3175–3183.

Luker, K. E., J. L. Collier, E. W. Kolodziej, G. R. Marshall, and W. E. Goldman. 1993. *Bordetella pertussis* tracheal cytotoxin and other muarmyl peptides: distinct structure-activity relationships for respiratory epithelial cytopathology. *Proc. Natl. Acad. Sci. USA* **90**:2365–2369.

Masure, H., R. Shattuck, and D. Storm. 1987. Mechanisms of bacterial pathogenicity that involve production of calmodulin-sensitive adenylate cyclases. *Microbiol. Rev.* **51**:60–65.

McCleary, W. R., and D. R. Zusman. 1990. Purification and characterization of the *Myxococcus xanthus* FrzE protein shows that it has autophosphorylation activity. *J. Bacteriol.* **172**:6661–6668.

Mekalanos, J. J. 1992. Environmental signals controlling expression of virulence determinants in bacteria. *J. Bacteriol.* **174**:1–7.

Melton, A. R., and A. A. Weiss. 1989. Environmental regulation of expression of virulence determinants in *Bordetella pertussis. J. Bacteriol.* **171**:6206–6212.

Melton, A. R., and A. A. Weiss. 1993. Characterization of environmental regulators of *Bordetella pertussis. Infect. Immun.* **61**:807–815.

Miller, J. F., S. A. Johnson, W. J. Black, D. T. Beattie, J. J. Mekalanos, and S. Falkow. 1992. Isolation and analysis of constitutive sensory transduction mutations in the *Bordetella pertussis bvgS* gene. *J. Bacteriol.* **174**:970–979.

Miller, J. F., J. J. Mekalanos, and S. Falkow. 1989a. Coordinate regulation and sensory transduction in the control of bacterial virulence. *Science* **243**:916–922.

Miller, J. F., C. R. Roy, and S. Falkow. 1989b. Analysis of *Bordetella pertussis* virulence gene regulation by use of transcriptional fusions in *Escherichia coli. J. Bacteriol.* **171**:6345–6348.

Monack, D. M., B. Aricò, R. Rappuoli, and S. Falkow. 1989. Phase variants of *Bordetella bronchiseptica* arise by spontaneous deletions in the *vir* locus. *Mol. Microbiol.* **3**:1719–1728.

Mooi, F. R., W. H. Jansen, H. Brunings, H. Gielen, H. G. J. van der Heide, H. C. Walvoort, and P. A. M. Guinee. 1992. Construction and analysis of *Bordetella pertussis* mutants defective in the production of fimbriae. *Microb. Pathog.* **12**:127–135.

Ota, I. M., and A. Varshavsky. 1993. A yeast protein similar to two-component regulators. *Science* **262**:566–569.

Parkinson, J. S., and E. C. Kofoid. 1992. Communication modules in bacterial signaling proteins. *Annu. Rev. Genet.* **26**:71–112.

Pazour, G. J., C. N. Ta, and A. Das. 1991. Mutants of *Agrobacterium tumefaciens* with elevated *vir* expression. *Proc. Natl. Acad. Sci. USA* **88**:6941–6945.

Pittman, M. 1984. The concept of pertussis as a toxin-mediated disease. *Pediatr. Infect. Dis.* **3**:467–486.

Radford, S. E., E. D. Laue, R. N. Perham, S. R. Martin, and E. Appella. 1989. Conformational flexibility and folding of synthetic peptides representing an interdomain segment of polypeptide chain in the pyruvate dehydrogenase multienzyme complex of *Escherichia coli. J. Biol. Chem.* **264**:767–775.

Relman, D., E. Tuomanen, S. Falkow, D. T. Golenbock, K. Saukkonen, and S. D. Wright. 1990. Recognition of a bacterial adhesin by an intergrin: macrophage CR3 ($\alpha_M\beta_2$, CD11b/CD18) binds filamentous hemagglutinin of *Bordetella pertussis. Cell* **61**:1375–1382.

Roberts, M., N. F. Fairweather, E. Leininger, D. Pickard, E. L. Hewlett, A. Robinson, C. Hayward, G. Dougan, and I. G. Charles. 1991. Construction and characterization of *Bordetella pertussis* mutants lacking the *vir*-regulated P.69 outer membrane protein. *Mol. Microbiol.* **5**:1393–1404.

Roy, C. R., and S. Falkow. 1991. Identification of *Bordetella pertussis* regulatory sequences required for transcriptional activation of the *fhaB* gene and autoregulation of the *bvgAS* operon. *J. Bacteriol.* **173**:2385–2392.

Roy, C. R., J. F. Miller, and S. Falkow. 1989. The *bvgA* gene of *Bordetella pertussis* encodes a transcriptional activator required for coordinate regulation of several virulence genes. *J. Bacteriol.* **171**:6338–6344.

Roy, C. R., J. F. Miller, and S. Falkow. 1990. Autogenous regulation of the *Bordetella pertussis bvgABC* operon. *Proc. Natl. Acad. Sci. USA* **87**:3763–3767.

Scarlato, V., B. Aricò, A. Prugnola, and R. Rappuoli. 1991. Sequential activation and environmental regulation of virulence genes in *Bordetella pertussis. EMBO J.* **10**:3971–3975.

Scarlato, V., B. Aricò, and R. Rappuoli. 1993. DNA topology affects transcriptional regulation of the pertussis toxin gene of *Bordetella pertussis* in *Escherichia coli* and in vitro. *J. Bacteriol.* **175**:4764–4771.

Scarlato, V., A. Prugnola, B. Aricò, and R. Rappuoli. 1990. Positive transcriptional feedback at the *bvg* locus controls expression of virulence factors in *Bordetella pertussis. Proc. Natl. Acad. Sci. USA* **87**:6753–6757.

Scarlato, V., and R. Rappuoli. 1991. Differential response of the *bvg* virulence regulon of *Bordetella pertussis* to MgSO$_4$ modulation. *J. Bacteriol.* **173:** 7401–7404.

Schneider, D., and C. Parker. 1982. Effect of pyridines on phenotypic properties of *Bordetella pertussis. Infect. Immun.* **38:**548–553.

Stibitz, S. 1994. Mutations in the *bvgA* gene of *Bordetella pertussis* that differentially affect regulation of virulence determinants. *J. Bacteriol.* **176:**5615–5621.

Stibitz, S. Complementation analysis of the *vir* locus of *Bordetella pertussis.* Submitted for publication.

Stibitz, S., W. Aaronson, D. Monack, and S. Falkow. 1989. Phase variation in *Bordetella pertussis* by a frameshift in a gene for a novel two component system. *Nature* (London) **338:**266–269.

Stibitz, S., W. Black, and S. Falkow. 1986. The construction of a cloning vector designed for gene replacement in *Bordetella pertussis. Gene* **50:**133–140.

Stibitz, S., A. A. Weiss, and S. Falkow. 1988. Genetic analysis of a region of the *Bordetella pertussis* chromosome encoding filamentous hemagglutinin and the pleiotropic regulatory locus *vir. J. Bacteriol.* **170:**2904–2913.

Stibitz, S., and M. S. Yang. 1991. Subcellular localization and immunological detection of proteins encoded by the *vir* locus of *Bordetella pertussis. J. Bacteriol.* **173:**4288–4296.

Stock, J., A. J. Ninfa, and A. M. Stock. 1989. Protein phosphorylation and regulation of adaptive responses in bacteria. *Microbiol. Rev.* **53:**450–490.

Swanson, R. V., R. B. Bourret, and M. Simon. 1993. Intermolecular complementation of the kinase activity of CheA. *Mol. Microbiol.* **8:**435–441.

Tuomanen, E., H. Towbin, G. Rosenfelder, D. Braun, G. Larson, G. Hansson, and R. Hill. 1988. Receptor analogs and monoclonal antibodies that inhibit adherence of *Bordetella pertussis* for human ciliated respiratory-epithelial cells. *J. Exp. Med.* **168:**267–277.

Tuomanen, E., and A. A. Weiss. 1985. Characterization of two adhesins of *Bordetella pertussis* for human ciliated respiratory-epithelial cells. *J. Infect. Dis.* **152:**118–125.

Uhl, M. A., and J. F. Miller. 1994. Autophosphorylation and phosphotransfer in the *Bordetella pertussis* BvgAS signal transduction cascade. *Proc. Natl. Acad. Sci. USA* **91:**1163–1167.

Uhl, M. A., and J. F. Miller. Comparative analysis of Bvg-dependent activation of the *Bordetella pertussis*

ptx and *fha* promoters in *E. coli.* Submitted for publication.

Uhl, M. A., and J. F. Miller. Unpublished data.

Ui, M. 1988. The multiple biological activities of pertussis toxin, p. 121–146. *In* A. Wardlaw and R. Parton (ed.), *Pathogenesis and Immunity in Pertussis.* John Wiley & Sons, Inc., New York.

Utsumi, R., S. Katayama, M. Taniguchi, T. Horie, M. Ikeda, S. Igaki, H. Nakagawa, A. Miwa, H. Tanabe, and M. Noda. 1994. Newly identified genes involved in the signal transduction of *Escherichia coli* K-12. *Gene* **140:**73–77.

Weiss, A., E. L. Hewlett, G. A. Myers, and S. Falkow. 1983. Tn*5*-induced mutations affecting virulence factors of *Bordetella pertussis. Infect. Immun.* **42:**33–41.

Weiss, A. A., and S. Falkow. 1982. Plasmid transfer to *Bordetella pertussis:* conjugation and transformation. *J. Bacteriol.* **152:**549–552.

Weiss, A. A., and S. Falkow. 1984. Genetic analysis of phase variation in *Bordetella pertussis. Infect. Immun.* **43:**263–269.

Weiss, A. A., and M. S. M. Goodwin. 1989. Lethal infection by *Bordetella pertussis* mutants in the infant mouse model. *Infect. Immun.* **57:**3757–3764.

Weiss, A. A., and E. L. Hewlett. 1986. Virulence factors of *Bordetella pertussis. Annu. Rev. Microbiol.* **40:**661–686.

Weiss, A. A., A. R. Melton, K. E. Walker, C. Andraos-Selim, and J. J. Meidl. 1989. Use of the promoter fusion transposon Tn*5 lac* to identify mutations in *Bordetella pertussis vir*-regulated genes. *Infect. Immun.* **57:**2674–2682.

Willems, R. J. L., C. Geuijen, H. G. J. van der Heide, G. Renauld, P. Bertin, W. M. R. van den Akker, C. Locht, and F. R. Mooi. 1994. Mutational analysis of the *Bordetella pertussis fim/fha* gene cluster: identification of a gene with sequence similarities to haemolysin accessory genes involved in export of FHA. *Mol. Microbiol.* **11:**387–397.

Willems, R. J. L., H. G. J. van der Heide, and F. R. Mooi. 1992. Characterization of a *Bordetella pertussis* fimbrial gene cluster which is located directly downstream of the filamentous hemagglutinin gene. *Mol. Microbiol.* **6:**2661–2671.

Wootton, J. C., and M. H. Drummond. 1989. The Q-linker: a class of interdomain sequences found in bacterial multidomain regulatory proteins. *Protein Eng.* **2:**535–543.

Three-Component Regulatory System Controlling Virulence in *Vibrio cholerae*

Victor J. DiRita

22

Studies of virulence and its regulation in *Vibrio cholerae* have classically been devoted to understanding the mechanism and consequences of toxin production by this human pathogen. The cholera toxin is a very well-studied molecule and is the virulence factor of primary importance for pathogenicity of *V. cholerae*. Early work, driven largely by the goal of overproducing the toxin for the purpose of making toxoid vaccines, led to the discovery that expression of cholera toxin is a regulated phenotype in *V. cholerae* (Richardson, 1969). More recent work identified several other gene products required for full virulence and showed that expression of these genes, which include those encoding a pilus colonization factor and an accessory colonization factor, is under coordinate control with cholera toxin expression (Taylor et al., 1987; Peterson and Mekalanos, 1988). This phenomenon of coordinate expression of virulence traits is observed in many other pathogens, some of which are addressed in this volume.

As is also often the case with other pathogenic bacteria and indeed for a broad range of physiological responses by both pathogenic and nonpathogenic bacteria, the products of a single regulatory locus control expression of cholera toxin and the genes with which it is coordinately expressed in *V. cholerae* (Miller and Mekalanos, 1984; Peterson and Mekalanos, 1988). This locus, the *toxR* locus, is made up of two genes, *toxR* and *toxS,* which encode membrane regulatory proteins (Miller et al., 1987; Miller et al., 1989; DiRita and Mekalanos, 1991). The mechanism by which these two proteins control expression of the ToxR regulon is the subject of this review. Although ToxR and ToxS represent a "two-component" regulatory system by definition, the relationship they share is not what has become conventionally understood by the term. Furthermore, because of the identification of the ToxT activator, whose expression is controlled by ToxR (DiRita et al., 1991), the regulatory pathway controlling virulence in *V. cholerae* is more accurately described as a "three-component" system (a model for this system is shown in Fig. 1; the details will be described). In the discussion that follows, I will not address the regulatory pathway controlled by available iron that plays a role in cholera virulence but will refer the reader to a recent comprehensive review (Litwin and Calderwood, 1993).

Victor J. DiRita, Unit for Laboratory Animal Medicine and Department of Microbiology and Immunology, University of Michigan Medical School, Ann Arbor, Michigan 48109.

Two-Component Signal Transduction, Edited by James A. Hoch and Thomas J. Silhavy, © 1995 American Society for Microbiology, Washington, DC 20005

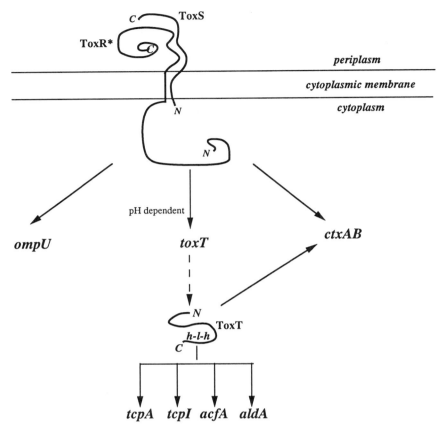

FIGURE 1 Model for the ToxR regulon in *V. cholerae.* The relative orientations of ToxR and ToxS with respect to the cytoplasmic membrane are indictated with N and C for the amino and carboxyl termini, respectively. ToxR* refers to the active form of ToxR, resulting from interaction with ToxS in the membrane. The putative carboxyl-terminal helix-loop-helix DNA binding domain of ToxT is indicated by h-l-h. See text for further details.

OVERVIEW OF PATHOGENICITY OF *VIBRIO CHOLERAE*

V. cholerae colonizes the small bowel, where it produces a potent exotoxin, the cholera toxin. Cholera toxin ADP-ribosylates a regulatory GTP binding protein in intestinal epithelial cells, thereby disrupting its function (Gill and Meren, 1978). The result is intracellular accumulation of cyclic AMP and consequent secretion of chloride ions from the cells, which leads to osmotically driven secretion of water and other nutrients from the cells and the characteristic rice water stools of infected individuals.

The question of why cholera toxin is produced during infection with *V. cholerae,* that is to say, what advantage it offers the microbe in the interaction between *V. cholerae* and its unfortunate host, remains open. Possible reasons range from acquisition of nutrients within the secretions of intoxicated epithelial cells to convenient passage for the bacteria out of the afflicted host back into the environment. Whatever the precise rationale for producing cholera toxin, the sine qua non of pathogenicity by *V. cholerae* is elaboration of toxin, and administration of toxin in the absence of the microbe is sufficient to mimic the disease (Finkelstein, 1973). Thus, mucosal colonization and toxin production are the main features of the pathogenesis of *V. cholerae* infection, and invasion of host cells with subsequent dissemination into other body sites is not part of the approach this pathogen uses to survive within a host.

Although colonization is the final destination for infecting vibrios, before reaching the cell surface the organism has to negotiate several environments. The first is the stomach, which is typically at a pH that *V. cholerae,* which is quite acid-sensitive, cannot survive. A large percentage of the inoculum is therefore probably killed in the first host environment encountered, which likely accounts for the high doses of organisms required to infect and for the fact that achlorhydric people (those with decreased stomach acidity) are more sensitive to cholera infection (Mims et al., 1993). This acidic environment may also induce a stress response that has been proposed to down-regulate transcription of *toxR,* as is discussed below (Parsot and Mekalanos, 1990). The organisms that survive the stomach and enter into the small bowel must get through the mucin layer that covers the epithelial cells of the bowel. Chemotaxis and motility of vibrios are virulence factors that may be required for them to get through this mucin layer to the epithelial cell surface (Freter et al., 1981; Richardson, 1991), as is the production of secreted proteases that degrade mucin (Schneider and Parker, 1978). Eventually, the cells reach the surface of the mucosal epithelial cells, where colonization factors and cholera toxin come into play. Given this sequence of events, it seems probable that the organism is inundated with environmental cues to which it must respond if it is to reach its goal, and this brings up the question of regulation.

The production of cholera toxin and coordinately expressed virulence factors is not a constitutive phenotype of *V. cholerae.* For in vitro production of virulence factors by many strains of *V. cholerae,* the optimum growth conditions are rich media of slightly decreased pH (6.5) and aerated growth at low temperature (25 to 30°C) (Richardson, 1969). Some strains of *V. cholerae,* as discussed below, require different conditions than these, and the reason for the strain differences may be due to alterations in the signal transduction pathway. Although these conditions are optimum for *V. cholerae* virulence expression in vitro, whether they reflect the natural signals perceived and responded to in vivo has not been demonstrated. It is unlikely, for example, that the temperature optimum of 25 to 30°C in vitro is an accurate representation of the temperature requirements during infection, which takes place at 37°C, but the dramatic difference in virulence factor production by cells grown at the two different temperatures has enabled investigators to ask general questions about virulence regulation.

ToxR REGULON

Virulence factors coordinately expressed with cholera toxin include the toxin-coregulated pilus (TCP) (Taylor et al., 1987) and the accessory colonization factor (ACF) (Peterson and Mekalanos, 1988). Several gene products are required for both TCP and ACF expression, and they are all regulated by the ToxR protein (Fig. 1).

Expression of TCP is a requirement for colonization of *V. cholerae* in humans (Herrington et al., 1988). In addition to the *tcpA* gene, which encodes the major subunit of TCP, the *tcp* gene cluster includes several genes and open reading frames whose products are involved (or predicted to be involved) in expression and assembly of the pilus structure (Kaufman et al., 1991; Kaufman et al., 1993; Peek and Taylor, 1992; Ogierman et al., 1993). Some of these are homologous to the family of proteins referred to as the general export pathway for extracellular protein deposition or secretion (Kaufman et al., 1993; Hobbs and Mattick, 1993). The specific nature of ACF has not yet been determined, but it is required for wild-type colonization in an animal model (Peterson and Mekalanos, 1988). The *acf* gene cluster is made up of four genes: *acfA, acfB, acfC,* and *acfD.*

All of the *tcp* and *acf* genes, as well as the cholera toxin operon, *ctxAB,* and the outer membrane protein OmpU require the ToxR regulatory protein for their expression. Two topics are addressed in the remainder of this chapter. The first deals with the structure and function of ToxR, which is an unusual member of the response regulator class of transcriptional activators (Miller et al., 1987), and of ToxS, an effector protein required by ToxR (Miller et al., 1989; DiRita and Mekalanos, 1991). Much of the work devoted to these two proteins to date has been concerned with how they interact with each other and how that interaction may result in a form of ToxR that can activate transcription. The second topic is the molecular basis of coordinate gene expression in the ToxR regulon. The work in this area reviewed below is aimed mainly at understanding how ToxR controls expression of the gene encoding the ToxT protein, which is the direct activator for many of the genes of the ToxR regulon (DiRita et al., 1991; Higgins et al., 1992; Higgins and DiRita, 1994) (Fig. 1).

TRANSMEMBRANE NATURE OF ToxR

ToxR was originally identified by virtue of its ability to activate expression of a *ctx-lacZ* transcriptional fusion in *Escherichia coli* cells lacking any other *V. cholerae* genes (Miller and Mekalanos, 1984). Nucleotide sequence analysis revealed an unusual domain structure for a protein that acts as a transcriptional activator (Miller et al., 1987). The amino-terminal portion, comprising approximately two-thirds of the protein, is similar to the carboxy-terminal domain of the OmpR protein, which is a response regulator of the two-component family of sensor-regulator pairs. ToxR does not share homology with OmpR in the region of the latter containing aspartic acid residues critical for phosphorylation by the sensor kinase EnvZ. This lack of an evident site of phosphorylation distinguishes ToxR from the typical two-component response regulator and suggests that the mechanism leading to active ToxR involves a molecular process other than phosphorylation.

The remaining primary structure of ToxR includes a short transmembrane domain and a periplasmic domain comprising the rest of the protein (approximately one-third). This atypical structure for a bacterial transcriptional activator is also seen in the CadC protein, an *E. coli* activator that activates expression of the *cadA* gene, encoding lysine decarboxylase, at low external pH (Watson et al., 1992). CadC is similar across its amino terminus to ToxR and other OmpR-like activators and, like ToxR, has putative transmembrane and periplasmic domains. The periplasmic region of CadC is proposed to be able to sense pH changes that lead to expression of the *cad* operon (Watson et al., 1992).

That ToxR contains a transmembrane domain has allowed investigators to use fusions to the *E. coli* alkaline phosphatase gene (*phoA*) (Manoil and Beckwith, 1986) to study its structure and activity. Fusing a signal sequence-dependent allele of *phoA* to the portion of *toxR* encoded downstream of the membrane domain results in a protein that has the ToxR DNA binding domain at its amino terminus and active alkaline phosphatase at its carboxy terminus. This observation confirmed that the hydrophobic domain in ToxR predicted to be a transmembrane domain actually is, because alkaline phosphatase must be translocated to the periplasm to have activity (Manoil and Beckwith, 1986). Analysis of alkaline phosphatase fusion proteins also confirmed that there is a single transmembrane domain in ToxR (Miller et al., 1987).

SIGNALING BY ToxR

Analysis of ToxR-PhoA fusion proteins led to the conclusion that environmental signals influencing *ctx* expression are transmitted through ToxR. A ToxR-PhoA fusion protein with most of the ToxR periplasmic domain replaced by alkaline phosphatase was able to bind to *ctx* promoter DNA and to activate *ctx* expression (Miller et al., 1987). Also, when expressed from the *tet* gene promoter on a multicopy plasmid, the gene encoding this fusion protein complemented a *toxR⁻* strain of *V. cholerae* for toxin production. *toxR⁻* cells complemented with the *toxR-phoA* gene produced high levels of toxin in media of elevated osmolarity (tryptone broth + 500 mM NaCl), a growth condition in which wild-type cells do not produce toxin. This constitutive phenotype did not hold for other environmental signals tested nor was it exhibited by *toxR⁻* cells complemented with wild-type *toxR* (Miller et al., 1987). The authors suggested that the C-terminal domain of ToxR may be required for sensing osmolarity changes in the environment but that other signals influencing ToxR-regulated gene expression can still be recognized when this domain is replaced by alkaline phosphatase.

Two other aspects of regulation in this system were brought out by these observations. The first regards whether transcriptional control over *toxR* expression plays a role in virulence expression. The plasmid encoding wild-type *toxR* used for complementation experiments expressed the gene under control of the *tet* promoter. Nevertheless, the wild-type regulation pattern was observed in these cells with respect to cholera toxin synthesis. This suggests that, at least for the conditions tested, transcriptional control over *toxR* expression is not a feature of the signaling pathway in *V. cholerae*. This is not the case for all signals, as *toxR* expression was later shown to be modulated by growth at different temperatures (Parsot and Mekalanos, 1990). These authors presented evidence that the level of *toxR-lacZ* expressed from a reporter plasmid is inversely related to the amount of the heat shock σ-factor, σ^{32}, in cells. This is presumably due to the divergently expressed, σ^{32}-dependent *htpG* gene upstream of *toxR* on the *V. cholerae* chromosome. These observations led to the hypothesis that the heat shock regulon and the ToxR regulon are expressed at different stages in the pathophysiology of *V. cholerae* infection (Parsot and Mekalanos, 1990). According to this hypothesis, the molecular mechanism for this involves access to the divergent *htpG/toxR* promoters by RNA polymerase charged with either σ^{32} or σ^{70},

the vegetative σ-factor (or the *Vibrio* equivalents of these σs). RNA polymerase(σ^{32}) that arises due to a heat shock response when the cells go from environmental temperatures to host temperature would result in transcription of *htpG* and, due to the proximity of the divergent promoters, would preclude access to the *toxR* promoter by RNA polymerase(σ^{70}). In this way, the cells could keep *toxR* transcription low in an environment where it would not be advantageous to synthesize ToxR–activated gene products (Parsot and Mekalanos, 1990).

The other aspect of ToxR-dependent regulation brought out by the *phoA* fusion studies (Miller et al., 1987) is the structure of active ToxR. Replacing nearly the entire C terminus of ToxR with alkaline phosphatase did not disrupt the ability of the resulting fusion protein to bind DNA and activate transcription. The authors suggested that in these fusion proteins, dimerization of alkaline phosphatase, a normal feature of this protein (Schlesinger, 1967), drives the dimerization of the DNA binding domain of ToxR. In fact, alkaline phosphatase activity of ToxR-PhoA comigrated with the shifted *ctx* promoter in gel shift assays (Miller et al., 1987), providing strong evidence that a dimerized fusion protein was associated with DNA binding activity. It was also proposed that dimerization of the DNA binding domain may be the result of a conformational change of the ToxR C terminus in response to an environmental signal (Miller et al., 1987). If so, then dimerization of the alkaline phosphatase moiety of ToxR-PhoA fusion proteins might be expected to lead to ToxR activity under conditions in which wild-type ToxR is inactive. This condition may be established, the authors argued, in high osmolarity, where cells expressing ToxR-PhoA produced toxin and those expressing ToxR did not. They proposed that dimerization of the C terminus of wild-type ToxR may occur in response to low osmolarity and that, at elevated osmolarity, this is somehow prohibited. Other workers noted similarities between the periplasmic domains of ToxR, PilB, a putative membrane sensor protein involved in pilin expression in *Neisseria gonorrhoeae,* and EnvZ, the membrane sensor kinase controlling expression of OmpR regulated proteins in *E. coli* in response to osmolarity, suggesting that the periplasmic domains of all three of these proteins may be sensitive to osmotic pressure (Taha et al., 1991). This is an interesting possibility, but these similarities are not compelling (15% over 95 residues at the C terminus of ToxR), and to date no evidence has been offered to support their biological relevance.

TRANSCRIPTIONAL ACTIVATION AS A CONSEQUENCE OF ToxR/ToxS PERIPLASMIC INTERACTION

Sequence analysis of the cloned *toxR* locus identified a second open reading frame encoded directly downstream of *toxR* (Miller et al., 1989). Initial characterization of *toxS* demonstrated that when ToxR is expressed from its own promoter in *E. coli,* the *toxS* gene product was required in *trans* for full ToxR activity. ToxS does not activate the *toxR* promoter, suggesting that the requirement for ToxS under conditions of low *toxR* expression is post-transcriptional (Miller et al., 1989). This suggests that whatever role ToxS plays in ToxR activity, if enough ToxR is present that role is obviated.

ToxS is a membrane-associated protein of approximately 19 kDa (DiRita and Mekalanos, 1991). A search of the data base of protein sequences did not identify any proteins with significant similarity to ToxS. Of note in particular is the lack of homology of ToxR and ToxS with response regulator and sensor kinase proteins, respectively, outside of the aforementioned similarity between ToxR and OmpR in their DNA binding domains. Given any lack of homology with sensor kinases, ToxS must have some other effect on ToxR activity besides phosphorylation. The domain structure of ToxS also suggests a novel role in ToxR activity. Hydrophobic analysis of the predicted amino acid sequence of ToxS suggested that the only transmembrane domain in the protein is at the amino terminus, with the remaining portion of the protein predicted to be in the periplasmic

space. This was confirmed by demonstrating enzymatic activity, and hence periplasmic localization, of alkaline phosphatase fused just six residues from the end of the ToxS (DiRita and Mekalanos, 1991).

The periplasmic localization of both ToxS and the C terminus of ToxR implies that any interaction between these two regulatory proteins takes place in the periplasmic space. Two deletion mutants of *toxR* were used to analyze this possibility. A *toxR* derivative deleted at a site that would leave only eight wild-type residues after the transmembrane domain, essentially removing the entire periplasmic domain of the protein, was ToxR⁻ irrespective of the presence of ToxS. However, a *toxR* derivative predicted to encode a protein truncated beyond residue 262 (leaving about two-thirds of the periplasmic domain intact) retained ToxS-dependent ToxR activity.

To investigate further the interaction between ToxR and ToxS in the periplasmic space, alkaline phosphatase fusion proteins of different lengths were used (DiRita and Mekalanos, 1991). When alkaline phosphatase was fused to ToxR residue Phe-206, which is just after the transmembrane domain (ToxR–PhoA-S), thereby replacing most of the periplasmic domain, the resulting fusion protein could activate *ctx* expression and did so in the absence or the presence of ToxS. This type of short fusion was used in the studies of ToxR structure described above to show that ToxR-PhoA still activated *ctx* expression (Miller et al., 1989). However, when the ToxR–PhoA fusion junction was eight residues from the end of ToxR, at residue Gln-262 (ToxR–PhoA-L), the ability to activate *ctx* required ToxS. These observations led to the conclusion that ToxR periplasmic residues between Phe-206 and Gln-262 are required for direct interaction between ToxR and ToxS (DiRita and Mekalanos, 1991).

The ToxR-PhoA-L fusion protein was used to identify interesting mutants of ToxR. For reasons that are not clear, this fusion expresses different levels of alkaline phosphatase activity in *E. coli* in the presence of ToxS than in its

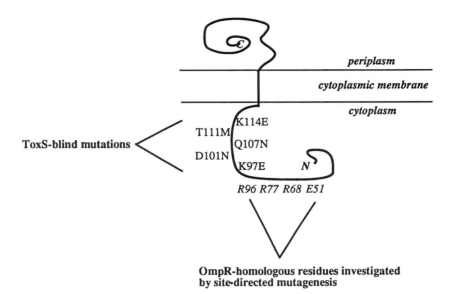

FIGURE 2 Schematic representation of the ToxR mutants described in the text. The one-letter amino acid code is used, and the number represents the position in ToxR of each residue. The ToxS blind mutants are referred to with the wild-type residue followed by the mutant residue isolated in the screen described in the text. The OmpR-homologous residues that were studied by in vitro mutagenesis were changed to various different residues, as described in the text.

absence. Also, the presence of ToxS in the cell protects this fusion protein from proteolytic cleavage in *E. coli,* perhaps because of a direct interaction between the two proteins that hinders protease access to the ToxR-PhoA-L. The different alkaline phosphatase activities of this fusion protein in the presence and absence of ToxS enabled a screen to be set up to identify mutants of ToxR-PhoA-L that had lost the ability to interact with ToxS. The rationale behind this screen was to identify ToxR residues that are critical for ToxR-ToxS interaction.

Mutants were identified that expressed elevated alkaline phosphatase activity in the presence of ToxS but that did not simultaneously acquire ToxR activity. It was necessary to screen for the latter phenotype to rule out promoter mutations, because overexpression of ToxR confers a ToxS-independent phenotype (Miller et al., 1987; Miller et al., 1989). Consistent with the behavior of the wild-type fusion protein in the absence of ToxS, the "ToxS-blind" mutants

were subject to proteolytic cleavage in *E. coli* in the presence of ToxS. Based on these criteria, the mutants behaved as if ToxS were not in the cell. Nucleotide sequence analysis determined that the lesions in these mutants were clustered within a region of 18 residues in the cytoplasmic domain of ToxR (Fig. 2). Given that the interaction between ToxR and ToxS had been determined to take place within the periplasm, the cytoplasmic lesions in ToxR-PhoA-L mutants identified by their inability to interact with ToxS led DiRita and Mekalanos to propose that the cytoplasmic domain identified by the mutations is important for maintaining the activated, dimerized state of ToxR. In this model, the interaction between ToxR and ToxS is required to stabilize dimers of ToxR that can form spontaneously. A critical concentration of ToxR dimers is required for activation of *ctx* expression, and at low levels of expression, ToxS enables this concentration to be maintained. The cytoplasmic domain identified by the

ToxS-blind mutations (Fig. 2) was proposed to be critical for locking ToxR homodimers subsequent to dimerization initiated in the periplasm, thus allowing the dimerized amino-terminal domain to bind to DNA (DiRita and Mekalanos, 1991). Hence, mutations in this domain disrupt the activated state of ToxR and do so in a way, according to this model, that also abolishes the ability of ToxR to interact with ToxS in the periplasm. This region of the protein was termed the *linker domain* to signify its location between the periplasmic ToxS responsive portion and the cytoplasmic DNA binding region. The choice of name for this domain was also influenced by mutations of the Tsr methyl-accepting chemotaxis protein of *E. coli* isolated by Ames and Parkinson (1988). These mutations, located between two functional domains of this signal transducing protein, were identified by virtue of their locking Tsr into an active state in the absence of the normal signaling molecule. "Locked output" mutations have also been described in the BvgS sensory protein controlling virulence gene expression in *B. pertussis* (Miller et al., 1992; reviewed in Chapter 21). Mutations in a region of this protein between a transmembrane domain and cytoplasmic signaling domains result in insensitivity to environmental signals that negatively modulate BvgS-dependent gene expression, thereby leading to a constitutive phenotype (Knapp and Mekalanos, 1988; Miller et al., 1992). The role of the linker region in BvgS was proposed to be similar to that of the ToxR linker, with the added feature that modulating signals were proposed to inhibit the activation function of the BvgS periplasmic domain (Miller et al., 1992).

ANALYZING THE DIMERIZATION MODEL OF ToxR FUNCTION

The dimerization model of ToxR structure makes two predictions. The first is that a domain within the C-terminal portion of the protein can dimerize, resulting in dimerization of sequences fused to that domain. The second prediction is that the DNA binding region of

ToxR requires dimerization to function. Preliminary results of experiments designed to test these predictions are mixed.

To investigate whether the ToxR C terminus acts as a dimerization domain, it was fused to the amino-terminal domain of the λ-repressor protein encoded by the *cI* gene. To bind to DNA and repress transcription of the P_R promoter of λ, the amino-terminal portion of the repressor must dimerize. This function is supplied by the normal C-terminal portion of the protein. If the C-terminal portion of ToxR can dimerize sequences N-terminal to it, then the amino terminus of λ-repressor should repress P_R transcription when fused to ToxR. This can be assayed by monitoring β-galactosidase activity of a P_R-*lacZ* transcriptional fusion. Other workers used this approach to analyze the yeast *GCN4* leucine zipper domain (Hu et al., 1990). When the amino terminus of the λ-repressor was fused to ToxR sequences beginning at ToxR residue Ile-11, thereby creating the fusion protein λ-ToxR, P_R-*lacZ* expression in *E. coli* was repressed nearly fourfold more than when the λ-repressor amino terminus alone was used (Dziejman and Mekalanos, 1994). The level of repression increased slightly (to fivefold over λ-amino terminus alone) when ToxS was expressed in the same cell. Western blot analysis indicated that ToxS was able to protect λ-ToxR from degradation in *E. coli,* as it could the ToxR-PhoA-L fusion protein discussed above. A λ-ToxR protein with the repressor amino terminus fused at ToxR residue Val-180 showed ToxS-dependent repression of P_R-*lacZ* that was fourfold over λ-amino terminus alone. This indicates that the linker domain proposed to be important for ToxS-dependent ToxR activity (DiRita and Mekalanos, 1991) is not required by λ-repressor for DNA binding. Finally, when a fusion was tested having the λ-amino terminus followed by ToxR cytoplasmic and transmembrane residues and periplasmic alkaline phosphatase, P_R repression was high and ToxS-independent (Dziejman and Mekalanos, 1994). The simple interpretation of these preliminary data is that the C-terminal, periplasmic domain of ToxR can dimerize the λ-repressor domain of this fusion protein. The result with λ–ToxR–PhoA fusion suggests that cytoplasmic sequences of ToxR are able to transmit a dimerization signal from the periplasmic space. These data also offer another example of how replacing the C terminus of ToxR with alkaline phosphatase relieves a ToxS requirement (DiRita and Mekalanos, 1991).

To test whether transcriptional activation by the N-terminal domain of ToxR requires its association with a dimerization domain, preliminary studies have been performed in which portions of ToxR were fused to the monomeric β-lactamase (Bla) protein and to the leucine zipper from *GCN4*, which, as noted above, has dimerization activity (Ottemann and Mekalanos, 1994). When the periplasmic domain of ToxR was replaced with either one of these proteins, ToxR activity was observed in *E. coli*. This argues against dimerization playing a role in this activity, because the ToxR-Bla fusion is predicted to be a monomer. These fusion proteins could also partially suppress the phenotype of a *toxR⁻* strain of *V. cholerae* in vitro but not in an infant mouse in vivo colonization model (Ottemann and Mekalanos, unpublished data).

A cytoplasmic fusion protein, created by fusing only the amino terminus of ToxR—not including the transmembrane domain—to *GCN4*, had no ToxR activity, either in *E. coli* or in *V. cholerae*. This implies that membrane insertion of ToxR is required for it to be able to activate gene expression. Whether this fusion protein does not activate transcription because it cannot bind to DNA or for some other reason has yet to be determined.

From these data, it appears, somewhat paradoxically, that while the C-terminal domain of ToxR can act as a dimerization domain, the amino-terminal DNA binding domain appears not to require dimerization for its activity. The role of the transmembrane domain is evidently critical, given that the soluble form of ToxR-GCN4 is inactive. Preliminary cross-linking experiments indicate that ToxR monomers can

be cross-linked to one another, but given the absence of evidence that transcriptional activation requires dimerization, it is unclear what this result implies. From all the available data, it appears that the interaction between ToxR and ToxS is stoichiometric rather than enzymatic, but given the mixed results using fusion proteins, purification of ToxR and ToxS may be required to analyze the structure of the active complex more completely.

DNA BINDING BY ToxR AT THE *ctx* PROMOTER

The *ctx* genes from different strains of *V. cholerae* harbor an upstream element, TTTTGAT, that is directly repeated from three to eight times depending on the strain. A *ctx* operon with eight such repeats was about 20 times more active in *V. cholerae* than one with only three (Miller and Mekalanos, 1984), and only *ctx* promoter fragments that included the tandem repeats were specifically recognized by ToxR (Miller et al., 1987). Precisely what ToxR is recognizing within the tandem repeats is not clear. The repeats are actually a component of a hyphenated inverted repeat element upstream of *ctx* (Mekalanos et al., 1983), raising the possibility that a structure containing the repeated heptad actually constitutes the ToxR recognition signal. The only other promoter DNA demonstrated to date to be recognized by ToxR in vitro, that of the *toxT* gene, differs from the *ctx* promoter in primary sequence but does have inverted repeat DNA within it (Higgins and DiRita, 1994; see below).

ANALYZING THE ToxR DNA BINDING DOMAIN BY MUTAGENESIS

Using the similarity between the amino terminus of ToxR and the carboxy terminus of OmpR as a guide, mutational analysis of the DNA binding domain of ToxR was undertaken (Ottemann et al., 1992) (Fig. 2). Different aspects of ToxR function were assessed with the mutants, including the ability to bind and activate the *ctx* promoter in *E. coli* and the ability of the mutants to complement a *toxR* null mutant of *V. cholerae* for toxin and OmpU production.

Three arginine (R) residues, R68, R77, and R96, were identified as critical for DNA binding and transcriptional activation. Replacing these residues with leucine or, in the case of R96, with the chemically similar amino acid lysine resulted in dramatic decrease in transcriptional activation. One of the mutants, ToxR R77L, retained a low level of DNA binding capacity but exhibited a drastic reduction in ToxR activity as measured by the ability to activate a *ctx-lacZ* fusion in *E. coli*. Nevertheless, this mutant was able to complement a *toxR* null mutant of *V. cholerae*, resulting in 60% of wild-type Ctx production and partial restoration of OmpU production.

The ToxR mutant with R96 replaced by leucine (L) had the phenotype of a positive control mutant (i.e., one that can bind DNA but is incapable of activating transcription). This protein had 2% wild-type ToxR activity in *E. coli* but bound *ctx* promoter DNA to the same extent as wild-type ToxR. When introduced into wild-type *V. cholerae*, the negative phenotype of ToxR R96L was not dominant. This observation is difficult to reconcile with the dimer model of ToxR function, unless it is assumed that ToxR-ToxR R96L heterodimers do not form frequently enough in the cell to inhibit steady-state levels of ToxR activity.

One other mutant, with glutamate 51 (E51) substituted by lysine (K), was also notable in this study. This protein showed less than 2% wild-type activity in *E. coli* and, when introduced into wild-type (*toxR*+) *V. cholerae*, led to 10-fold reduction in Ctx levels. The apparent transdominance of this mutant is evidence supporting a dimer structure of active ToxR; if homodimers of ToxR are required to activate transcription, heterodimers between a mutant and wild-type monomer would be expected to have decreased activity. ToxR E51K, alone among all the mutants characterized by Ottemann et al., was almost undetectable on Western blots using proteins prepared from *E. coli*.

Because membranes from *E. coli* expressing ToxR E51K exhibit residual DNA binding activity (Ottemann et al., 1992), the main defect leading to the ToxR⁻ phenotype of this mutant may be in its stability rather than in any intrinsic ToxR function, although it was more stable in *V. cholerae* than in *E. coli*.

EFFECT OF ToxS ON DNA BINDING BY ToxR

Whatever the precise role of ToxS in transcription activation may be, membrane preparations of *E. coli* cells expressing both ToxR and ToxS bound *ctx* promoter DNA at lower protein concentrations than did membrane preparations containing ToxR alone. This observation demonstrates that although ToxR can bind DNA in the absence of ToxS, the presence of ToxS enhances the ToxR-promoter interaction. This alone does not support or refute the dimerization model, because the level at which ToxR requires ToxS for DNA binding is not obvious from this experiment. It may be that having ToxS in the membranes keeps ToxR monomers stable until they find each other or until they find DNA. It may also be that ToxS specifically enhances dimer formation by ToxR. The nature of the complex between active ToxR and ToxS is currently being investigated using cross-linking reagents.

COORDINATE GENE EXPRESSION CONTROLLED BY ToxR AND ToxT

The ToxR-ToxS interaction clearly leads to the ability of ToxR to activate transcription from the *ctx* promoter. However, coordinate expression of *tcp* and *acf* genes with that of cholera toxin is effected through the activation by ToxR of *toxT* expression. ToxT is an AraC-like transcriptional activator protein encoded within the cluster of genes responsible for TCP production (Higgins et al., 1992; Kaufman et al., 1993; Ogierman and Manning, 1992).

Unlike ToxR, ToxT can directly activate expression of several genes in the ToxR regulon, including the *ctx* operon (see Fig. 1). Also, cloned *toxT* under control of the isopropyl-β-

D-thiogalactopyranoside (IPTG)-inducible *tac* promoter can suppress the null phenotype of a *V. cholerae toxR* mutant for *ctx* and *tcp*, but not *ompU*, when the cultures are grown with IPTG. Consistent with this observation is that transcription of *toxT* in the *toxR* mutant background is undetectable by Northern blot. These results suggest a model for coordinate expression in *V. cholerae* in which ToxR activates *toxT* expression and ToxT activates expression of virulence genes (see Fig. 1). They also suggest that *ompU* expression is independent of ToxT, which has been confirmed by analyzing a *toxT* null mutant of *V. cholerae* (DiRita and Higgins, unpublished data).

If this model is accurate, one prediction might be that overexpression of *toxT* from an inducible promoter would be expected to overcome signal requirements of the cell. When *toxT* was expressed in a *toxR* mutant of *V. cholerae* under control of the *tet* promoter of pBR327, cells grown at pH 8.5, a condition normally not permissive for ToxR-regulated gene expression, produced both Ctx and TcpA, the major subunit of TCP (DiRita et al., 1991) (see Fig. 1). So it would appear that, at least for pH, ToxR-dependent gene expression in response to the growth environment may be a function of activation of the *toxT* promoter.

V. cholerae strains are divided into two biotypes, classical and El Tor. Much of the work performed with ToxR-regulated gene expression has been done with classical strains principally because these strains express ToxR-regulated virulence determinants in standard laboratory media under the growth conditions outlined earlier in this chapter. El Tor strains, which carry the *toxRS* operon and also have the *tcp* cluster, including *toxT*, do not express the ToxR regulon under the same conditions. A special medium and growth conditions have been developed specifically to stimulate cholera toxin production by El Tor strains (Iwanga et al., 1986). These include growth in static culture with decreased oxygen tension, as well as at 37°C rather than 30°C, the optimum in vitro temperature for the classical strains. Growth under these conditions has demonstrated that El Tor strains express the ToxR regulon (Jonson et al., 1992; Waldor and Mekalanos, 1994).

As the signals required for ToxR-regulated gene expression by El Tor strains and classical strains are so different, the simplest conclusion is that the signal transduction pathway leading to gene expression is different in the two biotypes. Preliminary evidence supports this, as expression of *toxT* from the *tac* promoter was sufficient to allow expression of *ctx* and *tcpA* by El Tor strains in standard laboratory media supplemented with IPTG (Bruss and DiRita, unpublished data). Along with the observation of virulence gene expression in the classical strain at pH 8.5 when *toxT* was overproduced, we infer that the decision by the cell to express *toxT* is a critical event in the signal transduction pathway leading to virulence expression. The remainder of the work described in this chapter is aimed at understanding how this switch works.

ToxR-DEPENDENT EXPRESSION OF *toxT*

Northern blot experiments alluded to earlier showed that overnight cultures of wild-type *V. cholerae* classical strain O395 contain low levels of detectable *toxT*-specific RNA whereas the *toxR* mutant JJM43 did not. Primer extension analysis of RNA isolated from log-phase cultures of these same strains using a *toxT*-specific primer identified a ToxR-dependent product that maps to a site just upstream of the *toxT* open reading frame. In addition to the ToxR-dependent product, a less abundant ToxR-independent product (i.e., one generated with RNA from both wild-type and *toxR* mutant cells), mapped to a site 100 nucleotides farther upstream (Higgins and DiRita, 1994). Preliminary studies showed that the ToxR-dependent start site of transcription appeared rapidly after shifting cells from ToxR nonpermissive to ToxR permissive conditions and that a transcript from this site was detectable for approximately 2 h after the shift, after which time it was no longer detected. A lower level of expression from this promoter was then again observed once the cells were in stationary phase (Quinn, et al., unpublished data).

When DNA harboring sequences upstream of these start sites was fused to a promoterless *lacZ* on a multicopy plasmid, β-galactosidase activity levels were roughly eightfold higher in wild-type *V. cholerae* than in a *toxR* mutant, indicating the presence of a ToxR-dependent promoter on the DNA. Deletion of the *toxT* upstream DNA narrowed down the location of the region responsible for ToxR-dependent activity to between −73 and −114 nucleotides relative to the start site of transcription. Deletions to −73 resulted in a basal level of β-galactosidase activity that is unaffected by the presence of ToxR, whereas those at −114 retain full ToxR-dependent activation. Gel shift analysis of this region using membranes from *E. coli* cells expressing ToxR and ToxS demonstrated a tight correlation between ToxR-dependent expression from a particular promoter deletion and the ability of it to be specifically bound by ToxR (Higgins and DiRita, 1994).

Hence it appears that ToxR controls its regulon by controlling transcription of the *toxT* gene. Given the strong correlation between ToxR binding and ToxR-dependent promoter activity, the most straightforward molecular mechanism to explain this conclusion is that ToxR binds to and activates the *toxT* promoter. The *toxT* promoter, although it is AT-rich, does not harbor the TTTTGAT tandem repeat element discussed above. The DNA in the region bound by ToxR upstream of *toxT* is predicted by computer analysis to be rich in inverted repeat elements, with three of significant length, but there is no primary sequence in this DNA similar to the ToxR binding element within the *ctx* promoter (Higgins and DiRita, 1994). Mekalanos et al. (1983) pointed out that the TTTTGAT repeat unit in the *ctx* promoter is itself part of an element of hyphenated dyad symmetry, so the "consensus" binding sequence for ToxR may simply consist of inverted repeats that are AT-rich.

Although ToxR is sufficient for activation of *ctx-lacZ* in *E. coli* (Miller and Mekalanos, 1984), it did not activate *toxT-lacZ* in *E. coli,* irrespective of whether the fusion was plasmid or chromosomally located (Higgins and DiRita, 1994). Given that ToxR binds to the *toxT* promoter and is required for *toxT* expression in *V. cholerae* yet is insufficient in *E. coli,* it is possible that ToxR may interact with another factor to control *toxT* expression in *V. cholerae.* This opens up the possibility that this putative factor may confer the signaling aspect to ToxR-dependent *toxT* transcription. For example, perhaps a *V. cholerae* encoded factor interacts with ToxR to control whether ToxR activates *toxT* promoter. A model of this sort allows for amplification or modulation of whatever signal(s) may trigger ToxR-dependent *toxT* transcription in the cell.

CONCLUSIONS

The combination of its sequence similarity with other transcriptional activators and the novelty of its membrane location and interaction with ToxS makes ToxR a worthwhile subject in the study of transmembrane signal transduction. Several intriguing questions remain unanswered in this system. Those of more general appeal relate to how a membrane protein interacts with the transcription machinery. Membrane association for a transcriptional activator is not unique to ToxR since the identification of the aforementioned CadC regulator of pH-dependent gene expression in *E. coli* (Watson et al., 1992), as well as the more recent demonstration that the LuxR activator of luminescence genes in *Vibrio fischeri* also resides in the membrane (Kolibachuk and Greenberg, 1993). As transcription is a soluble process, how do these membrane activators, which presumably work through direct interaction with polymerase, become associated with the process? Gene fusion analysis in *E. coli* has shown that ToxR can interact with its binding site in both the *ctx* promoter and in the *toxT* promoter whether they are on the chromosome or on multicopy plasmids. When and where does

ToxR find these promoters? Related to this is the question of what ToxR recognizes in DNA to which it binds. Generating a consensus sequence for DNA binding by ToxR may be difficult with only two samples. Nevertheless, that these two promoters are so different from each other in primary sequence indicates either that a structure in the DNA is important or that the code for ToxR is multidegenerate, as has been proposed for that of the OxyR regulator of *E. coli* (Tartaglia et al., 1992).

The periplasmic domains of CadC and ToxR are logical places to begin analyzing the signaling capacity of these proteins. (Preliminary evidence concerning the domain structure of LuxR suggests that it does not have a large periplasmic domain, which makes it a slight variation on the theme.) Interaction between ToxR and its effector protein, ToxS, is evidently a periplasmic event. Unlike other signaling systems reviewed in this volume, the ToxR-ToxS interaction is probably stoichiometric rather than enzymatic. Does the presence or absence of specific environmental signals affect this interaction? ToxR protein fusion experiments reviewed above suggest that to activate gene expression, ToxR must be in the membrane. Is that due to some role of the periplasmic domain in DNA binding that requires a noncytoplasmic environment?

Questions of more specific interest to the matter of virulence regulation by ToxR also remain. Given that motility and pilus-dependent colonization, processes that appear mutually exclusive, are both important for virulence by *V. cholerae,* there must be some regulatory link between them. Biphasic regulation of complex processes controlled by a single regulatory locus is observed in *Bordetella* spp. and in *Salmonella* pathogenesis (Akerley et al., 1992; Behlau and Miller, 1993). Pertinent to this, at least one gene product, the major outer membrane protein OmpT, is negatively controlled by ToxR (Miller and Mekalanos, 1988). How does the swimming organism sense that it has reached a surface to be colonized? What are the molecular events that differentiate swimming

and colonization? One of the *tcp* gene products, TcpI, shares similarity with methyl-accepting chemotaxis proteins of *E. coli* and *S. typhimurium* (Peterson, personal communication), and this may shed some light on this question. ToxR-dependent gene expression requires activation of *toxT* expression. What is the purpose of having one activator control expression of another? What signals does ToxR recognize in vivo to initiate the virulence cascade? Once expressed, does ToxT have sensing requirements that must be met before its activation of gene expression? Finally, one of the more remarkable features of regulation of virulence in *V. cholerae* as analyzed to date is that a more standard signal kinase-response regulator pair of gene products has not been associated with any aspect of pathogenicity. It seems likely that as we continue to tie together the different phenotypes that are related to virulence, genes of this type will eventually be identified.

The output for ToxR activity in *V. cholerae* is virulence, but what is clear from investigations into this system so far is that along the way toward answering specific questions of virulence we are learning more about fundamental processes in molecular biology as well.

ACKNOWLEDGMENTS

I thank Karen Ottemann, Michelle Dziejman, John Mekalanos, and Ken Peterson for communicating data before publication and Paula Bruss for comments on the manuscript. I also thank Ron Taylor and Ken Peterson for thought-provoking discussions, Steve Winans for pointing out the OxyR work of B. N. Ames and colleagues as a way to think about ToxR DNA recognition, and Bobolly Brennan for assistance with the figures.

REFERENCES

Akerley, B. J., D. M. Monack, S. Falkow, and J. F. Miller. 1992. The *bvgAS* locus negatively controls motility and synthesis of flagella in *Bordetella bronchiseptica. J. Bacteriol.* **174:**980–990.

Ames, P., and J. S. Parkinson. 1988. Transmembrane signaling by bacterial chemoreceptors: *E. coli* transducers with locked signal output. *Cell* **55:**817–826.

Behlau, I., and S. I. Miller. 1993. A PhoP-repressed gene promotes *Salmonella typhimurium* invasion of epithelial cells. *J. Bacteriol.* **175:**4475–4484.

Bruss, P. M., and V. J. DiRita. Unpublished data.

DiRita, V. J., and D. E. Higgins. Unpublished data.

DiRita, V. J., and J. J. Mekalanos. 1991. Periplasmic interaction between two membrane regulatory proteins, ToxR and ToxS, results in signal transduction and transcriptional activation. *Cell* **64:**29–37.

DiRita, V. J., C. Parsot, G. Jander, and J. J. Mekalanos. 1991. Regulatory cascade controls virulence in *Vibrio cholerae*. *Proc. Natl. Acad. Sci. USA* **88:**5403–5407.

Dziejman, M., and J. J. Mekalanos. 1994. Analysis of membrane protein interactions: ToxR can dimerize the amino terminus of phage lambda repressor. *Mol. Microbiol.* **13:**485–494

Finkelstein, R. A. 1973. Cholera. *CRC Crit. Rev. Microbiol.* **2:**553–623.

Freter, R., P. C. M. O'Brien, and M. S. Macsai. 1981. Role of chemotaxis in the association of motile bacteria with intestinal mucosa: in vivo studies. *Infect. Immun.* **34:**234–240.

Gill, D. M., and R. Meren. 1978. ADP-ribosylation of membrane proteins catalyzed by cholera toxin: basis of activation of adenylate cyclase. *Proc. Natl. Acad. Sci. USA* **75:**3050–3054.

Herrington, D. A., R. H. Hall, G. Losonsky, J. J. Mekalanos, R. K. Taylor, and M. M. Levine. 1988. Toxin, toxin-coregulated pili, and the *toxR* regulon are essential for *Vibrio cholerae* pathogenesis in humans. *J. Exp. Med.* **168:**1487–1492.

Higgins, D. E., and V. J. DiRita. 1994. Transcriptional control of *toxT,* regulatory gene in the ToxR regulon of *Vibrio cholerae. Mol. Microbiol.* **14:**17–29.

Higgins, D. E., E. Nazareno, and V. J. DiRita. 1992. The virulence gene activator ToxT from *Vibrio cholerae* is a member of the AraC family of transcriptional activators. *J. Bacteriol.* **174:**6974–6980.

Hobbs, M., and J. S. Mattick. 1993. Common components in the assembly of type 4 fimbriae, DNA transfer systems, filamentous phage and protein-secretion apparatus: a general system for the formation of surface-associated protein complexes. *Mol. Microbiol.* **10:**233–243.

Hu, J. C., E. K. O'Shea, P. S. Kim, and R. T. Sauer. 1990. Sequence requirements for coiled-coils: analysis of λ repressor-*GCN4* leucine zipper fusions. *Science.* **250:**1400–1403.

Iwanga, M., K. Yamamoto, N. Higa, Y. Ichinose, N. Nakasone, and M. Tanabe. 1986. Culture conditions for stimulating cholera toxin production by *Vibrio cholerae* O1 El Tor. *Microbiol. Immunol.* **30:**1075–1083.

Jonson, G., J. Holmgren, and A.-M. Svennerholm. 1992. Analysis of expression of toxin-coregulated pili in classical and El Tor *Vibrio cholerae*

O1 in vitro and in vivo. *Infect. Immun.* **60:**4278–4284.

Kaufman, M. R., J. M. Seyer, and R. K. Taylor. 1991. Processing of TCP pilin by TcpJ typifies a common step intrinsic to a newly recognized pathway of extracellular protein secretion by gram-negative bacteria. *Genes Dev.* **5:**1834–1846.

Kaufman, M. R., C. E. Shaw, I. D. Jones, and R. K. Taylor. 1993. Biogenesis and regulation of the *Vibrio cholerae* toxin-coregulated pilus: analogies to other virulence factor secretory systems and localization of the *toxT* virulence regulator to the TCP cluster. *Gene* **126:**43–49.

Knapp, S., and J. J. Mekalanos. 1988. Two trans-acting regulatory genes (*vir* and *mod*) control antigenic modulation in *Bordetella pertussis. J. Bacteriol.* **170:**5059–5066.

Kolibachuk, D., and E. P. Greenberg. 1993. The *Vibrio fischeri* luminescence gene activator LuxR is a membrane-associated protein. *J. Bacteriol.* **175:**7307–7312.

Litwin, C. M., and S. B. Calderwood. 1993. Role of iron in regulation of virulence genes. *Clin. Microbiol. Rev.* **6:**137–149.

Manoil, C., and J. Beckwith. 1986. A genetic approach to analyzing membrane protein topology. *Science* **233:**1403–1408.

Mekalanos, J. J., D. J. Swartz, G. D. N. Pearson, N. Harford, F. Groyne, and M. de Wilde. 1983. Cholera toxin genes: nucleotide sequence, deletion analysis and vaccine development. *Nature* (London) **306:**551–557.

Miller, J. F., S. A. Johnson, W. J. Black, D. T. Beattie, J. J. Mekalanos, and S. Falkow. 1992. Constitutive sensory transduction mutations in the *Bordetella pertussis bvgS* gene. *J. Bacteriol.* **174:**970–979.

Miller, V. L., V. J. DiRita, and J. J. Mekalanos. 1989. Identification of *toxS,* a regulatory gene whose product enhances ToxR-mediated activation of the cholera toxin promoter. *J. Bacteriol.* **171:**1288–1293.

Miller, V. L., and J. J. Mekalanos. 1984. Synthesis of cholera toxin is positively regulated at the transcriptional level by *toxR. Proc. Natl. Acad. Sci. USA* **81:**3471–3475.

Miller, V. L., and J. J. Mekalanos. 1988. A novel suicide vector and its use in construction of insertion mutations: osmoregulation of outer membrane proteins and virulence determinants in *Vibrio cholerae* requires *toxR. J. Bacteriol.* **170:**2575–2583.

Miller, V. L., R. K. Taylor, and J. J. Mekalanos. 1987. Cholera toxin transcriptional activator ToxR is a transmembrane DNA binding protein. *Cell* **48:**271–279.

Mims, C. A., J. H. L. Playfair, I. M. Roitt, D. Wakelin, R. Williams, and R. M. Anderson.

1993. *Medical Microbiology.* Mosby Europe Limited, London.

Ogierman, M. A., and P. A. Manning. 1992. Homology of TcpN, a putative regulatory protein of *Vibrio cholerae,* to the AraC family of transcriptional activators. *Gene* **116:**93–97.

Ogierman, M. A., S. Zabihi, L. Mourtzios, and P. A. Manning. 1993. Genetic organization and sequence of the promoter-distal region of the *tcp* gene cluster of *Vibrio cholerae. Gene* **126:**51–60.

Ottemann, K. M., V. J. DiRita, and J. J. Mekalanos. 1992. ToxR proteins with substitutions in residues conserved with OmpR fail to activate transcription from the cholera toxin promoter. *J. Bacteriol.* **174:**6807–6814.

Ottemann, K. M., and J. J. Mekalanos. 1994. Regulation of cholera toxin expression, p. 177–185. *In* I. K. Wachsmuth, P. A. Blake, and Ø. Olsvik (ed.), *Vibrio cholerae and Cholera.* American Society for Microbiology, Washington, D.C.

Ottemann, K. M., and J. J. Mekalanos. Unpublished data.

Parsot, C., and J. J. Mekalanos. 1990. Expression of ToxR, the transcriptional activator of the virulence factors in *Vibrio cholerae,* is modulated by the heat shock response. *Proc. Natl. Acad. Sci. USA* **87:**9898–9902.

Peek, J. A., and R. K. Taylor. 1992. Characterization of a periplasmic thiol:disulfide interchange protein required for the functional maturation of secreted virulence factors of *Vibrio cholerae. Proc. Natl. Acad. Sci. USA* **89:**6210–6214.

Peterson, K. M. Personal communication.

Peterson, K. M., and J. J. Mekalanos. 1988. Characterization of the *Vibrio cholerae* ToxR regulon: identification of novel genes involved in intestinal colonization. *Infect. Immun.* **56:**2822–2829.

Quinn, R. H., D. E. Higgins, and V. J. DiRita. Unpublished data.

Richardson, K. 1991. Roles of motility and flagellar structure in pathogenicity of *Vibrio cholerae:* analysis of motility mutants in three animal models. *Infect. Immun.* **59:**2727–2736.

Richardson, S. H. 1969. Factors influencing in vivo skin permeability factor production by *Vibrio cholerae. J. Bacteriol.* **100:**27–34.

Schlesinger, M. J. 1967. Formation of a defective alkaline phosphatase subunit by a mutant of *Escherichia coli. J. Biol. Chem.* **242:**1604–1611.

Schneider, D. R., and C. D. Parker. 1978. Isolation and characterization of protease-deficient mutants of *Vibrio cholerae. J. Infect. Dis.* **138:**143–151.

Taha, M. K., B. Dupuy, W. Saurin, M. So, and C. Marchal. 1991. Control of pilus expression in *Neisseria gonorrhoeae* as an original system in the family of two-component regulators. *Mol. Microbiol.* **5:**137–148.

Tartaglia, L. A., C. J. Gimeno, G. Storz, and B. N. Ames. 1992. Multidegenerate DNA recognition by the OxyR transcriptional regulator. *J. Biol. Chem.* **267:**2038–2045.

Taylor, R. K., V. L. Miller, D. B. Furlong, and J. J. Mekalanos. 1987. Use of *phoA* gene fusions to identify a pilus colonization factor coordinately regulated with cholera toxin. *Proc. Natl. Acad. Sci. USA* **84:**2833–2837.

Waldor, M. K., and J. J. Mekalanos. 1994. ToxR regulates virulence gene expression in Non-O1 strains of *Vibrio cholerae* that cause epidemic cholera. *Infect. Immun.* **62:**72–78.

Watson, N., D. S. Dunyak, E. L. Rosey, J. L. Slonczewski, and E. R. Olson. 1992. Identification of elements involved in transcriptional regulation of the *Escherichia coli cad* operon by external pH. *J. Bacteriol.* **174:**530–540.

Ti Plasmid and Chromosomally Encoded Two-Component Systems Important in Plant Cell Transformation by *Agrobacterium* Species

Joe Don Heath, Trevor C. Charles, and Eugene W. Nester

23

Members of the genus *Agrobacterium* are common components of soil and rhizosphere populations. The best-characterized members of this genus are the virulent members, *Agrobacterium tumefaciens, A. vitis,* and *A. rhizogenes.* The first two cause the plant hypertrophic disease crown gall, and the latter causes hairy root. Plasmid-encoded virulence traits have historically had undue influence on the classification within this genus, especially considering that these plasmids are readily transferable between strains. A recent study of chromosomal 16S rRNA gene sequences suggested that this classification should be reevaluated (Sawada et al., 1993). This sequence analysis revealed that *A. tumefaciens* and the avirulent strains previously grouped within the species *A. radiobacter* are indeed members of a single species. Thus, the well-known tumorigenic strains previously known as *A. tumefaciens* can now be considered to be *A. radiobacter* strains that possess a tumor-inducing (Ti) plasmid.

OVERALL FEATURES OF *AGROBACTERIUM*-MEDIATED PLANT CELL TRANSFORMATION

Tumorigenic *Agrobacterium* strains incite the formation of crown gall tumors at wound sites on a wide variety of dicotyledonous plants as well as some monocotyledonous species. (For recent reviews, see Kado, 1991; Winans, 1992; Ankenbauer and Nester, 1993; Winans et al., 1994.) Tumor formation results after the transfer of a piece of DNA (T-DNA) of a large Ti plasmid into plant cells where the T-DNA becomes integrated into the plant chromosome through nonhomologous recombination and is subsequently expressed. The T-DNA, approximately 20 kb, codes for enzymes for the biosynthesis of auxin and cytokinin and also for control of endogenous levels of these plant growth regulators. Other genes on the T-DNA code for enzymes involved in the synthesis of a class of compounds termed opines, many of which are derivatives of basic amino acids. These opines, which include octopine and nopaline, serve as a source of carbon, nitrogen, and energy for the inducing strain of *Agrobacterium.*

The second region of the Ti plasmid essential for tumor formation is the virulence (*vir*) region. This 35-kb region of DNA, which is

Joe Don Heath and Eugene W. Nester, Department of Microbiology, University of Washington, Seattle, Washington 98195. *Trevor C. Charles,* Department of Natural Resource Sciences, McGill University, Macdonald Campus, Ste. Anne de Bellevue, Quebec, Canada H9X 3V9.

Two-Component Signal Transduction, Edited by James A. Hoch and Thomas J. Silhavy,
© 1995 American Society for Microbiology, Washington, DC 20005

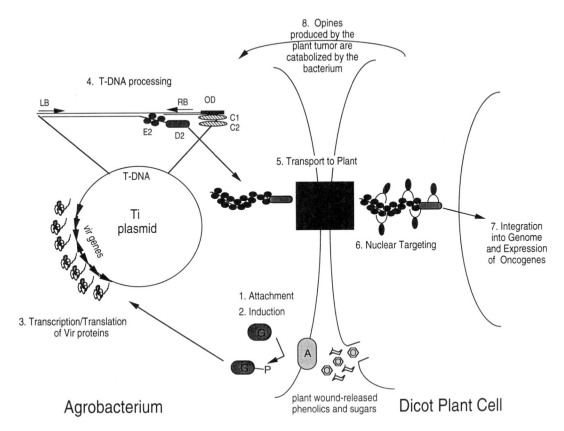

FIGURE 1 Overview of crown gall tumorigenesis. Schematic representation of the steps involved in the interaction of *Agrobacterium* with its plant host. (1) In response to plant wound-released sugars and a variety of other substituents of plant wound exudate, *Agrobacterium* moves toward the wound site and attaches to a plant cell. (2) In the acidic environment of the wound, *Agrobacterium* induces expression of its *vir* genes via the VirA/VirG regulatory system. (3) Transcription and translation of the *vir* genes lead to (4) T-DNA processing and (5) T-DNA transfer to the plant cell. (6) The T-DNA and bound *vir* gene products are targeted to the plant nucleus where (7) the T-DNA is integrated into the genome and the T-DNA oncogenes are expressed. The expression of the T-DNA encoded oncogenes leads to axenic tumor proliferation and (8) the production of opines that are used as carbon and nitrogen sources by the infecting *Agrobacterium*.

not transferred to the plant cell, codes for proteins that are required for the sensing of plant wound metabolites as well as the processing, transfer, nuclear targeting, and perhaps integration of the T-DNA into the plant chromosome. Four *vir* operons (*virA, virB, virD,* and *virG*) have been identified as being essential, and two others (*virC* and *virE*) greatly stimulate tumor formation. *virF* and *virH* (*pinF*) are found in some but not all strains of *Agrobacterium* and are not essential for tumor formation under labo-

ratory conditions. Sequencing of the *vir* region has revealed additional open reading frames that may or may not play a role in tumor formation.

The *vir* operons constitute a regulon that is induced in cells growing under acidic conditions by two classes of plant signal molecules: phenolic compounds and specific monosaccharides. The plant signals are recognized and transduced by the products of two *vir* genes, *virA* and *virG*. These two genes are members of the highly conserved class of two-component sensory transduction systems, *virA* coding for the sensor protein and *virG* for the response regulator.

In addition to the Ti plasmid-encoded virulence genes, several chromosomal loci are important for tumor formation. These include loci such as *chvA/B* and *att*, which are required for attachment of *Agrobacterium* to plant cells (Douglas et al., 1982; Douglas et al., 1985; Puvanesarajah et al., 1985; Matthysse, 1987; Zorreguieta et al., 1988); *chvE*, coding for a periplasmic galactose binding protein to which monosaccharides bind before interacting with the VirA protein (Cangelosi et al., 1990; Huang et al., 1990); the *ivr* loci, mutations in which result in avirulence (Metts et al., 1991); and another two-component regulatory system, *chvG* and *chvI* (Charles and Nester, 1993; Mantis and Winans, 1993). Thus, genes on both the chromosome and the Ti plasmid are required for tumorigenesis, and two-component regulatory systems that are involved in virulence are located on each replicon. However, it seems likely that most, if not all, of the genes on the chromosome that have been implicated in virulence also play a role in the general physiology of *Agrobacterium*. Thus, in most cases, their role in tumor formation may be indirect, whereas the Ti plasmid-encoded virulence genes generally play a direct role.

The salient features of crown gall tumor formation are diagrammed in Fig. 1.

TWO-COMPONENT REGULATORY SYSTEM ON THE Ti PLASMID: VirA/VirG

VirA, the Sensor Component

The key molecule that allows *Agrobacterium* to sense environmental conditions favorable for T-DNA transfer is the VirA protein. This protein is induced up to eightfold by *vir* gene inducers (Rogowsky et al., 1987; Winans et al., 1988; Turk et al., 1993a). The VirA protein of octopine-utilizing strains is 829 amino acids long with a molecular mass of 91 kDa (Leroux et al., 1987). VirA is anchored in the cytoplas-

mic membrane by two hydrophobic transmembrane regions, such that the N and C termini are in the cytoplasm along with the C-terminal portion of the protein. Most of the N-terminal region is in the periplasm. This topology has been confirmed by Tn*phoA* fusion analysis as well as by protease sensitivity analysis of the periplasmic domain of the protein following digestion of the cell wall (Melchers et al., 1989a; Winans et al., 1989). Recently, physical and genetic evidence has indicated that VirA exists as a homodimer in its native configuration, and this homodimer state is required for its functioning (Howitz and Binns, personal communication; Pan et al., 1993).

According to the nomenclature of Parkinson and Kofoid (1992), VirA can be conveniently divided into three domains or modules, each of which has a distinct function (Fig. 2). These are the input domain, the transmitter domain, which contains the phosphorylatable histidine, and the receiver domain, so termed because it bears some similarity to the region of the response regulator (VirG), which contains the phosphorylatable aspartate residue. Chang and Winans (1992) identified four domains of VirA that closely parallel the functions of the above three domains. These are the periplasmic and the linker domains (input), the kinase domain (transmitter), and the receiver domain (receiver). Chang and Winans (1992) constructed a series of in-frame deletions encompassing one or more of the domains. They assigned functional roles to each domain by studying the ability of the mutant proteins to respond to three different signal molecules: (i) the phenolic inducer, acetosyringone (AS), (ii) glucose, a monosaccharide that potentiates the induction by AS, and (iii) acid pH. The periplasmic domain is required for sensing monosaccharides but not phenolic compounds or acidic conditions. The linker domain is required for sensing both phenolic compounds and acidity.

The kinase domain (transmitter) contains the conserved phosphorylatable histidine (His-474), which is required for signal transduction

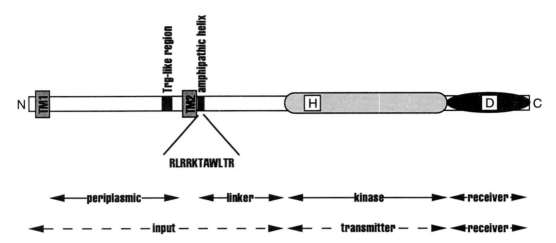

FIGURE 2 Functional domains of the sensor molecule VirA. The VirA protein can be divided into three major functional domains with a body plan of ITR, according to the nomenclature of Parkinson and Kofoid (1992) (the input domain, the transmitter domain, and the receiver domain). The input domain is further divided into the periplasmic domain, harboring the ChvE-responsive region, and the linker domain, which senses phenolic compounds and acidity. The transmitter (or kinase) domain contains the conserved histidine that is phosphorylated. The receiver domain is similar in sequence to the VirG protein, contains a conserved aspartate residue, and may play a regulatory role.

in all sensor molecules. Jin et al. (1990b) demonstrated that changing His-474 to glutamine (H474Q) results in a protein that is no longer phosphorylated in vivo, suggesting that, as in other two-component systems, this conserved histidine is the site of phosphorylation. Further, the H474Q mutant is avirulent and is unable to induce *vir* gene expression in the presence of plant signal molecules.

The function of the receiver domain is uncertain. It may play an inhibitory role in signal transduction because, when this domain was deleted, monosaccharides alone induced *vir* gene expression in the absence of phenolic compounds (Chang and Winans, 1992). However, a very similar deletion of the receiver domain of VirA from *A. rhizogenes* resulted in a protein that was nonfunctional in vivo (Endoh and Oka, 1993). The reason for this discrepancy is not clear but may be due to a slight variation in the region truncated. The receiver domain is somewhat similar to the region of VirG that is phosphorylated by VirA. The aspartate residue in VirG that is phosphorylated by VirA is conserved in the receiver domain of VirA. When

this aspartate in VirA was changed to asparagine, the inducing activity of VirA was severely reduced (Chang and Winans, 1992). Further, Melchers et al. (1989a) reported that small truncations of the extreme C-terminal region increased the inhibitory function of the domain. Consistent with these data, Turk (1993) observed that a deletion of the receiver domain reduced the autophosphorylating activity of VirA and phosphate transfer to VirG in vitro. However, the in vitro reactions involving phosphorylation of VirA and VirG do not require or respond to plant signal molecules (Jin et al., 1990b; Jin and Nester, unpublished data). Therefore, it is difficult to compare in vitro and in vivo data.

Very recently, a highly amphipathic helix sequence of 11 amino acids was identified that is located immediately C-terminal to the second transmembrane domain of VirA (Fig. 2). This motif (amino acids 278 to 288), which is highly conserved in most chemoreceptor proteins (Seligman and Manoil, 1994), is important for the insertion of the *Escherichia coli* chemoreceptor protein Tsr into the membrane. Mutations that greatly reduce the amphipathicity of the sequence in Tsr cause improper translocation of the cytoplasmic domain of Tsr to the periplasm. This sequence in VirA, as well as TM1, may serve to promote the normal membrane insertion of the VirA protein or play a role in dimer formation. Whether this region of the linker domain plays a role in the sensing of AS also remains an intriguing possibility.

SENSING OF PHENOLIC COMPOUNDS

The VirA/VirG system of *Agrobacterium* is one of the few two-component systems in which the signal compounds are known. The first natural compounds identified as possessing *vir*-inducing activity were the phenolic compounds, AS and hydroxy-acetosyringone, both isolated from tobacco (Stachel et al., 1985; Stachel et al., 1986). These compounds are derivatives of precursors of the plant cell wall component lignin and appear in the plant wound exudate soon after wounding at levels sufficient to induce the *vir* genes of *Agrobacterium* in a VirA/VirG-dependent fashion. The compounds themselves serve as inducers because AS does not appear to be metabolized by *Agrobacterium* in studies using either unlabeled AS or [³H]AS (Culianez-Macia and Hepburn, 1988; Lee et al., unpublished data a). The compounds are synthesized de novo as their appearance following wounding is inhibited by cycloheximide (Stachel et al., 1985). Other natural compounds with structures similar to AS that have *vir*-inducing activity include ethyl ferulate, ferulic acid, and feruloyl–substituted polysaccharide fragments, all isolated from wheat (Usami et al., 1988; Machida et al., 1989; Messens et al., 1990). Methyl syringate isolated from grape vines exhibits strong *vir*-inducing activity (Spencer et al., 1990). Compounds from extracts of petunia have *vir*-inducing activity, but their structures are strikingly different from that of AS (Zerback et al., 1989). However, all these compounds have a phenolic structure that is crucial for activity. Tobacco has been reported to produce dehydroconiferyl

acid, a potent *vir* inducer (Hess et al., 1991). Also, Douglas fir produces a compound, coniferin, that is likely converted to the *vir* inducer coniferyl alcohol by the action of an *Agrobacterium*-produced β-glucosidase (Morris and Morris, 1990). Only those *Agrobacterium* strains that are capable of producing tumors on Douglas fir have been shown to have high levels of β-glucosidase activity. The β-glucosidase gene has been cloned (Castle et al., 1992), but it has not been reported whether mutations in this gene lead to the predicted avirulent phenotype.

In several studies, the inducing activities of many phenolic compounds have been investigated (Stachel et al., 1985; Bolton et al., 1986; Spencer and Towers, 1988; Melchers et al., 1989b; Hess et al., 1991). The results of these studies suggest that the *vir*-inducing ability of a given compound is determined by its hydrophobicity and the pK$_a$ of the phenolic hydroxyl group. The more hydrophobic the compound and the lower the pK$_a$ of the phenolic hydroxyl group, the greater the inducing activity. A phenolic hydroxyl group with a low pK$_a$ may be required if the transfer of the phenolic proton to the phenolic binding site (either VirA or a phenolic binding protein) is a necessary component in signal transduction (Hess et al., 1991; Duban et al., 1993).

The requirement for the extreme hydrophobicity of an inducing compound supports a mechanism for plant-bacterial signaling that has been suggested in the *nod* system of *Rhizobium leguminosarum*. Naringenin is a hydrophobic flavonoid that induces expression of the *nod* genes. It has been shown that naringenin is rapidly partitioned to the cytoplasmic membrane where it may be detected directly by NodD (Recourt et al., 1989), a transmembrane sensor protein of the LysR family. The rapid membrane partitioning of naringenin is due to its highly hydrophobic nature. Likewise, *vir*-inducing compounds are probably partitioned to the cytoplasmic membrane, where they are detected by VirA, either directly or indirectly. Therefore, the more hydrophobic compounds will have a higher membrane partitioning coefficient and be more readily detected by the appropriate receptor. In support of this model, AS has been shown to be rapidly partitioned to the membrane of *Agrobacterium* (Lee et al., unpublished data c).

One of the many questions that remains unresolved in this signal transduction system is whether AS is sensed directly by VirA or is detected by another receptor that then interacts with VirA. In the *nod* system of *Rhizobium*, genetic evidence indicates that NodD senses inducing flavonoids directly (Györgypal et al., 1991). By analogy, it seems reasonable to expect that VirA would also sense phenolic compounds directly. However, data have been presented that suggest a role for one or both of a pair of chromosomally encoded cytoplasmic proteins in sensing AS (Lee et al., 1992). A 10-kDa and a 21-kDa protein were reported to be specifically labeled in *Agrobacterium* by an affinity label, ^{125}I-α-bromoacetosyringone. The label did not detect VirA. α-Bromoacetosyringone has been reported to be a specific inhibitor of the VirA/VirG signal transduction system (Hess et al., 1991) and, thus, should label the protein(s) responsible for sensing AS. However, another group has thus far been unable to demonstrate that α-bromoacetosyringone specifically inhibits *vir* induction (Lee et al., unpublished data b). As yet, no mutants lacking the putative AS binding protein(s) have been reported, so the biological significance of the 10-kDa and 21-kDa proteins is unclear. Thus, the question of what protein senses phenolic compounds remains unresolved.

MONOSACCHARIDES IN *vir* GENE INDUCTION

A variety of monosaccharides plays a very important role in potentiating the effect of the phenolic compounds in *vir* gene induction (Cangelosi et al., 1990; Shimoda et al., 1990). These include arabinose, galactose, galacturonic acid, glucose, glucuronic acid, mannose, fucose, cellobiose, and xylose (Ankenbauer and Nester, 1990). Most of the monosaccharides that po-

tentiate induction are components of polymers in typical cell walls of dicotyledonous plants. Galacturonic acid oligosaccharides (2 to 10 residues) were inactive in *vir* regulon induction, suggesting that sugar-mediated induction is specific for monosaccharides (Ankenbauer et al., unpublished data). The monosaccharides act synergistically with the phenolic compounds to induce *vir* genes. At low levels of AS, these sugars are absolutely essential for *vir* gene induction, whereas at higher levels of AS, their presence is not absolutely required. However, even at the optimum level of AS, monosaccharides increase the level of induction 5- to 10-fold (Cangelosi et al., 1990). Degradation of the plant cell wall components into monosaccharide subunits apparently occurs in response to wounding and the subsequent activation of plant glycosidases. Also, *Agrobacterium* cells contain a gene, *pgl*, that encodes a predicted protein with homology to known polygalacturonases that may play a role in degrading plant cell wall components. The gene is induced by a complex pectic polysaccharide from carrot root extracts (Rong et al., 1994). It is also conceivable that other bacteria in the wound environment could degrade plant cell walls.

None of the monosaccharides interacts directly with the VirA protein. Rather, a galactose binding protein, the product of the *chvE* gene, interacts with one of the various inducing monosaccharides and then the ChvE:: monosaccharide complex interacts with the periplasmic region of VirA. As with AS sensing, no *vir* induction in response to monosaccharides is observed in *virA* mutants, indicating that monosaccharide sensing occurs via the VirA/VirG pathway. *chvE* mutants were first identified as mutants that exhibited a very limited host range (Garfinkel and Nester, 1980; Huang et al., 1990). The *chvE* locus was mapped to the *Agrobacterium* chromosome (Garfinkel and Nester, 1980; Huang et al., 1990). The defect in *vir* gene induction observed in *chvE* mutants is presumably responsible for their limited host range.

The *chvE* gene product is very similar to two *E. coli* periplasmic proteins; a galactose binding protein (GBP) and a ribose binding protein. Each protein binds its specific monosaccharide ligand and then interacts with specific membrane-bound uptake and chemosensory complexes, resulting in the transport of the ligand and chemotaxis (Furlong, 1987; Macnab, 1987). A GBP with monosaccharide binding ability and transport activity has been isolated from an avirulent strain of *A. radiobacter* and is virtually identical to ChvE (Cornish et al., 1989).

Evidence indicates that ChvE functions on at least three levels: (i) to mediate monosaccharide uptake via a high-affinity monosaccharide transport system, (ii) to mediate chemotaxis toward sugars by interacting with *Agrobacterium* chemosensory proteins, and (iii) to promote *vir* regulon induction in a synergistic fashion with phenolic inducers by interacting with the periplasmic domain of VirA (Cornish et al., 1989; Cangelosi et al., 1990). As predicted, mutants with lesions in *chvE* are defective in all three functions. The similarities between the *E. coli* sugar/GBP/Trg and *Agrobacterium* sugar/ChvE/VirA signal transduction systems are striking. Both Trg and VirA have a conserved 15-amino-acid domain that is exposed to the periplasm (Cangelosi et al., 1990). Mutation of this sequence in Trg abolished chemotaxis to galactose and ribose (Park and Hazelbauer, 1986), and deletion of the sequence from VirA eliminated the ability of VirA to respond to inducing monosaccharides (Wu and Nester, unpublished data).

The direct interaction of ChvE with VirA has recently been demonstrated by a genetic analysis (Shimoda et al., 1993). A point mutant was isolated with an amino acid substitution in the periplasmic domain of VirA, which eliminated the monosaccharide-enhancing effect on *vir* gene induction. Two suppressor mutations in *chvE* were isolated in the *virA* mutant background, both of which exhibited restored monosaccharide-enhancing activity while one mutation restored virulence.

REGULATION OF *chvE* EXPRESSION

Upstream of *chvE* but divergent in transcription is a locus termed *gbpR,* the galactose binding protein regulator. Insertions in this locus have no effect on either virulence, uptake of glucose, or any other inducing sugar or chemotaxis toward the sugar (Doty et al., 1993). However, sequence analysis revealed that the gene product is a member of the LysR family of transcriptional regulators. *chvE::lacZ* expression was induced by three of the inducing sugars, arabinose, galactose, and fucose, when GbpR was present. In the absence of the inducing sugars, GbpR repressed *chvE::lacZ* expression (Doty et al., 1993). Thus, GbpR can function as both an activator and a repressor, depending on whether certain sugars are present in the environment. However, the levels of activation and repression by GbpR are apparently not great enough to effect a measurable change in cell physiology or tumor-forming ability when GbpR is absent.

SENSING OF ACIDITY

Early experiments that showed that the expression of *vir* genes was induced by plant cell metabolites were carried out in acidic plant cell culture medium. This observation led to the discovery that induction only occurred when the pH of the medium was less than 6.0 (Stachel et al., 1985). Other investigations confirmed the acidic pH optimum for *vir* regulon induction (Bolton et al., 1986; Stachel and Zambryski, 1986; Rogowsky et al., 1987; Culianez-Macia and Hepburn, 1988; Vernade et al., 1988; Turk et al., 1991). These studies reinforced the conclusion that the highest levels of induction occur at an acidic pH that still allows cell growth (4.8 to 5.5) and that induction does not occur at pH 7.0. The acidic requirement for expression is not surprising, considering the acidic environment of a plant wound site (Kahl, 1982; Grignon and Sentenac, 1991).

Melchers et al. (1989a) demonstrated that VirA-Tar hybrids in which the periplasmic domain of VirA was replaced by that of Tar, a transmembrane receptor protein, yielded high levels of induction at pH 6.8, in contrast to the lack of significant induction observed with wild-type VirA at pH values greater than 6.5. They suggested that the periplasmic domain of VirA must be able to sense acidic conditions to signal *vir* regulon induction. However, other investigators showed that *virA* deletion mutants that lack residues N-terminal to and including TM2 are still able to induce the *vir* regulon in response to low pH (Chang and Winans, 1992). These investigators localized the acid-responsive region of VirA to the linker domain between TM2 and the histidine protein kinase domain. As mentioned earlier, the linker domain is also responsible for detecting AS. It has been proposed that acidic pH is involved in the interaction of AS with VirA or a receptor molecule (Hess et al., 1991). Since the pK_a of the phenolic hydroxyl of AS is 7.4, at neutral pH, a large percentage of the AS molecules would be ionized. If only protonated phenolics are active inducers, as is hypothesized, this may account for some but certainly not all the pH effect on inducing activity.

The acid-sensing ability of *Agrobacterium* for *vir* gene induction involves *virG* as well as *virA* (see below).

SENSING OF TEMPERATURE

The effect of temperature on the formation of crown gall tumors has been known for almost 70 years. An early study of the influence of environmental factors in crown gall disease showed that gall development was substantially inhibited at 28°C and abolished at temperatures greater than 31°C (Riker, 1926). Subsequent studies confirmed that *Agrobacterium* was unable to form tumors at temperatures greater than 30°C (Braun, 1947; Rogler, 1980). These studies also determined that the thermosensitive step in tumor formation is the infective interaction between *Agrobacterium* and the plant host.

By assaying for the product of the *virD2* gene, Alt-Moerbe et al. (1988) showed that the optimal temperature for induction of *virD2* expression is 20° to 25°C. At temperatures of 30°C or greater, induction of *virD2* was not observed. In another study, a *virB::lacZ* fusion was expressed optimally at 29°C and was not expressed at temperatures greater than 30°C (Melchers et al., 1989a).

Further studies demonstrated that the autophosphorylation of VirA and the subsequent transfer of phosphate to VirG are both sensitive to elevated temperatures (Jin et al., 1993a). At temperatures greater than 32°C, the autophosphorylation activity of the VirA molecule is reversibly inactivated, whereas the VirG protein does not appear to be affected. However, this temperature effect on tumor formation cannot be overcome by infecting plants at 32°C with a strain of *Agrobacterium* harboring a VirA-independent *virG* locus that continually induces *vir* gene expression. This implies that at temperatures greater than 32°C, either the plant becomes more resistant to infection by *Agrobacterium* or some other required *vir* gene products no longer function.

VirG, the Response Regulator Component

The *virG* gene encodes a 25-kDa protein that has sequence similarity to a family of response regulator proteins (Winans et al., 1986). The VirG protein has neither a hydrophobic region nor a signal sequence, suggesting it is a cytoplasmic protein. However, cell fractionation experiments indicated that most VirG protein is associated with the inner membrane (Jin and Nester, unpublished data). It was suggested that this inner membrane localization resulted from the unphosphorylated VirG protein being associated with the VirA protein, whereas phosphorylated VirG is released into the cytoplasm where it promotes transcription of the *vir* genes.

VirG is a key molecule in *vir* gene induction, and its own expression is controlled by a complex regulatory system. Two promoters affect the expression of *virG*. P_1, the stronger of the two promoters, is induced by both AS in a VirA/VirG-dependent fashion (Stachel and Zambryski, 1986; Winans, 1990) and by phosphate starvation independently of VirA/VirG (Winans, 1990). The transcription of *virG* was also found to be inducible by acidic conditions (Veluthambi et al., 1987; Winans et al., 1988). This enhanced induction by acidic pH involves the second *virG* promoter, P_2 (Winans, 1990; Mantis and Winans, 1992). The acid pH induction of P_2 may contribute to but is certainly not the sole factor responsible for the acid pH optimum for *vir* gene induction (see Sensing of Acidity). Indeed, Chen and Winans (1991) showed that replacing the natural promoters P_1 and P_2 with the *lac* promoter does not abolish the acid pH optimum of *vir* gene induction. The fact that VirA-independent mutants of *virG* still demonstrate an enhancement of *vir* gene induction at acid pH indicates that an acid pH–dependent step of *vir* gene induction occurs downstream of VirA (Han et al., 1992; Pazour et al., 1992; Jin et al., 1993b). Other data indicate that the acidic pH requirement can be partially overcome by the provision of multiple copies of *virG* (Liu et al., 1993). These data suggest that induction steps both upstream and downstream of VirG activation by VirA are affected by acid pH.

Taken together, these observations lead to the hypothesis that the acidic pH requirement for *vir* regulon induction is an effect of both enhanced *virG* transcription in response to acidic conditions and a low pH optimum for induction conferred by VirA (see Sensing of Acidity). At neutral pH, VirA autophosphorylation or VirA-mediated phosphorylation of VirG may be inefficient. Also, the intracellular concentration of VirG might be too low for VirA-mediated phosphorylation to induce the *vir* regulon effectively. Since *vir* regulon induction requires VirG, the cell would benefit from increased levels of VirG protein prior to exposure to the inducing plant signals. Acidic conditions may serve to increase the levels of VirG. Following the increase of VirG concentration in the cell, AS-activated VirA would phosphorylate VirG, leading to further *virG* transcription and induction of the remaining members of the *vir* regulon.

The complex control of *virG* expression and the acid pH optimum for *vir* gene induction conferred by VirA indicate that very specific environmental conditions are required before a virulence program is initiated. These strict conditions likely exist only in close proximity to the wound site of a susceptible plant host.

The VirG protein belongs to the RO_{II} subfamily of receivers, which, according to the nomenclature of Parkinson and Kofoid (1992), are composed of a receiver domain and an output domain, the latter containing a helix-turn-helix DNA binding motif. The N-terminal receiver domain shares similarity with many response regulators and contains the conserved aspartate residue in an acid pocket. This residue (Asp-52) is phosphorylated in vitro by the sensor component, VirA (Jin et al., 1990a). Changing the conserved Asp-52 of VirG to an asparagine residue abolished *vir* gene expression in vivo and phosphorylation of VirG in vitro. The evidence indicates that this is the only amino acid phosphorylated by VirA.

As stated above, the C-terminal output domain of VirG contains a sequence-specific DNA binding domain, a feature characteristic of transcriptional activators. Footprinting analysis showed that the DNA binding domain of VirG protects a 12-bp consensus sequence called the *vir* box (Jin et al., 1990c). The *vir* box is present in all *vir* gene promoter regions in one or more copies and is required for VirA/VirG-mediated induction of the *vir* genes. However, recent data have shown that the 12-bp *vir* box, although required for VirG protein binding, is not sufficient for this sequence-specific binding. Roitsch et al. (1994) showed that nonconserved specific sequences within 19 bp 3′ to the *vir* box of the *virE* promoter are also required for VirG binding.

The effect of phosphorylation on VirG activity has recently been explored by two groups. Phosphorylation of VirG by VirA may increase its affinity for binding *vir* boxes or may promote its interaction with RNA polymerase. One group has examined the *vir* box binding affinity of $VirG_{N54D}$, a constitutively activated mutant derivative of VirG that is not phosphorylated (Han and Winans, 1994). Han and Winans found that $VirG_{N54D}$, serving as a model for phosphorylated wild-type VirG, bound the *vir* boxes of the VirG promoter with 10-fold higher affinity than the unphosphorylated wild-type protein. This agrees with several observations that phosphorylation increases the DNA binding affinity of other response regulators (Weiss and Magasanik, 1988; Forst et al., 1989; Makino et al., 1989; Feng et al., 1992). Han and Winans (1994) hypothesized that the increased *vir* box binding affinity of the $VirG_{N54D}$ mutant proteins results from the increased affinity of their receiver domains for each other, leading to cooperative binding of the operator by $VirG_{N54D}$ dimers. Presumably, phosphorylation of the wild-type VirG protein has the same effect. A second group recently showed that phosphorylated VirG was notably more active in in vitro transcription of *vir* genes than nonphosphorylated VirG. Also, the pH optimum of transcription was lower for phosphorylated VirG (pH 6.9) than for nonphosphorylated VirG (pH 8.1) (Endoh et al., 1993).

When inducing compounds are removed from the *Agrobacterium* growth medium, *vir* gene expression is quickly terminated (Alt-Moerbe et al., 1989; Hess et al., 1991). This suggests that activated VirG molecules are quickly eliminated, either by a phosphatase that dephosphorylates activated VirG or by a rapid turnover of VirG molecules in vivo. The latter possibility has been suggested by Chen and Winans, who found that the half-life of VirG protein under inducing conditions is about 45 to 60 min, less than one-third the half-life of bulk *Agrobacterium* protein (Winans, personal communication).

VirA/VirG-ChvE Model of *vir* Gene Regulation

In Fig. 3 (A and B), we present a model of signal transduction via VirA and VirG that attempts to incorporate much of the published and unpublished data. Although evidence has been presented that VirA exists as a homodimer in its functional state (Howitz and Binns, personal communication; Pan et al., 1993), we have illustrated VirA as a monomer for simplicity.

As previously stated, the sensor molecule VirA can be divided into three domains based on the nomenclature of Parkinson and Kofoid (1992): input, transmitter, and receiver domains (see Fig. 2). The activity of the transmitter domain is modulated by conformational changes induced by the input domain on recognition of environmental conditions. We believe that the data available for VirA/VirG suggest a three-state model in which the input domain in conjunction with the receiver domain modulates transmitter activity. Both inducing monosaccharides with ChvE and phenolic compounds play a role in *vir* induction via VirA/VirG. The two signals work synergistically to strongly induce *vir* gene expression.

The model diagrammed in Fig. 3 proposes three states of VirA activity: Off, Standby, and On. In this model, we propose that the transmitter domain of VirA is autophosphorylated in response to acid pH independently of ChvE or AS. The monosaccharide-bound ChvE interacts with the C-terminal half of the periplasmic domain of VirA at a site that includes the Trg-like 15-amino-acid motif (see Fig. 2). This interaction alters the conformation of VirA and converts it to the Standby mode. The Standby mode is a conformation in which a repressive or inhibitor-type function of the periplasmic domain is reversed or suppressed.

FIGURE 3 Models for *vir* gene induction in *Agrobacterium*. (A) Model for the sensing and transmission of signals by VirA in conjunction with ChvE. The model is consistent with phenolic compounds detected directly by VirA or through the action of a phenolic binding protein. For simplicity, the model is illustrated with VirA as the direct phenolic sensor. (B) Model for the activation of VirG by VirA and activation of *vir* gene transcription by activated VirG.

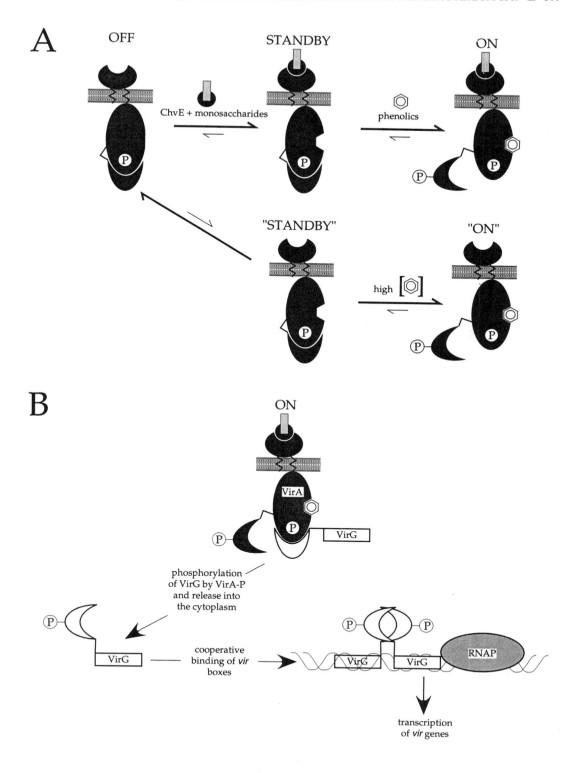

As a result, a region C-terminal to the second transmembrane domain of VirA is made available for interaction with phenolic compounds, either directly or indirectly through a phenolic binding protein. The region involved in recognizing the phenolic compound may be the amphipathic helix domain (see Fig. 2), which is believed to interact closely with the inner membrane.

Phenolic compounds such as AS convert the VirA molecule to the On state. The On state is achieved by VirA transferring a phosphate from the histidine residue of its transmitter domain to the aspartate residue of its receiver domain. The resulting conformational change reverses an inhibitory function of the receiver domain and allows access to the transmitter domain by VirG. The transmitter domain of VirA again autophosphorylates its histidine and in response to AS now transfers that phosphate to VirG, which is docked with VirA in the inner membrane. This transfer of phosphate to Asp-52 of VirG induces a conformational change in VirG that encourages its release from VirA and subsequent induction of *vir* gene expression, possibly via favorable interaction with RNA polymerase (Fig. 3B). The phosphorylation of VirG may lead to self-association of the VirG receiver domains, resulting in cooperative binding of the *vir* boxes and adjacent bases, which, in turn, leads to favorable interaction of the VirG dimer with RNA polymerase.

This model imposes an order of interaction of ChvE and AS with VirA. The model proposes that the interaction of ChvE with the periplasmic domain alters the conformation of the input domain of VirA and must occur first. The conformational change exposes the AS binding site, which defines the Standby mode. Several mutations of VirA have been described that mimic the ChvE-bound state (Standby) consistent with this model. A deletion of the periplasmic domain of VirA results in a molecule that is exquisitely sensitive to AS and no longer requires inducing monosaccharides for induction by low levels of AS (Melchers et al., 1989a; Cangelosi et al., 1990; Chang and Wi-

nans, 1992; Banta et al., 1994). A similar phenotype is observed when the periplasmic domain of VirA is replaced by the periplasmic domain of Trg (Melchers et al., 1989a; Turk, 1993). These results suggest that in the wild-type situation a repressive effect of the periplasmic domain is overcome on binding of the ChvE::monosaccharide, a hypothesis that has been suggested by several investigators (Cangelosi et al., 1990; Shimoda et al., 1990; Turk, 1993; Banta et al., 1994).

The model also accounts for the low level of *vir* gene induction by AS observed in the absence of monosaccharides. Due to the dynamic nature of sensor molecules, we propose that in the absence of monosaccharides a fraction of the VirA molecules will be in a Standby mode in which the phenolic binding site is unmasked. The action of ChvE::monosaccharide is proposed to strongly favor and drive this conformational change. In the presence of high levels of AS, the rare VirA molecule that spontaneously converts to the Standby mode will move quickly to the On state. According to the model, low levels of AS do not induce in the absence of inducing monosaccharides because those few molecules in the Standby mode at any point in time are not in close proximity to AS.

The strict requirement for acidic conditions, phenolic compounds, and inducing monosaccharides to achieve maximum *vir* induction ensures that *Agrobacterium* does not commit itself to *vir* gene expression and T-DNA production until it is in the environment of a plant wound site.

PREDICTIONS OF THE VirA/VirG-ChvE MODEL

The above model predicts that several types of *virA* mutants affected in each of the three states of activation could be obtained. One class is the ChvE::monosaccharide-independent mutants. The first identified members of this class are the periplasmic deletion mutants and periplasmic domain replacement mutants described by several investigators (Melchers et al., 1989a;

Cangelosi et al., 1990; Shimoda et al., 1990; Chang and Winans, 1992; Turk et al., 1993b; Banta et al., 1994). These mutants no longer require inducing monosaccharides to reach high levels of *vir* induction. Thus, these mutants effectively exist in a Standby mode in the absence of any signal. Random mutagenesis of *virA* clones from both octopine-utilizing (A6) and nopaline-utilizing (C58) strains has yielded mutants that are ChvE::monosaccharide-independent (Heath et al., unpublished data b).

A second class of mutants is predicted by this model. Clones of both A6 and C58 *virA* have been isolated that have mutations conferring an AS-independent phenotype (Ankenbauer et al., 1991; McLean et al., 1994; Heath et al., unpublished data a). These mutants induce *vir* genes to high levels in the absence of AS but still require monosaccharides for full induction. This class of mutant (AS-independent) effectively bypasses the Standby mode in the above model.

A third class of mutants that is predicted to be isolatable is both ChvE::monosaccharide and AS independent. For two reasons, this class of constitutively *vir*-expressing *virA* mutant may be difficult to isolate. First, according to the model (Fig. 3), this mutant VirA would have to harbor mutations that bypass both the Off and the Standby modes to become constitutively activated. Second, a mutant strain that is constitutively expressing its *vir* genes is likely to be less fit, and therefore revertants would arise at a high frequency. The *virA* mutants described as constitutive *vir*-inducing mutants by Pazour et al. (1991) were tested for induction in the presence of glucose, an inducing monosaccharide, and therefore it is not known whether they are constitutive in the absence of monosaccharides. A *virA* mutant was recently reported that is induced to high levels in the absence of AS and monosaccharides (McLean et al., 1994). However, this mutant does respond to AS addition in the absence of inducing monosaccharides and is induced to highest levels in the presence of inducing monosaccharides, indicating a dependence on signals for maximum induction.

Mutations in the amphipathic helix domain of VirA would be expected to have a severely debilitating effect on *vir* gene induction if this region is indeed involved in sensing phenolic compounds. Recently, several independent point mutations that affect *vir*-inducing activity have been isolated and mapped (Doty et al., unpublished data). The only mutations that completely abolished *vir* induction by AS mapped to the amphipathic helix, whereas the others that only affected sugar enhancement of *vir* induction mapped to the periplasmic domain. Since the amphipathic helix has been demonstrated in another system to be important in orientation of the protein in the membrane, it remains a possibility that these mutations are nonfunctional in *vir* induction simply because they are not inserted correctly in the membrane. Nevertheless, the observation is consistent with a model in which the amphipathic helix plays a role in phenolic sensing.

According to the model, *chvE* mutants that exhibit a sugar-independent *vir*-inducing phenotype should also be obtainable. This type of mutant has not yet been reported. The isolation and analysis of new mutants will lead to a more refined and accurate picture of VirA/VirG-ChvE-mediated *vir* gene regulation.

TWO-COMPONENT REGULATORY SYSTEM ON THE CHROMOSOME: ChvG/ChvI

A chromosomally encoded virulence locus was recently identified in *Agrobacterium* by independent approaches. During tumorigenesis, many of the most important events, from detection of plant signals to T-DNA transfer, take place at the level of the bacterial cell envelope. This realization led to a search for cell envelope protein-encoding genes that are involved in virulence (Cangelosi et al., 1991). Analysis of Tn*phoA*-generated avirulent mutants resulted in the identification of chromosomally located virulence loci. One of these has recently been the subject of more extensive characterization (Charles and Nester, 1993). The stimulatory effect of phosphate starvation in *virG* expression suggests that a PhoB-like protein may be involved in *virG* regulation. Surprisingly, functional complementation of an *E. coli phoB* mutant resulted in the cloning of the same chromosomal virulence locus as was identified by Tn*phoA* mutagenesis above (Mantis and Winans, 1993). Mutants of this locus exhibit a pleiotropic phenotype, being ultrasensitive to several antibiotics, acidic pH, and plant wound exudate and unable to grow on complex media. Complementation analysis identified two separate transcriptional units. Sequence analysis and subsequent data base homology searches indicated that each of these transcriptional units encodes one member of a novel two-component regulatory system, which was designated ChvG/ChvI. Both of these genes are transcribed in the same direction, with *chvI* upstream of *chvG*. ChvG is a 76-kDa (690-amino-acid) sensor protein containing a cytoplasmic membrane-spanning domain. ChvG is predicted to have protein kinase activity, which autophosphorylates the conserved histidine 352 and subsequently transfers the phosphate to an aspartic acid residue in ChvI, a transcriptional regulator protein of 241 amino acids.

Similar to many sensor histidine kinase proteins, ChvG is predicted to span the cytoplasmic membrane, with its carboxyl terminus extending into the cytoplasm. This protein is unusual, however, in that it contains a Walker type-A purine nucleotide binding site near its carboxyl terminus. This highly conserved domain has, in other proteins, been shown to be involved directly in interaction with the phosphate groups of purine nucleotides. The type A domain has been found in many proteins, both eukaryotic and prokaryotic, and several of these proteins are involved in the transport of substances either into or out of the cell (Higgins et al., 1986; Blight and Holland, 1990). In many cases, this domain has been shown to be necessary for nucleoside triphosphate hydrolysis, which presumably drives the transport process. This raises questions regarding the possible involvement of ChvG in some transport process

as well as its apparent involvement in activation of the transcriptional regulator ChvI by phosphorylation. At present, nothing is known regarding the function of this domain in ChvG, but its existence raises the possibility of a regulatory (as opposed to energetic) role for nucleoside triphosphate in this signal transduction system.

Although clones carrying *chvI* were able to functionally complement a *phoB* mutant of *E. coli,* the ChvG/ChvI system does not appear to be involved in *Agrobacterium* phosphate regulation. For example, *chvI* mutations did not abolish phosphate starvation-inducible phosphatase activity, as *phoB* mutations do in other bacteria. Also, *chvI* mutations did not affect *virG* expression resulting from phosphate starvation (Mantis and Winans, 1993). However, mutation of *chvI* did abolish induction of *virG* expression in response to acidic pH and *virB* and *virG* expression in response to AS. Similarly, *chvG* mutants generated by non-Tn*phoA* mutagenesis did not exhibit *vir* gene expression in response to AS (Charles and Nester, 1993). In Tn*phoA*-generated mutants of *chvG,* the *vir* genes are still inducible by AS in the presence of monosaccharides. However, these mutants are avirulent. This suggests that their inability to induce *vir* genes does not fully account for their inability to form tumors (Charles and Nester, 1993).

Mutation of *chvG* or *chvI* results in an increase in the sensitivity to cell envelope-perturbing compounds such as detergents and to acidic pH. Mutations also result in the inability of the cells to grow in the presence of an as-yet-unidentified component found in casamino acids. This unidentified component does not appear to be one of the 20 amino acids. Also, mutants are more sensitive than wild-type cells to leaf exudate from *Bryophyllum diagremontiana.* Does this complex pleiotropic phenotype result from alterations in the expression of a single regulated gene or several genes? Many two-component regulatory systems, including VirA/VirG, control regulons consisting of several transcriptional units, each with a corresponding promoter. We have seen already how expression of the Ti plasmid *vir* regulon is controlled by the VirA/VirG system in *Agrobacterium*. The pleiotropy of the phenotype may suggest that the ChvG/ChvI system is involved in the regulation of multiple genes.

Neither the virulence genes that are regulated by this system nor the inducing signal(s) have been identified yet. The gene *pckA*, encoding the gluconeogenic enzyme phosphoenolpyruvate carboxykinase, is located immediately upstream of *chvI*, but its relationship to *chvG/chvI* is unknown at this time. The identification of ChvG/ChvI-regulated genes is a priority for future studies. Many questions remain with regard to the role of *chvG/chvI* in tumorigenesis. Whether the ChvG/ChvI-regulated loci are directly or indirectly involved in the virulence of *Agrobacterium* is unknown. The pleiotropic nature of the phenotype suggests that any relation to virulence may be indirect. Simple explanations do not lend themselves well to this newly discovered two-component regulatory system.

EPILOGUE

Bacteria that interact with plants, like other microorganisms, respond to environmental changes by altering gene expression in characteristic and appropriate ways. This allows the cell to adapt favorably to changing conditions and especially to detect and take advantage of the presence of susceptible host plants. Members of the family of two-component regulatory systems play a major role in the monitoring of environmental changes (Charles et al., 1992), and in *Agrobacterium,* both plasmid- and chromosomally encoded members of this family affect virulence. The plasmid-borne VirA/VirG system is well characterized and indeed has served as a paradigm for the regulation of virulence determinants in other bacteria (Mekalanos, 1992), whereas study of the chromosomally encoded ChvG/ChvI system has suggested several intriguing possibilities, without revealing either the target gene(s) or the signal(s) that is detected by ChvG. This

emphasizes that although we have made tremendous progress in understanding *Agrobacterium* virulence, many gaps still remain in our understanding of *Agrobacterium* as a complete organism.

ACKNOWLEDGMENTS

We gratefully acknowledge Andrew Binns, Vicky Howitz, Colin Manoil, and Steve Winans for communicating unpublished data. We thank Sharon Doty, Karla Fullner, Kevin Piers, and Kathy Stephens for a critical reading of the manuscript and many valuable discussions. We especially thank Karla Fullner for providing Fig. 1.

The work from E.W.N.'s laboratory was supported by NIH grant GM32618. J. D. H. was supported by NIH fellowship GM15097.

REFERENCES

Alt-Moerbe, J., H. Kuhlmann, and J. Schroder. 1989. Differences in induction of Ti plasmid virulence genes *virG* and *virD*, and continued control of *virD* expression by four external factors. *Mol. Plant-Microbe Int.* **2**:301–308.

Alt-Moerbe, J., P. Nedderman, J. von Lintig, E. W. Weiler, and J. Schroder. 1988. Temperature-sensitive step in Ti plasmid *vir*-region induction and correlation with cytokinin secretion by *Agrobacteria*. *Mol. Gen. Genet.* **213**:1–8.

Ankenbauer, R., P. Albersheim, and E. Nester. Unpublished data.

Ankenbauer, R., and E. Nester. 1993. The *Agrobacterium* Ti plasmid and crown gall tumorigenesis: a model for signal transduction. Host–pathogen interactions, p. 67–104. *In* J. Kurjan and B. Taylor (ed.), *Signal Transduction, Prokaryotic and Simple Eukaryotic Systems.* Academic Press, San Diego, Calif.

Ankenbauer, R. G., E. A. Best, C. A. Palanca, and E. W. Nester. 1991. Mutants of the *Agrobacterium tumefaciens virA* gene exhibiting acetosyringone-independent expression of the *vir* regulon. *Mol. Plant-Microbe Int.* **4**:400–406.

Ankenbauer, R. G., and E. W. Nester. 1990. Sugar-mediated induction of *Agrobacterium tumefaciens* virulence genes: structural specificity and activities of monosaccharides. *J. Bacteriol.* **172**: 6442–6446.

Banta, L. M., R. D. Joerger, V. R. Howitz, A. M. Campbell, and A. N. Binns. 1994. Glu-225 outside the predicted ChvE binding site in VirA is crucial for sugar enhancement of acetosyringone perception by *Agrobacterium tumefaciens. J. Bacteriol.* **176**:3242–3249.

Blight, M. A., and I. B. Holland. 1990. Structure and function of haemolysin B, P-glycoprotein and other members of a novel family of membrane translocators. *Mol. Microbiol.* **4:**873–880.

Bolton, G. W., E. W. Nester, and M. P. Gordon. 1986. Plant phenolic compounds induce expression of the *Agrobacterium tumefaciens* loci needed for virulence. *Science* **232:**983–985.

Braun, A. C. 1947. Thermal studies on the factors responsible for tumor initiation in crown gall. *Am. J. Bot.* **34:**234–240.

Cangelosi, G. A., R. G. Ankenbauer, and E. W. Nester. 1990. Sugars induce the *Agrobacterium* virulence genes through a periplasmic binding protein and a transmembrane signal protein. *Proc. Natl. Acad. Sci. USA* **87:**6708–6712.

Cangelosi, G. A., E. A. Best, G. Martinetti, and E. W. Nester. 1991. Genetic analysis of *Agrobacterium. Methods Enzymol.* **204:**384–397.

Castle, L. A., K. D. Smith, and R. O. Morris. 1992. Cloning and sequencing of an *Agrobacterium tumefaciens* beta-glucosidase gene involved in modifying a *vir*-inducing plant signal molecule. *J. Bacteriol.* **174:**1478–1486.

Chang, C.-H., and S. C. Winans. 1992. Functional roles assigned to the periplasmic, linker, and receiver domains of the *Agrobacterium tumefaciens* VirA protein. *J. Bacteriol.* **174:**7033–7039.

Charles, T., S. Jin, and E. Nester. 1992. Two-component sensory transduction systems in phytobacteria. *Annu. Rev. Phytopathol.* **30:**463–484.

Charles, T. C., and E. W. Nester. 1993. A chromosomally encoded two-component sensory transduction system is required for virulence of *Agrobacterium tumefaciens. J. Bacteriol.* **175:**6614–6625.

Chen, C. Y., and S. C. Winans. 1991. Controlled expression of the transcriptional activator gene *virG* in *Agrobacterium tumefaciens* by using the *Escherichia coli lac* promoter. *J. Bacteriol.* **173:**1139–1144.

Cornish, A., J. A. Greenwood, and C. W. Jones. 1989. Binding-protein-dependent sugar transport by *Agrobacterium radiobacter* and *A. tumefaciens* grown in continuous culture. *J. Gen. Microbiol.* **135:**3001–3031.

Culianez-Macia, F. A., and A. G. Hepburn. 1988. The kinetics of T-strand production in a nopaline-type helper strain of *Agrobacterium tumefaciens. Mol. Plant-Microbe Int.* **1:**207–214.

Doty, S., I. Lundin, and E. Nester. Unpublished data.

Doty, S. L., M. Chang, and E. W. Nester. 1993. The chromosomal virulence gene, *chvE*, of *Agrobacterium tumefaciens* is regulated by a LysR family member. *J. Bacteriol.* **175:**7880–7886.

Douglas, C. J., W. Halperin, and E. W. Nester. 1982. *Agrobacterium tumefaciens* mutants affected in attachment to plant cells. *J. Bacteriol.* **152:**1265–1275.

Douglas, C. J., R. J. Staneloni, R. A. Rubin, and E. W. Nester. 1985. Identification and genetic analysis of an *Agrobacterium tumefaciens* chromosomal virulence region. *J. Bacteriol.* **161:**850–860.

Duban, M. E., K. Lee, and D. G. Lynn. 1993. Strategies in pathogenesis: mechanistic specificity in the detection of generic signals. *Mol. Microbiol.* **7:**637–645.

Endoh, H., T. Aoyama, and A. Oka. 1993. Transcription in vitro promoted by the *Agrobacterium* VirG protein. *FEBS Lett.* **334:**277–280.

Endoh, H., and A. Oka. 1993. Functional analysis of the VirG-like domain contained in the *Agrobacterium* VirA protein that senses plant factors. *Plant Cell Physiol.* **34:**227–235.

Feng, J., M. R. Atkinson, W. McCleary, J. B. Stock, B. L. Wanner, and A. J. Ninfa. 1992. Role of phosphorylated metabolic intermediates in the regulation of glutamine synthetase synthesis in *Escherichia coli. J. Bacteriol.* **174:**6061–6070.

Forst, S., J. Delgado, and M. Inouye. 1989. Phosphorylation of OmpR by the osmosensor EnvZ modulates expression of the *ompF* and *ompC* genes in *Escherichia coli. Proc. Natl. Acad. Sci. USA* **86:**6052–6056.

Furlong, C. E. 1987. Osmotic shock-sensitive transport systems, p. 768–796. *In* F. C. Neidhardt, J. L. Ingraham, K. B. Low, B. Magasanik, M. Schaechter, and H. E. Umbarger (ed.), *Escherichia coli and Salmonella typhimurium: Cellular and Molecular Biology.* American Society for Microbiology, Washington, D.C.

Garfinkel, D. J., and E. W. Nester. 1980. *Agrobacterium tumefaciens* mutants affected in crown gall tumorigenesis and octopine catabolism. *J. Bacteriol.* **144:**732–743.

Grignon, C., and H. Sentenac. 1991. pH and ionic conditions in the apoplast. *Annu. Rev. Plant Physiol. Plant Mol. Biol.* **42:**103–128.

Györgypal, Z., G. B. Kiss, and A. Konderosi. 1991. Transduction of plant signal molecules by the *Rhizobium* NodD proteins. *BioAssays* **13:**575–581.

Han, D. C., C.-Y. Chen, Y.-F. Chen, and S. C. Winans. 1992. Altered function mutations of the transcriptional regulatory gene *virG* of *Agrobacterium tumefaciens. J. Bacteriol.* **174:**7040–7043.

Han, D. C., and S. C. Winans. 1994. A mutation in the receiver domain of the *Agrobacterium tumefaciens* transcriptional regulator VirG increases its affinity for operator DNA. *Mol. Microbiol.* **12:**23–30.

Heath, J. D., M. Bartell, and E. Nester. Unpublished data a.

Heath, J. D., I. Lundin, and E. Nester. Unpublished data b.

Hess, K. M., M. W. Dudley, D. G. Lynn, R. D. Joerger, and A. N. Binns. 1991. Mechanism of phenolic activation of *Agrobacterium* virulence genes: development of a specific inhibitor of bacterial sensor/response systems. *Proc. Natl. Acad. Sci. USA* **88:**7854–7858.

Higgins, C. F., I. D. Hiles, G. P. C. Salmond, D. R. Gill, J. A. Downie, I. J. Evan, B. Holland, L. Gray, S. D. Buckel, A. W. Bell, and M. A. Hermondson. 1986. A family of related ATP-binding subunits coupled to many distant biological processes in bacteria. *Nature* (London) **323:**448–450.

Howitz, V. R., and A. N. Binns (University of Pennsylvania). 1993. Personal communication.

Huang, M. L., G. A. Cangelosi, W. Halperin, and E. W. Nester. 1990. A chromosomal *Agrobacterium tumefaciens* gene required for effective plant signal transduction. *J. Bacteriol.* **172:**1814–1822.

Jin, S., and E. Nester. Unpublished data.

Jin, S. G., R. K. Prusti, T. Roitsch, R. G. Ankenbauer, and E. W. Nester. 1990a. Phosphorylation of the VirG protein of *Agrobacterium tumefaciens* by the autophosphorylated VirA protein: essential role in biological activity of VirG. *J. Bacteriol.* **172:**4945–4950.

Jin, S., T. Roitsch, R. G. Ankenbauer, M. P. Gordon, and E. W. Nester. 1990b. The VirA protein of *Agrobacterium tumefaciens* is autophosphorylated and is essential for *vir* gene regulation. *J. Bacteriol.* **172:**525–530.

Jin, S. G., T. Roitsch, P. J. Christie, and E. W. Nester. 1990c. The regulatory VirG protein specifically binds to a *cis*-acting regulatory sequence involved in transcriptional activation of *Agrobacterium tumefaciens* virulence genes. *J. Bacteriol.* **172:**531–537.

Jin, S., Y.-N. Song, W.-Y. Deng, M. Gordon, and E. W. Nester. 1993a. The regulatory VirA protein of *Agrobacterium tumefaciens* does not function at elevated temperatures. *J. Bacteriol.* **175:**6830–6835.

Jin, S., Y.-N. Song, S. Q. Pan, and E. W. Nester. 1993b. Characterization of a *virG* mutation that confers constitutive virulence gene expression in *Agrobacterium*. *Mol. Microbiol.* **7:**555–562.

Kado, C. 1991. Molecular mechanisms of crown gall tumorigenesis. *Crit. Rev. Plant Sci.* **101:**1–32.

Kahl, G. 1982. Molecular biology of wound healing: the conditioning phenomenon, p. 211–267. *In* G. Kahl and J. S. Schell (ed.), *Molecular Biology of Plant Tumors.* Academic Press, London.

Lee, K., M. W. Dudley, K. M. Hess, D. G. Lynn, R. D. Joerger, and A. N. Binns. 1992. Mechanism of activation of *Agrobacterium* virulence genes: identification of phenol-binding proteins. *Proc. Natl. Acad. Sci. USA* **89:**8666–8670.

Lee, S., R. Ankenbauer, and E. Nester. Unpublished data a.

Lee, S., H. Floss, and E. Nester. Unpublished data b.

Lee, S., S. Jin, and E. Nester. Unpublished data c.

Leroux, B., M. F. Yanofsky, S. C. Winans, J. E. Ward, S. F. Ziegler, and E. W. Nester. 1987. Characterization of the *virA* locus of *Agrobacterium tumefaciens:* a transcriptional regulator and host range determinant. *Embo J.* **6:**849–856.

Liu, C.-N., T. R. Steck, L. L. Habeck, J. A. Meyer, and S. B. Gelvin. 1993. Multiple copies of *virG* allow induction of *Agrobacterium tumefaciens vir* genes and T-DNA processing at alkaline pH. *Mol. Plant-Microbe Int.* **6:**144–156.

Machida, Y., S. Okamoto, S. Matsumoto, S. Usami, A. Yamamoto, Y. Niwa, S. D. Jeong, J. Nagamine, N. Shimoda, C. Machida, and M. Iwahashi. 1989. Mechanisms of crown gall formation: T-DNA transfer from *Agrobacterium tumefaciens* to plant cells. *Bot. Mag. Tokyo* **102:**331–350.

Macnab, R. M. 1987. Motility and chemotaxis, p. 732–759. *In* F. C. Neidhardt, J. L. Ingraham, K. B. Low, B. Magasanik, M. Schaechter, and H. E. Umbarger (ed.), *Escherichia coli and Salmonella typhimurium: Cellular and Molecular Biology.* American Society for Microbiology. Washington, D.C.

Makino, K., H. Shinagawa, M. Amemura, T. Kawamoto, M. Yamada, and A. Nakata. 1989. Signal transduction in the phosphate regulon of *Escherichia coli* involves phosphotransfer between PhoR and PhoB proteins. *J. Mol. Biol.* **210:**551–559.

Mantis, N. J., and S. C. Winans. 1992. The *Agrobacterium tumefaciens vir* gene transcriptional activator *virG* is transcriptionally induced by acid pH and other stress stimuli. *J. Bacteriol.* **174:**1189–1196.

Mantis, N. J., and S. C. Winans. 1993. The chromosomal response regulatory gene *chvI* of *Agrobacterium tumefaciens* complements an *Escherichia coli phoB* mutation and is required for virulence. *J. Bacteriol.* **175:**6626–6636.

Matthysse, A. G. 1987. Characterization of non-attaching mutants of *Agrobacterium tumefaciens. J. Bacteriol.* **169:**313–323.

McLean, B. G., E. A. Greene, and P. Zambryski. 1994. Mutants of *Agrobacterium* VirA that activate *vir* gene expression in the absence of the inducer acetosyringone. *J. Biol. Chem.* **269:**2645–2651.

Mekalanos, J. J. 1992. Environmental signals controlling expression of virulence determinants in bacteria. *J. Bacteriol.* **174:**1–7.

Melchers, L. S., T. T. J. Regensburg, R. B. Bourret, N. J. Sedee, R. A. Schilperoort, and P. J. Hooykaas. 1989a. Membrane topology and functional analysis of the sensory protein VirA of *Agrobacterium tumefaciens. EMBO J.* **8:**1919–1925.

Melchers, L. S., A. J. Regensburg-Tuink, R. A. Schilperoort, and P. J. Hooykaas. 1989b. Specificity of signal molecules in the activation of *Agrobacterium* virulence gene expression. *Mol. Microbiol.* **7:**969–977.

Messens, E., R. Dekeyser, and S. Stachel. 1990. A nontransformable *Triticum monococcum* monocotyledenous culture produces the potent *Agrobacterium* *vir*-inducing compound ethyl ferulate. *Proc. Natl. Acad. Sci. USA* **87**:4368–4372.

Metts, J., J. West, S. H. Doares, and A. G. Matthysse. 1991. Characterization of three *Agrobacterium tumefaciens* avirulent mutants with chromosomal mutations that affect induction of *vir* genes. *J. Bacteriol.* **173**:1080–1087.

Morris, J. W., and R. O. Morris. 1990. Identification of an *Agrobacterium tumefaciens* virulence gene inducer from the pinaceous gymnosperm *Pseudotsuga menziesii*. *Proc. Natl. Acad. Sci. USA* **87**:3614–3618.

Pan, S. Q., T. Charles, S. Jin, Z.-L. Wu, and E. W. Nester. 1993. Preformed dimeric state of the sensor protein VirA is involved in plant-*Agrobacterium* signal transduction. *Proc. Natl. Acad. Sci. USA* **90**:9939–9943.

Park, C., and G. L. Hazelbauer. 1986. Mutations specifically affecting ligand interaction of the Trg chemosensory transducer. *J. Bacteriol.* **167**:101–109.

Parkinson, J. S., and E. C. Kofoid. 1992. Communication modules in bacterial signaling proteins. *Annu. Rev. Genet.* **26**:71–112.

Pazour, G. J., C. N. Ta, and A. Das. 1991. Mutants of *Agrobacterium tumefaciens* with elevated *vir* gene expression. *Proc. Natl. Acad. Sci. USA* **88**:6941–6945.

Pazour, G. J., C. N. Ta, and A. Das. 1992. Constitutive mutations of *Agrobacterium tumefaciens* transcriptional activator *virG*. *J. Bacteriol.* **174**:4169–4174.

Puvanesarajah, V., F. M. Schell, G. Stacey, C. J. Douglas, and E. W. Nester. 1985. Role for 2-linked-beta-D-glucan in the virulence of *Agrobacterium tumefaciens*. *J. Bacteriol.* **164**:102–106.

Recourt, K., A. A. N. van Brussel, A. J. M. Driessen, and B. J. J. Lugtenberg. 1989. Accumulation of a *nod* gene inducer, the flavonoid naringenin, in the cytoplasmic membrane of *Rhizobium leguminosarum* biovar *viciae* is caused by the pH-dependent hydrophobicity of naringenin. *J. Bacteriol.* **171**:4370–4377.

Riker, A. J. 1926. Studies on the influence of some environmental factors on the development of crown gall. *J. Agric. Res.* **32**:83–96.

Rogler, C. E. 1980. Plasmid-dependent temperature-sensitive phase in crown gall tumorigenesis. *Proc. Natl. Acad. Sci. USA* **77**:2688–2692.

Rogowsky, P. M., T. J. Close, J. A. Chimera, J. J. Shaw, and C. I. Kado. 1987. Regulation of the *vir* genes of *Agrobacterium tumefaciens* plasmid pTiC58. *J. Bacteriol.* **169**:5101–5112.

Roitsch, T., S. Jin, and E. W. Nester. 1994. The binding site of the transcriptional activator VirG from *Agrobacterium* comprises both conserved and specific nonconserved sequences. *FEBS Lett.* **338**:127–132.

Rong, L., N. C. Carpita, A. Mort, and S. B. Gelvin. 1994. Soluble cell wall compounds from carrot roots induce the *picA* and *pgl* loci of *Agrobacterium tumefaciens*. *Mol. Plant-Microbe Int.* **7**:6–14.

Sawada, H., H. Ieki, H. Oyaizu, and S. Matsumoto. 1993. Proposal for rejection of *Agrobacterium tumefaciens* and revised descriptions for the genus *Agrobacterium* and for *Agrobacterium radiobacter* and *Agrobacterium rhizogenes*. *Int. J. System. Bacteriol.* **43**:694–702.

Seligman, L., and C. Manoil. 1994. An amphipathic sequence determinant of membrane protein topology. *J. Biol. Chem.* **269**:19888–19896.

Shimoda, N., A. Toyoda-Yamamoto, S. Aoki, and Y. Machida. 1993. Genetic evidence for an interaction between the VirA sensor protein and the ChvE sugar-binding protein of *Agrobacterium*. *J. Biol. Chem.* **268**:26552–26558.

Shimoda, N., A. Toyoda-Yamamoto, J. Nagamine, S. Usami, M. Katayama, Y. Sakagami, and Y. Machida. 1990. Control of expression of *Agrobacterium vir* genes by synergistic actions of phenolic signal molecules and monosaccharides. *Proc. Natl. Acad. Sci. USA* **87**:6684–6688.

Spencer, P., A. Tanaka, and G. H. N. Towers. 1990. An *Agrobacterium* signal compound from grapevine cultivars. *Phytochemistry* **29**:3786–3790.

Spencer, P. A., and G. H. N. Towers. 1988. Specificity of signal compounds detected by *Agrobacterium tumefaciens*. *Phytochemistry* **27**:2781–2785.

Stachel, S. E., E. Messens, M. Van Montagu, and P. Zambryski. 1985. Identification of the signal molecules produced by wounded plant cells that activate T-DNA transfer in *Agrobacterium tumefaciens*. *Nature* (London) **318**:624–629.

Stachel, S. E., E. W. Nester, and P. C. Zambryski. 1986. A plant cell factor induces *Agrobacterium tumefaciens vir* gene expression. *Proc. Natl. Acad. Sci. USA* **83**:379–383.

Stachel, S. E., and P. C. Zambryski. 1986. *virA* and *virG* control the plant-induced activation of the T-DNA transfer process of *A. tumefaciens*. *Cell* **46**:325–333.

Turk, S. C., L. S. Melchers, H. den Dulk-Ras, T. A. J. Regensburg, and P. J. Hooykaas. 1991. Environmental conditions differentially affect *vir* gene induction in different *Agrobacterium* strains. Role of the VirA sensor protein. *Plant Mol. Biol.* **16**:1051–1059.

Turk, S. C. H. J. 1993. Characterization of the VirA receptor protein from *Agrobacterium tumefaciens*. Ph.D. dissertation. Leiden University, The Netherlands.

Turk, S. C. H. J., E. W. Nester, and P. J. J. Hooykaas. 1993a. The *virA* promoter is a host-range determinant in *Agrobacterium tumefaciens. Mol. Microbiol.* **7:**719–724.

Turk, S. C. H. J., R. P. vanLange, E. Sonneveld, and P. J. J. Hooykaas. 1993b. The chimeric VirA-Tar receptor is locked into a highly responsive state. *J. Bacteriol.* **175:**5706–5709.

Usami, S., S. Okamoto, I. Takebe, and Y. Machida. 1988. Factor inducing *Agrobacterium tumefaciens vir* gene expression is present in monocotyledonous plants. *Proc. Natl. Acad. Sci. USA* **85:**3748–3752.

Veluthambi, K., R. K. Jayaswal, and S. B. Gelvin. 1987. Virulence genes *A, G,* and *D* mediate the double-stranded border cleavage of T-DNA from the *Agrobacterium* Ti plasmid. *Proc. Natl. Acad. Sci. USA* **84:**1881–1885.

Vernade, D., E. A. Herrera, K. Wang, and M. M. Van. 1988. Glycine betaine allows enhanced induction of the *Agrobacterium tumefaciens vir* genes by acetosyringone at low pH. *J. Bacteriol.* **170:**5822–5829.

Weiss, V., and B. Magasanik. 1988. Phosphorylation of nitrogen regulator I (NRI) of *Escherichia coli. Proc. Natl. Acad. Sci. USA* **85:**8919–8923.

Winans, S. C. 1990. Transcriptional induction of an *Agrobacterium* regulatory gene at tandem promoters by plant-released phenolic compounds, phosphate starvation, and acidic growth media. *J. Bacteriol.* **172:**2433–2438.

Winans, S. C. 1992. Two way chemical signalling in *Agrobacterium*-plant interactions. *Microbiol. Rev.* **56:** 12–31.

Winans, S. C. (Cornell University). 1994. Personal communication.

Winans, S. C., P. R. Ebert, S. E. Stachel, M. P. Gordon, and E. W. Nester. 1986. A gene essential for *Agrobacterium* virulence is homologous to a family of positive regulatory loci. *Proc. Natl. Acad. Sci. USA* **83:**8278–8282.

Winans, S. C., R. A. Kerstetter, and E. W. Nester. 1988. Transcriptional regulation of the *virA* and *virG* genes of *Agrobacterium tumefaciens. J. Bacteriol.* **170:**4047–4054.

Winans, S. C., R. A. Kerstetter, J. E. Ward, and E. W. Nester. 1989. A protein required for transcriptional regulation of *Agrobacterium* virulence genes spans the cytoplasmic membrane. *J. Bacteriol.* **171:**1616–1622.

Winans, S. C., N. J. Mantis, C.-Y. Chen, C.-H. Chang, and D. C. Han. 1994. Host recognition by the VirA, VirG two-component regulatory proteins of *Agrobacterium tumefaciens. Res. Microbiol.* **145:**461–473.

Wu, L., and E. Nester. Unpublished data.

Zerback, R., K. Dressler, and D. Hess. 1989. Flavonoid compounds from pollen and stigma of *Petunia hybrida:* inducers of the *vir* region of the *Agrobacterium tumefaciens* Ti-plasmid. *Plant Sci.* **62:** 83–91.

Zorreguieta, A., R. Geremia, S. Cavaignac, G. Cangelosi, E. Nester, and R. Ugalde. 1988. Identification of the product of an *Agrobacterium tumefaciens* chromosomal virulence gene. *Mol. Plant-Microbe Int.* **1:**121–127.

Regulation of Glycopeptide Resistance Genes of Enterococcal Transposon Tn*1546* by the VanR–VanS Two-Component Regulatory System

Michel Arthur, Florence Depardieu, Theodore Holman, Zhen Wu, Gerard Wright, Christopher T. Walsh, and Patrice Courvalin

24

Glycopeptide antibiotics vancomycin and teicoplanin are used to treat severe infections caused by gram-positive cocci. Emergence in 1986 of glycopeptide-resistant enterococci resulted in therapeutic difficulty because these organisms are frequently coresistant to all other antibiotics of proven efficacy (Johnson et al., 1990; Murray, 1990). Glycopeptide-resistant enterococci are phenotypically and genotypically heterogeneous (Arthur and Courvalin, 1993). Strains displaying the so-called VanA resistance phenotype are inducibly resistant to high levels of vancomycin and teicoplanin. Spread of this type of resistance is due to conjugal transfer of plasmids that have acquired Tn*1546* or closely related elements by transposition (Arthur et al., 1993). Tn*1546*, 10,851 bp in size, encodes nine polypeptides (Fig. 1). The putative transposase (ORF1) and resolvase (ORF2) and the 38-bp terminal inverted repeats of Tn*1546* are structurally related to those of transposons of the Tn*3* family. The VanH dehydrogenase, the VanA ligase, and VanX, a protein of unknown function, are re-

quired for vancomycin resistance, whereas VanY and VanZ are not (Arthur et al., 1992a,b). VanA and VanH synthesize the depsipeptide D-alanyl-D-lactate (D-Ala-D-Lac), which is incorporated into peptidoglycan precursors instead of the dipeptide D-alanyl-D-alanine (D-Ala-D-Ala) found in glycopeptide-susceptible bacteria (Bugg et al., 1991). Glycopeptides form complexes with peptidoglycan precursors terminating in D-Ala-D-Ala at the cell surface, thereby inhibiting the transglycosylation and transpeptidation steps of peptidoglycan assembly (Reynolds, 1989). By contrast, D-Lac-containing precursors bind vancomycin with greatly reduced affinity and allow cell wall synthesis in the presence of the antibiotic (Bugg et al., 1991). Production of the depsipeptide D-Ala-D-Lac is controlled by the VanR-VanS two-component regulatory system that activates transcription of vancomycin resistance genes in response to the presence of glycopeptides in the culture medium (Arthur et al., 1992b).

PHENOTYPE

Analysis of growth curves of glycopeptide-resistant clinical isolates showed that exposure to subinhibitory concentrations of vancomycin resulted in a lag phase of several hours (Leclercq et al., 1989; Shlaes et al., 1989; William-

Michel Arthur, Florence Depardieu, and Patrice Courvalin, Unité des Agents Antibactériens, Centre National de la Recherche Scientifique, Institut Pasteur, Paris, France. *Theodore Holman, Zhen Wu, Gerard Wright, Christopher T. Walsh,* Department of Biological Chemistry and Molecular Pharmacology, Harvard Medical School, Boston, Massachusetts 02115.

Two-Component Signal Transduction, Edited by James A. Hoch and Thomas J. Silhavy,
© 1995 American Society for Microbiology, Washington, DC 20005

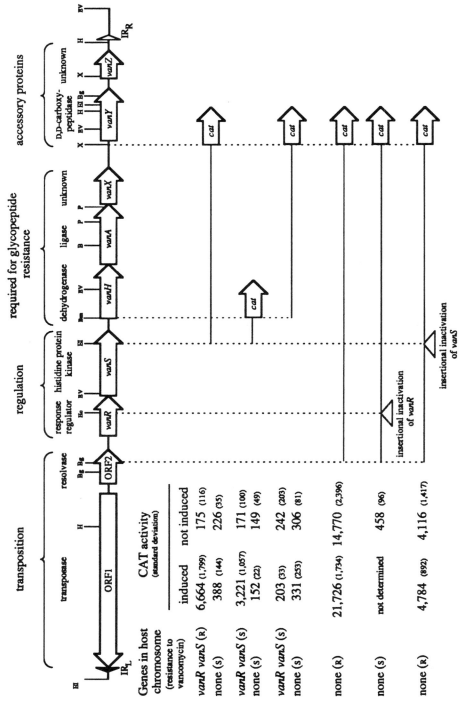

FIGURE 1 Map of Tn1546 and analysis of transcriptional fusions. Open arrows represent coding sequences. Closed and open arrowheads labeled IR$_L$ and IR$_R$ indicate the left and right inverted repeats of the transposon, respectively. Genes vanH, vanA, and vanX are cotranscribed from a promoter located in the vanS-vanH intergenic region (Arthur et al., 1992b). Restriction sites: B, BamHI; Bg, BglII; Bm, BsmI; EI, EcoRI; EV, EcoRV; H, HindIII; Hc, HincII; P, PstI; X, XbaI. The inserts cloned into the multicopy promoter probing vector pAT78 are indicated by solid bars. CAT activity is expressed in nanomoles per minute per milligram of protein. Expression of the cat reporter gene was studied in E. faecalis JH2-2 and in a derivative of JH2-2 harboring a chromosomal copy of vanR-vanS. Induction of vancomycin-resistant and -susceptible strains was performed with 50 and 1 μg of vancomycin per milliliter, respectively. R, resistant; S, susceptible; triangles, insertional inactivation of vanR or vanS.

son et al., 1989). This delay was not observed if enterococci were pregrown in the presence of low concentrations of vancomycin, indicating that expression of resistance was inducible. Induction of resistance is associated with increased production of a 40-kDa protein (Al-Obeid et al., 1990) corresponding to the VanA ligase (Dutka-Malen et al., 1990) and of a D,D-carboxypeptidase (Gutmann et al., 1992) subsequently identified as the product of the *vanY* gene of Tn*1546* (Arthur et al., 1992a). All glycopeptides induce resistance but to various extents (Shlaes et al., 1989). Induction was also observed with moenomycin, a competitive inhibitor of transglycosylases, but not with antibiotics that inhibit other steps of cell wall synthesis (Handwerger and Kolokathis, 1990). Inhibition of transglycosylation may therefore be part of the regulatory signal (Handwerger and Kolokathis, 1990).

SEQUENCE ANALYSIS OF VanR AND VanS

Amino acid sequence similarity was detected between VanR and response regulators (RR) of the OmpR-PhoB subclass, both in the effector and DNA binding domains (Arthur et al., 1992b). RRs of this subclass regulate transcription at specific promoters thought to be recognized by the main form of RNA polymerase holoenzyme, corresponding to $E\sigma^{70}$ in *Escherichia coli* (Stock et al., 1989). The aspartic acid residue at position 53 of VanR is aligned with Asp-57 of CheY, which is phosphorylated by the associated histidine protein kinase (HPK) CheA and corresponds to an invariant position in other RRs.

The C-terminal portion of VanS displays similarity with the kinase domain of HPK including three conserved amino acid motifs identified by Stock et al. (1989) and a presumed autophosphorylation site corresponding to His-164 (Arthur et al., 1992b). The first 122 amino acids at the N terminus of VanS are not related in sequence to other HPKs. This region of VanS contains two clusters of hydrophobic amino acids that could correspond to membrane-spanning regions. VanS was thus proposed to have the same membrane topology as HPK that transduce environmental signals. This feature is consistent with the fact that glycopeptide antibiotics do not enter the cytoplasm but inhibit peptidoglycan polymerization at the cell surface (Reynolds, 1989).

PHOSPHOTRANSFER REACTIONS

Validation of the predicted roles of VanS and VanR in sequential phosphoryl group transfer was obtained by overproduction, purification, and assay of the two proteins (Wright et al., 1993). The predicted cystolic domain of VanS (residues 95 to 374) was fused to the maltose binding protein, the 76-kDa fusion protein purified to homogeneity, and observed to undergo autophosphorylation with $[\gamma^{-32}P]ATP$ ($K_m = 0.62$ mM) on a histidine residue at a rate of 0.17 min^{-1}. About 10 to 15% stoichiometry of phosphorylation of the maltose binding protein-VanS fusion protein occurred, and all the ^{32}P label could be subsequently transferred to VanR, purified as a full-length, soluble 27-kDa protein. The ^{32}P-VanR was stable for 8 to 13 h, compared with 90 min for phospho-OmpR (Igo et al., 1989) and 6 to 15 s for phospho-CheY (Hess et al., 1988). The ^{32}P-VanR released radioactivity in alkali or with hydroxylamine, consistent with phosphorylation of the conserved Asp-53. Thus, the VanS-VanR system is indeed an ATP-using two-component system that sequentially transfers a phosphoryl group to the transmembrane sensor protein and then to the RR.

IDENTIFICATION OF THE TARGET PROMOTER

Transcriptional fusions with a chloramphenicol acetyltransferase (*cat*) reporter gene were used to study the role of VanR and VanS in the regulation of vancomycin resistance genes (Arthur et al., 1992b). The *vanR* and vanS genes were introduced into the chromosome of a susceptible strain of *Enterococcus faecalis* using an integrative vector. *trans*-Activation of transcriptional fusions carried by plasmids were analyzed based on determination of CAT activity (Fig. 1). VanR and VanS were found to be required for transcription of *vanH*, *vanA*, and *vanX*. A 505-bp *Eco*RI-*Bsm*I fragment comprising the *vanS-vanH* intergenic region carries a VanR-VanS-dependent vancomycin-inducible promoter. Deletion of this fragment abolishes transcription of the *cat* reporter gene located downstream from *vanX*, indicating that *vanH*, *vanA*, and *vanX* are cotranscribed from this promoter. Mapping of the 5′ end of mRNA by S1 nuclease protection and by primer extension assays identified one transcriptional start site in the *vanS-vanH* intergenic region.

The 3′ end of the mRNA encoding *vanH*, *vanA*, and *vanX* was not mapped. The location of *vanY* (Fig. 1) indicates that VanR and VanS may regulate synthesis of VanY D,D-carboxypeptidase through cotranscription with *vanH*, *vanA*, and *vanX* (Arthur et al., 1992a). This would account for coregulation of vancomycin resistance and D,D-carboxypeptidase production (Gutmann et al., 1992).

PROMOTER ACTIVATION BY VanR

Insertional inactivation of *vanR* in the *vanR*, *vanS*, *vanH*, *vanA*, and *vanX* gene cluster carried by a multicopy plasmid abolishes expression of vancomycin resistance, transcription of the *cat* reporter gene downstream from *vanX*, and transcription initiation at the promoter located in the *vanS-vanH* intergenic region (Arthur et al., 1992b). Thus, VanR is an activator required for transcriptional initiation at the promoter for *vanH*, *vanA*, and *vanX*.

Binding studies of VanR and phospho-VanR to a 254-bp DNA fragment containing the promoter for *vanH*, *vanA*, and *vanX* analyzed by gel shift revealed low-affinity binding for VanR (50% effective concentration of 20 μM), although phospho-VanR was estimated to bind with 500-fold tighter affinity (50% effective concentration of 40 nM) (Holman et al., 1994). A second DNA fragment of 197 bp upstream from *vanR* was also gel shifted by VanR (100 μM) and phospho-VanR (1.6 μM)

but with weaker 50% effective concentrations. DNase footprinting showed protection of an 80-bp fragment (−116 to −35) in the promoter upstream from *vanH* and a 40-bp fragment (−63 to −23) in the proposed *vanR* promoter. A 12-bp palindromic sequence, present once in the 40-bp sequence and twice in the 80-bp sequence, may be a consensus recognition sequence for phospho-VanR binding, transcriptional activation, and expression of the vancomycin resistance phenotype.

MODULATION BY VanS OF VanR ACTIVITY

Cloning of *vanR*, *vanS*, *vanH*, *vanA*, and *vanX* upstream from the *cat* gene in a multicopy vector resulted in high-level transcription of the reporter gene (Arthur et al., 1992b) (Fig. 1). Expression of *cat* was constitutive or nearly so. Although insertional inactivation of *vanS* led to a fourfold decrease in the level of *cat* expression, transcription of *vanH*, *vanA*, and *vanX* was sufficient for high-level resistance to vancomycin. Thus, VanS enhances the activity of VanR as transcriptional activator, although transcription occurs in the absence of the protein. In this expression system, high-level production of VanR encoded by a multicopy plasmid may have altered the regulation process so that the promoter is active in the absence of both the inducer and phosphorylation of VanR by VanS. Inefficient promoter activation by the nonphosphorylated form of VanR or inefficient phosphorylation of VanR by chromosomally encoded HPK could account for this observation. Such cross-reactivity (crosstalk) occurs between various two-component regulatory systems despite substantial amino acid sequence diversity among RRs and HPKs (Stock et al., 1989).

ACKNOWLEDGMENTS

We thank R. Leclercq and R. Quintiliani, Jr., for critical reading of the manuscript.

REFERENCES

Al-Obeid, S., L. Gutmann, D. M. Shlaes, R. Williamson, and E. Collatz. 1990. Comparison of vancomycin-inducible proteins from four strains of *Enterococci. FEMS Microbiol. Lett.* **70**:101–106.

Arthur, M., and P. Courvalin. 1993. Genetics and mechanisms of glycopeptide resistance in enterococci. *Antimicrob. Agents Chemother.* **37**:1563–1571.

Arthur, M., C. Molinas, and P. Courvalin. 1992a. Sequence of the *vanY* gene required for production of a vancomycin-inducible D,D-carboxypeptidase in *Enterococcus faecium* BM4147. *Gene* **120**:111–114.

Arthur, M., C. Molinas, and P. Courvalin. 1992b. The VanS-VanR two-component regulatory system controls synthesis of depsipeptide peptidoglycan precursors in *Enterococcus faecium* BM4147. *J. Bacteriol.* **174**:2582–2591.

Arthur, M., C. Molinas, F. Depardieu, and P. Courvalin. 1993. Characterization of Tn*1546*, a Tn*3*-related transposon conferring glycopeptide resistance by synthesis of depsipeptide peptidoglycan precursors in *Enterococcus faecium* BM4147. *J. Bacteriol.* **175**:117–127.

Bugg, T. D. H., G. D. Wright, S. Dutka-Malen, M. Arthur, P. Courvalin, and C. T. Walsh. 1991. Molecular basis for vancomycin resistance in *Enterococcus faecium* BM4147: biosynthesis of a depsipeptide peptidoglycan precursor by vancomycin resistance proteins VanH and VanA. *Biochemistry* **30**:10408–10415.

Dutka-Malen, S., C. Molinas, M. Arthur, and P. Courvalin. 1990. The VANA glycopeptide resistance protein is related to D-alanyl-D-alanine ligase cell wall biosynthesis enzymes. *Mol. Gen. Genet.* **224**:364–372.

Gutmann, L., D. Billot-Klein, S. Al-Obeid, I. Klare, S. Francoual, E. Collatz, and J. Van Heijenoort. 1992. Inducible carboxypeptidase activity in vancomycin-resistant enterococci. *Antimicrob. Agents Chemother.* **36**:77–80.

Handwerger, S., and A. Kolokathis. 1990. Induction of vancomycin resistance in *Enterococcus faecium* by inhibition of transglycosylation. *FEMS Microbiol. Lett.* **70**:167–170.

Hess, J. F., R. B. Bourret, and M. I. Simon. 1988. Histidine phosphorylation and phosphoryl group transfer in bacterial chemotaxis. *Nature* (London) **336**:139–143.

Holman, T. R., Z. Wu, B. L. Wanner, and C. T. Walsh. 1994. Identification of the DNA-binding site for the phosphorylated VanR protein required for vancomycin resistance in *Entococcus faecium*. *Biochemistry* **33**:4625–4631.

Igo, M. M., A. J. Ninfa, J. B. Stock, and T. J. Silhavy. 1989. Phosphorylation and dephosphorylation of a bacterial transcriptional activator by a transmembrane receptor. *Genes Dev.* **3**:1725–1734.

Johnson, A. P., A. H. C. Uttley, N. Woodford, and R. C. George. 1990. Resistance to vancomycin and teicoplanin: an emerging clinical problem. *Clin. Microbiol. Rev.* **3**:280–291.

Leclercq, R., E. Derlot, M. Weber, J. Duval, and P. Courvalin. 1989. Transferable vancomycin and teicoplanin resistance in *Enterococcus faecium*. *Antimicrob. Agents Chemother.* **33**:10–15.

Murray, B. E. 1990. The life and times of the *Enterococcus*. *Clin. Microbiol. Rev.* **3**:46–65.

Reynolds, P. E. 1989. Structure, biochemistry and mechanism of action of glycopeptide antibiotics. *Eur. J. Clin. Microbiol. Infect. Dis.* **8**:943–950.

Shlaes, D. M., A. Bouvet, C. Devine, J. H. Shlaes, S. Al-Obeid, and R. Williamson. 1989. Inducible, transferable resistance to vancomycin in *Enterococcus faecalis* A256. *Antimicrob. Agents Chemother.* **33**:198–203.

Stock, J. B., A. J. Ninfa, and A. M. Stock. 1989. Protein phosphorylation and regulation of adaptive responses in bacteria. *Microbiol. Rev.* **53**:450–490.

Williamson, R. C., S. Al-Obeid, J. H. Shlaes, F. W. Goldstein, and D. M. Shlaes. 1989. Inducible resistance to vancomycin in *Enterococcus faecium* D366. *J. Infect. Dis.* **159**:1095–1104.

Wright, G. D., T. R. Holman, and C. T. Walsh. 1993. Purification and characterization of VanR and the cytosolic domain of VanS: a two-component regulatory system required for vancomycin resistance in *Enterococcus faecium* BM4147. *Biochemistry* **32**:5057–5063.

Tetracycline Regulation of Conjugal Transfer Genes

Abigail A. Salyers, Nadja B. Shoemaker, and Ann M. Stevens

25

CONJUGATIVE TRANSPOSONS: A NOVEL TYPE OF INTEGRATED GENE TRANSFER ELEMENT

For a long time, transfer of DNA by conjugation was considered to be synonymous with plasmid transfer. Within the past 2 decades, however, nonplasmid conjugal transfer elements called conjugative transposons have been found in several different types of bacteria (Speer et al., 1992; Clewell and Flannagan, 1993; Scott, 1993). Conjugative transposons are DNA segments that are normally integrated in the bacterial genome and transfer themselves by excising from the genome of the donor and transferring by conjugation to the recipient, where they integrate into the recipient's genome. The best-studied conjugative transposons are a family of gram-positive conjugative transposons, exemplified by Tn916 and Tn1545 (Scott, 1993), and a family of *Bacteroides* transposons, which have also been called tetracycline resistance (Tc^r) elements (Bedzyk et al., 1992; Speer et al., 1992).

Members of the Tn916 family range in size from 18 to 25 kbp and carry the Tc^r determinant *tetM*. The *Bacteroides* conjugative transposons are much larger, ranging in size from 60 to 150 kbp in size, and most carry the Tc^r determinant *tetQ* (Bedzyk et al., 1992; Nikolich et al., 1992). Both *tetM* and *tetQ* encode a ribosome protection type of Tc^r protein (Burdett, 1991; Nikolich et al., 1992). Aside from the fact that the Tn916-type and the *Bacteroides*-type conjugative transposon carry a similar type of Tc^r gene, they differ considerably in most other ways (Bedzyk et al., 1992; Scott, 1993). For example, they have different integration mechanisms, and the *Bacteroides* conjugative transposons integrate relatively site-specifically, whereas the Tn916-type conjugative transposons integrate more randomly. The *Bacteroides* conjugative transposons can transfer plasmids in *cis* as well as in *trans*, whereas Tn916 does not transfer plasmids in *cis* (Shoemaker and Salyers, 1990; Scott, 1993). The *Bacteroides* conjugative transposons can also excise and mobilize unlinked chromosomal elements, a type of activity that has not been reported for Tn916 and its relatives. Finally, transfer genes of most *Bacteroides* conjugative transposons are regulated by tetracycline via a complex network of regulatory genes (Stevens et al., 1993), whereas this appears not to be the case for the Tn916-type elements. The tetracycline-responsive regulatory system

Abigail A. Salyers and Nadja B. Shoemaker, Department of Microbiology, University of Illinois, Urbana, Illinois 61801. *Ann M. Stevens,* Department of Microbiology, University of Iowa, Iowa City, Iowa 52242.

Two-Component Signal Transduction, Edited by James A. Hoch and Thomas J. Silhavy, © 1995 American Society for Microbiology, Washington, DC 20005

of the *Bacteroides* conjugative transposons is the subject of this chapter.

TETRACYCLINE-REGULATED ACTIVITIES OF THE *BACTEROIDES* CONJUGATIVE TRANSPOSONS

If a *Bacteroides* strain carrying a conjugative transposon is exposed to tetracycline, the frequency with which the conjugative transposon is transferred increases by at least 1,000-fold. Only a brief period of exposure to tetracycline (<30 min) is required for this enhancement of transfer frequency, and concentrations of tetracycline as low as 1 μg/ml are effective. All derivatives of tetracycline so far tested have the same effect as tetracycline, as does the nontoxic tetracycline analog, autoclaved chlortetracycline. At least four regulatory genes appear to be involved in tetracycline enhancement of transfer frequency (Stevens et al., 1992; Li et al., submitted). These regulatory genes have been designated *rte* (regulation of *Tc*ʳ element functions). The deduced amino acid sequences of two these genes, *rteA* and *rteB,* exhibit sequence similarity to known two-component regulatory systems (Stevens et al., 1992).

There are two other cases in which two-component regulatory systems have been connected with antibiotics. YecB of *Pseudomonas fluorescens* is the putative response regulator of a two-component system that controls genes necessary for synthesis of an antibiotic (Laville et al., 1992). The VanS-VanR two-component system of *Enterococcus faecium* controls expression of vancomycin resistance genes (Arthur et al., 1992; Chapter 24). The *Bacteroides* two-component system differs from these in that it regulates transfer rather than resistance or antibiotic production genes. An odd feature of the *Bacteroides* two-component system, which is described in more detail in the next section of this chapter, is that although expression of the resistance gene *tetQ* is regulated by tetracycline, this regulation is not mediated by the two-component system. That is, the action of the two-component system seems to be confined to control of transfer genes.

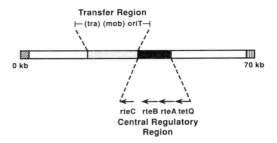

FIGURE 1 Schematic diagram of location of transfer and regulatory genes on *Bacteroides* conjugative transposon, Tcʳ Emʳ DOT. Three regulatory genes (*rteA, rteB,* and *rteC*) are located near the center of the 70-kbp element. The *oriT* of the conjugative transposon is located downstream of *rteC*. Adjacent to the *oriT* region is a 15-kbp region that contains all the genes necessary for transfer of the element. The mobilization gene(s) of Tcʳ DOT, which nicks at *oriT* to initiate the transfer process, is presumably located within this region but is indicated by (mob) because their precise location is not known.

The relative locations of known regulatory and transfer genes on the *Bacteroides* conjugative transposon TcʳEmʳ DOT are shown in Fig. 1 (Stevens et al., 1992; Li et al., submitted). The first step in the transfer process is excision of the element, which appears to occur by precise or nearly precise excision to form a double-stranded circular intermediate (Bedzyk et al., 1992). The genes involved in excision and integration have not yet been located and characterized and are thus not shown in Fig. 1, but they are probably located near one of the ends of the element. All the genes necessary for transfer of the circular intermediate are located within a 15-kbp region that lies adjacent to the regulatory genes (Fig. 1). Also located in this region is the transfer origin (*oriT*) of the conjugative transposon.

The *Bacteroides* conjugative transposons not only transfer themselves from a donor to a recipient cell but are also capable of mobilizing coresident plasmids and excising and mobilizing unlinked elements called NBUs (*non-replicating Bacteroides units*). NBUs are small (10 to 12 kbp) elements that are completely unrelated at the sequence level to the larger conjugative transposons and have a different mechanism of integration (Bedzyk et al., 1992; Shoemaker et al., 1993). Thus, NBUs are not simply small transfer-defective versions of the conjugative transposons. NBUs are usually located at a distance from the conjugative transposon on the chromosome and are induced to excise and transfer by *trans* action of proteins produced by the conjugative transposon. Excision of NBU DNA, like excision of the conjugative transposon, results in production of a double-stranded circular DNA intermediate (Shoemaker and Salyers, 1988; Shoemaker et al., 1993). The circular intermediate is then transferred by conjugation. The *oriT* of one NBU, NBU1, has been characterized and is located near the middle of the NBU (Li et al., 1993). Because the *oriT* of NBU1 is internal and not at the ends of the element, the NBU must recircularize in the recipient and probably becomes a double-stranded circle again before it integrates into the recipient's chromosome.

Both mobilization of coresident plasmids and transfer of NBUs are stimulated 100- to 1,000-fold by exposure of the donor cell to tetracycline, and this stimulation requires RteA-RteB (Valentine et al., 1988; Stevens et al., 1993; Li et al., 1993). Mobilization of coresident plasmids and NBU circle forms is mediated by the same transfer region that mediates self-transfer of the conjugative transposon. Presumably the conjugative transposon provides the proteins that form the mating pore, and the plasmid provides one or more mobilization proteins that nick the plasmid at its *oriT* and direct the DNA strand to be transferred to the mating pore. To date the mobilization genes for only one *Bacteroides* plasmid, pBFTM10, have been characterized (Hecht et al., 1991). It is still not clear whether the mobilization genes provided by the plasmid are regulated by tetracycline or whether the tetracycline induction effect results entirely from increased expression of transfer genes found on the conjugative transposon.

Both NBU excision and mobilization of the NBU circular form require proteins provided

in *trans* by the conjugative transposon (Stevens et al., 1992; Li et al., 1993) (Fig. 2). In the first stage, RteB provided by the conjugative transposon activates expression excision genes on the NBU and triggers formation of the excised circular intermediate. In the second stage, a single *mob* gene on the NBU provides a protein that presumably nicks the NBU *oriT* and initiates transfer of a single-stranded copy of the NBU. As in the case of plasmid mobilization, the conjugative transposon provides the mating pore proteins. Thus, NBUs not only parasitize the transfer machinery of the conjugative transposon but also its regulatory system. Integration of the NBU in a recipient is the only step in NBU transfer that appears not to require *trans* action by the conjugative transposon (Shoemaker et al., 1993).

TWO-COMPONENT REGULATORY SYSTEM: RteA AND RteB

rteA and *rteB* are located in the same operon, which also contains the Tc^r gene *tetQ* (Stevens et al., 1992) (Fig. 3). The gene order is *tetQ-rteA-rteB*. Insertional disruption of *rteA* or *rteB* abolishes element self-transfer of the conjugative transposon, mobilization of coresident plasmids, and excision-circularization and mobilization of NBUs. From this, it is clear that RteB is required for expression of essential transfer genes as well as for NBU excision-circularization. Because insertions in *rteA* are polar on *rteB,* it is not clear whether RteA is also essential. The deduced amino acid sequence of *rteA* exhibits sequence similarity to the sensor components of known two-component regulatory systems (Stevens et al., 1992). At the structural level, RteA is most similar to BvgS and LemA (Parkinson and Kofoid, 1992). These three putative sensors all have the unusual feature of containing potential receiver and output domains, similar to those found on response regulators, as well as the input and transmitter modules normally found in sensor components (Fig. 4). This raises the possibility that RteA might act as a transcriptional activator as well as a sensor. *rteA* alone does not

A Excision of NBUs

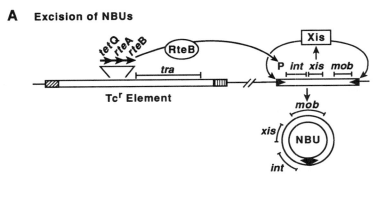

B Mobilization of NBUs

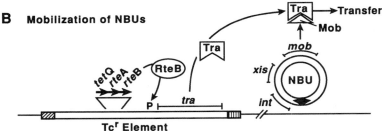

FIGURE 2 Interaction of NBUs with the conjugative transposon (Tc^r element). (A) RteB provided by the conjugative transposon activates expression of excision-integration genes (*int, xis*) on the NBU. Products of the excision gene(s), indicated here by Xis, act on the ends of the NBU to produce the excised circular form. (B) RteB, acting through RteC (not shown), activates expression of genes on the conjugative transposon that are needed for mobilization of the NBU (indicated by Tra). The NBU produces a protein, Mob, that presumably nicks at the NBU *oriT* and directs the single-stranded copy of the NBU to the mating pore provided by the conjugative transposon.

restore transfer activity to a conjugative transposon with a disruption in *rteB* (Stevens et al., 1992). Thus, if RteA is directly controlling expression of some genes on the conjugative transposon, it is controlling only a subset of the genes required for transfer. RteA has two regions that could be *trans*-membrane segments, but there is as yet no direct proof that RteA is located in the cytoplasmic membrane. The deduced amino acid sequence of *rteB* had closest similarity to the NtrC family of response regulators. Many members of this family of transcriptional activators recognize the sigma-54 form of RNA polymerase. It is not known whether RteB also recognizes a sigma-54 ho-

molog in *Bacteroides* or even whether *Bacteroides* has such a sigma factor.

At first, it seemed reasonable to assume that RteA was sensing tetracycline, but recent evidence indicates that this is not the case (Shoemaker and Stevens, unpublished data). Instead, tetracycline seems to be sensed through the promoter region of *tetQ*. Two lines of evidence support this contention. First, expression of the *tetQ-rteA-rteB* operon is increased about 20-fold when the bacteria are exposed to tetracycline and this stimulation does not require RteA or RteB. That is, when a 200-bp segment containing the putative promoter region of *tetQ* is fused to the β-glucuronidase gene, *uidA*,

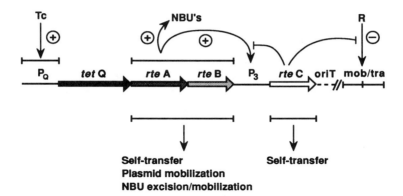

FIGURE 3 Model for tetracycline induction of transfer gene expression. *rteA* and *rteB* encode a two-component system, in which *rteA* is the putative sensor and *rteB* the putative response regulator. Tetracycline is actually sensed not through RteA but through the promoter of *tetQ*, the first gene in the operon that contains *rteA* and *rteB*. RteB activates expression of excision genes on the NBU and of *rteC*. RteC counters the effect of a repressor, which normally represses synthesis of genes needed for transfer (indicated as *mob/tra*), and interacts with RteB to suppress its own synthesis. The repressor gene has not yet been located but lies outside the transfer and regulatory areas shown in Fig. 1.

β-glucuronidase expression is still enhanced 20-fold by exposure to low levels of tetracycline in a strain that contains no copy of *rteA* or *rteB* (Nikolich, unpublished data). This finding suggests that the increased production of *rteA-rteB* message rather than tetracycline stimulation of RteA mediates the tetracycline induction effect. A second line of evidence supporting this hypothesis comes from experiments in which *rteA* and *rteB* were placed under control of a heterologous promoter, thus allowing them to be expressed independently of tetracy-

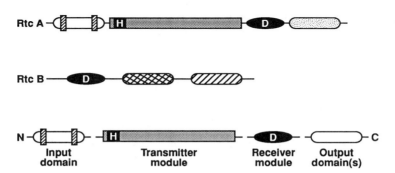

FIGURE 4 Modular structure of RteA and RteB. The output domains of RteA and RteB are shaded differently to indicate that they differ in amino acid sequence. H and D indicate histidine and aspartate residues, respectively, which are involved in phosphorylation reactions. Redrawn from Parkinson and Kofoid (1992).

cline. Under these conditions, there was no longer any tetracycline stimulation of element transfer functions (Shoemaker, unpublished data). If RteA was sensing tetracycline, the tetracycline stimulation should still have been seen.

SECOND LEVEL OF REGULATION: RteC AND A PUTATIVE REPRESSOR PROTEIN

Although the transfer region of the *Bacteroides* conjugative transposon Tc^r Em^r DOT has been localized (see Fig. 1), none of the genes encoded in this region has been characterized in any detail. Thus, it is not yet clear whether all transfer genes or only some of them are under control of RteB. The same can be said of genes involved in excision and integration. Nonetheless, there is some evidence to suggest that although RteB acts directly on the promoter regions of NBU excision genes, RteB does not act directly on the promoter regions of the Tc^r Em^r DOT transfer genes. Instead, RteB stimulates expression of another regulatory protein, RteC, which in turn acts to increase expression of transfer genes (Stevens et al., 1993). *rteC* is located downstream of *rteB* in a separate transcriptional unit (see Fig. 3). Disruption of *rteC* completely eliminates self-transfer of the conjugative transposon but has little effect on mobilization of coresident plasmids or excision and mobilization of NBUs (Stevens et al., 1993). This result suggests that RteC controls only genes that are specific to transfer of Tc^r Em^r DOT and not genes needed to construct the transfer pore, because transfer pore proteins are essential for mobilization of plasmids and NBU transfer as well as self-transfer of the conjugative transposon. Two obvious candidates for genes that might be controlled by RteC are the mobilization gene(s) of the conjugative transposon, which nick the element's own *oriT* and initiate transfer, and the genes involved in excision. At first, RteC was assumed to be a transcriptional activator, although there was no evidence that RteC binds DNA nor was any DNA binding motif apparent in the deduced amino acid sequence of RteC. New evidence suggests that RteC is actually an antirepressor (Li et al., submitted). When the 15-kbp transfer region was cloned away from the rest of Tc^r Em^r DOT, the plasmid carrying this region was now capable of constitutive high-level self-transfer. This result suggests that there is a repressor protein, which normally represses transfer gene expression but which lies outside the cloned transfer region. If such a repressor exists, the effect of RteC in allowing transfer gene expression in the presence of the repressor could be to bind the repressor and prevent it from binding to the promoter region of the transfer genes it normally represses (see Fig. 3). RteC appears to repress its own synthesis, because expression of an *rteC-uidA* fusion is much higher in a strain containing a conjugative transposon with a disruption in *rteC* than in a strain containing a wild-type conjugative transposon (Shoemaker and Stevens, unpublished data). Thus, RteC may also interact with RteB.

CONCLUSIONS

The regulatory machinery of the *Bacteroides* conjugative transposons is turning out to be much more complex than expected. The current model, illustrated in Fig. 3, can be summarized as follows. Tetracycline is sensed by the promoter of the *tetQ* operon, not by RteA. Elevated expression of *rteB* message then leads to an increase in the amount of RteB protein. RteA may still be needed to phosphorylate RteB, although this has not been proved. RteB in turn activates expression of RteC, which counters the action of a repressor that normally prevents essential excision or mobilization genes from being expressed. To prevent excess buildup of RteC, RteC suppresses its own synthesis, possibly by binding to RteB and interfering with RteB's efficacy as a transcriptional activator. RteB also activates genes on the NBUs that are necessary for NBU excision and circularization.

Why is such a complex regulatory system needed to regulate transfer, and why is tetracycline the signal it senses? Tetracycline has been in widespread use for only a relatively short time. It seems unlikely that such a complex signal transduction system would have evolved within the past several decades. One possible explanation is that the system evolved in soil bacteria, which are exposed to antibiotic-producing fungi and bacteria, and then moved into the *Bacteroides* group. Antibiotic-producing microorganisms are not present in the colon, where *Bacteroides* is normally found. *Flavobacterium,* a genus in the *Bacteroides* phylogenetic group, is found widely in soil, and the colonic *Bacteroides* could have picked up the conjugative transposons from flavobacteria transiently colonizing the intestine. So far, no information is available about gene transfer elements in *Flavobacterium* spp.

Another possible explanation for the role of tetracycline is that tetracycline is only an accidental inducer. Tetracycline consists of four fused phenolic rings. The fact that all derivatives of tetracycline so far tested activate the signal transduction system indicates that only a portion of the tetracycline molecule, the portion conserved in all these derivatives, is recognized by the system. Compounds with structures that mimic part of the tetracycline structure are abundant in nature (e.g., the plant flavones and flavinoids, lignin precursors, and tannins). Because the *Bacteroides* conjugative transposons transfer only when bacteria are grown on solid surfaces and because bacteria in the colon, where *Bacteroides* is normally found, encounter an abundance of plant surfaces, the sensory system could actually be intended to sense the presence of a surface that is conducive to successful mating. The complex regulatory system could be explained by the need to limit transfer frequency, possibly to save energy by reducing to the absolute minimum the number of transfer proteins actually produced. The only problem with this scenario is that a screen of commercially available plant phenolics has so far not revealed any compounds that stimulate the RteA-RteB-RteC system (Bonheyo, unpublished data). Whatever the actual inducer

turns out to be, it is clearly important to the bacteria or at least to the conjugative transposons themselves to limit the extent to which the conjugal transfer machinery is expressed.

The fact that an antibiotic is capable of stimulating transfer of a conjugative element, which carries antibiotic resistance genes and can mobilize other elements that also carry resistance genes, shows that antibiotics can do more than simply select for bacteria that have acquired a resistance gene: they can stimulate its transfer in the first place. The widespread use of tetracycline in animal feed as well as for treatment of human infections may explain why the *Bacteroides* conjugative transposons and the elements they mobilize are found so widely in *Bacteroides* clinical isolates. Antibiotic resistance genes have also been found on plasmids and NBUs that rely on the *Bacteroides* conjugative transposons for their transfer (Hecht et al., 1991; Li et al., 1993; Smith and Parker, 1993). Thus, the role of *Bacteroides* conjugative transposons in antibiotic resistance transfer extends beyond the conjugative transposons themselves.

It is not yet clear whether the phenomenon of antibiotic-induced transfer genes is restricted to the *Bacteroides* or will be found in other groups of bacteria. It is important to bear in mind the fact that transfer of *tetQ* would not have been detected if tetracycline had not been incorporated in the medium used to grow the donors. Thus, some chromosomally encoded genes that appear to be nontransmissible could actually be carried on a conjugal element, whose transfer functions must be stimulated by some inducer not normally included in the medium. This possibility needs to be kept in mind when screening environmental and clinical isolates for gene transfer capability.

ACKNOWLEDGMENT

Work described in this chapter was supported by grant AI22383 from the National Institutes of Health.

REFERENCES

Arthur, M., C. Molinas, and P. Courvalin. 1992. The VanS-VanR two component regulatory system controls synthesis of depsipeptide peptidoglycan precursors in *Enterococcus faecium* BM4147. *J. Bacteriol.* **174:**2582–2591.

Bedzyk, L. A., N. B. Shoemaker, K. E. Young, and A. A. Salyers. 1992. Insertion and excision of *Bacteroides* conjugative chromosomal elements. *J. Bacteriol.* **174:**166–172.

Bonheyo, G. Unpublished data.

Burdett, V. 1991. Purification and characterization of Tet(M), a protein that renders ribosomes resistant to tetracycline. *J. Biol. Chem.* **266:**2872–2877.

Clewell, D. B., and S. E. Flannagan. 1993. The conjugative transposons of gram-positive bacteria, p. 369–393. *In* A. L. Sonenshein, J. A. Hoch, and R. Losick (ed.), *Bacillus subtilis and Other Gram Positive Bacteria.* American Society for Microbiology, Washington, D.C.

Hecht, D. W., T. J. Jagielo, and M. H. Malamy. 1991. Conjugal transfer of antibiotic resistance factors in *Bacteroides fragilis:* the *btgA* and *btgB* genes of plasmid pBFTM10 are required for its transfer from *Bacteroides fragilis* and for its mobilization by IncPβ plasmid R751 in *Escherichia coli. J. Bacteriol.* **173:**7471–7480.

Laville, J., C. Voisart, C. Keel, M. Maurhofer, G. Defago, and D. Haas. 1992. Global control of *Pseudomonas fluorescens* mediates antibiotic synthesis and suppression of black root rot of tobacco. *Proc. Natl. Acad. Sci. USA* **89:**1562–1566.

Li, L. Y., N. B. Shoemaker, and A. A. Salyers. 1993. Characterization of the mobilization region of a *Bacteroides* insertion element (NBU1) that is excised and transferred by *Bacteroides* conjugative transposons. *J. Bacteriol.* **175:**6588–6598.

Li, L. Y., N. B. Shoemaker, and A. A. Salyers. Localization of the transfer origin (oriT) of a Bacteroides conjugative transposon and evidence for an additional layer of transfer genes. Submitted for publication.

Nikolich, M. Unpublished data.

Nikolich, M. P., N. B. Shoemaker, and A. A. Salyers. 1992. A *Bacteroides* tetracycline resistance gene represents a new class of ribosome protection tetracycline resistance. *Antimicrob. Agents Chemother.* **36:**1005–1012.

Parkinson, J. S., and E. C. Kofoid. 1992. Communications modules in bacterial signalling proteins. *Annu. Rev. Genet.* **26:**71–112.

Scott, J. R. 1993. Conjugative transposons, p. 597–614. *In* A. L. Sonenshein, J. A. Hoch, and R. Losick (ed.), *Bacillus subtilis and Other Gram Positive Bacteria.* American Society for Microbiology, Washington, D.C.

Shoemaker, N. Unpublished data.

Shoemaker, N., and A. Stevens. Unpublished data.

Shoemaker, N. B., and A. A. Salyers. 1988. Tetracycline-dependent appearance of plasmidlike forms in *Bacteroides uniformis* 0061 mediated by conjugal *Bacteroides* tetracycline resistance elements. *J. Bacteriol.* **170:**1651–1657.

Shoemaker, N. B., and A. A. Salyers. 1990. A cryptic 65 kb transposon-like element isolated from *Bacteroides uniformis* has homology to the *Bacteroides* conjugal tetracycline resistance elements. *J. Bacteriol.* **172:**1694–1702.

Shoemaker, N. B., G.-R. Wang, A. M. Stevens, and A. A. Salyers. 1993. Excision, transfer and integration of NBU1, a mobilizable site-selective insertion element. *J. Bacteriol.* **175:**6578–6587.

Smith, C. J., and A. C. Parker. 1993. Identification of a circular intermediate in the transfer and transposition of Tn*4555,* a mobilizable transposon from *Bacteroides* species. *J. Bacteriol.* **175:**2682–2691.

Speer, B. S., N. B. Shoemaker, and A. A. Salyers. 1992. Resistance to tetracycline: mechanisms and transmission. *Clin. Microbiol. Rev.* **5:**387–399.

Stevens, A. M., J. M. Sanders, N. B. Shoemaker, and A. A. Salyers. 1992. Genes involved in production of plasmidlike forms by a *Bacteroides* conjugal chromosomal element share significant amino acid homology with two component regulatory systems. *J. Bacteriol.* **174:**2935–2942.

Stevens, A. M., N. B. Shoemaker, L.-Y. Li, and A. A. Salyers. 1993. Tetracycline regulation of genes on *Bacteroides* conjugative transposons. *J. Bacteriol.* **175:**6134–6141.

Stevens, A. M., N. B. Shoemaker, and A. A. Salyers. 1990. Genes on a *Bacteroides* conjugal tetracycline resistance element which mediate production of plasmid-like forms from unlinked chromosomal DNA may be involved in transfer of the resistance element. *J. Bacteriol.* **172:**4271–4279.

Valentine, P. J., N. B. Shoemaker, and A. A. Salyers. 1988. Mobilization of *Bacteroides* plasmids by *Bacteroides* conjugal elements. *J. Bacteriol.* **170:**1319–1324.

CELLULAR
COMMUNICATION
AND DEVELOPMENT

VI

Switches and Signal Transduction Networks in the *Caulobacter crescentus* Cell Cycle

Todd Lane, Andrew Benson, Gregory B. Hecht, George J. Burton, and Austin Newton

26

Cell differentiation in the gram-negative bacterium *Caulobacter crescentus* results from asymmetric cell division that produces two morphologically distinct progeny, a nonmotile stalked cell and a motile swarmer cell. The stalked cell divides repeatedly like a stem cell to yield the same nonmotile stalked cell and a new motile swarmer cell. The pole-specific swarmer cell structures, which include the flagellum, DNA bacteriophage φCbK receptors, and pili, are assembled in a tightly ordered sequence during the course of the cell cycle (Fig. 1). Flagellar rotation begins immediately before cell separation, and at the completion of division, these new polar structures segregate with the motile swarmer cell. Each of the daughter cells also inherits its own distinct developmental program and pattern of chromosome replication (Degnen and Newton, 1972). The stalked cell initiates DNA synthesis immediately after division, whereas the swarmer cell enters a presynthetic gap, or G1 period, during which it retracts the pili, loses motility, sheds the flagellum, and forms a stalk before initiating chromosome replication (Newton

and Ohta, 1990; Brun et al., 1994) (reviewed in Fig. 1).

Complex developmental programs in many prokaryotes, including heterocyst formation in *Anabaena* (Haselkorn, 1992; Wolk, 1991), bacteroid development in *Rhizobium* spp. (Brewin, 1993; Gottfert, 1993), sporulation in *Bacillus* spp. (Chapter 8), and fruiting body formation in myxobacteria (Kim et al., 1992; Shimkets, 1990), are initiated in response to external stimuli such as nutrient deprivation or signals from neighboring cells. The temporal and spatial patterns of developmental events in *C. crescentus,* by contrast, are invariant in the cell division cycle and do not appear to depend on external signals. Genetic analysis has shown that the developmental programs in these cells are triggered instead by internal signals originating from the cell division cycle, namely, DNA replication and cell division (reviewed in Huguenel and Newton, 1982; Newton and Ohta, 1990; Horvitz and Herskowitz, 1992). There is now direct evidence that histidine kinases and response regulator proteins are key components in the regulation of *C. crescentus* differentiation. Two examples are considered in this chapter. In the first, evidence is discussed that developmental events are coupled to the cell division cycle by

Todd Lane, Andrew Benson, Gregory B. Hecht, George J. Burton, and Austin Newton, Department of Molecular Biology, Princeton University, Princeton, New Jersey 08544.

Two-Component Signal Transduction, Edited by James A. Hoch and Thomas J. Silhavy,
© 1995 American Society for Microbiology, Washington, DC 20005

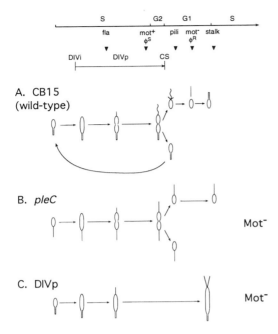

FIGURE 1 *C. crescentus* cell cycle. (A) The sequence of developmental events in the wild-type strain CB15 includes flagellum formation (fla), appearance of polar bacteriophage receptors (ϕ^S), activation of flagellum rotation (mot^+), pili formation (pili), loss of motility (mot^-), loss of bacteriophage receptors (ϕ^R), and stalk formation. The periods corresponding to DNA synthesis (S), postsynthetic gap (G2), presynthetic gap (G1), as well as division initiation (DIVi), division progression (DIVp), and cell separation (CS) are indicated. (B) Nonmotile *pleC* mutants assemble inactive flagella (designated as straight lines), are bacteriophage φCbK resistant, and fail to form stalks, but they divide normally (Sommer and Newton, 1989). (C) Cells blocked before the completion of DIVp form long unpinched filaments. Development is blocked, but flagellum biosynthesis continues; consequently, these cells do not form stalks but accumulate multiple inactive flagella at a single pole after several generations (Sheffery and Newton, 1981).

a complex signal transduction pathway mediated by sensor histidine kinases and effector proteins. In the second, the role of the response regulator FlbD is examined in flagellum biosynthesis, where it functions as both a transcription activator and repressor to regulate the timing of flagellar (*fla*) gene expression in the cell cycle.

CELL DIVISION CYCLE CHECKPOINTS AND THE REGULATION OF DEVELOPMENT

Cell division in *C. crescentus* can be visualized by the progressive pinching at the future site of cell separation, and genetic analysis shows that the process can be divided into three morphological stages: DIVi (initiation of division), DIVp (division progression), and CS (cell separation) (Fig. 1). Temperature-sensitive cell division mutants blocked in DIVi form long unpinched filaments, those blocked in DIVp are arrested at various stages of pinching, and those blocked at CS form chains of cells with highly constricted division sites (reviewed in Newton and Ohta, 1990). The first indication that cell division was required for development in *C. crescentus* was the observation that low concentrations of the β-lactam penicillin G, which blocks cells early in division at DIVi, also block motility and stalk formation (Terrana and Newton, 1976). A similar developmental phenotype was subsequently observed in mutants with temperature-sensitive defects in either DIVi or DIVp (Huguenel and Newton, 1982). Filamentous mutants blocked later in division at CS were shown to be motile but arrested in development before the assembly of pili (Sommer and Newton, 1988). As discussed in the second part of this chapter, there is also experimental evidence that DNA synthesis is required for initiation of the *fla* gene transcription cascade. Thus, successive cell division cycle events, DNA synthesis, DIVp, and CS, act as morphological or functional checkpoints required for specific events in the developmental program.

Pseudoreversion Analysis of Mutants in the Developmental Gene *pleC*

An approach to identifying genes acting at cell division checkpoints to control development was suggested by the phenotype of the pleiotropic gene *pleC*. *pleC* mutants divide normally, but their developmental program is arrested at the same stage as mutants blocked in the early DIVi or DIVp stages of division, namely, after the assembly of the flagellum at the cell pole but before the onset of flagellar rotation and stalk formation (Fig. 1). Consequently, both the cell division and *pleC* mutants retain a paralyzed flagellum and never form stalks. These observations suggested that *pleC* might be a developmental gene that responds directly to cell division checkpoints (Sommer and Newton, 1991).

Pseudoreversion analysis of a temperature-sensitive *pleC* mutation identified cold-sensitive suppressors that map to three new cell division genes, *divJ, divK,* and *divL.* Suppressors in these genes compensate for the original motility defect at 37°C, but they also display cell division phenotypes at 24°C (Sommer and Newton, 1991). The *divK* allele behaves as a bypass suppressor, suppressing the motility phenotype of all *pleC* alleles tested and conferring a severe division defect in either a *pleC* or wild-type genetic background. By contrast, the effectiveness of *divJ* and *divL* alleles in suppression depend strongly on the *pleC* mutations examined.

Sequence Analysis of *pleC, divJ,* and *divK*

DNA sequence analysis of *divJ* (Ohta et al., 1992) and *pleC* (Wang et al., 1993) show that both genes encode proteins with carboxy-terminal domains homologous to the histidine kinases of the bacterial sensor proteins (reviewed in J. B. Stock et al., 1989; Parkinson and Kofoid, 1992). The predicted *pleC* product contains 842 amino acids with a carboxy terminus containing the conserved or invariant residues including the histidine at position 610, an asparagine at residue 729, and the G1 and G2 blocks Asp-Xaa-Gly-Xaa-Gly and Gly-Xaa-Gly-Xaa-Gly located 30 to 60 residues downstream of residue 729 (Wang et al., 1993) (Fig. 2) (see Chapter 3). The amino terminus of PleC is predicted to contain two membrane-spanning regions on either side of a 230-residue periplasmic loop, a protein architecture typical of this protein family. Although the periplasmic domain is generally thought to sense environmental signals that modulate the

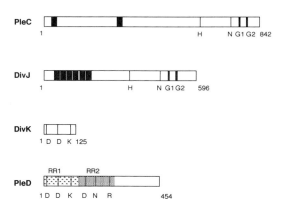

FIGURE 2 Predicted domain organization of histidine kinases PleC (Wang et al., 1993) and DivJ (Ohta et al., 1992) and response regulators DivK (Hecht et al., submitted) and PleD (Hecht and Newton, unpublished data) deduced from translated DNA sequences. Closed boxes indicate transmembrane domains; H, N, G1, and G2 indicate conserved His, Asn, Asp-Xaa-Gly-Xaa-Gly, and Gly-Xaa-Gly-Xaa-Gly motifs, respectively; RR1 and RR2 are the two response regulator domains in PleD; D, K, and R indicate conserved Asp, Lys, and Arg residues, respectively.

activity of the catalytic domain, the possibility that PleC may respond instead to autogenous signals generated during the cell cycle is discussed below.

The predicted DivJ protein contains 596 amino acids including a carboxy-terminal region with nearly 50% identity to the histidine kinase PleC (Ohta et al., 1992). The amino-terminal 150 residues of the protein are strongly hydrophobic; they may contain six membrane-spanning regions and are unlikely to include a periplasmic loop (Ohta et al., 1992) (Fig. 2). This amino-terminal architecture is unusual but not unique. The histidine kinase FixL of *Rhizobium meliloti* also contains a strongly hydrophobic amino terminus containing multiple transmembrane regions and a linker containing a heme binding domain (Gilles-Gonzalez et al., 1991). The histidine kinase UhpB of the *Escherichia coli* and *Salmonella typhimurium* sugar transport systems contains a hydrophobic amino terminus with 6 to 10 transmembrane domains and, like DivJ, appears to lack a large periplasmic loop. Along with the response regulator UhpA, this system requires a third regulatory protein, UhpC, which consists of 12 transmembrane domains and physically interacts with UhpB to form a signaling complex (Chapter 16).

The predicted DivK protein contains 125 amino acids and shares 25 to 35% homology with the phosphoacceptor or regulatory domains of several bacterial response regulator proteins (Hecht et al., submitted) (Fig. 2). In particular, the Glu-9, Asp-10, Asp-53, and Lys-105 residues of DivK correspond to the four typically invariant residues found in these proteins. Also, DivK contains the hydrophobic sequence motifs conserved in most of these proteins that are believed necessary for proper folding (A. M. Stock et al., 1989; J. B. Stock et al., 1989; Parkinson and Kofoid, 1992; Volz, 1993) and a primary sequence favorable for the formation of a γ-turn loop that is characteristic of response regulators (Volz, 1993).

Although response regulators often function to control transcription of cognate target genes via their carboxy-terminal DNA binding domains in response to phosphorylation, DivK belongs to a subfamily of response regulators that consists entirely of the conserved amino-terminal domain (Hecht et al., submitted). This group of proteins includes the *Bacillus subtilis* SpoOF and the enteric CheY proteins. Thus, DivK could function as a phosphorelay protein similar to SpoOF (Chapter 8), by interacting with a target protein such as CheY (Chapter 6), or both.

Catalytic Activities of PleC, DivJ, and DivK

Phosphotransfer experiments have been carried out in parallel with the genetic and molecular analyses described above. The purified carboxy-terminal sequences of PleC (PleC★) and DivJ (DivJ★) encoding the kinase domains of these proteins have been overexpressed and purified (Newton et al., 1994). The full-length DivK protein was purified by a similar strategy. Both PleC★ and DivJ★ display the expected autophosphorylation in the presence of ATP

and Mg²⁺ (Newton et al., 1994). Wang et al. (1993) demonstrated previously the autophosphorylation of a LacZ-PleC fusion protein by ATP.

The PleC★ and DivJ★ proteins also catalyze the rapid phosphotransfer from ATP to the purified DivK protein, but the initial rates of phosphorylation of the two kinases differ significantly in the presence and absence of DivK. The initial rate of DivK phosphorylation in the reaction with PleC★ is 25- to 100-fold higher than can be accounted for by the rate of PleC★ autophosphorylation in the absence of DivK. This result suggests a strong synergy between the PleC and DivK proteins. By contrast, addition of DivK to the reaction with DivJ★ increased the apparent initial rate of autophosphorylation only threefold (Newton et al., 1994). Preliminary results also indicate both PleC★ and DivJ★ have phospho-DivK phosphatase activity but that phospho-DivK is less stable in the presence of DivJ★ than in the presence of PleC★ (Lane et al., unpublished data). We can only speculate that the kinase and phosphatase activities measured in these experiments reflect activities of the full-length proteins. The enzymatic activities of the truncated PleC★ and DivJ★ kinases, along with those demonstrated for the response regulator DivK, are consistent, however, with the roles proposed for PleC, DivJ, and DivK in the regulation of *C. crescentus* division and development discussed below.

Is the PleC-DivK Signal Transduction Pathway Controlled by a Cell Cycle Checkpoint?

Several lines of evidence indicate that DivK is the cognate response regulator of PleC in its regulation of motility and stalk formation. The most compelling of these is genetic and comes from the isolation of *divK* alleles as bypass suppressors of the nonmotile phenotype of both *pleC* point mutations (Sommer and Newton, 1991) and null alleles (Hecht and Newton, unpublished data). Also consistent with the action of PleC and DivK in the same pathway is the observation that *divK* transcription is restricted to a period in the cell cycle during late S period when the PleC kinase is known to be required for activation of motility (Sommer and Newton, 1989; Hecht et al., submitted).

Given the coincidence of PleC function (Sommer and Newton, 1989) with the completion of DIVp, both of which are required for motility, it is tempting to speculate that successful completion of the DIVp checkpoint activates the PleC-DivK signal transduction pathway that is required in turn for motility and stalk formation. Thus, the PleC-DivK pathway may correspond to one of the couples between cell division and morphogenesis proposed originally by Huguenel and Newton (1982). Current evidence in support of this conclusion is circumstantial, and the response of a sensor histidine kinase to autogenous signals would constitute a novel form of regulation. However, if PleC does function as a conventional, environmentally regulated sensor, then it must respond to an unanticipated and as-yet-unidentified environmental stimulus.

Role of DivJ and DivK in Cell Division

DivK is unusual among response regulators and may be the first member of this protein family described in bacteria to play an essential role in regulating the cell division cycle. Conditional *divK* mutants at the nonpermissive temperature form straight unpinched filaments similar in phenotype to DIVi mutants, which suggests that DivK is responsible for an early step in the cell division pathway. Moreover, construction of a *divK* null allele in the *C. crescentus* chromosome has been successful only in strains carrying a complementing *divK* plasmid, strongly suggesting that physically disrupting the *divK* gene is a lethal event unless another functional copy is present in the cell (Hecht et al., submitted).

Although these results indicate that DivK plays a key role in regulating some aspect of cell division, the cognate histidine kinase PleC is not required for division. Null mutants of *pleC*

constructed by several different methods, including replacement of the entire gene with an antibiotic-resistance cassette, are nonmotile and resistant to bacteriophage φCbK, but they divide normally (Burton and Newton, unpublished data). Thus, *pleC,* unlike *divK,* is not essential for viability, and the function of this gene may be restricted to the regulation of polar morphogenesis. It seems likely that another histidine kinase is responsible for controlling the regulation of cell division by DivK.

One candidate for the gene encoding the proposed second kinase is *divJ.* Genetic results suggest that *divJ* alleles behave as gain of function mutations in their suppression of the *pleC* motility defect by altering the level of DivK phosphorylation, thus switching motility on. At the nonpermissive temperature, however, conditional mutations in the *divJ* kinase gene produce filamentous cells that are largely unpinched and similar in phenotype to *divK* mutants (Hecht et al., submitted). This early block in cell division coincides with the time during late G1 and early S period when *divJ* is maximally transcribed (Ohta et al., 1992). Although these results raise the possibility that DivJ and DivK function as a cognate pair to regulate cell division, direct genetic evidence for this conclusion is still lacking (see below).

The regulation of motility during late S period and cell division during late G1 and early S period by *divK* suggests that the DivK protein has separate functional targets. This conclusion is strongly supported by the observation that *divK* mutations are recessive for their cell division phenotype but dominant for their ability to activate motility (Hecht and Newton, unpublished data). The function of DivK at different times in the cell cycle to regulate motility and cell division, respectively, could depend on the temporal regulation of kinase or phosphatase activities in the cell cycle (Fig. 3). In this way, *divK* would act as a genetic switch to initiate the developmental pathway to motility and stalk formation in response to PleC activity and as a positive regulator to

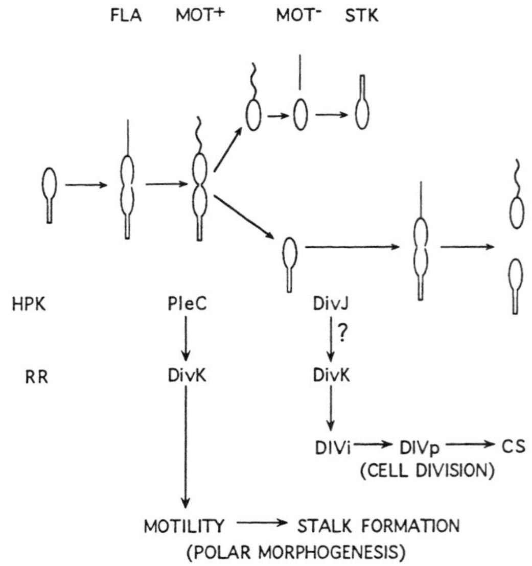

FIGURE 3 Roles of PleC, DivK, and DivJ in cell division and polar morphogenesis. This model proposes that the response regulator (RR) DivK acts at two different times in the cell cycle to control polar morphogenesis and cell division. Late in the cell cycle, DivK responds to the PleC histidine protein kinase (HPK) to initiate motility and stalk formation, and early in the subsequent cell cycle it responds to another sensor HPK, shown here as DivJ, to regulate cell division. For details, see text. Abbreviations: FLA, flagellum biosynthesis; MOT, motility; STK, stalk formation.

control cell division in response to the activity of DivJ or some other histidine kinase.

If the DivK protein is, in fact, a temporally regulated developmental switch, then the timing of PleC and DivJ activities would need to be precisely regulated to maintain the tightly ordered expression of developmental and division events in the *C. crescentus* cell cycle. In this context, it is not surprising that development in these cells is largely independent of environmental conditions and appears to depend instead on signals generated by the cell cycle (Newton and Ohta, 1990; Horvitz and Herskowitz, 1992). We know that flagellum biosynthesis, activation of motility, and pili formation require the completion of successive cell division cycle checkpoints, as discussed above in

this chapter and elsewhere (Huguenel and Newton, 1982).

The observation that mutations in *divJ* suppress the motility defect of *pleC* mutations raises the question of the normal functional relationship between DivJ and DivK. Systems with multiple kinases include the *E. coli* histidine kinases NarX and NarQ, both of which interact with the response regulators NarL and NarP to control nitrate- and nitrite-regulated gene expression (Schroder et al., 1994) (Chapter 14). We describe below a second response regulator, PleD, that may function in addition to DivK in *C. crescentus* to regulate motility. An alternative explanation of *divJ* suppression of *pleC* mutations is a mutationally enhanced form of cross regulation, like that described in the *E. coli* Pho system. PhoR is an environmentally responsive histidine kinase that along with its cognate response regulator, PhoB, regulates phosphorus uptake. Mutations in *phoR* that completely abolish its function still display a basal level of PhoB activity that can be accounted for by the interaction of PhoB with an unrelated kinase, CreC (Wanner, 1992) (Chapter 12). Genetic screens for bypass suppressors of various *divJ* mutations may resolve the issue of whether DivJ is the cognate kinase of DivK.

Other Players in Developmental Regulation: *pleD* and *divL*

Detailed models of developmental regulation in *C. crescentus* must at some point take into account additional genes that have been identified in pseudoreversion studies of *pleC*. One of these is *divL*. Temperature-sensitive and cold-sensitive alleles in this gene, like the motility suppressors in *divJ*, confer dramatic conditional cell division defects, the extent of which depends on the *pleC* mutation examined (Sommer and Newton, 1991). The possibility that the *divL* gene product is also involved in phosphotransfer must await DNA sequence analysis.

Another gene identified in the same study is *pleD*, which encodes a compound response regulator (see Fig. 2) and lies in the *divK* operon (Hecht and Newton, unpublished data). Suppressors in this gene bypass the motility defect of *pleC* mutations, and they are also defective in other functions, including turning off motility, shedding the flagellum, and forming the stalk (Sommer and Newton, 1989). Thus, "supermotile" *pleD* strains, which divide normally, retain their motility throughout the cell cycle (Sommer and Newton, 1989).

A *pleD::kan* null mutation confers the same supermotile phenotype as the original suppressors, which suggests that *pleD* normally acts as a negative regulator of motility in a *pleC*-dependent fashion. The PleD protein may play key roles in the gain of motility late in the cell cycle and in differentiation of the motile swarmer cell into a stalked cell during the G1 period. The semidominant *pleD301* point mutation also confers a supermotile phenotype in both wild-type and *pleC* strains (Hecht and Newton, unpublished data). One interpretation of this observation is that the conformation of the mutant PleD301 protein allows the onset of motility after flagellum biosynthesis, but it is unable to function later in development to effect the loss of motility and stalk formation. If these interpretations of the *pleD::kan* and *pleD301* phenotypes are correct, we speculate that PleD could be similar to DivK in that it responds to two temporally distinct cell cycle signals mediated by histidine kinases.

Prospects

The structures and activities of the histidine kinases and response regulators discussed above are typical of those described in other bacterial two-component systems. What sets them apart, however, is that they do not respond to any identified environmental conditions and are instead involved in the regulation of essential division and developmental functions. The roles played by the *C. crescentus* two-component systems may not be unique, and it is possible that similar two-component switches are a general feature of cell cycle and developmental regulation in prokaryotes. They could also fulfill related roles in other organisms. Recently, two-component systems have been discovered in eukaryotic cells and implicated in the regulation of the MAP (mitogen activated protein) kinases, which are thought to be involved in the control of mammalian cell division and differentiation along with other processes. In *Saccharomyces cerevisiae,* the SLN1 histidine kinase (Ota and Varshavsky, 1993) and the SSK1 response regulator (Maeda et al., 1994) osmoregulate the HOG1 MAP kinase pathway necessary for growth on various media. In *Arabidopsis thaliana,* the ETR1 histidine kinase (Chang et al., 1993), which mediates responses to both gaseous ethylene and to wounding and pathogen invasion (Abeles, 1992), may also regulate a MAP kinase activator (Chang et al., 1993). The two-component systems in both of these organisms are thought to respond to external stimuli in a manner similar to their counterparts in most bacterial systems.

Our working model in Fig. 3 proposes that PleC, DivJ, and DivK are members of signal transduction pathways that regulate polar morphogenesis and motility on the one hand and cell division on the other. Outstanding questions about the role of these proteins include the following: (i) Does DivK act at two different times in the cell cycle, and if so, does its differential activity depend on alternative states or levels of phosphorylation? (ii) Are the kinase and/or phosphatase activities of the PleC and DivJ proteins restricted to different times in the cell cycle, as suggested in Fig. 3? (iii) Is DivK or another protein(s) the cognate response regulator of DivJ? (iv) How do the activities of the PleD and DivL proteins contribute to this signal transduction system? (v) What proteins or genes are the targets of DivK regulation? It should be possible to address these questions using many of the genetic and biochemical approaches used originally to identify these signal transduction proteins.

ROLE OF FlbD AS A SWITCH PROTEIN IN FLAGELLAR GENE TRANSCRIPTION

Biosynthesis of the polar flagellum is at once the most dramatic and best-studied morphological change during *C. crescentus* differentia-

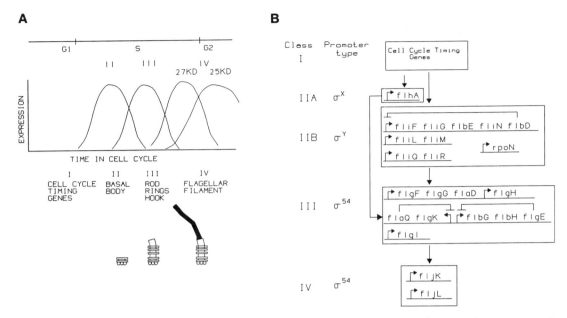

FIGURE 4 (A) Sequence of *fla* gene expression during the cell cycle. The graph illustrates the sequence and timing of class II, III, and IV *fla* gene transcription during the *C. crescentus* cell cycle. The periods of the presynthetic gap (G1), synthesis (S), and postsynthetic gap (G2) are indicated above the graph. (B) The flagellar hierarchy. Genes encoding proteins that are necessary for *fla* gene synthesis are arranged in hierarchical order based on epistasis experiments. The promoter structures of genes at each level are listed at the left, along with the class to which they belong. σ^X and σ^Y are hypothetical σ-factors that direct transcription from the unique promoters of the class IIA and class IIB genes.

tion (see Fig. 1). It is synthesized during a discrete portion of the cell cycle, and its function depends on the PleC and DivK proteins in the cell cycle, as discussed above. The flagellum consists of a basal body, which is embedded in the membrane and acts as the motor for flagellum rotation, the hook, and the filament, which is attached to the basal body by the hook and extends outside the cell. Although this structure is similar to that in the enteric bacteria, the mechanisms that regulate expression of the *C. crescentus* flagellar (*fla*) genes are very different, being driven by internal cell cycle signals rather than nutritional or environmental signals that modulate *fla* gene expression in *E. coli* and *S. typhimurium* (reviewed in Macnab, 1992). The *C. crescentus* flagellum is assembled sequentially during the cell cycle from flagellar proteins that are synthesized in generally the same order that they assemble, namely, from the

inside of the cell to the outside (Newton and Ohta, 1990; Brun et al., 1994). Thus, the basal-body proteins are expressed earliest in the cell cycle, followed by synthesis of the hook and, finally, the flagellins (Fig. 4A).

The *fla* genes are organized into a four-tiered regulatory hierarchy in which genes at each level (class I to IV) are required for expression of genes at the lower levels. Class II genes near the top of the hierarchy encode the innermost components, including the switch proteins, MS ring, and flagellar protein transport apparatus (Ramakrishnan et al., 1991; Dingwall et al., 1992b; Yu and Shapiro, 1992; Ramakrishnan et al., 1994). These genes are expressed earliest in the cell cycle, and they are required for expression of class III and IV genes lower in the hierarchy. Class III and IV genes, which are expressed later in the cell cycle, encode the remaining components of the basal body, the external hook, and the flagellar filament (Barrett et al., 1982; Minnich and Newton, 1987; Minnich et al., 1988; Dingwall et al., 1990; Dingwall et al., 1992a). Although the hierarchical organization of *C. crescentus fla* genes is similar to that in enteric bacteria (Macnab, 1992), the *C. crescentus* hierarchy has four, rather than three, classes of genes, and the positions of several genes in the hierarchies differ (summarized in Fig. 4B).

The exact nature of the cell cycle "clock" that governs *fla* gene expression in *C. crescentus* is unknown, but experiments with several class II, III, and IV genes have shown that interruption of ongoing DNA synthesis prevents their expression (Osley et al., 1977; Sheffery and Newton, 1981; Dingwall et al., 1992b). These results have led to the conclusion that *fla* gene transcription is triggered in some way by DNA replication or the replication machinery. Although it has not been definitively ruled out that an SOS-like function activated by DNA synthesis inhibition could account for these results, it has been shown that *fla* gene expression in *C. crescentus* strains with *rec* mutations deficient in Weigle reactivation display the same dependence on DNA synthesis as wild-type cells (Ohta et al., 1985). Irrespective of the mechanism, once the cascade has been initiated, the time at which a particular *fla* gene is expressed in the cell cycle is determined primarily by its position in the hierarchy. Thus, delaying transcription of certain class II genes until after their normal time of expression in the cell cycle leads to a similar displacement of the time at which the class III genes are expressed (Ohta et al., 1991).

Specialized Nature of the *fla* Gene Promoters

Insights into the nature of the machinery regulating the *fla* gene transcription cascade have come from analyses of promoter sequences. The class II *fla* genes are transcribed from one of two promoter types, class IIA or IIB. The class IIA promoter of the *flhA* gene does show some resemblance to the σ^{28}-consensus sequence of *fla* gene promoters from enteric bacteria (Helmann and Chamberlin, 1987), but the four class IIB promoters are unlike any previously characterized prokaryotic promoter type (Dingwall et al., 1992b; Yu and Shapiro, 1992; Stephens and Shapiro, 1993; VanWay et al., 1993; Anderson et al., 1995). The noncanonical structures of the class II promoters therefore imply that their cell cycle-specific transcription patterns are conferred by secondary forms of RNA polymerase carrying novel σ-factor subunits. A highly attractive but speculative hypothesis is that the class I *fla* genes, which have yet to be identified, encode specialized σ-factors, proteins that modulate the activity of RNA polymerases at these specialized level II promoters, or both.

The involvement of secondary RNA polymerase holoenzymes in *fla* gene transcription has been clearly demonstrated for the class III and IV genes that are expressed later in the cell cycle. DNA sequences positioned at −12 and −24 from the transcription start sites of the class III *flbG* operon and from the class IV flagellin genes were shown to be highly homologous to promoters of enteric bacteria that are transcribed by σ^{54}-RNA polymerase holoenzyme (Eσ^{54}) (Chen et al., 1986; Minnich and Newton, 1987; Mullin et al., 1987). Subsequent experiments showed that these promoters are recognized in vitro by reconstituted Eσ^{54} from *E. coli* (Ninfa et al., 1989) and that mutations in the highly conserved GG and GC residues at −12 and −24 respectively reduce or eliminate transcription from the templates both in vitro and in vivo (Ninfa et al., 1989; Mullin and Newton, 1989, 1993). The other class III genes now identified have also been shown to contain Eσ^{54} sequence elements (Dingwall et al., 1990; Dingwall et al., 1992a; Khambaty and Ely, 1992). Definitive evidence of the involvement of Eσ^{54} has recently come from the identification of the *rpoN* gene of *C. crescentus,* the demonstration that it is required for expression of class III and IV *fla* genes (Brun and Shapiro, 1992), and purification of an *rpoN* gene product with the expected transcriptional specificity for σ^{54}-dependent promoters (Anderson et al., 1995).

cis-Acting Sequences That Govern Class II, III, and IV *fla* Gene Transcription

Studies of Eσ^{54} in enteric bacteria have shown that, unlike most other prokaryotic RNA polymerase holoenzymes, it is incapable of isomerizing from closed promoter complexes to transcription-competent open complexes (reviewed in Kustu et al., 1989). Overcoming this energy barrier in the Eσ^{54} transcription cycle is facilitated by a specialized activator protein that is typically bound nearby to specific DNA sequence elements and uses the energy of ATP hydrolysis to catalyze isomerization (Kustu et al., 1989). Identification of analogous cis-acting sequences from which such an activator protein might function in *C. crescentus* was originally based on sequence analysis of several Eσ^{54}-dependent *fla* gene promoters. Conserved 19-bp sequences with inverted repeats were located ca. 100 bp upstream or downstream of the class III *flbG* and *flgK* (formerly *flaN*) promoters, respectively (Mullin et al., 1987; Mullin and Newton, 1989, 1993), as well as the class IV *fljK* and *fljL* flagellin gene promoters (Minnich and Newton, 1987). The requirement of these sequences, termed *ftr* elements (*flagellar transcription regulator*), for full transcriptional activation from *flbG* and *flgK* has recently been demonstrated by extensive mutational analysis (Mullin and Newton, 1989, 1993; Gober et al., 1991; Gober and Shapiro, 1992). An interesting aspect in the biology of Eσ^{54}-dependent *fla* gene transcription is that in at least two instances (the class III *flbG* and class IV *fljK* promoters), the minimal promoter elements, consisting of the Eσ^{54} binding site, an integration host factor binding site, and *ftr* sequences, have been shown to confer swarmer pole-specific transcription in the differentiating predivisional cell (Gober et al., 1991).

In addition to their role as positive control elements for transcription from class III and IV promoters, an *ftr* sequence (*ftr4*) has also been identified as a negative control element in transcription of the class IIB *fliF* operon (VanWay et al., 1993). The *ftr4* sequence overlaps the transcription start site of the *fliF* promoter, and mutations in this sequence derepress and extend *fliF* transcription well beyond its normal period of expression in the cell cycle (VanWay et al., 1993). The similarity of the *ftr* sequence elements indicated that they were recognized by a single transcription factor that could function as both a positive regulator of Eσ^{54}-dependent transcription of class III and IV promoters and a repressor of class IIB *fliF* operon transcription.

FlbD Protein Functions at *ftr* Sequence Elements

A candidate for the *ftr*-specific transcription factor was originally suggested by DNA sequence analysis of the class IIB *flbD* gene (Ramakrishnan and Newton, 1990), which is essential for transcription of the σ^{54}-dependent *fla* genes (Newton et al., 1989). The FlbD protein is homologous over its entire length to a family of response regulators that are characterized by the well-studied NtrC protein (also called NR I) of enteric bacteria (Ramakrishnan and Newton, 1990). Conserved regions include a central domain that is present in members of the NtrC response regulator family and is believed to contain the active site for ATP hydrolysis, which is required for open complex formation (Kustu et al., 1989).

That FlbD is, in fact, the protein that functions at the *ftr* sequence elements was supported by the finding that the *flbD* gene is necessary for both activation of class III and IV gene transcription and repression of transcription from the *fliF* operon at class II (Newton et al., 1989). Furthermore, when expressed in *E. coli*, the *flbD* gene is capable of activating transcription of the cloned *C. crescentus flbG* promoter by Eσ^{54} (Ramakrishnan and Newton, 1990). Definitive evidence for these conclusions has now been obtained by reconstituting in vitro transcription systems that recognize the class III *flbG* promoter and the class IIB *fliF* promoter, respectively (Benson et al., 1994a). These studies demonstrated that purified FlbD activates transcription from *flbG* and represses transcription from *fliF*. FlbD activity in vitro also depends on the presence of intact *ftr* sequences that are necessary for regulation in vivo (Benson et al., 1994a).

The mode of FlbD binding to the *ftr* sequences has been examined by DNase I, dimethylsulfate, and methylation interference footprinting assays (Benson et al., 1994a; Wingrove and Gober, 1994). FlbD contacts conserved guanine residues in the *ftr* sequences on both strands of the DNA that are positioned one helical turn apart within inverted repeats of the consensus half-site CCCGGCARRN (where R = purine and N = any nucleotide) (Benson et al., 1994a). When binding between FlbD and the *ftr1* and *ftr1** sequences, which are upstream of the class III *flbG* promoter, and the negative-acting *ftr4* sequence, which overlaps the start site of the class IIB *fliF* operon promoter, were compared, it was found that FlbD contacts G residues at conserved positions in all three sequences (Benson et al., 1994a). The symmetry of this sequence suggests that FlbD binds to the *ftr* elements as a dimer, with one monomer responsible for the contacts within each half-site.

Regulation of FlbD Activity

Although no FlbD kinase has been identified to date, two pieces of evidence strongly suggest that FlbD activity must be governed in the cell cycle. First, FlbD is synthesized early in the cell cycle, before compartmentalization that would limit its diffusion throughout the dividing cell. Its activity at the class III and IV promoters later in the cell cycle, however, is restricted to the swarmer cell pole, apparently as a consequence of a compartmentalized regulatory event (Gober et al., 1991; Wingrove et al., 1993). Second, transcription of the class III genes depends not only on synthesis of σ^{54} and FlbD but also on all other class II genes that have been identified (Newton et al., 1989; Xu et al., 1989; Ramakrishnan et al., 1994). Mutations in

any of the class II genes do not prevent FlbD synthesis, but they somehow block or prevent its function. Whether these two observations are the consequence of a single regulatory event, such as phosphorylation, remains unclear.

In the absence of an identified FlbD kinase, evidence that phosphorylation of FlbD controls its activity is indirect. Recent experiments have shown that a FlbD-polyhistidine fusion protein can be phosphorylated by crude cell extracts from *C. crescentus* and that a phosphate-labeled, epitope-tagged FlbD protein can be immunoprecipitated from cells labeled with orthophosphate (Wingrove et al., 1993). Also, genetic experiments indicate that the Asp residue in FlbD, corresponding to the likely site of phosphorylation, is critical for its function in vivo (Wingrove et al., 1993). Attempts to phosphorylate purified FlbD using either purified phosphokinases from other systems (phospho-CheA or phospho-NtrB) or the small-molecule acetyl phosphate have been unsuccessful (Ramakrishnan and Newton, unpublished data). Although acetyl phosphate also fails to stimulate FlbD activity in the transcription assay, it has been reported to enhance cooperative DNA binding of the FlbD-polyhistidine fusion protein, but by only 15% (Wingrove and Gober, 1994). Phosphoramidate does stimulate FlbD activity in the transcription assay, but its effect is modest (about two- to threefold), and the results are therefore difficult to interpret (Benson et al., 1994b).

The atypical behavior of FlbD in biochemical experiments may be related to the unusual structure of the amino-terminal regulatory domain. The highly conserved Asp-57 residue, which is the site of phosphorylation of CheY (Sanders et al., 1989), is present in FlbD, but several other residues that are virtually invariant in the response regulator family are lacking (Ramakrishnan and Newton, 1990). These include acidic residues (Asp and Glu) typically found at positions 12 and 13, which are Gly-12 and Lys-13 in FlbD, as well as Leu-109 instead of Lys. Most striking of these variations in FlbD is the presence of Lys-13 instead of Asp, a substitution that has been shown to confer constitutive activity on CheY (Bourret et al., 1990). Whether this indicates that the unmodified protein has some function in directing *fla* gene transcription or whether it reflects the uniqueness of regulatory protein(s) that controls FlbD function remains to be determined. Results using the reconstituted in vitro system do indicate that FlbD purified from *E. coli,* which is presumably unmodified, will activate transcription at relatively high concentrations from *C. crescentus* σ^{54}-*fla* gene promoters (Benson et al., 1994a; Benson et al., 1994b).

FlbD as a Global Regulator of Class III and IV *fla* Gene Transcription

In the absence of a method for "activating" FlbD, insights into the mechanism of its function come primarily from a comparative analysis of the well-characterized NtrC protein. Both DNase I and dimethylsulfate footprinting studies identified a pair of *ftr* sequence elements (*ftr1* and *ftr1**) upstream of the *flbG* promoter and showed that the centers of the *ftr1* and *ftr1** elements are positioned three helical turns apart (Benson et al., 1994a; Benson et al., 1994b). Activation of transcription from *glnAp2* in enteric bacteria depends on at least two NtrC binding sites, where oligomerization of NtrC dimers at these binding sites is essential for driving Eσ^{54} into the open complex (Weiss et al., 1992; Porter et al., 1993) (Chapter 9). This then suggests that activation of Eσ^{54}-dependent *fla* gene transcription may likewise depend on formation of FlbD oligomers in proximity to Eσ^{54} closed complexes.

In support of this conclusion, our laboratory recently identified pairs of sequences that are homologous to the *ftr* consensus sequence upstream of all identified promoters of *fla* genes at class III and IV of the hierarchy (Benson and Newton, unpublished data) (Fig. 4B). Reconstitution of transcription from these promoters in vitro has shown that FlbD functions as an activator and that its function in vitro requires highly conserved guanine residues of the newly identified *ftr* elements that correspond to those contacted by the FlbD protein in *ftr1, ftr1**, and *ftr4*. Genetic experiments have also demonstrated that these newly identified *ftr* elements are necessary for transcription in vivo (Wu et al., submitted). This work suggests that FlbD is a global activator of class III and IV *fla* gene transcription, but it also raises another interesting issue: if these genes share the same transcriptional machinery, what regulates the transition from expression of the class III to the class IV genes (see Fig. 4B)?

Prospects

Although the organization of the *C. crescentus fla* gene hierarchy is well defined, major questions remain to be answered about regulation of the *fla* genes within the hierarchy: (i) What are the class I genes, and how do they respond to cell cycle signals to initiate the *fla* gene cascade? (ii) If phosphorylation regulates the activity of FlbD, is the kinase synthesis or activity localized to one compartment of the predivisional cell, and to what signal does it respond? (iii) Does the native unmodified FlbD have a function in the cell as its unique amino terminus might suggest? (iv) What mediates the transition from class III to class IV gene transcription? Clearly the answers to these questions will be greatly facilitated by identification of the protein(s) involved in modulation of FlbD activity.

CONCLUSIONS

The developmental program of *C. crescentus* may be unique among those characterized in prokaryotic cells in its regulation by autogenous signals originating from events within the cell cycle itself. Evidence has been discussed implicating a multicomponent network of histidine kinases and response regulators in signal transduction pathways linking cell division cycle checkpoints to various developmental events. Several members of this network, including the histidine kinases PleC and DivJ and the response regulators DivK and PleD, have been identified and their roles in division and development partially characterized. The precise relationships between all members of these pathways have not been determined, however, and additional as yet uncharacterized proteins are also involved. Nevertheless, there is strong evidence that PleC and DivK function as a cognate kinase-response regulator pair to couple motility and stalk formation to a cell division cycle checkpoint. DivK plays a second essential role in cell division in response to DivJ or some other histidine kinase. One fascinating question is the precise nature of the cell cycle signals to which these sensor kinases respond, as well as the targets of these signal transduction pathways.

The nature of the regulated target genes in flagellum biosynthesis is much better understood, but nothing is known of the class I genes that respond to the cell cycle signal and initiate the *fla* gene cascade. Once initiated, the subsequent timing of *fla* gene expression relies in large part on the function of specialized σ-factors and the response regulator FlbD, which are themselves members of the transcription cascade. It will be interesting to determine if FlbD activity, like the device that initiates the *fla* gene cascade, is also subject to cell cycle regulation.

ACKNOWLEDGMENTS

Research from this laboratory was supported by Public Health Service grant GM-22299 from the National Institutes of Health and American Cancer Society grant MV-386. Todd Lane was supported by postdoctoral fellowship 1 F32 AI08488 from the National Institutes of Health, and Gregory B. Hecht was supported by National Institutes of Health training grant T32 GM07388.

REFERENCES

Abeles, F. B. 1992. *Ethylene in Plant Biology,* 2nd ed. Academic Press, New York.

Anderson, D. K., N. Ohta, J. Wu, and A. Newton. 1995. Regulation of the *Caulobacter crescentus rpoN* gene and function of the purified σ^{54} in flagellar gene transcription. *Mol. Gen. Genet.* **246**:697–706.

Barrett, J. T., C. S. Rhodes, D. M. Ferber, B. Jenkins, S. A. Kuhl, and B. Ely. 1982. Construction of a genetic map for *Caulobacter crescentus. J. Bacteriol.* **149**:889–896.

Benson, A., and A. Newton. Unpublished data.

Benson, A. K., G. Ramakrishnan, N. Ohta, J. Feng, A. Ninfa, and A. Newton. 1994a. The *Caulobacter crescentus* FlbD protein acts at *ftr* sequence elements both to activate and repress transcription of cell cycle regulated flagellar genes. *Proc. Natl. Acad. Sci. USA* **91**:4989-4993.

Benson, A. K., J. Wu, and A. Newton. 1994b. The role of FlbD in regulation of flagellar gene transcription in *Caulobacter crescentus. Res. Microbiol.* **12**:420–430.

Bourret, R. B., J. F. Hess, and M. I. Simon. 1990. Conserved aspartate residues and phosphorylation in signal transduction by the chemotaxis protein CheY. *Proc. Natl. Acad. Sci. USA* **87**:41–45.

Brewin, N. J. 1993. The *Rhizobium*-legume symbiosis: plant morphogenesis in a nodule. *Semin. Cell Biol.* **4**:149–156.

Brun, Y. V., G. Marczynski, and L. Shapiro. 1994. The expression of asymmetry during *Caulobacter* cell differentiation. *Annu. Rev. Biochem.* **63**:419–450.

Brun, Y. V., and L. Shapiro. 1992. A temporally controlled σ-factor is required for polar morphogenesis and normal cell division in *Caulobacter. Genes Dev.* **6**:2395–2408.

Burton, G., and A. Newton. Unpublished data.

Chang, C., S. F. Kwok, A. B. Bleecker, and E. M. Meyerowitz. 1993. *Arabidopsis* ethylene-response gene *ETR1*: similarity of product to two-component regulators. *Science* **262**:539–544.

Chen, L.-S., D. Mullin, and A. Newton. 1986. Identification, nucleotide sequence, and control of developmentally regulated promoters in the hook operon region of *Caulobacter crescentus. Proc. Natl. Acad. Sci. USA* **83**:2860–2864.

Degnen, S. T., and A. Newton. 1972. Chromosome replication during development in *Caulobacter crescentus. J. Mol. Biol.* **64**:671–680.

Dingwall, A., J. D. Garman, and L. Shapiro. 1992a. Organization and ordered expression of *Caulobacter* genes encoding flagellar basal body rod and ring proteins. *J. Mol. Biol.* **228**:1147–1162.

Dingwall, A., J. W. Gober, and L. Shapiro. 1990. Identification of a *Caulobacter* basal body structural gene and a *cis*-acting site required for activation of transcription. *J. Bacteriol.* **172**:6066–6076.

Dingwall, A., W. Y. Zhuang, K. Quon, and L. Shapiro. 1992b. Expression of an early gene in the flagellar regulatory hierarchy is sensitive to an interruption in DNA replication. *J. Bacteriol.* **174**:1760–1768.

Gilles-Gonzalez, M. A., G. S. Ditta, and D. R. Helinski. 1991. A haemoprotein with kinase activity encoded by the oxygen sensor of *Rhizobium meliloti. Nature* (London) **350**:170–172.

Gober, J. W., R. Champer, S. Reuter, and L. Shapiro. 1991. Expression of positional information during cell differentiation in *Caulobacter*. *Cell* **64:**381–391.

Gober, J. W., and L. Shapiro. 1992. A developmentally regulated *Caulobacter* flagellar promoter is activated by 3' enhancer and IHF binding elements. *Mol. Biol. Cell* **3:**913–926.

Gottfert, M. 1993. Regulation and function of rhizobial nodulation genes. *FEMS Microbiol. Rev.* **10:**39–63.

Haselkorn, R. 1992. Developmentally regulated gene rearrangements in prokaryotes. *Annu. Rev. Genet.* **26:**113–130.

Hecht, G. B., T. Lane, J. Sommer, and A. Newton. Submitted for publication.

Hecht, G. B., and A. Newton. Unpublished data.

Helmann, J. D., and M. J. Chamberlin. 1987. DNA sequence analysis suggests that expression of flagellar and chemotaxis genes in *Escherichia coli* and *Salmonella typhimurium* is controlled by an alternative σ factor. *Proc. Natl. Acad. Sci. USA* **84:**6422–6424.

Horvitz, H. R., and I. Herskowitz. 1992. Mechanisms of asymmetric cell division: two B's or not two B's, that is the question. *Cell* **68:**237–255.

Huguenel, E. D., and A. Newton. 1982. Localization of surface structures during procaryotic differentiation: role of cell division in *Caulobacter crescentus*. *Differentiation* **21:**71–78.

Khambaty, F. M., and B. Ely. 1992. Molecular genetics of the *flgI* region and its role in flagellum biosynthesis in *Caulobacter crescentus*. *J. Bacteriol.* **174:**4101–4109.

Kim, S. K., D. Kaiser, and A. Kuspa. 1992. Control of cell density and pattern by intercellular signaling in *Myxococcus* development. *Annu. Rev. Microbiol.* **46:**117–139.

Kustu, S., E. Santero, J. Keener, D. Popham, and D. Weiss. 1989. Expression of σ⁵⁴ (*ntrA*)-dependent genes is probably united by a common mechanism. *Microbiol. Rev.* **53:**367–376.

Lane, T., N. Ohta, and A. Newton. Unpublished data.

Macnab, R. M. 1992. Genetics and biogenesis of bacterial flagella. *Annu. Rev. Genet.* **26:**131–158.

Maeda, T., S. M. Wurler-Murphy, and H. Saito. 1994. A two-component system that regulates osmosensing MAP kinase cascade in yeast. *Nature* (London) **369:**242–245.

Minnich, S. A., and A. Newton. 1987. Promoter mapping and cell cycle regulation of flagellin gene transcription in *Caulobacter crescentus*. *Proc. Natl. Acad. Sci. USA* **84:**1142–1146.

Minnich, S. A., N. Ohta, N. Taylor, and A. Newton. 1988. Role of the 25-, 27-, and 29-kilodalton flagellins in *Caulobacter crescentus* cell motility: method for construction of deletion and Tn5 insertion mutants by gene replacement. *J. Bacteriol.* **170:**3953–3960.

Mullin, D., S. Minnich, L. S. Chen, and A. Newton. 1987. A set of positively regulated flagellar gene promoters in *Caulobacter crescentus* with sequence homology to the *nif* gene promoters of *Klebsiella pneumoniae*. *J. Mol. Biol.* **195:**939–943.

Mullin, D. A., and A. Newton. 1989. Ntr-like promoters and upstream regulatory sequence *ftr* are required for transcription of a developmentally regulated *Caulobacter crescentus* flagellar gene. *J. Bacteriol.* **171:**3218–3227.

Mullin, D. A., and A. Newton. 1993. A σ⁵⁴ promoter and downstream sequence elements *ftr2* and *ftr3* are required for regulated expression of divergent transcription units *flaN* and *flbG* in *Caulobacter crescentus*. *J. Bacteriol.* **175:**2067–2076.

Newton, A., G. Hecht, T. Lane, and N. Ohta. 1994. Role of histidine protein kinases and response regulators in cell division and polar morphogenesis in *Caulobacter crescentus*, p. 296–301. *In* A. M. Torriani-Gorini, E. Yagil, and S. Silver (ed.), *Cellular and Molecular Biology of Phosphate and Phosphorylated Compounds in Microorganisms*. American Society for Microbiology, Washington, D.C.

Newton, A., and N. Ohta. 1990. Regulation of the cell division cycle and differentiation in bacteria. *Annu. Rev. Microbiol.* **44:**689–719.

Newton, A., N. Ohta, G. Ramakrishnan, D. Mullin, and G. Raymond. 1989. Genetic switching in the flagellar gene hierarchy of *Caulobacter* requires negative as well as positive regulation of transcription. *Proc. Natl. Acad. Sci. USA* **86:**6651–6655.

Ninfa, A. J., D. A. Mullin, G. Ramakrishnan, and A. Newton. 1989. *Escherichia coli* σ-54 RNA polymerase recognizes *Caulobacter crescentus flaK* and *flaN* flagellar gene promoters in vitro. *J. Bacteriol.* **171:**383–391.

Ohta, N., L.-S. Chen, D. Mullin, and A. Newton. 1991. Timing of flagellar gene expression in the *Caulobacter* cell cycle is determined by a transcriptional cascade of positive regulatory genes. *J. Bacteriol.* **173:**1514–1522.

Ohta, N., L.-S. Chen, E. Swanson, and A. Newton. 1985. Transcriptional regulation of a periodically controlled flagellar gene operon in *Caulobacter crescentus*. *J. Mol. Biol.* **186:**107–115.

Ohta, N., T. Lane, E. G. Ninfa, J. M. Sommer, and A. Newton. 1992. A histidine protein kinase homologue required for regulation of bacterial cell division and differentiation. *Proc. Natl. Acad. Sci. USA* **89:**10297–10301.

Osley, M. A., M. Sheffery, and A. Newton. 1977. Regulation of flagellin synthesis in the cell cycle of *Caulobacter*: dependence on DNA replication. *Cell* **12:**393–400.

Ota, I. M., and A. Varshavsky. 1993. A yeast protein similar to bacterial two-component regulators. *Science* **262**:566–569.

Parkinson, J. S., and E. C. Kofoid. 1992. Communication modules in bacterial signaling proteins. *Annu. Rev. Genet.* **26**:71–112.

Porter, S. C., A. K. North, A. B. Wedel, and S. Kustu. 1993. Oligomerization of NTRC at the *glnA* enhancer is required for transcriptional activation. *Genes Dev.* **7**:2258–2273.

Ramakrishnan, G., and A. Newton. 1990. FlbD of *Caulobacter crescentus* is a homologue of NtrC (NR$_I$) and activates sigma-54 dependent flagellar gene promoters. *Proc. Natl. Acad. Sci. USA* **87**:2369–2373.

Ramakrishnan, G., and A. Newton. Unpublished data.

Ramakrishnan, G., J.-L. Zhao, and A. Newton. 1991. The cell-cycle regulated flagellar gene *flbF* of *Caulobacter crescentus* is homologous to the virulence locus (*lcrD*) of *Yersinia pestis. J. Bacteriol.* **173**:7283–7292.

Ramakrishnan, G., J.-L. Zhao, and A. Newton. 1994. Multiple structural proteins are required for both transcriptional activation and negative autoregulation of *Caulobacter* flagellar genes. *J. Bacteriol.* **176**:7587–7600.

Sanders, D. A., B. L. Gillece-Castro, A. M. Stock, A. L. Burlingame, and D. E. Koshland. 1989. Identification of the site of phosphorylation of the chemotaxis response regulator protein CheY. *J. Biol. Chem.* **264**:21770–21778.

Schroder, I., C. D. Wolin, R. Cavicchioli, and R. P. Gunsalus. 1994. Phosphorylation and dephosphorylation of the NarQ, NarX, and NarL proteins of the nitrate-dependent two-component regulatory system of *Escherichia coli. J. Bacteriol.* **176**:4985–4992.

Sheffery, M., and A. Newton. 1981. Regulation of periodic protein synthesis in the cell cycle: control of initiation and termination of flagellar gene expression. *Cell* **24**:49–57.

Shimkets, L. J. 1990. Social and developmental biology of the myxobacteria. *Microbiol. Rev.* **54**:473–501.

Sommer, J. M., and A. Newton. 1988. Sequential regulation of developmental events during polar morphogenesis in *Caulobacter crescentus*: assembly of pili on swarmer cells requires cell separation. *J. Bacteriol.* **170**:409–415.

Sommer, J. M., and A. Newton. 1989. Turning off flagellum rotation requires the pleiotropic gene *pleD: pleA, pleC,* and *pleD* define two morphogenic pathways in *Caulobacter crescentus. J. Bacteriol.* **171**:392–401.

Sommer, J. M., and A. Newton. 1991. Pseudoreversion analysis indicates a direct role of cell division

genes in polar morphogenesis and differentiation in *Caulobacter crescentus. Genetics* **129**:623–630.

Stephens, C. M., and L. Shapiro. 1993. An unusual promoter controls cell-cycle regulation and dependence on DNA replication of the *Caulobacter fliLM* early flagellar operon. *Mol. Microbiol.* **9**:1169–1179.

Stock, A. M., J. M. Mottonen, J. B. Stock, and C. E. Schutt. 1989. Three-dimensional structure of CheY, the response regulator of bacterial chemotaxis. *Nature* (London) **337**:745–749.

Stock, J. B., A. J. Ninfa, and A. M. Stock. 1989. Protein phosphorylation and regulation of adaptive responses in bacteria. *Microbiol. Rev.* **53**:450–490.

Terrana, B., and A. Newton. 1976. Requirement of cell division step for stalk formation in *Caulobacter crescentus. J. Bacteriol.* **128**:456–462.

VanWay, S. M., A. Newton, A. H. Mullin, and D. A. Mullin. 1993. Identification of the promoter and a negative regulatory element, *ftr4* that is needed for cell cycle timing of *fliF* operon expression in *Caulobacter crescentus. J. Bacteriol.* **175**:367–376.

Volz, K. 1993. Structural conservation in the CheY superfamily. *Biochemistry* **32**:11741–11753.

Wang, S. P., P. L. Sharma, P. V. Schoenlein, and B. Ely. 1993. A histidine protein kinase is involved in polar organelle development in *Caulobacter crescentus. Proc. Natl. Acad. Sci. USA* **90**:630–634.

Wanner, B. L. 1992. Is cross regulation by phosphorylation of two-component response regulator proteins important in bacteria? *J. Bacteriol.* **174**:2053–2058.

Weiss, V., F. Claverie-Martin, and B. Magasanik. 1992. Phosphorylation of nitrogen regulator I of *Escherichia coli* induces strong cooperative binding to DNA essential for activation of transcription. *Proc. Natl. Acad. Sci. USA* **89**:5088–5092.

Wingrove, J. A., E. K. Mangan, and J. W. Gober. 1993. Spatial and temporal phosphorylation of a transcriptional activator regulates pole-specific gene expression in *Caulobacter. Genes Dev.* **7**:1979–1992.

Wingrove, J. A., and J. W. Gober. 1994. A σ^{54} transcriptional activator also functions as a pole-specific repressor in *Caulobacter. Genes Dev.* **8**:1839–1852.

Wolk, C. P. 1991. Genetic analysis of cyanobacterial development. *Curr. Opin. Genet. Dev.* **1**:336–341.

Wu, J., A. Benson, and A. Newton. Submitted for publication.

Xu, H., A. Dingwall, and L. Shapiro. 1989. Negative transcriptional regulation in the *Caulobacter* flagellar hierarchy. *Proc. Natl. Acad. Sci. USA* **86**:6656–6660.

Yu, J., and L. Shapiro. 1992. Early *Caulobacter crescentus* genes *fliL* and *fliM* are required for flagellar gene expression and normal cell divison. *J. Bacteriol.* **174**:3327–3338.

The *frz* Signal Transduction System Controls Multicellular Behavior in *Myxococcus xanthus*

Wenyuan Shi and David R. Zusman

27

Myxococcus xanthus is an unusual gram-negative bacterium in that it exhibits a variety of social behaviors and has a complex life cycle (for a recent review, see Dworkin and Kaiser, 1993). When nutrients are abundant, the bacteria grow vegetatively. On a solid surface containing nutrients, cells swarm as a thin spreading colony, expanding to new areas that contain additional food. When deprived of nutrients, the cells aggregate to form fruiting bodies that contain approximately 100,000 cells. With continued starvation, the aggregated cells develop into metabolically dormant spherical myxospores, which are resting cells that are resistant to prolonged periods of starvation, desiccation, and high temperature. The spherical myxospores germinate and become rod-shaped vegetative cells in the presence of high levels of nutrients. The developmental cycle is dependent on cell motility. *M. xanthus* is one of many diverse gram-negative bacteria that move on a solid surface by a mechanism called *gliding*. Gliding motility is still a poorly understood method of locomotion, defined as the slow movement of the bacteria in the direction of the long axis of cells without the aid of flagella (McBride et al., 1993). The mechanism of gliding motility remains a mystery, although many hypotheses have been suggested (McBride et al., 1993).

Because of the complexity of fruiting body formation, *M. xanthus* has become recognized as an excellent prokaryotic model system for studying cellular development (Dworkin and Kaiser, 1993). To show vegetative swarming and developmental aggregation, cells must sense many signals from the environment, both favorable and unfavorable, process these signals, and then direct their motility to the appropriate response: swarming or aggregation. The cells must also be able to sense the signals from potentially catastrophic environmental conditions such as prolonged starvation or hazardous conditions. Under these conditions, the cells trigger the sporulation response, which allows cells to protect themselves for extensive periods of time (years). A lot of research has recently been directed at the molecular mechanisms of these cellular processes (Dworkin and Kaiser, 1993). This chapter focuses on the *frz* signal transduction system, which is a two-component signal transduction system involved in the social behavior of *M. xanthus* (McBride et al., 1993; Shi et al., 1993).

Wenyuan Shi and David R. Zusman, Department of Molecular and Cell Biology, University of California, Berkeley, Berkeley, California 94720.

Two-Component Signal Transduction, Edited by James A. Hoch and Thomas J. Silhavy,
© 1995 American Society for Microbiology, Washington, DC 20005

ISOLATION AND CHARACTERIZATION OF THE *frz* MUTANTS

Our laboratory has been interested in the signal transduction systems required for fruiting body formation. We began this investigation with the isolation of mutants defective in development. About 1,800 non-fruiting mutants of *M. xanthus* were isolated and characterized (Morrison and Zusman, 1979). Some of them did not show any aggregation or sporulation; some aggregated but showed very low or undetectable levels of sporulation (translucent-mound mutants); some showed defective aggregation but normal sporulation. We became particularly interested in one group of these mutants, called *frizzy* or *frz*, which sporulated normally but formed tangled, frizzy filaments under fruiting conditions instead of the normal fruiting bodies (Zusman, 1982) (Fig. 1). These cells were fully motile but were unable to direct themselves to form fruiting bodies. In later studies,

we found that these mutants were also defective in other forms of directed cell movement: for example, they were no longer able to form large swarming colonies on rich medium containing soft agar (Shi et al., 1993).

Using the transposon Tn5 tagging method for genetic analysis of *M. xanthus* developed by Kuner and Kaiser (1981), 36 frizzy mutants were analyzed and found to be linked to the same Tn5 transposon insertion site (Zusman, 1982). A wide range of different cotransduction frequencies suggested a cluster of linked genes (Zusman, 1982). Using the linked Tn5 insertion as a selectable marker (it confers kanamycin resistance), a 7.5-kbp DNA fragment containing the *frz* genes was cloned (Blackhart and Zusman, 1985a). The cloned DNA was analyzed by isolating new Tn5 insertions at frequent intervals within the *M. xanthus* DNA and by constructing deletion mutations in vitro; the mutated DNA was then transduced into *M. xanthus* for recombinational and com-

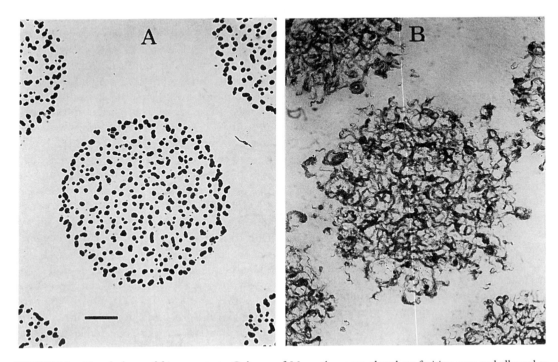

FIGURE 1 Morphology of frizzy mutants. Cultures of *M. xanthus* were placed on fruiting agar and allowed to incubate for 7 days at 28°C. (A) strain DZF1 (wild type); (B) strain DZF1227 (a *frzE* mutant). From Zusman (1982).

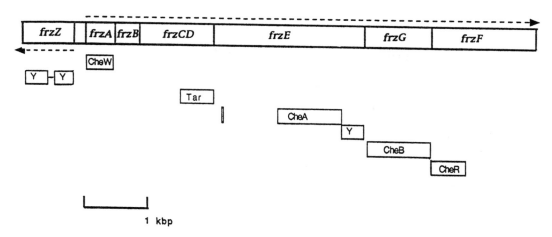

FIGURE 2 Sequence similarities between *frz* genes of *M. xanthus* and enteric chemotaxis (*che*) genes.

plementation analyses. Based on these studies, the *frz* genes were grouped into at least five complementation groups: *frzA, frzB, frzC, frzE,* and *frzF*. Also, two Tn5 mutations defined the *frzD* locus, which was dominant and caused a non-spreading colony phenotype (Blackhart and Zusman, 1985a). Sequence analysis later showed that *frzD* was located within the C-terminal region of *frzC;* the *frzC* and *frzD* loci are now referred to as parts of a single *frzCD* gene (McBride et al., 1989) (Fig. 2). Sequence analysis also showed that a 1-kbp gap between *frzE* and *frzF* constituted a new gene called *frzG* (McCleary et al., 1990) (Fig. 2). Recently the DNA upstream of the *frzA* was sequenced and an additional open reading frame, called *frzZ,* was identified. This gene, when mutated, also caused a frizzy phenotype (Trudeau and Zusman, unpublished data). The *frz* genes were subcloned into expression vectors, and the gene products of *frzA, frzB, frzCD, frzE, frzF,* and *frzZ* were purified and used for the preparation of polyclonal antibodies. Using the Western immunoblot technique and other methods (e.g., maxicell experiments), the Frz proteins were each identified in cell extracts of *M. xanthus* (Blackhart and Zusman, 1986; McCleary et al., 1990; McCleary and Zusman, 1990b; Trudeau et al., unpublished data).

frz GENES ARE CHEMOTAXIS GENES IN *M. XANTHUS*

As shown in Fig. 2, sequence analysis comparisons revealed that most of the *frz* gene products were homologous to chemotaxis gene products from the enteric bacteria (McBride et al., 1989; McCleary et al., 1990; McCleary and Zusman, 1990a). For example, FrzA was homologous to CheW, FrzE was homologous to both CheA and CheY, FrzF was homologous to CheR, FrzG was homologous to CheB, FrzCD was homologous to the methyl-accepting chemotaxis proteins (MCPs), and FrzZ had two domains that both were homologous to CheY. FrzB had no homolog in the enteric system, and no homolog of CheZ was found in *M. xanthus*.

The chemotaxis signal transduction system in *Escherichia coli* enables the bacteria to detect the chemicals or other signals in the environment, process the sensory signals, and direct themselves toward attractants or away from repellents. Because the *frz* genes were required for directed cell movement, it seemed reasonable to hypothesize that they were responsible for chemotaxis in *M. xanthus*. However, chemotaxis had never been clearly demonstrated in *M. xanthus*. The homologies did not prove that the *frz* genes were chemotaxis genes, because the conserved protein motifs could have evolved new functions. The problem with characterizing chemotaxis in *M. xanthus* was caused by the rates of cell movement. Gliding motility in *M. xanthus* is very slow: the rate of cell movement on 1.5% agar (the standard medium for culturing this organism) is about 2 to 4 µm/min, about 500 times slower than *E. coli* on swarm plates (0.3% agar). With this slow rate of movement, the usual methods for studying motility and chemotaxis in the enteric bacteria do not apply (Dworkin and Eide, 1983). To find out the biological functions of the *frz* genes, we developed several new methods that were particularly suitable for the study of motility and chemotaxis in slow-moving cells (Blackhart and Zusman, 1985b; Shi et al., 1993, 1994b; Shi and Zusman, 1994). For example, the motility of *E. coli* is so rapid that it is studied by strobe-illuminated microscopy to track cells. By contrast, the movement of individual *M. xanthus* cells was so slow that it was studied by time-lapse video microscopy (Blackhart and Zusman, 1985b). Using these methods, it was noticed that wild-type cells reversed their direction of gliding about every 6 to 8 min, but the mutants defective in *frzA, frzB, frzC, frzE,* and *frzF* loci reversed their direction of movement very rarely, about once every 2 h. By contrast, cells containing the *frzD* alleles of *frzCD* reversed their direction of movement very frequently, about once every 1 to 2 min. The altered reversal patterns found in the various *frz* mutants were similar in scope to some of the mutation phenotypes found in known chemotaxis mutants in enteric bacteria and suggested that the *frz* genes were involved in the control of directed cell movement.

Chemotaxis in enteric bacteria is routinely measured using the classical capillary assay (Adler, 1973). However, due to the slow gliding motility of *M. xanthus* (which is comparable with the diffusion rate of small molecules), the chemical gradient generated in and around the capillary will diffuse away before it can be recognized by the cells (Dworkin and Eide, 1983). Furthermore, *M. xanthus* cells are non-motile in liquid medium and require a solid surface for movement. To show spatial chemotactic movement in *M. xanthus,* it was necessary to set up agar plates, which maintain steep and stable chemical gradients. We were able to set up these gradients using petri plates that contain multiple compartments (Shi et al., 1993) (Fig. 3). The compartments were filled with the chemicals to be tested in 0.3% agar (wild-type cells move up to five times faster on soft agar) (Shi and Zusman, 1993) and nalidixic acid (to inhibit growth) and were then bridged with a very thin (1 mm thick) agar overlay (Fig. 3). In these compartmentalized plates, each compartment is isolated, except for the thin wick between them. This greatly limits diffusion of the chemical being tested. The chemicals or dyes (in the left compartment) that have crossed the border area are quickly dispersed by vertical diffusion in the right compartment. Thus, a sharp gradient is maintained. The empirical data showed that sharp chemical gradients were maintained for 20 h (Fig. 3). Using this assay, we observed that *M. xanthus* cells exhibited clear-cut tactic movements to many chemicals (Shi et al., 1993) (Fig. 4). For example, the bacteria spread into compartments with abundant nutrients like yeast extract or Casitone and avoided compartments with no nutrients, used (spent) medium, or repellents (e.g., short-chain alcohols or dimethyl sulfoxide). When individual cells were followed by time-lapse video microscopy, their movements were found to be consistent with the spatial chemotaxis behavior. For example, *M. xanthus* cells increased the frequency of cell reversal in response to repellents; in the presence of attractants, large groups of cells were observed to align with each other and move forward with few reversals (McBride et al., 1992; Shi et al., 1993). We found that most *frz* mutants were no longer able to respond to the spatial or temporal chemical gradients and did not exhibit any chemotactic movements (Fig. 4). These results strongly suggested that the *frz* genes control the chemotactic movement of *M. xanthus* in a manner similar to that of the *che* genes in enteric bacteria.

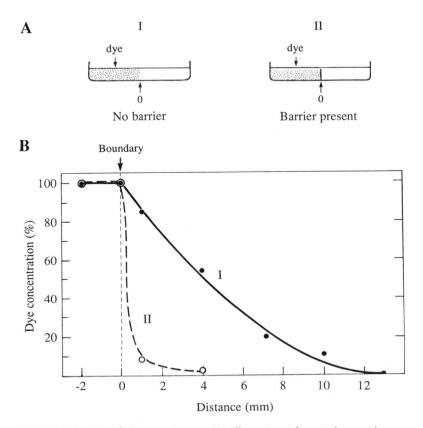

FIGURE 3 Spatial chemotaxis assay. (A) Illustration of petri plates without a barrier (I) and with a barrier plus a thin overlay (II). See text for detailed description. (B) Chemical gradients in both petri plates after 20 h of incubation. The solid line represents the case of a petri plate without a barrier (I), and the broken line, the case of a petri plate with a barrier and a thin overlay of agar (II). The gradients were analyzed by adding a dye (Congo red) to the left compartments at 0 h, incubating the plates for 20 h at 28°C, collecting 10-μl samples at different locations on the petri plates, resuspending them in 2 ml of water, and measuring the dye concentrations by absorbance at optical density at 499 nm. From Shi et al. (1993).

Sensory adaptation is essential for chemotaxis because it allows cells to remember past chemical concentrations and make temporal comparisons; cells can then be directed to the most favorable environment. Once again, sensory adaptation in *M. xanthus* was very difficult to study because of the very slow rate of cell motility and the difficulty in following the changes in the frequency of cell reversal. Fortunately, we were able to find an unusual method for studying adaptation based on the rate of

dispersal of cell clumps (Shi and Zusman, 1994). Unlike enteric bacteria, wild-type *M. xanthus* cells grown in liquid media contain many clumps that contain several hundred to tens of thousands of cells adhering to each other. Cell clumping is associated with the S-motility system and the presence of pili (Kaiser, 1979). When cell clumps were placed on a solid surface without chemical stimuli, individual cells began to move out from the clumps without a detectable lag. This dispersal pattern,

0 h   10 h   20 h

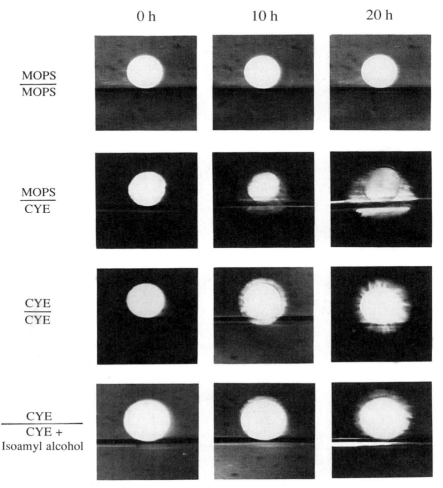

MOPS / MOPS

MOPS / CYE

CYE / CYE

CYE / CYE + Isoamyl alcohol

FIGURE 4 Chemotactic behavior of *M. xanthus* in compartmentalized petri plate assay. Wild-type DZ2 cells spread into the area containing attractants CYE (yeast extract plus Casitone) and avoid the area containing repellent isoamyl alcohol. From Shi et al. (1993).

when played back at accelerated time, appeared as a "blast" (Fig. 5). However, when cell clumps were placed on a surface containing repellents, a different pattern was observed. Individual cells showed an immediate increased frequency of cell reversals and therefore no dispersal of the cell clumps. With continued incubation, sensory adaptation did take place and the blast dispersal pattern was observed (Fig. 5). Adaptation time could therefore be measured directly from the delay in onset of the blast dispersal pattern.

BIOCHEMICAL ANALYSIS OF THE *frz* GENE PRODUCTS SHOWS THAT TWO ARE MEMBERS OF THE FAMILY OF TWO-COMPONENT SYSTEMS

In enteric bacteria, CheA is an autokinase that phosphorylates itself at a histidine residue after chemical stimulation and then transfers the phosphate to CheY at an aspartate residue. The phosphorylated CheY then interacts with the motor to control the rotation direction of the flagella (Matsumura et al., 1990). FrzE of *M. xanthus* is homologous to both CheA and

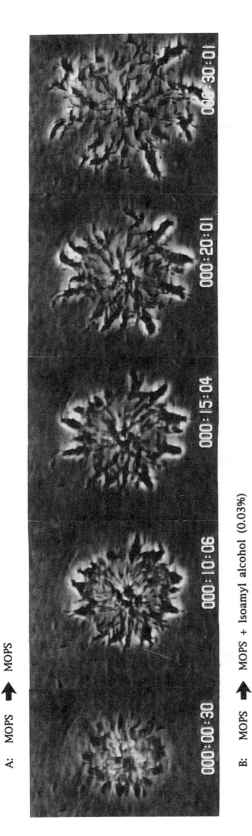

A: MOPS → MOPS

000:00:30   000:10:06   000:15:04   000:20:01   000:30:01

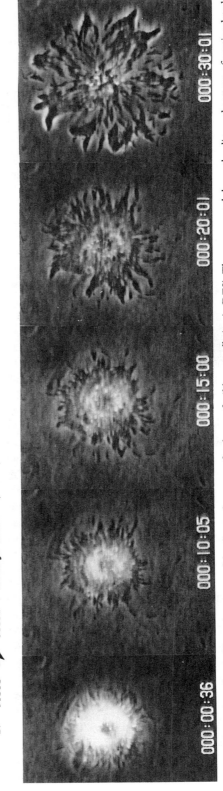

B: MOPS → MOPS + Isoamyl alcohol (0.03%)

000:00:36   000:10:05   000:15:00   000:20:01   000:30:01

FIGURE 5 Time-lapse video microscopy of the dispersal pattern of a clump of wild-type cells (strain DZ2). The upper panel shows the dispersal pattern of unstimulated cells. The lower panel shows the dispersal pattern in the presence of a negative stimulus (0.03% isoamyl alcohol). In this experiment, cell dispersal lag time is about 15 min. From Shi and Zusman (1994).

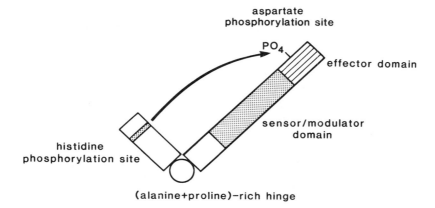

FIGURE 6 Model of FrzE autophosphorylation. FrzE has two domains that are homologous to CheA (shaded regions) and CheY (striped region). The center circle consists of an alanine- and proline-rich segment that is hypothesized to act as a flexible hinge. The CheA domain of FrzE was found to be autophosphorylated, and the phosphate was transferred to the CheY domain of FrzE (Acuna et al., 1995). Adapted from McCleary and Zusman (1990b).

CheY (McCleary and Zusman, 1990a). The N terminus of FrzE contains a conserved stretch of 20 amino acids that is homologous to the putative histidine phosphorylation site of CheA. Also, there is a large central region of 360 amino acids that shows 33% amino acid identity to CheA. Between the small and large regions of homology to CheA, there is an unusual 68-amino-acid domain that is 38% alanine and 34% proline. The C-terminal region of FrzE contains a 124-amino-acid domain that shows 31% amino acid identity to CheY and contains a region homologous to the aspartate phosphorylation site (Fig. 6). The purified FrzE protein, like CheA, was also autophosphorylated in the presence of ATP (McCleary and Zusman, 1990b). Analysis of the phosphorylated FrzE protein indicated that it contained an acylphosphate, probably phospho-aspartate (McCleary and Zusman, 1990b). It was suggested that autophosphorylation of FrzE initially occurs at a conserved histidine residue within the "CheA" domain and then that the phosphate is transferred to a conserved aspartate residue within the "CheY" domain (Fig. 6). To test this hypothesis, two genetically engineered proteins were produced: a protein

containing only the CheA domain of FrzE and a protein containing only the CheY domain of FrzE. The CheA domain polypeptide but not the CheY domain polypeptide was autophosphorylated (Acuna et al., 1995). The phosphorylated form of the CheA domain polypeptide was much more stable than phosphorylated FrzE; the CheY domain of FrzE apparently has phosphatase activity. When the CheA and CheY domain polypeptides were incubated together with ATP, both polypeptides were phosphorylated. These data indicate that FrzE indeed contains the biochemical activities of both CheA and CheY (Fig. 6). Although the CheY protein in enteric bacteria is the mediator between the membrane reception complex (MCP-CheA-CheW) and motor/switch (FliG-FliM-FliN), FrzE has only been shown to be associated with FrzCD (McBride and Zusman, unpublished data). Future work is needed to show the nature of the switch proteins that must interact with FrzE.

FrzZ is a novel protein consisting of two CheY-like domains with conserved aspartate phosphorylation sites. Between the two domains, there is a flexible region (hinge) rich in alanine and proline (Trudeau and Zusman, unpublished data). To our knowledge, this is the first example of a protein with two response domains in a single protein. Western immunoblot and primer extension analysis showed that FrzZ is indeed expressed in vivo during both vegetative growth and development. Mutations in *frzZ* caused the frizzy phenotype under developmental conditions and defective swarming under vegetative conditions (Trudeau and Zusman, unpublished data). Thus, the ability of FrzZ to translate chemosensory information correctly was somehow impaired in the *frzZ* mutants. The *frzZ* cells are still able to respond to sharp gradients of chemoattractants and repellents, indicating that FrzZ may play a fine-tuning function in signal transduction. Some clue to the role of FrzZ comes from biochemical studies. The FrzZ protein was overexpressed in *E. coli* and purified. The protein was found to be phosphorylated by FrzE in the presence of ATP and Mn^{2+} (Trudeau and Zusman, unpublished data), suggesting that it may play a role in modulating levels of phosphorylation of FrzE.

FrzCD is homologous to the MCPs of enteric bacteria, especially Tar, the receptor for aspartate in *E. coli* (McBride et al., 1989). The methylation of these receptor proteins in enteric bacteria is catalyzed by a specific methyltransferase, CheR, which modifies the MCPs at specific glutamate residues using *S*-adenosylmethionine as a methyl donor (Hazelbauer et al., 1990). The carboxymethylation of the glutamate residues of the MCPs is easily identifiable because this modification changes protein mobility on sodium dodecyl sulfate (SDS)-polyacrylamide gel electrophoresis (because of complex interactions between SDS and the modified proteins) and because the methyl groups are base-labile (Springer et al., 1979; McBride et al., 1992). FrzCD was found to be methylated at the homologous glutamate residues, and *S*-adenosylmethionine was found to be the methyl donor. The methylation was catalyzed by the CheR homolog FrzF (McCleary et al., 1990). The methylated form of FrzCD was base-labile (McCleary et al., 1990).

These results indicated that FrzCD is modified in a manner consistent with an MCP. However, unlike the MCPs in enteric bacteria, FrzCD appears to be a soluble cytoplasmic protein. It is the only MCP identified to date in *M. xanthus*, and it is possible that it may be the only one present, unlike enteric bacteria, which contain at least four different kinds of MCPs. It is likely that FrzCD interacts with various receptors in the membrane and transduces the signals to the gliding motility apparatus.

THE *frz* SENSORY SYSTEM IS REQUIRED FOR THE SOCIAL BEHAVIORS OF *M. XANTHUS*

Movement toward attractants or away from repellents requires the *frz* genes and is correlated with methylation or demethylation of FrzCD (Shi et al., 1993). Attractants such as Casitone and yeast extract caused methylation of FrzCD, and repellents such as isoamyl alcohol or isopropanol caused FrzCD demethylation. Using the methylation or demethylation of FrzCD as an indicator, we studied how the *frz* genes were involved in swarming on nutrient-rich medium and in fruiting body formation during starvation (Shi et al., 1993). When nutrients were abundant, cells grew rapidly, and the cells moved away from the colony center toward the colony edge, where more nutrients were present. Wild-type cells collected from the edge of swarming colonies contained primarily the methylated form of FrzCD, whereas the cells taken from the center of swarming colonies contained mainly demethylated FrzCD (Fig. 7). This suggests that cells recognized the colony center as a repellent area (with the resultant demethylation of FrzCD) and the colony edge as an attractant area (increased methylation of FrzCD). This may be the cause of colony expansion. Indeed, mutants that could not recognize used growth medium as a repellent failed to demethylate FrzCD in the presence of used medium and also were defective in colony swarming. Cells subjected to severe starvation contained largely demethylated FrzCD. During fruiting body formation, which

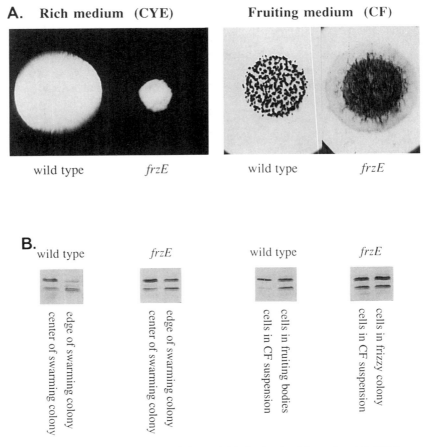

FIGURE 7 *frz* genes are required for social behavior in *M. xanthus*. (A) Wild type (strain DZ2) formed large swarming colony on CYE medium plus 0.3% agar and fruiting bodies on fruiting medium; the frzE mutants (strain DZ4148) failed to swarm or fruit. (B) Correlation between modification of FrzCD and directed cell movements in social behavior. See text for detailed description. From Shi et al. (1993).

occurs after an extended period of starvation, cells aggregate; at this time, FrzCD became more methylated even though the cells were still starving (McBride and Zusman, 1993; Shi et al., 1993) (Fig. 7). These results suggest that there are self-generated attractants for cellular aggregation during fruiting body formation and that these originate at aggregation centers and cause the methylation of FrzCD. The *frzA, frzB, frzCD, frzE,* and *frzF* mutants failed to show colony swarming and cellular aggregation during fruiting. Furthermore, they no longer showed changes in the methylation or

demethylation of FrzCD under these conditions (Shi et al., 1993). These results indicate the *frz* genes and chemotaxis play a very important role in the social behavior of *M. xanthus*.

What are the putative self-generated attractants or repellents for *M. xanthus,* and how are they sensed by the *frz* signal transduction system? The answers to these questions are currently under investigation. Recently, we isolated several new mutants that are not linked to the *frz* genes but are involved in signal transduction. For example, the *negA* mutants can no longer detect the repellents generated in used medium and therefore can no longer respond to this medium (Shi et al., 1994a). Similarly, we found that many non-fruiting mutants did not respond to starvation as a repellent signal (Shi et al., 1994a). Further characterization of these mutants should help us to have a better understanding of the molecular mechanisms behind the social behaviors of *M. xanthus*.

MANY QUESTIONS REMAIN

Genetic, biochemical, behavioral, and molecular biology studies all indicate that the *frz* genes are the chemotaxis genes of *M. xanthus*. They play very important roles in sensing the signals, processing the signals, and directing cell movement. Like other chemotaxis systems, the *frz* signal transduction system contains some of the conserved features of a two-component system such as a sensor and response regulator. However, the two-component system of *M. xanthus* has several unique features. For example, FrzE contains both sensor (kinase) and response regulator (phosphatase) on the same protein. Why does this occur? Also, FrzZ is a novel protein that contains two response regulators of unknown function. What proteins does this interact with? Obviously, there are a lot of missing elements in this sensory pathway. We do not know how the signals are initially sensed by the cells and how the signals are transduced to FrzCD. It will be necessary to further identify and characterize the components that interact with the *frz* signal transduction system to fully understand its function.

REFERENCES

Acuna, G., W. Shi, K. Trudeau, and D. R. Zusman. 1995. The "CheA" and "CheY" domains of *Myxococcus xanthus* FrzE function in vitro as an autokinase and a phosphate acceptor, respectively. *FEBS Lett.* **358:**31–33.

Adler, J. 1973. A method for measuring chemotaxis and use of the method to determine optimum conditions for chemotaxis by *Escherichia coli*. *J. Gen. Microbiol.* **74:**77–91.

Blackhart, B. D., and D. R. Zusman. 1985a. Cloning and complementation analysis of the "frizzy" genes of *Myxococcus xanthus. Mol. Gen. Genet.* **198:** 243–254.

Blackhart, B. D., and D. R. Zusman. 1985b. "Frizzy" genes of *Myxococcus xanthus* are involved in control of frequency of reversal of gliding motility. *Proc. Natl. Acad. Sci. USA* **82:**8767–8770.

Blackhart, B. D., and D. R. Zusman. 1986. Analysis of the products of the *Myxococcus xanthus frz* genes. *J. Bacteriol.* **166:**673–678.

Dworkin, M., and D. Eide. 1983. *Myxococcus xanthus* does not respond chemotactically to moderate concentration gradients. *J. Bacteriol.* **154:**437–442.

Dworkin, M., and D. Kaiser. 1993. *Myxobacteria II.* American Society for Microbiology, Washington, D.C.

Hazelbauer, G. L., R. Yaghmai, G. G. Burrows, J. W. Baumgartner, D. P. Dutton, and D. G. Morgan. 1990. Transducers: transmembrane receptor proteins involved in bacterial chemotaxis, p. 107–134. *In* J. P. Armitage and J. M. Lackie (ed.), *Biology of the Chemotactic Response.* Cambridge University Press, Cambridge.

Kaiser, D. 1979. Social gliding is correlated with the presence of pili in *Myxococcus xanthus. Proc. Natl. Acad. Sci. USA* **76:**5952–5956.

Kuner, J., and D. Kaiser. 1981. Introduction of transposon Tn5 into *Myxococcus* for analysis of developmental and other nonselectable mutants. *Proc. Natl. Acad. Sci. USA* **78:**425–429.

Matsumura, P., S. Roman, K. Volz, and D. McNally. 1990. Signalling complexes in bacterial chemotaxis, p. 135–154. *In* J. P. Armitage and J. M. Lackie (ed.), *Biology of the Chemotactic Response.* Cambridge University Press, Cambridge.

McBride, M. J., P. Hartzell, and D. R. Zusman. 1993. Motility and tactic behavior of *Myxococcus xanthus*, p. 285–305. *In* M. Dworkin and D. Kaiser (ed.), *Myxobacteria II.* American Society for Microbiology, Washington, D.C.

McBride, M. J., T. Köhler, and D. R. Zusman. 1992. Methylation of FrzCD, a methyl-accepting taxis protein of *Myxococcus xanthus*, is correlated with factors affecting cell behavior. *J. Bacteriol.* **174:**4246–4257.

McBride, M. J., R. A. Weinberg, and D. R. Zusman. 1989. "Frizzy" aggregation genes of the gliding bacterium *Myxococcus xanthus* show sequence similarities to the chemotaxis genes of enteric bacteria. *Proc. Natl. Acad. Sci. USA* **86:**424–428.

McBride, M. J., and D. R. Zusman. 1993. FrzCD, a methyl-accepting taxis protein from *Myxococcus xanthus*, shows modulated methylation during fruiting body formation. *J. Bacteriol.* **175:**4936–4940.

McBride, M. J., and D. R. Zusman. Unpublished data.

McCleary, W. R., M. J. McBride, and D. R. Zusman. 1990. Developmental sensory transduction in *Myxococcus xanthus* involves methylation and demethylation of FrzCD. *J. Bacteriol.* **172:**4877–4887.

McCleary, W. R., and D. R. Zusman. 1990a. FrzE of *Myxococcus xanthus* is homologous to both CheA and CheY of *Salmonella typhimurium. Proc. Natl. Acad. Sci. USA* **87:**5898–5902.

McCleary, W. R., and D. R. Zusman. 1990b. Purification and characterization of the *Myxococcus xanthus* FrzE protein shows that it has autophosphorylation activity. *J. Bacteriol.* **172:**6661–6668.

Morrison, C. E., and D. R. Zusman. 1979. *Myxococcus xanthus* mutants with temperature-sensitive stage-specific defects: evidence for independent pathways in development. *J. Bacteriol.* **140:**1036–1042.

Shi, W., T. Köhler, and D. R. Zusman. 1993. Chemotaxis plays a role in the social behavior of *Myxococcus xanthus. Mol. Microbiol.* **9:**601–611.

Shi, W., T. Köhler, and D. R. Zusman. 1994a. Isolation and phenotypic characterization of *Myxococcus xanthus* mutants which are defective in sensing negative stimuli. *J. Bacteriol.* **176:**696–701.

Shi, W., T. Köhler, and D. R. Zusman. 1994b. Motility and chemotaxis in *Myxococcus xanthus. Methods Mol. Genet.* **3:**258–269.

Shi, W., and D. R. Zusman. 1993. The two motility systems of *Myxococcus xanthus* show different selective advantages on various surfaces. *Proc. Natl. Acad. Sci. USA* **90:**3378–3382.

Shi, W., and D. R. Zusman. 1994. Sensory adaptation during negative chemotaxis in *Myxococcus xanthus. J. Bacteriol.* **176:**1517–1520.

Springer, M. S., M. F. Goy, and J. Adler. 1979. Protein methylation in behavioral control mechanisms and in signal transduction. *Nature* (London) **280:**279–284.

Trudeau, K., and D. R. Zusman. Unpublished data.

Trudeau, K., K. Smith, M. J. McBride, W. Shi, and D. R. Zusman. Unpublished data.

Zusman, D. R. 1982. Frizzy mutants, a new class of aggregation-defective developmental mutants of *Myxococcus xanthus. J. Bacteriol.* **150:**1430–1437.

Intercellular Communication in Marine *Vibrio* Species: Density-Dependent Regulation of the Expression of Bioluminescence

Bonnie L. Bassler and Michael R. Silverman

28

BIOLUMINESCENT MARINE BACTERIA

Luminescent bacteria are widespread in the marine environment, where they exist in a variety of diverse habitats: as free-living organisms, gut symbionts, saprophytes, parasites, or in specialized light organs of certain fish and squid (Nealson and Hastings, 1979). In the symbiotic associations, the ecological benefit for the host has been established (Morin et al., 1975). For example, the host organism can use bacterial luminescence for attraction of prey, escape from predators, and intraspecies communication. However, we do not yet understand what advantage the bacteria, either free living or as symbionts, derive from producing light. To begin to understand the adaptive benefit of light production, both the biochemistry of bacterial bioluminescence and the genetic mechanisms that regulate expression of the genes for luminescence (*lux*) have been explored. This chapter focuses on the different molecular mechanisms two model luminous bacteria, *Vibrio fischeri* (a symbiont) and *Vibrio harveyi* (a free-living microbe), use for regulating *lux* expression. Definition of the genes, proteins, in-

teractions, chemical modifications, and signals involved in the regulation of luminescence is leading to a molecular understanding of some of the mechanisms these bacteria use for intercellular communication and gene regulation. And by studying regulation of luminescence, we are learning how marine bacteria interact in a community, how they detect stimuli, and how they integrate, process, and transduce information to control expression of *lux* genes and other genes in the same control network or regulon.

Light production by luminous bacteria is catalyzed by the enzyme luciferase, a mixed function oxidase consisting of two different subunits (α and β), each approximately 40,000 molecular weight (Ziegler and Baldwin, 1981). In the generation of light, luciferase catalyzes the oxidation of a reduced flavin and a long-chain aldehyde, producing oxidized flavin and the corresponding long-chain fatty acid. A multienzyme complex containing three proteins, a reductase, a transferase, and a synthetase is involved in recycling the fatty acid to the aldehyde (Meighen, 1991). It is estimated for *V. harveyi* and *V. fischeri* that as much as 20% of the total oxygen consumed by the bacteria is used in the luciferase reaction (Makemson, 1986).

The *lux* genes have been cloned from several luminescent bacteria, including *Photobacter-*

Bonnie L. Bassler, Department of Molecular Biology, Princeton University, Princeton, New Jersey 08544. *Michael R. Silverman,* The Agouron Institute, La Jolla, California 92037.

Two-Component Signal Transduction, Edited by James A. Hoch and Thomas J. Silhavy,
© 1995 American Society for Microbiology, Washington, DC 20005

ium phosphoreum, Photobacterium leiognathi, V. fischeri, V. harveyi, and *Xenorhabdus luminescens,* and in every case, the sequence and organization of the *lux* genes encoding the luminescence enzymes are very similar. The genes are arranged in a single operon and transcribed in the order *luxCDABE.* The *luxA* and *luxB* genes encode the α- and β-subunits of luciferase, and *luxC, luxD,* and *luxE* encode components of the fatty acid reductase complex. In *V. fischeri,* an additional gene, *luxG,* and in *V. harveyi,* two additional genes, *luxG* and *luxH,* follow immediately downstream of *luxE.* These genes could encode proteins involved in the synthesis of the reduced flavin (Meighen, 1991).

DENSITY-DEPENDENT REGULATION OF LIGHT PRODUCTION: AUTOINDUCTION

Expression of luminescence in most bacteria is tightly regulated by the density of the population. As the density of a growing culture increases, the light each cell produces can increase as much as 1,000-fold. The mechanism of density-dependent control has been well characterized in the light organ symbiont *V. fischeri.* This bacterium synthesizes a small signal molecule, termed *autoinducer,* which diffuses across the cell membrane. As the population density increases, the autoinducer accumulates in the external environment and induces the expression of luminescence (Nealson et al., 1970). Therefore, *V. fischeri* monitors cell density by using the concentration of the autoinducer as an indicator. The *V. fischeri* autoinducer is β–ketocaproyl homoserine lactone (Eberhard et al., 1981). *V. harveyi,* a free-living marine bacterium, also uses an autoinducer to regulate density-dependent expression of luminescence. The autoinducer from *V. harveyi* has been determined to be β-hydroxybutryl homoserine lactone (Cao and Meighen, 1989). Recently, an alternative nomenclature for autoinducers was introduced; these designations are *N*-(3-oxohexanoyl)-L-homoserine lactone (VAI) for the *V. fischeri* autoinducer and *N*-(3-hy-

droxybutanoyl)-L-homoserine lactone (HAI) for the *V. harveyi* autoinducer (Fuqua et al., 1994). Studies described below suggest that *V. harveyi* produces an additional autoinducer substance (Bassler et al., 1993). Although the autoinducers produced by *V. fischeri* and *V. harveyi* have similar structures, neither species expresses luminescence after exposure to the autoinducer of the other.

REGULATION OF *lux* EXPRESSION IN *V. FISCHERI:* QUORUM SENSING

In *V. fischeri,* the regulatory genes involved in density-dependent control of luminescence are adjacent to the *luxCDABEG* operon encoding the luciferase enzymes (Engebrecht and Silverman, 1984). In this bacterium, the luminescence genes are organized into two divergently transcribed operons called operon R (right) and operon L (left) (Fig. 1). Operon R contains the *luxCDABEG* genes for luminescence and also the *luxI* gene, which encodes a protein that is required for the synthesis of autoinducer. The order of the genes in operon R is *luxICDABEG.* Operon L contains one gene, *luxR* (*R* for regulation), and this gene encodes a protein that is believed to activate transcription of the *luxICDABEG* operon in response to the presence of autoinducer. The *luxR* gene is adjacent to *luxI* but is transcribed in the opposite direction. Engebrecht et al. (1983) demonstrated that regulation of light production in recombinant *Escherichia coli* containing the *luxRICDABEG* genes resembles that observed in *V. fischeri* (i.e., expression of luminescence is dependent on cell density and can be stimulated by the exogenous addition of autoinducer). It is apparent that a positive feedback circuit controls *lux* expression in *V. fischeri* (see model in Fig. 1). In dilute cultures of *V. fischeri* (preinduction condition), weak transcription of the *luxICDABEG* operon occurs (1/100 to 1/1,000 of the induced level), and a relatively low level of autoinducer synthesis results. When a critical concentration of autoinducer has accumulated, transcription of the right operon is induced. Expression of *luxICDABEG*

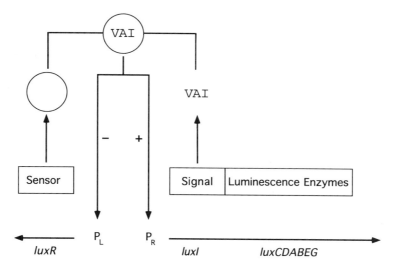

FIGURE 1 Model for regulation of bioluminescence in *V. fischeri*. Synthesis of the enzymes for light production in *V. fischeri* is dependent on transcription of operon R, which contains *luxICDABEG*. The product of the *luxG* gene is not required for luminescence. Transcription of operon R is regulated by the interaction of autoinducer (the production of which is a function of the *luxI* gene) and the protein LuxR, encoded by the *luxR* gene. In dilute cell suspensions, weak constitutive transcription of operon R occurs, resulting in a low concentration of autoinducer. As the cell density increases, a concentration of autoinducer is reached that is sufficient to induce transcription of operon R. Because *luxI* is in operon R, autoinducer synthesis is also increased on induction of operon R. The functions regulating operon R form a positive feedback circuit, and induction of operon R results in an exponential increase in light production per cell. A compensatory negative feedback circuit operates to indirectly modulate the expression of operon R (and light production) after induction. Negative regulation is achieved by limiting the expression of operon L containing *luxR* and requires the interaction of autoinducer and LuxR. VAI, *V. fischeri* autoinducer; P_L and P_R, left- and rightward promoters, respectively.

(post-induction condition) then increases exponentially as a result of the positive feedback control circuit (i.e., increased expression of *luxI* results in increased synthesis of autoinducer), which further amplifies expression of *luxI* and the genes encoding the enzymes for luminescence. At high concentrations of autoinducer, expression of the *luxICDABEG* operon reaches a plateau level. This modulation is thought to result from repression (negative feedback) of *luxR* expression by a LuxR-autoinducer complex (Engebrecht and Silverman, 1987). Recent evidence suggests that LuxR is membrane-associated (Kolibachuk and Greenberg, 1993). Binding of autoinducer by LuxR is thought to

facilitate multimerization of LuxR, and in this conformation, LuxR binds to the *luxICDA-BEG* promoter and activates transcription. Deletion analysis and dominance studies indicate that the C-terminal region of LuxR is sufficient for DNA binding and transcriptional activation and that the N terminus of LuxR is involved in autoinducer binding and regulates formation of multimers (Choi and Greenberg, 1992a,b).

The *luxR-luxI* signal response system, or the "Quorum Sensing System" as it has recently been termed (Fuqua et al., 1994), is not restricted to regulation of luminescence in *V. fischeri*. Homologs of this *luxR-luxI* signal response system are used for intercellular communication and control of different functions in other genera of bacteria. LuxR-like proteins and LuxI-like homoserine lactone autoinducers have been implicated in the regulation of the production of virulence factors in *Pseudomonas aeruginosa* (Passador et al., 1993), conjugal transfer of the Ti plasmid in *Agrobacterium tumefaciens* (Piper et al., 1993), antibiotic synthesis in *Erwinia carotovora* (Bainton et al., 1992), and nodulation in *Rhizobium leguminosarum* (Cubo et al., 1992). The analogous genes in *P. aeruginosa* are called *lasR* and *lasI* and apparently regulate the *lasB, lasA, aprA,* and *toxA* genes. The *lux* gene homologs are called *traR* and *traI* in *A. tumefaciens* and activate two *tra* operons as well as the *traR* and *traI* genes themselves. A *luxI* homolog *expI* has been identified in *E. carotovora*, and a *luxR* homolog called *rhiR* in *R. leguminosarum*. Properties such as luminescence, production of virulence factors, Ti plasmid transfer, antibiotic synthesis, and nodulation could be beneficial only when the organism is in a dense population or community, and linking the genes encoding such properties to the *luxR-luxI* sensor effectively couples expression to the appropriate environmental conditions.

REGULATION IN *V. HARVEYI*: A MULTICHANNEL SENSOR CIRCUIT

The regulatory genes that control luminescence in *V. harveyi* are different from those of *V. fischeri*. Sequencing of the promoter proximal region adjacent to the *luxCDABEGH* operon of *V. harveyi* demonstrated that neither a *V. fischeri luxR* nor a *luxI* regulatory gene homolog was located in that position, and probing the *V. harveyi* chromosome with the cloned *V. fischeri luxR* and *luxI* genes detected no candidates for these regulatory genes (Miyamoto et al., 1988). Transposon mini-Mu*lac* was used for in vivo mutagenesis of the *V. harveyi* genome to locate genes required for expression of luminescence (Martin et al., 1989). Transposon insertions that caused a Lux⁻ (nonluminous) phenotype were identified and mapped to two noncontiguous regions of the *V. harveyi*

chromosome. The *luxCDABEGH* operon was located in one region, and the second region contained a regulatory gene because mutants with defects in this locus were incapable of transcription of the *luxCDABEGH* operon. This second locus was cloned and sequenced and found to contain a single gene, which was named *luxR*. The *V. harveyi* LuxR protein is similar in amino acid sequence to some DNA binding proteins, but it is not similar to LuxR of *V. fischeri*. In recombinant *E. coli*, transcription of the *luxCDABEGH* operon was very strongly stimulated by providing *luxR* in *trans*. However, *luxR* alone was not sufficient for regulatory control because luminescence in the *luxCDABEGH-luxR* recombinant was not dependent on cell density or exogenous addition of *V. harveyi* autoinducer (Showalter et al., 1990; Swartzman et al., 1992).

Regulatory functions other than LuxR must exist in *V. harveyi* because no gene responsible for autoinducer production had been identified and because the LuxR regulatory protein did not appear to be directly involved in the response to autoinducer. Transposon mutagenesis might not have revealed these regulatory functions because they were not individually obligatory for expression of luminescence. For example, multiple independent sensory channels could regulate expression of *luxCDABEGH*, and eliminating one channel would not prevent induction of luminescence. Mutants impaired in one of these hypothetical sensory channels could be partially defective and might have a dim rather than a dark phenotype. And genes encoding these new sensory functions might be identified by isolating recombinant clones, which restored full light production to such dim mutants. Therefore, a recombinant library containing fragments of the *V. harveyi* genome cloned into a broad-host-range cosmid vector was transferred into a collection of spontaneous dim mutants that produced 1 to 10% of the light of the wild type, and individual clones that restored light production (i.e., complemented the defective genes) were recovered for further analysis.

One complementation group of *V. harveyi* dim mutants could be restored to full light production by a family of recombinant cosmids containing a subset of common restriction fragments. The positions and regulatory functions of the *lux* genes contained on these cosmids were determined by mutagenizing the recombinant cosmids with transposon Tn*5* or Tn*5lac* in *E. coli* followed by transferring the transposon insertion mutations into the chromosome of wild-type *V. harveyi* for phenotype analysis. All the gene-replacement strains of *V. harveyi* containing transposon insertions in the putative *lux* regulatory region produced wild-type levels of light, and the expression of luminescence in these strains was still dependent on the density of the culture. However, transposon insertions in one region of the DNA caused a defect in autoinducer production, and transposon insertions in an adjacent region of the DNA resulted in a defect in autoinducer response. Specifically, mutants with transposon insertions in one locus, which was subsequently shown to contain the *luxL* and *luxM* genes, produced significantly less autoinducer activity (Bassler et al., 1993). This observation was interpreted to mean that the wild-type strain must produce two structurally distinct autoinducers (HAI-1 and HAI-2) that specifically stimulate different response pathways (Sensor 1 and Sensor 2) and that LuxL and LuxM mutants are defective in the production of one autoinducer (HAI-1). Mutants with transposon insertions in the adjacent locus, which was subsequently shown to contain the *luxN* gene, were incapable of responding to the exogenous addition of synthetic autoinducer, β-hydroxybutryl homoserine lactone, but were still stimulated to express luminescence on exposure to supernatant from the wild-type strain or supernatants from LuxL and LuxM mutants. This pattern of responses indicated that LuxN mutants were defective in the specific response to HAI-1 (Sensor 1 function) but were capable of response to HAI-2 (Sensor 2 function). It was also inferred that HAI-1 is identical to β-hydroxybutryl homoserine lac-

tone. The structure of HAI-2 remains to be determined. A model for multiple systems (i.e., signalling system 1 with HAI-1 and Sensor 1 and signalling system 2 with HAI-2 and Sensor 2) that control density-dependent expression of luminescence is shown in Fig. 2.

Mutants with defects in signalling system 2 were required to isolate and characterize the components of the second regulatory pathway. Because either signalling system alone appeared to be sufficient to activate expression of luminescence, system 2 mutants were expected to be bright because signalling system 1 would still be operational. However, double mutants, containing a defect in signalling system 1 and a defect in signalling system 2, were predicted to be dark or very dim and would therefore have a phenotype that could be conveniently recognized. The strategy used to obtain such mutants consisted of two sequential mutageneses (Bassler et al., 1994b). The first mutagenesis consisted of transposon Tn*5* mutagenesis of the cloned *luxL* gene in recombinant *E. coli* followed by transfer of the transposon insertion into the chromosome of *V. harveyi* using a gene replacement technique (Bassler et al., 1993). The resulting LuxL mutant produced approximately the same intensity of luminescence as wild-type *V. harveyi* but did not produce the system 1 autoinducer, HAI-1 (β-hydroxybutryl homoserine lactone). The LuxL mutant could still be stimulated to express luminescence via signalling system 1 by the exogenous addition of HAI-1 because the signalling system 1 sensor protein (LuxN) was functional. The hypothetical signalling system 2 pathway, composed of components for the synthesis of the second unidentified autoinducer (HAI-2) and components mediating the response to HAI-2 (Sensor 2), was expected to be operational in this strain.

The LuxL mutant was then mutagenized with nitrosoguanidine, and the resulting dark or dim (less than 10% of wild-type luminescence) double mutants were isolated. An autoinducer cross-stimulation test was used to differentiate between double mutants containing an additional defect in signalling system 2

Signalling System 1   Signalling System 2

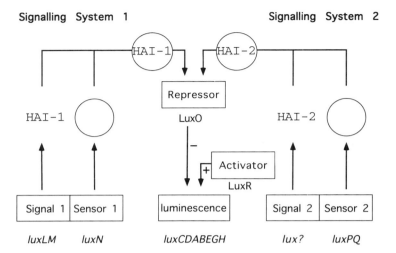

luxLM   *luxN*   *luxCDABEGH*   *lux?*   *luxPQ*

FIGURE 2 Model for genetic control of luminescence in *V. harveyi*. Expression of the operon (*luxCDABEGH*) encoding luminescence enzymes is controlled by two signalling channels, signalling systems 1 and 2. Each signalling system consists of functions that produce an extracellular signal substance (Signal 1 or 2) and functions that regulate the response to the corresponding signal substance (Sensor 1 or 2). Each signal substance (autoinducer 1 or 2 denoted by HAI-1 or HAI-2) interacts with the corresponding sensor (O). Signal 1 function is encoded by *luxL* and *luxM,* and Sensor 1 function is encoded by *luxN.* The Sensor 2 function is the product of the *luxP* and *luxQ* genes, and the gene(s) encoding Signal 2 synthesis function has not been identified. Sensor proteins regulate expression of luminescence indirectly by inactivating LuxO, a repressor of the *luxCDABEGH* operon. Expression of *luxCDABEGH* also requires LuxR, a positive transcription factor. The interactions and activities of the sensor and repressor proteins are probably controlled by phosphorylation reactions (see text).

genes and those containing an additional defect in either *luxCDABEGH* or *luxR*. The double mutant strains (recipients) were grown on agar plates in close proximity to a dark *luxA*::Tn*5* strain (autoinducer HAI-1, HAI-2 donor). Autoinducers diffusing from the donor strain cannot stimulate luminescence in double mutants with defects in *luxR* or *luxCDABEGH* but should be capable of stimulating light production in double mutants in which the additional defect is in signalling system 2. Several independent dim double mutants (less than 5% of wild-type luminescence) were responsive to cross-stimulation by the donor. These mutants were HAI-1⁻ (due to introduction of the *luxL* mutation in the first step of the mutagenesis) and were predicted to have acquired either a defect in HAI-2 synthesis or in the HAI-2

response (Sensor 2) in the second mutagenesis. Thus, these double mutants could be HAI-1⁻, Sensor 1⁺, HAI-2⁻, Sensor 2⁺, or HAI-1⁻, Sensor 1⁺, HAI-2⁺, Sensor 2⁻. Using more refined tests for the production of and response to specific autoinducers, all cross-stimulatable mutants were found to have a HAI-1⁻, Sensor 1⁺, HAI-2⁺, Sensor 2⁻ phenotype.

A library of recombinant cosmids containing fragments of the wild-type *V. harveyi* genome was transferred into a double mutant with the putative Sensor 2 defect to isolate complementing *lux* genes. Recombinant cosmids with the *luxLMN* locus, which presumably complemented the initial *luxL* mutation, as well as cosmids containing a new locus, which apparently complemented the putative Sensor 2 gene defect, were isolated. Exhaustive transposon mutagenesis was performed on the latter class of recombinant cosmid to identify and characterize the region encoding the Sensor 2 function. The mutated cosmids were mobilized into the system 2 mutant strain to identify those incapable of complementing the system 2 signalling defect. The transposon insertions that eliminated the Sensor 2 function were localized to one region of the recombinant cosmid that corresponded to two genes called *luxP* and *luxQ*. Each of these Tn5 insertions was then transferred to the genome of the wild-type *V. harveyi* to analyze the phenotype of a putative sensor 2 mutant. All the gene replacement mutant strains produced wild-type levels of light. This was expected because, as discussed earlier, a defect in a system 2 function alone might not be sufficient to prevent Lux expression. The capacity of the gene replacement mutants to produce specific autoinducer activity, or to respond to autoinducers, was examined.

V. harveyi strains with transposon insertions in the *luxPQ* locus were not deficient in production of HAI-1 or HAI-2 autoinducer activity, but LuxP or LuxQ mutants were defective in their response to HAI-2. The data in Fig. 3 show the autoinducer responses of three strains: wild-type *V. harveyi,* a LuxN mutant (Sensor 1⁻), and a LuxQ mutant (Sensor 2⁻). The strains were assayed for stimulation of expression of luminescence after the addition of cell-free supernatant from the wild-type strain (representing HAI-1 + HAI-2), supernatant from a LuxM mutant (representing HAI-2), and synthetic β-hydroxybutryl homoserine lactone (representing HAI-1). During initial growth (preinduction phase) of the diluted cultures in the absence of added autoinducers, all three strains exhibited a precipitous decline in relative light production. When the cell density reached a critical level, light production then increased abruptly as a result of accumulation of endogenously produced autoinducer (postinduction phase). These data indicate that the wild-type strain and the LuxN and LuxQ mutants are all subject to density-dependent regulation and that at least one signal response

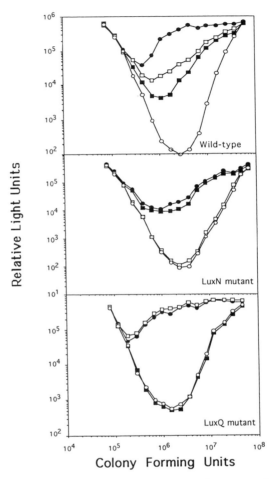

FIGURE 3 Response phenotypes of LuxN and LuxQ transposon insertion mutants of *V. harveyi*. The wild-type *V. harveyi* (top panel) is a control strain with a Tn5*lac* insertion outside the *lux* regulatory region. Response curves for transposon insertion mutants with the LuxN (middle panel) and LuxQ (bottom panel) phenotypes are also shown. Bright cultures of these strains were diluted 1:5,000, and light production was then measured during growth of the cultures. Cell-free supernatant (10%) or synthetic autoinducer (1 μg ml⁻¹) was added at the time of the first measurement. No addition (O), wild-type supernatant (●), LuxM supernatant (i.e., HAI-2 [■]), and synthetic autoinducer (i.e., HAI-1 [□]). Relative light units are defined as light emission per cell (i.e., counts min⁻¹ ml⁻¹ × 10³/colony-forming units ml⁻¹) and are plotted as a function of cell density (colony-forming units/ml). Data were taken from Bassler et al. (1994b).

system is operating in each strain. However, these strains had distinct differences in receptivities to exogenously added autoinducers. The wild-type strain responded to wild-type supernatant (HAI-1 + HAI-2) and maintained a level of light production that was more than 1,000-fold higher than the no-addition control culture late in the preinduction phase. Additions representing HAI-1 or HAI-2 alone also stimulated expression of luminescence in the wild type but were considerably less effective. Therefore, the wild type was capable of responding to both autoinducers and was Sensor 1[+] Sensor 2[+]. Light production in the LuxN mutant was stimulated by wild-type supernatant (HAI-1 + HAI-2) and by LuxM mutant supernatant (HAI-2), but no response was elicited by the synthetic autoinducer (HAI-1). Therefore, the LuxN mutant was capable of responding only to HAI-2 and was Sensor 1[-] Sensor 2[+]. Luminescence in the LuxQ mutant (or a LuxP mutant) was stimulated to the same extent by wild-type supernatant (HAI-1 + HAI-2) and synthetic autoinducer (HAI-1), but no stimulation was observed by supernatant from the LuxM mutant (HAI-2). Therefore, the LuxQ mutant was capable of responding only to HAI-1 and was Sensor 1[+] Sensor 2[-]. The signalling phenotypes of LuxN and LuxPQ mutants support the model shown in Fig. 2. LuxN and LuxPQ define two independent signalling pathways that respond to different stimuli. In the wild type, both pathways operate and converge to control the density-dependent expression of the luxCDABEGH operon (Bassler et al., 1994b). The gene(s) encoding HAI-2 function remains to be identified.

TWO-COMPONENT SIGNAL RELAY

The DNA sequences of the lux regulatory genes luxL, M, N, P, and Q were determined, and the functions encoded by each were inferred from the deduced amino acid sequences. The luxL and luxM gene sequences and the derived amino acid sequences of the gene products showed no correspondence to those of luxI from V. fischeri or to other gene or gene

product sequences in the data base, so the roles of these genes in the synthesis of HAI-1, β-hydroxybutryl homoserine lactone, are not clear. Regions of amino acid sequence in both LuxN and LuxQ are very similar to both the sensor kinase and response regulator domains of proteins of the "two-component" family of adaptive response regulators. This family of regulators is found in many bacteria and controls a diversity of functions such as chemotactic behavior, metabolic activities, and cell differentiation. The two components involved are a sensor kinase, which is a histidine protein kinase, and a response regulator. Recognition of a specific stimulus causes autophosphorylation of a histidine residue in the sensor kinase protein, and the phosphoryl group is subsequently transferred to an aspartate residue in the response regulator protein. Phosphorylation of the response regulator component results in an alteration in that protein's regulatory activity. In general, the sensor kinase and response regulator domains are located in different proteins, but in LuxN and LuxQ both domains are located in the same protein. This unusual motif has been observed previously; for example, the FrzE protein of Myxococcus, which has a role in controlling motility and development (McCleary and Zusman, 1990; Chapter 27), the ArcB protein of E. coli, which is involved in regulating the switch from aerobic to anaerobic growth (Iuchi et al., 1990; Chapter 13), and the VirA protein of A. tumefaciens, which has a role in virulence (Leroux et al., 1987; Chapter 23).

The relationship between LuxN and LuxQ and the family of two-component response regulators suggests that Lux signal transduction probably uses a phosphorelay process, but what specific mechanistic interactions occur? The initial event in signalling could be the binding of autoinducers to their cognate sensors (i.e., HAI-1 to Sensor 1 [LuxN] and HAI-2 to Sensor 2 [LuxQ]). Both LuxN and LuxQ have N-terminal sequences that could function as membrane-spanning segments (Bassler et al., 1993). Regions of LuxN or LuxQ exposed to the periplasm could interact directly with autoinducers or interact with other proteins that actually bind autoinducers. The latter possibility could apply to Lux signalling system 2 because Sensor 2 function, in addition to LuxQ, also requires LuxP, which is similar to the periplasmic ribose binding protein of *E. coli* and *Salmonella typhimurium* (Groarke et al., 1983). Thus, signal relay in system 2 could be initiated by the binding of HAI-2 to LuxP, which then interacts with LuxQ. Signal reception and relay in *Agrobacterium tumefaciens* has been shown to require the interaction of a periplasmic protein and a membrane-spanning sensor protein (Cangelosi et al., 1990; Huang et al., 1990; Chapter 23). ChvE, which resembles the periplasmic ribose binding protein of *E. coli* and *S. typhimurium,* is thought to bind galactose and subsequently to interact with VirA, a signalling protein similar to LuxN and LuxQ.

SIGNAL INTEGRATION: THE LuxO PROTEIN

Initial HAI-1 and HAI-2 signal recognition by LuxN and LuxQ could activate a series of phosphotransfer reactions. However, because neither of these Lux proteins contains an obvious DNA binding domain, it was not apparent how this information is transduced to affect a change in transcription of the *luxCDABEGH* operon. Is there another Lux regulatory protein involved in signal transduction? In studies similar to those described for *luxN*, a recombinant clone that restored Lux function to a second dim mutant complementation group was found, and the cloned DNA encoded an additional function required for density-dependent control of luminescence (Bassler et al., 1994a). Transposon Tn*5* insertions in this recombinant clone were isolated, and the mutations were transferred to the genome of *V. harveyi* for examination of mutant phenotypes. The locus defined by the Tn*5* insertions was designated *luxO* and the product of this locus, the LuxO protein. Expression of luminescence in *V. harveyi* strains with transposon insertions in this locus was independent of the density of the culture or of the addition of autoinducers and

was similar in intensity to the maximal level observed in wild-type bacteria. One interpretation of the constitutive phenotype of LuxO mutants is that LuxO acts negatively to control expression of luminescence and that relief of repression by LuxO in the wild type could result from interactions with other components in the Lux signalling system (Bassler et al., 1994a). It appears that both signalling systems 1 and 2 exert their respective autoinducer-mediated, density-dependent influence on *luxCDA-BEGH* expression by interaction with LuxO because elimination of this function abolished dependence of luminescence expression on both autoinducers. One hypothesis is that when the cells are dilute and there is a low concentration of autoinducer, LuxO is an active repressor of expression of *lux* genes, whereas at high cell density and high autoinducer concentration, the repressor activity of LuxO is inactivated (see Fig. 2).

The *luxO* sequence was determined and revealed a single open reading frame encoding the LuxO protein. The amino acid sequence of LuxO is similar to the response regulator domain of the family of two-component signal transduction proteins (Bassler et al., 1994a). Figure 4 shows a partial amino acid sequence alignment of the LuxN, LuxQ, and LuxO proteins as well as the ArcB and ArcA proteins of *E. coli*. Short blocks of sequence common to the sensor kinase domain (Parkinson and Kofoid, 1992; Chapter 3) are present in LuxN, LuxQ, and ArcB. The locations of the H block, which includes the site of histidine phosphorylation, and the N, G1, F, and G2 homology blocks are shown in Fig. 4. Blocks of sequence common to the response regulator domain (Stock et al., 1989) are also present in LuxN, LuxQ, LuxO, ArcB, and ArcA. The locations of the Asp-12,13 block, the Asp-57 block, which includes the site of aspartate phosphorylation, and the Lys-109 block are also shown (amino acid positions refer to those of CheY). The LuxN and LuxQ proteins clearly have roles in signal reception because LuxN and LuxQ mutants each fail to respond to one of the auto-

FIGURE 4

H Block

```
LuxQ  394  EDDHEIHHPVFQGFKEKLTPHLKMAAQGATLTGVNVPIGNKIYRWNLSPIRVDGDISGIIVQGQDITT-LIEAEK----QSNIARREAEKSAQ----ARADFLAK-[MSHEIRTP]INGIIGVAQLLKDS
           :.        :.::    ....:                                                                            .:::.   ...  ::
LuxN  378  DELSEELEYKVSAKRSMKALYDKMSSNNTALVMPLFGQGKSVTHLLISPHKSNNQMPSNEEISAVQTL-LTRVQS----TIEADRRIRQSRAL-ANS----- LAHEMRNPLAQVLQFEALKQH
           :.  : ::  :   :.                ...       :                             .: .:.                     ..:::  .        :::
ArcB  215  TDEKVFRHNVSLITYEQWLDYP----DGRKACFEIRKVPYYDRVGKRH------------GLMGFGRDIT----ERKRYQALERASR--DKITFIST-ISHELRTPLNGIVGLSRILLDT
```

N Block

```
LuxQ  513  VDTQEQRNQIDVLCHSGEHLLAVLNDILDFSKIEQGKFNIQKHPFSF[TDTMRTLENIYRPICTNKGVELVIENELDPNV-EIFTDQVR-[LNQILFNLVSNAVKFT]---PIG---SIRLHAEL
           .                :  .:  .:   :  .             :. :.:                 ::.                  ::::..:::::::::::
LuxN  492  IENHAPVEQITLDIENGQAAIQRGRQLIDIILREVSDSSPEHEPIAWTSIHKAVDQAVSHYGFEN--EKIIERIRLPQH-TDFVAKLNETLFNFVIFNLIRNAIYYFDSYPDS-----QIEISTKT
           .         :.:    .        :::..    :         :::  :  . ::        . ::.:::.:::.::                 ::. ::.:::::
ArcB  313  ELTAEQEKYLKTIHVSAVTLGNIFNDIIDMDKMERRKVQLDNQPVDFSFLADLENLSALQAQQKGLRPNLEPTLPLPH-QVITDGTR---LRQILWNLISNAVKFT---QQG----QVTVRVRY
```

G1 Block   F Block   G2 Block

```
LuxQ  628  EQFYGAENSVI[V--VELLTDTGIGIESDKLDQ][MFEPFVQ-EESTTTREV][GSSGLGL]TIVKNLVDMLEGDVQVRSSKGGGTTFVITLPVKDRERVLRPLEVS-----QRIKPFALFD-ES
                 .    :. .::::::. :  ::        :::..        :::::::       :. .:                                    :
LuxN  611  GPY------ENTLI----FRDTGPGIDETISHKIFDDFFS-YQKS-----GGSSGLGLGYCQRVMRSFGGRIECKSKLGTFTEFHLYYPVVPNAPKADTLRTPYFNDWKQNKRSNEHKVAPNVQINNQS
                     :. :      ::::: :.   :            :      ::::::::                                    ::
ArcB  427  D------EGDMLH---FEVEDSGIGIPDELDKIFAMYYQVKDSHGGKPATGTGLAVSRRLAKNMGDITVTSEQGKGSTFTLTIHAPSVAE-----EVD---------DAFDEDDMPL-PA
ArcA    1  MQTPH-------
```

Lys109 Block

Asp12,13 Block   Asp57 Block

```
LuxQ  738  L[RVLLIVEDN]FTNAFILQAPCKKYKMQVDWAKDGLDAMELLSDT[TFDLILLMDNQ]PHLGGIETHEI-------RQNLRLGTPIYACTADTAKETSDAFMAAGANYV[MLKPIKENALHEAFVDFKQ]
            :.:.::    .:         :  . . :          .::  : :  .::.::                    .    ::  .                       ::  . :.  :  .
LuxN  724  PTVLIVDDKEVQRALVQMYLNQLGVNSLQANNGENAVEVFKANHVDLIIMDVQMPVMNGFDASQRI----KE-LSPQTPIVALSGESGERELD-MINKLMDGRLEKPTTLNALRHVLGNWLN
            :.:.::    .:    :  . . :    . .:      ::::::.:::.: : :::.:::::::: ::.:      .  ::    :  .::.::.::  .::::   .:  :.: ..
ArcB  529  LNVLLVEDIELNVIVARSVLEKLGNSVDVAMTGKAALEMFKPGEYDLVLLDIQLPDMTGLDISRELTKRYPREDLP--PLVALTANVLKDKQE-YLNAGMDDVLSKPLSVPALTAMIKKFWD
             :. ::    ::        :  :  .       .::.::.:::.:::: ::.: : ::.::       ::        .     .   :  :::    ::   :
ArcA    7  --ILIVEDELVTRNTLKSIFEAEGYDVFEATDGAEMHQILSEYDINLVIMDINLPGKNGLLLAREL------RE-QANVALMFLTGRDNEVDKILGLEIGADDYIT-KP-----FN-
             :.:.::      .         :  .  . :  .::  . :.:::::.:::.: : :::.:::::::::: ::.:       :    .        .          :
LuxO    1  MVEDTASVAALYRSYLTPIGIDINIVGRDALE SLNERIPDLILLDLRLPDMTGMDVLHAV------KKSHPDVPIIFMTAHGSIDTAVEAMRHGSQDF-LIKPCEADRLRV-----
```

FIGURE 4 Amino acid sequence comparisons between Lux regulatory proteins and members of Arc bacterial two-component sensory transduction system. The protein comparisons were aligned to LuxQ. Amino acid sequence similarities between the histidine protein kinase and response regulator domains of *V.harveyi* LuxQ, and those domains in *V.harveyi* LuxN and LuxO and *E. coli* ArcB and ArcA are shown. The symbols "." and ":" denote amino acid similarities and identities, respectively. The highly conserved H, N, G1, F, and G2 blocks in the histidine protein kinase domain and the Asp-12,13, Asp-57, and Lys-109 blocks in the response regulator domain are boxed in LuxQ. Numbers next to protein designations refer to the position in the protein of the first amino acid residue used in each line of the sequence comparison. The full lengths of the proteins are LuxN, 849; LuxQ, 859; LuxO, 453; ArcB, 778; and ArcA, 238 amino acids.

inducers of *V. harveyi*. However, as mentioned, LuxN and LuxQ do not contain DNA binding motifs that are common to many response regulator proteins, so it is not evident that LuxN or LuxQ is capable of directly affecting gene transcription. The LuxO protein does contain DNA binding motifs so it could directly interact with *lux* DNA (such as the *luxCDABEGH* promoter) and control its transcription. The amino acid sequence of LuxO in Fig. 4 is not extended to include the helix-turn-helix DNA binding domains, but these regions correspond to amino acids 203 to 214 and 315 to 326 (Bassler et al., 1994a).

Other two-component circuits have been characterized in which a single protein contains both a sensor kinase and a response regulator domain (similar to LuxN and LuxQ) and a second protein contains both a response regulator domain and a DNA binding motif (similar to LuxO). Examples are the ArcB, ArcA system of *E. coli* (Iuchi and Lin, 1992; Chapter 13) and the VirA, VirG system of *Agrobacter-*

ium (Winans et al., 1986; Leroux et al., 1987; Chapter 23). Iuchi and Lin (1992) showed, using mutant ArcB proteins, that phosphoryl transfer from the histidine residue in the sensor kinase domain of ArcB to the aspartate residue in the response regulator domain of ArcB facilitates the subsequent intermolecular phosphoryl transfer from histidine in ArcB to the aspartate residue in the response regulator domain of ArcA. So, directly linking the response regulator domain to the sensor kinase domain could result in additional modulatory interactions controlling signal-to-noise discrimination and signal amplification.

We propose an analogous phosphorelay mechanism for LuxN, LuxQ, and LuxO (Fig. 5). The presence of autoinducer (HAI-1) stimulates an initial autophosphorylation of LuxN followed by phosphoryl transfer to aspartate in the response regulator domain of LuxN. The phosphotransfer reaction could be intramolecular, or the transfer could occur in *trans* (i.e., phosphoryl transfer from histidine on one

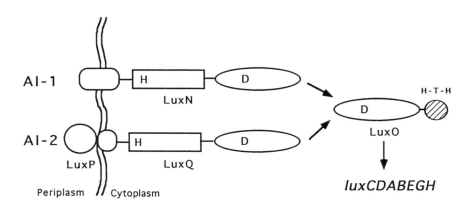

FIGURE 5 Two-component signal relay in density-sensing. Density-dependent regulation of luminescence in *V. harveyi* begins with binding of the signalling system 1 and 2 autoinducers (HAI-1 and HAI-2) by their cognate sensors LuxN and LuxP,Q, respectively. The LuxP protein is similar to the periplasmic ribose binding protein of *E. coli* and *S. typhimurium*. LuxN and LuxQ are members of the bacterial "two-component" family of adaptive response regulators, and each contains a histidine-protein kinase sensor domain and a response regulator domain. Subsequent signal transduction is proposed to occur via a series of phosphorelay reactions (see text). Coupling and integration of the signals from the two pathways could be accomplished by LuxO, a negative regulator of luminescence. LuxO is similar to the response regulator class of two-component regulatory proteins, and it contains a DNA binding domain so it could directly interact with *lux* DNA. H, histidine; D, aspartate; and H-T-H, helix-turn-helix DNA binding domain.

LuxN molecule to aspartate in another LuxN molecule). Phosphorylation at aspartate in LuxN facilitates a second, subsequent, auto-inducer-dependent autophosphorylation of LuxN, which is followed by an intermolecular phosphoryl group transfer to aspartate in LuxO. The consequence of phosphorylation of LuxO is inactivation of its repressor activity. An analogous series of phosphoryl transfer reactions results from HAI-2 stimulation of LuxQ. LuxO functions to integrate sensory inputs from signalling systems 1 and 2. LuxR is a positive transcription factor that is necessary and sufficient for activation of *luxCDABEGH* transcription. LuxO could directly repress *luxCDABEGH* transcription, or it could regulate transcription of the operon indirectly by controlling the expression of *luxR*. In Fig. 2, we have suggested that LuxO and LuxR act independently on *luxCDABEGH* transcription, but these terminal interactions in *lux* regulation have not been characterized.

DISCUSSION: SENSORY CHANNELS AND CONNECTIONS

The differences between the regulatory circuits controlling density-dependent expression of luminescence in *V. fischeri* and *V. harveyi* are striking. Density-sensing in *V. fischeri* is controlled by the *luxR-luxI* signal response system. LuxI is involved in the synthesis of a single autoinducer molecule, and LuxR regulates *luxICDABEG* operon transcription in response to autoinducer. Similar LuxR–LuxI signalling systems control density-dependent functions in a variety of nonluminous bacteria in other genera (Fuqua et al., 1994). However, *V. harveyi* (a member of the same genus) uses different proteins and a different signal transduction mechanism to accomplish a similar task. *V. harveyi* produces two distinct autoinducer signals. Two sensor proteins, LuxN and LuxQ, coupled with the signal integrator protein, LuxO, control the specific response of *V. harveyi* to each of the autoinducers. The proteins LuxN, LuxQ, and LuxO are all related to members of the two-component family of adaptive response regula-

tors, suggesting that Lux signal transduction in *V. harveyi* is accomplished by a phosphorelay cascade.

Why are the density-sensing systems of *V. fischeri* and *V. harveyi* different? *V. fischeri* grows as a symbiont in pure culture in the light organs of particular fish and squid (Nealson and Hastings, 1979), whereas *V. harveyi* is commonly found as a free-living bacterium in diverse marine environments or on the surfaces or in the gut tracts of various marine animals. The *V. harveyi* autoinducers do not stimulate expression of luminescence in *V. fischeri,* and the *V. fischeri* autoinducer does not stimulate *V. harveyi*. However, other nonluminous marine *Vibrio* do produce an autoinducer activity, which stimulates expression of luminescence in *V. harveyi* (Greenberg et al., 1979). So, *V. harveyi* lives in mixed populations and in a diversity of habitats and might communicate with other bacteria as well. Intercellular communication in such complex and varying circumstances could require a complicated sensory apparatus capable of detecting, discriminating, and interpreting many signals. In contrast to *V. fischeri,* *V. harveyi* has two autoinducer receptors, LuxN (Sensor 1) for HAI-1 and LuxP,Q (Sensor 2) for HAI-2. LuxP,Q function is required for sensing HAI-2, but it is possible that this signalling pathway can function to detect additional signals. As we argued earlier, LuxQ could detect HAI-2 indirectly via an interaction with LuxP, which is a periplasmic receptor for HAI-2. If LuxQ can interact with one periplasmic receptor, other as-yet-unidentified interactions might exist, and the range of receptivities of the Lux sensory system could be even larger.

Use of a two-component signalling mechanism could facilitate the construction of connections between signalling channels. The connection of LuxN and LuxP,Q sensors at LuxO has already been suggested, and connections to heterologous sensors are a possibility. Such connections or "crosstalk" between different two-component systems (e.g., interaction of the sensor kinase of one system with the response regulator of another system) is thought to occur

(Ninfa et al., 1988). And the metabolic or redox state of a cell could be connected to a two-component signalling process as a result of the direct interaction of small-molecule phosphoryl donors such as acetyl phosphate (Wanner and Wilmes-Riesenberg, 1992). Intercellular communication is but one of many factors known to affect *lux* expression. Environmental factors such as iron availability, oxygen tension, catabolite concentration, and genotoxic agents also affect *lux* expression. The reception of and response to such signals could require a complex interconnected sensory apparatus, and the two-component mechanism could have the plasticity to evolve the multiple channels and connections required for sensing and integrating diverse signals. Specifically, sensor kinases receptive to the varied signals mentioned above could interact with LuxO to couple perception of the environment to expression of luminescence.

Lux expression in *V. fischeri* is mediated by feedback circuits: a positive feedback circuit amplifies the response to autoinducer, and a negative feedback circuit represses the response at high autoinducer concentration. Possibly greater versatility in signal modulation is achieved by using the two-component mechanism in *V. harveyi*. The possibility that the composite structure of the LuxN and LuxQ sensors (i.e., the presence of both a sensor kinase and a response regulator domain in the same protein) has a function in signal amplification and signal-noise discrimination was discussed earlier. It is also possible that feedback relationships could modulate signal production and sensitivity of reception. For example, LuxO could control transcription of the *luxL,M,N,P,Q* genes encoding signal production and reception: if dephosphoryl-LuxO activated transcription of these genes, signal production and signal reception would be maximal at low concentrations of autoinducers, and if phosphoryl-LuxO repressed transcription, signal production and receptivity to signal would be limited at high autoinducer concentrations. Many possibilities for modulating signal transduction can be imagined.

The contrasts between Lux signalling in *V. fischeri* and *V. harveyi* have been emphasized, but both systems function in density-sensing and use similar chemical signals. What is the evolutionary relationship between the two signalling systems? One speculation is that recombination between the *luxR* gene and a sensor kinase gene in a progenitor of *V. fischeri* resulted in linkage of the autoinducer binding domain of LuxR to a sensor kinase protein. Receptiveness to the autoinducer would then be coupled to a different regulatory pathway. Subsequent mutations and gene duplications and rearrangements generated new and multiple autoinducers, receptivities, and regulatory connections, finally resulting in a bacterium with the properties of *V. harveyi*.

REFERENCES

Bainton, N. J., B. W. Bycroft, S. Chhabra, P. Stead, L. Gledhill, P. J. Hill, C. E. D. Rees, M. K. Winson, G. P. C. Salmond, G. S. A. B. Stewart, and P. Williams. 1992. A general role for the *lux* autoinducer in bacterial cell signalling: control of antibiotic biosynthesis in *Erwinia. Gene* **116:**87–91.

Bassler, B., M. Wright, and M. Silverman. 1994a. Sequence and function of *luxO,* a negative regulator of luminescence in *Vibrio harveyi. Mol. Microbiol.* **12:**403–412.

Bassler, B. L., M. Wright, R. E. Showalter, and M. R. Silverman. 1993. Intercellular signalling in *Vibrio harveyi:* sequence and function of genes regulating expression of luminescence. *Mol. Microbiol.* **9:**773–786.

Bassler, B. L., M. Wright, and M. R. Silverman. 1994b. Multiple signalling systems controlling expression of luminescence in *Vibrio harveyi:* sequence and function of genes encoding a second sensory pathway. *Mol. Microbiol.* **13:**273–286.

Cangelosi, G. A., R. G. Ankenbauer, and E. W. Nester. 1990. Sugars induce the *Agrobacterium* virulence genes through a periplasmic binding protein and a transmembrane signal protein. *Proc. Natl. Acad. Sci. USA* **87:**6708–6712.

Cao, J., and E. A. Meighen. 1989. Purification and structural identification of an autoinducer for the luminescence system of *Vibrio harveyi. J. Biol. Chem.* **264:**21670–21676.

Choi, S. H., and E. P. Greenberg. 1992a. Genetic dissection of DNA binding and luminescence gene activation by the *Vibrio fischeri* LuxR protein. *J. Bacteriol.* **174:**4064–4069.

Choi, S. H., and E. P. Greenberg. 1992b. Genetic evidence for multimerization of LuxR, the transcriptional activator of *Vibrio fischeri* luminescence. *Mol. Marine Biol. Biotech.* **1:**408–413.

Cubo, M. T., A. Economou, G. Murphy, A. W. B. Johnston, and J. A. Downie. 1992. Molecular characterization and regulation of the rhizosphere-expressed genes *rhiABCR* that can influence nodulation of *Rhizobium leguminosarum* biovar *vicae*. *J. Bacteriol.* **174:**4026–4035.

Eberhard, A., A. L. Burlingame, C. Eberhard, G. L. Kenyon, K. H. Nealson, and N. J. Oppenheimer. 1981. Structural identification of autoinducer of *Photobacterium fischeri* luciferase. *Biochemistry* **20:**2444–2449.

Engebrecht, J., K. Nealson, and M. Silverman. 1983. Bacterial bioluminescence: isolation and genetic analysis of functions from *Vibrio fischeri*. *Cell* **32:**773–781.

Engebrecht, J., and M. Silverman. 1984. Identification of genes and gene products necessary for bacterial bioluminescence. *Proc. Natl. Acad. Sci. USA* **81:**4154–4158.

Engebrecht, J., and M. Silverman. 1987. Nucleotide sequence of the regulatory locus controlling expression of bacterial genes for bioluminescence. *Nucleic Acids Res.* **15:**10455–10467.

Fuqua, W. C., S. C. Winans, and E. P. Greenberg. 1994. Quorum sensing in bacteria: the LuxR-LuxI family of cell density–responsive transcriptional regulators. *J. Bacteriol.* **176:**269–275.

Greenberg, E. P., J. W. Hastings, and S. Ulitzur. 1979. Induction of luciferase synthesis in *Beneckea harveyi* by other marine bacteria. *Arch. Microbiol.* **120:**87–91.

Groarke, J. M., W. C. Mahoney, J. N. Hope, C. E. Furlong, F. T. Robb, H. Zalkin, and M. A. Hermodson. 1983. The amino acid sequence of the D-ribose binding protein from *Escherichia coli* K12. *J. Biol. Chem.* **258:**12952–12956.

Huang, M.-L. W., G. A. Cangelosi, W. Halperin, and E. W. Nester. 1990. A chromosomal *Agrobacterium tumefaciens* gene required for effective plant signal transduction. *J. Bacteriol.* **172:**1814–1822.

Iuchi, S., and E. C. C. Lin. 1992. Mutational analysis of signal transduction by ArcB, a membrane sensor protein responsible for anaerobic repression of operons involved in the central aerobic pathways in *Escherichia coli*. *J. Bacteriol.* **174:**3972–3980.

Iuchi, S., Z. Matsuda, T. Fujiwara, and E. C. C. Lin. 1990. The *arcB* gene of *Escherichia coli* encodes a sensor-regulator protein for anaerobic repression of the *arc* modulon. *Mol. Microbiol.* **4:**715–727.

Kolibachuk, D., and E. P. Greenberg. 1993. The *Vibrio fischeri* luminescence gene activator LuxR is a membrane-associated protein. *J. Bacteriol.* **175:**7307–7312.

Leroux, B., M. F. Yanofsky, S. C. Winans, J. E. Ward, S. F. Ziegler, and E. W. Nester. 1987. Characterization of the *virA* locus of *Agrobacterium tumefaciens*: a transcriptional regulator and host range determinant. *EMBO J.* **6:**849–856.

Makemson, J. C. 1986. Luciferase-dependent oxygen consumption by bioluminescent *Vibrios*. *J. Bacteriol.* **165:**461–466.

Martin, M., R. Showalter, and M. Silverman. 1989. Identification of a locus controlling expression of luminescence genes in *Vibrio harveyi*. *J. Bacteriol.* **171:**2406–2414.

McCleary, W. R., and D. R. Zusman. 1990. FrzE of *Myxococcus xanthus* is homologous to both CheA and CheY of *Salmonella typhimurium*. *Proc. Natl. Acad. Sci. USA* **87:**5898–5902.

Meighen, E. A. 1991. Molecular biology of bacterial bioluminescence. *Microbiol. Rev.* **55:**123–142.

Miyamoto, C. M., A. F. Graham, and E. A. Meighen. 1988. Nucleotide sequence of the *luxC* gene and the upstream DNA from the bioluminescent system of *Vibrio harveyi*. *Nucleic Acids Res.* **16:**1551–1562.

Morin, J. G., A. Harrington, K. Nealson, N. Krieger, T. O. Baldwin, and J. W. Hastings. 1975. Light for all reasons: versatility in the behavioral repertoire of the flashlight fish. *Science* **190:**74–76.

Nealson, K. H., and J. W. Hastings. 1979. Bacterial bioluminescence: its control and ecological significance. *Microbiol. Rev.* **43:**496–518.

Nealson, K. H., T. Platt, and J. W. Hastings. 1970. Cellular control of the synthesis and activity of the bacterial luminescent system. *J. Bacteriol.* **104:**313–322.

Ninfa, A. J., E. G. Ninfa, A. N. Lupas, A. Stock, B. Magasanik, and J. Stock. 1988. Crosstalk between bacterial chemotaxis signal transduction proteins and regulators of transcription of the Ntr regulon: evidence that nitrogen assimilation and chemotaxis are controlled by a common phosphotransfer mechanism. *Proc. Natl. Acad. Sci. USA* **85:**5492–5496.

Parkinson, J. S., and E. C. Kofoid. 1992. Communication modules in bacterial signaling proteins. *Annu. Rev. Genet.* **26:**71–112.

Passador, L., J. M. Cook, M. J. Gambello, L. Rust, and B. H. Iglewski. 1993. Expression of *Pseudomonas aeruginosa* virulence genes requires cell to cell communication. *Science* **260:**1127–1130.

Piper, K. R., S. Beck von Bodman, and S. K. Farrand. 1993. Conjugation factor of *Agrobacterium tumefaciens* regulates Ti plasmid transfer by autoinduction. *Nature* (London) **362:**448–450.

Showalter, R. E., M. O. Martin, and M. R. Silverman. 1990. Cloning and nucleotide sequence of *luxR,* a regulatory gene controlling luminescence in *Vibrio harveyi. J. Bacteriol.* **172:** 2946–2954.

Stock, J. B., A. J. Ninfa, and A. M. Stock. 1989. Protein phosphorylation and regulation of adaptive responses in bacteria. *Microbiol. Rev.* **53:**450–490.

Swartzman, E., M. Silverman, and E. A. Meighen. 1992. The *luxR* gene product of *Vibrio harveyi* is a transcriptional activator of the *lux* promoter. *J. Bacteriol.* **174:**7490–7493.

Wanner, B. L., and M. R. Wilmes-Riesenberg. 1992. Involvement of phosphotransacetylase, acetate kinase, and acetyl phosphate synthesis in control of the phosphate regulon in *Escherichia coli. J. Bacteriol.* **174:**2124–2130.

Winans, S. C., P. R. Ebert, S. E. Stachel, M. P. Gordon, and E. W. Nester. 1986. A gene essential for *Agrobacterium* virulence is homologous to a family of positive regulatory loci. *Proc. Natl. Acad. Sci. USA* **83:**8278–8282.

Ziegler, M. M., and T. O. Baldwin. 1981. Biochemistry of bacterial bioluminescence. *Curr. Top. Bioenerg.* **12:**65–113.

A Signal Transduction Network in *Bacillus subtilis* Includes the DegS/DegU and ComP/ComA Two-Component Systems

Tarek Msadek, Frank Kunst, and Georges Rapoport

29

TWO-COMPONENT PARADIGM

Optimization of cell survival requires that gene expression be continually adapted in response to signals reflecting changes in the environment. As illustrated by the chapters of this book, this signal transduction process involves the interaction of two regulatory proteins, one of which controls the activity of the other (Nixon et al., 1986; Ronson et al., 1987; Kofoid and Parkinson, 1988), giving bacteria the flexibility to adapt to a wide variety of stimuli. The first component, a histidine protein kinase, is autophosphorylated at a conserved histidine residue in an ATP-dependent reaction. In a second step, the phosphoryl group is transferred to an aspartate residue in the conserved amino-terminal domain of the second component, the response regulator. Response regulators have been shown to catalyze their own phosphorylation, either from their cognate phosphorylated histidine kinase or from low-molecular-weight phosphodonors such as acetyl phosphate (Hess et al., 1988; Lukat et al., 1992; Mc-

Cleary et al., 1993). Several members of the histidine protein kinase family also act as phosphatases, catalyzing the dephosphorylation of the associated response regulator. Regulation by environmental signals may therefore take place by modulating either the kinase activity or the phosphatase activity of the protein.

These paired families of highly related proteins, designated "two-component systems," appear to be ubiquitous among bacteria and have been the subject of several reviews (Ronson et al., 1987; Kofoid and Parkinson, 1988; Stock et al., 1989; Gross et al., 1989; Albright et al., 1989; Bourret et al., 1989; Miller et al., 1989; Stock et al., 1990; Bourret et al., 1991; Parkinson and Kofoid, 1992; Parkinson, 1993; Volz, 1993; Alex and Simon, 1994). With more than 70 members reported to date and new examples continuing to emerge, two-component systems constitute the largest known family of bacterial regulators. Similar proteins have now been reported among eukaryotes such as *Arabidopsis thaliana* (Chang et al., 1993) and *Saccharomyces cerevisiae* (Brown et al., 1993; Ota and Varshavsky, 1993; Maeda et al., 1994), and less related proteins also exist in *Trypanosoma brucei* and rat (Popov et al., 1992; Chang and Meyerowitz, 1994).

Tarek Msadek, Frank Kunst, and Georges Rapoport, Unité de Biochimie Microbienne, Centre National de la Recherche Scientifique, Département des Biotechnologies, Institut Pasteur, Paris, France. *Present address for Tarek Msadek:* Division of Cellular Biology, The Scripps Research Institute, La Jolla, California 92037.

Two-Component Signal Transduction, Edited by James A. Hoch and Thomas J. Silhavy,
© 1995 American Society for Microbiology, Washington, DC 20005

TWO-COMPONENT REGULATORY SYSTEMS OF *BACILLUS SUBTILIS* FORM A SIGNAL TRANSDUCTION NETWORK

Soil bacteria such as *Bacillus subtilis* are subject to drastic variations of environmental conditions such as temperature, humidity, and nutrient source availability. At the onset of the stationary phase, faced with a depletion of essential nutrients, *B. subtilis* can adopt several responses, including synthesis of macromolecule-degrading enzymes, competence for genetic transformation, increased motility and chemotaxis, antibiotic production, and finally, sporulation (Sonenshein, 1989; Smith et al., 1991). Each of these post-exponential-phase responses is controlled by at least one two-component regulatory system. To date, there have been 9 proved or potential histidine protein kinase/response regulator systems identified in *B. subtilis* (DegS/DegU, ComP/ComA, KinA-KinB/Spo0F-Spo0A, PhoR/PhoP, CheA/CheY-CheB, SpaK/SpaR, DegM, URF-1/URF-2, and ORFX-18/ORFX-17) (Trach et al., 1985; Ferrari et al., 1985; Yoshikawa et al., 1986; Seki et al., 1987, 1988; Kunst et al., 1988b; Henner et al., 1988c; Tanaka and Kawata, 1988; Weinrauch et al., 1989; Perego et al., 1989; Antoniewski et al., 1990; Weinrauch et al., 1990; Bischoff and Ordal, 1991; Moir and Kemp, personal communication; Fuhrer and Ordal, 1991; Masui et al., 1992; Sorokin et al., 1993; Trach and Hoch, 1993; Kirsch et al., 1993; Klein et al., 1993).

Regulation by the *B. subtilis* two-component systems presents several original features, as described below.

First among these is the considerable amount of overlap that exists between the regulatory pathways controlling post-exponential-phase responses in *B. subtilis*. DegS/DegU, ComP/ComA, and KinA-KinB/Spo0F-Spo0A are all involved in controlling competence gene expression. DegS/DegU and KinA-KinB/Spo0F-Spo0A control protease gene expression, and the ComP/ComA system also influences degradative enzyme production under certain conditions. Mutations affecting ComP, DegS, or DegU can also modify the sporulation process, albeit to a minor extent. The ComP/ComA and DegS/DegU systems appear to be closely related, as they share several common targets, including *degQ, srfA,* and late competence genes. Production of alkaline phosphatase involves regulation by PhoR/PhoP, ORFX-18/ORFX-17, and KinA-KinB/Spo0F-Spo0A, and the ORFX-18/ORFX-17 pair also affects the initiation of sporulation (see Chapter 18). Apparently, few of these systems are dedicated to a single response but instead behave as global regulators that interact to form a signal transduction network.

Another unique feature of this type of regulation in *B. subtilis* is the notion that the unphosphorylated form of the response regulator can be active and control specific functions. At first glance, this appears to contradict the established dogma that phosphorylation is the sine qua non requirement for activating the response regulator. However, biochemical and genetic evidence has shown that the DegU response regulator acts as a molecular switch, controlling two alternative functions: the phosphorylated form of the protein is essential for degradative enzyme production, whereas the unphosphorylated form allows the expression of late competence genes (Msadek et al., 1990; Dahl et al., 1992). Furthermore, it has recently been shown that although phosphorylation of the Spo0A response regulator enhances binding to weak-affinity binding sites, unphosphorylated Spo0A can bind to high-affinity binding sites and activate transcription of *spoIIG* in vivo (Baldus et al., 1994).

Finally, it appears that several *B. subtilis* histidine protein kinase and response regulator proteins can associate to form a multicomponent phosphorelay. Indeed, several kinases (KinA, KinB) are thought to phosphorylate the Spo0F protein, which would thus integrate different environmental signals. Spo0B is a unique protein that acts as a phosphotransferase, catalyzing phosphotransfer between two response regulators, Spo0F and Spo0A (see Chapter 8).

Some of these original aspects are discussed in this review, within the framework of the DegS/DegU and ComP/ComA signal transduction network.

DEGRADATIVE ENZYME SYNTHESIS

Regulatory Genes

Synthesis of several degradative enzymes in *B. subtilis,* including an intracellular protease and several secreted enzymes (levansucrase, alkaline and metalloproteases, α-amylase, xylanase, β-glucanase), is positively controlled at the transcriptional level by the products of the *degS* and *degU* genes (Kunst et al., 1974; Ayusawa et al., 1975; Shimotsu and Henner, 1986; Aymerich et al., 1986; Amory et al., 1987; Ruppen et al., 1988; Henner et al., 1988a; Stülke, personal communication) and by at least two accessory regulatory genes, *degQ* and *degR,* which encode small polypeptides of 46 and 60 amino acids, respectively (Nagami and Tanaka, 1986; Yang et al., 1986; Yang et al., 1987; Amory et al., 1987; Tanaka et al., 1987). The *degS* and *degU* genes have been identified in several *Bacillus* species including *B. brevis* (Louw et al., 1994) and an alkalophilic *Bacillus* species (Hastrup, personal communication). The *degR* gene was originally isolated from *B. natto* (Nagami and Tanaka, 1986), and *degQ* has also been found in *B. amyloliquefaciens* and *B. licheniformis* (Tomioka et al., 1985; Amory et al., 1987).

The *degS* and *degU* genes form an operon encoding a two-component system (Tanaka and Kawata, 1988; Henner et al., 1988c; Kunst et al., 1988b; Msadek et al., 1990). Just upstream from the conserved carboxy-terminal histidine protein kinase domain in DegS lies a region sharing strong similarities with the catalytic domain of eukaryotic protein kinases (Hanks et al., 1988; Taylor, 1989; Msadek et al., 1990). DegS appears to be a cytoplasmic protein, as it does not contain any significant hydrophobic domains. Carboxy-terminal amino acid sequence similarities place DegU within the UhpA response regulator subfamily (Stock et al., 1989; Msadek et al., 1990).

The domain shared by members of the UhpA subfamily contains a potential helix-turn-helix DNA binding motif (Friedrich and Kadner, 1987) and is present in a wide spectrum of bacterial regulatory proteins that are not members of the response regulator family (Henner et al., 1988c; Kunst et al., 1988b; Henikoff et al., 1990; Msadek et al., 1990; Kahn and Ditta, 1991; Stout et al., 1991). Included among these proteins are MalT of *Escherichia coli* (Cole and Raibaud, 1986; Richet and Raibaud, 1989), GerE of *B. subtilis* (Cutting and Mandelstam, 1986; Holland et al., 1987), RcsA of *Klebsiella pneumoniae, E. coli, Erwinia amylovora,* and *Erwinia stewartii* (Allen et al., 1987; Coleman et al., 1990; Bernhard et al., 1990; Poetter and Coplin, 1991; Stout et al., 1991), RmpA of *K. pneumoniae* (Nassif et al., 1989; Vasselon et al., 1991), LasR of *Pseudomonas aeruginosa* (Gambello and Iglewski, 1991), LuxR of *Vibrio fischeri* (Engebrecht and Silverman, 1987; Devine et al., 1988), and BrpA of *Streptomyces hygroscopicus* (Raibaud et al., 1991), as well as several bacterial σ-factors (Kahn and Ditta, 1991; Lonetto et al., 1992).

The *B. subtilis* GerE protein is particularly interesting because it consists entirely of the conserved carboxy-terminal domain of the UhpA response regulator subfamily, suggesting an independent function for this region (Cutting and Mandelstam, 1986; Holland et al., 1987; Gross et al., 1989; Kahn and Ditta, 1991). GerE binds to the promoter regions of two spore coat protein genes, *cotB* and *cotC,* and stimulates transcription by σ^K-RNA polymerase holoenzyme (Zheng et al., 1992). A truncated protein consisting of only the conserved carboxy-terminal domain of the *Rhizobium meliloti* FixJ response regulator was shown to be capable of activating transcription from the *nifA* promoter (Kahn and Ditta, 1991; Da Re et al., 1994). The independent activity of the conserved carboxy-terminal domain was also demonstrated for the *V. fischeri* LuxR protein, which is not a member of the response regulator family (Choi and Greenberg, 1991, 1992; Fuqua et al., 1994). Finally, for three bacterial σ-factors,

DNA binding by this conserved carboxy-terminal domain was shown to be inhibited by the amino-terminal region (Dombroski et al., 1993). It thus appears that the activity of this conserved carboxy-terminal domain can be modulated through an intramolecular negative interaction with the amino-terminal domain of the protein.

In addition to the DegS/DegU two-component system, several accessory regulatory genes such as *degQ* and *degR* are also involved in controlling degradative enzyme production (Kunst et al., 1988a; Smith, 1993). The *degQ36-*(Hy) mutation, a single base change at position −10, leads to overexpression of the *degQ* gene and a strong increase in degradative enzyme production (Lepesant et al., 1972; Kunst et al., 1974; Yang et al., 1986; Amory et al., 1987). Overexpression of *degQ* or *degR* using a multicopy plasmid leads to the same phenotype, yet single or simultaneous deletions of the two genes have no detectable effect on degradative enzyme synthesis (Nagami and Tanaka, 1986; Yang et al., 1987; Tanaka et al., 1987). Furthermore, activation of transcription by DegQ or DegR does not take place in the absence of DegS or DegU (Amory et al., 1987; Msadek et al., 1991; Mukai et al., 1992). Thus, although both DegS and DegU are required for degradative enzyme production (Tanaka and Kawata, 1988; Henner et al., 1988c; Kunst et al., 1988b; Mukai et al., 1990; Msadek et al., 1991), DegQ and DegR appear to be dispensable under the conditions that have been tested (Yang et al., 1986; Yang et al., 1987; Klier et al., 1992). The mechanism by which DegQ acts on the transcription of genes encoding degradative enzymes remains to be determined. The possible role of DegR is discussed below (see Phosphorylation of DegU).

Target Genes

The DegS/DegU two-component system controls the expression of a wide variety of genes, encoding degradative enzymes such as proteases (*aprE, nprE, isp*), levansucrase (*sacB*), α-amylase (*amyE*), β-glucanase (*licS*), and β-xylanase, as well as regulatory genes (*degQ,*

degR), surfactin and polyketide biosynthesis genes (*srfA, pksX*), and the *comK* regulatory gene, which in turn controls both late competence gene expression and competence-specific induction of *recA* expression (see Competence Gene Expression) (Kunst et al., 1974; Ayusawa et al., 1975; Shimotsu and Henner, 1986; Aymerich et al., 1986; Henner et al., 1987; Amory et al., 1987; Ruppen et al., 1988; Henner et al., 1988a; Msadek et al., 1990; Hahn and Dubnau, 1991; Msadek et al., 1991; Msadek and Kunst, unpublished data; Mukai et al., 1992; Albertini et al., personal communication; Cheo et al., 1993; Scotti et al., 1993; Hahn et al., 1994; Stülke, personal communication; van Sinderen et al., 1994; van Sinderen and Venema, 1994). The *mpr, bpr,* and *vpr* genes encoding minor proteases also appear to be regulated by DegQ (Sloma et al., 1990a; Sloma et al., 1990b; Sloma et al., 1991). Most of the degradative enzymes whose synthesis is controlled by DegS/DegU can be regarded as scavenging enzymes that degrade polymers, providing bacteria with amino acids or sugars to replenish depleted intracellular pools.

Several reports have shown that Hy mutations in the *degU* or *degQ* genes lead to an increase in the mRNA level of *sacB* and *aprE* (Shimotsu and Henner, 1986; Aymerich et al., 1986; Henner et al., 1988a). DegR has been shown to increase the mRNA level of *aprE* and *nprE* (Tanaka et al., 1987). Thus, in every case examined, these pleiotropic regulators increase the rate of transcription initiation of their target genes. The DegU response regulator controls *sacB* at an additional level by increasing the expression of the *sacXY* operon, which encodes a pair of regulatory proteins mediating sucrose induction of levansucrase synthesis through antitermination (Crutz and Steinmetz, 1992). Regions essential for transcriptional activation by both DegS/DegU and DegQ were located between positions −117 and −96 and positions −164 and −141 with respect to the transcriptional start sites of *sacB* and *aprE*, respectively (Henner et al., 1987; Henner et al., 1988a,b). Although comparison of the corresponding regions of the *sacB* and *aprE* genes showed that 27 bases of 48 are identical in both of these upstream promoter regions, no obvious consensus sequence could be established (Henner et al., 1988a,b). Also, no biochemical evidence yet exists for specific binding of DegU to these target sites.

Phosphorylation of DegU: A Molecular Switch

The DegS and DegU proteins have been purified. DegS has been shown to be autophosphorylated when incubated in the presence of $[\gamma\text{-}^{32}P]$ATP and to transfer its phosphate group to the DegU response regulator (Mukai et al., 1990; Dahl et al., 1991). The pH stabilities of the phosphorylated proteins are consistent with the existence of a histidinyl-phosphate group for DegS and an aspartyl-phosphate group for DegU (Mukai et al., 1990). Sequence similarities with other two-component systems suggest the conserved His-189 residue of the DegS protein kinase and Asp-56 residue of the DegU response regulator as likely candidates for the respective phosphorylation sites of the two proteins (Mukai et al., 1990; Msadek et al.,

1990). Finally, by comparing the stability of the purified phosphorylated DegU response regulator in the presence or absence of DegS, we demonstrated that DegS acts as a DegU phosphatase (Dahl et al., 1992).

Two classes of mutations have been identified in both the *degS* and *degU* genes, leading to either deficiency of degradative enzyme synthesis or to a pleiotropic Hy phenotype, which includes hyperproduction of degradative enzymes, the ability to sporulate in the presence of glucose, decreased genetic competence, and the absence of flagella (Kunst et al., 1974; Ayusawa et al., 1975; Henner et al., 1988c; Msadek et al., 1990). Several of these mutations have been characterized at the molecular level, and the associated amino acid modifications in DegS or DegU are shown in Table 1, along with their effects, both in vivo and in vitro.

Mutations that lead to the accumulation of the unphosphorylated form of DegU abolish the expression of genes encoding degradative enzymes but do not affect the competence pathway. Mutations of this type either prevent autophosphorylation of DegS (DegS E300K), allow DegS autophosphorylation but strongly

TABLE 1 Mutations in the *degS* and *degU* genes and their associated phenotypes

Mutations	Associated modification	In vivo phenotypes	In vitro phenotypes	References
degS100(Hy) *degS200*(Hy)	DegS V236M DegS G218E	Degradative enzyme hyperproduction, competence deficiency	Reduced DegU phosphatase activity	Henner et al., 1988c; Tanaka et al., 1991; Dahl et al., 1992
degS42	DegS E300K	Deficiency in degradative enzyme production	Loss of autophosphorylation activity	Msadek et al., 1990; Tanaka et al., 1991
degS220	DegS A193V	Deficiency in degradative enzyme production	Deficiency in phosphotransfer to DegU	Msadek et al., 1990; Dahl et al., 1992
degU24(Hy) *degU31*(Hy)	DegU T98I DegU V131L	Degradative enzyme hyperproduction, competence deficiency	Increased DegU phosphorylation	Msadek et al., 1990; Dahl et al., 1991
degU32(Hy) *degU9*(Hy)	DegU H12L DegU E107K	Degradative enzyme hyperproduction, competence deficiency	Decreased DegU dephosphorylation	Henner et al., 1988c; Dahl et al., 1992; Tanaka and Kawata-Mukai, 1994
degU146	DegU D56N	Deficiency in degradative enzyme production	Deficiency in DegU phosphorylation	Msadek et al., 1990; Dahl et al., 1991, 1992

diminish phosphorylation of DegU (DegS A193V), or inactivate the presumed phosphorylation site of DegU (DegU D56N) (Tanaka et al., 1991; Dahl et al., 1992) (Table 1).

In a reciprocal fashion, mutations leading to a Hy phenotype increase degradative enzyme production and prevent competence gene expression by favoring an increase in the phosphorylated form of DegU. Mutations of this class that modify DegS result in the loss of DegU phosphatase activity (DegS V236M, DegS G218E) (Tanaka et al., 1991; Dahl et al., 1992). Mutations modifying DegU were also identified that either lead to a stronger DegU phosphorylation signal (DegU T98I, DegU V131L) or strongly decrease the rate of dephosphorylation of DegU (DegU H12L, DegU E107K) (Dahl et al., 1991, 1992; Tanaka and Kawata-Mukai, 1994). The phosphorylated form of the DegU H12L protein has a half-life of 120 min, as compared with 18 min for the native DegU response regulator. This strongly increased stability is consistent with the associated Hy phenotype of DegU H12L in vivo (Dahl et al., 1992). Regulation by DegR is also thought to involve stabilization of the phosphorylated form of DegU, because the addition of purified DegR increased the apparent half-life of phosphorylated DegU approximately fourfold (Mukai et al., 1992). Strong amino acid sequence similarities were noted between DegR and the amino-terminal domain of DegS (Kunst, unpublished data; Ogura et al., 1994). It has been suggested that the amino-terminal domain of DegS might be involved in DegU dephosphorylation and that DegR may inhibit this reaction by competition (Ogura et al.,1994).

Missense mutations in *degS* and *degU* that abolish degradative enzyme production do not affect the competence pathway, whereas those that lead to hyperproduction of degradative enzymes result in a lowered transformation frequency. Apparently, transformation frequency is high when degradative enzyme production is low, and vice versa (Msadek et al., 1990).

The interpretation of these phenotypes led to the hypothesis that the DegU response regulator could act as a molecular switch controlling the expression of alternate sets of genes, as shown in Fig. 1. Degradative enzyme production requires both DegS and DegU, indicating that the phosphorylated form of DegU is necessary for this process (Tanaka and Kawata, 1988; Henner et al., 1988c; Kunst et al., 1988b; Msadek et al., 1990).

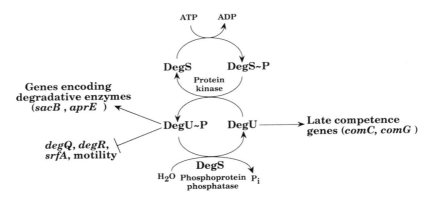

FIGURE 1 Pleiotropic regulation by DegS/DegU signal transduction pathway controlling degradative enzyme synthesis and competence gene expression in *B. subtilis*. Arrows indicate positive regulation, and perpendicular bars indicate negative regulation. Regulation by DegS/DegU has not been shown to be direct and may therefore involve possible intermediate genes.

DegU is also required for the development of genetic competence, whereas DegS is not (Tanaka and Kawata, 1988; Roggiani et al., 1990; Msadek et al., 1990; Msadek et al., 1991). Roggiani et al. (1990) showed that Hy mutations in *degS* or *degU* (DegS G218E, DegU H12L, DegU E107K) or a disruption of the *degU* gene result in decreased expression of late competence genes such as *comG*, whereas a disruption of the *degS* gene has little or no effect on competence gene expression. Taken together, these results strongly suggest that the unphosphorylated form of DegU allows late competence gene expression but that an excess of the phosphorylated form of DegU prevents competence development. Indeed, although DegU is not required for expression of *srfA*, an essential component of the competence pathway (see Competence Gene Expression), this expression is markedly reduced in a DegU H12L strain (Hahn and Dubnau, 1991; van Sinderen, personal communication).

A *comG'-'lacZ* fusion is expressed at approximately the same level in strains carrying wild-type *degS* and *degU* genes, a deletion of the *degS* gene, or a mutation preventing DegU phosphorylation (DegU D56N) (Fig. 2). This expression is abolished, however, in strains in which the *degU* gene is deleted or in the presence of a mutation stabilizing the phosphorylated form of DegU (DegU H12L) (Fig. 2). These results confirm the essential role of DegU and indicate that the unphosphorylated form of the DegU response regulator can act as a positive regulator of late competence gene expression. Identical results were obtained using a *comK'-'lacZ* fusion (van Sinderen and Venema, 1994). DegU thus appears to be the first response regulator described as having two active conformations: a phosphorylated form, which is necessary for degradative enzyme production, and a nonphosphorylated form, which allows the expression of genetic competence (see Fig. 1). Support for this hypothesis has been provided by the recent demonstration that although phosphorylation of the *B. subtilis* Spo0A protein enhances binding to weak-

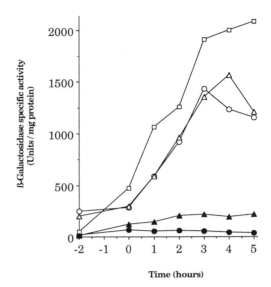

FIGURE 2 Expression of *comG'-'lacZ* during growth in competence minimal medium in strains carrying mutations in *degS* or *degU* genes. Time is expressed in hours before or after the time of transition from the exponential growth phase to the stationary phase. □: *degU146* D56N; △: Δ*degS*; ○: wild-type *degS* and *degU* genes; ▲: Δ(*degSdegU*); ●: *degU32*(Hy) H12L.

affinity binding sites, the unphosphorylated Spo0A response regulator can bind to high-affinity binding sites and activate transcription of *spoIIG* in vivo (Baldus et al., 1994). However, although we could not detect any phosphorylation of DegU D56N in vitro (Dahl et al., 1992), it cannot be ruled out that the DegU D56N protein may be phosphorylated in vivo at a low level on some alternate residue, perhaps in a DegS-independent fashion. Secondary phosphorylation sites have previously been reported for the CheY, OmpR, and FixJ response regulators (Bourret et al., 1990; Delgado et al., 1993; Reyrat et al., 1994).

Phosphorylation of response regulators may simply be regarded as a means to an end, allowing an increased affinity for protein-protein or protein-DNA interaction by either relieving a negative interaction between the two halves of the protein or by inducing a transient conformational change. For several two-component systems, kinase-independent mutations have been described, leading to modifications that appear to mimic phosphorylation of the response regulator (Popham et al., 1989; Bourret et al., 1990; Dixon et al., 1991; Green et al., 1991a; Green et al., 1991b; Brissette et al., 1991; Pazour et al., 1992; Kanamaru and Mizuno, 1992; Klose et al., 1993; Jin et al., 1993; Moore et al., 1993). In several cases, evidence has also been obtained that it is not phosphorylation per se, but rather the multimerization of the phosphorylated effector that induces a conformational change to the active conformation (Scarlato et al., 1990; Weiss et al., 1992; Porter et al., 1993). Results obtained by Podvin and Steinmetz (1992) could suggest that transcriptional activation by DegU may also involve oligomerization. These authors introduced a recombinant SPβ phage carrying a wild-type degU allele into a strain with the degU32(Hy) mutation (DegU H12L). The resulting strain carrying two different degU alleles had a lower level of sacB'-'lacZ expression as compared with the reference strain containing the single degU32-(Hy) allele, indicating partial suppression of the hyperproduction phenotype associated with the degU32 mutation. This phenotype could be due to the formation of heterologous multimers. An alternative hypothesis is that this suppression could be the result of competition between wild-type and mutant DegU proteins, either for phosphorylation by DegS or for DNA binding at the target site.

Little is known about the putative environmental signal(s) regulating the levels of phosphorylated DegU. Genes encoding degradative enzymes are often subject to multiple redundant regulatory pathways, often involving more than one two-component system (Klier et al., 1992; Jarnagin and Ferrari, 1992; Strauch and Hoch, 1992; Msadek et al., 1993). Phosphorylated DegU is required for the expression of genes expressed during the late exponential phase and the beginning of the stationary phase. One may therefore hypothesize that DegU is present in its phosphorylated form during this time interval and that regulation could involve a signal increasing the phospha-

tase activity of DegS. This would lead to a decrease in phosphorylated DegU and a corresponding arrest in degradative enzyme synthesis under growth conditions such as those leading to competence gene expression.

Perception of environmental signals may involve additional regulatory proteins that could modulate the activity of the kinase in response to molecular signals, as shown for the chemotaxis signal transduction pathway in E. coli (Borkovich et al., 1989; Borkovich and Simon, 1990; Gegner and Dahlquist, 1991; Gegner et al., 1992). Although sacB expression is also regulated by components of the phosphotransferase system, these proteins do not appear to affect regulation by DegS/DegU (Msadek, unpublished data). Regulation by the BvgS/BvgA two-component system in Bordetella pertussis has been shown to be modulated by several environmental signals including nicotinic acid (Melton and Weiss, 1989, 1993; Coote, 1991). Growth of a B. subtilis degQ(Hy) strain in the presence of 40 mM nicotinic acid led to an 18-fold decrease in sacB'-'lacZ expression, although the physiological significance of this effect is not clear (Msadek, unpublished data). Based on present data, the DegS/DegU system is thought to be involved in controlling secondary metabolism. Ogura et al. (1994) reported that the B. subtilis proB gene, encoding γ-glutamyl kinase, plays a role in enhancing alkaline protease synthesis in a DegS-dependent fashion. The authors suggested that the accumulation of γ-glutamyl phosphate in the cell may lead to higher levels of phosphorylated DegU, either by directly acting as a high-energy acylphosphate phosphodonor for DegS or by regulating the phosphatase activity of DegS (Ogura et al., 1994). However, the precise environmental signals to which the DegS/DegU system responds remain to be determined.

COMPETENCE GENE EXPRESSION

Regulatory Genes

Competence in B. subtilis is a natural physiological state allowing the uptake of exogenous DNA molecules. The products of more than 24

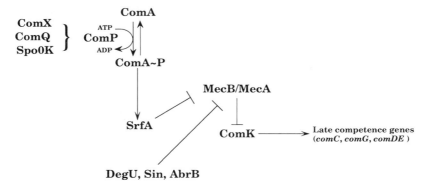

FIGURE 3 Regulation of competence gene expression in *B. subtilis*. Arrows indicate positive regulation, and perpendicular bars indicate negative regulation. The ComP/ComA and DegS/DegU two-component systems comprise two parallel pathways controlling competence gene expression (Dubnau, 1993).

loci have been identified as playing a part in competence development (Dubnau, 1993). These loci can be divided into two classes, one of which encodes regulatory proteins that are required for the expression of genes in the second class. This second class of genes, referred to as late competence genes because they are only expressed during the post-exponential growth phase, encodes products involved in DNA binding and uptake (see Target Genes). Specific regulators controlling the expression of these genes include ComP, ComA, ComQ, ComK, ComX, and ComS, a product of the *srfA* operon (Nakano et al., 1988; Weinrauch et al., 1989; Jaacks et al., 1989; van Sinderen et al., 1990; Weinrauch et al., 1990; Nakano et al., 1991a; Weinrauch et al., 1991; Msadek et al., 1991; Cosmina et al., 1993; Fuma et al., 1993; Magnuson et al., 1994; van Sinderen et al., 1994; D'Souza et al., 1994; Hamoen et al., 1995). Other proteins controlling the expression of late competence genes include DegS, DegU, Spo0A, Sin, AbrB, and the Spo0K oligopeptide permease (Dubnau, 1993). The regulatory pathway controlling competence gene expression is shown in Fig. 3 and is discussed below.

The *comP* and *comA* genes encode histidine kinase and response regulator proteins, respectively (Weinrauch et al., 1989; Weinrauch et al.,

1990). The amino acid sequence of ComP contains eight potential transmembrane segments in the amino-terminal domain (Weinrauch et al., 1990), and preliminary results suggest that ComP is an integral membrane protein (Dubnau et al., 1994). The carboxy-terminal domain of ComA shares amino acid sequence similarities with DegU and other response regulators of the UhpA subfamily and contains a potential helix-turn-helix DNA binding motif (Stock et al., 1989; Weinrauch et al., 1989; Msadek et al., 1990).

ComQ and ComX are the products of two genes lying just upstream from the *comP* and *comA* genes, which are thought to play a role in the phosphorylation of ComA (see Phosphorylation of ComA) (Msadek et al., 1991; Weinrauch et al., 1991; Magnuson et al., 1994).

Target Genes

Two sets of genes are controlled by ComP/ComA, either directly or indirectly. Genes whose expression directly requires ComP/ComA include the *degQ* regulatory gene, *gsiA*, a glucose starvation-inducible gene thought to play a role in early sporulation gene expression, and the *srfA* operon encoding surfactin synthetase subunits (Nakano and Zuber, 1989, 1991; Nakano et al., 1991b; Msadek et al., 1991; Hahn and Dubnau, 1991; Mueller et al., 1992).

DNA sequences required for regulation by ComP/ComA were located by deletion analysis upstream from *degQ* (positions −78 to −40) and *srfA* (positions −160 to −75) (Msadek et al., 1991; Nakano et al., 1991b). Comparison of these sequences allowed the identification of a 16-bp region of imperfect dyad symmetry (TTGCGGNNTCCCGCAN) thought to be the ComA target site ("ComA box"), which is also present upstream from *gsiA* (positions −75 to −60) (Nakano et al., 1991b; Mueller et al., 1992). This sequence is repeated twice upstream from the *srfA* promoter (box 1: −118 to −103; box 2: −74 to −59) (Nakano et al., 1991b). Site-directed mutagenesis studies revealed that alteration of positions 3 and 4 of the CCGCAA motif prevents ComA-dependent transcription of *srfA* (Nakano et al., 1991b; Nakano and Zuber, 1993). It was also shown that the box 1 sequence is no longer required for ComP/ComA-dependent activation when the box 2 sequence is modified to form perfect dyad symmetry (TTGCGGNNNNCCGC-AA) (Nakano and Zuber, 1993). Positioning of the two ComA boxes upstream from *srfA* is shown to be helix face-dependent. A 5-bp insertion between the two boxes, leading to their positioning on opposite faces of the DNA helix, strongly reduced *srfA* expression, whereas a 10-bp insertion led to a twofold increase in expression (Nakano and Zuber, 1993). These results led to the suggestion that ComA may bind DNA as a dimer, with cooperative binding first to box 1 and then to box 2 and loop formation by the intervening DNA (Nakano and Zuber, 1993).

ComA has been shown to bind directly to the *srfA* promoter regulatory sequences in vitro, and the regions protected by DNase I footprinting analysis encompass the ComA boxes (Roggiani and Dubnau, 1993). It has been suggested that the primary if not only role of ComA in competence development is to allow expression of the *srfA* operon. Indeed, when expression of *srfA* was placed under the control of the isopropyl-β-D-thiogalacto-pyranoside (IPTG)-inducible *spac* promoter, ComA was no longer required for competence

development (Hahn and Dubnau, 1991; Nakano and Zuber, 1991).

Genes controlled indirectly by ComP/ComA are dependent on ComS, a product of the *srfA* operon (D'Souza et al., 1994; Hamoen et al., 1995). Expression of *comS* is required in turn for the synthesis of ComK (see Fig. 3), a DNA binding protein previously referred to as competence transcription factor, which positively regulates the expression of late competence genes involved in binding and uptake of exogenous DNA (Mohan and Dubnau, 1990; van Sinderen et al., 1990; van Sinderen et al., 1994; van Sinderen et al., 1995; van Sinderen and Venema, 1994). These genes include *comC* and the *comG* operon, whose products share significant similarities to proteins required for protein secretion in *Klebsiella* and pilin processing and assembly in *Pseudomonas, comE,* and *comF,* which encodes a protein with similarities to ATP-dependent RNA/DNA helicases (Albano et al., 1989; Mohan et al., 1989; Breitling and Dubnau, 1990; Mohan and Dubnau, 1990; Dubnau, 1991; Londoño-Vallejo and Dubnau, 1993; Hahn et al., 1993). Expression of late competence genes mirrors the development of competence and responds to signals that are at least in part nutritional: they are expressed in glucose-minimal salt medium but not in complex medium and are turned on after the time of transition from the exponential growth phase to the stationary phase (Dubnau, 1991). Competence development is strongly diminished when glucose is replaced by glycerol or when glutamine is added to the culture medium (Albano et al., 1987).

Regulatory sequences required for expression of *comC* were located by deletion analysis between positions −97 and −79 (Mohan and Dubnau, 1990). A comparison of this sequence with regions upstream from *comG* revealed the existence of a partial palindrome whose presence on a multicopy plasmid prevents expression of *comC, comE,* and *comG,* presumably by titrating the positive regulator ComK (Mohan and Dubnau, 1990). This sequence was shown to be required for DNA binding of competence transcription factor (now known to be ComK) to the *comC* promoter region (Mohan and Dubnau, 1990; van Sinderen et al., 1995), and a similar sequence is also found upstream from the *comE* operon (Hahn et al., 1993). ComK is also required for its own synthesis (van Sinderen and Venema, 1994) and has been shown to be a DNA binding protein that binds to the promoter regions of *comC, comE, comF, comG,* and *comK* itself (van Sinderen et al., 1995). Although no clear-cut consensus "ComK box" has yet been established, the partial palindromes that act as ComK targets upstream from some of the late competence genes contain a conserved CAGAAA motif (Hahn et al., 1993).

Phosphorylation of ComA

ComA has been shown to be phosphorylated in vitro using either the heterologous *E. coli* NtrB histidine protein kinase or acetyl phosphate, and phosphorylation of ComA has been shown to enhance binding affinity for the ComA boxes located upstream from the *srfA* promoter (Roggiani and Dubnau, 1993; Dubnau, 1993). Mutations that modify the conserved aspartate at position 55 of ComA (D55E or D55N) do not allow competence development, suggesting this residue may be the phosphorylation site of ComA, as shown for several other response regulators (Parkinson and Kofoid, 1992; Dubnau et al., 1994).

ComQ, ComX, and the oligopeptide permease Spo0K are all required for *srfA* expression, and are thought to play a role in the ComP-dependent phosphorylation of ComA (Nakano and Zuber, 1989, 1991; van Sinderen et al., 1990; Nakano et al., 1991b; Weinrauch et al., 1991; Grossman et al., 1991; Hahn and Dubnau, 1991; Magnuson et al., 1994). Indeed, ComQ, Spo0K, ComP, and ComA are no longer required for competence development when *srfA* is placed under the control of the IPTG-inducible *spac* promoter (Nakano and Zuber, 1991; Hahn and Dubnau, 1991). Overexpression of ComA also bypasses the ComP, ComQ, and Spo0K requirements, and competence development no longer requires glucose

and becomes insensitive to repression by glutamine (Weinrauch et al., 1990; Weinrauch et al., 1991; Dubnau et al., 1994). Thus, the ComP/ComA signal transduction system is thought to respond to nutritional signals as well as to signals relayed via the ComQ-ComX-Spo0K pathway.

The *spo0K* operon, also known as *opp,* encodes the subunits of an oligopeptide permease of the ABC transporter family, required for both the initiation of sporulation and the development of competence (Perego et al., 1991; Rudner et al., 1991; Higgins, 1992; Fath and Kolter, 1993). ComX was originally defined as a competence extracellular signaling factor accumulated in the culture medium as cells grow to high density (Magnuson et al., 1994). The addition of conditioned medium containing the extracellular factor induced expression of *srfA-lacZ* in cells at low density (Magnuson et al., 1994). The extracellular factor was purified and shown to be a peptide of 9 to 10 amino acids, corresponding to the carboxy-terminal portion of a 55-amino-acid polypeptide, ComX, encoded by an open reading frame partially overlapping the end of the *comQ* gene (Magnuson et al., 1994). ComQ is required for the production of the mature pheromone derived from ComX, and this is suggested to be the main role of ComQ in competence development (Magnuson et al., 1994). ComP, ComA, and the Spo0K oligopeptide permease are all required for induction of *srfA-lacZ* in response to the extracellular peptide pheromone. It was therefore suggested that Spo0K might be responsible for sensing, binding or transporting the pheromone into the cell, and signaling this event to the membrane-bound ComP kinase, leading to phosphorylation of ComA and expression of the *srfA* operon (Dubnau et al., 1994; Magnuson et al., 1994) (see Fig. 3).

SIGNAL TRANSDUCTION NETWORK

MecA and MecB: A Pair of Negative Regulatory Proteins

The link between expression of *srfA* and that of *comK* has remained a gray area in the competence signal transduction pathway. In addition to *srfA,* expression of *comK* also requires DegU, AbrB, and Sin, whereas expression of *srfA* does not, suggesting that these regulators act downstream from *srfA,* as discussed below (Hahn and Dubnau, 1991; Hahn et al., 1994; van Sinderen and Venema, 1994; Dubnau et al., 1994). Further complicating the story is the requirement of ComK for its own expression (van Sinderen and Venema, 1994).

The *srfA* operon encodes the three subunits of surfactin synthetase, with domains responsible for binding and activating each of the seven amino acids of the surfactin peptide (Nakano et al., 1991a; Cosmina et al., 1993; Fuma et al., 1993). The region coding for the valine-activating domain has been shown to be the only part of the *srfA* operon that is required for competence development (van Sinderen et al., 1993). However, a point mutation within the amino-acylation site of this domain, which effectively abolishes surfactin biosynthesis, had no effect on competence development or on *comG'-'lacZ* expression, suggesting this domain may have some activity other than surfactin synthesis that is required for competence development (D'Souza et al., 1993). It has since been shown that it is not surfactin synthetase per se that is required for competence development but instead the product of *comS,* a small open reading frame embedded within the coding sequence of *srfAB,* the second gene of the *srfA* operon, and that encodes a 46-amino-acid polypeptide (D'Souza et al., 1994; Hamoen et al., 1995). Together with ComX (see above), this is the second example of a small polypeptide involved in competence development and encoded by an open reading frame overlapping that of another gene (Magnuson et al., 1994; D'Souza et al., 1994). Some similarities have been reported between ComS and the homeodomain region of the POU domain class of eukaryotic transcriptional regulators (D'Souza et al., 1994). Some insight into how ComS controls *comK* expression and thus that of late competence genes may be provided by the recently characterized *mec* genes (Kong et al., 1993; Msadek et al., 1994).

Dubnau and Roggiani (1990) isolated mutants capable of expressing late competence genes in complex medium, which were called *mec* (for *m*edium-independent *e*xpression of *c*ompetence). The *mec* mutations were mapped on the *B. subtilis* chromosome at two distinct loci: *mecA* was located between *spo0K* and *comK* (Dubnau, 1993), and *mecB* was linked by transformation to the *cysA* and *rpoB* genes (Dubnau and Roggiani, 1990). The *mec* mutations allow a complete bypass of null mutations in *comP, comA, srfA, sin,* and *degU* for late competence gene expression (Roggiani et al., 1990). These results suggested that the corresponding regulatory proteins could exert their effects on the expression of late competence genes before the action of the *mecA* and *mecB* gene products (Dubnau, 1993).

The *mecA* and *mecB* genes were cloned and their nucleotide sequences determined (Kong et al., 1993; Msadek et al., 1994). The predicted MecB protein shows very high similarity over its entire length with members of the ClpC family of ATPases (60% identity) (Msadek et al., 1994). The *mecB* gene is part of a putative operon containing five other genes whose function is unknown (Ogasawara et al., 1994; Msadek et al., 1994). Computer analysis suggests that the products of two of these genes could contain potential zinc-finger DNA binding domains and that one of the deduced proteins has a domain that is highly conserved among ATP:guanidino phosphotransferases such as creatine kinases, whereas another shares 46% amino acid sequence identity with the *E. coli sms* gene product, whose absence confers sensitivity to methyl methane sulfonate (Neuwald et al., 1992; Msadek, unpublished data).

Clp ATPases are highly conserved among both eukaryotes and prokaryotes and contain two nucleotide binding sites (Gottesman et al., 1990; Squires and Squires, 1992). The *E. coli* ClpA ATPase acts as the regulatory subunit for the ClpP protease (Gottesman and Maurizi, 1992). MecA shares some amino acid sequence similarities with ClpP but lacks the conserved catalytic serine and histidine residues, suggest-

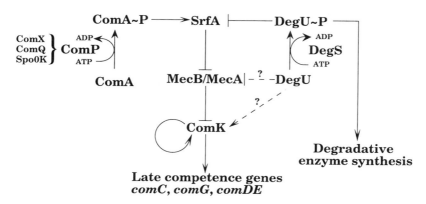

FIGURE 4 Provisional model of signal transduction network controlling competence gene expression and degradative enzyme synthesis. The phosphorylated form of ComA controls *srfA* expression, and the phosphorylated form of DegU is required for degradative enzyme synthesis. Negative regulation of *comK* expression by MecB/MecA may be relieved by an *srfA*-generated signal, sensed by MecB, thus allowing expression of late competence genes. DegU is required for expression of *comK*, but whether it acts directly or by relieving negative regulation by MecB/MecA remains to be determined (dotted lines). Arrows and perpendicular bars indicate positive and negative regulation, respectively.

ing that it may not act as a protease (Kong and Dubnau, 1994). However, these sequence similarities may reflect residues required for the interaction of the two proteins, suggesting that MecB could act as the regulatory subunit of MecA, as demonstrated for ClpA/ClpP (Kong and Dubnau, 1994; Msadek et al., 1994).

Both MecA and MecB appear to act as negative regulators, because deletions of either of these genes bypass the requirement of the early regulatory genes *comP*, *comA*, and *degU* for the expression of late competence genes (Kong et al., 1993; Msadek et al., 1994). However, they do not bypass the need for ComK (Msadek et al., 1994; van Sinderen and Venema, 1994; Hahn et al., 1994). These results seem to indicate that MecA and MecB may act downstream from the ComP, ComA, and DegU regulatory proteins but before ComK (Fig. 4). Both MecA and MecB have indeed been shown to be negative regulators of *comK* expression (Msadek et al., 1994; Hahn et al., 1994; van Sinderen and Venema, 1994).

Because both MecA and MecB act as negative regulators of *comK* expression, MecB could

act through MecA, either by activating it in some way or by binding to it to form an active complex. Support for this hypothesis has been provided by Kong and Dubnau (1994), who have shown that MecB is apparently required to activate MecA. It has also been shown that MecB and MecA can associate to form a complex (Dubnau and Kong, personal communication). MecB and MecA may thus interact to antagonize, sequester, or otherwise inactivate ComK under inappropriate environmental conditions, as indicated in Fig. 4, thus preventing ComK from positively autoregulating its own synthesis. Indeed, protein-protein interaction has been shown between MecA and ComK (Kong and Dubnau, 1994), and ComK is still required for its own synthesis in a *mecA* or *mecB* mutant (Msadek et al., 1994; van Sinderen and Venema, 1994; Hahn et al., 1994). Under conditions leading to expression of competence genes, MecB may respond to some intracellular signal, presumably generated through the ComP/ComA/ComS pathway, preventing MecA activation. Amounts of free active ComK thus produced would rapidly accumulate through a positive feedback loop, allowing expression of late competence genes.

MecB and MecA are thus part of a signal transduction pathway linking *comS* expression to that of *comK,* although the nature of the signal releasing ComK synthesis from negative regulation through MecB/MecA remains to be determined.

Deg/Com/Mec Network

ComP/ComA and DegS/DegU form part of a pleiotropic signal transduction regulatory network controlling competence and degradative enzyme synthesis, with links to several other post-exponential-phase phenomena such as sporulation, motility, and antibiotic synthesis.

ComA, like DegU, is a pleiotropic regulator, because it also controls the expression of genes that are not required for competence development, such as *degQ,* a regulator of degradative enzyme synthesis, or *gsiA,* which is involved in the regulation of early sporulation gene expression (Msadek et al., 1991; Mueller and Sonenshein, 1992; Mueller et al., 1992). Thus, the ComP/ComA regulatory pair can also control degradative enzyme production through *degQ* gene expression and has been shown to be required for expression of the *degQ36*(Hy) phenotype (Msadek et al., 1991).

Expression of *degQ* was shown to be subject to growth phase regulation (Yang et al., 1986). This expression increased under conditions of carbon or phosphate source limitations or during amino acid deprivation (Msadek et al., 1991). Expression was reduced under all the conditions tested in a strain deleted for the *degS* and *degU* genes or carrying the DegU H12L (Hy) substitution (Henner et al., 1987; Msadek et al., 1990; Msadek et al., 1991). Separate regulatory targets for DegS/DegU and ComP/ComA were localized by deletion analysis upstream from the *degQ* gene, between positions −393 and −186 and between positions −78 and −40, respectively (Msadek et al., 1991). Other examples of gene regulation by dual two-component systems include the Vsr system of *Pseudomonas solanacearum* and the Nar system of *E. coli* (Schell et al., 1993; Rabin and Stewart, 1993).

Both the ComP/ComA and DegS/DegU two-component systems control the expression of late competence genes; however, they seem to act through two different branches in the competence regulatory pathway (Fig. 4) that intersect to allow expression of *comK* (van Sinderen et al., 1990; Hahn et al., 1994; van Sinderen et al., 1994; van Sinderen and Venema, 1994). Indeed, the ComP/ComA branch involves at least one intermediate regulatory locus, *srfA/comS,* whereas the DegS/DegU regulatory pair is not required for the expression of *srfA* (Nakano and Zuber, 1990; Hahn and Dubnau, 1991; van Sinderen, personal communication). The *mec* mutations allow a bypass of both ComP/ComA and DegU for competence gene expression (Roggiani et al., 1990).

We have shown that the *mec* mutations also lead to overexpression of *sacB,* encoding levansucrase, bypassing the DegS/DegU requirement, and thus providing a further link between the ComP/ComA and DegS/DegU systems. Mutations selected as allowing expression of *sacB* in the absence of DegS/DegU were isolated and found to map at the *mecB* locus. The increased expression of *sacB* in *mec* mutants was shown to be entirely dependent on ComK (Msadek et al., 1994; Kunst et al., 1994). ComK has been shown to bind to the *sacB* promoter region, albeit with an apparently lower affinity than to promoters of late competence genes (Hamoen and van Sinderen, personal communication).

The Spo0A phosphorelay is also implicated in this regulatory network. Indeed, *gsiA* expression is negatively controlled by Spo0A and positively regulated by ComP/ComA (Mueller and Sonenshein, 1992; Mueller et al., 1992). The *gsiA* gene is now known to be identical to *spo0L,* encoding a response regulator aspartate phosphatase that specifically dephosphorylates Spo0F, a component of the Spo0A phosphorelay, thus lowering the levels of phosphorylated Spo0A within the cell (Perego et al., 1994) (see Chapter 8).

Both the DegS/DegU two-component system and the Spo0A phosphorelay control the expression of the *aprE* and *nprE* genes encoding the major *B. subtilis* proteases. A *spo0A* null mutation is epistatic to the *degU32*(Hy) mutation (H12L) for expression of these genes, because it leads to repression of protease gene expression through AbrB. However, in a strain carrying both these mutations, *sacB* was still expressed at a high level (Msadek and Kunst, unpublished data).

DegS/DegU, ComP/ComA, the Spo0A phosphorelay, and AbrB are all required for competence gene expression (Dubnau, 1993). Expression of *comK* requires DegU, ComS, and a low level of AbrB; however, an excess of AbrB negatively regulates *comK* transcription (Hahn et al., 1994). Thus, the Spo0A phosphorelay and ComP/ComA are both involved in establishing the choice between competence development and sporulation, by respectively controlling the levels of AbrB and ComS in the cell (Hahn et al., 1994; Dubnau et al., 1994; Strauch and Hoch, 1992, 1993). The ComP/ComA system can also control AbrB levels indirectly through the Spo0L phosphatase (now called RapA), which lowers the levels of phosphorylated Spo0F and Spo0A in the cell (Perego et al., 1994). Spo0A, ComP/ComA, and DegS/DegU also control expression of the polyketide synthase gene *pksX* (Albertini et al., 1993; Scotti et al., 1993).

This network is also involved in controlling DNA repair and recombination. Competence-specific induction of the *recA* gene requires Spo0A, ComA, and DegU (Lovett et al., 1989; Yasbin et al., 1992; Cheo et al., 1992, 1993). Sequences required for the competence-specific induction of the *recA* gene were defined by deletion analysis, and this region (centered at position −126) contains a putative "ComK box," suggesting that *recA* may be regulated by ComK in the same manner as late competence genes (Cheo et al., 1993; Dubnau et al., 1994). ComK has been shown to bind to the *recA* and *dinR* promoter regions and also controls expression of the *addAB*-encoded ATP-dependent deoxyribonuclease (Haijema et al., 1995, in press).

Both ComP and DegS have been proposed to also affect sporulation to a minor extent,

especially in the absence of the SpoIIJ (KinA) kinase, possibly through crosstalk with the Spo0A phosphorelay (Weinrauch et al., 1990; Smith et al., 1992). Expression of *srfA* is also required for efficient sporulation (Nakano et al., 1991a). Signals involved in the initiation of sporulation include the depletion of carbon, nitrogen, or phosphate sources (Schaeffer et al., 1965; Sonenshein, 1989). Information is channeled to the Spo0A phosphorelay by multiple kinases that may include ComP and DegS under certain environmental conditions that trigger sporulation. As suggested for the *E. coli* Che system (Lukat et al., 1992), intermediary metabolites such as acetyl phosphate or carbamoyl phosphate may also be involved in regulating inputs to the Com/Deg/Spo signal transduction network, providing further links between these systems and the nutritional status of the cell. In strains carrying *degS*(Hy) or *degU*(Hy) mutations, which lead to the accumulation of the phosphorylated DegU response regulator, sporulation is no longer repressed by glucose (Kunst et al., 1974; Henner et al., 1988c; Msadek et al., 1990; Tanaka et al., 1991; Dahl et al., 1992). These mutations also affect several σ^D-dependent functions in *B. subtilis,* such as flagellar synthesis and expression of the *degR* gene, which encodes a 60-amino-acid accessory regulatory peptide (Nagami and Tanaka, 1986; Tanaka et al., 1987; Yang et al., 1987; Helmann et al., 1988; Márquez et al., 1990; Helmann, 1991). Indeed, although deletion of the *degS* and *degU* genes has no effect on these functions, *degR* is no longer expressed in a strain carrying a DegU H12L (Hy) substitution, and the cells are devoid of flagella and grow as long filaments (Kunst et al., 1974; Henner et al., 1987; Msadek and Kunst, unpublished data; Mukai et al., 1992). It has been shown that the *degS*(Hy) and *degU*(Hy) mutations, in fact, abolish expression of *sigD,* encoding an alternative σ-factor, thus strongly reducing expression of the σ^D-regulon (Tokunaga et al., 1994). Why Hy mutations in *degS* and *degU* lead to glucose-resistant sporulation is not clear. Mutations in several other regulators (*gsiA, pai, hpr*) also lead to the same phenotype (Perego and Hoch, 1988; Honjo et al., 1990; Mueller et al., 1992; Mueller and Sonenshein, 1992).

MecB may provide links between this regulatory network and several stress response mechanisms. Indeed, a mutation in the closely related *E. coli clpB* ATPase gene leads to reduced cell survival at 50°C (Squires et al., 1991). Investigation of *mecB* mutant phenotypes revealed that MecB is essential for growth of *B. subtilis* at high temperature (54°C) (Msadek et al., 1994). Expression of *mecB* has also been shown to be strongly induced by several stress conditions including heat shock as well as salt and ethanol stress (Krüger et al., 1994). The heat shock role of MecB might involve interaction with the ClpP proteolytic subunit, recently identified in *B. subtilis* (Völker et al., 1992). The thermosensitive phenotype of the Δ*mecB* mutant was not reversed in a Δ*mecB* Δ*comK* double mutant, indicating that this phenotype is not due to *comK* overexpression, unlike the *mec* effect on competence genes (Msadek et al., 1994). Furthermore, the Δ*mecA* mutant grew normally at 54°C, indicating that the heat shock role of MecB appears to be distinct from its role in controlling competence gene expression (Msadek et al., 1994). Heat shock induction of MecB synthesis may therefore first act to shut down competence gene expression, after which MecB could possibly switch partners, releasing MecA to associate with ClpP. Indeed, *B. subtilis* 168 *trpC2* cells did not develop competence when grown at 50°C.

CONCLUDING REMARKS

In *B. subtilis,* the two-component systems controlling post-exponential-phase responses appear to interconnect to a large extent to form a sensory transduction network. This network appears to be involved in a hierarchy of environmental signal responses, involving a choice between competence gene expression, DNA repair and recombination, degradative enzyme production, and finally, sporulation. At the onset of the stationary phase, when growth is limited by the depletion of nitrogen sources such as amino acids but glucose is still present as a carbon source, the ComP/ComA system would allow competence gene expression (Dubnau, 1993), and the DegS/DegU regulatory pair could be involved in the choice between producing degradative enzymes or expressing competence genes. Finally, when glucose is no longer present and no alternative nutrient sources are available as substrates for degradative enzymes, the Spo0A phosphorelay would trigger sporulation initiation. Degradative enzyme synthesis, competence development, and sporulation thus appear to constitute three mutually exclusive post-exponential growth phase responses, with the Deg, Com, and Spo two-component relays acting as consecutive switches, directing the flow of environmental signals toward one response or another. An exciting area of future research will be to identify the types of signals involved in regulation by each of these two-component systems and by the other regulators such as MecB/MecA and the ComQ-ComX-Spo0K pathway, as well as determining how these regulators interact within the signal transduction network. Biochemical experiments will be required to complement the available genetic data and help provide an answer to several remaining questions. Binding of DegU to its target genes, such as *sacB* or *aprE*, remains to be demonstrated. Additional experimentation will be needed to understand how DegU controls *comK* expression. In particular, does DegU bind to a DNA region upstream from *comK*, or is the main effect of DegU to abolish negative regulation by MecB/ MecA? Finally, what is the precise role of ComS in regulation of *comK*? Does ComS interact with MecA/MecB, and if so, what is the nature of this interaction?

ACKNOWLEDGMENTS

We thank A. Albertini, D. Dubnau, A. Galizzi, A. D. Grossman, J. Hahn, L. Hamoen, S. Hastrup, J. A. Hoch, F. M. Hulett, E. H. Kemp, M. Louw, R. Magnuson, A. Moir, J. P. Mueller, M. Nakano, C. Scotti, I. Smith, D. van Sinderen, J. Solomon, Y. Weinrauch, N. Ogasawara, and P. Zuber for helpful discussion and for providing us with unpublished information.

The work from our laboratory reported in this review was supported by funds from Institut Pasteur, Centre National de la Recherche Scientifique, Institut National de la Recherche Agronomique, and Université Paris 7.

REFERENCES

Albano, M., R. Breitling, and D. A. Dubnau. 1989. Nucleotide sequence and genetic organization of the *Bacillus subtilis comG* operon. *J. Bacteriol.* **171:** 5386–5404.

Albano, M., J. Hahn, and D. Dubnau. 1987. Expression of competence genes in *Bacillus subtilis. J. Bacteriol.* **169:**3110–3117.

Albertini, A., C. Scotti, and A. Galizzi (University of Pavia, Pavia, Italy). 1993. Personal communication.

Albright, L. M., E. Huala, and F. M. Ausubel. 1989. Prokaryotic signal transduction mediated by sensor and regulator pairs. *Annu. Rev. Genet.* **23:** 311–336.

Alex, L. A., and M. I. Simon. 1994. Protein histidine kinases and signal transduction in prokaryotes and eukaryotes. *Trends Genet.* **10:**133–138.

Allen, P., C. A. Hart, and J. R. Saunders. 1987. Isolation from *Klebsiella* and characterization of two *rcs* genes that activate colanic acid capsular biosynthesis in *Escherichia coli. J. Gen. Microbiol.* **133:**331–340.

Amory, A., F. Kunst, E. Aubert, A. Klier, and G. Rapoport. 1987. Characterization of the *sacQ* genes from *Bacillus licheniformis* and *Bacillus subtilis. J. Bacteriol.* **169:**324–333.

Antoniewski, C., B. Savelli, and P. Stragier. 1990. The *spoIIJ* gene, which regulates early development steps in *Bacillus subtilis,* belongs to a class of environmentally responsive genes. *J. Bacteriol.* **172:**86–93.

Aymerich, S., G. Gonzy-Tréboul, and M. Steinmetz. 1986. 5′-noncoding region *sacR* is the target of all identified regulation affecting the levansucrase gene in *Bacillus subtilis. J. Bacteriol.* **166:**993–998.

Ayusawa, D., Y. Yoneda, K. Yamane, and B. Maruo. 1975. Pleiotropic phenomena in autolytic enzyme(s) content, flagellation and simultaneous hyperproduction of extracellular α-amylase and protease in a *Bacillus subtilis* mutant. *J. Bacteriol.* **124:**459–469.

Baldus, J. M., B. D. Green, P. Youngman, and C. P. Moran, Jr. 1994. Phosphorylation of *Bacillus subtilis* transcription factor SpoOA stimulates transcription from the *spoIIG* promoter by enhancing binding to weak 0A boxes. *J. Bacteriol.* **176:**296–306.

Bernhard, F., K. Poetter, K. Geider, and D. L. Coplin. 1990. The *rcsA* gene from *Erwinia amylovora:* identification, nucleotide sequence, and

regulation of exopolysaccharide biosynthesis. *Mol. Plant-Microb. Int.* **3:**429–437.

Bischoff, D. S., and G. W. Ordal. 1991. Sequence and characterization of *Bacillus subtilis* CheB, a homolog of *Escherichia coli* CheY, and its role in a different mechanism of chemotaxis. *J. Biol. Chem.* **266:**12301–12305.

Borkovich, K. A., N. Kaplan, J. F. Hess, and M. I. Simon. 1989. Transmembrane signal transduction in bacterial chemotaxis involves ligand-dependent activation of phosphate group transfer. *Proc. Natl. Acad. Sci. USA* **86:**1208–1212.

Borkovich, K. A., and M. I. Simon. 1990. The dynamics of protein phosphorylation in bacterial chemotaxis. *Cell* **63:**1339–1348.

Bourret, R. B., K. A. Borkovich, and M. I. Simon. 1991. Signal transduction pathways involving protein phosphorylation in prokaryotes. *Annu. Rev. Biochem.* **60:**401–441.

Bourret, R. B., J. F. Hess, K. A. Borkovich, A. A. Pakula, and M. I. Simon. 1989. Protein phosphorylation in chemotaxis and two-component regulatory systems of bacteria. *J. Biol. Chem.* **264:**7085–7088.

Bourret, R. B., J. F. Hess, and M. I. Simon. 1990. Conserved aspartate residues and phosphorylation in signal transduction by the chemotaxis protein CheY. *Proc. Natl. Acad. Sci. USA* **87:**41–45.

Breitling, R., and D. Dubnau. 1990. A membrane protein with similarity to *N*-methylphenylalanine pilins is essential for DNA binding by competent *Bacillus subtilis. J. Bacteriol.* **172:**1499–1508.

Brissette, R. E., K. Tsung, and M. Inouye. 1991. Intramolecular second-site revertants to the phosphorylation site mutation in OmpR, a kinase-dependent transcriptional activator in *Escherichia coli. J. Bacteriol.* **173:**3749–3755.

Brown, J. L., S. North, and H. Bussey. 1993. *SKN7,* a yeast multicopy suppressor of a mutation affecting cell wall β-glucan assembly, encodes a product with domains homologous to prokaryotic two-component regulators and to heat shock transcription factors. *J. Bacteriol.* **175:**6908–6915.

Chang, C., S. F. Kwok, A. B. Bleecker, and E. M. Meyerowitz. 1993. *Arabidopsis* ethylene-response gene *ETR1:* similarity of product to two-component regulators. *Science* **262:**539–544.

Chang, C., and E. M. Meyerowitz. 1994. Eukaryotes have "two-component" signal transducers. *Res. Microbiol.* **145:**481–486.

Cheo, D. L., K. W. Bayles, and R. E. Yasbin. 1992. Molecular characterization of regulatory elements controlling expression of the *Bacillus subtilis recA*[+] gene. *Biochimie* **74:**755–762.

Cheo, D. L., K. W. Bayles, and R. E. Yasbin. 1993. Elucidation of regulatory elements that control damage induction and competence induction of the *Bacillus subtilis* SOS system. *J. Bacteriol.* **175:** 5907–5915.

Choi, S. H., and E. P. Greenberg. 1991. The C-terminal region of the *Vibrio fischeri* LuxR protein contains an inducer-independent *lux* gene activating domain. *Proc. Natl. Acad. Sci. USA* **88:**11115–11119.

Choi, S. H., and E. P. Greenberg. 1992. Genetic dissection of DNA binding and luminescence gene activation by the *Vibrio fischeri* LuxR protein. *J. Bacteriol.* **174:**4064–4069.

Cole, S. T., and O. Raibaud. 1986. The nucleotide sequence of the *malT* gene encoding the positive regulator of the *Escherichia coli* maltose regulon. *Gene* **42:**201–208.

Coleman, M., R. Pearce, E. Hitchin, F. Busfield, J. W. Mansfield, and I. S. Roberts. 1990. Molecular cloning, expression and nucleotide sequence of the *rcsA* gene of *Erwinia amylovora*, encoding a positive regulator of capsule expression: evidence for a family of related capsule activator proteins. *J. Gen. Microbiol.* **136:**1799–1806.

Coote, J. G. 1991. Antigenic switching and pathogenicity: environmental effects on virulence gene expression in *Bordetella pertussis*. *J. Gen. Microbiol.* **137:** 2493–2503.

Cosmina, P., F. Rodriguez, F. de Ferra, G. Grandi, M. Perego, G. Venema, and D. van Sinderen. 1993. Sequence and analysis of the genetic locus responsible for surfactin synthesis in *Bacillus subtilis*. *Mol. Microbiol.* **8:**821–831.

Crutz, A. M., and M. Steinmetz. 1992. Transcription of the *Bacillus subtilis* sacX and sacY genes, encoding regulators of sucrose metabolism, is both inducible by sucrose and controlled by the DegS-DegU signalling system. *J. Bacteriol.* **174:**6087–6095.

Cutting, S., and J. Mandelstam. 1986. The nucleotide sequence and the transcription during sporulation of the *gerE* gene of *Bacillus subtilis*. *J. Gen. Microbiol.* **132:**3013–3024.

Dahl, M. K., T. Msadek, F. Kunst, and G. Rapoport. 1991. Mutational analysis of the *Bacillus subtilis* DegU regulator and its phosphorylation by the DegS protein kinase. *J. Bacteriol.* **173:**2539–2547.

Dahl, M. K., T. Msadek, F. Kunst, and G. Rapoport. 1992. The phosphorylation state of the DegU response regulator acts as a molecular switch allowing either degradative enzyme synthesis or expression of genetic competence in *Bacillus subtilis*. *J. Biol. Chem.* **267:**14509–14514.

Da Re, S., S. Bertagnoli, J. Fourment, J.-M. Reyrat, and D. Kahn. 1994. Intramolecular signal transduction within the FixJ transcriptional activator: in vitro evidence for the inhibitory effect of the

phosphorylatable regulatory domain. *Nucleic Acids Res.* **22:**1555–1561.

Delgado, J., S. Forst, S. Harlocker, and M. Inouye. 1993. Identification of a phosphorylation site and functional analysis of conserved aspartic acid residues of OmpR, a transcriptional activator for *ompF* and *ompC* in *Escherichia coli*. *Mol. Microbiol.* **10:**1037–1047.

Devine, J. H., C. Countryman, and T. O. Baldwin. 1988. Nucleotide sequence of the *luxR* and *luxI* genes and structure of the primary regulatory region of the *lux* regulon of *Vibrio fischeri* ATCC 7744. *Biochemistry* **27:**837–842.

Dixon, R., T. Eydmann, N. Henderson, and S. Austin. 1991. Substitutions at a single amino acid residue in the nitrogen-regulated activator protein NTRC differentially influence its activity in response to phosphorylation. *Mol. Microbiol.* **5:**1657–1667.

Dombroski, A. J., W. A. Walter, and C. A. Gross. 1993. Amino-terminal amino acids modulate σ-factor DNA-binding activity. *Genes Dev.* **7:**2446–2455.

D'Souza, C., M. M. Nakano, N. Corbell, and P. Zuber. 1993. Amino-acylation site mutations in amino acid-activating domains of surfactin synthetase: effects on surfactin production and competence development in *Bacillus subtilis*. *J. Bacteriol.* **175:**3502–3510.

D'Souza, C., M. M. Nakano, and P. Zuber. 1994. Identification of *comS*, a gene of the *srfA* operon that regulates the establishment of genetic competence in *Bacillus subtilis*. *Proc. Natl. Acad. Sci. USA* **91:**9397–9401.

Dubnau, D. 1991. Genetic competence in *Bacillus subtilis*. *Microbiol. Rev.* **55:**395–424.

Dubnau, D. 1993. Genetic exchange and homologous recombination, p. 555–584. *In* A. L. Sonenshein, J. A. Hoch, and R. Losick (ed.), *Bacillus subtilis and Other Gram-Positive Bacteria: Biochemistry, Physiology, and Molecular Genetics.* American Society for Microbiology, Washington, D.C.

Dubnau, D., J. Hahn, M. Roggiani, F. Piazza, and Y. Weinrauch. 1994. Two-component regulators and genetic competence in *Bacillus subtilis*. *Res. Microbiol.* **145:**403–411.

Dubnau, D., and L. Kong (Public Health Research Institute, New York). 1994. Personal communication.

Dubnau, D., and M. Roggiani. 1990. Growth medium-independent genetic competence mutants of *Bacillus subtilis*. *J. Bacteriol.* **172:**4048–4055.

Engebrecht, J., and M. Silverman. 1987. Nucleotide sequence of the regulatory locus controlling expression of bacterial genes for bioluminescence. *Nucleic Acids Res.* **15:**10455–10467.

Fath, M. J., and R. Kolter. 1993. ABC transporters: bacterial exporters. *Microbiol. Rev.* **57**:995–1017.

Ferrari, F. A., K. Trach, D. LeCoq, J. Spence, E. Ferrari, and J. A. Hoch. 1985. Characterization of the *spo0A* locus and its deduced product. *Proc. Natl. Acad. Sci. USA* **82**:2647–2651.

Friedrich, M. J., and R. J. Kadner. 1987. Nucleotide sequence of the *uhp* region of *Escherichia coli. J. Bacteriol.* **169**:3556–3563.

Fuhrer, D. K., and G. W. Ordal. 1991. *Bacillus subtilis* CheN, a homolog of CheA, the central regulator of chemotaxis in *Escherichia coli. J. Bacteriol.* **173**:7443–7448.

Fuma, S., Y. Fujishima, N. Corbell, C. D'Souza, M. M. Nakano, P. Zuber, and K. Yamane. 1993. Nucleotide sequence of 5′ portion of *srfA* that contains the region required for competence establishment in *Bacillus subtilis. Nucleic Acids Res.* **21**:93–97.

Fuqua, W. C., S. C. Winans, and E. P. Greenberg. 1994. Quorum sensing in bacteria: the LuxR-LuxI family of cell density-responsive transcriptional regulators. *J. Bacteriol.* **176**:269–275.

Gambello, M. J., and B. H. Iglewski. 1991. Cloning and characterization of the *Pseudomonas aeruginosa lasR* gene, a transcriptional activator of elastase expression. *J. Bacteriol.* **173**:3000–3009.

Gegner, J. A., and F. W. Dahlquist. 1991. Signal transduction in bacteria: CheW forms a reversible complex with the protein kinase CheA. *Proc. Natl. Acad. Sci. USA* **88**:750–754.

Gegner, J. A., D. R. Graham, A. F. Roth, and F. W. Dahlquist. 1992. Assembly of an MCP receptor, CheW, and kinase CheA complex in the bacterial chemotaxis signal transduction pathway. *Cell* **70**:975–982.

Gottesman, S., and M. R. Maurizi. 1992. Regulation by proteolysis: energy-dependent proteases and their targets. *Microbiol. Rev.* **56**:592–621.

Gottesman, S., C. Squires, E. Pichersky, M. Carrington, M. Hobbs, J. S. Mattick, B. Dalrymple, H. Kuramitsu, T. Shiroza, T. Foster, W. P. Clark, B. Ross, C. L. Squires, and M. R. Maurizi. 1990. Conservation of the regulatory subunit for the Clp ATP-dependent protease in prokaryotes and eukaryotes. *Proc. Natl. Acad. Sci. USA* **87**:3513–3517.

Green, B. D., M. G. Bramucci, and P. Youngman. 1991a. Mutant forms of Spo0A that affect sporulation initiation: a general model for phosphorylation-mediated activation of bacterial signal transduction proteins. *Semin. Dev. Biol.* **2**:21–29.

Green, B. D., G. Olmedo, and P. Youngman. 1991b. A genetic analysis of Spo0A structure and function. *Res. Microbiol.* **142**:825–830.

Gross, R., B. Aricó, and R. Rappuoli. 1989. Families of bacterial signal-transducing proteins. *Mol. Microbiol.* **3**:1661–1667.

Grossman, A. D., K. Ireton, E. F. Hoff, J. R. LeDeaux, D. Z. Rudner, R. Magnuson, and K. A. Hicks. 1991. Signal transduction and the initiation of sporulation in *Bacillus subtilis. Semin. Dev. Biol.* **2**:31–36.

Hahn, J., and D. Dubnau. 1991. Growth stage signal transduction and the requirements for *srfA* induction in development of competence. *J. Bacteriol.* **173**:7275–7282.

Hahn, J., G. Inamine, Y. Kozlov, and D. Dubnau. 1993. Characterization of *comE*, a late competence operon of *Bacillus subtilis* required for the binding and uptake of transforming DNA. *Mol. Microbiol.* **10**:99–111.

Hahn, J., L. Kong, and D. Dubnau. 1994. The regulation of competence transcription factor synthesis constitutes a critical control point in the regulation of competence in *Bacillus subtilis. J. Bacteriol.* **176**:5753–5761.

Haijema, B.-J., L. W. Hamoen, J. Kooistra, G. Venema, and D. van Sinderen. 1995. Expression of the ATP-dependent deoxyribonuclease of *Bacillus subtilis* is under competence-mediated control. *Mol. Microbiol.* **15**:203–211.

Haijema, B.-J., L. W. Hamoen, J. Kooistra, G. Venema, and D. van Sinderen. Regulated expression of the *Bacillus subtilis dinR* and *recA* genes during competence development and SOS induction. *Mol. Microbiol.,* in press.

Hamoen, L., and D. van Sinderen (University of Groningen, The Netherlands). 1993. Personal communication.

Hamoen, L. W., H. Eshuis, J. Jongbloed, G. Venema, and D. van Sinderen. 1995. A small gene, designated *comS*, located within the coding region of the fourth amino acid-activation domain of *srfA*, is required for competence development in *Bacillus subtilis. Mol. Microbiol.* **15**:55–63.

Hanks, S. K., A. M. Quinn, and T. Hunter. 1988. The protein kinase family: conserved features and deduced phylogeny of the catalytic domains. *Science* **241**:42–52.

Hastrup, S. (Novo Nordisk, Denmark). 1992. Personal communication.

Helmann, J. D. 1991. Alternative sigma factors and the regulation of flagellar gene expression. *Mol. Microbiol.* **5**:2875–2882.

Helmann, J. D., L. M. Márquez, V. L. Singer, and M. J. Chamberlin. 1988. Cloning and characterization of the *Bacillus subtilis* sigma-28 gene, p. 189–193. *In* A. T. Ganesan and J. A. Hoch (ed.), *Genetics and Biotechnology of Bacilli,* vol. 2. Academic Press, San Diego, Calif.

Henikoff, S., J. C. Wallace, and J. P. Brown. 1990. Finding protein similarities with nucleotide sequence databases. *Methods Enzymol.* **183**:111–132.

Henner, D. J., E. Ferrari, M. Perego, and J. A. Hoch. 1988a. Location of the targets of the *hpr-97, sacU32*(Hy), and *sacQ36*(Hy) mutations in upstream regions of the subtilisin promoter. *J. Bacteriol.* **170:**296–300.

Henner, D. J., E. Ferrari, M. Perego, and J. A. Hoch. 1988b. Upstream activating sequences in *Bacillus subtilis,* p. 3–9. *In* A. T. Ganesan and J. A. Hoch (ed.), *Genetics and Biotechnology of Bacilli,* vol. 2. Academic Press, San Diego, Calif.

Henner, D. J., M. Yang, L. Band, H. Shimotsu, M. Ruppen, and E. Ferrari. 1987. Genes of *Bacillus subtilis* that regulate the expression of degradative enzymes, p. 81–90. *In* M. Alacevic, D. Hranueli, and Z. Toman (ed.), *Genetics of Industrial Microorganisms.* Proceedings of the 5th International Symposium on the Genetics of Industrial Microorganisms. Pliva, Zagreb, Yugoslavia.

Henner, D. J., M. Yang, and E. Ferrari. 1988c. Localization of *Bacillus subtilis sacU*(Hy) mutations to two linked genes with similarities to the conserved procaryotic family of two-component signalling systems. *J. Bacteriol.* **170:**5102–5109.

Hess, J. F., R. B. Bourret, and M. I. Simon. 1988. Histidine phosphorylation and phosphoryl group transfer in bacterial chemotaxis. *Nature* (London) **336:**139–143.

Higgins, C. F. 1992. ABC transporters: from microorganisms to man. *Annu. Rev. Cell Biol.* **8:**67–113.

Holland, S. K., S. Cutting, and J. Mandelstam. 1987. The possible DNA-binding nature of the regulatory proteins, encoded by *spoIID* and *gerE,* involved in the sporulation of *Bacillus subtilis. J. Gen. Microbiol.* **133:**2381–2391.

Honjo, M., A. Nakayama, K. Fukazawa, K. Kawamura, K. Ando, M. Hori, and Y. Furutani. 1990. A novel *Bacillus subtilis* gene involved in negative control of sporulation and degradative-enzyme production. *J. Bacteriol.* **172:**1783–1790.

Jaacks, K. J., J. Healy, R. Losick, and A. D. Grossman. 1989. Identification and characterization of genes controlled by the sporulation-regulatory gene *spo0H* in *Bacillus subtilis. J. Bacteriol.* **171:**4121–4129.

Jarnagin, A. S., and E. Ferrari. 1992. Extracellular enzymes: gene regulation and structure function relationship studies, p. 189–217. *In* R. H. Doi and M. McGloughlin (ed.), *Biology of Bacilli: Applications to Industry. Biotechnology,* vol. 22. Butterworth-Heinemann, Stoneham, Mass.

Jin, S., Y.-N. Song, S. Q. Pan, and E. W. Nester. 1993. Characterization of a *virG* mutation that confers constitutive virulence gene expression in *Agrobacterium. Mol. Microbiol.* **7:**555–562.

Kahn, D., and G. Ditta. 1991. Modular structure of FixJ: homology of the transcriptional activator domain with the −35 binding domain of sigma factors. *Mol. Microbiol.* **5:**987–997.

Kanamaru, K., and T. Mizuno. 1992. Signal transduction and osmoregulation in *Escherichia coli:* a novel mutant of the positive regulator, OmpR, that functions in a phosphorylation-independent manner. *J. Biochem.* **111:**425–430.

Kirsch, M. L., P. D. Peters, D. W. Hanlon, J. R. Kirby, and G. W. Ordal. 1993. Chemotactic methylesterase brings about adaptation to attractants in *Bacillus subtilis. J. Biol. Chem.* **268:**18610–18616.

Klein, C., C. Kaletta, and K.-D. Entian. 1993. Biosynthesis of the antibiotic subtilin is regulated by a histidine kinase/response regulator system. *Appl. Environ. Microbiol.* **59:**296–303.

Klier, A., T. Msadek, and G. Rapoport. 1992. Positive regulation in the Gram-positive bacterium: *Bacillus subtilis. Annu. Rev. Microbiol.* **46:**429–459.

Klose, K. E., D. S. Weiss, and S. Kustu. 1993. Glutamate at the site of phosphorylation of nitrogen-regulatory protein NTRC mimics aspartyl-phosphate and activates the protein. *J. Mol. Biol.* **232:**67–78.

Kofoid, E. C., and J. S. Parkinson. 1988. Transmitter and receiver modules in bacterial signaling proteins. *Proc. Natl. Acad. Sci. USA* **85:**4981–4985.

Kong, L., and D. Dubnau. 1994. Regulation of competence-specific gene expression by Mec-mediated protein-protein interaction in *Bacillus subtilis. Proc. Natl. Acad. Sci. USA* **91:**5793–5797.

Kong, L., K. J. Siranosian, A. D. Grossman, and D. Dubnau. 1993. Sequence and properties of *mecA:* a negative regulator of genetic competence in *Bacillus subtilis. Mol. Microbiol.* **9:**365–373.

Krüger, E., U. Völker, and M. Hecker. 1994. Stress induction of *clpC* in *Bacillus subtilis* and involvement in stress tolerance. *J. Bacteriol.* **176:**3360–3367.

Kunst, F. Unpublished data.

Kunst, F., A. Amory, M. Débarbouillé, I. Martin, A. Klier, and G. Rapoport. 1988a. Polypeptides activating the synthesis of secreted enzymes, p. 27–31. *In* A. T. Ganesan and J. A. Hoch (ed.), *Genetics and Biotechnology of Bacilli,* vol. 2. Academic Press, San Diego, Calif.

Kunst, F., M. Débarbouillé, T. Msadek, M. Young, C. Mauël, D. Karamata, A. Klier, G. Rapoport, and R. Dedonder. 1988b. Deduced polypeptides encoded by the *Bacillus subtilis sacU* locus share homology with two-component sensor-regulator systems. *J. Bacteriol.* **170:**5093–5101.

Kunst, F., T. Msadek, and G. Rapoport. 1994. Signal transduction network controlling degradative enzyme synthesis and competence in *Bacillus subtilis,* p. 1–20. *In* P. J. Piggot, C. P. Moran, Jr., and P. Youngman (ed.), *Regulation of Bacterial Differentia-*

tion. American Society for Microbiology, Washington, D.C.

Kunst, F., M. Pascal, J. Lepesant-Kejzlarová, J.-A. Lepesant, A. Billault, and R. Dedonder. 1974. Pleiotropic mutations affecting sporulation conditions and the synthesis of extracellular enzymes in *Bacillus subtilis* 168. *Biochimie* **56:**1481–1489.

Lepesant, J.-A., F. Kunst, J. Lepesant-Kejzlarová, and R. Dedonder. 1972. Chromosomal location of mutations affecting sucrose metabolism in *Bacillus subtilis* Marburg. *Mol. Gen. Genet.* **118:**135–160.

Londoño-Vallejo, J. A., and D. Dubnau. 1993. *comF*, a *Bacillus subtilis* late competence locus, encodes a protein similar to ATP-dependent RNA/DNA helicases. *Mol. Microbiol.* **9:**119–131.

Lonetto, M., M. Gribskov, and C. A. Gross. 1992. The σ^{70} family: sequence conservation and evolutionary relationships. *J. Bacteriol.* **174:**3843–3849.

Louw, M. E., S. J. Reid, D. M. James, and T. G. Watson. 1994. Cloning and sequencing the *degS-degU* operon from an alkalophilic *Bacillus brevis.* *Appl. Microbiol. Biotechnol.* **42:**78–84.

Lovett, C. M., Jr., P. E. Love, and R. E. Yasbin. 1989. Competence-specific induction of the *Bacillus subtilis* RecA protein analog: evidence for dual regulation of a recombination protein. *J. Bacteriol.* **171:**2318–2322.

Lukat, G. S., W. R. McCleary, A. M. Stock, and J. B. Stock. 1992. Phosphorylation of bacterial response regulator proteins by low molecular weight phospho-donors. *Proc. Natl. Acad. Sci. USA* **89:**718–722.

Maeda, T., S. M. Wurgler-Murphy, and H. Saito. 1994. A two-component system that regulates an osmosensing MAP kinase cascade in yeast. *Nature* (London) **369:**242–245.

Magnuson, R., J. Solomon, and A. D. Grossman. 1994. Biochemical and genetic characterization of a competence pheromone from *Bacillus subtilis.* *Cell* **77:**207–216.

Márquez, L. M., J. D. Helmann, E. Ferrari, H. M. Parker, G. W. Ordal, and M. J. Chamberlin. 1990. Studies of σ^D-dependent functions in *Bacillus subtilis.* *J. Bacteriol.* **172:**3435–3443.

Masui, A., N. Fujiwara, M. Takagi, and T. Imanaka. 1992. Cloning and nucleotide sequence of the regulatory gene, *degM,* for minor serine protease in *Bacillus subtilis.* *J. Ferment. Bioeng.* **74:**230–233.

McCleary, W. R., J. B. Stock, and A. J. Ninfa. 1993. Is acetyl phosphate a global signal in *Escherichia coli*? *J. Bacteriol.* **175:**2793–2798.

Melton, A. R., and A. A. Weiss. 1989. Environmental regulation of expression of virulence determinants in *Bordetella pertussis.* *J. Bacteriol.* **171:** 6206–6212.

Melton, A. R., and A. A. Weiss. 1993. Characterization of environmental regulators of *Bordetella pertussis.* *Infect. Immun.* **61:**807–815.

Miller, J. F., J. J. Mekalanos, and S. Falkow. 1989. Coordinate regulation and sensory transduction in the control of bacterial virulence. *Science* **243:**916–922.

Mohan, S., J. Aghion, N. Guillen, and D. Dubnau. 1989. Molecular cloning and characterization of *comC,* a late competence gene of *Bacillus subtilis.* *J. Bacteriol.* **171:**6043–6051.

Mohan, S., and D. Dubnau. 1990. Transcriptional regulation of *comC*: evidence for a competence-specific transcription factor in *Bacillus subtilis.* *J. Bacteriol.* **172:**4064–4071.

Moir, A., and E. H. Kemp (University of Sheffield, United Kingdom). 1991. Personal communication.

Moore, J. B., S.-P. Shiau, and L. J. Reitzer. 1993. Alterations of highly conserved residues in the regulatory domain of nitrogen regulator I (NtrC) of *Escherichia coli.* *J. Bacteriol.* **175:**2692–2701.

Msadek, T. Unpublished data.

Msadek, T., and F. Kunst. Unpublished data.

Msadek, T., F. Kunst, D. Henner, A. Klier, G. Rapoport, and R. Dedonder. 1990. Signal transduction pathway controlling synthesis of a class of degradative enzymes in *Bacillus subtilis*: expression of the regulatory genes and analysis of mutations in *degS* and *degU.* *J. Bacteriol.* **172:**824–834.

Msadek, T., F. Kunst, A. Klier, and G. Rapoport. 1991. DegS-DegU and ComP-ComA modulator-effector pairs control expression of the *Bacillus subtilis* pleiotropic regulatory gene *degQ.* *J. Bacteriol.* **173:**2366–2377.

Msadek, T., F. Kunst, and G. Rapoport. 1993. Two-component regulatory systems, p. 729–745. *In* A. L. Sonenshein, J. A. Hoch, and R. Losick (ed.), *Bacillus subtilis and Other Gram-Positive Bacteria: Biochemistry, Physiology, and Molecular Genetics.* American Society for Microbiology, Washington, D.C.

Msadek, T., F. Kunst, and G. Rapoport. 1994. MecB of *Bacillus subtilis,* a member of the ClpC ATPase family, is a pleiotropic regulator controlling competence gene expression and growth at high temperature. *Proc. Natl. Acad. Sci. USA* **91:**5788–5792.

Mueller, J. P., G. Bukusoglu, and A. L. Sonenshein. 1992. Transcriptional regulation of *Bacillus subtilis* glucose starvation-inducible genes: control of *gsiA* by the ComP-ComA signal transduction system. *J. Bacteriol.* **174:**4361–4373.

Mueller, J. P., and A. L. Sonenshein. 1992. Role of the *Bacillus subtilis gsiA* gene in regulation of early sporulation gene expression. *J. Bacteriol.* **174:**4374–4383.

Mukai, K., M. Kawata, and T. Tanaka. 1990. Isolation and phosphorylation of the *Bacillus subtilis degS* and *degU* gene products. *J. Biol. Chem.* **265:**20000–20006.

Mukai, K., M. Kawata-Mukai, and T. Tanaka. 1992. Stabilization of phosphorylated *Bacillus subtilis* DegU by DegR. *J. Bacteriol.* **174**:7954–7962.

Nagami, Y., and T. Tanaka. 1986. Molecular cloning and nucleotide sequence of a DNA fragment from *Bacillus natto* that enhances production of extracellular proteases and levansucrase in *Bacillus subtilis. J. Bacteriol.* **166**:20–28.

Nakano, M. M., R. Magnuson, A. Myers, J. Curry, A. D. Grossman, and P. Zuber. 1991a. *srfA* is an operon required for surfactin production, competence development, and efficient sporulation in *Bacillus subtilis. J. Bacteriol.* **173**:1770–1778.

Nakano, M. M., M. A. Marahiel, and P. Zuber. 1988. Identification of a genetic locus required for biosynthesis of the lipopeptide antibiotic surfactin in *Bacillus subtilis. J. Bacteriol.* **170**:5662–5668.

Nakano, M. M., L. Xia, and P. Zuber. 1991b. Transcription initiation region of the *srfA* operon, which is controlled by the *comP-comA* signal transduction system in *Bacillus subtilis. J. Bacteriol.* **173**:5487–5493.

Nakano, M. M., and P. Zuber. 1989. Cloning and characterization of *srfB*, a regulatory gene involved in surfactin production and competence in *Bacillus subtilis. J. Bacteriol.* **171**:5347–5353.

Nakano, M. M., and P. Zuber. 1990. Identification of genes required for the biosynthesis of the lipopeptide antibiotic surfactin in *Bacillus subtilis*, p. 397–405. *In* M. M. Zukowski, A. T. Ganesan, and J. A. Hoch (ed.), *Genetics and Biotechnology of Bacilli*, vol. 3. Academic Press, San Diego, Calif.

Nakano, M. M., and P. Zuber. 1991. The primary role of ComA in establishment of the competent state in *Bacillus subtilis* is to activate expression of *srfA. J. Bacteriol.* **173**:7269–7274.

Nakano, M. M., and P. Zuber. 1993. Mutational analysis of the regulatory region of the *srfA* operon in *Bacillus subtilis. J. Bacteriol.* **175**:3188–3191.

Nassif, X., N. Honoré, T. Vasselon, S. T. Cole, and P. J. Sansonetti. 1989. Positive control of colanic acid synthesis in *Escherichia coli* by *rmpA* and *rmpB*, two virulence-plasmid genes of *Klebsiella pneumoniae. Mol. Microbiol.* **3**:1349–1359.

Neuwald, A. F., D. E. Berg, and G. V. Stauffer. 1992. Mutational analysis of the *Escherichia coli serB* promoter region reveals transcriptional linkage to a downstream gene. *Gene* **120**:1–9.

Nixon, B. T., C. W. Ronson, and F. M. Ausubel. 1986. Two-component regulatory systems responsive to environmental stimuli share strongly conserved domains with the nitrogen assimilation regulatory genes *ntrB* and *ntrC. Proc. Natl. Acad. Sci. USA* **83**:7850–7854.

Ogasawara, N., S. Nakai, and H. Yoshikawa. 1994. Systematic sequencing of the 180 kilo bases region of the *Bacillus subtilis* chromosome containing the replication origin. *DNA Res.* **1**:1–14.

Ogura, M., M. Kawata-Mukai, M. Itaya, K. Takio, and T. Tanaka. 1994. Multiple copies of the *proB* gene enhance *degS*-dependent extracellular protease production in *Bacillus subtilis. J. Bacteriol.* **176**:5673–5680.

Ota, I. M., and A. Varshavsky. 1993. A yeast protein similar to bacterial two-component regulators. *Science* **262**:566–569.

Parkinson, J. S. 1993. Signal transduction schemes of bacteria. *Cell* **73**:857–871.

Parkinson, J. S., and E. C. Kofoid. 1992. Communication modules in bacterial signaling proteins. *Annu. Rev. Genet.* **26**:71–112.

Pazour, G. J., C. N. Ta, and A. Das. 1992. Constitutive mutations of *Agrobacterium tumefaciens* transcriptional activator *virG. J. Bacteriol.* **174**:4169–4174.

Perego, M., S. P. Cole, D. Burbulys, K. Trach, and J. A. Hoch. 1989. Characterization of the gene for a protein kinase which phosphorylates the sporulation-regulatory proteins Spo0A and Spo0F of *Bacillus subtilis. J. Bacteriol.* **171**:6187–6196.

Perego, M., C. Hanstein, K. M. Welsh, T. Djavakhishvili, P. Glaser, and J. A. Hoch. 1994. Multiple protein aspartate phosphatases provide a mechanism for the integration of diverse signals in the control of development in *B. subtilis. Cell* **79**:1047–1055.

Perego, M., C. F. Higgins, S. R. Pearce, M. P. Gallagher, and J. A. Hoch. 1991. The oligopeptide transport system of *Bacillus subtilis* plays a role in the initiation of sporulation. *Mol. Microbiol.* **5**:173–185.

Perego, M., and J. A. Hoch. 1988. Sequence analysis and regulation of the *hpr* locus, a regulatory gene for protease production and sporulation in *Bacillus subtilis. J. Bacteriol.* **170**:2560–2567.

Podvin, L., and M. Steinmetz. 1992. A *degU*-containing SPβ prophage complements superactivator mutations affecting the *Bacillus subtilis degSU* operon. *Res. Microbiol.* **143**:559–567.

Poetter, K., and D. L. Coplin. 1991. Structural and functional analysis of the *rcsA* gene from *Erwinia stewartii. Mol. Gen. Genet.* **229**:155–160.

Popham, D. L., D. Szeto, J. Keener, and S. Kustu. 1989. Function of a bacterial activator protein that binds to transcriptional enhancers. *Science* **243**:629–635.

Popov, K. M., Y. Zhao, Y. Shimomura, M. J. Kuntz, and R. A. Harris. 1992. Branched chain α-ketoacid dehydrogenase kinase: molecular cloning, expression, and sequence similarity with histidine protein kinases. *J. Biol. Chem.* **267**:13127–13130.

Porter, S. C., A. K. North, A. B. Wedel, and S. Kustu. 1993. Oligomerization of NTRC at the *glnA* enhancer is required for transcriptional activation. *Genes Dev.* **7:**2258–2273.

Rabin, R. S., and V. Stewart. 1993. Dual response regulators (NarL and NarP) interact with dual sensors (NarX and NarQ) to control nitrate- and nitrite-regulated gene expression in *Escherichia coli* K-12. *J. Bacteriol.* **175:**3259–3268.

Raibaud, A., M. Zalacain, T. G. Holt, R. Tizard, and C. J. Thompson. 1991. Nucleotide sequence analysis reveals linked *N*-acetyl hydrolase, thioesterase, transport, and regulatory genes encoded by the bialaphos biosynthetic gene cluster of *Streptomyces hygroscopicus. J. Bacteriol.* **173:**4454–4463.

Reyrat, J.-M., M. David, J. Batut, and P. Boistard. 1994. FixL of *Rhizobium meliloti* enhances the transcriptional activity of a mutant FixJD54N protein by phosphorylation of an alternate residue. *J. Bacteriol.* **176:**1969–1976.

Richet, E., and O. Raibaud. 1989. MalT, the regulatory protein of the *Escherichia coli* maltose system, is an ATP-dependent transcriptional activator. *EMBO J.* **8:**981–987.

Roggiani, M., and D. Dubnau. 1993. ComA, a phosphorylated response regulator protein of *Bacillus subtilis,* binds to the promoter region of *srfA. J. Bacteriol.* **175:**3182–3187.

Roggiani, M., J. Hahn, and D. Dubnau. 1990. Suppression of early competence mutations in *Bacillus subtilis* by *mec* mutations. *J. Bacteriol.* **172:**4056–4063.

Ronson, C. W., B. T. Nixon, and F. M. Ausubel. 1987. Conserved domains in bacterial regulatory proteins that respond to environmental stimuli. *Cell* **49:**579–581.

Rudner, D. Z., J. R. LeDeaux, K. Ireton, and A. D. Grossman. 1991. The *spo0K* locus of *Bacillus subtilis* is homologous to the oligopeptide permease locus and is required for sporulation and competence. *J. Bacteriol.* **173:**1388–1398.

Ruppen, M. E., G. L. Van Alstine, and L. Band. 1988. Control of intracellular serine protease expression in *Bacillus subtilis. J. Bacteriol.* **170:**136–140.

Scarlato, V., A. Prugnola, B. Aricó, and R. Rappuoli. 1990. Positive transcriptional feedback at the *bvg* locus controls expression of virulence factors in *Bordetella pertussis. Proc. Natl. Acad. Sci. USA* **87:**6753–6757.

Schaeffer, P., J. Millet, and J.-P. Aubert. 1965. Catabolic repression of bacterial sporulation. *Proc. Natl. Acad. Sci. USA* **54:**704–711.

Schell, M. A., T. P. Denny, and J. Huang. 1993. VsrA, a second two-component sensor regulating virulence genes of *Pseudomonas solanacearum. Mol. Microbiol.* **11:**489–500.

Scotti, C., M. Piatti, A. Cuzzoni, P. Perani, A. Tognoni, G. Grandi, A. Galizzi, and A. M.

Albertini. 1993. A *Bacillus subtilis* large ORF coding for a polypeptide highly similar to polyketide synthases. *Gene* **130:**65–71.

Seki, T., H. Yoshikawa, H. Takahashi, and H. Saito. 1987. Cloning and nucleotide sequence of *phoP,* the regulatory gene for alkaline phosphatase and phosphodiesterase in *Bacillus subtilis. J. Bacteriol.* **169:**2913–2916.

Seki, T., H. Yoshikawa, H. Takahashi, and H. Saito. 1988. Nucleotide sequence of the *Bacillus subtilis phoR* gene. *J. Bacteriol.* **170:**5935–5938.

Shimotsu, H., and D. J. Henner. 1986. Modulation of *Bacillus subtilis* levansucrase gene expression by sucrose and regulation of the steady-state mRNA level by *sacU* and *sacQ* genes. *J. Bacteriol.* **168:**380–388.

Sloma, A., C. F. Rudolph, G. A. Rufo, Jr., B. J. Sullivan, K. A. Theriault, D. Ally, and J. Pero. 1990a. Gene encoding a novel extracellular metalloprotease in *Bacillus subtilis. J. Bacteriol.* **172:**1024–1029.

Sloma, A., G. A. Rufo, Jr., C. F. Rudolph, B. J. Sullivan, K. A. Theriault, and J. Pero. 1990b. Bacillopeptidase F of *Bacillus subtilis:* purification of the protein and cloning of the gene. *J. Bacteriol.* **172:**1470–1477.

Sloma, A., G. A. Rufo, Jr., K. A. Theriault, M. Dwyer, S. W. Wilson, and J. Pero. 1991. Cloning and characterization of the gene for an additional extracellular serine protease of *Bacillus subtilis. J. Bacteriol.* **173:**6889–6895.

Smith, I. 1993. Regulatory proteins that control late-growth development, p. 785–800. *In* A. L. Sonenshein, J. A. Hoch, and R. Losick (ed.), *Bacillus subtilis and Other Gram-Positive Bacteria: Biochemistry, Physiology, and Molecular Genetics.* American Society for Microbiology, Washington, D.C.

Smith, I., E. Dubnau, M. Predich, U. Bai, and R. Rudner. 1992. Early *spo* gene expression in *Bacillus subtilis:* the role of interrelated signal transduction systems. *Biochimie* **74:**669–678.

Smith, I., I. Mandic-Mulec, and N. Gaur. 1991. The role of negative control in sporulation. *Res. Microbiol.* **142:**831–839.

Sonenshein, A. L. 1989. Metabolic regulation of sporulation and other stationary-phase phenomena, p. 109–130. *In* I. Smith, R. A. Slepecky, and P. Setlow (ed.), *Regulation of Procaryotic Development: Structural and Functional Analysis of Bacterial Sporulation and Germination.* American Society for Microbiology, Washington, D.C.

Sorokin, A., E. Zumstein, V. Azevedo, S. D. Ehrlich, and P. Serror. 1993. The organization of the *Bacillus subtilis* 168 chromosome region between the *spoVA* and *serA* genetic loci, based on sequence data. *Mol. Microbiol.* **10:**385–395.

Squires, C., and C. L. Squires. 1992. The Clp proteins: proteolysis regulators or molecular chaperones? *J. Bacteriol.* **174:**1081–1085.

Squires, C. L., S. Pedersen, B. M. Ross, and C. Squires. 1991. ClpB is the *Escherichia coli* heat shock protein F84.1. *J. Bacteriol.* **173:**4254–4262.

Stock, J. B., A. J. Ninfa, and A. M. Stock. 1989. Protein phosphorylation and regulation of adaptive responses in bacteria. *Microbiol. Rev.* **53:**450–490.

Stock, J. B., A. M. Stock, and J. M. Mottonen. 1990. Signal transduction in bacteria. *Nature* (London) **344:**395–400.

Stout, V., A. Torres-Cabassa, M. R. Maurizi, D. Gutnick, and S. Gottesman. 1991. RcsA, an unstable positive regulator of capsular polysaccharide synthesis. *J. Bacteriol.* **173:**1738–1747.

Strauch, M. A., and J. A. Hoch. 1992. Control of postexponential gene expression by transition state regulators, p. 105–121. *In* R. H. Doi and M. McGloughlin (ed.), *Biology of Bacilli: Applications to Industry. Biotechnology,* vol. 22. Butterworth-Heinemann, Stoneham, Masss.

Strauch, M. A., and J. A. Hoch. 1993. Transition-state regulators: sentinels of *Bacillus subtilis* post-exponential gene expression. *Mol. Microbiol.* **7:**337–342.

Stülke, J. (University of Greifswald, Germany). 1994. Personal communication.

Tanaka, T., and M. Kawata. 1988. Cloning and characterization of *Bacillus subtilis iep,* which has positive and negative effects on production of extracellular proteases. *J. Bacteriol.* **170:**3593–3600.

Tanaka, T., M. Kawata, and K. Mukai. 1991. Altered phosphorylation of *Bacillus subtilis* DegU caused by single amino acid changes in DegS. *J. Bacteriol.* **173:**5507–5515.

Tanaka, T., M. Kawata, Y. Nagami, and H. Uchiyama. 1987. *prtR* enhances the mRNA level of the *Bacillus subtilis* extracellular proteases. *J. Bacteriol.* **169:**3044–3050.

Tanaka, T., and M. Kawata-Mukai. 1994. Stabilization of the phosphorylated form of *Bacillus subtilis* DegU caused by *degU9* mutation. *FEMS Microbiol. Lett.* **115:**93–96.

Taylor, S. S. 1989. cAMP-dependent protein kinase. *J. Biol. Chem.* **264:**8443–8446.

Tokunaga, T., M. H. Rashid, A. Kuroda, and J. Sekiguchi. 1994. Effect of *degS-degU* mutations on the expression of *sigD,* encoding an alternative sigma factor, and autolysin operon of *Bacillus subtilis. J. Bacteriol.* **176:**5177–5180.

Tomioka, N., M. Honjo, K. Funahashi, K. Manabe, A. Akaoka, I. Mita, and Y. Furutani. 1985. Cloning, sequencing, and some properties of a novel *Bacillus amyloliquefaciens* gene involved in the increase of extracellular protease activities. *J. Biotechnol.* **3:**85–96.

Trach, K. A., J. W. Chapman, P. J. Piggot, and J. A. Hoch. 1985. Deduced product of the stage 0 sporulation gene *spoOF* shares homology with SpoOA, OmpR, and SfrA proteins. *Proc. Natl. Acad. Sci. USA* **82:**7260–7264.

Trach, K. A., and J. A. Hoch. 1993. Multisensory activation of the phosphorelay initiating sporulation in *Bacillus subtilis:* identification and sequence of the protein kinase of the alternate pathway. *Mol. Microbiol.* **8:**69–79.

van Sinderen, D. (University of Groningen, The Netherlands). 1993. Personal communication.

van Sinderen, D., G. Gali, P. Cosmina, F. de Ferra, S. Withoff, G. Venema, and G. Grandi. 1993. Characterization of the *srfA* locus of *Bacillus subtilis:* only the valine-activating domain of *srfA* is involved in the establishment of genetic competence. *Mol. Microbiol.* **8:**833–841.

van Sinderen, D., A. Luttinger, L. Kong, D. Dubnau, G. Venema, and L. Hamoen. 1995. *comK* encodes the competence transcription factor, the key regulatory protein for competence development in *Bacillus subtilis. Mol. Microbiol.* **15:**455–462.

van Sinderen, D., A. ten Berge, B.-J. Haijema, L. Hamoen, and G. Venema. 1994. Molecular cloning and sequence of *comK,* a gene required for genetic competence in *Bacillus subtilis. Mol. Microbiol.* **11:**695–703.

van Sinderen, D., and G. Venema. 1994. *comK* acts as an autoregulatory switch in the signal transduction route to competence in *Bacillus subtilis. J. Bacteriol.* **176:**5762–5770.

van Sinderen, D., S. Withoff, H. Boels, and G. Venema. 1990. Isolation and characterization of *comL,* a transcription unit involved in competence development of *Bacillus subtilis. Mol. Gen. Genet.* **224:**396–404.

Vasselon, T., P. J. Sansonetti, and X. Nassif. 1991. Nucleotide sequence of *rmpB,* a *Klebsiella pneumoniae* gene that positively controls colanic acid biosynthesis in *Escherichia coli. Res. Microbiol.* **142:**47–54.

Völker, U., H. Mach, R. Schmid, and M. Hecker. 1992. Stress proteins and cross-protection by heat shock and salt stress in *Bacillus subtilis. J. Gen. Microbiol.* **138:**2125–2135.

Volz, K. 1993. Structural conservation in the CheY superfamily. *Biochemistry* **32:**11741–11753.

Weinrauch, Y., N. Guillen, and D. A. Dubnau. 1989. Sequence and transcription mapping of *Bacillus subtilis* competence genes *comB* and *comA,* one of which is related to a family of bacterial regulatory determinants. *J. Bacteriol.* **171:**5362–5375.

Weinrauch, Y., T. Msadek, F. Kunst, and D. Dubnau. 1991. Sequence and properties of *comQ,* a new competence regulatory gene of *Bacillus subtilis. J. Bacteriol.* **173**:5685–5693.

Weinrauch, Y., R. Penchev, E. Dubnau, I. Smith, and D. Dubnau. 1990. A *Bacillus subtilis* regulatory gene product for genetic competence and sporulation resembles sensor protein members of the bacterial two-component signal-transduction systems. *Genes Dev.* **4**:860–872.

Weiss, V., F. Claverie-Martin, and B. Magasanik. 1992. Phosphorylation of nitrogen regulator-I of *Escherichia coli* induces strong cooperative binding to DNA essential for activation of transcription. *Proc. Natl. Acad. Sci. USA* **89**:5088–5092.

Yang, M., E. Ferrari, E. Chen, and D. J. Henner. 1986. Identification of the pleiotropic *sacQ* gene of *Bacillus subtilis. J. Bacteriol.* **166**:113–119.

Yang, M., H. Shimotsu, E. Ferrari, and D. J. Henner. 1987. Characterization and mapping of the *Bacillus subtilis prtR* gene. *J. Bacteriol.* **169**: 434–437.

Yasbin, R. E., D. L. Cheo, and K. W. Bayles. 1992. Inducible DNA repair and differentiation in *Bacillus subtilis:* interactions between global regulons. *Mol. Microbiol.* **6**:1263–1270.

Yoshikawa, H., J. Kazami, S. Yamashita, T. Chibazakura, H. Sone, F. Kawamura, M. Oda, M. Isaka, Y. Kobayashi, and H. Saito. 1986. Revised assignment for the *Bacillus subtilis spo0F* gene and its homology with *spo0A* and with two *Escherichia coli* genes. *Nucleic Acids Res.* **14**:1063–1072.

Zheng, L., R. Halberg, S. Roels, H. Ichikawa, L. Kroos, and R. Losick. 1992. Sporulation regulatory protein GerE from *Bacillus subtilis* binds to and can activate or repress transcription from promoters for mother-cell-specific genes. *J. Mol. Biol.* **226**: 1037–1050.

Index

The page numbers for entries occurring in figures are followed by an f; those for entries occurring in tables, by a t.

0A boxes, 164, 165f, 167–168, 169
  reverse, 164, 166

ABC protein, 206
Ablations, 15, 16
abrB gene, 160
  promotion repression in, 166–167
  Spo0A~P repression in, 293
  transcriptional activation in, 168
  transcriptional repression in, 167, 168
AbrB protein, 139, 160, 293, 455, 458, 461
abrBp gene, 160, 162, 167
abrB1p gene, 168f
abrB2p gene, 168f
Accessory colonization factor (ACF), 353–354
Acetate, 226
Acetosyringone (AS), 369, 371, 372, 374, 376, 378, 379, 380
Acetyl coenzyme A (CoA), 73, 75
Acetyl phosphate, 3, 42–43, 73, 210, 212–213
Acetyladenylate, 191
Acetylation, 191
acf gene, 360
acfA gene, 353
acfB gene, 353
acfC gene, 353
acfD gene, 353
Acidity, 373–374, 375
ackA gene, 73, 75, 213
Acyl phosphates, 133–134
addAB gene, 461
Adenosine diphosphate (ADP)
  bacterial virulence factors and, 352
  histidine protein kinase and, 30, 31, 32

magnesium, 32
  nitrogen fixation genes and, 280
  nitrogen regulator I and, 80
  nitrogen regulator II and, 75
  protein II and, 83
  thermodynamics and, 42
Adenosine triphosphate (ATP), *see also* Magnesium adenosine triphosphate
  ArcA and, 226
  ArcB and, 225, 226
  BvgS and, 338
  DegU and, 451
  EnvZ and, 109
  FrzE and, 426
  histidine protein kinases and, 25, 30, 31, 32
  nitrogen fixation genes and, 275, 280, 282–283
  nitrogen regulator I and, 70, 74, 82, 147, 148, 150, 153–154, 156, 173
  nitrogen regulator II and, 75–76, 80
  phosphatase regulation and, 44
  phosphate incorporation into, 203, 215
  protein II and, 83–84
  thermodynamics and, 42
  Uhp system and, 264
  uridylyltransferase/uridylyl and, 81
  VanR/VanS and, 389
Adenylyltransferase (ATase), 3, 68–69, 81
adhE gene, 237
ADP, *see* Adenosine diphosphate
aeg-46.5 locus, 235, 236–237, 238, 242
Aerobic respiratory enzymes, 223–229
AgrA protein, 309, 310, 312
AgrB protein, 309

Agrobacterium
    Ti plasmid of, 367–381
    virulence factors of, 307
Agrobacterium radiobacter, 367, 373
Agrobacterium rhizogenes, 367
Agrobacterium tumefaciens
    Ti plasmid of, 308, 367, 434
    virulence factors of, 306t, 307t, 308, 438
Agrobacterium vitis, 367
Ail protein, 325, 326
Alcaligenes eutrophus, 237
AlgR protein, 312
AlgR1 protein, 307
AlgR2 protein, 307
Alkaline phosphatase genes, 290
Amino acid sequence analysis
    of *divJ,* 405–406
    of *divK,* 405–406
    of *pleC,* 405–406
    of Spo0A, 161f
    of transmitters and receivers, 19–21
    of VanR/VanS, 389
amyE gene, 450
Anabaena, 403
Antimicrobials, 62
APase, 297–299
APase A, 289, 290, 293
APase B, 289, 290, 293
Apo-CheY, 56–57
aprA gene, 434
aprE gene, 139, 450–451, 461, 462
Arabidopsis, 30–31
Arabidopsis thaliana, 409, 447
AraC protein, 307
Arc system, 223–229
arcA gene, 213, 223, 227–228
ArcA protein, 35, 228
    bacterial virulence factors and, 320, 321
    Lux compared with, 439, 441
    phosphorylation by ArcB-P, 226
    phosphotransfer to, 346
    primary structures of, 225
ArcA-P protein, 227, 228
arcB gene, 223, 227–228
ArcB protein, 34–35, 227, 229f
    autophosphorylation of, 225–226
    bacterial virulence factors and, 320, 321, 338, 339, 342
    Bvg compared with, 339
    compounds controlling activity of, 226
    Lux compared with, 438, 439, 441
    phosphorylation of ArcA and, 226
    phosphotransfer from, 346
    primary structures of, 225
ArcB-P protein, 226, 227
Association-dissociation mechanism, 209–210
ATase, *see* Adenylyltransferase
ATP, *see* Adenosine triphosphate

att gene, 369
atx gene, 313
aut gene, 71
Autophosphatase, 40–41
Autophosphorylation, *see also* Phosphorylation
    of ArcB, 225–226
    of BvgS, 338–342
    of CheA, 31–33, 110
    of EnvZ, 33–34, 109, 110–111
    of FixL★, 279–280
    of FrzE, 426
    of histidine protein kinases, 31–34
    of nitrogen regulator II, 32, 33, 75–76, 80
Azorhizobium caulinodans, 284, 285

Bacillus, 135, 403
Bacillus anthracis, 307t, 313
Bacillus brevis, 449
Bacillus subtilis, 39, 206, 271, 406, 447–462
    competence gene expression in, 454–457
    degradative enzyme synthesis in, 449–454
    Pho regulon of, 289–300
    phosphatase regulation and, 44
    phosphorelay and, 137
    phosphorylation and, 30
    transcriptional regulation in, 159–176
Bacterial alkaline phosphatase (Bap), 203, 211, 212–213
Bacterial virulence factors, 305–314, 434, 438
    Bordetella pertussis, 306, 307, 308, 333–346
    Clostridium perfringens, 306t, 311–312
    Erwinia amylovora, 306t, 312
    group A streptococci, 312–313
    Klebsiella pneumoniae, 306t, 311, 312
    Neisseria gonorrhoeae, 306t, 310–311, 356
    Salmonella, 307, 308, 311, 319–330
    Shigella flexneri, 306, 307t, 308–309, 328
    Staphylococcus aureus, 309–310, 312
    Vibrio cholerae, 306, 307, 308, 319, 351–363
Bacteroides fragilis, 306t
Bacteroides transposons, 393–399
Bap, *see* Bacterial alkaline phosphatase
BarA protein, 342
Bioluminescence, 431–443
    autoinduction of, 432
    quorum sensing in, 432–434, 442
Bordetella, 307, 363
Bordetella avium, 334
Bordetella bronchiseptica, 334, 344–345
Bordetella parapertussis, 334
Bordetella pertussis, 306, 307, 308, 333–346
bpr gene, 450
Bradyrhizobium japonicum, 279, 285
α-Bromoacetosyringone, 372
BrpA protein, 449
Brute-force screens, 9, 13
Bryophyllum diagremontiana, 380
bvg gene, 334
    repression by, 344–345

transcriptional regulation by operon, 335
BvgA protein, 307, 333–346
    activation of virulence factors by, 343–344
    domain organization of, 334–335
    mutational analysis of, 337–338
    phosphorylation of, 342
    phosphotransfer to, 342, 346
BvgS protein, 307, 333–346, 358
    autophosphorylation of, 338–342
    domain organization of, 334–335
    intramolecular complementation of mutations, 337
    requirements for domains in vivo, 336–337
    signal-insensitive mutations in, 336
Bypass suppression, 22

C ring, 185, 186
cadA gene, 354
CadC protein, 354, 362, 363
Calcium
    bacterial virulence factors and, 306, 308
    flagellar switch and, 191–192
cAMP, see Cyclic adenosine-3′,5′-cyclic monophosphate
Capsule synthesis, see Colanic acid capsule synthesis
Carbamyl phosphate, 73
Carbon sufficiency, 83–84
cat gene, 389–390
Catabolite control
    by acetyl phosphate, 212–213
    by CreC, 210–212
Catabolite gene activator protein (CAP), 266
Catabolite repression protein, 175
Caulobacter crescentus
    cell cycle of, 403–415
    response regulator of, 39
φCbK receptors, 403, 407
Cecropins, 324
Cell division cycle checkpoints, 404–409
Cell separation (CS), 404
cheA gene, 2, 94, 95
CheA protein, 3, 270, 448
    autophosphorylation of, 31–33, 110
    Bvg compared with, 334, 339
    catabolite control by CreC and, 212
    chemotaxis and, 93, 94–95, 96–97, 98
    CheY and, 39, 389
    crosstalk and, 214
    dimerization and, 29–30, 31–32
    failure to function as phosphatase, 44
    flagellar switch and, 189, 190
    Frz compared with, 421, 424–426
    phosphorylation of, 29–30, 38
    phosphotransfer and, 43–44
CheA_L protein, 94–95, 96, 97
CheA_S protein, 30, 94–95, 96, 97
CheB protein, 34, 39, 41, 448
    chemotaxis and, 93, 94, 96–97, 98
    Frz compared with, 421
    phosphorylation of, 38, 42, 283

phosphotransfer and, 43
CheB-P protein, 96, 97
Chemotaxis, 13, 89–98, 454
    flagellar regulon motility and expression in, 89–91
    frz in, 421–424
    membrane receptors in, 93–94
CheR protein, 93, 96–97, 421, 427
CheW protein, 270
    chemotaxis and, 93, 94–95, 96–97, 98
    CheY autophosphorylation and, 33
    flagellar switch and, 190
    Frz compared with, 421, 426
    phosphotransfer and, 44
cheY gene, 2, 95, 98, 184, 187
CheY protein, 4, 20, 34, 41, 115, 448
    active site structure of, 56–57
    amino acid sequence analysis of, 406
    chemotaxis and, 93, 94, 95–96, 97, 98
    core and tertiary structure conservation in, 57
    flagellar switch and, 183, 184, 187, 188, 189–191, 192, 193, 196
    FlbD compared with, 413
    Frz compared with, 421, 424–426
    historical perspective on, 53–54
    Lux compared with, 439
    magnesium puzzle and, 60–61
    molecular structure of, 54–57
    phosphatase regulation and, 44
    phosphorelay and, 133, 134
    phosphorylation of, 35, 38–40, 61–62, 283, 389, 453
    phosphotransfer and, 43–44, 110
    possible activation mechanism for, 61–62
    secondary structure conservation in, 57–58
    Spo0A compared with, 129, 163
    structural roles of conserved residues in, 59t
    structure-function relationships in, 35–38
    as tertiary template, 61
    VanR and, 389
CheY~P (CheY-P) protein, 133
    chemotaxis and, 95–96, 97, 98
    flagellar switch and, 181, 182, 188, 190, 192, 193
cheZ gene, 40, 96, 184, 187
CheZ protein
    chemotaxis and, 93, 96, 97, 98
    CheY and, 38, 40, 44
    flagellar switch and, 187, 188, 189–191
    Frz compared with, 421
    as phosphatase, 44
    phosphorelay and, 136
chvA gene, 369
chvB gene, 369
chvE gene, 369, 372–373
ChvE protein, 373
    Lux compared with, 439
    vir regulation by, 376–379
chvG gene, 369, 380–381
ChvG protein, 379–381

chvI gene, 369, 380–381
ChvI protein, 379–381
Citrobacter freundii, 255, 306t
Clostridium, 135
Clostridium perfringens, 306t, 311–312
clpB gene, 462
ClpC protein, 458–459
ClpP protein, 462
CoA, *see* Acetyl coenzyme A
coi gene, 163
Colanic acid capsule synthesis, 253–260
comA gene, 133, 455, 458, 459
ComA protein, 271, 300, 448, 449, 461, 462
    competence genes and, 455–456
    Mec and, 459, 460
    phosphorelay and, 136, 137
    phosphorylation of, 457
comC gene, 456–457
comE gene, 456–457
comF gene, 456–457
comG gene, 456
comK gene, 450, 457–458, 460, 461, 462
ComK protein, 455, 456–457, 459–460, 461
Communication modules, 19–22
comP gene, 455, 458, 459
ComP protein, 300, 448, 449, 461, 462
    ComA phosphorylation and, 457
    competence genes and, 455–456
    Mec and, 459, 460
    phosphorelay and, 136, 137
Competence genes, 454–457
Complementation analysis, 14
ComQ protein, 455, 457, 462
comS gene, 456, 458
ComS protein, 455, 456, 458, 460, 462
comX gene, 141
ComX protein, 455, 457, 458, 462
Conformational change mechanism, 209–210
Conformational suppression, 22
Coniferin, 371
Conjugal transfer genes, 393–399
Conservation, 53–62
CopR protein, 307
CopS protein, 307
Core structure conservation, 57
Corynebacterium diphtheriae, 306, 307
cotB gene, 449
cotC gene, 449
cps gene, 255, 258
CrC protein, 204, 290
creA gene, 211, 213
CreA protein, 211
creABCD gene, 210–212
creB gene, 211, 213
CreB protein, 210, 211, 214, 215, 217, 218
creC gene, 211, 212–213
CreC protein, 30, 203, 211, 213, 214, 215, 408

catabolite control of Pho regulon by, 210–212
cross regulation and, 216, 217, 218
creD gene, 211, 213
Cross regulation, 203, 213–218
Crosstalk, 17, 213–214, 442
crp gene, 1, 213
Crp protein, 233, 277, 319, 322
Cryptococcus neoformans, 313
ctaA gene, 295
CtaA protein, 295
ctx gene, 359, 360–361
Ctx protein, 360, 361
cya gene, 1, 213, 338, 343–344
cyc gene, 235
Cyclic adenosine-3′,5′-cyclic monophosphate (cAMP),
    1, 212, 352
Cycloheximide, 371
cydAB gene, 228
cysA gene, 458

Defensins, 324–325
DegM protein, 448
degQ gene, 448, 449, 450, 455–456, 460
DegQ protein, 450
degR gene, 449, 450, 461
DegR protein, 450
Degradative enzyme synthesis, 449–454
degS gene, 449, 451, 452, 453, 460
DegS protein, 300, 448, 449, 462
    competence genes and, 455
    degradative enzyme synthesis and, 450–451
    DegU phosphorylation and, 451–454
    functions and properties of, 460–461
degS(Hy) mutants, 461–462
degU gene, 458, 459
    degradative enzyme synthesis and, 449
    phosphorylation of, 451–454
DegU protein, 271, 295, 300, 448, 449, 458, 462
    competence genes and, 455
    degradative enzyme synthesis and, 450–451
    functions and properties of, 460–461
    Mec and, 459
    phosphorylation of, 451–454
degU(Hy) mutants, 300, 461–462
Dehydroconiferyl acid, 371
2-Deoxy-glucose-6-phosphate, 265–266
Dephosphorylation, *see also* Phosphorylation
    of CheY-P, 190
    of FixJ, 280–281
    of nitrogen regulator II, 32, 80
Deuridylylation, 81
Dimerization
    of histidine protein kinases, 29–32
    of nitrogen regulator I, 147, 150–151
    of ToxR, 358–359
dinR gene, 461
divJ gene, 405–406, 407–408
DivJ protein, 409, 414

amino acid sequence analysis of, 405–406
catalytic activities of, 406
cell division and, 407–408
divK gene, 405–406, 407
DivK protein, 409, 414–415
amino acid sequence analysis of, 405–406
catalytic activities of, 406
cell cycle checkpoints and, 406–408
flagellar switch and, 410
PleD compared with, 409
divL gene, 405, 408–409
DivL protein, 409
dmsA gene, 237, 248
DNA binding
by Arc, 225
by FixJ, 284
by FlbD, 413
by Lux O, 441
by nitrogen regulator I, 149, 150, 151, 154, 156–157
by OmpR, 112–114, 115–117, 119, 120
receivers and, 11
by response regulators, 34
by RteC, 398
by Spo0A, 135, 160–162, 164, 167–168, 175
by ToxR, 354, 355, 358, 359–360, 363
by UhpA, 449
by VirG, 375–376
dnaJ gene, 259
DnaJ protein, 98, 259
dnaK gene, 259
DnaK protein, 98, 259
Domain liberation, 17–19
dsbA gene, 257
dsbB gene, 257
dsrA gene, 260
DsrA protein, 260

emm6 gene, 313
Enteric bacteria, 67–85; *see also* Nitrogen assimilation;
    specific types
Enterobacter cloacae, 326
Enterococcus faecium, 306t, 394
envZ gene, 2
bacterial virulence factors and, 308–309, 330
porin regulation and, 105, 109–110, 114
EnvZ protein, 164
autophosphorylation of, 33–34, 109, 110–111
bacterial virulence factors and, 308, 310, 320, 321,
    354, 356
CreC compared with, 212
crosstalk and, 214
domain structure of, 108f
kinase:phosphatase ratio regulated by, 243
phosphatase regulation by, 44, 45
phosphorylation of, 29
porin regulation and, 105–112, 114, 115, 307
quenching and, 17–18

transplantations and, 17
Epistasis, 14–15
Erwinia amylovora, 255, 306t, 312, 449
Erwinia carotovora, 434
Erwinia stewartii, 255, 449
Escherichia coli, 28, 39, 162, 173–174, 370–371, 373,
    380, 405, 412, 427, 438, 439, 441, 449, 458,
    460, 461, 462
aerobic respiratory enzymes in, 223, 226
bioluminescence in, 432, 434, 435
catabolite repression protein in, 175
chemotaxis and, 13, 89–98, 421, 454
CheY in, 55
colanic acid capsule synthesis in, 253, 255, 259
flagellar switch and, 181–196
glutamine synthetase in, 3
jamming and, 18
mutations in genes of, 1
nifA regulation by, 281–282
nitrate/nitrite gene expression in, 233–249, 408
nitrogen assimilation in, 2, 67, 69, 71, 72f
nitrogen fixation genes in, 277, 279, 281–282, 284
Pho regulon of, 203–218, 290–292
porin regulon of, 105–124, 174, 307
response regulator of, 54
RNA polymerase of, 69, 169, 173, 282–283, 284, 389, 411
succinyl-CoA synthetase of, 29
Uhp system of, 263–273
virulence factors of, 306, 308, 310, 311, 312, 321, 325,
    328, 329, 336, 338, 343, 344, 354, 356–357,
    358, 359, 362, 363
Ethyl ferulate, 371
ETR1, 409
expI gene, 434

Factor Z, 294, 295
FapR protein, 307
fdnG gene, 233, 234, 236, 238, 239, 240, 242, 243, 249
narX null mutants and, 244
operon control region of, 245, 247
Ferulic acid, 371
fha gene, 335, 338, 343, 344
fhaB gene, 338, 344
fim gene, 335, 343
FIS protein, 54
fixGHIS gene, 277
fixJ gene, 277–278, 281–282, 283
FixJ protein, 34, 54, 271, 280–281, 282–286, 449
FixL★ and, 278, 279
phosphorylation of, 283, 453
principal domains of, 283–284
RcsB compared with, 255, 258
transcriptional activation and, 282–283
fixK gene, 281, 284
fixL/fixJ and, 277–278
promoters of, 284
transcriptional activation and, 282

FixK protein, 277
fixL gene, 277–278, 281–282
FixL protein, 242, 282
    amino acid sequence analysis of, 405
    principal domains of, 279
    properties of, 277–278
    transcriptional activation and, 282
fixL★ gene, 282
FixL★ protein, 278
    autophosphorylation of, 279–280
    FixJ dephosphorylation and, 280–281
    transcriptional activation and, 282–283
fixLJ gene, 277
fixNOQP gene, 277
fla gene, 404, 415
    cis-acting sequences in transcription of, 412
    flagellar switch and, 410–411
    FlbD in transcription of, 414
    specialized nature of promoters, 411–412
flaA gene, 344
Flagellar protein transport apparatus, 410
Flagellar regulon, 89–91
Flagellar switch, 96, 181–196
    in *Caulobacter crescentus,* 409–414
    components of, 183–184
    functional analysis of, 187–188
    interaction among, 194–195
    location of, 184–185
    model for, 192–193, 194f
    proteins in, 95, 184
    rotor versus stator of, 186–187
    stoichiometric composition of, 186
flaN gene, *see flgK* gene
Flavobacterium, 399
flbD gene, 412
FlbD protein, 39, 404, 415
    in *fla* transcription, 414
    as flagellar switch protein, 409–414
    at *ftr* sequence elements, 412–413
    regulation of activity, 413–414
flbG gene, 412–413
flgK gene, 412
flgL gene, 412
flhC gene, 90
FlhC protein, 90
flhCD gene, 344
flhD gene, 90, 97–98
FlhD protein, 90
fliA gene, 90
fliC gene, 90
fliF gene, 185, 412–413
FliF protein, 185, 186
fliG gene, 95, 183, 184–185
FliG protein, 95, 97, 183–184, 190, 192, 193
    biochemical properties of, 184
    CheY and, 426
    description of, 189

    functional analysis of, 187, 188
    location of, 184–185
    stoichiometric composition of, 186
fliM gene, 95, 183
FliM protein, 34, 95, 97, 183–184, 190, 192, 193
    biochemical properties of, 184
    CheY and, 39, 196, 426
    description of, 189
    functional analysis of, 187, 188
    location of, 185
    stoichiometric composition of, 186
fliN gene, 183
FliN protein, 95, 183–184, 190, 192, 193
    biochemical properties of, 184
    CheY and, 426
    description of, 189
    functional analysis of, 187, 188
    location of, 185
    stoichiometric composition of, 186
FN516 protein, 227
FN517 protein, 227
fnr gene, 227
Fnr protein, 233, 245, 247, 248, 277
Fosfomycin, 265
frdA gene, 233, 234, 236, 238, 239, 240, 242, 243, 249
    operon control region of, 248
frlAB gene, 344–345
FruR protein, 319, 322
frz gene, 419–429
    biochemical analysis of, 424–427
    chemotaxis and, 421–424
    social behaviors and, 427–429
frzA gene, 421, 422, 428
FrzA protein, 421
frzB gene, 421, 422, 428
frzC gene, 421, 422
frzCD gene, 421, 422, 428
FrzCD protein, 421, 427–428, 429
frzE gene, 421, 422, 428
FrzE protein, 424–426, 429
    autophosphorylation of, 426
    Bvg compared with, 334, 338, 339
    chemotaxis and, 421
    FrzZ phosphorylation and, 427
    Lux compared with, 438
frzF gene, 421, 422, 428
FrzF protein, 421
FrzG protein, 39, 421
frzZ gene, 421, 427
FrzZ protein, 421, 426–427, 429
ftr gene, 412–413
FtsY protein, 310
ftsZ gene, 257
Fumarate, 191
Functional analysis, 187–188

Gain-of-function mutants, 14, 15

β-Galactosidase, 3, 13
gbpR gene, 373
GbpR protein, 373
GerE protein, 449
glnA gene, 2, 67, 153, 154, 156
  in absence of nitrogen regulator II, 72
  derivatives of, 150
  glnL mutations and, 78, 79
  mutations in histidine kinase C-terminal, 76, 78
  nitrogen regulator I dimer binding to, 150–151
  protein II regulation of transcription, 81–82
  regulation of, 70–71
glnALG gene, 71
glnAp1 gene, 69, 70
glnAp2 gene, 2, 4, 67, 69, 70
  in absence of nitrogen regulator II, 71, 73–74, 75
glnB gene, 2, 69, 79, 85
glnD gene, 68, 81
glnD::Tn*10,* 84–85
glnE gene, 68
glnG (*ntrC*) gene, 2, 67, 71, 276
glnH gene, 153
glnHp2 gene, 71
glnL (*ntrB*) gene, 2, 67, 71, 276
    mutations in histidine kinase C-terminal domain
      and, 76–78
    nitrogen regulator II/protein II interaction and,
      78–79
glnL::Tn*10,* 78–79
GlpR protein, 265
GlpT protein, 265
Glucose, 1, 73, 211, 272–273
Glucose 6-phosphate (Glu6P), 263, 265–266, 268,
      272, 273
Glutamate synthase, 75
Glutamine, 68, 69
  nitrogen regulator I phosphorylation and, 82–83
  uridylyltransferase/uridylyl and, 81
Glutamine synthetase, 3, 85
  in absence of nitrogen regulator II, 71, 72–73, 75
  mechanisms of regulation of, 67–69
  regulation of expression, 70–71
Glutamine synthetase-adenosine monophosphate
      (AMP), 69
Glycerol, 73
Glycopeptide resistance genes, 387–390
Group A streptococci, 312–313
GrpE protein, 98
gsiA gene, 136, 137, 455–456, 460–461, 462

HAI, 432
HAI-1, 435–439, 441
HAI-2, 435–439, 442
Heparin, 171
hil gene, 326
his gene, 253
HisP protein, 207

Histidine protein kinases, 25–34
  dimerization of, 29–32
  mutations in C-terminal domain of, 76–78
  phosphatase regulation and, 44–45
  phosphotransfer regulation and, 43–44
  as response regulators, 34–43
  structure-function relationships between domains,
      26–29
HOG1 protein, 409
hpr gene, 462
Hpr protein, 139–140
hrp gene, 312
HrpR protein, 307
HrpS protein, 307, 312
hut gene, 71
hutC gene, 1
hutU gene, 1
Hydroxy-acetosyringone, 371
N-(3-hydroxybutanoyl)-L-homoserine lactone, *see*
      HAI

icd gene, 213
In vitro transcription regulation, 167–173
Initiation of division (DIVi), 404, 407
Input-output communication, 15–17
Integration host factor (IHF), 121, 245–247
isp gene, 450
ivr gene, 369

Jamming, 18

2-Ketoglutarate, 68, 69, 75, 80
  nitrogen regulator I phosphorylation and, 82–83
  protein II as carbon sufficiency sensor and, 83–84
  uridylyltransferase/uridylyl and, 81
3-Ketoglutarate, 83
kinA gene, 133
KinA protein, 129, 132–133, 135, 136, 448
KinA~P protein, 135
Kinase-phosphatase antagonism, 137–139
Kinase-phosphatase ratio
  EnvZ regulation of, 243
  in porin regulation, 110–111
kinB gene, 132
KinB protein, 129, 132–133, 448
Klebsiella, 255, 307, 312, 456
Klebsiella aerogenes, 1, 71, 72f
Klebsiella pneumoniae, 2, 4, 449
  colanic acid capsule synthesis in, 259, 260
  nitrogen assimilation in, 71, 72f
  nitrogen fixation genes in, 276, 277
  virulence factors of, 306t, 311, 312

D-Lactate, 226
LacZ protein, 266
lasA gene, 434
lasB gene, 434

lasI gene, 434
lasR gene, 434
LasR protein, 449
licS gene, 450
Listeria monocytogenes, 307t, 308
Lom protein, 325–326
lon gene, 255, 257, 259
Lon protease, 253
Loss-of-function mutants, 15
lux gene, 431, 432–434
luxA gene, 432
luxB gene, 432
luxC gene, 432
luxCDABEG gene, 432
luxCDABEGH gene, 434, 435–436, 438, 439, 441,
    442
luxD gene, 432
luxE gene, 432
luxG gene, 432
luxH gene, 432
luxI gene, 434
LuxI protein, 308, 434, 442
luxICDABEG gene, 432–433
luxL gene, 435, 436, 438, 443
LuxL protein, 435
luxM gene, 435, 438, 443
LuxM protein, 435, 438
luxN gene, 435, 438, 439, 443
LuxN protein, 437, 438–442, 443
LuxO protein, 439–443
luxP gene, 437, 438, 443
LuxP protein, 437, 438, 439
luxPQ gene, 437
LuxPQ protein, 438
luxQ gene, 437, 438, 443
LuxQ protein, 437, 438–442, 443
luxR gene, 432, 433, 434, 435–436
LuxR protein, 271, 434, 442, 449
    bacterial virulence factors and, 308, 362, 363
    quorum sensing and, 433–434
    RcsB compared with, 255, 258
luxRICDABEG gene, 432
luxR-luxI signal response system, *see* Quorum sensing
    system
LysR protein, 311, 371–372

M protein, 313
M ring, 186
Magainins, 324–325, 328–329
Magnesium
    CheY and, 36–37, 40, 41, 53, 60–61
    glutamine synthetase and, 68
    protein II and, 84
Magnesium adenosine diphosphate (MgADP), 32
Magnesium adenosine triphosphate (MgATP)
    CheA and, 31, 32
    histidine protein kinases and, 26, 28, 45

phosphatase regulation and, 44
    thermodynamics and, 41
MalT protein, 255, 270, 449
Manganese
    FixL★ and, 279–280
    glutamine synthetase and, 68
MAP protein, 409
Mastoparan, 324–325
MCPs, *see* Methyl-accepting chemotaxis proteins
mdh gene, 213
mecA gene, 458, 462
MecA protein, 457–460, 462
mecB gene, 458, 460, 462
MecB protein, 457–460, 462
Melittin, 324–325
Membrane-derived oligosaccharides (MDO), 107
Membrane receptors, 93–94
MerR protein, 173
Methyl-accepting chemotaxis proteins (MCPs), 93–
    95, 98
    Frz compared with, 421, 426, 427
    methylation of, 96–97
    nar compared with, 235
Methyl syringate, 371
Methylation, 96–97
micF gene, 122–123
modA (*chlD*) gene, 237
Monosaccharides, 372–373
motA gene, 90
MotA protein, 184, 185, 186, 187
MotB protein, 184, 185, 186
MoxX protein, 271
mpr gene, 450
Mry protein, 313
MS ring, 181–182, 183, 184–185, 186, 193, 410
MudJ protein, 326
Mutant screens, 13
Mutations, *see also* Null mutants
    BvgA, 337–338
    BvgS, 336, 337
    in communication modules, 21–22
    complementation analysis of, 14
    crp, 1
    in C-terminal histidine kinase domain, 76–78
    cya, 1
    frz, 421
    functional defects of, 13–14
    gain-of-function, 14, 15
    glnL, 78–79
    loss-of-function, 15
    narL, 239
    narX★, 241–242
    narX(H399Q), 240–241
    nitrogen regulator I, 150
    pleC, 404–405
    in response regulators, 58–60
    sporulation, 297–299

ToxR DNA binding analysis and, 359–360
Myxococcus, 13, 438
Myxococcus xanthus, 206, 334, 338, 419–429
    chemotaxis in, 421–424
    response regulator of, 39
    social behaviors of, 427–429

nac gene, 71
NAC protein, 311
NapAB protein, 237
nap-cyc fusion gene, 235, 236–237, 238, 242, 248
Nar protein, 233, 460
narG gene, 233, 236, 238, 240, 242, 243, 249
    narX null mutants and, 244
    operon control region of, 245–247
Narigenin, 371–372
narK gene, 236, 247
narL gene, 239
    effect of null mutants on nitrate/nitrite, 235–238
    identification and characterization of, 234–235
NarL protein, 156, 233, 234, 235–236, 237, 243, 244, 271, 408
    hierarchy of nitrate/nitrite regulation by, 242
    operon control region of, 234t, 245–247, 248
    phosphatase regulation and, 44
    phosphotransfer with NarX, 238–239
    structure of, 235
narP gene, 234–238
NarP protein, 233, 235–236, 237, 238, 243, 244–245, 271, 408
    operon control region of, 234t, 245, 247, 248
    structure of, 235
narQ gene, 236, 239, 242
    identification and characterization of, 234–235
    null mutations of, 238, 239–240
narQ narL double null strain, 242, 244
NarQ protein, 233, 238, 239, 240, 242, 408
    equilibrium model for functions of, 242–244
    negative regulation by, 244–245
    phosphatase regulation and, 44
    structure of, 235
narX gene, 236, 242
    identification and characterization of, 234–235
    null mutants of, 238, 239–240, 243, 244
narX narQ double null strain, 237, 239, 240, 244
NarX protein, 233, 234, 240, 242, 408
    equilibrium model for functions of, 242–244
    negative regulation by, 244–245
    phosphatase regulation and, 44
    phosphotransfer with NarL, 238–239
    signal-insensitive mutations in, 336
    structure of, 235
narX★ gene, 241–242, 244
NarX★ protein, 242
narX(H399Q) gene, 240–241, 244
NarX(H399Q) protein, 240, 241
narXL gene, 235, 247

NDPK, *see* Nucleotide diphosphate kinase
negA gene, 428–429
Neisseria gonorrhoeae, 306t, 310–311, 356
nif gene, 4, 71
nifA gene, 276, 277, 284, 449
    fixL/fixJ and, 277–278
    negative and positive regulation of, 281–282
    promoters of, 284
    transcriptional activation and, 282
NifA protein, 4, 312
nifHDK gene, 276, 277
NifL protein, 311
nifLA gene, 71, 153
nirB gene, 247–248, 249
Nitrate-nitrite responsive gene expression, 233–249, 408
    equilibrium model for, 242–244
    negative regulation of, 244–245
    target operon control regions for, 245–248
Nitrite-responsive gene expression, *see* Nitrate-nitrite responsive gene expression
Nitrogen assimilation, 2, 67–85
    in absence of nitrogen regulator II, 71–75
    redundancy of uridylyltransferase/uridylyl and protein II in, 84–85
Nitrogen fixation genes, 275–276
Nitrogen regulator I (NR$_I$/NtrC), 2, 3–4, 34, 54, 67, 69–70, 71, 73–74, 207
    autophosphorylation of, 33
    background information on, 147–150
    bacterial virulence factors and, 310, 311
    central activation domain of, 154–156
    dimerization of, 147, 150–151
    FlbD compared with, 412
    mode of action of, 284
    nitrogen regulator II/protein II regulated phosphatase and, 79–80
    oligomerization of, 147, 148, 151–156, 173
    phosphorylation of, 35, 39, 41, 74, 80, 82–83, 150–151, 180
    Spo0A compared with, 173, 174
    transcriptional activation by, 147–157
    UhpA distinguished from, 270
Nitrogen regulator II (NR$_{II}$/NtrB), 2, 3, 67, 70–71, 270
    autophosphorylation of, 32, 33, 75–76, 80
    bacterial virulence factors and, 311
    crosstalk and, 214
    dephosphorylation of, 32, 80
    dimerization of, 31
    glnL alteration of protein II interaction, 78–79
    nitrogen regulation in absence of, 71–75
    nitrogen regulator I phosphorylation and, 82, 83, 151
    phosphatase regulation and, 44, 45
    phosphorylation of, 29, 39
    protein II as carbon sufficiency sensor and, 84
    reconstitution of phosphatase regulated by, 79–80
    structure-function analysis of, 75–79

nod gene, 371–372
NodD protein, 371–372
Nonreplicating *Bacteroides* units (NBUs), 394–395, 398, 399
nprE gene, 139, 450, 461
nrfA gene, 237, 242–243, 244, 248, 249
NtrB, *see* Nitrogen regulator II
ntrB (*glnL*) gene, *see glnL* (*ntrB*) gene
NtrC, *see* Nitrogen regulator I
ntrC (*glnG*) gene, *see glnG* (*ntrC*) gene
Nucleotide diphosphate kinase (NDPK), 28
Null mutants, 15
   functional defects of, 14
   narL, 235–238
   narP, 235–238
   narQ, 238, 239–240
   narX, 238, 239–240, 243, 244

obg gene, 141
Obg protein, 141
Oligomerization, 147, 149, 151–156, 173
ompB gene, 105, 107, 309
ompC gene, 105, 106, 108, 112, 114, 116f, 119, 123, 162, 164
   bacterial virulence factors and, 309, 321
   DNA binding and, 115, 117
   promoter of, 121–122
   transcriptional activation of, 117, 118
   transcriptional repression of, 167
OmpC protein, 105, 106, 107, 123–124, 174
   bacterial virulence factors and, 308, 309, 321
ompF gene, 105, 114, 116f, 122, 123, 162, 164
   bacterial virulence factors and, 321
   DNA binding and, 115, 117
   promoter of, 120–121
   transcriptional activation of, 117
   transcriptional repression of, 118–119, 167
OmpF protein, 105, 106, 107, 122–124, 174
   bacterial virulence factors and, 308, 321
ompR gene, 2, 105, 109–110, 114
   bacterial virulence factors and, 308, 309, 321, 330
   Pho regulon and, 213
OmpR protein, 34, 35, 54, 105–106, 109–110, 112–119, 156, 174, 307
   bacterial virulence factors and, 308, 309, 310, 319, 320, 321, 322, 323, 354, 356, 359
   domain structure of, 108f
   EnvZ and, 243
   mode of action of, 284
   phosphatase regulation and, 44
   phosphorylation of, 453
   porin promoters and, 120–122
   quenching and, 17–18
   Spo0A compared with, 162, 164, 167, 174
   transcriptional activation in, 117–118
   transcriptional repression in, 118–119
   transplantations and, 17

   UhpA distinguished from, 270
   VanR compared with, 389
OmpR-P protein, 109, 110, 112, 117
OmpT protein, 363
ompU gene, 361
OmpU protein, 360
OmpX protein, 326
opp gene, *see spo0k* (*opp*) gene
ops gene, 213
ORF1 protein, 387
ORF2 protein, 387
ORFX-17 protein, 448
ORFX-18 protein, 448
orgA gene, 321
oriT gene, 394, 395, 398
Osmolarity, 105–108, 109, 112, 113f, 114–115, 118, 123–124
Oxaloacetate, 83
OxyR protein, 266

P_{II}, *see* Protein II
pagA gene, 323
pagB gene, 323
pagC gene, 325–326, 327, 329
PagC protein, 323, 325–326
pai gene, 462
Paracoccus denitrificans, 271
Penicillin G, 404
Petunia extracts, 371
PfeR protein, 307
PfeS protein, 307
PfixK protein, 284–285
pfl gene, 228, 237
pfoR gene, 312
pgl gene, 372
PgtP protein, 265
pH
   autophosphatase and, 40
   bacterial virulence factors and, 305, 306, 310, 323–324, 325, 327, 353, 361
   porin regulation and, 122
   Ti plasmid and, 373–374, 375, 376, 380
Phenolic compounds, 371–372
PhnCDE protein, 215
Pho box, 290–291
Pho regulon, 73, 203–218, 289–300
   alkaline phosphatase genes as reporters of, 290
   catabolite control by acetyl phosphate, 212–213
   catabolite control by CreC, 210–212
   cross regulation of, 203, 213–218
   regulatory network of, 292–295
   regulatory site mechanism for, 207, 208f, 210
   stoichiometric mechanism for, 207, 208f, 209–210
phoA gene, 203, 289, 290, 291
   bacterial virulence factors and, 327
   cross regulation and, 218
   induction of, 292

ToxR and, 354
PhoA protein, 266
phoB gene, 2, 289, 290, 291
   induction of, 292
   Spo0A~P repression of, 293
   sporulation mutants and, 297–298
   virG regulation and, 380
PhoB protein, 4, 34, 35, 54, 173–174, 204, 205, 214,
    291, 408
   bacterial virulence factors and, 327, 329
   catabolite control by acetyl phosphate and, 213
   catabolite control by CreC and, 210
   cross regulation and, 216, 217, 218
   discovery of, 322
   gene regulation by, 215
   VanR compared with, 389
   virG regulation and, 380
PhoB-P protein, 174
PhoE protein, 215
phoH gene, 215
PhoM protein, *see* CreC protein
phoN gene, 322, 325, 329
phoP gene, 290, 294
   bacterial virulence factors and, 322, 323–324, 326,
    327, 328–329, 330
   discovery of, 322
   induction of, 292
   phylogenetic distribution of, 328–329
PhoP protein, 54, 289, 290–292, 300, 448
   bacterial virulence factors and, 307, 319, 320, 322–
    330
   discovery of, 322
   phoPR induction and, 293
   sap sporulation mutants and, 297–299
   signal transduction during sporulation, 295–296
PhoP~P protein, 291, 294, 299
phoPQ protein, 323, 325–326, 330
phoPR gene
   autoregulated induction of, 292–293
   induction during spore development, 295–296
   sap sporulation mutants and, 297–299
phoPR protein, 293–294, 295, 300
phoQ gene, 322, 323–324, 326
PhoQ protein, 307, 320, 322–330
   discovery of, 322
   pathogenicity of, 322–323
   signals transmitted by, 327–328
phoR gene, 2, 290, 294
   catabolite control by acetyl phosphate and, 212–213
   catabolite control by CreC and, 210, 211
   cross regulation and, 217, 218
   induction of, 292
PhoR protein, 203, 214, 242, 270, 289, 290–292, 300,
    408, 448
   bacterial virulence factors and, 329
   catabolite control by acetyl phosphate and, 213
   catabolite control by CreC and, 211

cross regulation and, 218
discovery of, 322
gene regulation by, 215
phoPR induction and, 293
phosphate control of, 204–210
sap sporulation mutants and, 298, 299
signal transduction role during sporulation, 295–296
PhoR^A protein, 205, 207–208, 209–210
PhoR^R protein, 205, 207–208
   cross regulation and, 217, 218
   mechanisms of interconversion of, 209–210
phoS gene, 299
PhoS protein, *see* PstS protein
Phosphatases, 44–45
   kinase antagonism with, 137–139
   kinase ratio to, *see* Kinase-phosphatase ratio
   nitrogen regulator II regulated, 79–80
   phosphorelay phosphate regulation of, 135–137
   protein II regulated, 44, 79–80, 81–82
   in response regulators, 42–43
Phosphate, 203–204
   Bacillus subtilis, 289–300
   catabolite control by CreC and, 210–211
   detection of environmental, 207–209
   pathways of genes regulated by, 214–215
   Pst system control of, 204–210
   regulation of phosphorelay, 135–137
   repression complexes of, 205–207
Phosphate regulon, *see* Pho regulon
Phosphoenolpyruvate–sugar phosphotransferase
    (PTS) system, 92–93
Phosphoramidate, 43, 73
Phosphorelay system, 129–142
   enzymatic activities of proteins, 133–134
   kinase-phosphatase antagonism in, 137–139
   nature of signals in, 140–141
   quaternary structure of, 134–135
   repressor-activator antagonism in, 139–140
   signal input into, 132–133
   transcriptional regulation of, 130–132
Phosphorus assimilation, 214–215
Phosphoryl group transfer, *see* Phosphotransfer
Phosphorylation, *see also* Autophosphorylation;
    Dephosphorylation
   of ArcA, 226
   of BvgA, 342
   of CheY, 35, 38–40, 61–62, 283, 389, 453
   of ComA, 457
   of DegU, 451–454
   of FixJ, 283, 453
   of FrzZ, 427
   of histidine protein kinases, 25–26, 29–31
   of nitrogen regulator I, 35, 39, 41, 74, 80, 82–83, 150–
    151
   of RcsB, 257–258
   of response regulators, 35, 38–40, 41–43
   transmitters and receivers in, 11

Phosphotransacetylase-acetate kinase (Pta-AckA), 204, 216–217
Phosphotransfer, 238–239
    to BvgA, 342, 346
    CheY and, 43–44, 110
    NarL-NarX, 238–239
    in VanR/VanS, 389
Photobacterium leiognathi, 432
Photobacterium phosphoreum, 432
phoU gene, 207
PhoU protein, 203, 270
    catabolite control by CreC and, 211
    cross regulation and, 216, 218
    gene regulation by, 215
    phosphate control of, 204–210
pilA gene, 310–311
PilA protein, 310–311
pilB gene, 310–311
PilB protein, 310–311, 355–356
pilE gene, 310, 311
PilR protein, 307
PilS protein, 307
Piston model, 93
Pivot model, 93
pksX gene, 450
Plant cell transformation, 367–381
pleC gene, 409
    amino acid sequence analysis of, 405–406
    cell division and, 407–408
    pseudoreversion analysis of mutants, 404–405
PleC protein, 409, 414–415
    catalytic activities of, 406
    cell division cycle checkpoints and, 406–408
    flagellar switch and, 410
pleD gene, 408–409
PleD protein, 408, 409, 414
pmrA gene, 321
PmrA protein, 320–321
pmrB gene, 321
PmrB protein, 320–321
PnifA protein, 284–285
Porin promoters, 116f, 120–122
Porin regulation, 105–124, 174, 307
    response regulator in, 112–119
    sensor in, 107–112
    stimulus for, 106–107
    transmembrane signal transduction in, 111–112
ppk gene, 213
prgB gene, 327
prgH gene, 326, 327, 329
PrgH protein, 326
proB gene, 454
Progression of division (DIVp), 404, 407
Promoters
    activated by Spo0A, 164–166
    in nitrogen fixation genes, 284–285
    repressed by Spo0A, 166–167

    in VanR/VanS, 389–390
Protein II (P_{II}), 2, 3, 69, 70, 270
    glnL in nitrogen regulator II interaction, 78–79
    mutations in histidine kinase C-terminal domain and, 76, 78
    nitrogen regulator I phosphorylation and, 82
    nitrogen regulator II autophosphorylation and, 76
    phosphatase regulation and, 44, 79–80, 81–82
    PhoU compared with, 206–207
    purification and crystallization of, 79–80
    redundancy of, 84–85
    as sensor of carbon sufficiency, 83–84
    uridylylation/deuridylylation of, 81
Protein II-uridine monophosphate (P_{II}-UMP), 69, 71, 81–82
Protein D, 149, 152
Proton motive force, 191
Pseudomonas, 306t, 308, 456
Pseudomonas aeruginosa, 449
    virulence factors of, 306, 307–308, 312, 434
Pseudomonas fluorescens, 394
Pseudomonas solanacearum, 460
Pseudomonas syringae, 306t, 312
Pseudoreversion analysis, 404–405
psiD gene, 323, 326
psiE gene, 215
psiF gene, 215
psiH gene, *see phoH* gene
Pst system, 203, 204–210
    catabolite control by CreC and, 211
    cross regulation and, 218
PstA protein, 204, 206, 207, 211
pstA2 gene, 206
PstB protein, 204, 206, 207, 211
PstC protein, 204, 206, 207, 211
pstHI gene, 213
PstS protein, 204, 206, 207, 208, 209, 211
PstSCAB protein, 215, 216
pta gene, 73, 75
Pta-AckA, *see* Phosphotransacetylase-acetate kinase
PtsG protein, 310
ptx gene, 338, 343–344
pur gene, 213
put gene, 71
Pyruvate, 73, 211, 226

Quenching, 17–18
Quorum sensing (*luxR-luxI*) system, 432–434, 442

Random walk, 91
rapA gene, 137
RapA protein, 137, 461
RapB protein, 137
rcsA gene, 255, 258, 260
RcsA protein, 256–257, 260, 449
    bacterial virulence factors and, 311
    RcsB stimulation by, 258–259

rcsB gene, 256, 257, 259, 312
RcsB protein, 255–257, 259–260, 271
    bacterial virulence factors and, 311
    phosphorylation of, 257–258
    RcsA stimulation of, 258–259
rcsC gene, 257, 258, 259, 312
RcsC protein, 257–258, 259, 260, 311
rcsF gene, 258
RcsF protein, 256, 258
recA gene, 450, 461
Receivers, 10–11; *see also specific receivers*
    phosphorylation activities of, 11
    signaling properties of, 11–12
    signaling transactions between transmitters and, 17–19
    structure-function studies of, 19–22
Regulatory genes, 454–455
Regulatory site mechanism, 207, 208f, 210
resD gene, 295
ResD protein, 289, 294–295, 299, 300
resDE gene, 294–295, 299
resE gene, 295
ResE protein, 289, 294–295, 299, 300
Response regulators, 2, 10–11, 25, 34–43; *see also specific response regulators*
    autophosphatase activities in, 40–41
    conservation in, 53–62
    family relationships in, 54
    historical perspective on, 53–54
    modular design of, 54
    mutant sites in regulatory domains of, 58–60
    phosphatase kinetics in, 42–43
    phosphorylation of, 35, 38–40, 41–43
    in porin regulation, 112–119
    Spo0A as model for, 173–175
    structure-function relationships in, 35–38
    thermodynamics and, 41–42
Reverse 0A boxes, 164, 166
Reversion analysis, 22
rfb gene, 253
rhiR gene, 434
Rhizobium, 312, 403
Rhizobium leguminosarum, 371–372, 434
Rhizobium meliloti, 275–286, 405, 449
rmpA gene, 259
RmpA protein, 259, 449
rmpA2 gene, 311
RmpA2 protein, 311
RNA III, 309–310
RNA polymerase
    bacterial virulence factors and, 344, 355
    FixJ and, 282–283, 284
    fla and, 411
    nitrogen assimilation and, 69, 70, 73
    PhoB and, 204
    phosphorelay and, 135, 140
    porin regulation and, 117–118, 119

    RteA/RteB and, 396
    Spo0A and, 162, 166, 168–171, 173, 174, 175
    VanR/VanS and, 389
    VirG and, 375, 378
Rotors, 186–187
rpiA gene, 213
rpoA gene, 118, 119
RpoA protein, 344
rpoB gene, 458
rpoN gene, 411–412
rpoS gene, 123
RpoS protein, 322
rscA gene, 253
RscA protein, 253, 255
rscB gene, 253
RscB protein, 253
rscC gene, 253
RscC protein, 256
rteA gene, 394, 395–398
RteA protein, 395–398, 399
rteB gene, 394, 395–398
RteB protein, 395–398, 399
rteC gene, 398
RteC protein, 398, 399
rvtA gene, 163

S ring, 186
sacB gene, 300, 450–451, 454, 460, 461, 462
Saccharomyces cerevisiae, 30, 328, 409, 447
sacXY gene, 450
sad gene, 163, 164
Salmonella, 307, 308, 311, 319–330
    pathogenesis biology in, 319–320
    virulence phenotypes in, 323–325
Salmonella typhi, 320, 321, 324
Salmonella typhimurium, 150, 405, 439
    chemotaxis in, 89–98
    flagellar switch and, 181–196
    nitrogen assimilation in, 67, 71
    organophosphate transport systems of, 264, 265
    response regulator of, 54
    virulence factors of, 308, 319, 320, 321, 322, 323–324, 325, 328, 329, 363
sap gene, 297–299
sapA gene, 297, 299
sapB gene, 297
sasA gene, 206
Scissions, 15, 16–17
Φ(*sdh-lacZ*), 223, 226, 227
Secondary structure conservation, 57–58
Selection schemes, 9, 13
Sensors, 2, 10–11; *see also specific sensors*
    in porin regulation, 107–112
Sesbania rostrata, 284
Shielding, 19
Shigella flexneri, 306, 307t, 308–309, 328
sigD gene, 462

Sigma factor-6, 123
Sigma factor-32, 355
Sigma factor-54, 2, 3–4, 147, 149f, 152, 154, 156, 173
  bacterial virulence factors and, 311
  fla and, 411–412
  nitrogen assimilation and, 70, 73
  RteA/RteB and, 396
Sigma factor-70, 4, 156, 173
  bacterial virulence factors and, 311, 355
  FixJ and, 282, 284
  nitrogen assimilation and, 69
  PhoB and, 204
  VanR/VanS and, 389
Sigma factor-A, 130–132, 140, 162, 168
Sigma factor-F, 30
Sigma factor-H, 130–132, 140, 168, 169
Signaling pathways
  chemotactic, 92–98
  genetic analysis of, 13–15
  reconstruction of, 14–15
Signaling proteins, 15–19
sin gene, 458
Sin protein, 455, 458
sinI gene, 139
SinI protein, 139, 140
sinR gene, 140
SinR protein, 139–140
SLN1 protein, 409
Slow-switchers, 183
Smooth swimming, 91
sms gene, 458
sob gene, 163, 164
sodA gene, 228
sof gene, 163, 164
soxRS gene, 123
SoxRS protein, 123
spac gene, 456
SpaK protein, 448
SpaR protein, 448
spo0A gene, 159–160, 164, 175, 293, 299
  activation of, 166
  res not dependent on, 295
  transcriptional activation of phosphorelay components and, 131
Spo0A protein, 129, 133–134, 299, 300, 448, 460–461, 462
  competence genes and, 455
  domains of, 160–164
  kinase-phosphatase antagonism and, 137, 139
  phoPR during spore development and, 295–296
  phosphorylation of, 39, 453
  promoters activated by, 164–166
  promoters repressed by, 166–167
  quaternary structure of phosphorelay components and, 135
  repressor-activator antagonism and, 140
  signal input into phosphorelay and, 132

transcriptional activation of phosphorelay components and, 131, 132
  transcriptional regulation by, 159–176
Spo0ABD protein, 160–162, 167, 171, 172f, 175
spo0Ap gene, 168
Spo0A~P (Spo0A-P) protein, 129–130, 135, 136, 141–142, 160, 167, 174, 175, 299, 300
  conversion of Spo0F~P to, 138f
  kinase-phosphatase antagonism and, 137
  phoPR during spore development and, 295–296
  phoPR repression by, 293–294
  repressor-activator antagonism and, 139, 140
  ResD antagonism to, 295
  transcriptional activation and, 168, 169, 170–171, 172f
  transcriptional activation of phosphorelay components and, 130–132
  transcriptional repression and, 168
spo0B gene, 132, 141, 163, 293, 295
Spo0B protein, 39, 133, 137, 141, 142, 299
Spo0B~P protein, 133
spo0E gene, 136
Spo0E protein, 136, 137, 139
spo0F gene, 163, 293
  promotion activation in, 166
  promotion repression in, 166
  res not dependent on, 295
  signal input into phosphorelay and, 132
  transcriptional activation of phosphorelay components and, 131
Spo0F protein, 129, 133–134, 137, 141, 299, 448, 461
  amino acid sequence analysis of, 406
  kinase-phosphatase antagonism and, 138f
  phosphorylation and, 39
  quaternary structure of phosphorelay components and, 134
  signal input into phosphorelay and, 132
  transcriptional activation of phosphorelay components and, 131, 132
spo0Fp gene, 169
Spo0F~P protein, 133, 134, 136, 141, 142
  conversion to Spo0A~P, 138f
  kinase-phosphatase antagonism and, 137
spo0H gene, 168
Spo0K protein, 455, 457, 462
spo0K(opp) gene, 141, 457
spo0L gene, 136–137, 460–461
Spo0L protein, 136–137
spo0P gene, 136, 137
Spo0P protein, 136, 137
spoIIA gene, 130, 160, 162, 175
  promotion activation in, 166
  repressor-activator antagonism and, 139, 140
  signal input into phosphorelay and, 132
  sporulation APase and, 297
SpoIIAA protein, 30
SpoIIAB protein, 30
spoIIAp gene, 162, 166, 168, 169, 174

spoIIE gene, 139, 140, 166, 175, 297
spoIIG gene, 130, 160, 173, 175, 448
    promotion activation in, 166
    repressor-activator antagonism and, 139, 140
    signal input into phosphorelay and, 132
    Spo0A phosphorylation and, 453
    sporulation APase and, 297
    transcriptional repression in, 168
spoIIGp gene, 160, 174
    promotion activation in, 164–166
    transcriptional activation in, 169–173
SpoIIJ protein, 461
Sporulation
    PhoP/PhoR and, 295–296
    phosphorelay system in, 129–142
    regulatory mutants of, 297–299
Sporulation APase, 297–299
SpvR protein, 322
srfA gene, 448, 450, 455–456, 457–458, 460
srfAB gene, 458
SSK1 protein, 409
Staphylococcus aureus, 309–310, 312
Stators, 186–187
Stoichiometric mechanism, 207, 208f, 209–210
Streptococci, 306t
Streptococcus pyogenes, 312
Streptomyces griseus, 271
Streptomyces hygroscopicus, 449
Strong enhancers, 150, 154, 155f
Strong sites, 150, 155f
Structure-function relationships
    in CheY, 35–38
    in communication modules, 19–22
    in histidine protein kinase domains, 26–29
    in nitrogen regulator II, 75–79
    in response regulators, 35–38
Succinyl-CoA synthetase, 29
suv-3 gene, 162
suv-4 gene, 162

tac gene, 361
Tap protein, 93, 212
tar gene, 90
Tar protein, 212
    chemotaxis and, 93, 94
    porin regulation and, 107–108
    Ti plasmid and, 373–374
    transplantations and, 17
Taz1 protein, 107–108, 111–112
Taz1d1A protein, 112
tcp gene, 360, 361
tcpA gene, 353
TcpA protein, 361
TcpI protein, 363
T-DNA, 367–368, 369, 379
Teicoplanin, 387
Temperature
    bacterial virulence factors and, 305

    porin regulation and, 122, 123–124
    Ti plasmid and, 374
Tertiary structure conservation, 57
tetQ gene, 393, 394, 395, 396–397, 398, 399
Tetracycline, 393–399
Thermodynamics, 41–42
Ti plasmid, 308, 367–381, 434
    acidity and, 373–374, 375
    monsaccharides and, 372–373
    phenolic compounds and, 371–372
    temperature and, 374
Tn*916,* 393
Tn*1545,* 393
Tn*1546,* 387–390
Tobacco, 371
toxA gene, 434
Toxin-coregulated pilus (TCP), 353–354, 360
toxR gene, 351, 353, 354, 360, 361
ToxR protein, 307, 351, 362–363
    coordinate gene expression controlled by, 360–361
    dimerization of, 358–359
    mutagenesis analysis and, 359–360
    regulon of, 353–354
    signaling by, 354–356
    ToxS periplasmic interaction with, 356–358
    toxT expression and, 361–362
    transmembrane nature of, 354
toxR-phoA fusion gene, 355
ToxR-PhoA fusion protein, 354–355, 356, 359
toxRS gene, 361
toxS gene, 351
ToxS protein, 351, 363
    periplasmic interaction with ToxR, 356–358
    ToxR dimerization and, 358–359
    ToxR DNA binding and, 360
toxT gene, 360–362
ToxT protein, 307, 351, 360–361, 363
traI gene, 434
Transcriptional activation
    in FixJ, 282–283
    by nitrogen regulator I, 147–157
    by OmpR, 117–118
    by Spo0A, 168–173
    in ToxR/ToxS, 356–358
Transcriptional regulation
    by *bvg* operon, 335
    of phosphorelay components, 130–132
    by Spo0A, 159–176
Transcriptional repression
    by OmpR, 118–119
    by Spo0A, 167–168
Transmembrane signal transduction, 111–112
Transmitters, 10–11; *see also specific transmitters*
    phosphorylation activities of, 11
    signaling properties of, 11–12
    signaling transactions between receivers and, 17–19
    structure-function studies of, 19–22

Transplantations, of foreign domains, 15, 16f, 17
traR gene, 434
traY gene, 228
Trg protein, 17, 93, 94, 108, 212, 373
Trypanosoma brucei, 447
Trz1 protein, 108
tsr gene, 241
Tsr protein, 212, 370–371
    bacterial virulence factors and, 358
    chemotaxis and, 93, 94
    jamming and, 18
    shielding and, 19
    signal-insensitive mutations in, 336
Tumbling, 13, 91–92, 97, 98

ubiD gene, 226
ugpAB gene, 327
UgpBAEC protein, 215
Uhp system, 263–273
    complexity of signal transduction in, 270
    genes of, 265
    signal and response in, 265–266
uhpA gene, 265, 266, 267, 268, 271
UhpA protein, 265, 266, 268, 284
    degradative enzyme synthesis and, 449
    intracellular signal of, 270–271
    sequence analysis of, 405
    target of, 271–273
uhpABC gene, 268
uhpABCT gene, 265
uhpB gene, 267, 268, 270
UhpB protein, 242, 265, 266–267, 268, 270, 271
    sequence analysis of, 405–406
uhpBC gene, 268
uhpC gene, 267, 268
UhpC protein, 265, 266–267, 268, 270, 271, 406
uhpT gene, 265, 266, 268, 270, 271–273
UhpT protein, 264–266, 267, 268, 273
uidA gene, 396–397, 398
URF-1 protein, 448
URF-2 protein, 448
Uridylylation, 81
Uridylyltransferase, 3
Uridylyltransferase/uridylyl (UTase/UR), 68–69, 70
    nitrogen regulator I phosphorylation and, 82, 83
    protein II uridylylation/deuridylylation by, 81
    purification of, 80, 81
    redundancy of, 84–85
Urocanase, 1
Urocanate, 1
UvrC2, 271

vanA gene, 390
VanA protein, 387, 389
Vancomycin resistance, 387–390
vanH gene, 390
VanH protein, 387

VanR protein, 387–390, 394
VanS protein, 387–390, 394
vanX gene, 390
VanX protein, 387
VanY protein, 387
VanZ protein, 387
Vibrio
    bioluminescence in, 431–443
    virulence factors of, 308
Vibrio cholerae, 306, 307, 308, 319, 351–363
    pathogenicity in, 352–353
    transcriptional activation in, 356–358
Vibrio fischeri, 362, 442, 443, 449
    bioluminescence in, 431, 432–434
    virulence factors of, 308
Vibrio harveyi, 431, 432, 439, 441, 442, 443
    multichannel sensory circuit of, 434–438
Vibrio parahemolyticus, 306t
vir box, 375–376
vir gene, 344, *see also bvg* gene
    monosaccharides in induction of, 372–373
    VirA/VirG-ChvE model of, 376–379
virA gene, 368–369
VirA protein, 34
    bacterial virulence factors and, 307
    Lux compared with, 438, 439
    monosaccharides and, 372
    Ti plasmid and, 369–374, 375, 376–379, 380–381
    vir regulation by, 376–379
virB gene, 368, 374, 380
virC gene, 368
virD gene, 368
virE gene, 368
VirF protein, 307
virG gene, 368–369, 375, 380
VirG protein, 54
    bacterial virulence factors and, 307
    monosaccharides and, 372
    phosphorylation and, 35
    Ti plasmid and, 369, 370, 371, 374–379, 380–381
    vir regulation by, 376–379
virH gene, 368
virR gene, 312
VirR protein, 312
vpr gene, 450
Vsr protein, 460

Weak enhancers, 150

Xenorhabdus luminescens, 432

YecB protein, 394
Yersinia, 306, 307, 308, 325
Yersinia enterocolitica, 328
Yersinia pestis, 307, 328
YopN protein, 308